AF774054

NONLINEAR GRAVITODYNAMICS

The Lense-Thirring Effect

NONLINEAR GRAVITODYNAMICS

The Lense-Thirring Effect

A documentary introduction to current research

Editors

Remo Ruffini

Costantino Sigismondi

University of Rome "La Sapienza", Italy

World Scientific

New Jersey • London • Singapore • Hong Kong

Published by

World Scientific Publishing Co. Pte. Ltd.
5 Toh Tuck Link, Singapore 596224
USA office: 27 Warren Street, Suite 401-402, Hackensack, NJ 07601
UK office: 57 Shelton Street, Covent Garden, London WC2H 9HE

British Library Cataloguing-in-Publication Data
A catalogue record for this book is available from the British Library.

The editors and the publisher would like to thank the following for their permission to reproduce the articles found in this volume:

The American Physical Society, The American Astronomical Society, Elsevier Science, The Royal Society, Kluwer Academic/Plenum Publishers and the Government of The United States of America.

NONLINEAR GRAVITODYNAMICS
The Lense-Thirring Effect

Copyright © 2003 by World Scientific Publishing Co. Pte. Ltd.

All rights reserved. This book, or parts thereof, may not be reproduced in any form or by any means, electronic or mechanical, including photocopying, recording or any information storage and retrieval system now known or to be invented, without written permission from the publisher.

For photocopying of material in this volume, please pay a copying fee through the Copyright Clearance Center, Inc., 222 Rosewood Drive, Danvers, MA 01923, USA. In this case permission to photocopy is not required from the publisher.

ISBN-13 978-981-238-347-1
ISBN-10 981-238-347-6

FOREWORD

REMO J. RUFFINI AND COSTANTINO SIGISMONDI
ICRA, International Center for Relativistic Astrophysics and Physics Department, University of Rome "La Sapienza" , P.le A. Moro 2, I-00185 Rome, Italy

This book gives an up-to-date detailed account of knowledge of the Lense-Thirring effect and its implications for physics and astrophysics.

Einstein's theory of General Relativity, born in 1916 as a purely theoretical work with a splendid mathematical formalism and feeble experimental verifications, is today becoming more and more the theoretical driving force of one of the largest observational and experimental programs in the history of science. Involving discoveries in relativity, cosmology, astronomy and astroparticle physics, this great adventure reflects the enormous investments since the 1960s in a whole suite of new technologies especially, but not exclusively, in space.

One grand example of these broad-ranging developments is the Lense-Thirring effect. Starting from a profound intuition of Lense and Thirring in 1918, based on a simple solution of the linearized Einstein field equations, this has emerged in the past four decades as a phenomenon of extraordinary importance in cosmology, radio jets in quasars, and the physics of neutron stars and black holes, besides leading to some of the most sophisticated experiments ever performed in the space surrounding our planet Earth.

This book is dedicated to three persons:

Francis Everitt, who, starting from the complementary visions of William Fairbank, George Pugh and Leonard Schiff has, with many colleagues, brought the Gravity Probe B Relativity Mission to actuality, ready for launch in mid-2003.

Riccardo Giacconi, who, with the establishment of X-ray astronomy and the discovery of neutron stars and black holes in binary x-ray sources, has opened the way to the study of Lense-Thirring processes in very strong general relativistic regimes.

Joseph Taylor, whose radioastrometric timing observations have not only provided quantitative evidence for gravitational radiation effects in binary pulsar systems but have also opened yet another possibility of observing the Lense-Thirring effect in such systems.

This book has been motivated by the "Third William Fairbank Meeting" held in Rome and Pescara from June 29 to July 4, 1998. The book contains the contributions presented at that meeting, expanded by the addition of a complete set of classical and prominent contemporary papers on this subject and a general introduction by one of us (R. J. R.).

For their grants in support of the activities which have made this volume possible, we gratefully acknowledge the Italian Ministry of Foreign Affairs (MAE), the Italian Ministry of Education University and Research (MIUR), the City of Pescara and the University of Rome "La Sapienza".

Pescara, November 2002

Remo J. Ruffini and Costantino Sigismondi

CONTENTS

Foreword v

The Contributions to the WFM3 and INW1

Introduction to Nonlinear Gravitodynamics: The Lense-Thirring Effect
Remo Ruffini 3

Section A: Equations of Motion of Spinning Particles in Electrodynamics and General Relativity

Some considerations on the varieties of frame dragging
Kenneth Nordtvedt 35

Equations of Motion of Spinning Relativistic Particle in External Fields
I. B. Khriplovich and A. A. Pomeransky 46

Equations with Intrinsic Rotation in Gravitational Theory
Leopold Halpern 70

Spinning Particles in the Kerr Field
O. Semerák 74

Energy First Integral for Spinning Particles in the Schwarzschild Background
D. Bini, G. Gemelli and R. Ruffini 81

Nongeodesic Motion of Charged Spinning Test Particles
D. Bini, G. Gemelli and R. Ruffini 84

Section B: Inertial Forces and Gravitoelectromagnetism

Absolute and Relative Frenet-Serret Frames for Accelerated Black Hole Circular Orbits
D. Bini, R. T. Jantzen and A. Merloni 95

Centripetal Acceleration and Centrifugal Force in General Relativity
D. Bini, F. De Felice and R. T. Jantzen 119

On the (Non)Existence of a Gravitomagnetic Dynamo
J.-F. Pascual-Sánchez 128

Rotation of the Polarization Plane in the Gravitational Field of Rotating Objects
V. Perlick 135

Section C: The GP-B Mission: The Orbiting Gyroscope Experiment around the Earth

General Treatment of Geodetic and Lense-Thirring Effects on an Orbiting Gyroscope
R. J. Adler and A. S. Silbergleit 145

Classical Torque Errors in Gravity Probe B Experiment
A. S. Silbergleit, M. I. Heifetz and G. M. Keiser 155

Data Reduction in the Gravity Probe B Experiment: Optimal Estimation and Filtering
M. I. Heifetz, G. M. Keiser and A. S. Silbergleit 172

Section D: Probing the Lense-Thirring Effect around the Earth with Twin Satellites

The Gravitomagnetic Field and its Measurement with the Lageos Satellites
Ignazio Ciufolini and Erricos Pavlis 185

Lense-Thirring Precession Determination from Laser Ranging to Artificial Satellites
J. C. Ries, R. J. Eanes and B. D. Tapley 201

On Gravitomagnetic and Rotational Analogue of the Hall Effect: A Possibility to Detect Lense-Thirring Field of Earth
B. J. Ahmedov 212

Determination of the PPN Parameter γ Through Geodetic VLBI
M. Cataldo, L. Guerriero, G. Bianco, R. Lanotte and F. Dilettuso 219

Quantum Tests of Lense-Thirring Type Effects
C. Lämmerzahl 225

Section E: Probing the Gravitomagnetic Lense-Thirring effect with Neutron Stars and Black Holes

Relativistic Periastron and Nodal Precession and Quasi-Periodic Oscillations from Low Mass X-Ray Binaries
L. Stella 235

On Gravitomagnetic Precession and QPO in Black Hole Candidates
Wei Cui, Wan Chen and S. N. Zhang 248

Millisecond Phenomena in Accreting Neutron Stars – An Update
M. Van Der Klis 262

Variability of Black-Hole Binary Sources and Lense-Thirring Orbital Precession
V. Karas, O. Semerák and F. De Felice 282

Magnetospheres Around Rotating Black Holes
M. Dovčiak and V. Karas 288

Lense-Thirring Effect in the Superfluid Inside Neutron Stars
H. Casini and R. Montemayor 296

General Relativistic Calculations of Precession Around Rapidly Rotating Neutron Stars
Sharon M. Morsink 303

Relativistic Diskoseismology: C-Modes and the Lense-Thirring Effect
A.S. Silbergleit and R.V. Wagoner 316

Generation and Evolution of Magnetic Fields in the Gravitomagnetic Field of a Kerr Black Hole
Ramon Khanna 329

Lense-Thirring Precession of Accretion Disks and Quasi-Periodic Oscillations in X-Ray Binaries
D. Markovic and F. K. Lamb 336

Classical Papers

On the Gravitational Effects of Rotating Masses: The Thirring-Lense Papers
B. Mashhoon, F. W. Hehl and D. S. Theiss (1984) 349

An Example of A New Type of Cosmological Solutions of Einstein's Field Equations of Gravitation
K. Gödel (1949) 389

Spinning Test-Particles in General Relativity. I
A. Papapetrou (1951) 393

Spinning Test-Particles in General Relativity. II
E. Corinaldesi and A. Papapetrou (1951) 404

Proposal for a Satellite Test of the Coriolis Prediction of General Relativity
G. E. Pugh (1959) 414

Motion of a Gyroscope According to Einstein's Theory of Gravitation
L. I. Schiff (1960) 427

The Gyroscope Experiment I. General Description and Analysis of Gyroscope Performance
C. W. F. Everitt (1974) 439

Bound Geodesics in the Kerr Metric
D. C.Wilkins (1972) 469

Generalized Wilkins Effect and Selected Orbits in a Kerr-Newmann Geometry
M. Johnston and R. Ruffini (1974) 478

Sur Certaines Vérifications Nouvelles de la Relativité Générale Rendues Possibles par la Découverte d'un Pulsar Membre d'un Systéme Binaire
T. Damour and R. Ruffini (1974) 484

Possible Experiment with Two Counter-Orbiting Drag-Free Satellites to Obtain a New Test of Einstein's General Theory of Relativity and Improved Measurements in Geodesy
R. A. Van Patten and C. W. F. Everitt (1976) 487

Contemporary Papers

Measurement of the Lense-Thirring Drag on High-Altitude, Laser-Ranged Artificial Satellites
I. Ciufolini (1986) 493

New Test of General Relativity: Measurement of de Sitter Geodetic Precession Rate for Lunar Perigee
B. Bertotti, I. Ciufolini and P. L. Bender (1987) 497

Measurement of the de Sitter Precession of the Moon: A Relativistic Three-Body Effect
I. I. Shapiro, R. D. Reasenberg, J. F. Chandler and R. W. Babcock (1988) 501

Determination of the Geometry of the PSR B1913+16 System by Geodetic Precession
M. Kramer (1998) 505

The Contributions to the first ICRA Network Workshop and the Third William Fairbank Meeting on the Lense-Thirring Effect

INTRODUCTION TO NONLINEAR GRAVITODYNAMICS: THE LENSE-THIRRING EFFECT

REMO J. RUFFINI

Department of Physics, University of Rome "La Sapienza" and ICRA, International Center for Relativistic Astrophysics, P.le A. Moro 2, 00185 Rome and Piazza della Repubblica 10, 65100 Pescara, Italy

1 From electrodynamics to nonlinear gravitodynamics

1.1 From electricty and magnetism to electrodynamics

That electric charges have repulsive or attractive interactions decreasing with the square of their distances was the result of a heroic set of work begun by Henry Cavendish (1771–1773)[1] and followed by Charles Augustin Coulomb (1785).[2] Today, still using the Cavendish method, the validity of the inverse square law has reached the very impressive accuracy of one part in 10^{16}.[3]

Similarly the properties of magnetic interactions had been described even earlier in the classic book of William Gilbert *De Magnete* (1600).[4] In 1820 Hans Christian Oersted published the epochal book *Experimenta circa effectum conflictus electrici in acum magneticum.*[5] In this fundamental contribution to physics he showed how an electric current gives rise to a magnetic field, establishing the first link between the two concepts, the electric field and the magnetic field, which until then had been physically unrelated.

An additional step forward was taken by Ampère in his *Recueil d'Observations Electrodynamiques* of 1825 unifying both electric and magnetic effects under the single name of electrodynamics, relating their origin and their intensity to electric current distributions.[6] In parallel in 1831 Faraday (*Experimental Researches in Electricity*) reached a deeper understanding based on graphical representations of the lines of force of magnetic fields due to time varying electric currents.[7]

1.2 Classical and quantum electrodynamics

In 1862 with an elegant set of differential equations Maxwell established the constitutive equations of electromagnetic fields relating their propagation to the speed of light. We recall his truly remarkable words: "We can scarcely avoid the inference that *light consists in the transverse undulations of the same medium which is the cause of electric and magnetic phenomena.*"[8] This monumental work was confirmed in the years 1886–1888 by Hertz's discovery of transverse electromagnetic waves which indeed propagated at the speed of light, a result published in his book *Electric Waves.*[9] The first person to explicitly show that a rotating charged body generates a magnetic field was H. A. Rowland in 1876. In 1905 Albert Einstein showed that Maxwell's equations were the only field equations in the realm of physics consistent with the principle of special relativity and they were put at the very foundation of the development of relativistic field theories.[10]

After the work of Dirac, Heisenberg and Pauli, the attempt to quantize electrodynamics reached completion in the 1960s through the work of Feynman, Dyson, Schwinger and Tomonaga.[11] The quantum microphysics description (Quantum Electrodynamics, QED) as well as the macroscopic description (Classical Electrodynamics) were joined together in a unique framework. Classical electrodynamics is the limit of QED for small momentum and energy transfers and for large average numbers of photons.[12] The extremely impressive agreement between the theory and experimental tests in phenomena such as the Lamb shift of atomic levels,[13,14] the value of the fine structure constant α, and the anomalous magnetic moment of the electron represent the greatest achievements of this unified picture. A very impressive test of QED in the strong field regime was performed using the Zeeman splitting of a hydrogen-like atom under the influence of an external magnetic field. The energy shift due to this external field is $E_{mag} = <-\mu \cdot \mathbf{B}>$, where the magnetic moment μ of a particle with charge q and mass m is related to its angular momentum J by the g-factor: $\mu = g\frac{q}{2mc}\mathbf{J}$. The g factor of the spin of the free electron was measured to a relative precision of 4 parts in 10^{12} by Van Dyck et al.[15] in 1987. The value og g obtained from QED calculations[16] is one of the most precise agreements between theory and experiment in the entire realm of physics.

Today, as clearly expressed by Jackson,[12] classical electrodynamics lives in a sector of the unified description of particles and interactions known as the standard model. The standard model gives a coherent quantum-mechanical description of electromagnetic, weak, and strong interactions based on fundamental constituents—quarks and leptons—interacting via force carriers—photons, W and Z bosons and gluons. The unified theoretical framework is built upon continuous gauge symmetries and discrete symmetries of particle properties. QED is a consequence of a spontaneously broken symmetry in a theory in which initially the weak and electromagnetic interactions are unified and the force carriers of both are massless. The symmetry breaking leaves the electromagnetic force carrier (photon) massless with a Coulomb's law of infinite range, while the weak force carriers acquire masses of the order of 80–90 GeV/c^2 with a weak interaction at low energies of extremely short range ($2 \cdot 10^{-18}$ m).[17] Because of their origin in a unified theory the range and the strength of the weak interaction are related to the electromagnetic coupling.

This entire field starting from the work of Cavendish and Coulomb all the way to the current standard model of particle physics was corroborated by a continuous, extremely vaste and fertile interaction with experimental data. Each new step in high precision measurements prepared the way to new theoretical formulations, and each new theoretical idea could always been tested within a few years of its introduction with high precision experimental verifications. The theory could be proved or disproved in a most succesful and fertile interaction with the experimental framework through the developments of entire families of dedicated high energy accelerators specifically developed for this scope. This fortunate situation is at the very ground of the success of this scientific endeavour.

1.3 Newtonian gravitational theory

In this grand theoretical synthesis which spans from classical electrodynamics to particle physics to the regime of quantum-relativistic and classical electrodynamics, the gravitational interaction has been left out.

The case of gravitational interaction is indeed markedly different from the one of the electrodynamics we have just mentioned, and surprisingly presents some very interesting analogies. At the time in which Coulomb and Cavendish were struggling to unveil the fundaments of the electric forces, and Ampére was moving the first step toward the formulation of electromagnetism, the theory of gravity was standing already on a much firmer basis, due to the theoretical work of Newton and his formulation of the gravitational theory.[18] This theory had a phenomenal development through the work of Lagrange (who won the Paris Academy Prize in 1764 for a work on the libration of the Moon), Euler[22] (same prize in 1774), Laplace[23] (author of a treatise on Celestial Mechanics in 1799), followed by Hamilton[25] (1834-5) and Jacobi[24] (1843),[26] reaching a profound theoretical maturity for the physical and mathematical point of view and tremendous predicting power and explanation of all observed phenomena in Astronomy, especially in the explanation of the physics of the planetary systems[19,27] and binary stars. It is well known that this formidable theoretical framework at the turn of the nineteenth century presented only a minor disagreement with the so called anomalous precession of the perihelion of Mercury discovered by Le Verrier with calculations in 1845 and re-calculated by Newcomb.[21] The perihelion of Mercury precess of about 5000 arcseconds per century due mainly to the many body interactions, well explained in Newtonian theory, and only a minor quantity of 43 arcseconds per century was left unexplained.[20,21] It would have been hard to imagine that such a meagre discrepancy would fuel one of the greatest revolution in the entire scientific thinking.

1.4 The general theory of relativity

After 1905 a major conceptual difficulty surfaced. In the case of electromagnetism, the formulation of Maxwell equations preceded the introduction of special relativity, and were shown by Einstein[10] to be fully consistent with the relativity principle. In the case of gravitation the Newtonian equations were not relativistically invariant.

It is well known that the formulation of a relativistic theory of gravitation has not been as straightforward as in the case of electromagnetism.[35] The corresponding goal in the case of the gravitational interaction has implied the abandonment of the Newtonian theory in which the gravitational interaction is instantaneous, it is described by a scalar field and a Laplacian operator. It was necessary to move on to the theory of general relativity introduced by Albert Einstein in 1916. The gravitational interaction in Einstein theory is described by a rank-two symmetric tensor fulfilling ten nonlinear partial differential equations in four dimensions. Such a new theory surpassed in mathematical complexities not only the Maxwell equations, but all known field equations of physical interactions.

Such an extreme modification further introduced the geometrization procedure of the gravitational interaction which in turn has led to the reformulation of the known spacetime structure of our Universe.

The most outstanding difference between general relativity and the previous mentioned field theories consisted in overcoming three major obstacles:
1) The extremely small number of measurable new effects.
2) The outstanding mathematical difficulties inherent to the integration of the field equations.
3) The geometrical formulation of the interaction extremely challenging from a conceptual point of view and very far away from the every-day experience of a traditional physicist.

1.5 The birth of relativistic astrophysics

This situation has markedly changed since 1967 with the opening of the entire new field of knowledge which goes under the name of relativistic astrophysics. And the difference have mainly matured due to three different factors:
1) The introduction of computers, which have enormously facilitated the theoretical work in relativistic theories, both by the integration of the fundamental differential equations in cases of scientific interest, and more recently in the development of very powerful algebraic manipulators technologies software able to overcome an immense amount of very tedious mathematical work.
2) The development of space technology which has allowed unprecedented human activities both in the observational field and in the very high precision measurement. To recall the lifting of very large observatories in the optical, x-Ray, gamma-ray wavelength over the Earth atmosphere which have allowed for the first time to look with eye accuracy at the entire Universe in totally unexplored region of the electromagnetic spectrum, and
3) The very fertile wealth of scientific discoveries ranging from neutron stars to black holes to gravitational waves.

We are today in the middle of witnessing this enormous scientific revolution and this book wants to record some specific moments in the development of this new era. Paradoxically to gauge the success and the future perspectives of this entire new field of research it can be of interest to establish some correspondence between the field of gravitation including the profound novelties introduced by general relativity and the field of electrodynamics. There are indeed analogies but at the same time the field of gravitation presents a new realm of unsurpassed conceptual magnitude and richness of new physical regimes which were unimaginable in the previous theoretical framework.

1.6 The "Nonlinear gravitodynamics"

This book only refers to the basic classical non-quantum theory of gravity, and shows some profound analogies and great differences between gravitation and classical electrodynamics. Much of these differences are due to the tensorial nature of the gravitational interaction and to the nonlinearity of the Einstein field equations. We see how the effect of rotation in general relativity leads to some unexpected and profound analogies with the magnetic interactions. These effects have been expressed in the study of Gravitomagnetism, to which the majority of the contributions in this volume are dedicated. We indicate this field as the study of the "nonlinear gravitodynamics" in order to evidence the analogies with electromagnetism, but at the same time to stress the profound differences introduced by Einstein field equations. To illustrate these analogies and differences, we give three specific examples (sections 2, 3 and 4).

2 The gravitational analogue of Coulomb-like interactions

We first compare the classic work of Coulomb and Cavendish in electricity with the corresponding situation in the case of gravitation in trying to identify a "Coulomb-like" behavior in the gravitational interaction. We see that both at small and large distances there exist totally new and unexplored phenomena in the gravitational case. Looking first at small distances:

a) In the infinitely small there still exists the greatest challenges of physics and mathematics advanced by Riemann[37] that space does not conform to the hypothesis of a continuous geometry and a discrete geometrical description might be appropriate instead, see also Wheeler.[38]

b) In sharp contrast with electromagnetic interaction at submillimeter distances the behavior of the gravitational interaction is still very poorly known (for a review see Ohanian and Ruffini[36]).

c) At kilometer distances and in strong gravitational fields, the concept of an horizon and the issue of gravitational collapse occur. A very drastic departure from a Coulomb-like behavior is observed and through the horizon the very basic space-like structure is subverted.

Turning now to large distances, the determination of the Coulomb-like behavior also encounters some totally new challenges and interpretations:

d) At planetary distances the accurate analysis of motions of space trajectories outside our solar system[39] presents some yet unexplained feature, which may lead to a new understanding of gravitational interaction.

e) At galactic scales the verification of the Coulomb-like behaviour on distances of tens of kiloparsecs is hampered by the unknown contribution of dark matter to the rotation curves of galaxies.[40]

f) At extragalactic distances the simple Coulomb-like behavior is overcomed by the necessity of describing the gravitational interaction taking into account the space-time geometry of an expanding Friedmann Universe. Looking at large distances in the past light cone we reach the totally new concept of cosmological singularity where both space and time, as we know it today, loose their meaning. Looking at the very large distances in the future we do not yet know if the Universe in finite or infinite and what will be its predominant dynamics.

Already these six points elucidate the wealth of new information that the description of the gravitational interaction within general relativity has introduced linking the determination of the law of force, so straightforward in the electromagnetic case in the Coulomb-Cavendish work, to totally new regimes encompassing conceptually new phenomena ranging from discreteness of space-time structure in the infinitely small, to black holes and to the cosmological singularity in the infinetely large.

3 The gravitational analogue of Hertz-like wave solutions

As a second example we recall how the very structure of the Einstein field equations of general relativity led to the theoretical prediction of a totally new phenomena: the existence of gravitational waves (see e.g. the Landau and Lifshitz text[41]). Analogous to the work of Hertz, the experimental verification of the existence of such waves and their conformity to the energy flux predicted in the formula given by Landau and Lifshitz[41] has been splendidly confirmed by the observations of binary pulsars made by Hulse and Taylor.[42] This discovery also nicely represents the enormous conceptual and experimental differences between the electromagnetic and gravitational fields, as indicated by the following three points:
a) Due to the weakness of the gravitational coupling the very existence of gravitational waves had passed unnoticed in the traditional physics framework. We had to wait for the discovery of a total new class of stellar objects, the neutron stars, discovered following the development of radio telescopes.[43] It was using specifically the Arecibo radio telescope that the measurement of the very regular emission from two rotating neutron stars, a binary pulsar, made possible the discovery of gravitational radiation.[42]
b) The theoretical framework implied by general relativity is of enormous complexity in comparison to Maxwell's linear equations: the implications that the equation of motions of a relativistic system are a consequence of the Einstein field equations has led to an entirely new formulation of nonlinear effects encountered in the emission and the general energetic balance of the system, including the radiation reaction terms, as clearly pointed out by Damour.[44]
c) The previous two points stress the great astrophysical interest for the sources of the gravitational wave emission as well as the formidable difference in beauty and complexity between the theoretical treatment of electromagnetic and gravitational radiation emission. It is appropriate, nevertheless, to stress that the amount of information of any physical or astrophysical system obtainable from electromagnetic radiation is incomparably larger than the one obtainable from gravitational radiation. The reason is simply stated: the gravitational radiation emission is proportional to the time variability of the quadrupole moment of the entire emitting system, which a such is a global quantity which does not contain the detailed description of the physical or astrophysical system. The detailed description of an astrophysical system can only come from data obtained in the electromagnetic interaction in the optical, radio, x-ray and gamma ray wavelengths and data from particle physics process, due to the physical dimensions of the wavelength characterising their interaction.

4 The gravitational analogue of the Oersted-Ampère-like magnetic interaction

We turn now from the Cavendish-Coulomb-Hertz work on charges and radiation to the Oersted-Ampère work on the occurrence of magnetic fields originating from currents. We again look for gravitational analogue. In doing so we move on to the third example which refers to the topic of the present book. Since the earliest

days of general relativity it became clear, through the work of Lense and Thirring (here reproduced in the classic paper section in the translation by B. Mashhoon, F. W. Hehl, and D. S. Theiss[45] *On the Gravitational Effects of Rotating Masses: The Thirring-Lense Papers*) that a rotating mass would introduce a dragging of the inertial frames. Thanks to the work of R.L. Forward,[46] itself based on earlier work by Møller,[47] this concept matured over the years into a gravitational analogue of the concept of a magnetic field. Such a "gravitomagnetic" component had not been previously theoretically conceived nor had it ever manifested itself experimentally due to the smallness of its magnitude in earthbound experiments. Also in this case, as in the previous two examples, we expect an enormous conceptual difference in moving from the electrodynamic phenomenon to the description of the corresponding gravitational phenomenon in the framework of general relativity. Also particularly interesting will be the identification of physical and astrophysical settings under which such phenomenon will manifest itself. We see in this book that this condition will be fulfilled on three enormously different scales:
a) Around our planet Earth. Using extremely high precision measurements around the Earth, made possible by the most advanced space technologies available today, it is possible that in the forthcoming year the precise measurement of the gravito magnetic field of the Earth will be obtained due to the launch of the Gravity Probe B mission of NASA, and possibly by measurements by LAGEOS and future specially conceived space missions, e.g. LARES.
b) Around neutron stars and black holes. This possibility appears to be promising both in the explanation of quasi-periodic oscillations (QPO) observed in binary X-ray sources and in galactic and extragalactic relativistic jets and
c) On the largest possible scale imaginable: our Universe. Indeed to show the cosmological implications of rotating masses we have reproduced in the classic paper section *An Example of a New Type of Cosmological Solutions of Einstein's Field Equations of Gravitation* by Kurt Gödel.[48]

5 The goal of this volume

It is clear from the previous three examples that the vigorous coming of age of relativistic astrophysics, the many technical and computational progress, the development of very accurate space technologies around our planet are all participating in making fundamental physics and astrophysics the natural domain of applicability of general relativity. General relativity has become *the* theoretical framework for promoting one of the strongest develoments possibly of the entire scientific endeavour, assisted by an unprecedented, vigorous development of obervatories in space, on the ground as well as underground.

Returning to the analogy between electricity and gravitation the three examples we have considered are only on classical level. Although a quite large effort has been invested in the last twenty years by the scientific community not conclusive success have yet been reached in extending to gravitation the success obtained by Tomonaga, Schwinger, Feynman and Dyson on quantum electrodynamics nor in the standard model of elementary particles. Undoubtely the clarification of the relative role of general relativity on the quantum domain will have a still larger inpact on the understanding of fundamental laws of nature. The goal of this book is to reach a better understanding of the "nonlinear gravitodynamics" still at the classical level. This represents a significant step in reaching a better understanding and familiarity with a vast number of new physical and astrophysical processes. It may be as well propedeutical to the extension of these concepts to the microphysical world and promote the development of a complete approach of the unification of physical interaction.

This volume is organized in three parts as follows:

- The papers presented at the Third William Fairbank Meeting (WFM3) and the First ICRA Network Workshop (INW1) on Relativistic Gravitational Experiments in Space.

  The meeting was held in Rome at the University "La Sapienza" and in the newly restructured ICRA Network Center in Pescara, with the participation of leading experts both in theoretical and experimental fields.

- The reproduction of some classic papers.

- The reproduction of some basic contemporary papers.

6 The contributions to the WFM3/INW1

The contributions are divided into five sections:

- **Section A: Equations of Motion of Spinning Particles in Electrodynamics and General Relativity.**

  Some Considerations on the Varieties of Frame Dragging by **K. Nordtvedt** addresses the issue of the local Lorentz-invariance of gravity and its role in the interpretation of the geodetic precession versus gravito-magnetic effects in the Gravity-Probe-B mission (GP-B mission).

  Equations of Motion of Spinning Particles in External Fields, by **I. B. Khriplovich** and **A. A. Pomeransky** deals with the noncovariant description of spin in deriving the equation of motion of spinning relativistic particles. Classic results for the gravitational spin-orbit interaction are recovered.

  Equations with Intrinsic Rotation in Gravitational Theory by **L. Halpern** considers spinning particles within the anti-deSitter group $G = SO(3,2)$ instead of the traditional Poincaré group.

  Spinning Particles in the Kerr Field by **O. Semerák** describes the numerical solution of Mathisson-Papapetrou equations for particles moving around Kerr black holes.

  Finally *Energy First Integral for Spinning Particles in the Schwarzschild Background* and *Non Geodesic Motion of Charged Spinning Test Particles* by **D. Bini, G. Gemelli** and **R. Ruffini** addresses the issue of supplementary conditions for the motion of charged spinning test particles from a relative observer point of view in a Schwarzschild and a Reissner-Nordström black hole spacetime.

- **Section B: Inertial Forces and Gravitoelectromagnetism.**

  Absolute and Relative Frenet-Serret Frames for Accelerated Black Hole Circular Orbits by **D. Bini, R. T. Jantzen** and **A. Merloni** uses absolute and relative Frenet-Serret frames to study circular orbits in black hole spacetimes. The nice geometrical properties of these orbits are elucidated in the framework of gravitoelectromagnetism.

  Centripetal acceleration and Centrifugal Force in General Relativity, by **D. Bini, F. De Felice** and **R. T. Jantzen** uses comoving relative Frenet-Serret frames to define centrifugal forces in general relativity.

  On the (Non-)Existence of a Gravitomagnetic Dynamo by **J. F. Pascual-Sanchez** shows the impossibility to amplify and maintain an initial spin motion as it occurs with a magnetic dynamo and the nonexistence in gravity of the induction term as in Faraday's equation.

  Finally in *Rotation of the Polarization Plane in the Gravitational Field of Rotating Objects* by **V. Perlick** it is noted that in the Faraday effect the polarization direction of electromagnetic radiation undergoes a rotation if it passes through a magnetized medium. This paper discusses whether spacetime geometries can produce an additional rotation to the polarization direction.

- **Section C: The GP-B Mission: the Orbiting Gyroscope Experiment around the Earth.**

  Pugh (1959)[54] and Schiff (1960)[55] independently proposed to measure the gravitomagnetic field of the Earth by detecting the mechanical precession of gyroscopes carried by a satellite. (See the reproduction of both articles in the classic paper section.) The proposal of Schiff gave birth to the GP-B project. The paper of Everitt[56] (1974) reprinted in the classic papers section records the first stages of this project now finally scheduled for launch on October 30, 2002. The Gravity Probe B Mission, which owes so much to the inspiration of William Fairbank, to whose memory this meeting has been dedicated, is by far the most technically sophisticated experiment carried out in fundamental physics. It will probe the Einstein theory of general relativity to an unprecedented accuracy, with many new technologies that also have other potential applications in astrophysics and gravitational physics missions: for example in the Satellite Test of the Equivalence Principle (STEP)[57] and the Laser Interferometer Space Antenna (LISA),[58] where the space alignment and drag-free control of satellites is crucial.

  The GP-B experimental group at Stanford University has been preparing and testing equipment with the intention of detecting the gravitomagnetic field of the Earth and measuring it (Lense-Thirring and Schiff precessions) to about one percent accuracy. The GP-B project is nearing completion. Here we have the contributions of the GP-B team:

  General Treatment of Geodetic and Lense-Thirring Effects on an Orbiting Gyroscope by **R. J. Adler** and **A. S. Silbergleit** considers linearized general relativity theory from a flat background and shows that only the Eddington parameters α and γ are needed for the description of the Lense-Thirring effects. The possible influences of the Earth quadrupole moment on the GP-B mission are outlined.

Classical Torque Errors in Gravity Probe B Experiment by **A. S. Silbergleit, M. I. Heifetz** and **G. M. Keiser** presents the error tree spreadsheet of the mission GP-B. Contributions of different groups of torques to the total expected experimental error are compared in view of the latest developments of the mission.

Data Reduction in the Gravity Probe B Experiment: Optimal Estimation and Filtering by **M. I. Heifetz, G. M. Keiser** and **A. S. Silbergleit** describes from the point of view modern control theory three main problems addressed in the data reduction of the GP-B mission: 1) direct observation of the GP-B gyroscope's precession angle; 2) precise estimation of the gyroscope's geodetic and frame-dragging average drift rates; 3) gyroscope-telescope matching: elimination of the reference direction pointing error.

In the classic paper section on this topic we have reproduced also: *Proposal for a Satellite Test of the Coriolis Prediction of General Relativity* by G.E. Pugh (1959) and *Motion of a Gyroscope According to Einstein's Theory of Gravitation* by L.I. Schiff (1960).

- **Section D: Probing the Lense-Thirring Effect around the Earth with Twin Satellites.**

In a classic paper reproduced in our volume, Van Patten and Everitt[59] (1976) proposed to observe the spin-orbit and the Lense-Thirring effect with two counter-orbiting polar satellites. A later paper by Ciufolini[60] (1986) reproduced in the contemporary papers session proposed a different approach in the measurement of this effect by using a high-altitude laser-ranged artificial satellite: the Italy-US LAGEOS-III/LARES satellite mission.

The Van Patten-Everitt experiment, as originally proposed, would have done somewhere between a 1% and 5% measurement; the Ciufolini experiment is at least 10% in the most optimistic claim. The reason is that Van Patten-Everitt by measuring the distance between the two satellites directly by Doppler-ranging as they pass over the poles, eliminates the large error due to the uncertainty in the positions of the spacecraft with respect to the poles, occurring with Ciufolini's method.[60] Ciufolini indeed has ingeniously thought about how one might do an experiment with existing satellites.

The current status of the LAGEOS satellite experiments and observations has been critically reviewed and the merits of different approaches are discussed. Recent experimental evidence for the Lense-Thirring precession of the laser ranged satellites[61] LAGEOS and LAGEOS II around the Earth is described in the paper of Ciufolini et al.[62] At the present time, the latest improvements on this topic regard the understanding of the perturbations to the orbit of the satellites by lunar tides with the nutation period.[65] New insights about satellite gravitatiomal orbital perturbations and gravitomagnetic clock effect are given by Iorio.[66,67] Recent analyses[68] have almost doubled the period of observation with respect to the measurement described in these proceedings by Ciufolini and Pavlis and in Ciufolini et al.[62]

The Gravitomagnetic Field and its Measurements with the Lageos Satellite by **I. Ciufolini** and **E. Pavlis** shows the results of the measurement of the Lense-Thirring precession by analyzing the data of the LAGEOS and LAGEOS II satellites with the Earth's gravity field models JGM-3 and EGM-96. The Lense-Thirring effect is detected with a quoted accuracy of 1.1 ± 0.2, where 1 is the value in General Relativity.

Lense-Thirring Precession Determination from Laser Ranging to Artificial Satellites by **J. C. Ries, J. Eanes** and **B. D. Tapley** considers the uncertainties on the even zonal harmonics of the Earth's gravity field and in contrast with the previous paper concludes that they are still too large for measuring the Lense-Thirring precession with a precision better than 50 to 100%.

On Gravitomagnetic and Rotational Analog of the Hall Effect: a Possibility to Detect Lense-Thirring Field of Earth by **B. J. Ahmedov** advances the possibility of using the Coriolis force which generates an azimuthal potential difference in a spinning rotor carrying radial electric current in order to measure the Lense-Thirring effect on the Earth.

Determination of the PPN Parameter γ through Geodetic VLBI by **M. Cataldo, L. Guerriero, G. Bianco, R. Lanotte** and **F. Dilettuso** addresses the measurement of the parametrized post-Newtonian term γ using VLBI observations of distant quasars. After data reduction they obtain $\gamma = 1.0008 \pm 0.0012$.

Quantum Test on Lense-Thirring Type Effects by **C. Lämmerzahl** discusses the possibility of quantum testing the Lense-Thirring effect. The accuracy of present-day techniques only permits measurement of the rotation of the Earth with atomic interferometry by measuring the Coriolis force on the atoms.

- **Section E: Probing the Gravitomagnetic Lense-Thirring effect with Neutron Stars and Black Holes.**

In this section the Lense-Thirring effect is considered around neutron stars and black holes, since some evidence of gravitomagnetic effects may have been observed by analyzing the data coming from the region of strong gravitational field near neutron stars and black holes. Since many astrophysical compact objects like neutron stars and black holes have a large rotational velocity and a radius which is close to the gravitational one, general relativistic gravitomagnetic effects should have crucial importance for some astrophysical processes. Accretion processes around neutron stars and black hole candidates may provide another example of interaction between a rapidly rotating body and gravitating matter. The related contributions in this section are:

Relativistic Periastron and Nodal Precession and Quasi-Periodic Oscillations From Low Mass X-Ray Binaries by **L. Stella** considers the fundamental frequencies for geodesic motion in the gravitational field of the accreting compact objects in the relativistic precession model as an explanation for quasi-periodic oscillations (QPOs) in low mass X-ray binaries. The Lense-Thirring frequency is suggested to manifest itself when the orbit of the accreting material is tilted with respect to the equatorial plane.

On Gravitomagnetic Precession and QPOs in Black Hole Candidates by **W. Cui, W. Chen** and **S. N. Zhang**, addresses the alleged observational evidence for gravitomagnetic precession around spinning black holes candidates. Orbiting matter is shown to be a viable process for explaining QPOs.

Millisecond Phenomena in Accreting Neutron Stars, an Update by **M. Van Der Klis** reviews observations and theories in the beat-frequency interpretation of QPOs.

Variability of Black Hole Binary Sources, and Lense-Thirring Orbital Precession by **V. Karas, O. Semerák** and **F. De Felice** discusses possible observational signatures of Lense-Thirring precession of companion stars around a central galactic black hole in active galactic nuclei (AGN).

Magnetospheres around Rotating Black Holes by **M. Dovčiak** and **V. Karas** addresses influential and recent selected work concerning mutual interactions between electromagnetic fields and a rotating Kerr black hole.

Lense-Thirring Effect in the Superfluid Inside Neutron Stars by **H. Casini** and **R. Montemayor** refers to the frame-dragging forces in the superfluid inside neutron stars. This issue is also studied from a perspective of seeking to verify the existence of the Lense-Thirring effect in coherent quantum systems.

General Relativistic Calculations of Precession around Rapidly Rotating Neutron Stars by **S. M. Morsink** deals with general relativistic derivations of Lense-Thirring precession orbits which overcome the limitations of weak-field approximations and explores their behavior in the region close to a neutron star.

Relativistic Diskoseismology: C-Modes and the Lense-Thirring Effect by **A. S. Silbergleit** and **R. V. Wagoner** addresses the issue of the modes of oscillation trapped within the inner region of accretion disks induced by a rotating central body. The theory is compared with the data of two galactic 'micro-quasars'.

Generation and Evolution of Magnetic Fields in the Gravitomagnetic Field of a Kerr Black Hole by **R. Khanna** considers primary and induced sources for magnetic fields around rotating black holes, and the structure of a black-hole-driven current is presented.

Lense-Thirring Precession of Accretion Disks and Quasi-Periodic Oscillations in X-Ray Binaries by **D. Markovic** and **F. K. Lamb** presents a model for the accretion disk made of a family of rings in order to take into account warp-like deformations. High and low frequency gravitomagnetic modes are discussed with respect to their observability.

7 The Classic Papers

H. Thirring (1918) *Über die Wirkung rotierender ferner Massen in der Einstenschen Gravitationstheorie* [Physikalische Zeitschrift (1918), 19(3):33-39] *[On the Effect of Rotating Distant masses in Einstein's Theory of Gravitation]* translated into English in B. Mashhoon, F. W. Hehl, and D. S. Theiss (1984),[45] *On the Gravitational Effects of Rotating Masses: The Thirring-Lense Papers, General Relativity and Gravitation* **16**, 711.

J. Lense and H. Thirring (1918) *Über den Einfluß der Eigenrotation der Zentralkörper auf die Bewegung der Planeten und Monde nach der Einsteinschen Gravitationstheorie* [Physikalische Zeitschrift (1918), **19** (3), p.156-1163] *[On the Influence of the Proper Rotation of Central Bodies on the Motions of Planets and Moons According to Einstein's Theory of Gravitation]* translated into English in B. Mashhoon, F. W. Hehl, and D. S. Theiss (1984),[45] *On the Gravitational Effects of Rotating Masses: The Thirring-Lense Papers, General Relativity and Gravitation* **16**, 711.

The general relativistic effect of the rotation of a central object (the rotating Sun generates a mass current) on planetary orbits was first calculated by de Sitter (1916)[34] when the orbital plane is the equatorial plane of a central body. He found that the angular momentum of the Sun modifies the spacetime geometry by generating a precession of the planetary orbits. For Mercury he found an effect $\sim 2.62 \cdot 10^{-4}$ times weaker than the static general relativistic value for the anomalous precession, which was beyond any reasonable hope of being measured then or even now.

The work of de Sitter preceded that of 1918 of the Austrian physicist Hans Thirring (1888–1976) and of Josef Lense (1890–1985) and Thirring (1918)[45] and was done well before the introduction of the Kerr metric (1963).[71] Such ideas were already presented in the works of the Friedländer brothers (1896)[72] and Föppl (1904).[73] Their approach followed Mach's principle concerning the origin of centrifugal and inertial forces: no absolute (Newtonian) space exists and the effects of every rotation of a body are the same as if it were at rest and the entire universe were rotating around it. Föppl (with a gyroscope) and the Friedländer brothers (with a torsion balance) intended to measure how the angular momentum of the Earth acts on a body, but their accuracy was 10^{10} times below the value needed to observe such effect for a 10% measurement.[61] It is remarkable that their attempt was made well before the introduction of general relativity.

Primarily Thirring, in connection with Mach's principle, considered distant rotating masses in his paper H. Thirring *Über die Wirkung rotierender ferner Massen in der Einstenschen Gravitationstheorie* [Physikalische Zeitschrift (1918), 19(3):33–39] *[On the Effect of Rotating Distant masses in Einstein's Theory of Gravitation]* reproduced as first classic paper. He described the influence of a rotating hollow sphere upon a body orbiting inside, finding the dragging of inertial frames by the motion of nearby matter.

Working on a second paper: J. Lense and H. Thirring *Über den Einfluß der Eigenrotation der Zentralkörper auf die Bewegung der Planeten und Monde nach der Einsteinschen Gravitationstheorie* [Physikalische Zeitschrift (1918), 19(3):156-163] *[On the Influence of the Proper Rotation of Central Bodies on the Motions of Planets and Moons According to Einstein's Theory of Gravitation]*, reproduced as a second classical paper, Thirring asked Lense to assist him in finding a perturbative solution to the corresponding differential equations for the gravitational field around a rotating sphere because of his experience in astronomy.

Thirring and Lense observed that the Newtonian equivalence of the field outside a sphere with uniform density with the field of a material point of the same mass concentrated at its center no longer applies for rotating spheres in general relativity. They calculated the precession of the orbital plane of a body orbiting around a rotating mass (the secular rate of change of the longitude of the nodes, the intersection between the orbital plane and the equatorial plane of the central object) within Einstein's theory of general relativity. They found the perturbations on the osculating orbital elements of planets and moons, and in the case of Mercury, linearizing the Einstein equations, they calculated a value of 0.01 *arcsec* per century that is $\sim 2.3 \cdot 10^{-4}$ times smaller than the relativistic precession already calculated by Einstein,[33] in agreement with the numbers of de Sitter. This phenomenon is new with respect to Einstein's paper on Mercury.[33] A table of predictions of such an effect for the Moon and other natural satellites in the solar system in a time scale of a century is given. Thirring and Lense stressed the importance of testing their effect as an additional test of Einstein's general relativity. In fact the effect should really be called the Thirring-Lense effect because of Hans Thirring's primary initiative.

K. Gödel (1949), *An Example of a New Type of Cosmological Solutions of Einstein's Field Equations of Gravitation, Rev. Mod. Phys.* **21**, 447–450.

In 1949 in Princeton Kurt Gödel presented Albert Einstein with a very special gift for his seventieth birthday: an exact cosmological solution of the equations of general relativity modified by the presence of a cosmological term. The location and the audience for the presentation were also peculiar. It took place in the first lecture hall on the left on the ground floor of Fuld Hall at the Institute for Advanced Study. At the blackboard was Kurt Gödel, in the first row Albert Einstein, and in the last row after all the empty seats, as Chandra later recalled to me, were the only other two people present: Subramanyan Chandrasekhar and Martin Schwarzschild. The solution had at each point "a rotation of matter relative to the compass of inertia". The matter was assumed to rotate at each point relative to the compass of inertia with angular velocity $\omega^2 = 4\pi G\rho$, where ρ is the mean density and G the gravitational constant. This solution is stationary in time and spatially homogeneous. An increasing direction of time can be consistently introduced everywhere in the spacetime and a temporal orientation is defined along each world-line of matter or light. Every world-line of matter occurring in the solution is an open line of infinite length, which never approaches any of its preceding points again. However, there do exist closed time-like lines in this exact solution. As Gödel states, "In particular, if P, Q are any two points on a world-line of matter, and P precedes Q on this line, there exists a time-like line connecting P and Q on which Q precedes P; i.e., it is theoretically possible in these worlds to travel into the past, or otherwise influence the past."

Einstein was quite disturbed by the existence of the closed timelike lines and argued that this solution could not be physically realized: "God will not allow them" he stated to Gödel.

Twenty years later I had occasion to discuss this matter again with Kurt Gödel. At that time I was a young member of the Institute for Advanced Study, and every day I noticed Gödel walking in his black coat with the collar turned up and his head covered by a large black hat with the brim pulled down, walking at a precise time from his office to the shiny blue limousine which was carrying some members from the Institute to the town at 12 pm. To speak to him was impossible. Even to approach him was difficult: as one would physically approach him, he would smoothly or slowly rotate himself away not leaving any opportunity to see his face and have a verbal interaction. This total detachment from his surroundings had lasted at least twenty years, the previous members of the Institute told me. I was very determined to find a way to interact with him, so I decided to take the limousine every day and sit next to him. This was also convenient in order to reach my office at the University and have lunch at the University cafeteria. This also became a habit for me which I followed every week-day and I almost forgot about the real motivation for taking the limousine in order to interact with him. I was usually using the travel time to check some of the computations in my notebook. I was doing this one day when I heard the almost metallic voice of Gödel address me about the computation I was doing. "This is General Relativity" he said. I realized I had finally established contact, and I was extremely careful not to break that fateful opportunity. "Yes" I answered without taking my eyes away from my notebook. "I have also been working in General Relativity" Gödel added with some excitement. Again not to break this contact, without looking at him, I simply answered "Really?" with an apparently disinterested attitude. After a few minutes of deep silence treating him with apparent nonchalance I added "If you have written any articles on it, please put them in my mailbox at the Institute. I will be happy to glance at them." And so the next morning I found in my mailbox one of the original reprints of his classic paper which we reproduce here in this volume. Months later I asked Gödel to autograph it for me, which he did, and I keep this as one of my dearest documents.

This started a series of encounters I had with Gödel, each one following a complex ceremonial procedure, first alone and then I used the occasion to bring along some of my young Princeton undergraduate students including Mark Johnston, Robert Leach and Robert Jantzen. We were all extremely fascinated by the knowledge of Gödel: he was clearly very familiar with relativity, including the Lense and Thirring papers. This familiarity with Thirring's work was later explained by Jantzen who learned that indeed Gödel had been a student of Hans Thirring in Vienna.[74] Gödel was moving in the realm of a different culture reaching very powerful physical conclusions on the content of Einstein equations starting from mathematical formalisms like quaternions, Clifford algebras, and topological considerations on the nature of space which were quite foreign to us. Even the Bianchi work on the classification of three-dimensional positive-definite Riemannian spaces was taken for granted by him. Curiously enough those were the days in which Yevgeny Lifshitz, I. M. Khalatnikov and Vladimir Belinsky in Moscow were rediscovering independently the Bianchi geometries in their analysis of the initial cosmological singularity.

We were very surprised that Gödel was perfectly up to date on all the most recent works on the Kerr-Newmann solution including the then very recent paper by Brandon Carter (1968).[75] We spent much time during our meetings discussing the analogies and differences between the closed timelike lines in his solution and the ones found either inside the horizon of a Kerr-Newmann solution or in presence of a naked Kerr-Newmann singularity endowed with values of charge and rotation larger than the critical ones. Again Gödel insisted on the fact that those features which so profoundly disturbed Einstein were only a consequence of the large amount of angular momentum he had assumed corresponding to the extreme values of a Kerr-Newmann solution. As in that case a smaller value of the specific angular momentum he had assumed should erase the causalty-violating features.

Similarly he was quite aware that the only reason for having introduced a cosmological term in his solution was solely in order to obtain a mathematically simple analytic solution possibly describing a special temporal instant of the evolution of our Universe. We were very interested to have indications from him if one could find a more general Friedmann-like solution which was still spatially homogeneous and endowed with expansion, rotation and shear and with a subcritical amount of angular momentum. We discussed this matter for a few weeks for Bianchi Type IX, VII and I solutions. Then in an usual Friday afternoon meeting, following the traditional complex ceremonial procedure, Gödel came very radiant: "Yes, it is possible to obtain that more general solution!" By this time we had become aware of the difficulty of having concrete and detailed indications from him: once we were struggling in reconstructing assertion 8 in his first classic paper. To our request for some hints he answered with some joy "You are right to find that proof difficult. It took to me more then one month to prove it. I then threw away the proof, but I assure you is correct!" Such an attitude applied as well to the present circumstance. To us trying to advance a belated request of how to approach finding such a solution, he almost disdainfully answered "I was very interested to prove that the solution could indeed exist, now to express it is just a technicality!" It took almost thirteen years for Bob Jantzen, by then *the* world expert on Bianchi universes, to make some concrete progress in this general direction with Kjell Rosquist[76] by finding an explicit family of exact solutions of the Einstein field equations for a perfect fluid with expansion, rotation and shear, although in a very special class of Bianchi type VI Universes.

One Friday afternoon I asked Gödel about the physical motivations he had in order to formulate his rotating cosmological model. The discussion took place as usual in his office at the end corner of the then new library at the Institute for Advanced Study, overlooking the pond. A thick white curtain was covering almost the entire window. He did not answer and instead he slid down in his chair until his eyes could catch a glance outside through a narrow gap left by the curtain in the window. I was somewhat concerned that he might have fallen asleep. After four or five minutes of silence I repeated the question: "What motivated you to introduce rotation in the description of the Universe" and finally he answered "I was just looking outside to find an object not rotating ..." (I realized that he was figuratively looking out to Mars, Jupiter, the moon, the galaxy, Andromeda) ... null rotation is a set of measure zero!" I recall this as one of the most powerful and profound answers I have received in my entire life on any subject. I was particularly attracted also because in those days I had just computed that the specific angular momentum of our galaxy was indeed close to the critical value of a Kerr metric.

In one of those Friday afternoon meetings Mark Johnston asked Gödel to sign a copy of a book on the "Gödel proof". He was initially reluctant. We were somewhat concerned to have offended him recalling the remark by Einstein that asking for an autograph was a last sign of cannibalism still present in our culture! But then he expressed himself "In this book there are two parts, my article and an introduction. The introduction is very long and very difficult and moreover is wrong. If you do not mind I will just autograph my part of the book" and so he did.

In 1974 I left Princeton but before leaving I attempted almost the impossible. Having made so much progress in having discussions with him and one or two of my students, I boldly asked him to give a lecture in my class at the University to some twenty students on any topic he liked. I was also extremely interested in his thoughts on Spinoza. Surprisingly he did not reject the possibility. He asked me to give him some time to think about it. A few weeks later in our meeting he replied "I would have liked to speak about General Relativity, but I have not been working in this field in the last twenty years" and when I reminded him that he could speak on any topic he added "I would have liked to speak on philosophy, but I have been working in this field only in the last twenty years!" This was our last Friday afternoon meeting, following as usual a complex ceremonial procedure."

A. Papapetrou (1951), *Spinning Test-Particles in General Relativity. I*, *Proc. Roy. Soc. London* **A 209**, 248–58.

E. Corinaldesi, and A. Papapetrou (1951), *Spinning Test-Particles in general relativity. II*, *Proc. Roy. Soc. London* **A 209**, 259–68.

In 1951 Papapetrou introduced a model for treating spinning test particles in General Relativity which still today is considered fundamental. He developed an approximation procedure according to which a "small" extended body is represented by a single line (instead of a small tube in the spacetime) plus a multipole structure, namely a number of tensor fields defined on it and characterizing the body itself. In this context a test particle becomes a "single-pole", defined only by a timelike line and a scalar field (the mass m) defined on it. A spinning test particle instead is a "pole-dipole" particle characterized by a timelike line plus a scalar field (the mass m) and an antisymmetric tensor field (the spin $S_{\alpha\beta}$) defined along it. Of course the Papapetrou procedure equally applies to particles with higher multipole moments like quadrupole moments and it is at the same time simple and natural. The celebrated Papapetrou equations for a spinning test particle are the following

$$\frac{DP^\alpha}{ds} = \frac{1}{2} R^\alpha{}_{\beta\mu\nu} u^\beta S^{\mu\nu} , \qquad \frac{DS^{\alpha\beta}}{ds} = P^\alpha u^\beta - P^\beta u^\alpha ,$$

where u^α is the unit timelike tangent vector to the line of the particle parametrized by the proper time s, P^α is its linear momentum, i.e. a vector field defined along the world line of the particle but in general not coinciding with its tangent and containing (in its modulus) information about the mass of the particle, and $S^{\alpha\beta}$ its spin, i.e. a tensor field also defined along the world line of the particle. This system, however, has 13 unknowns (3 for u^α which is a unit vector, 4 for P^α and 6 for $S^{\alpha\beta}$ which is antisymmetric) but only 10 independent equations. This is the main problem with this model: one must choose three supplementary conditions in order for the model to be complete and this choice is not straightforward.

Papapetrou's second 1951 paper, in collaboration with E. Corinaldesi, apart from applications to the case of a Schwarzschild spacetime, discussed a set of supplementary conditions which bear his name: $S^{0\alpha} = 0$. This set appears to be noncovariant and coordinate dependent but still is a possible choice widely discussed by Papapetrou and Corinaldesi and many other authors in the following years. There are also other choices used now, two others being $S^{\alpha\beta}u_\beta = 0$ due to Pirani[77] and $S^{\alpha\beta}P_\beta = 0$ due to Tulczjyew.[78]

Finally it is worthwhile mentioning that alternative derivations and analysis of the Papapetrou equations of motion have also been given by Dixon, Souriau, Ehlers and others in the 1970s.[79]

G. E. Pugh (1959), *Proposal for a Satellite Test of the Coriolis Prediction of General Relativity*, WSEG Research Memorandum no. 11.

L. I. Schiff (1960), *Motion of a Gyroscope According to Einstein's Theory of Gravitation*, *Proc. Nat. Acad. Sci. Am.* **46**, 871–882.

C. W. F. Everitt (1974), *The Gyroscope Experiment I: General Description and Analysis of Gyroscope Performance.*, Proc. of the International School of Physics "Enrico Fermi" Course LVI, Edited by B. Bertotti, Academic Press, NY London, p. 331-360.

These three papers are the milestones of the GP-B mission.

The Everitt paper of 1974 gave an extensive account of the possible sources of error for the measurement of two relativistic effects: the frame-dragging and the geodetic precessions. An accompanying paper by Lipa, Fairbank and Everitt (1974)[81] gave an account of the technical status of the experiment at that time including the techniques for obtaining ultra low magnetic fields in superconductors invented by W.O. Hamilton and B. Cabrera.

The GP-B experiment consists of a reference telescope pointed at an appropriate selected guide star, four gyroscopes with the spin axes aligned parallel to the line of sight to the star, all mounted in a quartz block structure attached by silicate bonding to the telescope. The whole instrument is maintained at a temperature 1.8 K within a Dewar vessel containing 2300 liters of superfluid helium. The gyroscope is a 38 mm diameter fused quartz sphere coated with superconducting niobium spinning at $\sim$ 150 Hz. The changes in the spin axis of the gyroscopes are detected with a London-moment readout with great sensitivity and null stability, which come from the special properties of the Superconducting QUantum Interference Devices (SQUID). An angle of a milliarcsecond is detected in 7 hours of observation time.

Fundamentally the sources of error can be divided into three groups:

1. residual Newtonian drifts of the gyroscopes,
2. readout errors,
3. uncertainty in the proper motion of the reference star.

Originally the guide star selected was Rigel which is a 0.7 magnitude star in Orion. Now the improvements on the telescope allow the use of the 5.9 magnitude IM Peg, HR 8703 which is also active in the radio and close to a radio quasar.

In this way the VLBI uncertainty in its proper motion is 0.09 milliarcseconds per year in declination and right ascension.[82] The orbital ($\pm$5 arcseconds and period 90 minutes) and annual ($\pm$20.116 arcseconds) aberrations produce large, very accurately known, periodic displacements of the apparent position of the guide star which, as the experiment is mechanized, appear in the readout of each gyroscope, supplying a continuous absolute calibration of the gyro scale factor accurate to a few parts in 10^7. The aberration signals also provide an absolute phase reference for the rotation of the spacecraft essential for separating the Lense-Thirring effect from the geodetic effect. The telescope, moreover, is made from fused quartz and maintained at liquid-helium temperatures to eliminate thermal distortions. A beam-splitter system produce two images of the reference 5.9 magnitude star, one for each readout axis. The location of the center of the star image is limited only by the sharpness of the knife edge of the beam-splitter prism and by photon-counting statistics (including the effects of photon noise). Regarding the rotating sphere it should be homogeneous to better than one part in 10^6 and spherical to 3 parts in 10^7. An extremely detailed analysis of all known Newtonian torques on the gyroscope has been carried out. Under expected orbit conditions, the residual non-relativistic drift of each gyroscope has a worst-case sum below 0.08 milliarc seconds per year. The gyros are spun up by gas jets running at sonic velocity around the periphery of the ball; the optimum spin speed is $\sim$ 150 Hz at which speed the deviations from sphericity due to centrifugal distortion become comparable with the polishing errors. The characteristic spin down time at the expected operating pressure (10^{-11} torr) is 30,000 years.

Disturbances on the gyroscopes are reduced by making the satellite 'drag-free' using measurements from the gyro suspension system to provide control signals for thrusters using the boil-off gas from the dewar. The concept of the drag-free satellite was first suggested by Pugh in his classic paper. Through it, disturbances from air-drag, solar radiation pressure, solar wind and micrometeorite impact are greatly reduced. An important factor in the proper design of Gravity Probe B is the in-line configuration of the gyroscopes which minimizes the effect of tidal disturbances on them due to the finite distance between a gyroscope and the drag-free point in the spacecraft.

D. C. Wilkins (1972), *Bound Geodesics in the Kerr Metric*, *Phys. Rev.* **D 5**, 814–822.

An "effective potential" technique had been used very successfully by Carl Størmer in the 1930s in studying the trajectories of cosmic rays in the Earth's magnetic field.[83] In the fall of 1967 Brandon Carter visited Princeton and presented his remarkable mathematical work leading to the separability of the Hamilton-Jacobi equations for the trajectories of charged particles in the field of a Kerr-Newmann geometry.[75] This visit had a profound impact on our small group working with John Wheeler on the physics of gravitational collapse. Indeed it was Johnny who had the idea of exploiting the analogy between the trajectories of cosmic rays and spacetime trajectories in general relativity, using the Størmer "effective potential" technique in order to obtain physical consequences from the set of first order differential equations obtained by Carter. I still remember the excitement of preparing the $2m \times 2m$ grid plot of the effective potential for particles around a Kerr black hole which finally appeared later in print [Rees, Ruffini and Wheeler 1973,1974;[84] see Fig. (1)].

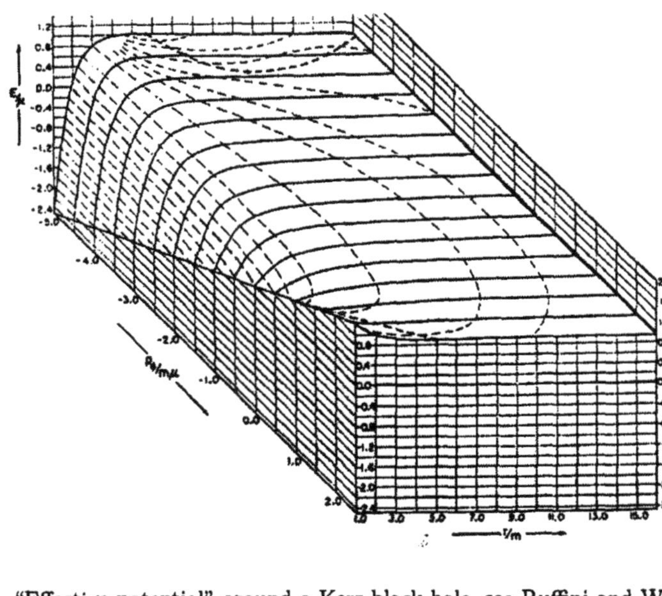

Figure 1. "Effective potential" around a Kerr black hole, see Ruffini and Wheeler 1971

Out of this work came the celebrated result for the maximum binding energy $1 - \frac{1}{\sqrt{3}} \sim 42\%$ for corotating orbits and $1 - \frac{5}{3\sqrt{3}} \sim 3.78\%$ for counter-rotating orbits in the Kerr geometry. We were very pleased to be later associated with Brandon Carter in a "gold medal" award for this work presented by Yevgeny Lifshitz: in the fourth and last edition of volume 2 of the Landau and Lifshitz series (*The Classical Theory of Fields* 1975[41]), both Brandon's work and my own work with Wheeler were proposed as named exercises for bright students! This was certainly a simple and fruitful analogy which led to a successful theoretical accomplishment, opening up a new window on very unexpected general relativistic effects. I find it interesting that in recent years it has become clear to me that the difference in the binding energies of the corotating and counterrotating orbits in a Kerr-Newman geometry have become the object of direct astrophysical observations in binary X-ray sources.

In 1970 a special collaboration was established between Francis Everitt, Bill Fairbank, Leonard Schiff and myself, then an assistant professor at Princeton. We discussed my moving to Stanford in order to create a theoretical activity on relativistic astrophysics. We then started an exchange program, while waiting for further developments. One of Francis Everitt's Ph.D. students, Daniel Wilkins, started to collaborate with me in Princeton, and it was during one of Daniel's visits to Princeton we learned about the untimely death of Leonard Schiff. Among the many topics discussed with Francis and Daniel, we decided to generalize the effective potential technique developed by Wheeler and myself for the equatorial plane of a Kerr black hole to nonequatorial orbits. This work was beautifully and carefully developed by Daniel and I decided to make him the only author of the paper as a tribute to his profound dedication. We returned later to numerical work on this general problem with Mark Johnston, see the following paper.

M. Johnston and R. Ruffini (1974), *Generalized Wilkins Effect and Selected Orbits in a Kerr-Newmann Geometry, Phys. Rev.* **D10**, 2324–2329.

After the work of Wilkins, I started an intense activity in Princeton with a group of undergraduate students including Robert Jantzen, Robert Leach, Mark Johnston, and Richard Hanni. All of them gave important contributions in the following years, Bob Jantzen developing into one of the world experts in fundamental general relativity, Bob Leach in contributing to the first identification of Cygnus X1 as a black hole,[86] Rick Hanni in studying the electrostatic behaviour of a black hole.[87] With Mark in an exhausting amount of work, at times staying up a day and night continously at the Princeton computer (at that time on the ground floor corner of the newly built Joseph Henry Laboratories), we explored the noncircular orbits around a Kerr black hole, both by the effective potential technique and direct numerical computation.

We treated the computer as an experimental machine by which we were simulating "scattering" and "trapping" of test particles from black holes in order to visualize their space-time geometry. This very intense work was manifested in the Figs. 5–11 in that paper. Particularly attractive was Fig. 10 also from an aesthetical point of view. I was so fascinated by the geometrical effects due to the periastron precession combined with the Moebius strip like behaviour, a combination uniquely weighted by the solving the Einstein equations, that a few years later I made this figure the symbol of our International Center for Relativistic Astrophysics (ICRA). Also later in collaboration with my future wife, Anna Imponente, we proposed the realization of this image in a silver sculpture which was created by the Italian sculptor Attilio Pierelli. This work of art and science then became the prize of the "Marcel Grossmann Awards". It is represented on the cover of this volume. The text written by Anna Imponente on this collaboration between art and science is an integral part of those awards and is reproduced in *http : //www.icra.it/MG/awards/award_grossmann.htm TEST: Traction of Events in Space-Time.*

T. Damour and R. Ruffini (1974), *Sur Certaines Vérifications Nouvelles de la Relativité Générale Rendues Possibles par la Découverte d'un Pulsar Membre d'un Système Binaire*, Académie de Sciences (Paris), *Comptes Rendus, Serie A – Sciences Mathematiques* **279**, 971–973 [in French].

This paper was written just after the discovery by Hulse and Taylor of their famous binary pulsar system. In our paper we pointed out that in principle this system could become an astrophysical analogue of the Stanford Gravity Probe-B experiment. GP-B would not represent a duplication of an astrophysical system but a way to obtain an independent measurement of the Lense-Thirring effect around the Earth under controlled conditions. This verification been acquired the Lense Thirring effect could then be used in order to gain additional information in the astrophysical regime. The paper was written a few weeks before the return to Paris of Thibault Damour to discuss his doctoral thesis developed under my supervision at Princeton during the preceding two years. We wanted to publish this result as quickly as possible, and we intended to stress the role of the research both of us had planned to start in the following years in Europe, so we decided to ask André Lichnerowicz to present our paper to the French Academy and it was then published in French in the *Comptes Rendues*. It was very gratifying to see at the recent Marcel Grossmann Meeting held in Rome (2000) that our predictions were indeed verified by the splendid observations of Michael Kramer (see the last contemporary paper) and the order of magnitude of the effect we had estimated was confirmed.

R. A. Van Patten and C. W. F. Everitt (1976), *Possible Experiment with Two Counter-Orbiting Drag-Free Satellites to Obtain a New Test of Einstein's General Theory of Relativity and Improved Measurements in Geodesy, Phys. Rev. Lett.* **36**, 629–32.

The proposal of Pugh and Schiff at the beginning of the 1960s, led to the forty-year project of Gravity Probe B, the first alternative project to measure relativistic effects with satellites around the Earth was presented by Van Patten and Everitt[59] in 1976. This paper opened the field of measurement of gravitomagnetic effects with twin satellites around the Earth. Van Patten and Everitt began by considering the accuracy at which the orbital parameters of TRIAD transit navigation satellites could be known after some years in orbit, taking into account that this was the first drag free satellite. This accurate knowledge of orbital parameters allows one to improve the knowledge on the classic perturbations of the orbit due to the quadrupole moment of the Earth gravitational field and to its higher order moments. In fact by using two satellites, with Doppler ranging beteween them, counter-orbiting along identical circular polar orbits (with a small displacement only to avoid collision) eliminates the uncertainty in the contribution of the quadrupole moment (proportional to the cosine of the inclination above the equatorial plane). The solar and lunar gravity-gradient perturbations yield secular and periodic perturbations equal in magnitude for the two satellites, but opposite in direction. Finally, other perturbations are at frequencies which are integral multiples of Earth's rotational frequency. They proposed two new satellites because the error in tracking uncertainties in the time history of the inclination angle of TRIAD should cause an error about six times larger than the relativistic nodal motion to be detected. The analysis of the errors is crucial in the extraction of such tiny relativistic effects in the orbital motions of the satellites: one satellite uses external data to fix the even zonal harmonics of the Earth's gravitational potential, but there are errors in their determination. The two satellite philosophy would overcome this problem. The most important point referred to this method is the elimination of the error due to uncertainty in the position of the Pole by Doppler-ranging between the two spacecraft. This error is not eliminated in the version of the twin satellite experiment by Ciufolini.[60]

8. Contemporary papers

I. Ciufolini (1986), *Measurement of the Lense-Thirring Drag on High-Altitude, Laser-Ranged Artificial Satellites, Phys. Rev. Lett.* **56**, 278–81.

This paper describes a new method of measuring the Lense-Thirring effect using *one passive, non-polar, laser-ranged satellite* to be coupled with the already orbiting satellite LAGEOS. The idea of this experiment (today called LARES) can be divided into two parts:

(a) Position measurements of laser-ranged satellites, of LAGEOS (1976) type, are accurate enough to detect the very tiny effect due to the gravitomagnetic field: the Lense-Thirring precession.

(b) To "cancel out" the enormous perturbations due to the non-sphericity of the Earth gravity field, it is suggested to launch a new *non-polar* satellite: LAGEOS III (LARES), with inclination supplementary to that of LAGEOS, and with the other orbital parameters, a and e, equal to those of LAGEOS, i.e. the inclination of LARES should be 70 degrees, whereas that of LAGEOS is 110 degrees. LAGEOS is a high-altitude, small cross-sectional-area to mass ratio, spherical, laser-ranged satellite whose position can today be measured with a precision of less than 1 cm over about 5900 km of altitude; the orbit of the satellite can be fit with an accuracy of about 1 cm (RMS of orbital residuals). The use of the new laser ranged satellite may allow a detection of the Lense-Thirring effect with an uncertainty, using the refined perturbation models to-day available, of less than 1 milliarcsec, i.e. with an uncertainty of about 2% of the gravitomagnetic effect. However Ries, et. al., in this volume, have quoted for the error due to uncertainty in co-inclination numbers of $\sim 4\%$;

Shapiro, in many discussions puts a factor from experience that in this general style of modeling, one needs to take the error given by the computer error model and put in a real-life correction factor of 2 to 3. This would lead a possible experiment reaching an accuracy of 10 – 15%.

B. Bertotti, I. Ciufolini and P.L. Bender (1987), *New Test of General Relativity: Measurement of de Sitter Geodetic Precession Rate for Lunar Perigee*, *Phys. Rev. Lett.* **58**, 1062–5.

According to general relativity, the calculated rate of motion of the lunar perigee should include a contribution of 19.2 milliarcsec/yr due to the geodetic (de Sitter) precession. In this paper the authors analyze the results of Lunar Laser Ranging (LLR) with respect to the planetary dynamical frame and compare the Earth-rotation result from LLR and from Very Long Baseline Interferometry (VLBI) by then confirming the geodetic precession rate for the lunar perigee, as predicted by general relativity, with an error of about 10%.

I. I. Shapiro, R. D. Reasenberg, J. F. Chandler and R. W. Babcock (1988), *Measurement of the de Sitter Precession of the Moon: A Relativistic Three-Body Effect*, *Phys. Rev. Lett.* **61**, 2643–6.

They improved the accuracy of the measurement of the geodesic precession of the orbital motion of the Moon up to 2%. In fact the measurement of the de Sitter precession for the lunar orbit has been performed by Bertotti et al.[63] (1987) and improved by Shapiro et al.[64] (1988) as a test of Einstein's general relativity. They overcome the uncertainty in the determination of the lunar motion using the four lasers left on the Moon during the Apollo missions for determining the libration and the motion of its center of mass.

M. Kramer (1998), *Determination of the Geometry of the PSR B1913+16 System by Geodetic Precession*, *Astrophys. J.* **509**, 856–860.

Damour and Ruffini (1974),[69] reproduced in the classic papers section, examined the binary pulsar PSR B1913+16 discovered by Hulse and Taylor[42] and predicted the possibility of observing the Lense-Thirring precession of the pulsar's orbit around the central object, as well as the spin-orbit coupling, manifested by a change in the emission angle of the pulsar signal. This binary system has become a classic in relativistic astrophysics having led to the discovery of gravitational radiation emission Hulse and Taylor (1975).[42] Kramer[70] (1998) has given statistical evidence for the geodetic precession of the pulsar with a secular change in the emission angle over a range of 25 years, in excellent agreement with the theoretical prediction by Damour and Ruffini. The possible disappearance of the pulsar signal in the year 2025 should give an additional information on the beaming angle of the pulsar emission.

Acknowledgments

It is a pleasure to thank Ignazio Ciufolini, Thibault Damour, Francis Everitt, Robert Jantzen and Costantino Sigismondi for their many suggestions and discussions.

Rome, October 2002
Remo Ruffini

References

1. See e.g. *The Electrical Researches of the Honourable Henry Cavendish F.R.S.* edited by J. Clerk Maxwell F.R.S., Cambridge: at the University Press, 1879, pp. 104–113. The Cavendish experiment was performed sometime between 1771 and 1773 and written up for publication in 1873. On p. xlv of his book, Maxwell speculates but cannot really explain why Cavendish never bothered to publish it. Knowledge of Cavendish's unpublished work floated around in a general way in the scientific community for many years, but his proof of the inverse square law was not explicitly stated until George Wilson's *Life of Cavendish* (London 1851). Cavendish proved the inverse square law to $r^{-2\pm0.04}$. Maxwell and his student, Donald MacAlister, extended the proof by the same method to 1 part in a million.
2. The work of Coulomb was read at a meeting of the French Académie on 9 September 1784, and published in the 1785 volume of the *Mémoires de l'Académie Royale des Sciences*; see also *Collection des mémoires relatifs à la physique , publiés par la Société française de Physique*, 5 vols (Paris, 1884–1891), vol. 1 (ed A. Potier) *Mémoires de Coulomb.*
3. Williams, E. R., J. E. Faller and H. A. Hill *New Experimental Test of Coulomb's Law: A Laboratory Upper Limit on the Photon Rest Mass*, Physical Review Letters **26**, 721–724 (1971). See also Jackson (1999),[12] p. 7–8.
4. Gilbert, William, *De Magnete, Magneticisque Coporibvs, et de Magno Magnete Tellure; Physiologia Nova, Plurimis & Argumentis, & Experimentis Demonstrata.* Londini, excudebat P. Short (1600). See also *De magnete: On the loadstone and magnetic bodies, and on the great magnet of the earth,* translated by P. Fleury Mottelay, 1893; reprinted New York, Dover 1958, or *De Magnete: On the magnet, magnetick bodies also, and on the great magnet the earth; a new physiology, demonstrated by many arguments & experiments,* translated S.P. Thompson and others, London: Chiswick Press (1900).
5. Oersted, Hans Christian, *Experimenta Circa Effectum Conflictus Electrici in Acum Magneticam. Neuere Electromagnetische Versuche*, Nurnberg, Schrag (1820).
6. Ampère, André Marie, *Recueil d'Observations Electrodynamiques*, Mem. de l'Acad. 6, 175 (1825).
7. The paper by Faraday was read to the Royal Society on November 24, 1831 and appears in *Phil. Trans.* for that year. It was reprinted in *Experimental Researches in Electricity; Reprinted from the Philosophical Transactions of 1831-1838*, London, R. and J.E. Taylor, 1839, which is the first of the three volumes of Faraday's collected *Experimental Researches in Electricity.* See also Faraday, Michael *Experimental Researches in Electricity*, London: Printed by Richard Taylor (1832).
8. see Part III of James Clerk Maxwell paper *"On Physical Lines of Force"* 1861–1862, published in *Phil. Mag.* and reprinted in *The Scientific Papers of James Clerk Maxwell* (Cambridge 1890, reprinted Dover Publications 1965, vol. 1, p. 500). The full development of his equations came in his 1865 paper *"A Dynamical Theory of the Electromagnetic Field"*, plus his important 1868 *"Note*

on the Electromagnetic Theory of Light" and in the *Treatise.* see also the excellent book by Francis Everitt : *James Clerk Maxwell Physicist and Natural Philosopher*, New York: Scribner (1975), chapters 7–8.
9. Hertz's first demonstration of electromagnetic waves with an approximate computation of their velocity as $\sim 3 \cdot 10^{10}$ centimeters per second was published in Annalen der Physik **34**, 551 (1888). *Electric Waves* translated by D.E. Jones was published in German in 1892 and in English in 1893. In Whittaker's *History of the Theories of Aether and Electricity* (2nd edition, vol. 1, p. 324) there is an interesting discussion of how Hertz got the right answer for the wrong reason by means of two cancelling mistakes, one theoretical and the other experimental, which were corrected by Poincaré and by Ernst Lecher, respectively, in 1890.
10. Einstein, Albert *Zur Electrodynamik bewegter Körper, On the Electrodynamics of Moving Bodies*, Ann. d. Physik 17, 891 (1905).
11. Schweber, S. S., *QED and the Men Who Made It : Dyson, Feynman, Schwinger, and Tomonaga* Princeton: Princeton University Press (1994).
12. Jackson, John D., *Classical Electrodynamics* (3rd Ed.), New York: Wiley (1999).
13. Lamb, W. E. and R. C. Retherford, Phys. Rev. **72**, 241 (1947).
14. Pal'chikov, V. G., Yu. L. Sokolov, and V. P. Yakovlev, *Interference of Atomic States in an External Field and the Lamb Shift in Hydrogen-like Helium*, Optics and Spectroscopy **87**, 708–712 (1999).
15. Van Dyck, R. S., Jr., P. B. Schwinberg and H. G. Dehmelt, *New High-Precision Comparison of Electron and Positron g Factors*, Phys. Rev. Lett. **59**, 26 (1987).
16. Hughes, V. M., and T. Kinoshita, *Anomalous g values of the electron and muon*, Reviews of Modern Physics **71**, 133 (1999).
17. The LEP Collaboration ALEPH, DELPHI, L3, OPAL, and LEP Electroweak Group, and the SLD Heavy Flavour and Electroweak Groups *A combination of Preliminary Electroweak Measurements and Constraints on the Standard Model* hep-ex/0112021 (2001).
18. Newton, Isaac, *Philosophiae Naturalis Principia Mathematica*, Streater, London, (1687).
19. Flammarion, Camille, *Astronomie Populaire*, C. Marpon et E. Flammarion éditeurs, Paris, (1880).
20. Baum, R. and W. Sheehan, *In Search of Planet Vulcan: The Ghost in Newton's Clockwork Universe* Plenum Trade, New York, (1997).
21. Newcomb, Simon, *Note on accounting for the secular variations of the orbits of Venus and Mercury*, Astronomical Journal, vol. 14, iss. 327, p. 117-118, (1894).
22. Euler, Leonhard *Theoria motuum lunae : nova methodo pertractata* Leo Courvoisier editor O. Fussli, Turici, (1958).
23. Laplace, Pierre Simon de *Celestial mechanics by the marquis de La Place*; translated, with a commentary, by Nathaniel Bowditch Hilliard, Gray, Little and Wilkins, Boston (1829).
24. Jacobi, Carl Gustav Jacob *Vorlesungen uber Dynamik* edited by A. Clebsch, Reimer, Berlin, (1884).

25. Hamilton, William, *Geometry, analysis, astronomy, probability and finite differences, miscellaneous* edited for the Royal Irish academy by B. K. P. Scaife, Cambridge university press, Cambridge, (2000).
26. O'Connor, J. J. and E. F. Robertson, http://www-gap.dcs.st-and.ac.uk/ history/HistTopics/Orbits.html School of Mathematics and Statistics, St. Andrews University, Scotland (1996).
27. Tisserand, F., Traité de Mecanique Céleste, 4 vol., Gauthier-Villars, Paris (1889-96), (1960^2). This work, an update of Pierre-Simon Laplace's work of 1799 on the same subject, it is still used as a sourcebook by authors writing on celestial mechanics.
28. Audretsch, J.; Dehnen, H.; Hnl, H. , *Crucial Test for the Existence of a Solar Quadrupole Moment*, Astrophysical Journal, vol. 150, p.L127, (1967).
29. Dicke, R. H., *The Solar Oblateness and the Gravitational Quadrupole Moment* Astrophysical Journal, vol. 159, p.1, (1970).
30. Eddington, A. S. *The total eclipse of 1919 May 29 and the influence of gravitation on light* , The Observatory, Vol. 42, p. 119-122, (1919)
31. Walsh, D., Carlswell, R. F. ans Weyman, R. J., *0957+561 A, B: Twin Quasistellar Objects or Gravitational Lens?*, Nature, 279, 381, (1979).
32. Pound, R. V., and G. A. Rebka, Jr., *Apparent Weight of Photons*, Phys. Rev. Lett., 4, 337, (1960).
33. Einstein, Albert, *Erklärung der Perihelbewegung des Merkur aus der allgemeinen Relativitätstheorie*, Preuss. Akad. Wiss. Berlin Sitzber 47, 831-39, (1915).
34. De Sitter, W., *On Einstein's Theory of Gravitation and its Astronomical Consequences*, Mon. Not. Roy. Astron. Soc. 76, 699-728, (1916).
35. Is it worth recalling that Maxwell in 1865 explicitly stated that the formulation of a *field* theory of gravitation was profoundly problematical. See paragraph 82, *Note on the Attraction of Gravitation* in his 1865 paper *"A Dynamical Theory of the Electromagnetic Field" Scientific Papers* vol. 1, pp. 570–571. His problem was that generating an attractive force between gravitating masses in Newtonian space would require infinite (or at least enormously great) energy density disseminated throughout the Universe.
36. Ruffini, R. and H. Ohanian, *Gravitazione e Spaziotempo*, Bologna: Zanichelli (1998).
37. Riemann, B. , *Über die Hypothesen welche der Geometrie zu Grunde liegen*, in *Gesammelte Mathematische Werke*, H. Weber, ed., Leipzig (1876).
38. Wheeler, J. A., *Beyond the End of Time*, in *Black Holes, Gravitational Waves and Cosmology*, M. Rees, R. Ruffini and J. A. Wheeler, New York: Gordon and Breach (1975), pp. 286–307.
39. Anderson, J. D., Laing, P. A., Lau, E. L., Liu, A. S., Nieto, M. M., Turyshev, S. G., *Study of the Anomalous Acceleration of Pioneer 10 and 11*, e-print gr-qc/0104064 (2001).
40. Sancisi R., and T.S. Van Albada, in J. Kormendy e G.R. Knapp, *Dark Matter in the Universe* Dordrecht: Reidel (1987).
41. Landau, L. D., E. M. Lifshitz, *The Classical Theory of Fields, Course of Theoretical Physics Volume 2*, Fourth Rev. English Edition, New York:

Butterworth-Heinemann (1997).
42. Hulse, R. A. and J. H. Taylor, *Discovery of a pulsar in a binary system* Astrophysical Journal **196**, L1 (1975).
43. Hewish, A., Bell, S. J., Pilkington, J. D. H., Scott, P. F. and R. A. Collins, Nature, **217**, 709, (1968).
44. Damour, T. and J. H. Taylor, *Strong-field Tests of Relativistic Gravity and Binary Pulsars*, Phys. Rev. **D 45**, 1840–1868 (1992).
45. Mashhoon, B., F. W. Hehl and D. S. Theiss, *On the Gravitational Effects of Rotating Masses: The Thirring-Lense Papers*, Gen. Relativ. Grav. **16**, 711 (1984).
46. Forward, Robert L., *Proc. IRE* **49**, 892 (1961).
47. Møller, C., *The Theory of Relativity*, First Edition Oxford: Oxford University Press (1952);
The Theory of Relativity, Second Edition Oxford: Oxford University Press (1972).
48. Gödel, Kurt, *An Example of a New Type of Cosmological Solutions of Einstein's Field Equations of Gravitation*, Reviews of Modern Physics **21**, 447–450 (1949).
49. Ruffini R., and H. Gursky editors, *Neutron Stars, Black Holes and Binaries Sources*, Dordrecht: D. Reidel (1975).
50. Papapetrou, A., *Spinning Test-Particles in general relativity. I, Proc. Roy. Soc. London* **A 209**, 248–258 (1951).
51. Corinaldesi, E., and A. Papapetrou, *Spinning Test-Particles in general relativity II*, Proc. Roy. Soc. London **A 209**, 259–268 (1951).
52. Wilkins, D. C., *Bound Geodesics in the Kerr Metric*, Phys. Rev. **D 5**, 814–822 (1972).
53. Johnston M., and R. Ruffini, *Generalized Wilkins Effect and Selected Orbits in a Kerr-Newmann Geometry*, Phys. Rev. **D 10**, 2324–2329 (1974).
54. Pugh, G. E., *Proposal for a Satellite Test of the Coriolis Prediction of General Relativity*, WSEG Research Memorandum no. 11 (1959).
55. Schiff, L.I., *Motion of a Gyroscope According to Einstein's Theory of Gravitation*, Proc. Nat. Acad. Sci. Am. **46**, 871 (1960).
56. Everitt, C. W. F., *The Gyroscope Experiment I. General Description and Analysis of Gyroscope Performance*, in Proc. of the International School of Physics "Enrico Fermi" Course LVI on Experimental Gravitation edited by B. Bertotti, pp. 331–360, New York/London: Academic Press (1974).
57. http://einstein.stanford.edu/STEP/
58. http://lisa.jpl.nasa.gov/
59. Van Patten, R. A. and C. W. F. Everitt, *Possible Experiment with Two Counter-Orbiting Drag-Free Satellites to Obtain a New Test of Einstein's General Theory of Relativity and Improved Measurements in Geodesy*, Physical Review Letters **36**, 629–632 (1976).
60. Ciufolini, Ignazio, *Measurement of the Lense-Thirring Drag on High-Altitude, Laser-Ranged Artificial Satellites*, Physical Review Letters **56**, 278–281 (1986).
61. Ciufolini, I. and J. A. Wheeler, *Gravitation and Inertia*, Princeton: Princeton University Press (1995).

62. Ciufolini I., E. Pavlis, et al., *Test of General Relativity and Measurement of the Lense-Thirring Effect with Two Earth Satellites*, Science **279**, 2100 (1998).
63. Bertotti, B., I. Ciufolini and P.L. Bender, *New Test of General Relativity: Measurement of de Sitter Geodetic Precession Rate for Lunar Perigee*, Physical Review Letters **58**, 1062–1065 (1987).
64. Shapiro, I. I.; Reasenberg, R. D.; Chandler, J. F.; Babcock, R. W., *Measurement of the de Sitter Precession of the Moon: A Relativistic Three-Body Effect*, Physical Review Letters **61**, 2643–2646, (1988).
65. Iorio, L. *Earth Tides and Lense-Thirring Effect*, Celestial Mechanics and Dynamical Astronomy **79** (3), 201–230 (2001).
66. Iorio, L. *Satellite Gravitational Orbital Perturbations and the Gravitomagnetic Clock Effect*, Int. J. Mod. Phys. **D 10**, 465–476 (2001).
67. Iorio, L. *Effects of Earth Tesseral and Sectorial Tides on the Determination of the Lense-Thirring Effect*, Class. Quantum Grav. **18**, 4303–4310 (2001).
68. Ciufolini, I. *The 1995–99 Measurements of the Lense-Thirring Effect Using Laser-Ranged Satellites*, Class. Quantum Grav. **17**, 2369–2380 (2000).
69. Damour, T. and R. Ruffini, *Sur Certaines Vérifications Nouvelles de la Relativité Générale Rendues Possibles par la Découverte d'un Pulsar Membre d'un Systéme Binaire*, Académie de Sciences (Paris), Comptes Rendus, Serie A: Sciences Mathematiques **279**, 971–973 (1974).
70. Kramer, M., *Determination of the Geometry of the PSR B1913+16 System by Geodetic Precession*, Astrophys. J. **509**, 856 (1998).
71. Kerr, R. P., *Gravitational Field of a Spinning Mass as an Example of Algebraically Special Metrics*, Phys. Rev. Lett. **11**, 237–38 (1963).
72. Friedländer, B. and I., *Absolut und Relativ Bewegung?* Berlin: Simion-Verlag (1896).
73. Föppl, A., *Über Absolut und Relativ Bewegung*, Sitzber Bayer. Akad. Wiss. **34**, 383–95 (1904).
74. *The Princeton Mathematics Community in the 1930s: An Oral History Project* (On-Line Version by R. T. Jantzen): `http://www.princeton.edu/mudd/math`, Princeton University Library (2000).
75. Carter, Brandon *Global Structure of the Kerr Family of Gravitational Fields*, Phys. Rev. **174**, 1559–1571 (1968).
76. Rosquist, K. and R. T. Jantzen, *Exact Power Law Solutions of the Einstein Equations*, Phys. Lett. **107A**, 29–32 (1985)
77. Pirani, F.A.E., *On the Physical Significance of the Riemann Tensor*, Acta Physica Polonica **15** 389–405 (1956).
78. Tulczjyew, W., *Motion of multipole particles in general relativity theory*, Acta Physica Polonica **18**, 393 (1959).
79. Dixon, W. G., et al., *Dynamics of extended bodies in general relativity I. Momentum and angular momentum*, Proc. Roy. Soc. Lond. A. **314**, 499–527 (1970).
80. Everitt, C. W. F., *The Gyroscope Experiment I: General Description and Analysis of Gyroscope Performance.*, Proc. of the International School of Physics "Enrico Fermi" Course LVI, Edited by B. Bertotti, New York/London: Academic Press, pp. 331–360 (1974).

81. Lipa, J. A., W. M. Fairbank and C. W. F. Everitt, *The Gyroscope Experiment II. Development of the London-Moment Gyroscope and of Cryogenic Technology for Space.*, in Proc. of the International School of Physics "Enrico Fermi" Course LVI on Experimental Gravitation edited by B. Bertotti, p. 361–380, New York/London: Academic Press (1974).
82. Ransom, R. R.; Bartel, N.; Bietenholz, M. F.; Lebach, D. E.; Ratner, M. I.; Shapiro, I. I.; Lestrade, J.-F., *VLBI Imaging of the Gravity Probe B Guide Star IM Pegasi* American Astronomical Society Meeting 194, n. 50.06 (1999).
83. Stφrmer, C., Astrophysica Norvegica **1**, 1 (1934).
84. Rees, M., Ruffini, R., Wheeler, J. A. *Black Holes, Gravitational Waves and Cosmology*, New York: Gordon and Breach (1974).
85. Ruffini, R., Wheeler, J. A., *Introducing the Black Hole*, Physics Today **24** (1), 30 (1971).
86. Leach, R. W. and R. Ruffini, *On the Masses of X-Ray Sources* Astrophysical Journal **180**, p. L15 (1973).
87. Hanni, R. S. and R. Ruffini, *Lines of Force of a Point Charge near a Schwarzschild Black Hole* Phys. Rev. **D 8**, 3259–3265 (1973).

Section A

Equations of Motion of Spinning Particles in Electrodynamics and General Relativity

SOME CONSIDERATIONS ON THE VARIETIES OF FRAME DRAGGING

KENNETH NORDTVEDT

Northwest Analysis 118 Sourdough Ridge, Bozeman MT 59715; e-mail: kennordtvedt@one800.net

A major consequence of General Relativity and related theories of gravity is that *all inertial frames are local.* These local frames are accelerated, warped and stretched, and rotated with respect to each other due to the surrounding mass-energy distributions. While only their relative rotations are typically called *frame dragging effects*, this phrase describes a broader range of gravitational influences on inertia. Discussing the rotational effects on inertial frames within the wider framework, two conclusions are reached. Concerning *gravito-magnetism*, the relativistic gravitational force between two sources of moving matter, this interaction is already well verified through its contributions to a number of experimentally measured phenomena, and its presence is essential to the local Lorentz invariance of gravity which is also well confirmed by observation. So explicit observation of the gravito-magnetic interaction between the spinning Earth and an orbitting gyroscope will, at the expected one percent level of precision, add little to our existing empirical knowledge concerning relativistic gravity. However, measurement to a part in 10^5 of the larger form of frame dragging on such a gyroscope — the *deSitter geodetic precession* which results from the motion of the gyroscope through the Earth's gravitational field — will result in a most significant quantitative improvement in our knowledge of relativistic gravity.

1 Introduction

The *Lense-Thirring effect*, the induced rotation of local inertial frames by nearby rotating (or more generally, simply moving) matter is just one of the many novel post-Newtonian phenomena emerging from the order $1/c^2$ interaction terms in General Relativity and other modern theories of gravity. This moving matter-moving matter interaction, called gravito-magnetism in analogy with electro-magnetism, is being explicitly sought observationally in a variety of experiments, though I argue here that this interaction is already well-measured (to about a part in a thousand) by its contributions to several other phenomena measured in gravitational physics. The entire package of experimental knowledge concerning relativistic gravity must be considered as a whole in order to make sense of what we know and do not know. So I discuss in a unified fashion the several forms of inertial frame draggings and rescalings which occur in modern theories, and both their experimental measurement and theoretical interconnections are emphasized. Geodetic precession of inertial frames, first studied by de Sitter, occurs when a body's co-moving frame moves non-radially through a gravitational field. The forthcoming experiment, Gravity Probe-B, will measure this precession to almost a part in 10^5 in the process of measuring the Lense-Thirring precessions to a part in 10^2, and we should consider this geodetic precession measurement as the most important science goal of GP-B, since it will improve knowledge of metric gravity's most significant parameterized post-Newtonian (PPN) coefficient parameter γ by more than an order of magnitude in precision.

2 General Considerations

If we wish to use the description *frame dragging* for effects in post-Newtonian gravitational physics, then that phrase either applies to or is closely related to several phenomena concerning inertia:

1. Both the acceleration as well as mere presence of nearby matter accelerates (local inertial frames) another body.

2. Presence of nearby matter alters the inertia of a body.

3. Motion past nearby matter induces rotation of (local inertial frames) a body.

4. Rotating nearby matter induces rotation of (local inertial frames) a body.

And other quite related 'frame scaling' effects are produced by nearby matter (simlar to the second point above) which should be considered as well, such as:

5. The 'redshift' of clock rates by nearby matter.

6. The production of non-Euclidean geometrical structure of extended systems of objects by presence of nearby matter.

Since we are especially concerned with what has been called gravito-magnetism, I will eventually focus on that interaction, but along the way I will imbed gravito-magnetic phenomena into its place in the broader picture of all relativistic, post-Newtonian effects in gravity, since I believe it makes most sense to be so viewed.

A. Most all forms of (inertial) frame draggings or scalings are of related theoretical origins, significance and novelty — and they are different aspects of the maxim "all inertial frames are local" which is now generally accepted.

B. Frame dragging is directly analagous and similar to relativistic corrections occuring in other field theories of interaction (hence the coining of the name gravito-magnetism from the prior name electro-magnetism). It is only because these relativistic effects in gravity occur in proportion to body masses, with the accelerations then being body-independent, that frame dragging and scaling are alleged to be effects of (or from) the space-time arena rather than effects of forces on matter, though the latter is, in my view, the more mainstream and natural way to represent physical interactions. So I suggest replacing the common textbook cliche "gravity is a manifestation of space-time geometry" with the view point that "space-time geometry is a manifestation of or consequent category emerging from the tensor gravitational interaction of general relativity", and this paper is produced from that perspective.

C. Mach's ideas are vindicated in many respects by the emergence of frame dragging and scaling in relativistic gravity, though we need not dwell on this since Mach's own limitations prevented him from seeing that dynamical fields were the mechanisms by which influences of surrounding and distant matter are brought into a local system.

My operational taste is to keep theoretical considerations close to experimental possibilities and actualities, so I proceed on this basis and first consider the frame dragging varieties 1) and 2) listed above. If one takes a metric (or near-metric) theory of matter and fields, including gravity, which is described by an action of the type

$$A = \int (\mathcal{L}(g_{\mu\nu}...) + \mathcal{L}(\eta_{\mu\nu} \rightarrow g_{\mu\nu})_{SM})\ d^4x \tag{1}$$

with ... indicating other possible metric or 'arena' fields, and SM stands for *Standard Model*, then it will reduce to an effective, lowest order N-body Lagrangian of localized particles

$$L = \sum_i \left(-m_i + \frac{1}{2}m_i v_i^2\right) - \frac{1}{2}\sum_{ij}\frac{e_i e_j}{r_{ij}} + \frac{1}{2}\sum_{ij}\frac{Gm_i m_j}{r_{ij}} \tag{2}$$

in which the kinetic energy and electrostatic Lagrangian terms stand as surrogates for all the fermionic particles and bosonic fields in the Standard Model of physics. If some surrounding *spectator* matter is then added to our total universe system as shown in Figure 1, the effective Lagrangian for the localized system becomes rescaled in each of its terms

$$L = \sum_i \left(-m_i\,(1-U) + \frac{1}{2}m_i v_i^2\,(1+(2\gamma+1)U)\right) \tag{3}$$

$$-\frac{1}{2}\sum_{ij}\frac{e_i e_j}{r_{ij}}\,(1-(\gamma+1)U) + \frac{1}{2}\sum_{ij}\frac{Gm_i m_j}{r_{ij}}\,(1-(4\beta-2)U)\,, \tag{4}$$

in which U is the Newtonian potential of the spectator matter

$$U \equiv \sum_s \frac{Gm_s}{c^2 R_s}\,. \tag{5}$$

γ is a PPN parameter within the space of all metric theories of gravity, most often described as the strength of the metric theory to generate non-Euclidean spatial geometry, but $\gamma - 1$ more basically being a measure of the participation of other fields in the metric gravity. On its face, this modified Lagrangian indicates that the surrounding spectator matter alters the inertia of matter (by the common scale factor $1 + (2\gamma + 1)U$ in metric theories), as Mach suggested should occur and as Einstein argued did occur in General Relativity in his 1920s semi-popular book: "The Meaning of Relativity". Carl Brans and others later argued that this alteration of inertia does not really occur in metric theories, that it is just a coordinate choice effect and can be "swept under the space-time rug" by local rescaling of space and time coordinates

$$d\tau = dt\,(1-U)\,, \tag{6}$$

$$d\vec{\rho} = d\vec{r}\,(1+\gamma U)\,. \tag{7}$$

The local Lagrangian then becomes ($Ldt \rightarrow L'd\tau$)

$$L' = \sum_i \left(-m_i + \frac{1}{2}m_i \dot{\rho}^2\right) - \frac{1}{2}\sum_{ij}\frac{e_i e_j}{r_{ij}} + \frac{1}{2}\sum_{ij}\frac{G'm_i m_j}{r_{ij}} \tag{8}$$

with

$$G' = G\,(1-(4\beta-3-\gamma)U)\,. \tag{9}$$

Here β is another PPN parameter in *the space of metric theories* connected with the non-linearity of gravity. Although the kinetic and electrostatic Lagrangian terms are now returned to their original forms when reexpressed in the new coordinates, even in metric theories not all can be swept into space-time recalibration.

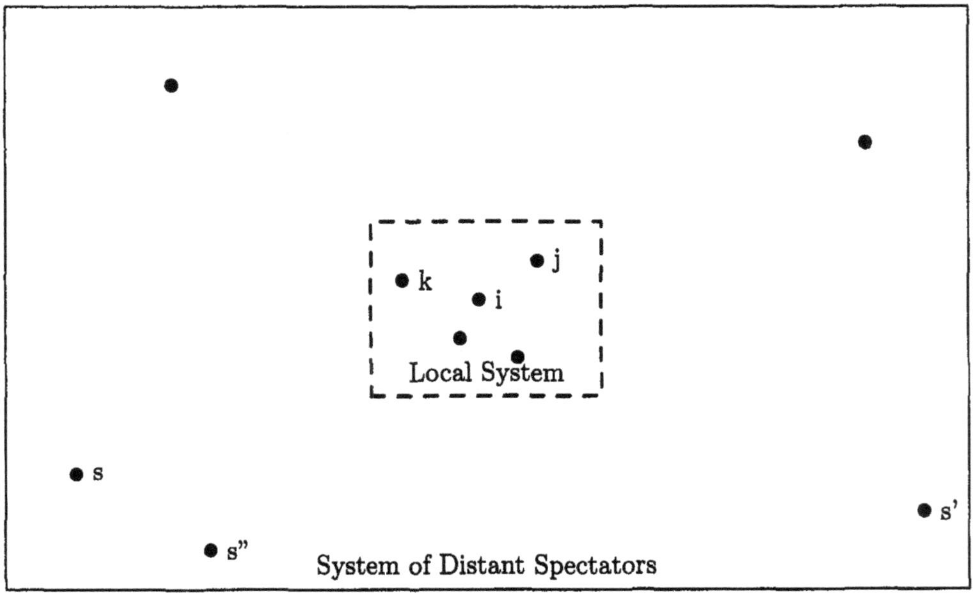

Figure 1. The Lagrangian for a local system of particles i,j,k,... shown within the dashed box is rescaled in metric theories of gravity by presence of other distant 'spectator' bodies s,s',s",... Locally viewed, the strength of Newton's gravitational coupling parameter is not necessarily invariant under this rescaling, and the non-gravitational physics is only invariant if in the metric case. Globally viewed, inertial mass and the strength of interactions are rescaled.

Depending on the specific gravitational theory, Newton's coupling parameter G will generally be rescaled, becoming dependent on location in space and time, and spawning a variety of interesting modern experiments. Even the local invariance of non-gravitational physics in presence of spectator matter after local rescaling of space-time coordinates, which occurs in metric theories but breaks down in the non-metric theories, still leaves prediction of novel phenomena which is today tested repeatedly by experiment, for example, the global comparison of clock rate 'red shifts' and the non-Euclidean spatial metrology found in solar system ranging experiments proportional to PPN γ, both manifestations of the location-dependence of the coordinate rescalings exhibited above. So not only is inertia affected by surrounding matter, but also are the coupling parameters on the force-side of matter's equations of motion. These rescalings on both sides of the equation of motion can only be discussed in their combination and entirety, but they represent real and observable, not merely coordinate, effects. One wonders why Mach and others who followed his ideas did not generally discuss the possible influence of the universe on both sides of $\vec{f} = m\vec{a}$, rather than solely on inertial side of these equations of physical law?

In metric theories the accelerative frame dragging terms of the post-Newtonian equation of motion have the particular form

$$\vec{A}_i = -(2\gamma + 1) \sum_j \frac{Gm_j}{c^2 r_{ij}} \vec{A}_i + \qquad (10)$$

$$+\ \frac{3+4\gamma+\alpha_1-\alpha_2}{2}\sum_j \frac{Gm_j}{c^2 r_{ij}}\vec{A}_j\ +\ \frac{1-\alpha_2}{2}\sum_j \frac{Gm_j}{c^2 r_{ij}}\hat{r}_{ij}\hat{r}_{ij}\cdot\vec{A}_j\,. \tag{11}$$

The terms of the first line beg the interpretation of *rescaling of the inertia of body i due to nearby presence of bodies j*, while the second line shows *induced acceleration of body i due to the acceleration of nearby bodies j*. The separate terms are not unique and vary with theory, but their strengths are related. For the case of binary pulsar systems these induced accelerations acting between the pair of bodies make a major contribution to the net periastron precession of the orbits, as I will discuss more fully later.

For a celestial body undergoing collective acceleration as a whole, these accelerative frame dragging terms, acting collectively within the body, provide a self-acceleration or net contribution to the body's total inertial mass proportional to the gravitational binding energy of the body

$$\delta M(I)\ =\ -\frac{1+\alpha_2-\alpha_1}{2}\sum_{ij}\frac{Gm_i m_j}{c^2 r_{ij}}\ -\ \frac{1-\alpha_2}{2}\sum_{ij}\frac{Gm_i m_j}{c^2 r_{ij}}\hat{r}_{ij}\hat{r}_{ij}\,. \tag{12}$$

The contributions proportional to PPN γ have canceled, and if the *preferred frame* PPN coeffients α_1 and α_2 vanish, this inertial mass contribution is consistent with the dictates of special relativity's Lorentz invariance. (This is one, but not the only, illustration of why $\alpha_{1,2}$ have been called *preferred frame* parameters.) The second spatial tensor inertial term in this contribution is also essential in canceling other tensor inertia contributions from the body's kinetic energy content, its electromagnetic energy content, etc. Though experiment does not in isolation measure the inertial mass of any celestial body, it does measure the gravitational to inertial mass ratio of such bodies; in lunar laser ranging (LLR), for example, the ratio of the Earth's two masses are measured to be one to precision of better than a part in a thousand of the mass equivalence of that body's gravitational binding energy. And this measurement should continue to improve in the future. Unless there are conspiring anomalies in the gravitational binding energy contributions to both gravitational and inertial masses, we confirm accelerative frame dragging as required in Lorentz-invariant, relativistic gravity.

As microscopic mechanism, these inductive interactions in gravity are quite identical in nature to inductive forces from other fields which contribute to the body's (inertial) mass-energy; for instance in electromagnetic theory the charges in an accelerated body produce induced electrical forces on their nearby charges of amount

$$\delta\vec{f}_i\ =\ -\frac{1}{2}\sum_j\frac{e_i e_j}{c^2 r_{ij}}\vec{A}_j\ -\frac{1}{2}\sum_j\frac{e_i e_j}{c^2 r_{ij}}\hat{r}_{ij}\hat{r}_{ij}\cdot\vec{A}_j\,. \tag{13}$$

Collectively these forces produce the needed electrostatic inertia to the whole body of

$$\delta M(I)\ =\ \frac{1}{2}\sum_{ij}\frac{e_i e_j}{c^2 r_{ij}}\ +\ \frac{1}{2}\sum_{ij}\frac{e_i e_j}{c^2 r_{ij}}\hat{r}_{ij}\hat{r}_{ij} \tag{14}$$

and similarly for all other field energies in the body. It is conceivable that all inertia results from such contributions from the force fields within bodies and there is no fundamental mechanical inertia of particles. A quite general slow motion, weak field phenomenological expression for the $1/c^2$ order relativistic and post-Newtonian gravity equation of motion for matter elements of a system can usefully be organized into the following block forms of acceleration terms

$$\begin{aligned}\vec{A}_i &= \sum_j \vec{g}_{ij} + Order\left(\sum_j \vec{g}_{ij}\vec{v}_i\vec{v}_i\right) + Order\left(\sum_j \vec{g}_{ij}\vec{v}_j\vec{v}_j\right) + \\ &+ Order\left(\sum_j \vec{g}_{ij}\vec{v}_i\vec{v}_j\right) + Order\left(\sum_j \frac{Gm_j}{r_{ij}}\vec{A}_i\right) + \\ &+ Order\left(\sum_j \frac{Gm_j}{r_{ij}}\vec{A}_j\right) + Order\left(\sum_{jk} Gm_k\left(\frac{1}{r_{jk}} or \frac{1}{r_{ik}}\right)\vec{g}_{ij}\right) \qquad (15)\end{aligned}$$

with the Newtonian accelerations being

$$\vec{g}_{ij} = \vec{\nabla}_i \frac{Gm_j}{r_{ij}}. \qquad (16)$$

Looking into the details of the gravito-magnetic block proportional to the velocities of both accelerated and accelerating bodies, $\vec{v}_i\vec{v}_j$, reveals an expression containing four possible forms of vector accelerations

$$\begin{aligned}\vec{A}_i(gravito-magnetic) &= C_1\sum_j \vec{g}_{ij}\vec{v}_i\cdot\vec{v}_j + C_2\sum_j \vec{g}_{ij}\cdot\vec{v}_i\vec{v}_j + \\ &+ C_3\sum_j \vec{g}_{ij}\cdot\vec{v}_j\vec{v}_i + C_4\sum_j \vec{g}_{ij}\vec{v}_i\cdot\hat{r}_{ij}\vec{v}_j\cdot\hat{r}_{ij} \qquad (17)\end{aligned}$$

with C_{1-4} being dimensionless, phenomenological coefficients to be measured by experiment. A complete phenomenological expansion for each of the other blocks can similarly be made. The space and time coordinates are assumed to have been chosen such that the speed of light function is given by an expansion with leading terms

$$c(\vec{r}) = 1 - 2\gamma^*\sum_i \frac{Gm_i}{r_i} + \cdots. \qquad (18)$$

Taking a very cautious approach and assuming the least, these equations of motion must be assumed valid in only one special *preferred inertial frame*, with any system under analysis referred to that frame of reference, rather than its rest frame, for calculation of experimental observables. Or with care, transformation rules for the position and velocity vectors in this phenomenological equation might be developed so that systems could be analyzed from the perspective of more convenient frames of reference, though this is not a necessary ingredient for scientific prediction. This equation will generally not be form-invariant under such changes in the asymptotic inertial frames. The goal of experiments is to measure the coefficients $C_{1,2,3,4}$ appearing in the gravito-magnetic block of the equation of motion, along with all the other coefficients in the other blocks of that equation. The post-Newtonian contributions to this equation of motion can be described as motional, accelerative, and non-linear. We are here most interested in the emergence of the various frame dragging effects from this equation. But, in fact, the entire collection of solar system relativistic gravity observations are calculated from these equations. If this equation is specialized to the case of General Relativity theory or its close cousins, most all possible observable effects turn out to be null effects, and this seems not to be widely realized. The totality of experiments over-constrains the general phenomenological equation of motion and to reasonable precision specifies its unique form. An extremely rich spectrum of possible oscillating perturbations and secular perturbations emerge from application of the general equation of motion to the near-Earth satellites, to the Moon, to the planets of the solar system, and even to binary pulsar systems. These observables have been cataloged here and there over the last 30 years and have been measured either actually or virtually by radar, laser and gps ranging during the same period. (By being "virtually" measured, I mean that tests for some null effects have not yet actually been performed by those who have published analyses of the experimental data, but such tests could straightforwardly be performed, and the results are already approximately known by comparison with the published test results). **To approximately the part in a thousand level of precision or better, there is no uncertainty left in the numerical coefficients of the phenomenological equation of motion outlined above. Through a collection of measurements of many so-called observables, perturbations or effects, each of which is given by some combination of the many parameters of the equation of motion, the total $1/c^2$ equation of motion has been systematically mapped out. This is just as true of the gravito-magnetic block as it is of the other blocks in the overall equation of motion.** I will illustrate in more detail how this mapping works, focusing on so-called frame dragging effects. These arguments are based on points I first outlined in papers of the late 1980s.[1,2]

3 Periastron Precession of Binary Pulsar Orbits.

Literally every post-Newtonian block type of accelerations appearing in the general, phenomenological equation of motion shown above contributes to the net periastron precession of a binary system's orbit. And when the two bodies have comparable masses, as they do in PSR 1913+16, all the post-Newtonian contributions to periastron precession are of the same characteristic magnitude, entering with both signs, and each being comparable to the net precession effect, itself, which is about 4.2 degrees per year.

4 Orbits of the Moon and Near-Earth Satellites.

The orbit of the Moon and its ranging distance from Earth used to model the LLR data is calculated at NASA's Jet Propulsion Laboratory and elsewhere using equations of motion formulated in the solar system's center of mass (barycentric) rest frame. Even supposing one neglects the several preferred frame accelerative terms and their resulting perturbations which will be proportional to W^2, $\vec{W} \cdot \vec{V}$,

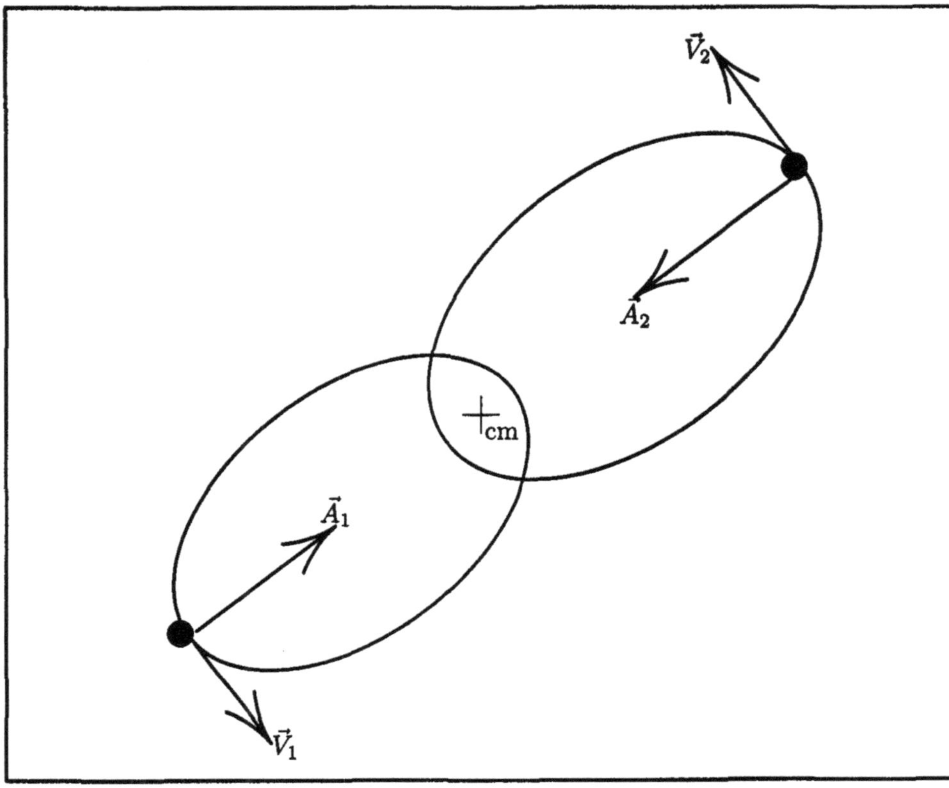

Figure 2. The periastron precession of the orbits of a binary system of comparable masses includes contributions from gravito-magnetism proportional to $\vec{V}_1 \vec{V}_2$ and from frame dragging interactions proportional to the accelerations $\vec{A}_1$ and $\vec{A}_2$ which are comparable in magnitude to the usual motional and non-linear post-Newtonian interactions.

and $\vec{W} \cdot \vec{U}$ (in fact such possible perturbations have been measured to be null in LLR and elsewhere), these three velocity vectors $\vec{W}$, $\vec{V}$, $\vec{U}$ being those of the solar system with respect to the preferred frame, the Earth relative to the solar system, and Moon relative to the Earth, respectively, the Moon's equation of motion relative to Earth will nevertheless still contain accelerations proportional to the product of factors $\vec{g}_{earth}$, $\vec{V}$, and $\vec{U}$. Such terms emerge from only two of the blocks in the equation of motion — from the Moon's own motional corrections proportional to V^2_{moon}, and from the gravito-magnetic block proportional to the product of Earth and Moon velocities relative to the solar system, which are

$$\vec{V}_{earth} \equiv \vec{V}\,, \tag{19}$$

$$\vec{V}_{moon} \equiv \vec{V} + \vec{U}\,. \tag{20}$$

The net perturbation of the Earth-Moon distance resulting from these accelerations, unless both sources contribute and precisely cancel, is a synodic month oscillation of characteristic magnitude

$$\delta r \cong \frac{Gm_e}{r_e^2}\frac{VU}{c^2}\frac{1}{\omega\Omega} \tag{21}$$

$$\cong 200\,cm \tag{22}$$

with Ω being the Earth's orbital frequency and ω being the Moon's (satellite's) orbital frequency. For application to near-Earth satellite orbits, this perturbation has even dramatically larger characteristic magnitude

$$\delta r \cong 100\,meters\,. \tag{23}$$

The JPL computer model could be used, if necessary, to calculate near-Earth satellite orbits as well as that of the Moon. The orbits of both the satellites and the Moon are presently measured and fit to precision of about a centimeter, and to this level no such anomalous perturbations are found. The gravito-magnetic accelerations must be there in just the predicted strength of the Lorentz-invariant PPN metric in order to nullify the contributions from the Moon or satellite's own motional acceleration corrections. This situation is a more complex dynamical analogy to Lorentz's demonstration, made even prior to Einstein's full special relativity theory, that the magnetic force between moving charges was intimately involved in producing the needed *Lorentz* contraction of the Michelson-Morley apparatus. Lorentz showed that an arbitrary crystal with atomic ions at equilibrium sites $\vec{r}_i$ would remain in equilibrium at the *Lorentz* transformed sites $\vec{r}_i'(t')$ under the force law from Maxwell's equations

$$if \qquad \vec{E}(\vec{r}_i) = 0 \qquad for\ all\ sites\ i \tag{24}$$

$$then \qquad \vec{E}'(\vec{r}_i') + \vec{v}/c \times \vec{B}'(\vec{r}_i') = 0 \tag{25}$$

neglecting the thermal motions of the atoms about their equilibrium sites (Lorentz did not yet have the correct mechanical laws of motion). The previously mentioned "own motional" relativistic gravity corrections, in partnership with the non-linear block of accelerations, produce the classic perihelion precession of Mercury which was General Relativity's initial great empirical achievement. One can adjust the equation of motion in one place to fit one observation, but such adjustment will then generally destroy agreement with experiment at other places, and the chain of fitting observation to the equation of motion soon spreads to the entire collection of observables. **You can not play with the equation of motion for the sole purpose of suggesting uncertainty in one effect's value without examining the consequences of claimed uncertainty on all the many other observations — both null and non-null — in gravitational physics. The collection of all observations is a united package which the assumed equations of physics must face as a whole.** Sceptics of my argument just given have pointed out that near-Earth satellite orbits are typically computed from an equation of motion formulated in the Earth's rest frame in which no gravito-magnetic interaction proportional to Earth motion exists. So how can gravito-magnetism be a factor in fitting the satellite range data? But I believe this illustrates my general point. To formulate the problem in this constantly accelerating geocentric frame and not include any gravitational potentials from the Earth's motion relative to the preferred frame or to the Sun assumes the very Lorentz invariance of the metric gravitational physics whose validity compels the gravito-magnetic interaction to exist in a certain proportion to other well established motional terms in the gravitational interaction. So I see GP-B's one percent measurement of its gyroscopes' Lense-Thirring gravito-magnetic precessions as adding very little to our present knowledge of the relativistic, post-Newtonian gravitational interaction. It will be GP-B's measurement of the geodetic *deSitter* precession which will add significantly to our knowledge. *Geodetic precession* of a local inertial frame is produced as consequence of the motion of that frame through the gravitational field of a body; it was first theoretically predicted and discussed in the context of the lunar orbit by deSitter within months of the publication of General Relativity. In fact, LLR presently confirms its existence through its effect on the lunar orbit to precision of half of one percent of its predicted value of 19.2 *mas/year*. It is interesting that geodetic precession can be predicted by a special relativistic generalization of the Equivalence Principle. Consider a non-rotating rod moving with constant velocity vector having both positive vertical and horizontal components. It moves through force-free inertial space, leaving some surface at $t = 0$. This surface accelerates upward at a constant rate and intersects the rod again at some future time as shown in Figure 2. Straightforward applications of the Lorentz transformation to bring the two events of rod and surface intersections to rest reveals that the rod's orientations have changed between these two events. Invoking the Equivalence Principle and thereby requiring that the same phenomena occur for a rod freely falling at velocity $\vec{v}$ through a gravitational field $\vec{g}$, then predicts a precession rate for such a rod of

$$\vec{\Omega} = \frac{1}{2}\,\frac{\vec{g} \times \vec{v}}{c^2}\,. \tag{26}$$

How dramatic it would have been if Einstein had predicted in 1907, a decade before he had a complete relativistic theory of gravity, that local inertial space not only accelerated, but also generally rotated, due to influence of surrounding matter.

The predicted total geodetic precession rate in metric gravity is actually found to be

$$\vec{\Omega} = \frac{2\gamma + 1}{2}\,\frac{\vec{g} \times \vec{v}}{c^2}\,. \tag{27}$$

The additional precession in this total results from the curvature (non-Euclidean structure) of the spatial geodesics in metric gravity and the assumption that the axes of any inertial frame moving along elements of these spatial geodesics will be parallel transported; a phenomena proportional to the fundamental PPN γ coefficient. For the GP-B mission, the gyroscope will experience a geodetic precession rate due to its motion through the Earth's gravitational field which is several hundred times stronger than the *Lense-Thirring* precession rate it should experience due to the spin of the Earth's matter. GP-B stands a chance to measure this geodetic precession to a part in 10^5 and consequently the most basic PPN γ coefficient to that same level. In my view, this will be the real science achievement of this

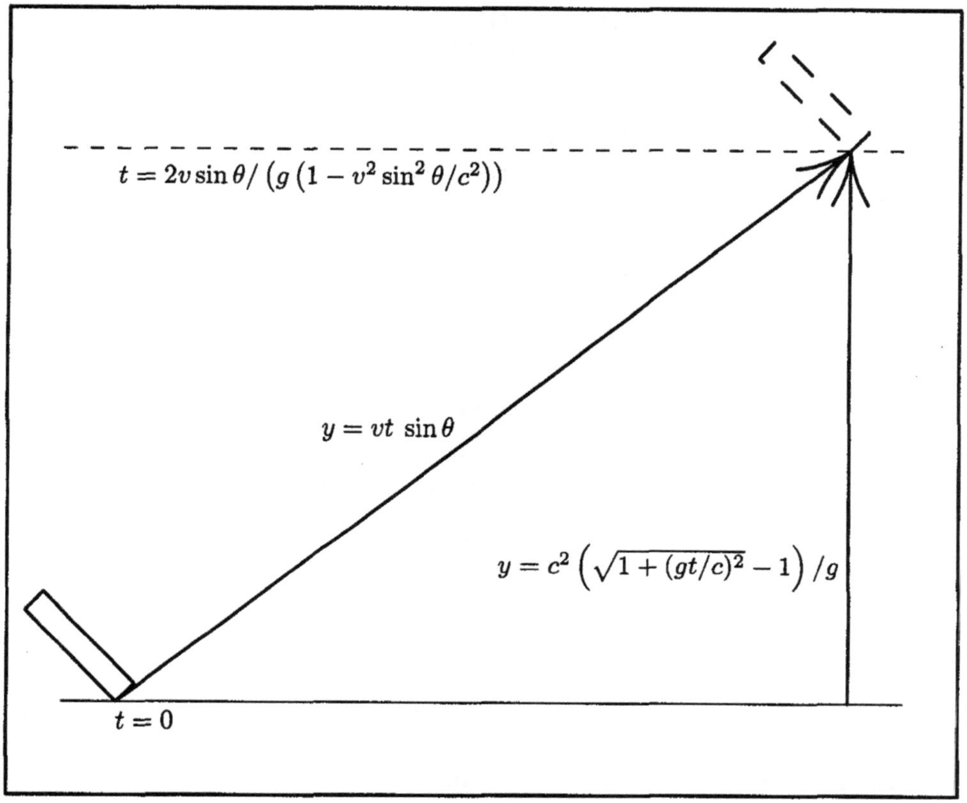

Figure 3. In force-free inertial space a rod travels at constant velocity and is intercepted twice by a constantly accelerated floor. As measured in the rest frame of the floor at the two interceptions, the rod is found to have rotated. By the Equivalence Principle one then concludes that motion through a gravitational field rotates local inertial frames.

experiment, as it will advance by perhaps two orders of magnitude our knowledge of whether the tensor gravity of General Relativity produces the whole story or, on the other hand, whether there are also additional very weak long-range interactions representing new physical law.

References

1. Gravitomagnetic Interaction and Laser Ranging to Earth Satellites. Phys. Rev. Lett. 61, No. 23, 2647 (1988).
2. Existence of the Gravitomagnetic Interaction, Int. J. Theor. Phys. 27, No. 11, 1395 (1988).

EQUATIONS OF MOTION OF SPINNING RELATIVISTIC PARTICLE IN EXTERNAL FIELDS

I.B. KHRIPLOVICH, A.A. POMERANSKY
Budker Institute of Nuclear Physics, 630090 Novosibirsk, Russia
and Novosibirsk University
E-mail: khriplovich@inp.nsk.su, pomeransky@vxinpz.inp.nsk.su

We consider the motion of a spinning relativistic particle in external electromagnetic and gravitational fields, to first order in the external field, but to an arbitrary order in spin. The noncovariant spin formalism is crucial for the correct description of the influence of the spin on the particle trajectory. We show that the true coordinate of a relativistic spinning particle is its naive, common coordinate $\vec{r}$. Concrete calculations are performed up to second order in spin included. A simple derivation is presented for the gravitational spin-orbit and spin-spin interactions of a relativistic particle. We discuss the gravimagnetic moment (GM), a specific spin effect in general relativity. It is shown that for the Kerr black hole the gravimagnetic ratio, i.e. the coefficient at the GM, equals unity (just as for the charged Kerr hole the gyromagnetic ratio equals two). The equations of motion obtained for relativistic spinning particle in external gravitational field differ essentially from the Papapetrou equations.

1 Introduction

The problem of the motion of a particle with internal angular momentum (spin) in an external field consists of two parts: the description of the spin precession and accounting for the spin influence on the trajectory of motion. To lowest nonvanishing order in c^{-2} the complete solution for the case of an external electromagnetic field was given more than 70 years ago.[1] The gyroscope precession in a centrally symmetric gravitational field had been considered to the same approximation even earlier.[2] Then, much later the spin precession was investigated in the case of the gravitational spin-spin interaction.[3] The fully relativistic problem of the spin precession in an external electromagnetic field was also solved more than 70 years ago,[4] and then in a more convenient formalism, using the covariant vector of spin, in Ref. 5.

The situation with the second part of the problem, which refers to how the spin influences the trajectory, is different. Covariant equations of motion for a relativistic spinning particle in an electromagnetic field were written in the same paper,[4] and for the case of a gravitational field in Ref. 6. These equations have been discussed repeatedly from various points of view in numerous papers (see, e.g. Refs. 7–18). The problem of the influence of the spin on the trajectory of a particle in external fields is not only of purely theoretical interest. It attracts attention being related to the description of the motion of ultrarelativistic particles in accelerators[19] (see also recent review: Ref. 20).

In fact, it is far from being obvious whether one can observe in practice the discussed spin corrections to the equations of motion of elementary particles, say, electron or proton. According to the well-known argument by Bohr (see Ref. 21), an additional Lorentz force due to the finite size of the wave packet of a charged particle and to the uncertainty relation, exceeds the corresponding component of the Stern–Gerlach force. However, this argument by itself does not exclude in principle the possibility to observe a regular Stern–Gerlach effect, let even a small one, in the presence of a larger background due to the uncertainty relation. This possibility is supported, in particular, by recent calculations.[22] Moreover, spin-dependent correlations certainly exist in differential cross sections of scattering processes. So, it was proposed long ago to separate by polarizations a beam of charged particles in a storage ring through the spin interaction with external fields.[23] Though this proposal is being discussed rather actively (see review: Ref. 20), it is not clear up to now whether it is feasible technically.

There are however macroscopic objects for which internal rotation certainly influences their trajectories. We mean the motion of Kerr black holes in external gravitational fields. This problem is of importance in particular for the calculation of the gravitational radiation of binary stars. In this connection it was considered in Refs. 24–27. However, when we turned to these calculations, we found[28] that the equations of motion with spin taken into account to the lowest nonvanishing order in c^{-2}, used in these papers, even in the simpler case of an external field lead to results which differ from the well-known gravitational spin-orbit interaction. The problem is essentially related to the correct definition of the center-of-mass coordinate. Moreover, it turned out that the widely used Papapetrou equations[6] also fail to reproduce in the same c^{-2} approximation the result for the gravitational spin-orbit interaction found in the classical work.[2] This discrepancy was pointed out long ago in Ref. 29; however the explanation suggested in Ref. 29 does not look satisfactory (see Ref. 28).

In our recent work[30] (the present talk is essentially based on it) we derived equations of motion of a relativistic particle with the noncovariant description of spin. These equations agree with well-known limiting cases. Though for external electromagnetic field such equations in the linear in spin approximation have been obtained previously[19] (see also Ref. 20), we would like to start with comments related to this approximation in electrodynamics.

2 Covariant and Noncovariant Equations of Motion of a Spinning Particle in an Electromagnetic Field

2.1 The Problem with Covariant Equations of Motion

The interaction of spin with external electromagnetic field is described, up to terms on the order of c^{-2} included, by the well-known Hamiltonian (see, e.g. Ref. 31):

$$H = -\frac{eg}{2m}\vec{s}\cdot\vec{B} + \frac{e(g-1)}{2m^2}\vec{s}\cdot[\vec{p}\times\vec{E}]\,. \tag{1}$$

Here $\vec{B}$ and $\vec{E}$ are external magnetic and electric fields; e, m, $\vec{s}$, and $\vec{p}$ are the particle charge, mass, spin, and momentum, respectively; g is its gyromagnetic ratio. Let us emphasize that the structure of the second (Thomas) term in this expression not only has been firmly established theoretically, but has also been confirmed with high accuracy experimentally, at any rate in atomic physics. To avoid misunderstandings, let us note that, generally speaking, the last term in formula (1) should be rewritten in a Hermitian form (see, e.g. Ref. 32):

$$[\vec{p}\times\vec{E}] \to \frac{1}{2}\left([\vec{p}\times\vec{E}] - [\vec{E}\times\vec{p}]\right) = [\vec{p}\times\vec{E}] + \frac{i}{2}\nabla\times\vec{E}\,.$$

We will be interested, however, in the semiclassical approximation mainly, when in the interaction linear in spin, field derivatives are neglected. (Besides, the correction with $\nabla\times\vec{E}$ vanishes in the case of potential electric field considered in Ref. 31.)

Let us try to construct a covariant equation of motion accounting for spin, which would reproduce in the same approximation the force

$$f_m = \frac{eg}{2m}(\vec{s}\cdot\vec{B})_{,m} + \frac{e(g-1)}{2m}\left(\frac{d}{dt}[\vec{E}\times\vec{s}]_m - (\vec{s}\cdot[\vec{v}\times\vec{E}])_{,m}\right), \tag{2}$$

corresponding to the Hamiltonian (1) (here and below, comma with a subscript denotes a partial derivative[a] and the scalar product between two vectors is written without dot in between). A covariant correction f^μ to the Lorentz force $eF^{\mu\nu}u_\nu$, linear in the tensor of spin $S_{\mu\nu}$ and in the gradient of the tensor of electromagnetic field $F_{\mu\nu,\lambda}$, may depend also on the 4-velocity u^μ. Since $u^\mu u_\mu = 1$, this correction must satisfy the condition $u_\mu f^\mu = 0$. From the mentioned tensors one can construct only two independent structures meeting the last condition. The first,

$$\eta^{\mu\kappa}F_{\nu\lambda,\kappa}S^{\nu\lambda} - F_{\lambda\nu,\kappa}u^\kappa S^{\lambda\nu}u^\mu, \tag{3}$$

reduces in the c^{-2} approximation to

$$(2\vec{s}(\vec{B} - [\vec{v}\times\vec{E}]))_{,m}$$

and the second,

$$u^\lambda F_{\lambda\nu,\kappa}u^\kappa S^{\nu\mu}, \tag{4}$$

reduces to

$$\frac{d}{dt}[\vec{s}\times\vec{E}]_m\,.$$

(Let us note that the structures with the contraction $F_{\nu\kappa,\lambda}S^{\kappa\lambda}$ reduce to these two expressions, due to the Maxwell equations and the antisymmetry of $S_{\kappa\lambda}$.)

Obviously, no linear combination of these two structures can reproduce the correct expression (2) for the spin-dependent force. In a somewhat less general way it was shown in Ref. 28.

But why the correct (in the c^{-2} approximation) formula (2) cannot be obtained from a covariant expression for the force? Obviously, one can easily reproduce those terms in (2) which are proportional to g by a linear combination of (3) and (4), i.e. there is no problem to present in a covariant form the terms which describe, so to say, direct interaction of a magnetic moment with external fields. It is the terms in (2) independent of g and corresponding to the Thomas precession which cannot be written covariantly. Certainly, the Thomas precession can be described beyond the c^{-2} approximation, for arbitrary velocities. But there are no reasons why this essentially noncovariant phenomenon should have a covariant description. This is the point.

[a]Note that partial derivatives of vectors on the LHS of Eq. (2) are scalar coefficients of the corresponding vectors.

2.2 What Is the Correct Definition of the Coordinate of a Spinning Particle?

It was noted in Ref. 28 that the covariant formalism can be reconciled with the correct results if the coordinate $\vec{x}$ entering the covariant equation is related to the usual one $\vec{r}$ in the c^{-2} approximation as follows:

$$\vec{x} = \vec{r} + \frac{1}{2m}\vec{s} \times \vec{v}. \tag{5}$$

The generalization of this substitution to the case of arbitrary velocities

$$\vec{x} = \vec{r} + \frac{\gamma}{m(\gamma+1)}\vec{s} \times \vec{v}, \quad \gamma = \frac{1}{\sqrt{1-v^2}}. \tag{6}$$

was recently pointed out in Ref. 20.

But why the spin precession itself (as distinct from the spin influence on the trajectory) can be described covariantly[4,5] without any concern for the coordinate definition? First of all, the covariant equations of the spin precession

$$\frac{dS_\mu}{d\tau} = \frac{e}{2m}\left[gF_{\mu\nu}S^\nu - (g-2)u_\mu F_{\lambda\nu}u^\lambda S^\nu\right] \tag{7}$$

(here S_μ is the covariant 4-vector of spin) are written in the semiclassical approximation, i.e., the coordinate dependence of external fields is neglected at all. Second, equations (7) are homogeneous and linear in spin. So, even if one went here beyond the semiclassical approach, but stayed within the approximation linear in spin, the use of the usual coordinate $\vec{r}$, which differs from $\vec{x}$ in terms proportional to $\vec{s}$ only, would be completely legitimate.

Of course, the choice of the variable, $\vec{r}$ or $\vec{x}$, is by itself a matter of convention. But still, which of them is the true center-of-mass coordinate of a relativistic spinning body?

We note first of all that relation (5) is valid for a free particle as well. So, to answer the question, we will consider a simple example of the free Dirac particle with the Hamiltonian

$$H = \vec{\alpha}\vec{p} + \beta m.$$

Here, the operator whose expectation value equals to $\vec{r}$, is not $\vec{r}$ itself, but[33]

$$\vec{x} = \vec{r} + \frac{i\beta\vec{\alpha}}{2\epsilon} - \frac{i\beta(\vec{\alpha}\vec{p})\vec{p} + [\vec{\Sigma} \times \vec{p}]\,\epsilon}{2\epsilon^2(\epsilon+m)}; \quad \epsilon = \sqrt{p^2+m^2}; \quad \vec{\Sigma} = \frac{1}{2i}[\vec{\alpha} \times \vec{\alpha}]. \tag{8}$$

To lowest nonvanishing order in c^{-2} expression (8) reduces to

$$\vec{x} = \vec{r} + \frac{1}{2m}\vec{s} \times \vec{v}, \quad \vec{s} = \frac{1}{2}\vec{\sigma}, \tag{9}$$

which might prompt indeed substitution (5). However, under the Foldy–Wouthuysen (FW) transformation, which separates positive-energy states from negative-energy ones, the relativistic operator $\vec{x}$ goes over into mere $\vec{r}$. And the transition from the exact Dirac equation in an external field to its approximate form containing only the first-order correction in c^{-2} is performed just by means of the FW transformation. Thus, in the arising c^{-2} Hamiltonian the coordinate of a spinning electron is the same $\vec{r}$ as in the completely nonrelativistic case. Nobody makes substitution (5) in the Coulomb potential when treating the spin-orbit interaction in the hydrogen atom.

As to a classical particle, it is in fact a well-localized wave packet constructed from positive-energy states, i.e. it is properly described in the FW representation. Therefore, it is $\vec{r}$ which is the true coordinate of a classical relativistic spinning particle.

2.3 The Noncovariant Formalism

The correct equations of motion in an electromagnetic field including spin to first order are known for a fairly long time.[19] Though being fully relativistic, they are noncovariant and based on the initial physical definition of spin. According to this definition, spin is the 3-dimensional vector $\vec{s}$ (or 3-dimensional antisymmetric tensor s_{mn}) of the internal angular momentum defined in the rest frame of the particle. The covariant vector of spin S_μ (or the covariant antisymmetric tensor $S_{\mu\nu}$) are obtained from $\vec{s}$ (or s_{mn}) merely by the Lorentz transformation. By the way, an advantage of this approach is that the constraints $u^\mu S_\mu = 0$ and $u^\mu S_{\mu\nu} = 0$ hold identically. The precession frequency for spin $\vec{s}$ at an arbitrary velocity is well-known (see, for instance, Ref. 31):

$$\vec{\Omega} = \frac{e}{2m}\left\{(g-2)\left[\vec{B} - \frac{\gamma}{\gamma+1}\vec{v}(\vec{v}\vec{B}) - \vec{v}\times\vec{E}\right] + 2\left[\frac{1}{\gamma}\vec{B} - \frac{1}{\gamma+1}\vec{v}\times\vec{E}\right]\right\}. \tag{10}$$

Naturally, the corresponding interaction Lagrangian (the Lagrangian description is here somewhat more convenient than the Hamiltonian one) equals

$$L_{1s} = \vec{\Omega}\vec{s} = \frac{e}{2m}\vec{s}\left\{(g-2)\left[\vec{B} - \frac{\gamma}{\gamma+1}\vec{v}(\vec{v}\vec{B}) - \vec{v}\times\vec{E}\right]\right.$$
$$\left. +2\left[\frac{1}{\gamma}\vec{B} - \frac{1}{\gamma+1}\vec{v}\times\vec{E}\right]\right\}. \tag{11}$$

The equation of motion for coordinate is the usual one:

$$\left(\nabla - \frac{d}{dt}\nabla_{\vec{v}}\right) L_{tot} = 0, \tag{12}$$

where L_{tot} is the total Lagrangian of the system. The equation of motion for spin in general form is

$$\dot{\vec{s}} = -\{L_{tot}, \vec{s}\}, \tag{13}$$

where $\{\dots,\dots\}$ is the Poisson bracket, or

$$\dot{\vec{s}} = -i[L_{tot}, \vec{s}] \tag{14}$$

in the quantum problem. This applicability of a common canonical formalism is one more advantage of the noncovariant approach. Meanwhile, in the covariant approach we have to deal in particular with higher time derivatives, which is obvious already from relation (9).

3 Equations of Motion of a Spinning Particle in an Electromagnetic Field

3.1 The General Approach

Our approach to the problem is based on the following physically obvious argument. As long as we do not consider excitations of internal degrees of freedom of a body moving in an external field, this body (even if it is macroscopic) can be treated as an elementary particle with spin.

Therefore, the Lagrangian of the spin interaction with an external field can be derived from the elastic scattering amplitude

$$-eJ^{\mu}A_{\mu} \tag{15}$$

of a particle with spin s by a vector potential A_{μ}. Due to the arguments presented in Section 1, the discussion of the effects nonlinear in spin (which are of primary interest to us) may be physically meaningful only in the classical limit $s \gg 1$. It is this approximation that is basically used below.

The matrix element J_{μ} of the electromagnetic current operator between states with momenta k and k' can be written (under P and T invariance) as follows (see, for instance, Refs. 34 and 35):

$$J_{\mu} = \frac{1}{2\epsilon}\,\bar{\psi}(k')\,\{p_{\mu}F_e + \Sigma_{\mu\nu}q^{\nu}F_m\}\,\psi(k). \tag{16}$$

Here $p_{\mu} = (k'+k)_{\mu}$, $q_{\mu} = (k'-k)_{\mu}$.

The wave function of a particle with an arbitrary spin ψ can be written (see, for instance, Refs. 31, §31) as

$$\psi = \frac{1}{\sqrt{2}}\begin{pmatrix}\xi\\ \eta\end{pmatrix}. \tag{17}$$

Both spinors,

$$\xi = \{\xi^{\alpha_1\,\alpha_2\,..\,\alpha_p}_{\dot\beta_1\,\dot\beta_2\,..\,\dot\beta_q}\}$$

and

$$\eta = \{\eta^{\beta_1\,\beta_2\,..\,\beta_q}_{\dot\alpha_1\,\dot\alpha_2\,..\,\dot\alpha_p}\},$$

are symmetric in the dotted and undotted indices separately, and

$$p + q = 2s.$$

For a particle of half-integer spin one can choose

$$p = s + \frac{1}{2}, \quad q = s - \frac{1}{2}.$$

In the case of integer spin it is convenient to use

$$p = q = s.$$

The spinors ξ and η are chosen in such a way that under reflection they transform into each other (to within a phase). For $p \neq q$ they are different objects which belong to different representations of the Lorentz group. For $p = q$, these two spinors coincide. Nevertheless, we will use the same expression (17) for the wave function of any spin, i.e., we will also formally introduce the object η for an integer spin, bearing in mind that it is expressed in terms of ξ. This will allow us to perform calculations in the same way for the integer and half-integer spins.

In the rest frame both ξ and η coincide with a nonrelativistic spinor ξ_0, which is symmetric in all indices; in this frame there is no difference between dotted and undotted indices. The spinors ξ and η are obtained from ξ_0 through the Lorentz transformation:

$$\xi = \exp\{\vec{\Sigma}\vec{\phi}/2\}\xi_0\,; \qquad \eta = \exp\{-\vec{\Sigma}\vec{\phi}/2\}\xi_0\,. \tag{18}$$

Here the vector $\vec{\phi}$ is directed along the velocity, $\tanh\phi = v$;

$$\vec{\Sigma} = \sum_{i=1}^{p}\vec{\sigma}_i - \sum_{i=p+1}^{p+q}\vec{\sigma}_i\,,$$

and $\vec{\sigma}_i$ acts on the ith index of the spinor ξ_0 as follows:

$$\vec{\sigma}_i\,\xi_0 = (\vec{\sigma}_i)_{\alpha_i\beta_i}\,(\xi_0)_{\dots\beta_i\dots}\,. \tag{19}$$

In the Lorentz transformation (18) for ξ, after the operator $\vec{\Sigma}$ has acted on ξ_0 the first p indices are identified with the upper undotted indices and the next q indices are identified with the lower dotted indices. The inverse situation takes place for η.

We note that in an external field the components of velocity $\vec{v}$ (and together with them the components of $\vec{\phi}$) do not commute, in general. However, to the adopted approximation, linear in the external field, one can ignore this noncommutativity which is itself proportional to the field. Moreover, we are mainly interested in the classical limit of the final result where such commutators are negligible since they are proportional to an extra power of $\hbar$. Therefore, $\vec{v}$ and $\vec{\phi}$ will be treated as ordinary numerical parameters.

Next,

$$\bar{\psi} = \psi^\dagger\gamma_0 = \psi^\dagger\begin{pmatrix} 0 & I \\ I & 0 \end{pmatrix};$$

here I is the sum of 2×2 unit matrices acting on all indices of the spinors ξ and η. The components of the matrix $\Sigma_{\mu\nu} = -\Sigma_{\nu\mu}$ are:

$$\Sigma_{0n} = \begin{pmatrix} -\Sigma_n & 0 \\ 0 & \Sigma_n \end{pmatrix}; \tag{20}$$

$$\Sigma_{mn} = -\,2i\epsilon_{mnk}\begin{pmatrix} s_k & 0 \\ 0 & s_k \end{pmatrix}; \tag{21}$$

$$\vec{s} = \frac{1}{2}\sum_{i=1}^{2s}\vec{\sigma}_i.$$

The scalar operators $F_{e,m}$ depend on two invariants, $t = q^2$ and $\tau = (S^\mu q_\mu)^2$. The covariant vector of spin S_μ is defined, e.g., for the state with momentum k_μ, and is obtained via the Lorentz transformation from the vector of spin $(0,\vec{s})$ in the rest frame:

$$S^{\mu}=(S_0,\vec{S}),\quad S_0=\frac{(\vec{s}\vec{k})}{m},\quad \vec{S}=\vec{s}+\frac{\vec{k}(\vec{k}\vec{s})}{m(\epsilon+m)}. \tag{22}$$

In the expansion in the electric multipoles

$$F_e(t,\tau)=\sum_{n=0}^{N_e} f_{e,2n}(t)\tau^n$$

the highest power N_e is obviously s and $s-1/2$ for integer and half-integer spin, respectively. In the magnetic multipole expansion

$$F_m(t,\tau)=\sum_{n=0}^{N_m} f_{m,2n}(t)\tau^n$$

the highest power N_m is $s-1$ and $s-1/2$ for integer and half-integer spin. The invariant form factors $f_{e,m}(t)$, being Fourier-transformed into the coordinate representation, describe the space distribution of the charge, magnetic moment, and higher multipoles, i.e., describe the finite size of the particle.

Clearly,

$$f_{e,0}(0)=1,\quad f_{m,0}(0)=\frac{g}{2}.$$

Let us note at last that we have chosen the noncovariant normalization for the amplitude (15), being interested in the Lagrangian referring to the world time t and not to the proper time τ.

3.2 Effects Linear in the Spin

As a warm-up exercise and check of our approach, let us reproduce now the well-known result (11) for the case of a constant external field. We start with the terms proportional to g-factor. The corresponding term in the scattering amplitude can be written as

$$\frac{eg}{4\epsilon}\,\xi_0'^{\dagger}\Big\{[\exp\{\vec{\Sigma}\vec{\phi}/2\}(\vec{s}\vec{B})\exp\{-\vec{\Sigma}\vec{\phi}/2\}+\exp\{-\vec{\Sigma}\vec{\phi}/2\}(\vec{s}\vec{B})\exp\{\vec{\Sigma}\vec{\phi}/2\}]$$

$$+\frac{i}{2}\,[\exp\{\vec{\Sigma}\vec{\phi}/2\}(\vec{\Sigma}\vec{E})\exp\{-\vec{\Sigma}\vec{\phi}/2\}-\exp\{-\vec{\Sigma}\vec{\phi}/2\}(\vec{\Sigma}\vec{E})\exp\{\vec{\Sigma}\vec{\phi}/2\}]\Big\}\,\xi_0. \tag{23}$$

It is essential that in the considered case of a constant external field, one may put $\vec{k}'=\vec{k}$, $\vec{v}'=\vec{v}$, $\vec{\phi}'=\vec{\phi}$, since $\vec{q}=\vec{k}'-\vec{k}$ corresponds to the field gradient.

In our further calculations we use the well-known identity

$$\exp\{\hat{A}\}\hat{B}\exp\{-\hat{A}\}=\hat{B}+\frac{1}{1!}\,[\hat{A},\,\hat{B}]+\frac{1}{2!}\,\Big[\hat{A},\,[\hat{A},\,\hat{B}]\Big]+\cdots,$$

and the following relationships:

$$[\Sigma_i,\,\Sigma_j]=4i\epsilon_{ijk}s_k\,,\quad [\Sigma_i,\,s_j]=i\epsilon_{ijk}\Sigma_k\,; \tag{24}$$

$$\cosh\phi = \gamma, \quad \sinh\phi = v\,\gamma. \tag{25}$$

After simple algebraic transformations, expression (23) reduces to

$$\frac{eg}{2m}\,\vec{s}\left[\vec{B} - \frac{\gamma}{\gamma+1}\,\vec{v}(\vec{v}\vec{B}) - \vec{v}\times\vec{E}\right]. \tag{26}$$

Let us now discuss the contribution of the convection term

$$-\frac{e}{2\epsilon}\,\bar{\psi}(k')\psi(k)\,p^{\mu}A_{\mu}. \tag{27}$$

We write the product of exponents in the expression

$$\bar{\psi}(k')\psi(k) = \frac{1}{2}\xi_0'^{\dagger}[\exp\{\vec{\Sigma}\vec{\phi}'/2\}\exp\{-\vec{\Sigma}\vec{\phi}/2\} + \exp\{-\vec{\Sigma}\vec{\phi}'/2\}\exp\{\vec{\Sigma}\vec{\phi}/2\}]\xi_0 \tag{28}$$

as

$$\begin{aligned}&\exp\{\vec{\Sigma}\vec{\phi}'/2\}\exp\{-\vec{\Sigma}\vec{\phi}/2\}\\ &\quad = \prod_p \exp\{\vec{\sigma}\vec{\phi}'/2\}\exp\{-\vec{\sigma}\vec{\phi}/2\}\prod_q \exp\{-\vec{\sigma}\vec{\phi}'/2\}\exp\{\vec{\sigma}\vec{\phi}/2\}.\end{aligned} \tag{29}$$

Let us consider a typical factor in this formula:

$$\begin{aligned}\exp\{\vec{\sigma}\vec{\phi}'/2\}\exp\{-\vec{\sigma}\vec{\phi}/2\} &= \cosh(\phi'/2)\cosh(\phi/2) - (\vec{n}'\vec{n})\sinh(\phi'/2)\sinh(\phi/2)\\ &\quad + \vec{\sigma}[\vec{n}'\sinh(\phi'/2)\cosh(\phi/2) - \vec{n}\cosh(\phi'/2)\sinh(\phi/2)]\\ &\quad - i(\vec{\sigma}[\vec{n}'\times\vec{n}])\sinh(\phi'/2)\sinh(\phi/2);\end{aligned} \tag{30}$$

here $\vec{n}' = \vec{v}'/v'$, $\vec{n} = \vec{v}/v$. Since we are interested in gradients only as long as they enter together with spin, in the first term, $\cosh(\phi'/2)\cosh(\phi/2) - (\vec{n}'\vec{n})\sinh(\phi'/2)\sinh(\phi/2)$, we put $\phi' = \phi/2$, $\vec{n}' = \vec{n}$, after which this term turns to unity. Then, we are discussing the interaction linear in spin, so that the product (29) reduces to

$$1 + \vec{\Sigma}\left[\vec{n}'\sinh(\phi'/2)\cosh(\phi/2) - \vec{n}\cosh(\phi'/2)\sinh(\phi/2)\right]$$

$$-2i\,(\vec{s}[\vec{n}'\times\vec{n}])\sinh(\phi'/2)\sinh(\phi/2).$$

When substituted into formula (28), the terms proportional to $\vec{\Sigma}$ cancel out. Now, limiting ourselves to the terms linear in $\vec{q}$, we reduce the spin-dependent part of (27) to

$$-e\,\frac{p^{\mu}}{2\epsilon}\,\frac{i(\vec{s}[\vec{k}\times\vec{q}])}{m(\epsilon+m)}\,A_{\mu}.$$

Let us note further that since $p^{\mu}q_{\mu} = 0$, the following identity holds

$$p^{\mu}q_{\alpha}A_{\mu} = p^{\mu}(q_{\alpha}A_{\mu} - q_{\mu}A_{\alpha}) = p^{\mu}iF_{\alpha\mu}. \tag{31}$$

Then, we can put $p_{\mu} = 2mu_{\mu}$, where u_{μ} is the 4-velocity. As a result we arrive at the following expression:

$$-\frac{e}{2m}\,\vec{s}\left[2\left(1-\frac{1}{\gamma}\right)\vec{B} - \frac{2\gamma}{\gamma+1}\,\vec{v}(\vec{v}\vec{B}) - \frac{2\gamma}{\gamma+1}\,\vec{v}\times\vec{E}\right]. \tag{32}$$

The sum of (26) and (32) yields (11). Thus, we have reproduced the well-known result for the interaction linear in the spin, starting from the relativistic wave equation for an arbitrary spin

Below we repeatedly use identities of the form (31). In the classical language such a transformation corresponds to discarding in a Lagrangian (or adding to it) a total time derivative Indeed,

$$u^\mu q_\mu \to u^\mu \partial_\mu = \gamma \left(\frac{\partial}{\partial t} + \vec{v}\nabla \right) = \gamma \frac{d}{dt}$$

3.3 Effects Quadratic in the Spin

Let us now investigate the interaction of second order in the spin. The "bare", explicitly quadrupole interaction present in expressions (15) and (16) is

$$-e \frac{p^\mu}{2\epsilon} f_{e,2} (S^\alpha q_\alpha)^2 A_\mu . \tag{33}$$

Using the identity (31) and relations (22), and then discarding the total time derivative $\partial/\partial t + \vec{v}\nabla$, we write this interaction as

$$L_{2s} = -e\, f_{e,2} \left[(\vec{s}\nabla) - \frac{\gamma}{\gamma+1} (\vec{v}\vec{s})(\vec{v}\nabla) \right] \left[(\vec{s}\vec{E}) - \frac{\gamma}{\gamma+1} (\vec{s}\vec{v})(\vec{v}\vec{E}) + (\vec{s}[\vec{v} \times \vec{B}]) \right] \tag{34}$$

Using the Maxwell equations and adding a total derivative with respect to t, one can show that the tensor $s_i s_j$ in (34) can be rewritten in the following irreducible form: $s_i s_j \to s_i s_j - (1/3)\delta_{ij}\vec{s}^{\,2}$. Now from the nonrelativistic limit of formula (34), it is clear that this formula describes indeed the interaction with an external field of the quadrupole moment

$$Q_{ij} = -2\, e\, f_{e,2}\, (3\, s_i s_j - \delta_{ij} \vec{s}^{\,2}); \quad Q = Q_{zz}|_{s_z = s} = -2\, e\, f_{e,2}\, s(2s-1). \tag{35}$$

In the asymptotics, as $\gamma \to \infty$, the interaction (34) tends to a constant

$$L_{2s} = -e\, f_{e,2} \left[(\vec{s}\nabla) - (\vec{v}\vec{s})(\vec{v}\nabla) \right] \left[(\vec{s}\vec{E}) - (\vec{s}\vec{v})(\vec{v}\vec{E}) + (\vec{s}[\vec{v} \times \vec{B}]) \right]. \tag{36}$$

It is well-known that even in the absence of the bare quadrupole term, i.e. at $f_{e,2} = 0$, a quadrupole interaction arises in the nonrelativistic limit due to the convection and magnetic terms in interaction (15). The value of this induced quadrupole moment at an arbitrary spin of the particle is Ref. 35 (in the formula below we corrected a misprint in the original paper[35]):

$$Q_1 = -e\,(g-1) \left(\frac{\hbar}{mc} \right)^2 \begin{cases} s & \text{integer spin,} \\ s - 1/2 & \text{half-integer spin.} \end{cases} \tag{37}$$

Here we have explicitly displayed the Planck constant to show that the induced quadrupole moment Q_1 vanishes in the classical limit $\hbar \to 0$, $s \to \infty$, $\hbar s \to$ const. Therefore, the interaction of second order in spin proportional to Q_1 does not influence in fact equations of motion of a classical particle (although it plays a role in atomic spectroscopy[35]).

Still, the convection and magnetic terms in expression (15) induce an interaction of second order in spin which has a classical limit and is therefore of interest for our problem. It is convenient here to start with the convection current interaction. Let us come back to formula (30). Again we put in it

$$\cosh\left(\frac{\phi'}{2}\right)\cosh\left(\frac{\phi}{2}\right)-(\vec{n}'\vec{n})\sinh\left(\frac{\phi'}{2}\right)\sinh\left(\frac{\phi}{2}\right)=1\,.$$

In the other terms, linear in $\vec{\sigma}$, we keep only the first power of $\vec{q}\rightarrow -i\hbar\nabla$, in the hope that in the final result (29) $\hbar$ will enter in the combination $\hbar s\rightarrow$ const. Nevertheless, these terms by themselves are small as compared to unity, so that in the classical limit expression (30) can be rewritten as

$$\exp\left\{\vec{\sigma}\left[\vec{n}'\sinh\left(\frac{\phi'}{2}\right)\cosh\left(\frac{\phi}{2}\right)-\vec{n}\cosh\left(\frac{\phi'}{2}\right)\sinh\left(\frac{\phi}{2}\right)\right]-i(\vec{\sigma}[\vec{n}'\times\vec{n}])\sinh^2\left(\frac{\phi}{2}\right)\right\}.$$

Clearly, in the product (29) the operators $\vec{\sigma}$ attached to

$$\vec{n}'\sinh\left(\frac{\phi'}{2}\right)\cosh\left(\frac{\phi}{2}\right)-\vec{n}\cosh\left(\frac{\phi'}{2}\right)\sinh\left(\frac{\phi}{2}\right),$$

combine in the resulting exponent into the operator $\vec{\Sigma}$ which vanishes in the classical limit. In this limit only those operators $\vec{\sigma}$ survive that are attached to $[\vec{n}'\times\vec{n}]\sinh^2(\phi/2)$; they combine into $2\vec{s}$. Thus, in the classical limit the product (29) reduces, with the account for the second identity (25), to

$$\exp\left\{\frac{1}{m}\frac{\gamma}{\gamma+1}\,(\vec{s}[\vec{v}\times\nabla])\right\}. \tag{38}$$

Let us note that the action of the operator (38) on any function of coordinates, whether it is a vector potential or field strength, amounts to the shift of its argument:

$$\vec{r}\rightarrow\vec{r}+\frac{1}{m}\frac{\gamma}{\gamma+1}\,\vec{s}\times\vec{v}\,.$$

Curiously, just this substitution was pointed out in Ref. 20 for the transition from covariant equations linear in spin to noncovariant equations (see (6)).

Now, taking into account the second term in the expansion of the exponential function (38) and using again the identity (31), we obtain the following expression for the quadratic in spin interaction arising from the convection current:

$$-\frac{e}{2m^2}\frac{\gamma}{\gamma+1}\,(\vec{s}[\vec{v}\times\nabla])\left[\left(1-\frac{1}{\gamma}\right)(\vec{s}\vec{B})-\frac{\gamma}{\gamma+1}(\vec{s}\vec{v})(\vec{v}\vec{B})-\frac{\gamma}{\gamma+1}\left(\vec{s}[\vec{v}\times\vec{E}]\right)\right]. \tag{39}$$

Let us discuss now the contribution to the discussed effect due to the magnetic moment. It is convenient to write the term in the scattering amplitude we are interested in (it is proportional to g-factor) as

$$\frac{eg}{4\epsilon}\,\xi_0'^{\dagger}\left\{\left[\exp\{\vec{\Sigma}\vec{\phi}'/2\}\exp\{-\vec{\Sigma}\vec{\phi}/2\}\exp\{\vec{\Sigma}\vec{\phi}/2\}(\vec{s}\vec{B})\exp\{-\vec{\Sigma}\vec{\phi}/2\}\right.\right.$$
$$+\exp\{-\vec{\Sigma}\vec{\phi}'/2\}\exp\{\vec{\Sigma}\vec{\phi}/2\}\exp\{-\vec{\Sigma}\vec{\phi}/2\}(\vec{s}\vec{B})\exp\{\vec{\Sigma}\vec{\phi}/2\}\Big]$$
$$+\frac{i}{2}\Big[\exp\{\vec{\Sigma}\vec{\phi}'/2\}\exp\{-\vec{\Sigma}\vec{\phi}/2\}\exp\{\vec{\Sigma}\vec{\phi}/2\}(\vec{\Sigma}\vec{E})\exp\{-\vec{\Sigma}\vec{\phi}/2\}$$

$$- \exp\{-\vec{\Sigma}\vec{\phi}/2\}\exp\{\vec{\Sigma}\vec{\phi}/2\}\exp\{-\vec{\Sigma}\vec{\phi}/2\}(\vec{\Sigma}\vec{E})\exp\{\vec{\Sigma}\vec{\phi}/2\}\Big]\Big\}\xi_0 . \quad (40)$$

Using in this case the first term in the expansion of the exponential function (38), we arrive at the following expression for the contribution proportional to the magnetic moment:

$$\frac{eg}{2m^2}\frac{\gamma}{\gamma+1}\,(\vec{s}\,[\vec{v}\times\nabla])\left[(\vec{s}\vec{B}) - \frac{\gamma}{\gamma+1}(\vec{s}\vec{v})(\vec{v}\vec{B}) - \left(\vec{s}\,[\vec{v}\times\vec{E}]\right)\right]. \quad (41)$$

The total result for the induced interaction, quadratic in the spin, is

$$L^i_{2s} = \frac{e}{2m^2}\frac{\gamma}{\gamma+1}\,(\vec{s}\,[\vec{v}\times\nabla])\left[\left(g-1+\frac{1}{\gamma}\right)(\vec{s}\vec{B}) - (g-1)\frac{\gamma}{\gamma+1}(\vec{s}\vec{v})(\vec{v}\vec{B})\right.$$
$$\left. - \left(g-\frac{\gamma}{\gamma+1}\right)\left(\vec{s}\,[\vec{v}\times\vec{E}]\right)\right]. \quad (42)$$

Let us note that in the nonrelativistic limit the induced interaction with magnetic field tends to zero as v/c, and that with electric field as $(v/c)^2$. Moreover, the part of interaction (42) that is not related to g-factor, is reducible in spin; in other words, $s_i s_j$ in it cannot be rewritten as an irreducible tensor $s_i s_j - (1/3)\delta_{ij}\vec{s}^{\,2}$. Therefore, the interaction (42) is not in fact a quadrupole one. However, its asymptotic behaviour for $\gamma \to \infty$ is of interest. In this limit

$$L^i_{2s} = \frac{e}{2m^2}(g-1)\,(\vec{s}\,[\vec{v}\times\nabla])\,[(\vec{s}\vec{B}) - (\vec{s}\vec{v})(\vec{v}\vec{B}) - \left(\vec{s}\,[\vec{v}\times\vec{E}]\right)]. \quad (43)$$

Surprisingly, the asymptotical formulae (36) and (43) coincide to within a factor and a total time derivative. To prove this, it is convenient to introduce three orthogonal unit vectors

$$\vec{v};\ \vec{\rho} = \frac{[\vec{v}\times\vec{s}]}{|[\vec{v}\times\vec{s}]|};\ \vec{\zeta} = [\vec{v}\times\vec{\rho}].$$

Using the completeness of this basis and the equation $\dot{\vec{E}} = [\nabla\times\vec{B}]$, and discarding a total derivative with respect to t, one can check that

$$[(\vec{s}\nabla) - (\vec{v}\vec{s})(\vec{v}\nabla)][(\vec{s}\vec{E}) - (\vec{s}\vec{v})(\vec{v}\vec{E}) + (\vec{s}[\vec{v}\times\vec{B}])] = [\vec{v}\times\vec{s}]^2(\vec{\zeta}\nabla)[(\vec{\zeta}\vec{E}) + (\vec{\rho}\vec{B})],$$

coincides indeed with

$$(\vec{s}\,[\vec{v}\times\nabla])[(\vec{s}\vec{B}) - (\vec{s}\vec{v})(\vec{v}\vec{B}) - (\vec{s}\,[\vec{v}\times\vec{E}])] = -[\vec{v}\times\vec{s}]^2\,(\vec{\rho}\nabla)[(\vec{\rho}[\vec{v}\times\vec{B}]) + (\vec{\rho}\vec{E})].$$

Thus, there is a special value of the bare quadrupole moment

$$Q = -2(g-1)\frac{es^2}{m^2} \qquad \text{or} \qquad f_{e,2} = (g-1)\frac{1}{2m^2} \quad (44)$$

(let us recall that we consider now a classical situation, when $s \gg 1$), at which the total interaction quadratic in spin, $L_{2s} + L^i_{2s}$, asymptotically decreases with energy.

The situation resembles that which takes place for the interaction linear in spin. It is well-known (see, for instance, Refs. 11, 36 and 37) that there is a special value of g-factor, $g = 2$, at which the interaction linear in spin decreases as $\gamma \to \infty$.

This follows immediately from formula (11) for the first-order Lagrangian. Thus, putting additionally $g = 2$, we obtain

$$Q = -2\,\frac{es^2}{m^2} \qquad \text{or} \qquad f_{e,2} = \frac{1}{2m^2}\,. \tag{45}$$

Let us note that the choice $g = 2$ for the bare magnetic moment is a necessary (but insufficient) condition of renormalizability in quantum electrodynamics.[11,36,37] It is satisfied not only for the electron, but also for the charged vector boson in the renormalizable electroweak theory.

In some respect, however, the situation with the special values (44), (45) of the quadrupole moment differs from the situation with g-factor. The conditions (44), (45), as distinct from the condition $g = 2$, are not universal, since they are valid only for large spins, $s \gg 1$; in other words, they refer only to classical objects with internal angular momentum. In particular, for the charged vector boson of the renormalizable electroweak theory the bare quadrupole interaction is absent at all, $f_{e,2} = 0$. The quadrupole moment of this particle is (in our language) of the induced nature, it is given by formula (37) at $s = 1$, $g = 2$.

4 A Simple-Minded Aside on the Spin Precession in a Gravitational Field

In this section we present a simple and general derivation of the equations of spin precession in a gravitational field. This approach not only allows us to easily reproduce and generalize known results for spin effects. Pointed out here remarkable analogy between gravitational and electromagnetic fields allows also to easily transform the results of the previous section to the case of an external gravitational field.

It follows from the angular momentum conservation in flat space-time taken together with the equivalence principle that the 4-vector of spin S^μ is parallel transported along the particle world-line. The parallel transport of a vector along a geodesic $x^\mu(\tau)$ means that its covariant derivative vanishes:

$$\frac{DS^\mu}{D\tau} = 0\,. \tag{46}$$

(In this section we restrict our discussion to the effects linear in the spin.) We will use the tetrad formalism natural for the description of spin. In view of relation (46), the equation for the tetrad components of spin $S^a = S^\mu e^a_\mu$ is

$$\frac{DS^a}{D\tau} = \frac{dS^a}{d\tau} = S^\mu e^a_{\mu;\nu} u^\nu = \eta^{ab}\gamma_{bcd} u^d S^c\,. \tag{47}$$

Here

$$\gamma_{abc} = e_{a\mu;\nu} e^\mu_b e^\nu_c = -\gamma_{bac} \tag{48}$$

are the Ricci rotation coefficients.[38] Certainly, the equation for the tetrad 4-velocity components is exactly the same:

$$\frac{du^a}{d\tau} = \eta^{ab}\gamma_{bcd} u^d u^c \tag{49}$$

The meaning of Eqs. (47), (49) is clear: the tetrad components of both vectors vary in the same way, due to the rotation of the local Lorentz vierbein only.

In exactly the same way, the 4-dimensional spin and velocity of a charged particle with the gyromagnetic ratio $g = 2$ precess with the same angular velocity in an external electromagnetic field, by virtue of equation (7) at $g = 2$ and the Lorentz equation:

$$\frac{dS_a}{d\tau} = \frac{e}{m}F_{ab}S^b; \qquad \frac{du_a}{d\tau} = \frac{e}{m}F_{ab}u^b.$$

Thus, the correspondence:

$$\frac{e}{m}F_{ab} \longleftrightarrow \gamma_{abc}u^c. \tag{50}$$

is obvious. This correspondence allows one to obtain the precession frequency $\vec{\omega}$ of the 3-dimensional vector of spin $\vec{s}$ in external gravitational field from expression (10) via the simple substitution

$$\frac{e}{m}B_i \longrightarrow -\frac{1}{2}\epsilon_{ikl}\gamma_{klc}u^c; \quad \frac{e}{m}E_i \longrightarrow \gamma_{0ic}u^c. \tag{51}$$

This frequency is

$$\omega_i = -\epsilon_{ikl}\left(\frac{1}{2}\gamma_{klc} + \frac{u^k}{u^0+1}\gamma_{0lc}\right)\frac{u^c}{u^0_w}. \tag{52}$$

The factor $1/u^0_w$ in this expression is related to the transition in the left-hand side of Eq. (47) to the differentiation with respect to the world time t:

$$\frac{d}{d\tau} = \frac{dt}{d\tau}\frac{d}{dt} = u^0_w\frac{d}{dt}.$$

The quantity u^0_w is supplied with the subscript w to emphasize that this is a world component of 4-velocity, but not a tetrad component. All other indices in expression (52) are tetrad ones, $c = 0, 1, 2, 3$; $i, k, l = 1, 2, 3$. The corresponding spin-dependent correction to the Lagrangian is

$$L_{1sg} = \vec{s}\vec{\omega}\,. \tag{53}$$

As an illustration of formulae (52) and (53), let us apply them to the cases of spin-orbit and spin-spin interactions. We restrict, as is common in the problems discussed, to the linear approximation in the gravitational field. However, in our approach, as distinct from the standard ones, both problems can be easily solved for arbitrary particle velocities.

The tetrads $e_{a\mu}$ are related to the metric as follows:

$$e_{a\mu}e_{b\nu}\eta^{ab} = g_{\mu\nu}.$$

To linear approximation we can put $g_{\mu\nu} = \eta_{\mu\nu} + h_{\mu\nu}$ and do not distinguish between the tetrad and world indices in $e_{a\mu}$. The ambiguity in the choice of tetrads will be fixed by choosing the symmetric gauge $e_{\mu\nu} = e_{\nu\mu}$. Then

$$e_{\mu\nu} = \eta_{\mu\nu} + \frac{1}{2}h_{\mu\nu}\,.$$

Using expression (48) for the Ricci coefficients, we find to linear approximation

$$\gamma_{abc} = \frac{1}{2}(h_{bc,a} - h_{ac,b})\,. \tag{54}$$

Let us start with the spin-orbit interaction. In the centrally symmetric field created by a mass M, the metric is

$$h_{00} = -\frac{2kM}{r}, \qquad h_{mn} = -\frac{2kM}{r}\delta_{mn}. \tag{55}$$

Here the nonvanishing Ricci coefficients are

$$\gamma_{ijk} = \frac{kM}{r^3}(\delta_{jk}r_i - \delta_{ik}r_j)\,, \qquad \gamma_{0i0} = -\frac{kM}{r^3}r_i\,. \tag{56}$$

Their substitution into formula (52) yields the following expression for the precession frequency:

$$\vec{\omega}_{ls} = \frac{2\gamma+1}{\gamma+1}\frac{kM}{r^3}\,\vec{v}\times\vec{r}\,. \tag{57}$$

In the limit of low velocities, $\gamma \to 1$, the answer goes over into the classical result.[2]

Now we consider the spin-spin interaction. Let the spin of the central body be $\vec{s}_0$. Linear in $\vec{s}_0$ components of metric, which are responsible for the spin-spin interaction, are:

$$h_{0i} = 2k\,\frac{[\vec{s}_0\times\vec{r}]_i}{r^3}\,.$$

Here the nonvanishing Ricci coefficients are

$$\gamma_{ij0} = k\left(\nabla_i\frac{[\vec{s}_0\times\vec{r}]_j}{r^3} - \nabla_j\frac{[\vec{s}_0\times\vec{r}]_i}{r^3}\right), \qquad \gamma_{0ij} = -k\nabla_i\frac{[\vec{s}_0\times\vec{r}]_j}{r^3}\,. \tag{58}$$

The frequency of the spin-spin precession is

$$\vec{\omega}_{ss} = -k\left(2-\frac{1}{\gamma}\right)(\vec{s}_0\nabla)\nabla\frac{1}{r} + k\frac{\gamma}{\gamma+1}\left[\vec{v}(\vec{s}_0\nabla) - \vec{s}_0(\vec{v}\nabla) + (\vec{v}\vec{s}_0)\nabla\right](\vec{v}\nabla)\frac{1}{r}\,. \tag{59}$$

In the limit of low velocities this formula also goes over into the corresponding classical result.[3]

In the conclusion of this section we note that in the case of an external gravitational field there is no covariant expression for the force linear in the particle spin. In other words, the deviation from geodesics of the trajectory of a spinning particle is not described by the Riemann tensor. If it were the case, there would be a unique possible covariant structure, to within a factor (in Ref. 6 it equals $-1/2m$): $R_{\mu\nu ab}u^\nu S^{ab}$. As mentioned already in Section 1, the covariant description (as distinct from our our formulae (52), (53)) contradicts the classical results in the limit of low velocities.

5 Equations of Motion of a Spinning Particle in a Gravitational Field

5.1 The General Approach

The equations of motion in an external gravitational field to any order in spin are constructed similarly to the equations of motion in the case of an electromagnetic field.

We start with the elastic scattering amplitude in a weak external gravitational field $h_{\mu\nu}$. We use it as an heuristic argument only, and afterwards will go beyond the linear approximation. This amplitude is

$$-\frac{1}{2}T_{\mu\nu}h^{\mu\nu}. \tag{60}$$

The matrix element $T_{\mu\nu}$ of the energy-momentum tensor between states of momenta k and k' can be written as

$$T_{\mu\nu} = \frac{1}{4\epsilon}\,\bar{\psi}(k')\Big\{p_\mu p_\nu\,F_1 + \frac{1}{2}\,(p_\mu\Sigma_{\nu\lambda} + p_\nu\Sigma_{\mu\lambda})\,q^\lambda\,F_2 + (\eta_{\mu\nu}q^2\ -\ q_\mu q_\nu)\,F_3$$
$$+\,[S_\mu S_\nu q^2 - \ (S_\mu q_\nu +\ S_\nu q_\mu)(Sq)\ +\ \eta_{\mu\nu}(Sq)^2]\,F_4\Big\}\,\psi(k). \tag{61}$$

The scalar operators F_i in this expression are also expanded in powers of $\tau = (Sq)^2$:

$$F_i(t,\tau) = \sum_{n=0}^{N_i} f_{i,2n}(t)\tau^n.$$

It can be easily proven that the total number of invariant form factors $f_{i,2n}$ is $4s+2$ and $4s+1$ for integer and half-integer spin, respectively. The independence of the four tensor structures in (61) is obvious. As for the completeness of the expansion, it can be proven, for instance, by demonstrating that the total number of invariant form factors, as calculated in the annihilation channel, coincides with the above result.

In the generally covariant form, the structure $(\eta_{\mu\nu}q^2\ -\ q_\mu q_\nu)\,h^{\mu\nu}$ corresponds to the scalar curvature R, and $[S_\mu S_\nu q^2 - (S_\mu q_\nu + S_\nu q_\mu)(Sq) + \eta_{\mu\nu}(Sq)^2]\,h^{\mu\nu}$ corresponds to the product $R_{\mu\nu}S^\mu S^\nu$, where $R_{\mu\nu}$ is the Ricci tensor. Since we are interested in the equations of motion in a sourceless field, the corresponding terms in the expansion (61) will be omitted.

Just as in electrodynamics charge conservation dictates the condition $f_{e,0}(0) = 1$, here energy conservation leads to $f_{1,0}(0) = 1$. As to the term in the amplitude (60) which contains $f_{2,0}$, it is convenient to write it in a different form, using the analogy (50) with electromagnetic field. Putting $g = 2$ and $(e/m)F_{ab} \rightarrow f_{ab} = \gamma_{abc}u^c$ in the corresponding electromagnetic term

$$i\,\frac{eg}{8\epsilon}\,\bar{\psi}(k')\,\Sigma^{ab}F_{ab}\,\psi(k),$$

we arrive at the following contribution to the Lagrangian of the gravitational interaction:

$$i\,\frac{1}{4u_w^0}\,\bar{\psi}(k')\,\Sigma^{ab} f_{ab}\,\psi(k); \tag{62}$$

here, as usual, $u_w^0 = \epsilon/m$. Using for γ_{abc} the linear approximation (54), one can easily see that the expression (62) corresponds indeed to the discussed contribution to the amplitude provided that $f_{2,0} = 1$. Thus, in gravity the value of one more form factor at zero momentum transfer t is fixed. It corresponds to conservation of angular momentum. This was pointed out earlier in Ref. 39 and 40.

Let us now come back to the convection term in formula (60). As in electrodynamics, when we go over here to spinors in the rest frame, the term of first order in spin is written as

$$-\frac{p^\mu p^\nu}{8\epsilon}\,\frac{1}{m}\,\frac{u^0}{u^0+1}\,(\vec{s}\,[\vec{v}\times\nabla])\;h_{\mu\nu}. \tag{63}$$

Using (31), (54), we obtain

$$p^\mu \nabla_k h_{\mu\nu} \to -\,p^\mu(-\partial_k h_{\mu\nu} + \partial_\mu h_{k\nu}) \to -\,2p^a\gamma_{ak\nu}.$$

Thus, the expression (63) can be written in terms of the Ricci coefficients:

$$\frac{1}{u_w^0}\,\frac{u^0}{u^0+1}\,\epsilon^{mnk} s^m v^n u^a u^c \gamma_{akc}. \tag{64}$$

As can be easily seen, the sum of (62) and (64) reproduces the Lagrangian (53).

5.2 *Effects Quadratic in the Spin*

Let us investigate now the effects of second order in the spin in the equations of motion in a gravitational field. In the case of a binary star these effects are of the same order of magnitude as the spin-spin interaction when the spins of the components of the system are comparable.[28] The influence of the latter on the characteristics of gravitational radiation becomes noticeable for a system of two extreme black holes.[25] Correspondingly, second-order spin effects in the equations of motion become substantial if at least one component of a binary is close to an extreme black hole.[28] Therefore, the investigation of these effects is not of a purely theoretical interest only. In principle they can be observed with the gravitational wave detectors under construction.

An obvious source of second-order spin effects is the term

$$L_{2sg} = -\,f_{1,2}\,\frac{1}{8\epsilon}\,p^\mu p^\nu (Sq)^2 h_{\mu\nu} \tag{65}$$

in the amplitude (60). Due to the relation

$$\begin{aligned} p^\mu p^\nu q_\alpha q_\beta h_{\mu\nu} &= p^\mu p^\nu (q_\alpha q_\beta h_{\mu\nu} + q_\mu q_\nu h_{\alpha\beta} - q_\alpha q_\nu h_{\mu\beta} - q_\beta q_\mu h_{\nu\alpha}) \\ &\to 2p^\mu p^\nu R_{\mu\alpha\nu\beta}, \end{aligned}$$

the Lagrangian (65) can be written in terms of the Riemann tensor:

$$L_{2sg} = -\,\frac{\kappa}{2\epsilon}\,u^a S^b u^c S^d R_{abcd}. \tag{66}$$

Instead of $f_{1,2}$, we have introduced a dimensionless parameter κ:

$$f_{1,2} = \frac{\kappa}{2m^2}.$$

Now, it is convenient to use the Petrov representation for the components of the Riemann tesor (see Ref. 38):

$$E_{kl} = R_{0k0l}, \quad E_{kl} = E_{lk}; \quad C_{kl} = \frac{1}{4}\epsilon_{kmn}\epsilon_{lrs}R_{mnrs}, \quad C_{kl} = C_{lk};$$

$$B_{kl} = \frac{1}{2}\epsilon_{lrs}R_{0krs}, \quad B_{kk} = 0. \tag{67}$$

We limit our discussion to the case of a sourceless gravitational field. Then, at $R_{ab} = 0$, further simplifications hold:

$$C_{kl} = -E_{kl}, \quad B_{kl} = B_{lk}, \quad E_{kk} = C_{kk} = 0. \tag{68}$$

Finally, we arrive at the following interaction Lagrangian quadratic in the spin:

$$L_{2sg} = -\frac{\kappa}{2\epsilon}\left[(2\vec{u}^2 + 1)\,E_{kl} - 2\left(2 - \frac{1}{u_0 + 1}\right)u_k u_m E_{lm} + \delta_{kl} u_m u_n E_{mn}\right.$$
$$\left. + \frac{1}{(u_0+1)^2}\,u_k u_l u_m u_n E_{mn} - 2\,u_0\,\epsilon_{kmn} u_m B_{nl} + \frac{2}{u_0+1}\,u_k u_m \epsilon_{lrn} u_r B_{mn}\right]$$
$$\times\left(s_k s_l - \frac{1}{3}\,\delta_{kl}\vec{s}^2\right). \tag{69}$$

To avoid misunderstandings, we note that all three-dimensional indices in this equation (and in (70), (71)) are in fact contravariant.

As in electrodynamics, along with the "bare" interaction (69), there is here an induced interaction quadratic in spin. Its explicit form can be obtained most easily by setting $g = 2$ in the electromagnetic formula (42) and by making the substitution (51). We also take into account the correspondence

$$q_i \gamma_{abc} u^c = (q_i \gamma_{abc} - q_c \gamma_{abi})\,u^c \to i\,(\partial_i \gamma_{abc} - \partial_c \gamma_{abi})\,u^c \to i\,R_{abci} u^c.$$

Finally, using (67) and (68), we obtain the following result for the induced interaction:

$$L^i_{2sg} = \frac{1}{2\epsilon}\left\{\left(2\vec{u}^2 - \frac{u^0 - 1}{u^0 + 1}\right)E_{kl} - 2\left[2 - \frac{1}{u_0 + 1} - \frac{1}{(u_0+1)^2}\right]u_k u_m E_{lm}\right.$$
$$+ \left[1 - \frac{1}{(u_0+1)^2}\right]\delta_{kl} u_m u_n E_{mn} + \frac{1}{(u_0+1)^2}\,u_k u_l u_m u_n E_{mn}$$
$$\left. - 2\left(u_0 - \frac{1}{u_0+1}\right)\epsilon_{kmn} u_m B_{nl} + \frac{2}{u_0+1}\,u_k u_m \epsilon_{lrn} u_r B_{mn}\right\} s_k s_l\,. \tag{70}$$

As in the electromagnetic case, the induced interaction tends to zero in the nonrelativistic limit $\sim v/c$, and the spin factor in it, $s_k s_l$, is not an irreducible tensor.

The asymptotic behaviour of L_{2sg} and L^i_{2sg} is the same: both Lagrangians increase linearly with energy. However, in this case too the coefficient in the "bare" interaction can be chosen in such a way, $\kappa = 1$, that the total Lagrangian of second order in the spin decreases (as well as the analogous interaction in electrodynamics) when the energy tends to infinity. At $\kappa = 1$

$$L_{2sg} + L^{i}_{2sg} = -\frac{1}{\epsilon(u^0+1)}\left(u^0 E_{kl} - \frac{1}{u_0+1} u_k u_m E_{lm} \right.$$
$$\left. + \frac{1}{2(u_0+1)} \delta_{kl} u_m u_n E_{mn} + \epsilon_{kmn} u_m B_{nl}\right) s_k s_l. \qquad (71)$$

6 The Gravimagnetic Moment

There is a profound analogy between the linear in the spin interaction of the magnetic moment with the electromagnetic field

$$\mathcal{L}_{em} = -\frac{eg}{4m} F_{ab} S^{ab} \qquad (72)$$

and the "bare" gravitational Lagrangian (66) which is quadratic in the spin.[11] (Here it is more convenient to write the gravitational Lagrangian, like $\mathcal{L}_{em}$, for the proper time τ, rather than the world time t, i.e. to multiply expression (66) by ϵ/m.) This analogy is based on the following observation.[11] It is well-known that the canonical momentum $i\partial_\mu$ enters relativistic wave equations for a particle in electromagnetic and gravitational external fields via the combination

$$\Pi_\mu = i\partial_\mu - eA_\mu - \frac{1}{2}\Sigma^{ab}\gamma_{ab\mu}.$$

It follows from the structure of the commutator (or Poisson brackets in the classical limit)

$$[\Pi_\mu, \Pi_\nu] = -i\left(eF_{\mu\nu} - \frac{1}{2}\Sigma^{ab}R_{ab\mu\nu}\right)$$

that in a sense $-\frac{1}{2}\Sigma^{ab}R_{ab\mu\nu}$ plays the same role in gravity as $eF_{\mu\nu}$ in electromagnetism. It is quite natural then that the gravitational analogue of the electromagnetic spin interaction (72) is

$$\mathcal{L}_{gm} = \frac{\kappa}{8m} R_{abcd} S^{ab} S^{cd}. \qquad (73)$$

One can easily show that expressions (73) and (66) coincide (to within a factor ϵ/m). It is sufficient to this end to take into account the relation $S^{ab} = \epsilon^{abcd} S_c u_d$, as well as the identity

$$\tilde{\tilde{R}}_{abcd} = \frac{1}{4}\epsilon_{ab}{}^{ef}\epsilon_{cd}{}^{gh}R_{efgh} = -R_{abcd},$$

which is valid for a sourceless gravitational field.

By analogy with the magnetic moment

$$\frac{eg}{2m} S^{\mu\nu},$$

it is natural to define the gravimagnetic moment

$$-\frac{\kappa}{2m}S^{ab}S^{cd}.$$

The gravimagnetic ratio κ, like the gyromagnetic ratio g in electrodynamics, may have in general any value. However, according to our semiclassical arguments, the value $\kappa = 1$ in gravity is as preferable as $g = 2$ in electrodynamics. In any case, at $g = 2$ and $\kappa = 1$ the spin equations of motion have the simplest form.

On the other hand, it has been shown in Ref. 11 that just this value of the gravimagnetic ratio, $\kappa = 1$, follows from the wave equations in the Feynman gauge for the photon and graviton in an external gravitational field. The same value, $\kappa = 1$, follows from the squared Rarita-Schwinger equation for $s = 3/2$ in a gravitational field.[11]

A second-order wave equation for an arbitrary spin in a gravitational background had been proposed long ago in Ref. 41. For integer spins, its form corresponds also to $\kappa = 1$. However, the value of κ proposed in Ref. 41 for half-integer spins is quite different, it does not tend to unity even in the semiclassical limit $s \to \infty$. Clearly, such a prescription does not look reasonable: in the semiclassical limit $s \to \infty$ the parameter κ should not change when the spin changes by 1/2.

The situation with spin 1/2 is rather special.[11] The properties of the spin matrices for $s = 1/2$,

$$S^{ab} = \frac{i}{4}(\gamma^a\gamma^b - \gamma^b\gamma^a),$$

are such that the gravimagnetic interaction (73) degenerates here into the mere scalar curvature

$$\frac{\kappa}{16m}R, \tag{74}$$

without any consequences whatsoever for the motion of the spin. Obviously, the coefficient at this spin-independent stucture cannot be fixed by the above arguments. Formula (74) would predict for $\kappa = 1$ the following wave equation for $s = 1/2$

$$\left(-g^{\mu\nu}D_\mu D_\nu - m^2 + \frac{1}{8}R\right)\psi = 0 \tag{75}$$

(the factor $1/2m$ should be deleted from the expressions (73) and (74) when going over from a Lagrangian in our normalization to a wave equation). Meanwhile, the squared Dirac equation in a gravitational field is

$$\left(-g^{\mu\nu}D_\mu D_\nu - m^2 + \frac{1}{4}R\right)\psi = 0. \tag{76}$$

The discrepancy between (75) and (76) can be removed, for instance, by adding

$$\frac{1}{16m}R$$

to the Lagrangian (73) for half-integer spins. This term in no way influences our semiclassical arguments, vanishing in the limit $\hbar \to 0$. As to the recent proposal[42] to cure the discussed discrepancy by ascribing to the electron (which has no gravimagnetic interaction at all) the gravimagnetic ratio $\kappa = 2$, we cannot see in it any real physical meaning.

7 Multipoles of Black Holes

But let us come back from elementary particles to macroscopic bodies. For a classical object the values of both parameters g and κ depend in general on the various properties of the body. However, for black holes the situation is different. It has been shown in Ref. 43 from an analysis of the Kerr-Newman solution that the gyromagnetic ratio of a charged rotating black hole is universal (and equal to that of the electron!): $g = 2$.

We will show that for the Kerr black hole the gravimagnetic ratio is $\kappa = 1$. This value follows in fact from the analysis of the motion of spin of a black hole in an external field done in Ref. 24 (though this statement was not explicitly formulated there). We will present here an independent and, in our opinion, simpler derivation of this important result.

At great distance from a Kerr hole, the hole can be considered as a point source of a weak gravitational field. To linear approximation in the field of a hole at rest, the Lagrangian density corresponding to the interaction (66) can be written as

$$L = \frac{\kappa}{4m}\,(\vec{s}\nabla)^2\,h_{00}\delta(\vec{r})\,. \tag{77}$$

The so induced correction to the energy-momentum tensor has a single component:

$$\delta T_{00} = -\frac{\kappa}{2m}\,(\vec{s}\nabla)^2\,\delta(\vec{r})\,. \tag{78}$$

In the gauge

$$\bar{h}^{\mu\nu}{}_{,\nu} = 0, \;\; \bar{h}_{\mu\nu} = h_{\mu\nu} - \frac{1}{2}\eta_{\mu\nu}h^{\alpha}_{\alpha} \tag{79}$$

the static Einstein equation for the corresponding correction h_{00} to the 00-component of the metric is

$$\Delta h_{00} = 8\pi k T_{00}\,.$$

The correction itself is

$$h_{00} = \kappa\frac{k}{m}\,(\vec{s}\nabla)^2\,\frac{1}{r}\,. \tag{80}$$

Let us compare h_{00} with the corresponding contribution to the Kerr metric. In the Boyer-Lindquist coordinates this metric is

$$ds^2 = \left(1 - \frac{r_g r}{\Sigma}\right)dt^2 - \frac{\Sigma}{\Delta}dr^2 - \Sigma d\theta^2 - \left(r^2 + a^2 + \frac{r_g r a^2}{\Sigma}\sin^2\theta\right)r^2\sin^2\theta$$
$$+ \frac{2r_g r a}{\Sigma}\sin^2\theta d\phi dt\,, \tag{81}$$

where $\Delta = r^2 - r_g r + a^2$, $\Sigma = r^2 + a^2\cos^2\theta$, $\vec{a} = \vec{s}/m$. At $r_g = 0$ the metric (81) describes a flat space in spheroidal coordinates.[38] Meanwhile, it is Cartesian coordinates which correspond in the flat space to the gauge (79). The transition from the spheroidal coordinates to Cartesian ones is carried out with the required accuracy by the substitution

$$\vec{r} \to \vec{r} + \frac{\vec{a}(\vec{a}\vec{r}) - \vec{r}a^2}{2r^2}\,.$$

In the Cartesian coordinates the spin-dependent part of the 00-component of the metric

$$g_{00} = 1 - \frac{r_g}{r} + \frac{r_g a^2}{2r^3}(3\cos^2\theta - 1)$$

obviously coincides with h_{00} from formula (80) at $\kappa = 1$. Somewhat more tedious consideration of the space components of the Kerr metric leads to the same result, $\kappa = 1$.

Let us note that the motion of the Kerr black hole in an external gravitational field is not described by the Papapetrou equation even if one leaves aside the problem of spin-orbit interaction linear in spin. The point is that this equation refers to the case $\kappa = 0$ (Ref. 14.)

It is proven in the same way that for a charged Kerr hole as well the gravimagnetic ratio $\kappa = 1$. Moreover, it can be proven that the electric quadrupole moment of a charged Kerr hole also equals

$$Q = -2\frac{es^2}{m^2},$$

the value, at which the interaction quadratic in spin decreases with energy. It has been shown Ref. 44 that other, higher multipoles of a charged Kerr hole as well possess just those values which guarantee that the interaction of any order in spin (but of course, linear in an external field) asymptotically decreases with increasing energy.

Acknowledgements

We are grateful to I.A. Koop, R.A. Sen'kov, A.N. Skrinsky, and Yu.M. Shatu- nov for useful discussions. We are also grateful to D. Bini, G. Gemelli, and R. Ruffini for the advices due to which, hopefully, the present text has become more comprehensible. We greatly appreciate the warm hospitality extended to us by ICRA. The work was supported by the Russian Foundation for Basic Research through Grant No. 98-02-17797 and by the Federal Program Integration-1998 through Project No. 274.

References

1. L.H. Thomas, *Nature* **117**, 514 (1926); *Phil. Mag.* **3**, 1 (1926).
2. W. de Sitter, *Mon. Not. R. Astron. Soc.* **77**, 155, 181 (1916).
3. L. Schiff, *Phys. Rev. Lett.* **4**, 435 (1959).
4. J. Frenkel, *Z. Phys.* **37**, 243 (1926).
5. V. Bargman, L. Michel, and V. Telegdi, *Phys. Rev. Lett.* **2**, 435 (1959).
6. A. Papapetrou, *Proc. Roy. Soc. London* A **209**, 248 (1951).
7. A. Barducci, R. Casalbuoni, and L. Lusanna, *Nuovo Cimento* A **35**, 389 (1976).
8. F. Ravndal, *Phys. Rev.* D **21**, 2823 (1980).
9. P.L. Nash, *J. Math. Phys.* **25**, 2104 (1984).
10. U. Heinz, *Phys. Lett.* B **144**, 228 (1984);
Ann. Phys. (N.Y.) **161**, 48 (1985).

11. I.B. Khriplovich, *Zh. Eksp. Teor. Fiz.* **96**, 385 (1989) [*Sov. Phys. JETP* **69**, 217 (1989)].
12. J.W. van Holten, *Nucl. Phys.* B **356**, 3 (1991).
13. R.H. Rietdijk and J.W. van Holten, *Class. Quantum Grav.* **9**, 575 (1992).
14. K. Yee and M. Bander, *Phys. Rev.* D **48**, 2797 (1993).
15. J.P. Costella and B.H.J. McKellar, *Int. J. Mod. Phys.* A **9**, 461 (1994).
16. M. Chaichian, R. Gonzales Felipe, and D. Louis Martinez, *Phys. Lett. A* **236**, 188 (1997); E-print archive hep-th/9601119.
17. Ya.I. Azimov and R.M. Ryndin, in *Proceedings of the XXXI PNPI Winter School* (St. Petersburg, 1997); E-print archive hep-ph/9710433, hep-ph/9707468.
18. D.Bini, G.Gemelli, and R.Ruffini, in these *Proceedings.*
19. Ya.S. Derbenev and A.M. Kondratenko, *Zh. Eksp. Teor. Fiz.* **64**, 1918 (1973) [*Sov. Phys. JETP* **37**, 968 (1973)].
20. K. Heinemann, DESY report DESY 96-229; E-print archive physics/9611001.
21. W. Pauli, *Collected Scientific Papers*, eds. R. Kronig and V.F. Weisskopf (John Wiley and Sons, 1964), v. 2, p. 544.
22. H. Batelaan, T.J. Gay, and J.J. Schwendiman, *Phys. Rev. Lett.* **79**, 4517 (1997).
23. T.O. Niinikoski and R. Rosmanith, Nucl.Instr.Meth. A **225**, 460 (1987).
24. K.S. Thorne and J.B. Hartle, *Phys. Rev.* D **31**, 1815 (1985).
25. L.E. Kidder, C.M. Will, and A.G. Wiseman, *Phys. Rev.* **D47**, R4183 (1993).
26. L. Blanchet, T. Damour, B.R. Iyer, C.M. Will, and A.G. Wiseman, *Phys. Rev. Lett.* **74**, 3515 (1995).
27. H.T. Cho, *Class. Quantum Grav.* **15**, 2465, (1998); E-Print archive gr-qc/9703071.
28. I.B. Khriplovich and A.A. Pomeransky, *Phys. Lett.* A **216**, 7 (1996); E-print archive gr-qc/9602004.
29. B.M. Barker and R.F. O'Connell, *Gen. Rel. Grav.* **5**, 539 (1974).
30. A.A. Pomeransky and I.B. Khriplovich, *Zh. Eksp. Teor. Fiz.* **113**, 1537 (1998) [*Sov. Phys. JETP* **86**, 839 (1998)]; E-print archive gr-qc/9710098.
31. V.B. Berestetskii, E.M. Lifshitz, and L.P. Pitaevskii, *Quantum Electrodynamics* (Pergamon Press, 1994).
32. J.D. Bjorken and S.D. Drell, *Relativistic Quantum Mechanics*, (McGraw-Hill, 1964).
33. L.L. Foldy and S.A. Wouthuysen, *Phys. Rev.* **78**, 248 (1951).
34. I.B. Khriplovich, *Parity Nonconservation in Atomic Phenomena* (Gordon and Breach, 1991).
35. I.B. Khriplovich, A.I. Milstein, and R.A. Sen'kov, *Phys. Lett.* A **221**, 370 (1996); *Zh. Eksp. Teor. Fiz.* **111**, 1935 (1997) [*Sov. Phys. JETP* **84**, 1054 (1997)].
36. S. Weinberg, in *Lectures on Elementary Particles and Quantum Field Theory*, eds. S. Deser, M.Grisaru, and H. Pendleton (MIT Press, 1970).
37. S. Ferrara, M. Porrati, and V.L. Telegdi, *Phys. Rev.* D **46**, 3529 (1992).

38. L.D. Landau and E.M. Lifshitz, *The Classical Theory of Fields* (Butterworth-Heinemann, 1975).
39. I.Yu. Kobzarev and L.B. Okun', *Zh. Eksp. Teor. Fiz.* **43**, 1904 (1962) [*Sov. Phys. JETP* **16**, 1343 (1962)].
40. F.W. Hehl, A. Macias, E.W. Mielke, and Yu.N. Obukhov, in *On Einstein's Path, Festschrift for E. Schucking on the Occasion of His 70th Birthday*, ed. A. Harvey (Springer, 1998); E-print archive gr-qc/9706009.
41. S.M. Christensen and M.J. Duff, *Nucl. Phys.* B **154**, 301 (1979).
42. R. Aldrovandi, V.C. de Andrade, and J.G. Pereira, E-print archive gr-qc/9804117 v2.
43. B. Carter, *Phys. Rev.* **174**, 1559 (1968).
44. A.A. Pomeransky, in these *Proceedings*.

EQUATIONS WITH INTRINSIC ROTATION IN GRAVITATIONAL THEORY

LEOPOLD HALPERN
Department of Physics, Florida State University
Tallahassee, FL 32306

The generalization of the gravitational field equations is explored that give rise to the orbits of structureless (geodesic) particle motion as well as the orbits of particles with spin. In the simplest case these equations are equivalent to the Einstein equations with cosmological member of the Cartan-Killing metric of the manifold of the Anti De Sitter group, projected on the Anti De Sitter universe as base space. These equations are generally Einstein's equations with a source nonlinear in the curvature, combined with the equations suggested by C.N. Yang for a gauge theory of gravitation with third derivatives of the metric. The general case is found to be geometrically more complex, with torsion in the higher dimensional space.

General relativity predicts that spinning particles in gravitational fields manifest deviations from a geodetic path.[1] This is true also for zero rest mass polarized particle wave packets. One can conclude that such polarized particles of different spin, like photons, gravitons and zero rest mass neutrinos, which originate within a limited space-time domain of considerable curvature (e.g., a supernova) will reach a very distant observer at different time intervals.[2] Their orbits which depend somewhat on polarization and spin will not be exactly alike. Even within the solar system such polarization dependent effects need not remain undetectable — they are gyroscope experiment with zero rest mass particles. (Helicity for a particle of zero rest mass is strictly conserved with propagation through gravitational fields according to general relativity.) Elementary particle wave and spin are of nonlocal character and can thus not rigorously adjust to the principle of equivalence which applies only to domains where curvature effects are negligible.[3] Spin is on the other hand so fundamental a notion in elementary particle physics that one can feel tempted to treat spinning particles (at least) on an equal footing with structureless particles. This is attempted in a modernized, relativistic version of the principle of inertia in which the simple Anti De Sitter group $G = SO(3,2)$ replaces the Galilei group. The particle orbits are also in this case group orbits but because G is simple, a nonsingular Cartan-Killing metric is defined on the group manifold and the content of the principle becomes much richer: Einstein's equations with a cosmological member are fulfilled by this Cartan-Killing metric[4] on G. Our habitate is now the Anti De Sitter universe, which is the space of right cosets: $B = G/H$, where $H = SO(3,1)$ is the Lorentz subgroup of G. (B is a generalized sphere which can be imagined as imbedded into a five-dimensional space of which two dimensions are imaginary.) All orbits of G on G are geodesis of the metric. The natural projection $\pi : G \to B$ projects the Cartan-Killing metric of G onto a metric of constant curvature — the Anti De Sitter metric on B. A six-dimensional subvariety of the geodesics on G, which belongs to H, project only on a point on B. We call these vertical. The remaining 4-dimenstional variety, which has tangent vectors perpendicular to the vertical, we call horizontal. The horizontal geodesics of G project on geodesics on B. Horizontal orbits with timelike and lightlike tangent vectors describe the motion of structureless particles; spinning particles have besides this suitable vertical components. We require that the horizontal and the vertical components of each tangent vector commute when they are expressed in terms of left invariant vectors on G. This is equivalent to the condition that the spin tensor S^{ik} of the particle fulfills on B:

$$S^{ik}u_k = 0 \qquad u^k u_k = 1 \tag{1}$$

(u^k is the horizontal component of the tangent vector, the four velocity.) Orbits that fulfill these conditions we call admissible. Our principle of inertia takes the form: *"A particle moves along the projection of an admissible orbit of the Anti De Sitter group into the Anti De Sitter universe unless forces act on it."* This principle is at the origin of our approach. We are aware that excellent work has already been done[5] to describe the motion of relativistic spinning particles on the manifold of the Poincaré group and generalized to gravitational fields[6] by the methods of Kibble[7] and Hehl.[8] The adaption of their results to our case is straightfoward, but it is not suited to serve our purpose (unless we adopt even Finsler geometries on G to achieve our goal). They search for Lagrangians to describe spin whereas we are studying alternative geometries which *a priori* admit spinning particle like motions. We proceed by first making use of the natural principal fibre bundle $P(G, H, \pi, B)$ of our group manifold to obtain a generalized metric γ on G which projects by π' on a metric g on B. γ generalizes the Cartan-Killing metric.

$$\gamma_{RS} = Tr(Ad\, A_R \cdot Ad\, As) \tag{2}$$

on G, which is a solution of

$$R_{st} = \frac{1}{4}\gamma_{st} \tag{3}$$

equivalent to the Einstein equations with cosmological member:

$$R_{st} - \frac{1}{2}\gamma_{st}R + \gamma_{st} = 0 \tag{3a}$$

We admit now more general solutions γ which however must still have six Killing vector fields which fulfill the commutation relations of the Lie algebra of the subgroup H. We call these vector fields still vertical. Horizontal vector fields are now perpendicular to the vertical fields with the new metric γ. What we have obtained is equivalent to a Kaluza-Klein gauge theory with the Lorentz group H as gauge group. The resulting gauge field is however truly unified with the metric field (which is in general not true for K-K theories). Depending on our restrictions the gauge field either includes the Riemannian curvature two-form or is equal to it. This K-K approach is the first step in our project. The spin is described in it as an inner quantum number — somewhat like charge — neglecting its space-time properties. As such it correctly reproduces the interaction of the spin with the curvature tensor which occurs in Papapetrou's equations;[a] it completely ignores the spin precession. The Einstein equations on the group manifold G now project on

[a]For an excellent short introduction to these equations see the paper of Bini et al.

a more sophisticated set of equations on the space-time of the base B. A subset of their solutions are all solutions of Einstein's vacuum equations (with cosmological member). Other solutions besides this must also exist. The equations are:

$$B_{ik} - \frac{1}{2}(B+1)g_{ik} = \frac{3}{2}G\left(H_{ijhl}\, H_k{}^{jhl} - \frac{1}{4}g_{ik}\, H_{abcd}\, H^{abcd}\right) \tag{3b}$$

$$H^k{}_{hjl;k} = 0 \tag{3c}$$

Here B_{ik}, B denote the Ricci tensor and curvature invariant on the base. (This is to distinguish them from the corresponding terms on the bundle (see eq. (3a)) which we denote by R, R_{st}, etc.) The tensor denoted by H is the curvature tensor obtained from the curvature 2-form of the Yang-Mills field. Whereas the curvature tensors denoted by B and R are Riemannian (the first in four, the second in ten dimensions) H can have for suitable solutions of eq. (3a) contortion terms of its connection. The covariant derivative of eq. (3c) is however Riemannian. H equals B if torsion vanishes. The eqs. (3c) are equivalent to the equations proposed by C.N. Yang as necessary to occur in a gauge theory of gravitation.[9] He did not propose the combination with eqs. (3b) and thus had difficulties with unphysical solutions. The constant G in eqs. (3b) is of dimension length squared. In the units used in these equations (with the cosmological member equal to unit) one would expect G to be extremely small in order to fit conditions in the solar system. It is suggestive to choose it equal to Newton's gravitational constant (for the value of which we have no explanation hitherto). There is however still another possibility: The constant may indeed be unity however only such solutions be admitted which deviate from general relativity by very little under conditions similar to those in the solar system. The situation may become different in more extreme situations. The right hand member of eq. (3b) leads in any case to some (minor) violations of the principle of equivalence. We meanwhile defer search for further solutions of eqs. (3b,c) because, as we shall see, the next step in our approach will modify even eqs. (3a) in the presence of stronger gravitational fields considerably. To understand the situation, we need to contemplate the commutation relations of the horizontal and vertical vector fields. For the Cartan-Killing metric these are the commutation relations of the Lie algebra of G at every point:

$$[A_R, A_S] = c_{RS}{}^T\, A_T \tag{4}$$

For a generalized metric γ which projects a metric g on the base without torsion, we need to relax only the commutation relations between horizontal vectors:

$$[A_E, A_F]^M = \mathcal{C}(x^i)_{EF}{}^M\, A_M \tag{4a}$$

where A_E, A_F are horizontal and A_M is vertical. $\mathcal{C}_{EF}{}^M$ is a function of the points of the base B. All the other commutation relations which involve vertical vectors remain as in eq. (4). The Jacobi relations determine symmetries between the components of the curvature two-form and the curvature tensor. If we relax eq. (4) further, to admit even horizontal vectors on the right side of eq. (4a), we obtain a geometry which is metric on the base but with a contortion term in the connection.

(Einstein-Cartan geometry) The eqs. (4a) are for every point x^i of B still those of the Lie algebra of a group which is in general not semisimple. Let us denote these groups by $G(x^i)$. Our approximation with the K-K formalism yielded the right force exerted from the curvature on the spin but, as expected, not the right spin motion. We can expect to obtain the right relation between the vertical spin component of the tangent vector and the horizontal orbital component if we replace the metric connection on the bundle at every point x^i by the connection formed there only out of left invariant vectors. This can formally be done[10] and results in contortion — *not* for the connection on the base manifold but for the linear connection of the ten-dimensional manifold, notably with vertical components. The contortion terms are simple functions of the curvature on the base. We see that we have to go beyond the Kaluza-Klein approach and introduce on the group manifold a non-Riemannian geometry[b] (This does not alter the orbits in case of the Cartan-Killing metric.) The spin precession resulting from any accelerated motion is taken into account in the same way. There is however still a back reaction from the acceleration of the vertical component on the horizontal orbital component. This effect can only be taken into account by a more complicated Lagrangian with mixed horizontal and vertical terms (like the third Lagrangian term of ref. 5). Such a term cannot serve our purpose unless we use Finsler geometries. The effect we are considering is negligibly small and overshadowed by uncertainty in quantum theory; yet we believe it should be taken into full consideration in the search for generalizations of gravitational theory. We try to avoid the introduction of Finslerian geometry and try to follow a more physical lead which is not yet ripe for presentation. From the experimental side refinement of all methods to measure intertial and spin motion in gravitational fields offer hope for progress.

Acknowledgments

I would like to thank the late Prof. W. Fairbank and Prof. F. Everitt for numerous stimulating discussions and encouragement.

References

1. A. Papapetrou, Proc. Roy. Soc. **209**, p. 248 (1951).
2. L. Halpern, Proc. 5th Marcel Grossmann Mtg., Part A, p. 1143, World Scientific (1989), D.J. Blair, M. Buckingham, eds.
3. L. Halpern, Nuovo Cim **25**, p. 1239 (1962).
4. L. Halpern, Int. J. Th. Phys. **33**, p. 401 (1994); Astrophys. & Space Sci. **244**, p. 263 (1996).
5. A. Hanson & T. Regge, Ann. Phys. **87**, p. 498 (1974).
6. A. Balachandran et al., Phys. Lett. **89**B, p. 199 (1980).
7. T.W.B. Kibble, J. Math. Phys. **2**, p. 212 (1960).
8. F. Hehl et al., Rev. Mod. Phys. **48**, p. 393 (1976).
9. C.N. Yang, Phys. Rev. Lett. **33**, p. 445 (1974).
10. L.P. Eisenhart, Cont. Groups of Transf., Ch. V, Princeton (1933).

[b] I thank Prof. R. Jantzen for a stimulating discussion on this problem.

SPINNING PARTICLES IN THE KERR FIELD

O. SEMERÁK
Department of Theoretical Physics, Faculty of Mathematics and Physics, Charles University, V Holešovičkách 2, CZ-180 00 Praha 8, Czech Republic;
E-mail: semerak@mbox.troja.mff.cuni.cz

Mathisson-Papapetrou equations are solved numerically to obtain trajectories of spinning test particles in (the meridional section of) the Kerr spacetime. The conditions $p_\sigma S^{\mu\sigma}=0$ are used to close the system of equations. A spin-curvature interaction may in principle lead to considerable deviations from geodesic motion. However, in astrophysical situations probably no large spin effects can be expected for the values of spin consistent with a pole-dipole test-particle approximation.

1 Introduction

We refer to a paper[1] for a full introductory survey and a thorough list of references, giving just a short summary here. Geometrized units ($c=G=1$) and the metric signature ($-+++$) are used, Greek indices running 0-3 and Latin indices 1-3. Index-posed brackets mean antisymmetrization in the outer of the enclosed indices, a semicolon stands for covariant differentiation and a dot for absolute differentiation with respect to the particle's proper time τ.

A spinning particle has to have finite extension, hence it feels tidal gravitational forces, or, spacetime curvature. The role of spin-curvature interaction ranges from possible (partial) support of multiple systems against attraction between masses of their components to modulation of gravitational waves emitted by an orbiting particle and to induction of chaos in the particle's motion. Particles with spin are usually treated in relativity within the approximation which describes a body whose self-field can be neglected (a *test* body) and which is so small compared to the background curvature length scale (thus a test *particle*) that all its multipoles beyond the dipole can be neglected (*pole-dipole* test particle). Free motion of such *pole-dipole test particle*, or *torque-free gyroscope*, is described by Mathisson-Papapetrou (MP) equations,

$$\dot{p}^\mu = -\frac{1}{2}R^\mu{}_{\nu\rho\sigma}u^\nu S^{\rho\sigma}, \qquad \dot{S}^{\mu\nu} = 2\,p^{[\mu}u^{\nu]}, \tag{1}$$

where $p^\mu = mu^\mu - u_\sigma \dot{S}^{\mu\sigma}$ is the total 4-momentum of the particle, $m\,(>0)$ is a scalar parameter (particle's "kinematical" or "monopole" rest mass) which is assumed to be positive, the skew tensor $S^{\mu\nu}$ represents internal angular momentum ("spin") of the particle, the tangent vector to the particle's worldline u^μ is referred to as "kinematical 4-velocity"; $\Gamma^\mu{}_{\rho\sigma}$ and $R^\mu{}_{\nu\rho\sigma} = 2\,(\Gamma^\mu{}_{\iota[\rho}\Gamma^\iota{}_{\sigma]\nu} - \Gamma^\mu{}_{\nu[\rho,\sigma]})$ are Christoffel symbols and Riemann tensor of the background spacetime. Another important quantities are the particle's "dynamical", "total" or "effective" rest mass $\mathcal{M} = \sqrt{-p_\sigma p^\sigma}$, and its "dynamical 4-velocity" $\mathcal{U}^\mu = p^\mu/\mathcal{M}$. If $\mathcal{U}^\mu = u^\mu$, then $\mathcal{M}=m$.

Neither of the mass parameters needs be a constant of motion, namely

$$\mathrm{d}m/\mathrm{d}\tau = \dot{m} = -\dot{u}_\sigma p^\sigma = \dot{u}_\rho u_\sigma \dot{S}^{\rho\sigma}, \tag{2}$$

$$\mathrm{d}\mathcal{M}/\mathrm{d}\tau = \dot{\mathcal{M}} = -\mathcal{U}_\sigma \dot{p}^\sigma = (m\mathcal{M})^{-1}\dot{p}_\rho p_\sigma \dot{S}^{\rho\sigma}. \tag{3}$$

However, the existence of integrals of motion is guaranteed when the background spacetime has some symmetries: for each Killing field ξ^μ one has

$$p_\sigma \xi^\sigma - \frac{1}{2}\,\xi_{\rho;\sigma} S^{\rho\sigma} = \text{const.} \tag{4}$$

(In certain spacetimes also other expressions may be conserved involving a Killing-Yano tensor field.)

Equation (1) are three too few to determine p^μ and $S^{\mu\nu}$ and one needs further conditions, specifying a reference point about which the particle's momentum and spin are calculated. The latter is usually taken to be the particle's centre of mass which, however, is not invariant (it is frame-dependent). Two rival conditions for its definition are mostly imposed as the side constraints —

$$u_\sigma S^{\mu\sigma} = 0, \tag{5}$$

or

$$\mathcal{U}_\sigma S^{\mu\sigma} = 0. \tag{6}$$

The constraints (5) imply $\dot{u}_\sigma \dot{S}^{\mu\sigma} = 0$ and $\mathrm{d}m/\mathrm{d}\tau = 0$; the spin 4-vector, defined by $s^\mu = \frac{1}{2}\,\epsilon^{\nu\mu}{}_{\rho\sigma} u_\nu S^{\rho\sigma}$ ($\epsilon^{\nu\mu}{}_{\rho\sigma}$ is the Levi-Civita tensor) and thus orthogonal to u^μ ($u_\sigma s^\sigma = 0$), is Fermi-transported along the particle's worldline: $\dot{s}^\mu = u^\mu \dot{u}_\sigma s^\sigma$. The other constraints (6) imply $\dot{\mathcal{U}}_\sigma \dot{S}^{\mu\sigma} = 0$ and $\mathrm{d}\mathcal{M}/\mathrm{d}\tau = 0$ (the latter assures that $\mathcal{U}^\mu$ remains timelike, $\mathcal{U}_\sigma \mathcal{U}^\sigma = -1$); the spin 4-vector, in this case defined by $\mathcal{S}^\mu = \frac{1}{2}\,\epsilon^{\nu\mu}{}_{\rho\sigma} \mathcal{U}_\nu S^{\rho\sigma}$ and thus orthogonal to $\mathcal{U}^\mu$ ($\mathcal{U}_\sigma \mathcal{S}^\sigma = 0$), is Fermi-transported along the integral curve of $\mathcal{U}^\mu$: $\dot{\mathcal{S}}^\mu = \mathcal{U}^\mu \dot{\mathcal{U}}_\sigma \mathcal{S}^\sigma$. Both definitions yield $2\,s_\sigma s^\sigma = 2\,\mathcal{S}_\sigma \mathcal{S}^\sigma = S_{\rho\sigma} S^{\rho\sigma}$ and $\mathrm{d}(s_\sigma s^\sigma)/\mathrm{d}\tau = \mathrm{d}(\mathcal{S}_\sigma \mathcal{S}^\sigma)/\mathrm{d}\tau = 0$.

To determine the particle's trajectory, one must know the tangent $u^\mu(\tau)$ which, however, can only be found indirectly from $p^\mu(\tau)$ and $S^{\mu\nu}(\tau)$ and only in the case when one employs the constraints (6):

$$u^\mu = \frac{m}{\mathcal{M}} \left(\mathcal{U}^\mu + \frac{2 S^{\mu\nu} R_{\nu\iota\kappa\lambda} \mathcal{U}^\iota S^{\kappa\lambda}}{4\mathcal{M}^2 + R_{\alpha\beta\gamma\delta} S^{\alpha\beta} S^{\gamma\delta}} \right). \tag{7}$$

Note that in contrast to $\mathcal{U}^\mu$ the kinematical 4-velocity u^μ may *not* everywhere be timelike. This appears to indicate the limits of validity of the MP equations: For sufficiently large spin or/and sufficiently non-homogeneous field, the certain minimum size, required for a spinning particle, may become incompatible with the pole-dipole approximation or with the very test-body approach. We therefore check the character of u^μ within the integration of MP equations and restrict to situations when u^μ can be treated as usual 4-velocity, with $u_\sigma u^\sigma = -1$ ($\Rightarrow \dot{u}_\sigma u^\sigma = 0$). This normalization formula must be used to fix the parameter m in (7), since the latter is *not* constant of motion in that case.

2 Integration of MP equations in the Kerr field

MP equations have often been considered to deduce *spin* behaviour along prescribed orbits. Simple types of particle *motion* have been studied (analytically) in Schwarzschild or Kerr fields.[2–5] However, *actual trajectories* have only been computed by Plyatsko[6] (his monograph contains a thorough list of references), Apostolatos,[7] and Suzuki & Maeda.[8] We will find, in the Kerr field, the trajectories of particles with different spins and compare them with those followed by non-spinning particles (with geodesics starting from the respective initial conditions). Whereas the references given above mostly concentrated on azimuthal motion in the equatorial plane, we are rather interested in motion in the (r,θ)-plane, in particular, in the influence of the outgoing particles' spin on collimation of their trajectories about the axis of symmetry (studied in Ref. 9).

In Boyer-Lindquist coordinates $x^\mu = (t, r, \theta, \phi)$, the Kerr metric reads[10]

$$ds^2 = -\frac{\Delta\Sigma}{\mathcal{A}}\,dt^2 + \frac{\mathcal{A}}{\Sigma}\sin^2\theta\,(d\phi - \omega_K dt)^2 + \frac{\Sigma}{\Delta}\,dr^2 + \Sigma d\theta^2, \tag{8}$$

where $\Delta = r^2 - 2Mr + a^2$, $\Sigma = r^2 + a^2\cos^2\theta$, $\mathcal{A} = (r^2+a^2)^2 - \Delta a^2 \sin^2\theta$, $\omega_K = 2Mar/\mathcal{A}$; M and a is mass and specific angular momentum of the source.

At an initial point $x^\mu_{\rm in}$ in a given Kerr spacetime, we set the dynamical-velocity and the specific-spin magnitudes ($\hat{\mathcal{V}}$, $\hat{\mathcal{S}}/\mathcal{M}$) and directions ($\hat{\alpha}^{\rm v}$, $\hat{\beta}^{\rm v}$; $\hat{\alpha}^{\rm s}$, $\hat{\beta}^{\rm s}$) in the locally non-rotating frame (LNRF) — the local orthonormal triad

$$e^\mu_{\hat{r}} = (0, \sqrt{\Delta/\Sigma}, 0, 0), \quad e^\mu_{\hat{\theta}} = (0, 0, 1/\sqrt{\Sigma}, 0), \quad e^\mu_{\hat{\phi}} = (0, 0, 0, \sqrt{\Sigma/\mathcal{A}}\,\sin^{-1}\theta) \tag{9}$$

of a zero-angular-momentum observer (ZAMO). Initial components of the particle's dynamical 4-velocity and of its specific-spin 4-vector are then

$$\mathcal{U}^\mu = \hat{\Gamma}\,(u^\mu_{\rm ZAMO} + e^\mu_{\hat{\imath}}\hat{\mathcal{V}}^{\hat{\imath}}), \qquad \tilde{\mathcal{S}}^\mu = \hat{\Gamma}\,(u^\mu_{\rm ZAMO}\hat{\mathcal{V}}_{\hat{\imath}}\hat{\mathcal{S}}^{\hat{\imath}}/\mathcal{M} + e^\mu_{\hat{\imath}}\hat{\mathcal{S}}^{\hat{\imath}}/\mathcal{M}), \tag{10}$$

where $u^\mu_{\rm ZAMO} = \sqrt{\Delta^{-1}\Sigma^{-1}\mathcal{A}}\,(1, 0, 0, \omega_K)$ is the 4-velocity of ZAMO,

$$\hat{\Gamma} = -\mathcal{U}_\sigma u^\sigma_{\rm ZAMO} = (1 - \hat{\mathcal{V}}^2)^{-1/2}$$

is the relative boost factor, $\hat{\mathcal{V}}^2 = \hat{\mathcal{V}}_{\hat{\imath}}\hat{\mathcal{V}}^{\hat{\imath}}$, and the LNRF components of relative velocity and of specific relative spin read

$$\hat{\mathcal{V}}^{\hat{\imath}} = \hat{\mathcal{V}}\,(\cos\hat{\alpha}^{\rm v}, \sin\hat{\alpha}^{\rm v}\cos\hat{\beta}^{\rm v}, \sin\hat{\alpha}^{\rm v}\sin\hat{\beta}^{\rm v}), \tag{11}$$

$$\hat{\mathcal{S}}^{\hat{\imath}}/\mathcal{M} = (\hat{\mathcal{S}}/\mathcal{M})\,(\cos\hat{\alpha}^{\rm s}, \sin\hat{\alpha}^{\rm s}\cos\hat{\beta}^{\rm s}, \sin\hat{\alpha}^{\rm s}\sin\hat{\beta}^{\rm s}). \tag{12}$$

The tensor of specific spin is obtained according to $\tilde{S}^{\mu\nu} = \epsilon^{\mu\nu}{}_{\rho\sigma}\mathcal{U}^\rho\tilde{\mathcal{S}}^\sigma$.

We choose the supplementary conditions (6) which were justified and/or used in most papers treating the subject in the past. With these constraints, the integration of the MP equations is straightforward, each integration step,

$$d\mathcal{U}^\mu = -\left(\frac{1}{2}\,R^\mu{}_{\nu\rho\sigma}\tilde{S}^{\rho\sigma} + \Gamma^\mu{}_{\nu\sigma}\mathcal{U}^\sigma\right)u^\nu d\tau\,, \tag{13}$$

$$d\tilde{S}^{\mu\nu} = 2\left(\mathcal{U}^{[\mu}u^{\nu]} + \Gamma^{[\mu}{}_{\rho\sigma}\tilde{S}^{\nu]\rho}u^\sigma\right)d\tau\,, \tag{14}$$

$$dx^\mu = u^\mu d\tau, \tag{15}$$

being "closed" first by calculation of the instantaneous kinematical 4-velocity,

$$u^\mu = \frac{m}{\mathcal{M}}\left(\mathcal{U}^\mu + \frac{2\tilde{S}^{\mu\nu}R_{\nu\iota\kappa\lambda}\mathcal{U}^\iota\tilde{S}^{\kappa\lambda}}{4 + R_{\alpha\beta\gamma\delta}\tilde{S}^{\alpha\beta}\tilde{S}^{\gamma\delta}}\right). \tag{16}$$

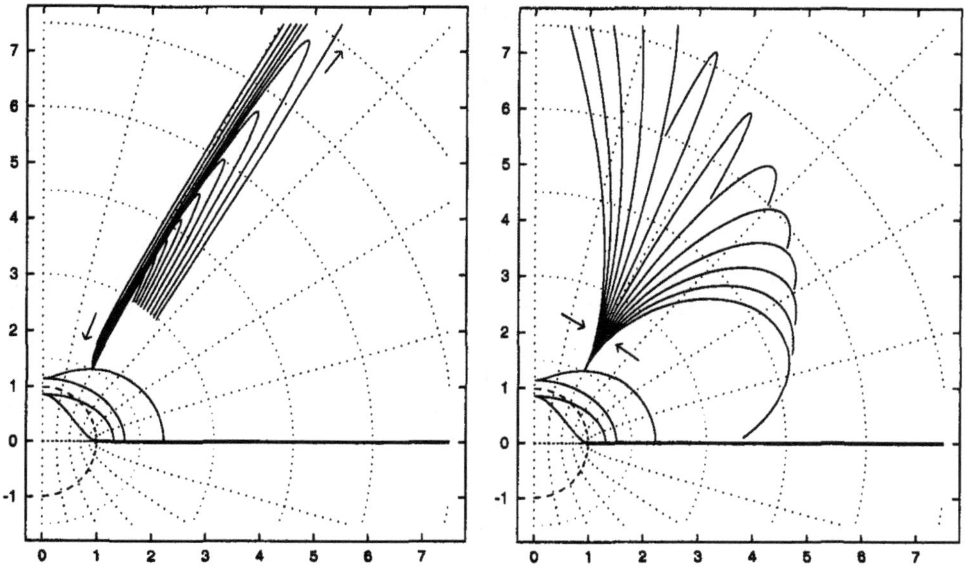

Figure 1. Trajectories of particles ejected radially outwards ($\hat{\alpha}^{\mathrm{v}}_{\mathrm{in}} = 0°$) from the LNRF at $r_{\mathrm{in}} = 1.5M$, $\theta_{\mathrm{in}} = 30°$, from a Kerr black hole with $a = 0.99M$. In the **left** plot, the particles have different initial spins parallel/anti-parallel to $\hat{\mathcal{V}}^{\mu}_{\mathrm{in}}$: going from left to right, you see respectively 6 paths starting with anti-parallel spins — $\hat{\alpha}^{\mathrm{s}}_{\mathrm{in}} = 180°$ and $(\mathcal{M}M)^{-1}\hat{\mathcal{S}}_{\mathrm{in}} = 0.6, 0.5, 0.4, \ldots, 0.1$, then one with zero spin (geodesic), and then 6 paths starting with parallel spins — $\hat{\alpha}^{\mathrm{s}}_{\mathrm{in}} = 0°$ and $(\mathcal{M}M)^{-1}\hat{\mathcal{S}}_{\mathrm{in}} = 0.1, 0.2, 0.3, \ldots, 0.6$. In the **right** plot, initial spins are perpendicular to $\hat{\mathcal{V}}^{\mu}_{\mathrm{in}}$ and lying in the plane of the sheet: going from left to right, you see respectively 6 paths starting with spins pointing to the lower right (towards the equatorial plane) — $\hat{\beta}^{\mathrm{s}}_{\mathrm{in}} = 0°$ and $(\mathcal{M}M)^{-1}\hat{\mathcal{S}}_{\mathrm{in}} = 0.6, 0.5, 0.4, \ldots, 0.1$, then one with zero spin (the same geodesic as in left plot), and then 6 paths starting with spins pointing to the upper left (towards the axis) — $\hat{\beta}^{\mathrm{s}}_{\mathrm{in}} = 180°$ and $(\mathcal{M}M)^{-1}\hat{\mathcal{S}}_{\mathrm{in}} = 0.1, 0.2, 0.3, \ldots, 0.6$. The figure may be compared with figure 2 of Ref. 9, where *spinless* particles were ejected radially with different velocities (and energies) from the same centre.

The metric (8) is stationary and axisymmetric, having Killing vectors $\xi^{\mu}_{(t)} = \delta^{\mu}_{t}$ and $\xi^{\mu}_{(\phi)} = \delta^{\mu}_{\phi}$. According to (4), particle's (minus) total energy $p_t - \frac{1}{2}\, g_{t\rho,\sigma} S^{\rho\sigma}$ and azimuthal angular momentum $p_\phi - \frac{1}{2}\, g_{\phi\rho,\sigma} S^{\rho\sigma}$ are constants of motion.[11] However, they are not sufficient to determine a generous trajectory and we will proceed according to the scheme (13)–(16).

3 Numerical results — trajectories

We used the 4th order Runge-Kutta method, with the step in τ chosen so that the proper spatial integration step $\sqrt{g_{ij}\mathrm{d}x^i\mathrm{d}x^j}$ be $M/100$ along the trajectories.

We present here just several figures thus obtained. They show trajectories projected onto the meridional plane of the Kerr-Schild coordinates, ($\rho = \sqrt{r^2 + a^2}\,\sin\theta$, $z = r\cos\theta$). The values are given in units of M along both axes. The Boyer-Lindquist mesh (the lines of $r =$ const and $\theta =$ const) is illustrated by dotted lines in all plots (the latitudes indicated are $\theta = 0°$, $15°$, ..., $90°$). The region $z > 0$ corresponds to the $(r > 0)$-sheet of the manifold; also a small piece of the $(r < 0)$-sheet is shown $(z < 0)$, where the gravitational field acts in an exactly reverse manner (thus there are neither static limits nor horizons there). The two regions can communicate nowhere but through the $(r = 0)$-surface $(\rho < a,\ z = 0)$; the thick lines going to the right from the very source (the Kerr ring singularity), which is just a point at $\rho = a$, $z = 0$ $(r = 0,\ \theta = 90°)$, indicate that there is no connection through the equatorial planes — the pictures end there from both sides. The source ring is spanned by a spherical region $\rho^2+z^2 < a^2$ $(r^2 < a^2\cos^2\theta)$, shown by dashed lines, which is the only part of Kerr spacetime where momentarily static particles (if they really can exist there) are pulled upwards. The static limit surfaces $(r = M \pm \sqrt{M^2 - a^2\cos^2\theta})$ are represented by solid lines; the inner static limits lead to the singularities. In Fig. 1 (with Kerr black hole of $a = 0.99M$), the horizons $(r = M \pm \sqrt{M^2 - a^2})$ are also shown by solid lines.

The figures confirm that the spin-curvature effect strengthens with spin and with non-homogeneity of the background field, and also with decreasing relative amount of the total energy E resulting from translational motion. With given spin, the *pole-dipole test-particle approximation* is better maintainable (u^μ remains timelike rather) for slow particles. The spin effect is best seen on low-energy particles outgoing from the vicinity of the source. For large spins $(\geq \mathcal{M}M)$, the orbits are often very complicated near the centre (though still in the region where u^μ is timelike). In the figures shown here we follow outgoing particles with $E \doteq (0.67 - 1.06)\mathcal{M}$ and with $\hat{\mathcal{S}}_{\rm in} \leq 0.6\mathcal{M}M$, whose trajectories still tend to have quite intuitive shapes.

The figures prove that the spin-curvature interaction may lead to considerable deviations from geodesic motion. One should ask, however, whether the values of spin which already imply distinct effect may actually occur in nature and whether they are compatible with the pole-dipole test-particle approximation. It seems that the answer is rather negative here. For an electron, for example, the specific spin is $\frac{\hbar/2}{m_e} = 1.93 \cdot 10^{-11}$cm. Hence, when expressed in units of mass of a compact stellar-size centre, with $M = 3M_\odot = 4.43 \cdot 10^5$cm say, we obtain $\hbar/2 = 0.44 \cdot 10^{-16} m_e M$. More promising than elementary particles appear to be the whole stars (orbiting around supermassive black holes in the nuclei of galaxies). Their specific angular momentum $J_{\rm star}/m_{\rm star}$ is considered to range from $10^5 m_{\rm star}$ for a protostar of solar mass to $\leq m_{\rm star}$ for a neutron star.[12] Than, for the most optimistic value and for the centre's mass $M = 10^8 M_\odot$, we arrive at $J_{\rm star} \sim 10^{-3} m_{\rm star} M$. This is *not* negligible, but it is still not more than just a lower limit when *some* spin effect may be expected. Though a normal star thus seems better than a pulsar, it is much harder to maintain the approximation for such a large and easily deformable body. One thus rather expects to find large spin effects in situations described not as a motion of a spinning "particle" in the field of a much more massive body, but rather as a two-particle problem with spins — as e.g. in binary pulsar.[7]

With large spins, u^μ easily becomes spacelike close to the source. If one does not care about it, very illogical motions may be obtained. Some paths already bend suspectedly *near* the region where u^μ turns spacelike. In such situations usually u^μ differs from $\mathcal{U}^\mu$ considerably. A question thus arises whether it would not be

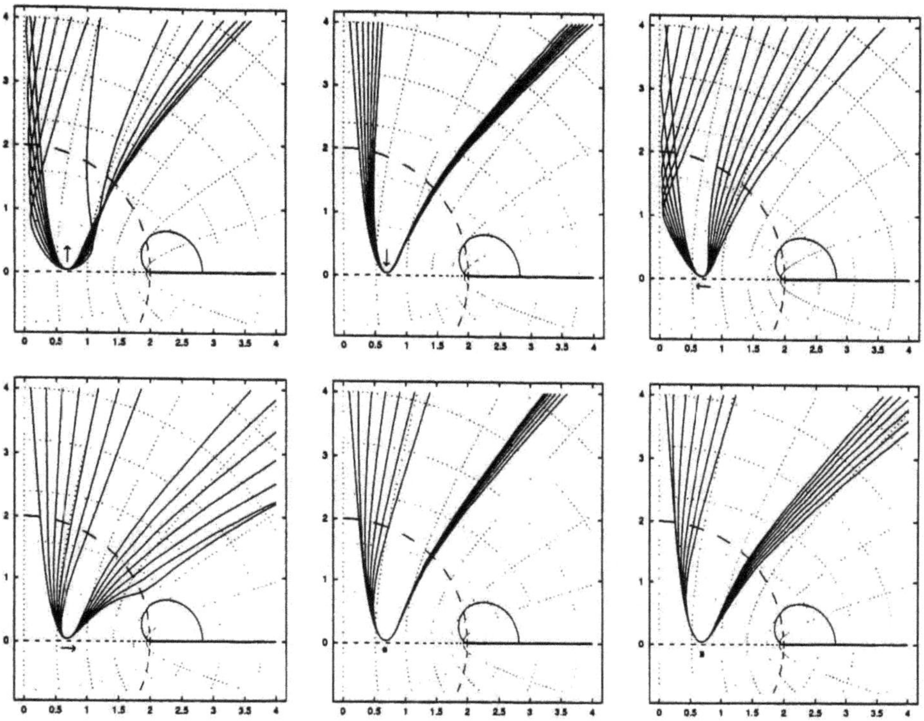

Figure 2. Trajectories of particles ejected in the local latitudinal directions ($\hat{\alpha}^{\rm v}_{\rm in} = 90°$, $\hat{\beta}^{\rm v}_{\rm in} = 0°, 180°$) from the LNRF at $r_{\rm in} = 0.04M$, $\theta_{\rm in} = 20°$, from a Kerr naked singularity with $a = 2M$. All the particles start with the same initial local dynamical speed, $\hat{\mathcal{V}}_{\rm in} = 0.150264$, having different initial spins pointing respectively in each of the main directions: upwards ($\hat{\alpha}^{\rm s}_{\rm in} = 0°$; 1st plot), downwards ($\hat{\alpha}^{\rm s}_{\rm in} = 180°$; 2nd plot), to the left ($\hat{\alpha}^{\rm s}_{\rm in} = 90°$, $\hat{\beta}^{\rm s}_{\rm in} = 180°$; 3rd plot), to the right ($\hat{\alpha}^{\rm s}_{\rm in} = 90°$, $\hat{\beta}^{\rm s}_{\rm in} = 0°$; 4th plot), towards you (i.e., against the source's rotation: $\hat{\alpha}^{\rm s}_{\rm in} = 90°$, $\hat{\beta}^{\rm s}_{\rm in} = 270°$; 5th plot), and away from you (i.e., along the source's rotation: $\hat{\alpha}^{\rm s}_{\rm in} = 90°$, $\hat{\beta}^{\rm s}_{\rm in} = 90°$; 6th plot). Going from left to right in all the plots (as taken where the trajectories leave them), you see respectively 7 paths starting to the left ($\hat{\beta}^{\rm v}_{\rm in} = 180°$) with $(\mathcal{M}M)^{-1}\hat{\mathcal{S}}_{\rm in} = 0.0$ (geodesic), $0.1, 0.2, \ldots, 0.6$, and then 7 paths starting to the right ($\hat{\beta}^{\rm v}_{\rm in} = 0°$) with $(\mathcal{M}M)^{-1}\hat{\mathcal{S}}_{\rm in} = 0.6, 0.5, 0.4, \ldots, 0.0$ (geodesic) in the 1st, 3rd and 5th plots, whereas with $(\mathcal{M}M)^{-1}\hat{\mathcal{S}}_{\rm in} = 0.0$ (geodesic), $0.1, \ldots, 0.6$ in the 2nd, 4th and 6th plots. The figure shows how the spin effects change Fig. 12 of Ref. 9, where a fan of parabolic spinless particles was ejected with $\hat{\mathcal{V}}_{\rm in} = 0.150264$ from the same point in the same spacetime. The trajectories, whose tangents spanned 180° initially, were accelerated in the vertical direction and collimated into a much less angle. Here we take two marginal paths of the fan (given by $\hat{\alpha}^{\rm v}_{\rm in} = 90°$) — that starting to the left ($\hat{\beta}^{\rm v}_{\rm in} = 180°$) and that starting to the right ($\hat{\beta}^{\rm v}_{\rm in} = 0°$) — and find their shapes for different initial spins.

better to fix, instead of the initial $\mathcal{U}^\mu$, the initial u^μ (which is the tangent to the trajectory), even if the integration logic requires, in case of our choice of the spin supplementary condition (6), to *calculate* u^μ from $\mathcal{U}^\mu$ and $S^{\mu\nu}$ in each step. And should not one rather adhere to the other supplementary condition, (5), as also suggested by Ref. 13, even though the relation between u^{μ} and p^{μ} [the counterpart of eq. (7)] is not known in this case?

It was demonstrated by Ref. 14 that even when keeping just second-order terms in the particle's spin or velocity or in the centre's gravitational radius, the different supplementary conditions lead to different non-geodesic equations of motion. It would therefore be interesting to compare the trajectories obtained with different supplementary conditions. This is our aim for the future.

Acknowledgements

I thank for support from Czech grants GACR-202/96/0206 and GAUK-230/96.

References

1. O. Semerák, "Spinning test particles in the Kerr field I.", submitted to *Mon. Not. R. Astron. Soc.*
2. E. Corinaldesi and A. Papapetrou, *Proc. Roy. Soc. A* **209**, 259 (1951).
3. S.N. Rasband, *Phys. Rev. Lett. 30*, 111 (1973).
4. K.P. Tod, F. de Felice and M. Calvani, *Nuovo Cim.* **34B**, 365 (1976).
5. M.A. Abramowicz and M. Calvani, *Mon. Not. R. Astron. Soc.* **189**, 621 (1979).
6. R.M. Plyatsko, The manifestations of the ultra-relativistic spin-orbital interaction (Naukova Dumka, Kiev 1988) (in Ukrainian).
7. T.A. Apostolatos, *Class. Quantum Grav.* **13**, 799 (1996).
8. S. Suzuki and K. Maeda, *Phys. Rev. D* **55**, 4848 (1997).
9. J. Bičák, O. Semerák and P. Hadrava, *Mon. Not. R. Astron. Soc.* **263**, 545 (1993).
10. C.W. Misner, K.S. Thorne and J.A. Wheeler, Gravitation (Freeman, San Francisco 1973).
11. R. Micoulaut, *Zeits. Phys.* **206**, 394 (1967) (in German).
12. F. de Felice and L. Di G. Sigalotti, *Astrophys. J.* **389**, 386 (1992).
13. S. Ragusa and M. Bailyn, *Gen. Rel. Grav.* **27**, 163 (1995).
14. B.M. Barker and R.F. O'Connell, *Gen. Rel. Grav.* **5**, 539 (1974).

ENERGY FIRST INTEGRAL FOR SPINNING PARTICLES IN THE SCHWARZSCHILD BACKGROUND

D. BINI
International Center for Relativistic Astrophysics,
Università di Roma, I–00185 Roma, Italy.
Istituto per Applicazioni della Matematica, C.N.R. , I–80131 Napoli, Italy.
e-mail: binid@icra.it

G. GEMELLI
Istituto Matematico G. Castelnuovo,
Università di Roma, Piazzale A. Moro 2, I-00185 Roma, Italy.
e-mail: gemelli@mat.uniroma1.it

R. RUFFINI
International Center for Relativistic Astrophysics,
Università di Roma, I–00185 Roma, Italy.
e-mail: ruffini@icra.it

The energy first integral for a spinning particle in the Papapetrou model, in the Schwarzschild background, is calculated and then transformed according to Barker & O'Connell center-of-mass shift law. Thus its expression reduces approximatively to that of non-spinning test particle.

1 Papapetrou equations in the Schwarzschild background

Papapetrou equations for a pole-dipole (spinning) particle, are the following[1]:

$$\begin{aligned} (D/d\tau_U)P^\alpha &= (1/2)R_{\rho\sigma\beta}{}^\alpha S^{\rho\sigma}U^\beta \\ (D/d\tau_U)S^{\alpha\beta} &= P^\alpha U^\beta - P^\beta U^\alpha \end{aligned} \tag{1}$$

Here $P^\alpha = M_0U^\alpha - U_\beta(D/d\tau_U)S^{\alpha\beta}$ is the (generalized) momentum vector; $M_0 = -P^\alpha U_\alpha$ is the proper mass; $S^{\alpha\beta}$ is the spin tensor ($S^{(\alpha\beta)} = 0$); U is the unit tangent vector ($U^\alpha U_\alpha = -1$) to the particle worldline ℓ_U. Let $U = (D/d\tau_U)X$, where $X = X(\tau_U)$ is the center point which describes ℓ_U.

The center point X in Papapetrou's model is a generic center for the multipole reduction of the stress-energy tensor of the full extended body[1–5]; therefore there are three degrees of freedom and the model must be completed by a choice of suitable supplementary conditions (SC). The different choices which can be found in the literature are the following:

(CP): $u_\alpha S^{\alpha\beta} = 0$ where u is a given reference frame (suggested by the background).

(P): $U_\alpha S^{\alpha\beta} = 0$

(T): $P_\alpha S^{\alpha\beta} = 0$

Once a choice of (SC) is made, ℓ_U, U, P and S are determined by the differential system (1); these fields therefore depend on SC,[5] i.e. a rigorous notation should be: $X^{(SC)}$, $U^{(SC)}$, $P^{(SC)}$, $S^{(SC)}$.

For a generic reference frame u we introduce the following splitting of U and P:

$$\begin{aligned} U &= \gamma_{(U,u)}[u + \nu_{(U,u)}] \\ P &= ||P||\bar{u} = E_{(P,u)}[u + \nu_{(\bar{u},u)}] \end{aligned} \tag{2}$$

where $E_{(P,u)} = \gamma_{(U,u)}||P||$ and $\gamma_{(U,u)}$ is the Lorentz factor.

In another paper[6] we studied the motion of a charged spinnning particle in the Reissner-Nordström spacetime. Let us now consider the background of Schwarzschild

$$ds^2 = -\Delta r^{-2} dt^2 + r^2 \sin^2\theta d\varphi^2 + r^2\Delta^{-1}dr^2 + r^2 d\theta^2 \tag{3}$$

where $\Delta = r^2 - 2Mr$.

Let us introduce the family of static observers, i.e. $u = m$, where $m^\alpha = r\Delta^{-1/2}\delta^\alpha{}_t$. Let us moreover introduce the unitary spatial vector n: $n^\alpha = \sqrt{\Delta} r^{-1}\delta^\alpha{}_r$.

2 Energy first integral

For equations (1), in the Schwarzschild background, under the supplementary condition CP, we have the energy first integral[7]:

$$E = -P_0 = cost. \tag{4}$$

which, in our variables, is:

$$E = \gamma_{(U,m)}\left[E_{(P,m)}\sqrt{1 - 2M/r} + \gamma_{(U,m)}\frac{M}{r^2}n \cdot [S_{(m)} \times_m \nu_{(U,m)}]\right], \tag{5}$$

where $r = |X_{(CP)}|$.

Let us assume "slow" motion, so that we can neglect squares of the dynamical variables (and therefore $\gamma \cong 1$) and "very large" distances, so that we can neglect factors of r^{-3} or more. In this approximation we can write:

$$E \cong M_0\Big(1 - \frac{M}{r} - \frac{M^2}{r^2}\Big) + \frac{M}{r^2}n \cdot (S \times v) \tag{6}$$

where $M_0 = -P \cdot U \cong E_{(P,m)}$.

If we then apply the following transformation[8]:

$$X_{(CP)} = X_{(T)} + \frac{1}{M_0}v_{(CP)} \times S, \tag{7}$$

which corresponds to the center of mass shift from $X_{(CP)}$ to $X_{(T)}$ (or, equivalently, $X_{(P)}$), we have:

$$E \cong M_0\Big(1 - \frac{M}{r'} - \frac{M^2}{r'^2}\Big), \tag{8}$$

with $r' = |X_{(T)}|$.

According to this transformation the expression of the Killing energy of the spinning particle therefore becomes (approximatively) that of non-spinning test particle:

$$\mathcal{E} = M_0\sqrt{1 - 2M/r},$$

provided one considers it at the proper center of mass $X_{(T)}$ rather than at $X_{(CP)}$, which is the centroid with respect to the observer congruence m.

References

1. A. Papapetrou, *Proc. Roy. Soc. London* **209**, 248 (1951).
2. W.G. Dixon, *Proc. Roy. Soc. London* A **314**, 499 (1970).
3. W.G. Dixon, *Proc. Roy. Soc. London* A **319**, 509 (1970).
4. W.G. Dixon, *Philos. Trans. R. Soc. London.* A **277**, 59 (1974).
5. J. Ehlers, E. Rudolph, *Gen. Rel. Grav.* **8**, 197 (1977).
6. D. Bini, G. Gemelli, R. Ruffini, "Nongeodesic motion of charged spinning test particles", in *The First ICRA Network and The Third William Fairbank Meeting on the Lense-Thirring effect*, (1998).
7. E. Corinaldesi, A. Papapetrou, *Proc. Roy. Soc. London* **209**, 259 (1951).
8. B.M. Barker, R.F. O'Connell, *Gen. Rel. Grav.* **5**, 539 (1974).

NONGEODESIC MOTION OF CHARGED SPINNING TEST PARTICLES

D. BINI
Istituto per Applicazioni della Matematica, C.N.R. , I–80131 Napoli, Italy and International Center for Relativistic Astrophysics, Università di Roma, I–00185 Roma, Italy.
e-mail: binid@icra.it

G. GEMELLI
Istituto Matematico G. Castelnuovo and International Center for Relativistic Astrophysics, Università di Roma, Piazzale A. Moro 2, I-00185 Roma, Italy
e-mail: gemelli@mat.uniroma1.it

R. RUFFINI
International Center for Relativistic Astrophysics, Università di Roma, I–00185 Roma, Italy.
e-mail: ruffini@icra.it

We study the Papapetrou-Dixon-Souriau model for the motion of a charged massive particle. We discuss supplementary conditions and the role of different families of observers. In this framework we consider spin-gravity and spin-electromagnetism coupling terms for an orbit in the Reissner-Nordström background, generalizing the picture given in the literature.

1 Introduction

In this paper we consider spin-electromagnetic and spin-gravitational coupling terms in the Papapetrou-Dixon-Souriau model for the orbit of a charged massive particle. Different supplementary conditions, with their corresponding meaning in terms of the center-of-mass worldline definition, are discussed. We provide the splitting of the equations of motion with respect to a frame associated with a generic 4-velocity field u. As an application, we consider the Reissner-Nordström background and study the evolution of a particle with respect to the family of static observers. We thus provide, in a unique framework, the exact generalization of earlier results belonging to the purely electromagnetic case[1–3] and the purely gravitational case.[2,4] We also discuss the weak field slow motion approximation generalizing the picture given in the literature. Complete details will be given in a forthcoming paper.

2 Equations of motion

The Dixon-Souriau equations of motion of a spinning test particle in a given gravitational and electromagnetic background[5–8] are given by:

$$\begin{aligned} Dp^\alpha/d\tau_U &= (1/2)R_{\rho\sigma\beta}{}^\alpha S^{\rho\sigma}U^\beta + eF^\alpha{}_\beta U^\beta - (1/2)\lambda S^{\mu\nu}\nabla^\alpha F_{\mu\nu}\ , \\ DS^{\alpha\beta}/d\tau_U &= p^\alpha U^\beta - p^\beta U^\alpha + \lambda[S^{\alpha\mu}F_\mu{}^\beta - S^{\beta\mu}F_\mu{}^\alpha]\ , \end{aligned} \qquad (1)$$

where $R_{\alpha\beta\rho\sigma}$ is the Riemann tensor, p^α is the (generalized) momentum vector, $S^{\alpha\beta}$ is a (antisymmetric) spin tensor, $F^{\alpha\beta}$ is the electromagnetic field, e is the electric charge, λ is a spin-electromagnetism coupling scalar and $U = DX/d\tau_U$ is the unit tangent vector ($U^\alpha U_\alpha = -1$) of the "center line" ℓ_U used to make the multipole reduction, and where $X = X(\tau_U)$ is the center point whose world line is ℓ_U. Units are chosen in order that the speed of light in empty space satisfies $c = 1$.

The classic Papapetrou's scheme is obtained from (1) by assuming $F = 0$, i.e. by neglecting the electromagnetic field.

It is convenient to introduce the following notation for the spin-gravity and spin-electromagnetism coupling terms: $\mathcal{R}_{\alpha\beta} = (1/2)R_{\alpha\beta\mu\nu}S^{\mu\nu}$, $Q_\alpha = \frac{1}{2}S^{\mu\nu}\nabla_\alpha F_{\mu\nu}$ and $N^{\alpha\beta} = S^{\alpha\mu}F_\mu{}^\beta - S^{\beta\mu}F_\mu{}^\alpha$.

Let us assume a) $p = ||p||\bar{u} = M_0\bar{u}$, defined along ℓ_U, is timelike: $\bar{u}^\alpha\bar{u}_\alpha = -1$ and b) U and $\bar{u}$ may be extended in a regular way in a neighbourhood of the line ℓ_U. We then have two timelike congruences, $\bar{u}$ and U, at our disposal. Both $\bar{u}$ and U are associated with the particle in a natural way and so are both candidates for defining the proper rest frame of the particle.

It is well known that the number of independent equations in (1) is less than that of the unknown quantities: 3 additional scalar supplementary conditions (SC) are needed for the scheme to be completed. Once a suitable choice has been made, ℓ_U, p and S are determined by the equations.

The different supplementary conditions used in the literature are:

(CP) $S_{\alpha\beta}u^\beta = 0$ (Corinaldesi-Papapetrou condition,[9,4]) where u is a (known) preferred family of observers usually suggested by the background;

(T) $S_{\alpha\beta}\bar{u}^\beta = 0$, where $\bar{u} = p/||p||$ (Tulkzyjew's condition[10–13]);

(P) $S_{\alpha\beta}U^\beta = 0$ (Pirani's condition[2,14,15]).

Clearly the fields ℓ_U, p and S all depend on the choice of supplementary conditions[13] so a more precise notation would be: $X_{(SC)}$, $U_{(SC)}$, $p_{(SC)}$, $S_{(SC)}$, where the index values $SC = CP, T, P$ correspond to these choices. This cumbersome notation will be avoided when possible.

3 Splitting with respect to a generic reference frame

In general relativity a reference frame is defined by a congruence of timelike curves, say the set of the world lines of a family of observers. We denote by u the unit tangent vector ($u \cdot u = -1$) of the world lines of a generic reference frame. The splitting of the spacetime along u and its orthogonal local rest space LRS_u gives the measurement relative to u of any tensor field defined on a domain of the spacetime; similarly one can obtain the formulation relative to u of any tensor equation (see Ref. 16 for more details).

Let u be a generic observer field. The splitting of U and p along u and LRS_u is:

$$\begin{aligned} U &= \gamma_{(U,u)}[u + \nu_{(U,u)}] \\ p &= ||p||\bar{u} = E_{(p,u)}u + p_{(u)} = M_0\gamma_{(\bar{u},u)}[u + \nu_{(\bar{u},u)}] \end{aligned} \tag{2}$$

where $E_{(p,u)} = \gamma_{(\bar{u},u)}||p|| = \gamma_{(\bar{u},u)}M_0$. Let us introduce the following notation for the electric and magnetic parts of the antisymmetric 2-tensor fields S, F, N and $\mathcal{R}$ which appear in the equations: $L_{(u)} = S^{(E)}_{(u)}$, $S_{(u)} = S^{(M)}_{(u)}$, $E_{(u)} = F^{(E)}_{(u)}$, $B_{(u)} = F^{(M)}_{(u)}$, $N_{(u)} = N^{(E)}_{(u)}$, $M_{(u)} = N^{(M)}_{(u)}$, $K_{(u)} = \mathcal{R}^{(E)}_{(u)}$, $T_{(u)} = \mathcal{R}^{(M)}_{(u)}$, $Q_{(u)} = Q - q_{(u)}u$, $q_{(u)} = -u \cdot Q$. The background is assumed to be completely known. With these definitions, the equations of motion (1) are equivalent to the following set:

$$\begin{aligned}
&\gamma[E_{(p,u)}\nu_{(\bar{u},u)}]^{\cdot} = E_{(p,u)}F^{(G)} + P_{(U,u)}[-K_{(U)} + eE_{(U)}] - \lambda Q_{(u)} \\
&\gamma\dot{E}_{(p,u)} = E_{(p,u)}\nu_{(\bar{u},u)} \cdot F^{(G)} - \lambda q_{(u)} + \nu_{(U,u)} \cdot P_{(U,u)}[eE_{(U)} - K_{(U)}] \\
&\gamma\dot{L}_{(u)} = S_{(u)} \times F^{(G)} + \gamma E_{(p,u)}[\nu_{(U,u)} - \nu_{(\bar{u},u)}] + \lambda N_{(u)} \\
&\gamma\dot{S}_{(u)} = F^{(G)} \times L_{(u)} + +\gamma E_{(p,u)}\nu_{(\bar{u},u)} \times \nu_{(U,u)} + \lambda M_{(u)}
\end{aligned} \tag{3}$$

where, for the sake of brevity, we have denoted $\gamma = \gamma_{(U,u)}$, $F^{(G)} = F^{(G)}_{(\mathrm{fw},U,u)}$ and $(\;)^{\cdot} = D_{(\mathrm{fw},U,u)}/d\tau_{(U,u)}$.

The acceleration of the particle relative to u can be described in principle by the Fermi-Walker spatial derivative of either $\nu_{(U,u)}$ or $\nu_{(\bar{u},u)}$, the two descriptions being not equivalent. The acceleration corresponding to $\nu_{(\bar{u},u)}$ is given by (3). On the right hand side of the first equation of (3) are terms corresponding to the gravitational field, the electromagnetic field, and a sum of spin-gravity-electromagnetism coupling terms. It is convenient to introduce the following (relative) tidal acceleration, due entirely to the spin of the particle, to represent this coupling:

$$\gamma_{(U,u)}A_{(\bar{u},u)} = -P_{(U,u)}K_{(U)} - \lambda Q_{(u)} \tag{4}$$

We also define an analogous term $A_{(U,u)}$:

$$\gamma A_{(U,u)} = \gamma[E_{(p,u)}\nu_{(U,u)}]^{\cdot} - E_{(p,u)}F^{(G)} - eP_{(U,u)}E_{(U)} \tag{5}$$

In (3) we have a law which allows us to shift from $\nu_{(\bar{u},u)}$ to $\nu_{(U,u)}$. Thus, by differentiation we have:

$$A_{(U,u)} = A_{(\bar{u},u)} - [\gamma^{-1}S_{(u)} \times_u F^{(G)}]^{\cdot} - \lambda[\gamma^{-1}N_{(u)}]^{\cdot} + \ddot{L}_{(u)} \tag{6}$$

Clearly the relative acceleration (6) leads to different spin-gravity and spin-electromagnetism coupling terms than the tidal acceleration (4).

The choice of the supplementary condition is fundamental: 1) it determines the world line ℓ_U where all the fields we are dealing with are defined; 2) the equations are not formally invariant for different choices, since they imply different relationships between $L_{(u)}$ and $S_{(u)}$. From (2) we have:-

$$\text{(P):} \qquad L_{(u)} = S_{(u)} \times \nu_{(U,u)} \;, \tag{7}$$

$$\text{(T):} \qquad L_{(u)} = S_{(u)} \times \nu_{(\bar{u},u)} \;, \tag{8}$$

$$\text{(CP):} \qquad L_{(u)} = 0 \;. \tag{9}$$

When substituted into the first equation of (3) or (6) these three conditions lead to a total of 6 different expressions for the sum of the spin-gravity and the spin-electromagnetic couplings.

4 Motion in the Reissner-Nordström background

Let us consider the background of a static black hole of mass M and charge Q, described by the Reissner-Nordström line element in Boyer-Lindquist coordinates:

$$ds^2 = -\Delta r^{-2}dt^2 + r^2\sin^2\theta d\varphi^2 + r^2\Delta^{-1}dr^2 + r^2 d\theta^2 \ , \tag{10}$$

where $\Delta = r^2 - 2Mr + Q^2$. The associated electromagnetic field is $F = (Q/r^2)dr\wedge dt$. Let us study the motion of a spinning particle satisfying equations (1), with respect to the family of static observers: $m^\alpha = r\Delta^{-1/2}\delta^\alpha{}_t$. Then $F^{(G)}_{(\mathrm{fw},U,m)} = -\gamma_{(U,m)}a_{(m)}$, where $a_{(m)} = (Mr - Q^2)/(r^2\sqrt{\Delta})n$ and $n = \sqrt{\Delta}r^{-1}\partial_r$.

Let us introduce the following approximations:

H1. Slow motion: squares of $\nu_{(U,m)}$, $\nu_{(\bar{u},m)}$, $L_{(m)}$, $S_{(m)}$ are negligible.

H2. Large distance: factors of r^{-4} are negligible.

Under these assumptions we have:

$$\left[\frac{Mr - Q^2}{r^4} + \frac{\lambda Q}{\gamma r^2}\frac{\sqrt{\Delta}}{r^2}\right] \cong \frac{1}{3}\left[\frac{3Mr - 4Q^2}{r^4} + 3\frac{\lambda Q}{\gamma r^2}\frac{\sqrt{\Delta}}{r^2}\right] \cong \frac{M + \lambda Q}{r^3}. \tag{11}$$

Let us denote, for the sake of brevity: $v \cong \nu_{(U,m)}$, $\bar{v} \cong \nu_{(\bar{u},m)}$, $M_0 \cong E_{(p,m)}$, $S \cong S_{(m)}$, $L \cong L_{(m)}$. We then have:

$$\begin{aligned} K_{(m)} &\cong \frac{3M}{r^3}\left[-\frac{1}{3}L + (L\cdot n)n\right], \\ T_{(m)} &\cong \frac{3M}{r^3}\left[-\frac{1}{3}S + (S\cdot n)n\right], \\ Q_{(m)} &\cong \frac{3Q}{r^3}\left[-\frac{1}{3}L + (L\cdot n)n\right]. \end{aligned} \tag{12}$$

The acceleration laws are then:

$$\begin{aligned} \dot{\bar{v}} &\cong -\frac{M + \lambda Q}{r^2}n - \frac{M^2 - Q^2}{r^3}n + \frac{1}{M_0}a_{(\bar{u},m)}, \\ \dot{v} &\cong -\frac{M + \lambda Q}{r^2}n - \frac{M^2 - Q^2}{r^3}n + \frac{1}{M_0}a_{(U,m)} \ ; \end{aligned} \tag{13}$$

where we have introduced the approximate spin-gravity-electromagnetism coupling terms:

$$\begin{aligned} a_{(\bar{u},m)} &= \frac{3M}{r^3}\left[-\frac{2}{3}v\times S+[S\cdot(n\times v)]n+(v\cdot n)n\times S+\frac{1}{3}L-(L\cdot n)n\right] \\ &+\frac{3\lambda Q}{r^3}\left[\frac{1}{3}L-(L\cdot n)n\right], \\ a_{(U,m)} &= \frac{3M}{r^3}\left[-v\times S+[S\cdot(n\times v)]n+2(v\cdot n)n\times S+\frac{1}{3}L-(L\cdot n)n\right] \\ &+\frac{3\lambda Q}{r^3}\left[-\frac{1}{3}v\times S+(v\cdot n)n\times S+\frac{1}{3}L-(L\cdot n)n\right] \\ &+\left[\frac{M+\lambda Q}{r^2}-\frac{Q^2}{r^3}\right]M_0(\bar{v}\times v)\times n+\ddot{L}, \end{aligned} \tag{14}$$

such that $a_{(\bar{u},m)}\cong A_{(\bar{u},m)}$, $a_{(U,m)}\cong A_{(U,m)}$; and where

$$\bar{v}\cong v+\frac{1}{M_0}\left[\frac{M+\lambda Q}{r^2}+\frac{M^2-Q^2}{r^3}\right]n\times S-\frac{1}{M_0}\dot{L}. \tag{15}$$

As for the factor λ, a natural choice[5] is to set: $\lambda=-e/M_0$.

As for the evolution laws of mass and spin we have:

$$\begin{aligned} \dot{M}_0 &\cong -M_0\left[\frac{M}{r^2}+\frac{M^2-Q^2}{r^3}\right](\bar{v}\cdot n)+\frac{eQ}{r^2}(v\cdot n) \\ &-\frac{3M}{r3}\left[-\frac{1}{3}(v\cdot L)+(L\cdot n)(v\cdot n)\right], \\ \dot{S} &\cong \left[\frac{M+\lambda Q}{r^2}+\frac{M^2-Q^2}{r^3}\right]L\times n+M_0\bar{v}\times v\,. \end{aligned} \tag{16}$$

The discussion so far holds for a generic center of mass worldline. It will now be interesting to specialize it to a particular worldline corresponding to one of the three choices of supplementary conditions. To distinguish the spin-gravity-electromagnetism coupling terms which occur in the relationship of the acceleration to the velocity and spin fields for different supplementary conditions (in correspondance with the different expressions for L as a function of S and v), the explicit SC subscript notation for $a_{(\bar{u},m)(SC)}$ and $a_{(U,m)(SC)}$ will be used.

The case (CP). Here $L\equiv 0$ so the picture is simplified:

$$\begin{aligned} a_{(\bar{u},m)(CP)} &= \frac{3M}{r^3}\left[-\frac{2}{3}v\times S+[S\cdot(n\times v)]n+(v\cdot n)n\times S\right], \\ a_{(U,m)(CP)} &= \frac{3M}{r^3}[-v\times S+[S\cdot(n\times v)]n+2(v\cdot n)n\times S] \\ &+\frac{3\lambda Q}{r^3}\left[-\frac{1}{3}v\times S+(v\cdot n)n\times S\right]. \end{aligned} \tag{17}$$

The case (T). In this case $L = S \times \bar{v}$ and consequently we have:

$$
\begin{aligned}
a_{(\bar{u},m)(T)} &= \frac{3M}{r^3}[-v \times S + 2[S \cdot (n \times v)]n + (v \cdot n)n \times S] \\
&+ \frac{3\lambda Q}{r^3}\left[-\frac{1}{3}v \times S + [S \cdot (n \times v)]n\right], \\
a_{(U,m)(T)} &= \frac{3M}{r^3}[-v \times S + 2[S \cdot (n \times v)]n + (v \cdot n)n \times S] \\
&+ \frac{3\lambda Q}{r^3}[-v \times S + [S \cdot (n \times v)]n + 2(v \cdot n)n \times S].
\end{aligned} \tag{18}
$$

The case (P). Here $L = S \times v$ and so we have:

$$
\begin{aligned}
a_{(\bar{u},m)(P)} &= \frac{3M}{r^3}[-v \times S + 2[S \cdot (n \times v)]n + (v \cdot n)n \times S] \\
&+ \frac{3\lambda Q}{r^3}\left[-\frac{1}{3}v \times S + [S \cdot (n \times v)]n\right].
\end{aligned} \tag{19}
$$

Thus both $a_{(\bar{u},m)(P)}$ and $a_{(\bar{u},m)(T)}$ have the same dependence on the fields S, n, r, v, but the latter fields are defined on different worldlines for the two cases.

An expression for the quantity $a_{(U,m)(P)}$ cannot be given without solving differential equations, a problem related to the fact that under the supplementary condition (P), the full system of equations is a second order system.[10,17] However, if by analogy with the (CP) and (T) cases, we "a priori" assume that $a_{(U,m)(P)}$ is proportional to S:

H3. $a_{(U,m)(P)} \times S \cong 0$;

then for $a_{(U,m)(P)}$ we find an expression which is formally identical to that of $a_{(U,m)(T)}$. Therefore the P and T models can be considered equivalent.

Only two of the acceleration vectors of our description, namely $a_{(U,m)(CP)}$ and $a_{(\bar{u},m)(P)}$, have been previously calculated, and only for the purely gravitational case (Papapetrou's equations: $Q = e = \lambda = 0$) in the paper by Barker and O'Connell[4] where they were denoted by $a_{S(CP)}$ and $a_{S(P)}$, respectively.

Again in the purely gravitational case, it is interesting to note that the same expression for $a_{(\bar{u},m)(P)}$ is the result of a different approach, that of the post-Minkowskian calculation of the spin-orbit interaction of the two-body problem, as carried out by Damour;[18] it suffices to consider the limiting case when one of the bodies is very heavy.

The classical interaction of spin with the gravitational field is usually obtained from the spin-orbit interaction potential: $V_{(G)} = \frac{3}{2}\frac{M}{r^2}S \cdot (n \times v)$ (see e.g. Refs. 2, 4 and 19); the corresponding acceleration is

$$
a_{(G)} = \frac{3M}{r^3}\left[-v \times S + \frac{3}{2}(n \cdot v)n \times S + \frac{3}{2}[S \cdot (n \times v)]n\right]. \tag{20}
$$

Similarly, for the interaction of spin with the electromagnetic field, using the Thomas interaction potential: $V_{(E)} = -\frac{1}{2}\frac{\lambda Q}{r^2}S \cdot (n \times v)$ we have the following expression for the acceleration:[2]

$$a_{(E)} = \frac{3\lambda Q}{r^3}\left[\frac{1}{3}v \times S - \frac{1}{2}(n\cdot v)n \times S - \frac{1}{2}[S\cdot(n\times v)]n\right]. \tag{21}$$

We are going to compare the classical results (20) and (21) with those coming from relativistic models.

In the Schwarzschild limit ($Q \to 0$) it is possible to define an auxiliary center point $X_{(NR)}$ which behaves like a non-relativistic center of mass, that is its evolution involves a spin-gravity coupling term formally identical to $a_{(G)}$.[4] To this end let us introduce

$$X_{(NR)} = X_{(CP)} - \frac{1}{2M_0}v_{(CP)} \times S \tag{22}$$

and the associated evolution equation

$$\ddot{X}_{(NR)} = \dot{v}_{(NR)} \cong -\frac{M+\lambda Q}{r^2_{(NR)}}n_{(NR)} - \frac{M^2-Q^2}{r^3_{(NR)}}n_{(NR)} + \frac{1}{M_0}a_{(NR)}\,, \tag{23}$$

where $r_{(NR)} = |X_{(NR)}|$. Equation (23) is indeed a definition for $a_{(NR)}$. It results $a_{(NR)} \cong a_{(G)}$, provided one replaces r and v in the expression (20) with $r_{(NR)}$ and $v_{(NR)}$ respectively.

Moreover, in the Schwarzschild case, the following "shift of the center of mass law" was found consistent with the Papapetrou equations by Barker and O'Connell in Ref. 4:

$$X_{(CP)} \cong X_{(T)} + \frac{1}{M_0}v_{(CP)} \times S. \tag{24}$$

Relation (24) permits to interpret $X_{(NR)}$ as the mean point between $X_{(CP)}$ and $X_{(T)}$: $X_{(NR)} \cong (1/2)[X_{(CP)} + X_{(T)}]$. Consequently equation (22) can be equivalently written as

$$X_{(NR)} = X_{(T)} + \frac{1}{2M_0}v_{(CP)} \times S. \tag{25}$$

Unfortunately in the Dixon-Souriau model relations (22), (24) and (25) cannot be generalized to the Reissner-Nordström case.

In the latter case it is worth noting that under assumptions H1 and H2 only, $a_{(\bar{u},m)(CP)}$ is formally identical (but for a numerical factor) to $a_{(G)} + a_{(E)}$, i.e. we have:

$$a_{(\bar{u},m)(CP)} = \frac{2M}{3M-\lambda Q}[a_{(G)} + a_{(E)}]\,, \tag{26}$$

provided one replaces r and v in the expression (20) and (21) with $r_{(CP)}$ and $v_{(CP)}$ respectively.

We thus note that the result of the nonrelativistic theory is also recovered at $X_{(CP)}$, modulo a factor determined by the background, without the necessity of introducing the auxiliary point $X_{(NR)}$; it suffices to follow the evolution of the momentum vector $\bar{u}$ instead of that of U. A problem with this alternative approach is that in the limit $M \to 0$ one obtains $a_{(\bar{u},m)(CP)} = 0$ instead of $a_{(E)}$ and that the singular case $3M = \lambda Q$ is excluded.

Acknowledgements

We are indebted to Prof. H. Ohanian, Prof. I.B. Khriplovich and to Prof. R.T. Jantzen for their very helpful suggestions.

References

1. V. Bargman, L. Michel, V. Telegdi *Phys. Rev. Lett.* **2**, 435 (1959).
2. I.B. Khriplovich, A.A. Pomeransky *Phys. Letters* A **216**, 7 (1996).
3. I.B. Khriplovich, A.A. Pomeransky gr-qc/9710098 e-print (1997).
4. B.M. Barker, R.F. O'Connell *Gen. Rel. Grav.* **5** 539, (1974).
5. J.M. Souriau *C.R. Acad. Sc. Paris* **271**, 751 (1970).
6. J.M. Souriau *C.R. Acad. Sc. Paris* **271**, 1086 (1970).
7. J.M. Souriau *C.R. Acad. Sc. Paris* **274**, 1082 (1972).
8. J.M. Souriau *Ann. Inst. H. Poincaré* **20**, 22 (1974).
9. E. Corinaldesi, A. Papapetrou *Proc. Roy. Soc. London* **209**, 259 (1951).
10. W. Tulczyjew *Acta Phys. Polon.* **18**, 393 (1959).
11. W.G. Dixon *Il Nuovo Cimento* **34**, 318 (1964).
12. W.G. Dixon *Philos. Trans. R. Soc. London* A **277**, 59 (1974).
13. J. Ehlers, E. Rudolph *Gen. Rel. Grav.* **8**, 197 (1977).
14. F. Pirani, *Acta Phys. Polon.* **15**, 389 (1956).
15. S. Ragusa, M. Bailyn *Gen. Rel. Grav.* **27**, 163 (1995).
16. R.T. Jantzen, P. Carini, D. Bini *Annals of Physics* **215**, 1 (1992).
17. G. Ferrarese *Acc. Sc. Torino-Atti Sc. Fis.* **131**, 41 (1997).
18. T. Damour *C. R. Acad. Sci. Paris* **294**, 1355 (1982).
19. B.M. Barker, R.F. O'Connell *Phys. Rev.* D **2**, 1428 (1970).

Section B

Inertial Forces and Gravitoelectromagnetism

ABSOLUTE AND RELATIVE FRENET-SERRET FRAMES FOR ACCELERATED BLACK HOLE CIRCULAR ORBITS

D. BINI

Istituto per Applicazioni della Matematica, C.N.R. , I-80131 Napoli, Italy and I.C.R.A., Università di Roma, I-00185 Roma, Italy.
e-mail: binid@icra.it

R. T. JANTZEN

Dept. of Math. Sciences, Villanova University, Villanova, PA 19085, USA and I.C.R.A., Università di Roma, I-00185 Roma, Italy.
e-mail: rjantzen@email.villanova.edu

A. MERLONI

I.C.R.A., Università di Roma, I-00185 Roma, Italy.
e-mail: merloni@icra.it

The Frenet-Serret approach is applied in several ways to a familiar but still not geometrically well understood example: circular orbits in black hole spacetimes. An invariant spacetime Frenet-Serret frame approach is useful in understanding the properties of these orbits and of Fermi-Walker transport along them, and provides a visual interpretation of the geometry of this family of orbits. Closely related to the spacetime frame for these special curves are the relative Frenet-Serret frames that may be defined with respect to a family of test observers on the spacetime. The latter connect more directly to our 3-dimensional intuition about the tangent, normal, and binormal to a curve in ordinary space. These absolute and relative frames together help interpret the effects of space curvature, and the gravitoelectric and gravitomagnetic effects on circular orbiting test particles and on their gyroscopic frames of reference.

1 Introduction

The geodesics of a black hole spacetime have been "known" for quite some time thanks to their complete integrability in terms of the four constants of the motion which exist due to the special symmetry of this class of spacetimes. However, "knowing" them and understanding what they tell us about the gravitational field of a black hole are two different things. Effective potentials provide a good tool for appreciating the qualitative behavior of the geodesics, but they don't tell us how the various parts of the gravitational field contribute to that behavior. In fact, our intuition about the kinematics of particle motion is rooted in our 3-dimensional world view, and indeed gravitoelectromagnetic terminology has naturally appeared to characterize an observer-based description of the gravitational field in many applications of general relativity, whether it be in investigating approximate solutions in linearized theory or in studying exact solutions with high symmetry. We relate to forces, and forces tell us how to interpret classical force fields.

It is therefore not a useless exercise to explore test particle motion in important known spacetimes like those describing black holes in order to bridge the gap between the Newtonian picture and the general relativistic one of how the gravitational field "works" in this context. Accelerated circular orbits are a nice case to study because of their high symmetry, because so many distinct families of such orbits characterize important aspects of the geometry of these spacetimes, and because we carry so much baggage about such orbits from Newtonian mechanics which needs to be sorted out in the relativistic regime. Examples are important in any theory in giving us mental pictures of how and why things happen the way they do. Since our mental pictures associated with circular orbits are still a bit cloudy, it is worthwhile devoting some effort to clearing up the situation a bit, even though the literature already has many discussions of this topic (for a review see Ref. 1).

In fact our mental image of a black hole depends on which family of circular orbits we choose to think of as "fixed in space." However, the usual static observers associated with the Boyer-Lindquist coordinate system at the heart of that image of an unchanging space geometry break down near the hole, while the usual ADM geometry associated with the zero-angular momentum observers which remain valid up to the horizon is referred to a family which we think of as being forcibly "dragged around the hole" relative to the static observers, and their geometry is distinct from the geometry of the static space of our mental image. Other circular orbits capture other features of our Newtonian intuition about points fixed in space around a gravitating object, and the acceleration of these various families tells us something about the geometry of spacetime in ways that we can relate to that Newtonian intuition.

The ultimate goal is to add clarity, not just muddy the waters with more formalism, so the present discussion must also place other work into perspective at the same time that it introduces new baggage, or it will be hard to justify. Hence the subtitle of this article.

2 General Frenet-Serret remarks

In classical differential geometry, given an arclength parametrized curve on a Riemannian 3-manifold, one can introduce the unit tangent, normal and binormal in a natural way, providing an orthonormal frame (the Frenet-Serret frame) with which to analyse the geometry of the curve independent of the parametrization as well as important aspects of parametrization-dependent effects. The special case of Euclidean 3-space is a fundamental example in elementary nonrelativistic mechanics, underpinning the concepts of longitudinal and transverse (centripetal) acceleration which still remain almost universal examples in calculus pedagogy. The curvature and torsion scalars characterize the geometry of the curve by giving an equivalent ("osculating") circle of best fit to the curve (or equivalently the angular velocity of the rotating tangent direction within this plane) and the angular velocity (with respect to the arc length parametrization) of the orientation of the plane containing that circle. These in turn help us interpret the components of the acceleration of the time-parametrized curve of a point particle in mechanics along (longitudinal) and perpendicular (transverse) to the direction of motion in terms of the time rate of change of speed and the centripetal acceleration due to the angular velocity about the center of the osculating circle.

One can generalize this procedure to a Riemanian or pseudo-Riemannian manifold any dimension, with one curvature scalar and multiple torsion scalars which together characterize the angular velocity or pseudo-angular velocity of the Frenet-Serret frame compared to one which is parallel transported along the given curve. For the case of spacetime, a Lorentzian 4-manifold, there are two torsions, but the geometry of the Frenet-Serret frame depends on the causal character of the curve: timelike, null, or spacelike. For a spacelike curve with a spacelike "first normal," the situation is similar to the Riemannian case with the curvature (and normal together) describing a circle of best fit to the curve in that spacelike plane (or equivalently the angular velocity of the rotating tangent direction within that plane), and the two torsions determine the rotation of the Frenet-Serret frame in a remaining spacelike plane and a boost of the frame in a single timelike plane formed by pairs of its member vectors. Such curves arise in analysing the time synchronization defect along a closed curve, which exists for a family of observers in spacetime with nonzero vorticity, for example.

However, by far the most interesting curves are the timelike and null curves which represent the motion of nonzero and zero rest mass test particles respectively. Here the timelike case will be considered, where the unit tangent vector is interpreted as the particle 4-velocity and the first normal is the (signed) direction of the acceleration. The curvature, which is the (signed) magnitude of the 4-acceleration, and the first normal then together describe a pseudo-circle (hyperbola) of best fit to the curve or equivalently a pseudo-angular velocity of the boost of the 4-velocity (boost rapidity derivative) compared to one which is parallel transported along the curve. The two torsions then determine the Fermi-Walker angular velocity of the spatial triad of the Frenet-Serret frame, namely its angular velocity compared to a triad which is orthogonal to the tangent vector and Fermi-Walker transported along the curve.

In a spacetime with a preferred family of observers with 4-velocity field u, one can also introduce a space-plus-time "relative" Frenet-Serret frame approach which reconnects with our Euclidean 3-space example so useful in nonrelativistic mechanics. The unit relative velocity of a spacetime curve then plays the role of the tangent, but the normal and binormal in this relative approach depend on the choice of the relative spatial derivative in the local rest space of the observer family where these spatial vectors live, and there are three "natural" choices. This leads to three possible relative centripetal acceleration vectors, for example.

However, one can also introduce a "comoving" relative Frenet-Serret frame along a timelike curve (world line of a nonzero rest mass test particle) adapted to its own local rest space by taking the unit relative velocity of the curve with respect to the observer family within that rest space as the first Frenet-Serret vector, its natural unique Fermi-Walker derivative along the curve as the normal, and their spatial cross-product as the bi-normal. This models the test particle's reconstruction of its relative motion as seen by the observer family, and corresponds in some sense to our notion of how one would see one's own motion in the "space" of the black hole spacetime. These relative frames are discussed in the companion article[2,3] in these proceedings.

3 General Gyro spin remarks

The precession of the spin of a gyroscope in a gravitational field is an exciting consequence of relativistic gravitational theory compared to the Newtonian case, but since test gyros actually define what it means for a direction to be "nonrotating," their precession only makes sense as a relative effect. For a generic spacetime it is problematic to decide what to compare the spin direction to locally, but in stationary axisymmetric spacetimes, the symmetry provides a rigid connection to nonrotating observers at infinity through the usual static observers following the Killing vector trajectories which are the Boyer-Lindquist time coordinate lines.

For observers at the same spacetime point but in relative motion, the orientation of a set of orthonormal spatial axes carried by the other observer is understood to be *the orientation of the axes which are boosted back to the observer doing the observing*. This assumption is inherent in the diagrams of primed and unprimed 3-frames in relative motion when we discuss inertial frames in elementary special relativity. All observers in a stationary axisymmetric spacetime in circular motion therefore agree on the orientation of any particular set of axes at a given spacetime point since their relative velocities are all "collinear" and the family of boosts along a fixed direction forms a group. (If three observers were not collinear, i.e., had linearly independent 4-velocities, boosting one's orthonormal frame to the local rest spaces of the other two, and then boosting either one of the latter to the other's local rest space would yield a frame differing by a rotation from the one boosted directly back to that local rest space from the original observer.) The precession rate of a set of gyro fixed axes along a given world line with respect to a fixed parametrization of the world line and a fixed set of reference axes is therefore observer-independent in this class of observers. (Observer proper time parametrizations then lead to differing rates.) The normalized Boyer-Lindquist spatial coordinate frame (boosted to the world line local rest space) provides a natural set of reference axes for circular motion. Since these frames rotate along a circular orbit with respect to nonrotating frames at spatial infinity with the orbital angular velocity, a subtraction must be performed to remove this additional rotation.

However, the question of the total precession during one revolution of a circular orbit does depend on which observer family the revolution is taken with respect to, since completion of a revolution is marked by the return to the original observer whose path is crossed at its start. Usually the static observers are used to mark revolutions with respect to nonrotating observers at spacelike infinity, although this only works outside the ergosphere of a black hole.

4 Circular orbits and Frenet-Serret formalism

The spacetime Frenet-Serret frame along the single timelike test particle world line with 4-velocity $U = e_0$, parametrized by the proper time τ, is described by the equations[5,6]

$$\frac{\mathrm{D}}{\mathrm{d}\tau}e_0 = \kappa\, e_1\ , \qquad \frac{\mathrm{D}}{\mathrm{d}\tau}e_2 = -\tau_1 e_1 + \tau_2 e_3\ ,$$
$$\frac{\mathrm{D}}{\mathrm{d}\tau}e_1 = \kappa\, e_0 + \tau_1\, e_2\ , \quad \frac{\mathrm{D}}{\mathrm{d}\tau}e_3 = -\tau_2 e_2\ , \tag{1}$$

namely

$$\mathrm{D}e_\alpha/\mathrm{d}\tau = L^\beta{}_\alpha\, e_\beta \ , \tag{2}$$

where the matrix $(L^\alpha{}_\beta)$ belongs to the Lie algebra of the Lorentz group. The curvature κ is allowed to be of either sign in order that the frame may be extended smoothly through isolated points where $DU/d\tau = 0$.Its absolute value is the magnitude of the acceleration $a = \kappa e_1$ and describes the rate of change of the rapidity parameter of the boost of the Frenet-Serret frame relative to an orthonormal frame parallel transported along the world line. The first and second torsions τ_1 and τ_2 are the components of the Frenet-Serret angular velocity vector

$$\omega_{(\mathrm{FS})} = \tau_2 e_1 + \tau_1 e_3 \tag{3}$$

of the spatial frame $\{e_a\}$ with respect to one Fermi-Walker transported along the curve,[4] i.e.,

$$\mathrm{D}_{(\mathrm{fw})} e_a/\mathrm{d}\tau = \omega^c_{(\mathrm{FS})a}\, e_c = \epsilon_{bca}\omega^b_{(\mathrm{FS})} e_c \ , \tag{4}$$

where $\mathrm{D}_{(\mathrm{fw})}/\mathrm{d}\tau$ is the projection orthogonal to U of the intrinsic derivative, namely the spatial Fermi-Walker derivative. The sign-reversed angular velocity $-\omega_{(\mathrm{FS})}$ therefore describes the precession angular velocity with respect to the Frenet-Serret frame of the Fermi-Walker transported spin vector of a test gyroscope moving along the world line of U.[4]

Consider the Boyer-Lindquist line element for the Kerr spacetime orthogonalized with respect to $n = N^{-1}(\partial_t - N^\phi \partial_\phi)$ (the unit normal to the Boyer-Lindquist time slicing of the Kerr spacetime and the 4-velocity of zero angular momentum observers or ZAMO's), leading to the lapse-shift form

$$\mathrm{d}s^2 = -N^2 \mathrm{d}t^2 + g_{\phi\phi}(\mathrm{d}\phi + N^\phi \mathrm{d}t)^2 + g_{rr}\mathrm{d}r^2 + g_{\theta\theta}\mathrm{d}\theta^2 \ . \tag{5}$$

The mutually orthogonal (t, ϕ) and (r, θ) subspaces of the tangent space reflect the $2 + 2$ splitting associated with the orthogonal transitivity symmetry of the spacetime. Most of the present remarks also apply to the wider class of orthogonally transitive stationary axisymmetric spacetimes. For definiteness, it is assumed that the black hole angular momentum parameter a satisfies $a \geq 0$ so that the black hole is rotating in the positive ϕ direction when $a \neq 0$. The context should always clearly distinguish the symbol a for the acceleration used below from the traditional angular momentum parameter a of the Kerr metric. The dimensionless parameter $\bar{a} = a/\mathcal{M}$ and radial variable $\bar{r} = r/\mathcal{M}$ are also useful, where $\mathcal{M}$ is the black hole mass parameter.

Furthermore, stationary axisymmetric quantities depend only on (r, θ), which are simplest to visualize by representing them graphically in the ρ-z coordinate plane of the corresponding flat spacetime cylindrical coordinates $(\rho, z) = (r \sin\theta, r\cos\theta)$;the $\pm\rho$ coordinate directions will be referred to as "horizontally outward" and "horizontally inward" respectively away from and towards the "vertical" symmetry axis $\rho = 0$.Similarly the z coordinate direction will be referred to as "vertical," either away from or towards the equatorial plane $z = 0$ or $\theta = \pi/2$.

World lines which correspond to circular orbits with constant speed $|\nu|$ (with respect to the ZAMO's) have 4-velocity

$$e_0 = U = \Gamma(\partial_t + \zeta\partial_\phi) = \gamma(n + \nu\partial_{\hat{\phi}}) = \cosh\alpha\, n + \sinh\alpha\, \partial_{\hat{\phi}} \ , \tag{6}$$

where the relative velocity $\nu\partial_{\hat\phi}$ has constant signed magnitude $\nu = \tanh\alpha \in (-1, 1)$ parametrized by the rapidity $\alpha \in (-\infty, \infty)$, in terms of which the following useful relations hold

$$[\gamma, \gamma\nu, \nu, \gamma^2(1+\nu^2), 2\gamma^2\nu] = [\cosh\alpha, \sinh\alpha, \tanh\alpha, \cosh 2\alpha, \sinh 2\alpha] \ . \qquad (7)$$

The rapidity is the natural pseudo-angular parameter for the hyperbola of such 4-velocities in the (t, ϕ) subspace of the tangent space, the relative observer plane for the set of all such observers at each spacetime point, and for more simply expressing certain naturally occurring functions of the relative velocity. Circular orbits with $\nu > 0$ (positive angular momentum) will be referred to as co-revolving, while those with $\nu < 0$ (negative angular momentum) will be referred to as counter-revolving, following the terminology of Wilkins.[7]

The acceleration of these orbits $a = DU/d\tau$ may be parametrized by polar coordinates in the $\partial_{\hat r}$-$\partial_{\hat\theta}$ plane in the tangent space to which they are confined (the "acceleration plane"),namely the signed magnitude κ of the acceleration and its orientation angle χ, both of which are constants along a given orbit and in turn functions of the rapidity α parametrizing all such orbits

$$a = \kappa(\cos\chi\,\partial_{\hat r} + \sin\chi\,\partial_{\hat\theta}) \ , \qquad |\kappa| = [(a^{\hat\theta})^2 + (a^{\hat r})^2]^{1/2} \ , \ \tan\chi = a^{\hat\theta}/a^{\hat r} \ . \qquad (8)$$

If we choose the convention $\kappa \geq 0$, $\chi \in [-\pi, \pi]$,the direction of the acceleration is

$$e_1 = \cos\chi\,\partial_{\hat r} + \sin\chi\,\partial_{\hat\theta} \ . \qquad (9)$$

Since e_1 and U are 1-parameter families of unit vectors, their angular/pseudo-angular derivatives must be orthogonal to them. In fact the final pair of Frenet-Serret vectors are

$$\begin{aligned} e_2 &= dU/d\alpha = \sinh\alpha\, n + \cosh\alpha\,\partial_{\hat\phi} \ , \\ e_3 &= -de_1/d\chi = \sin\chi\,\partial_{\hat r} - \cos\chi\,\partial_{\hat\theta} \ , \end{aligned} \qquad (10)$$

the latter vector being the clockwise rotation of e_1 by 90 degrees in the ∂_r-∂_θ plane, making the triad $\{e_a\}$ righthanded. Thus the e_0-e_2 relative velocity plane and the orthogonal e_1-e_3 acceleration plane (transverse to all such velocities, characterizing purely transverse relative acceleration for any observer of the same type) reflect the $2+2$ orthogonal transitivity splitting.

The outward $(+)$ and inward $(-)$ (with respect to the symmetry axis) "horizontal" directions along $\pm\partial_{\hat\rho}$ in the $\partial_{\hat r}$-$\partial_{\hat\theta}$ plane correspond to the values $\chi_\pm = \pm\pi/2-\theta$, while$\chi \in (-\pi/2, \pi/2)$ corresponds to outward acceleration$a^{\hat r} > 0$ with respect to the radial direction, and the complementary half-plane to inward such acceleration. Our convention that $\kappa \geq 0$ means that as $r \to \infty$, then $\chi \approx \chi_-$ for sufficiently large ν in region A, i.e., the acceleration is horizontally inward toward the symmetry axis far from the black hole at a high enough speed, corresponding to the flat spacetime inward centripetal acceleration along $-\partial_{\hat\rho}$. As one approaches the black hole above the equatorial plane, the ultrarelativistic acceleration vectors tilt farther up, rotating away from the symmetry axis towards the outward radial direction to resist the increasingly strong attraction toward the hole.

Our mental picture of a black hole corresponds to the projected spatial coordinate geometry on the (r, θ, ϕ) coordinate space which parametrizes the space of

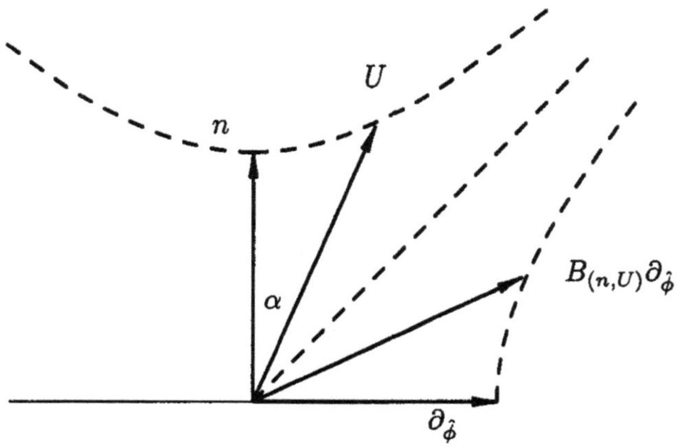

Figure 1. The hyperbola of unit 4-velocities U of the timelike circular orbits in the relative observer plane for circular motion, spanned by the orthogonal vectors n and $\partial_{\hat\phi}$. Spacelike circular orbits have unit tangents which are related by a boost to $\partial_{\hat\phi}$.

static (Killing) observers following the time coordinate lines with 4-velocity $m = \partial_{\hat t}$, a geometry which differs from the induced spatial coordinate geometry characterizing the space of the ZAMO observers. In the corresponding flat spatial geometry of our Newtonian picture on this same space, $-\partial_\rho$ is the normal to the spatial curve having unit tangent along the ϕ coordinate direction, while the binormal is along the z coordinate direction. These also correspond to the relative Frenet-Serret frame vectors, but differ by relativistic effects.[3]

As α varies from $-\infty$ to ∞, or ν from -1 to 1, the 4-velocity U traces out one branch of a hyperbola (pseudocircle) in the relative observer plane, while the acceleration vector a traces out one branch of a hyperbola (to be shown below) in the orthogonal transverse relative acceleration plane whose properties depend on the location of the tangent space, i.e., on r and θ.

The second branch of this hyperbola corresponds to superluminal motion (a spacelike circular orbit) with $\nu^{-1} \in (-1, 1)$, which includes the closed ϕ-coordinate circles. Figures 1 and 2 illustrate these curves.

The line element in the acceleration plane

$$ds_{(a)}{}^2 = d\kappa^2 + \kappa^2 d\chi^2 \tag{11}$$

measures the arclength along this curve of acceleration vectors in the tangent space. With the above choices for e_1 and e_3, the two torsions must be allowed to change sign. They are then simply equal to half the sign-reversed rate of change of the arclength along the two polar coordinate directions with respect to α

$$\tau_1 = -\frac{1}{2} d\kappa/d\alpha\,, \quad \tau_2 = -\frac{1}{2}\kappa d\chi/d\alpha\,. \tag{12}$$

The tangent to the acceleration hyperbola and consequently the Frenet-Serret angular velocity (3) can therefore be expressed as

$$\frac{da}{d\alpha} = \frac{d\kappa}{d\alpha} e_1 + \kappa \frac{de_1}{d\alpha} = 2(-\tau_1 e_1 + \tau_2 e_3)\,, \qquad \omega_{(FS)} = \frac{1}{2} e_2 \times_U \frac{da}{d\alpha}\,. \tag{13}$$

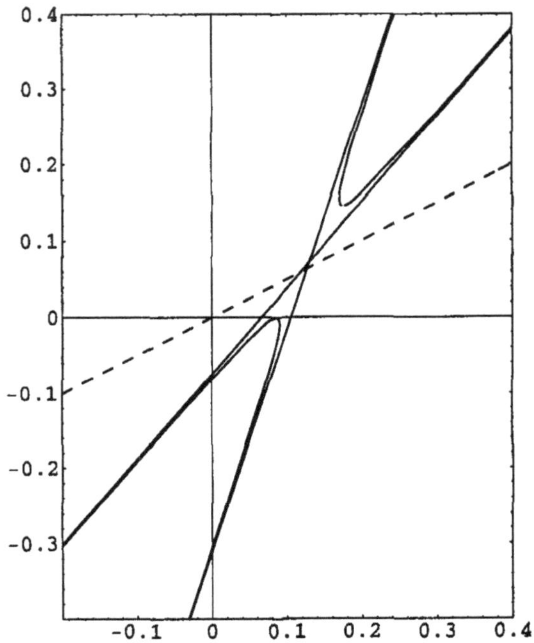

Figure 2. The hyperbola of acceleration vectors of the nonnull circular orbits in the acceleration plane for circular motion, spanned by $\partial_{\hat{r}}$ (horizontal axis) and $\partial_{\hat{\theta}}$ (vertical axis), shown for $\bar{a} = 0.5$, $\theta = \pi/3$, $\bar{r} = 4$. The dashed line is along the "horizontal" direction, with $\partial_{\hat{\rho}}$ pointing to the left along it towards the symmetry axis. The lower branch represents the timelike orbits whose right half corresponds to the counter-rotating orbits whose acceleration vectors must have a larger (outward) radial component to resist the greater attraction to the black hole. The upper branch represents the spacelike orbits, with the closed ϕ coordinate circle near its vertex.

Thus $\omega_{(\mathrm{FS})}$ is obtained from $\frac{1}{2}\mathrm{d}a/\mathrm{d}\alpha$ by a counterclockwise rotation of 90 degrees with respect to the ordered axes (e_3, e_1), which have the same orientation as $(\partial_{\hat{r}}, \partial_{\hat{\theta}})$, placing it along the inward normal to the hyperbola as will be seen below. Its magnitude

$$||\omega_{(\mathrm{FS})}|| = (\tau_1^2 + \tau_2^2)^{1/2} = \frac{1}{2}[(\mathrm{d}\kappa/\mathrm{d}\alpha)^2 + \kappa^2(\mathrm{d}\chi/\mathrm{d}\alpha)^2]^{1/2} = \frac{1}{2}|\mathrm{d}s_{(\mathrm{a})}/\mathrm{d}\alpha| \qquad (14)$$

is just half the speed with which the acceleration curve is traced out in its plane with respect to the α parametrization. The Frenet-Serret angular velocity traces out a curve in the acceleration plane which is also a hyperbola with its asymptotes rotated by 90° but lying in a complementary branch (the opening angles of the two hyperbolas sum to π). The derivation of these simple torsion formulas (12)on which all of this beautiful geometry rests starts from eqs. (49), (54) of Iyer and Vishveshwara.[4] They apply to the entire orthogonally transitive stationary axisymmetric symmetry class of spacetimes, enabling one to convert their lengthy formulas into concrete geometry nicely illustrated by 2-dimensional graphics.

5 Geometry of the transverse relative acceleration plane

The orthonormal components of the parametrized acceleration hyperbola in the $\partial_{\hat{r}}$-$\partial_{\hat{\theta}}$ plane have the fractional quadratic form

$$a^{\hat{r}} = \mathcal{F}(\nu; k^{\hat{r}}, \nu_+^{(r)}, \nu_-^{(r)}) \ , \ a^{\hat{\theta}} = \mathcal{F}(\nu; k^{\hat{\theta}}, \nu_+^{(\theta)}, \nu_-^{(\theta)}) \ , \qquad (15)$$

where $\nu = \tanh\alpha \in (-1, 1)$ and (using Eq. (7))

$$\begin{aligned} \mathcal{F}(\nu; k, \nu_+, \nu_-) &= k\frac{(\nu - \nu_-)(\nu - \nu_+)}{1 - \nu^2} \\ &= k[\sinh^2\alpha - (\nu_- + \nu_+)\sinh\alpha\cosh\alpha + \nu_-\nu_+\cosh^2\alpha] \\ &= \frac{1}{2}k[(\nu_-\nu_+ + 1)\cosh 2\alpha - (\nu_- + \nu_+)\sinh 2\alpha + (\nu_-\nu_+ - 1)] \ . \qquad (16) \end{aligned}$$

Thus with the natural rapidity parametrization, the acceleration curve takes the matrix form

$$\begin{bmatrix} a^{\hat{r}} \\ a^{\hat{\theta}} \end{bmatrix} = \begin{bmatrix} A & B \\ C & D \end{bmatrix} \begin{bmatrix} \cosh 2\alpha \\ \sinh 2\alpha \end{bmatrix} + \begin{bmatrix} E \\ F \end{bmatrix} \ . \qquad (17)$$

Solving this for $\cosh 2\alpha, \sinh 2\alpha$ and using the identity $\cosh^2 2\alpha - \sinh^2 2\alpha = 1$, one immediately finds the explicit form for the unparametrized equation of a general hyperbola with center $[E, F]$ displaced from the origin

$$\begin{bmatrix} a^{\hat{r}} - E & a^{\hat{\theta}} - F \end{bmatrix} \begin{bmatrix} A & B \\ C & D \end{bmatrix}^{-1\mathrm{T}} \begin{bmatrix} 1 & 0 \\ 0 & -1 \end{bmatrix} \begin{bmatrix} A & B \\ C & D \end{bmatrix}^{-1} \begin{bmatrix} a^{\hat{r}} - E \\ a^{\hat{\theta}} - F \end{bmatrix} = 1 \ . \qquad (18)$$

A change of observer corresponds to a translation in α which in turn leads to an $SL(2, R)$ transformation of the matrix of coefficients.

The Frenet-Serret angular velocity vector $\omega_{(\mathrm{FS})}$ is obtained from the acceleration vector by half the α-derivative, which interchanges $\cosh 2\alpha$ and $\sinh 2\alpha$, and a counterclockwise rotation by $\pi/2$

$$\begin{bmatrix} \omega^{\hat{r}}_{(\mathrm{FS})} \\ \omega^{\hat{\theta}}_{(\mathrm{FS})} \end{bmatrix} = \begin{bmatrix} 0 & -1 \\ 1 & 0 \end{bmatrix} \begin{bmatrix} A & B \\ C & D \end{bmatrix} \begin{bmatrix} 0 & 1 \\ 1 & 0 \end{bmatrix} \begin{bmatrix} \cosh 2\alpha \\ \sinh 2\alpha \end{bmatrix} \ . \qquad (19)$$

This is a hyperbola centered at the origin with complementary asymptotes to the acceleration hyperbola rotated by 90 degrees. The minimum value of the magnitude $||\omega_{(\mathrm{FS})}||$ clearly occurs at its vertex, which corresponds to the same relative velocity $\nu_{(\mathrm{vert})} = \tanh\alpha_{(\mathrm{vert})}$ describing the vertex of the acceleration hyperbola, defining a family of "extreme Fermi-Walker rotation" observers.

These general properties characterize the timelike circular orbits in any orthogonally transitive stationary axisymmetric spacetime. For the Kerr spacetime, one may use a computer algebra system to study the particular functions of r and θ which determine the behavior of the acceleration and Frenet-Serret hyperbolae as a function of position in order to give some insight into how the behavior differs from the Newtonian case. This has been done and discussed in detail in Ref. 8, so a complimentary discussion of their qualitative behavior will be given here.

The key difference with the picture of the static Newtonian gravitational field of an isolated stationary axially symmetric rotating source is that the natural (unaccelerated) stationary observers "at rest in space" in the Newtonian case correspond to the accelerated "static" observers in the relativistic case (those following the time coordinate lines in the Boyer-Lindquist coordinate system, along Killing trajectories, said to be nonrotating with respect to spatial infinity), so the acceleration vectors in the two cases differ by the acceleration needed to remain fixed in "space," which is itself an ambiguous concept in the relativistic regime. Thus the 4-acceleration of a circular orbit consists of the sum of a relative centripetal acceleration associated with the circular motion with respect to the observer family together with the acceleration needed by the observer family to resist the gravitational attraction, two pieces which individually remain ambiguous since there are a number of choices for the decomposition.

Furthermore, it is the locally nonrotating observers or ZAMO's which prove more useful than the static observers since the ZAMO's remain well defined right up to the horizon of a black hole spacetime, while the static observers are unable to remain "fixed in space" within the ergosphere. On the other hand the ZAMO's are "dragged around" the hole compared to the static observers, and while the "infinitesimal connecting vector" between nearby ZAMO's does not rotate with respect to Fermi-Walker propagated axes along their world lines because their vorticity vanishes, their acceleration vector does rotate with respect to such axes in the Kerr case. Since static observers follow Killing trajectories, their acceleration vector and nearby observers are locked together undergoing a rotation compared to Fermi-Walker propagated axes with an associated angular velocity which is the vorticity vector modulo sign.[4]

However, in the nonrotating Schwarzschild case, these two observer families coincide and also exhibit another feature of the Newtonian case: their 4-acceleration is extremal. In the Newtonian case, the circular orbiting observers at rest in space have the maximum radial acceleration, namely zero, compared to the negative (radially inward) radial acceleration for all objects in circular motion, due to their inward centripetal acceleration in the direction of the axis of symmetry. This is a minimum of the magnitude of the acceleration. In the relativistic nonrotating case this behavior of the relative centripetal acceleration is superimposed on the acceleration needed to resist the gravitational acceleration of the hole, but it is still a maximum of the radial component at least far enough from the hole, as well as a local maximum of the magnitude of the total 4-acceleration. In the rotating case, these "extremely accelerated observers"[10–12] are yet a third family, characterized by an extreme value of the magnitude of the total 4-acceleration (one of 3 distinct extremal values which exist for most radii), which in turn differs from the extreme value of the radial component.[13] This is typical of the way in which concepts which are unambiguous in the nonrelativistic case have multiple generalizations in the relativistic case, where properties which hold simultaneously in the nonrelativistic context no longer do so.

Figure 3 shows how the acceleration hyperbola above the equatorial plane at a fixed angle $\theta = \pi/3$ rotates upward (from the inward horizontal direction far from the hole) and outward (towards the outward radial direction) as one approaches

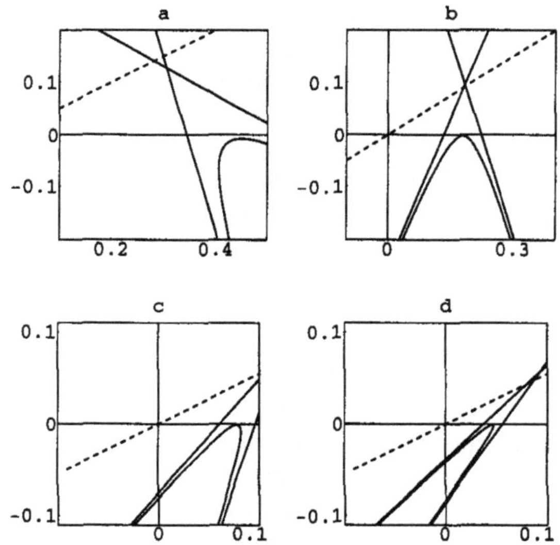

Figure 3. The acceleration hyperbolae in the tangent space (horizontal axis along $\partial_{\hat{r}}$, vertical axis along $\partial_{\hat{\theta}}$)for regions A, B, and C for $\tilde{a} = 0.5$,shown at a typical angle $\theta = \pi/3$ at radii (a) $\tilde{r} = 2.25$ (region C), (b) $\tilde{r} = 3$ (region B), (c) $\tilde{r} = 4$ (region A),(d) $\tilde{r} = 5$ (region A). The dashed line represents the "horizontal" direction in the ρ-z plane, with the symmetry axis to the left along it at the polar angle $\chi_- = -5\pi/6$, and the equatorial plane above it. Figure 9 shows the orientation of these diagrams in the ρ-z plane for the same angle. The hyperbola rotates inward as r increases (counterclockwise in this figure) and centripetal acceleration becomes more dominant, finally reaching a configuration like (d) where 3 local extrema of κ (the distance from the origin) are obvious.

the hole from far away, while the vertex of the hyperbola moves farther out in the radial direction, as more and more radially outward acceleration is needed to oppose the increasingly stronger nearly radial gravitational attraction towards the hole. Simultaneously the asymptotes of the hyperbola open up as the increasingly stronger dragging effect increasingly separates the acceleration directions of ultrarelativistic corevolving and counter-revolving orbits until a maximum opening angle is reached very near the hole beyond which they instead squeeze closer and closer together and to the outward radial direction, with the opening angle going to zero at the horizon. The directions of the two asymptotes themselves are the directions of the acceleration vectors of the counter and corotating photon orbits. The asymptote which is to the right (rotated further towards the outward radial direction) in Fig. 3 corresponds to $\nu \to -1$ (the counter-rotating photon), while the one to the left corresponds to $\nu \to 1$ (the corotating photon), which reflects the fact that counter-revolving (corevolving) orbits require a stronger (weaker) radial acceleration to resist the increased (decreased) gravitational attraction due to the contribution from the gravitomagnetic force, in the relative observer description of Ref. 14 (see Fig. 2 there). The next section will analyze the contributions of the various aspects of the spacetime geometry (space curvature, gravitoelectric and gravitomagnetic effects) to the behavior of the acceleration hyperbola, using the relative observer decomposition of the acceleration.

Figure 3 also reveals the geometry of the extremal acceleration points of the acceleration hyperbola[10,11,12] as a function of the radius, which occur when the position vector of the point on the hyperbola is orthogonal to its tangent vector (and therefore aligned with the Fermi-Walker angular velocity modulo sign), since the distance from the origin in the acceleration plane is the magnitude of the acceleration being extremized. Far from the hole as in Fig. 3(a), there is one global minimum on the origin side of the hyperbola and two local extrema, one maximum and one minimum, on the far side. As one approaches the hole (Figs. 3(b) and 3(c)), the symmetry axis of the hyperbola rotates towards the outward radial direction until these two local extrema coalesce and then disappear, leaving only the global minimum behind.

At each radius for fixed angle θ there is also a single global minimum $|\chi_{(\mathrm{min})}|$ of the absolute value of the acceleration angle, where the position vector is tangent to the hyperbola (and therefore orthogonal to the Fermi-Walker rotation), near its vertex. This angle is the minimum acute angle that the acceleration vector is tilted upward (downward) from the radial direction above (below) the equatorial plane. Examining this extreme value as a function of the radius for fixed angles, it turns out to be extremely small with a sharp peak near the black hole with a maximum value less than roughly 0.0025 (about 0.15 degrees) even for an extreme black hole. As the horizon is approached, even the photons are unable to resist the increasing radial attraction of the hole so the radial acceleration component overwhelms the tangential one for both corotating and counter-rotating photons, and the acceleration hyperbola branch squeezes to the outward radial direction, taking this angle back to zero. The relative velocity $\nu^{(\chi)} < 0$ at which the extreme value of ξ occurs is very small as is the relative velocity $\nu^{(\theta)} > 0$ at which $|a_{\hat\theta}|$ is minimized (horizontal tangent to the acceleration curve), both of which are sharply peaked at the same radius for a given θ very close to the hole with a maximum on the order of 10^{-3}. Thus the ZAMO's lie somewhere between the horizontal tangent and the point of tangency of the position vector on the acceleration hyperbola.

The orientation of the acceleration hyperbola can be characterized by whether the two photon acceleration asymptotes have a positive or negative radial component, i.e., are tilted radially outward or inward compared to the tangential θ direction. Far from the hole they are both initially tilted inward (region A), but first the counter-rotating photon acceleration rotates to an outward tilt (region B), and is then joined by the corotating photon acceleration in tilting outward (region C) near the hole, as illustrated respectively in Figs. 3(a), 3(b), and 3(c). Because of the orientation of the hyperbola in the acceleration plane, this is correlated with the number of zeros of the radial acceleration (where the hyperbola crosses the $\partial_{\hat\theta}$ axis), namely 2 (region A), 1 (region B), and 0 (region C). The two roots $\nu_{\pm}^{(r)}$ of the radial acceleration $a^{\hat r}$ are always real and of opposite sign and describe the circular orbits with vanishing radial acceleration, reducing to geodesics in the equatorial plane, where the acceleration is purely radial. Both velocities are subluminal at sufficiently large radius, leading to a pair of orbits with vanishing radial acceleration, but as one approaches the black hole first the counter-revolving velocity $\nu_{-}^{(r)}$ and then also the corevolving velocity $\nu_{+}^{(r)}$ become superluminal, defining three regions

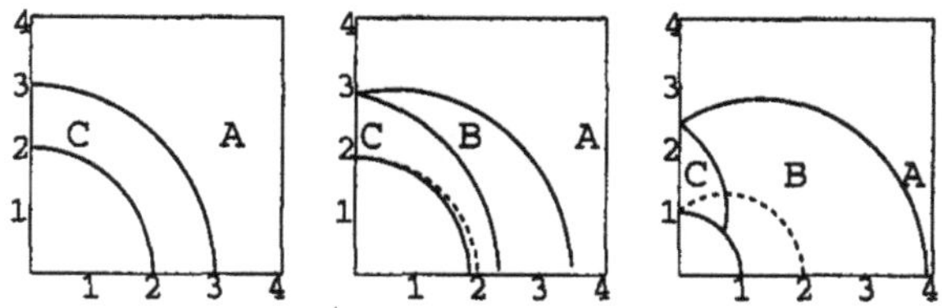

Figure 4. The profiles of the regions A, B, and C for a black hole with $\bar{a} = 0$ (left), $\bar{a} = 0.5$ (center) and $\bar{a} = 1$ (right) represented in the ρ-z coordinate plane (horizontal axis ρ, vertical axis z), showing the ergosphere boundary (dashed line) and the horizon (innermost curve). In the $\bar{a} = 0$ limit (see Fig. 3a), region B collapses to the sphere $\bar{r} = 3$, while the ergosphere collapses to the horizon $\bar{r} = 2$. In the $\bar{a} = 1$ limit (see Fig. 3c), region C collapses to the horizon for angles greater than the angle $\theta_* \simeq 47^0$. Region C continues to exist at smaller angles.

A (both subluminal, far from the hole), B (one subluminal, one superluminal), and C (both superluminal, close to the hole) exactly as discussed for the equatorial plane geodesics in Ref. 9 to which this discussion reduces in the equatorial plane case. In region C, the acceleration is always radially outward ($a^{\hat{r}} > 0$), while in region B the ultrarelativistic counter-revolving acceleration is radially outward but the ultrarelativistic corevolving acceleration is inward.

The interface radii $r_{\rm AB}(\theta)$ and $r_{\rm BC}(\theta)$ separating these three regions occur respectively where the counter-rotating and corotating photon orbits have zero radial acceleration (namely $\nu_-^{(r)} = -1$ and $\nu_+^{(r)} = 1$) and the corresponding asymptote of the acceleration hyperbola is vertical in the acceleration plane. The middle region B collapses to the sphere $\bar{r} = 3$ in the Schwarzschild limit $\bar{a} = 0$. Figure 4 illustrates these three regions in the ρ-z coordinate plane for $\bar{a} = 0, 0.5, 1$. In the extreme case $\bar{a} = 1$, the curve separating region C and B bifurcates on the horizon, while region B squeezes up against the horizon up until that bifurcation point for $\bar{a} \to 1$. The details are discussed in the appendix.

Figure 5 illustrates the behavior of the acceleration magnitude κ as a function of ν for selected radii in the three regions A, B, C at a typical angle of $\theta = \pi/3$. In most of region A extending out to infinity (lower curves), there are 3 local extrema of κ where the position vector in the 2-plane of the acceleration hyperbola is orthogonal to its tangent line, occurring at the ZAMO relative velocities $\nu_{-1}^{(\kappa)} < \nu_{-2}^{(\kappa)} < 0 < \nu_+^{(\kappa)}$ (local min, local max, global min respectively). As one approaches the hole the local max and min coallesce, leaving only a single global minimum near the hole (upper curves).

The angular opening of the pair of asymptotes of the acceleration hyperbola increases with $\bar{a}$ and decreases to zero when $\bar{a} = 0$. Figure 6 shows the limiting case $\bar{a} = 0$ of the Schwarzschild spacetime where the branch of the acceleration hyperbola degenerates to a half line terminating on the axis of purely radial acceleration at $\nu = 0$, traced once for each sign of ν. This half line intersects the zero radial acceleration ($e_{\hat{\theta}}$) axis in a single point in region A but as one moves towards the black hole, this half line rotates away from the symmetry axis towards the outward radial direction, aligning itself parallel to the zero radial acceleration axis ($e_{\hat{\theta}}$) at the interface with region C (region B shrinks to a surface), beyond which it is tilted

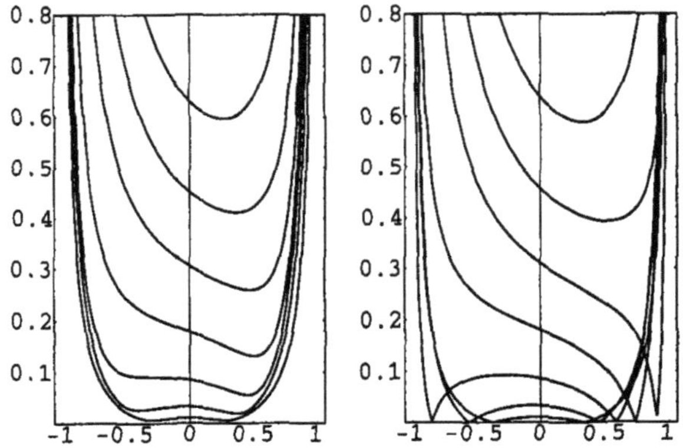

Figure 5. The magnitude κ of the acceleration versus ZAMO relative velocity $\nu \in (-1, 1)$ for selected radii in regions A (lower curves), B (intermediate curves), and C (upper curves) at a typical angle $\theta = \pi/3$ (left) and near the equatorial plane at for $\tilde{a} = 0.5$. The close proximity to the circular geodesics ($\kappa = 0$) in that plane is clearly seen at the kinks in the curves near the horizontal axis.

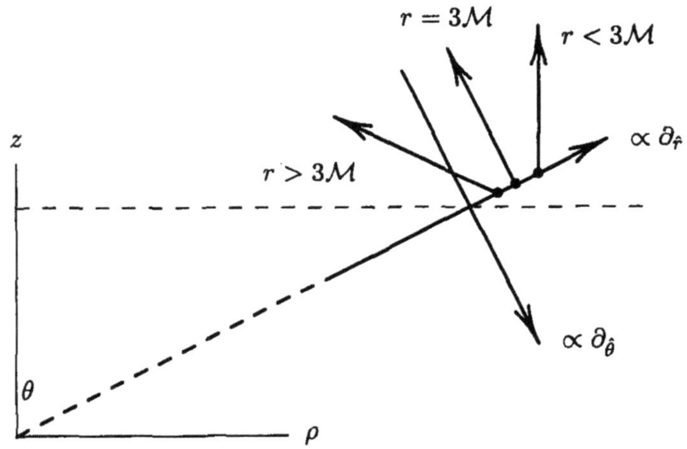

Figure 6. The degenerate acceleration hyperbola half lines for the Schwarzschild case for timelike circular orbits. Their directions are those of the acceleration of the photon orbits.

away from that axis and thus has no intersection. The angle of tilt reveals the competition between the outward radial acceleration necessary to resist falling into the hole and the centripetal acceleration towards the symmetry axis perpendicular to the circular orbit.

In the rotating case, the angular opening of the pair of asymptotes increases from zero far away from the hole to a maximum roughly in regions B and C and then decreases back to zero at the horizon. This is illustrated in Fig. 7 for selected angles between 0 and $\pi/2$ for $\tilde{a} = 0.5$. As the equatorial plane is approached, this angle approaches a step function with value π in region B and zero in regions A

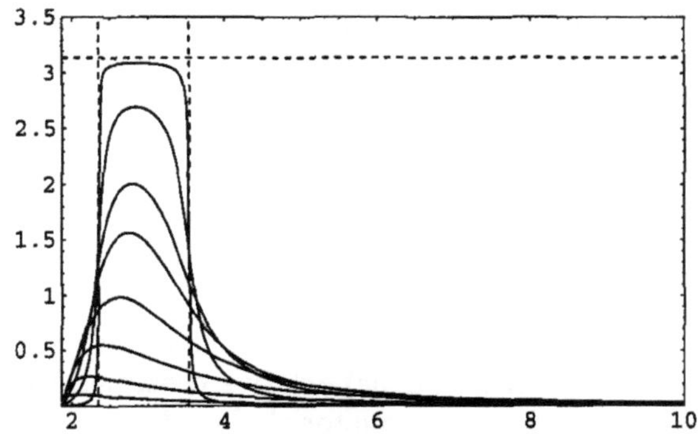

Figure 7. The maximum opening angle in radians for the acceleration hyperbola as a function of $\tilde{r}$ outside the horizon for $\tilde{a} = 0.5$ and $\theta = \pi/2.1, \pi/2.3, \pi/2.5, \pi/3, \pi/4, \pi/6, \pi/10$ (from top to bottom). The vertical dashed lines denote the equatorial radii of the interfaces between regions A, B, and C.

and C, reflecting the behavior of the purely radial component of the acceleration there as plotted in Fig. 6(c) of Ref. 9.

Figure 4 of Ref. 8 illustrates the velocities of the various special circular orbits with respect to the ZAMO's at selected values of θ, generalizing the equatorial plane case shown in Fig. 7(b) of Ref. 9. These include the extremal acceleration observers corresponding to the extremal values of the acceleration magnitude discussed above.

6 The Relative Observer Decomposition of the Acceleration

Given what we all know about Newtonian gravity and the very useful notions of centripetal acceleration / centrifugal force in the description of orbits in such fields, it is worth trying to understand how the picture may be stretched to fit the relativistic case. The gravitational field is sensed through the motion of test particles, and given that our mental picture of a black hole, like the corresponding nonrelativistic gravitational field, is based on an implicit family of test observers, it is only natural to analyze the gravitational field by calculating how these test observer families view test particle motion. This subject has a long and rich history[1] that gave rise to the terminology of "gravitoelectric" and "gravitomagnetic" fields, two of the three aspects of spacetime geometry which result from splitting spacetime into space-plus-time through an observer family decomposition, the remaining one being space curvature. It is also somewhat obscured by a Fermat's theorem approach to light ray paths in stationary spacetimes which leads to a conformal transformation of the spatial geometry used to discuss nonzero rest mass motion. This has its origins in old work by Møller, but has been analyzed geometrically in detail for black hole spacetimes by Abramowicz and many collaborators. While this is a nice geometrization of the relative motion of test particles and light rays in static spacetimes, it seems less useful in more general stationary spacetimes where the gravitomagnetic effect cannot be absorbed into the spatial geometry, and appears to be no longer useful in nonstationary spacetimes. However, it leads to some semantical controversy over claims for the "correct generalization" of centrifugal and

Coriolis forces to general relativity. Such debates can of course never be won since the richness of general relativity removes the degeneracy of many properties which hold simultaneously in nonrelativistic physics. Distinct mathematical objects one can introduce with the same nonrelativistic limits simply describe different aspects of the more complex relativistic landscape.

The qualitative behavior of the acceleration hyperbola for circular orbits as a function of position in the r-θ plane can be understood in terms of the geometry of the relative motion as seen by the ZAMO's[16,17] by starting with the nonrotating black hole case. In the Schwarzschild spacetime, this behavior is largely the result of the competition between the (unique) relative centripetal acceleration of the circular motion, tilted away from the equatorial plane by the spatial geometry, and the outward acceleration necessary to resist falling into the hole, scaled by a proper time dilation factor. Turning on the rotation of the hole then breaks the symmetry between corevolving and counter-revolving orbits, spreading apart the asymptotes from the symmetry axis of the degenerate acceleration half-line hyperbola branch, tilting the counter-revolving orbit half farther from the black hole symmetry axis and the corevolving half closer towards it due to an increased/decreased radial acceleration term, while moving the vertex slightly away from the equatorial plane off the $\partial_{\hat{r}}$ axis.

Since the circular orbit motion is a case of purely transverse relative acceleration, the acceleration vector already lies in the observer local rest space and so may itself be directly represented by its relative observer decomposition (given by $a(U) = \gamma A(U,u)$ in Eq. (9.3) of Ref. 17 with $u = n$). Since the acceleration lies in the $\partial_{\hat{r}}$-$\partial_{\hat{\theta}}$ plane, it is convenient to introduce the orthonormal 2-vector notation $\vec{X} = (X^{\hat{r}}, X^{\hat{\theta}})$. For the ZAMO decomposition, the two nonzero acceleration components have the form

$$\begin{aligned} \vec{a} &= \vec{k}\sinh^2\alpha + 2\vec{\theta}_{\hat{\phi}}\sinh\alpha\cosh\alpha + \vec{A}\cosh^2\alpha \\ &= (\vec{k}+\vec{A})\sinh^2\alpha + 2\vec{\theta}_{\hat{\phi}}\sinh\alpha\cosh\alpha + \vec{A} \\ &= \frac{1}{2}(\vec{k}+\vec{A})\cosh 2\alpha + \vec{\theta}_{\hat{\phi}}\sinh 2\alpha + \frac{1}{2}(-\vec{k}+\vec{A})\ . \end{aligned} \tag{20}$$

The first line is its relative observer decomposition into a relative centripetal acceleration term involving the relative Lie curvature vector $\vec{k}$ of the ϕ coordinate circles in the slicing geometry of the ZAMO's (see Eq. (13.3) of Ref. 16), a gravitomagnetic term involving the observer expansion tensor $\theta^{\alpha}{}_{\beta}$ (let $\vec{\theta}_{\hat{\phi}} = (\theta^{\hat{r}}{}_{\hat{\phi}}, \theta^{\hat{\theta}}{}_{\hat{\phi}})$), and a gravitoelectric term involving the observer acceleration $\vec{A}$. The second line is the corresponding optical metric decomposition, where the first term is the relative Lie optical centripetal acceleration and $\vec{k}+\vec{A} = N^{-1}\vec{\tilde{k}}$ is along the optical Lie relative curvature vector of the ϕ-coordinate circles in the optical slicing geometry of the ZAMO's (see equation (13.4) of Ref. 16 and 18) and the remaining two terms are the optical gravitomagnetic and gravitoelectric terms. The final line identifies the coefficients in the standard hyperbola parametrization in equation (17).

The ZAMO Lie Frenet-Serret frame is $\{\partial_{\hat{\phi}}, \hat{k}, \partial_{\hat{\phi}} \times_n \hat{k}\}$, where $\hat{k}$ is the unit vector direction of $\vec{k}$.[3] The ZAMO Fermi-Walker and corotating Fermi-Walker relative curvature vectors agree (zero vorticity condition) and are related to the Lie one by a speed-dependent relationship

$$\vec{k}_{(\mathrm{fw})} = \vec{k} + \vec{\theta}_{\hat{\phi}}/\tanh\alpha \ , \tag{21}$$

which absorbs half of the second term in the first line of (20) into the first.

For the static observer decomposition, the ZAMO gravitomagnetic expansion vector $\vec{\theta}_{\hat{\phi}} \sim \theta^{\alpha}{}_{\beta}\hat{\nu}^{\beta}$ term is instead replaced by the gravitomagnetic cross product $\hat{\nu} \times_m \vec{H}$, where $\vec{H} = 2\omega$ is twice the vorticity of the static observer 4-velocity m and $\hat{\nu}$ is the unit relative velocity with respect to the relevant observer. In the static observer case, the Lie and corotating Fermi-Walker relative curvature vectors agree (zero expansion condition) and are related to the Fermi-Walker relative curvature by an analogous relationship.

Each of these decompositions views the circular orbit from the point of view of an observer family also in circular motion. If instead one wishes to view this relative motion from the point of view of the circular orbit, one needs a comoving frame along that orbit, and one can perform a Frenet-Serret analysis of the derivative properties of the relative velocity vector of the test particle following the orbit with respect to either family but as seen by the test particle itself, as discussed in a companion article in these proceedings.[3] This comoving relative Frenet-Serret frame coincides with the spatial frame of the spacetime Frenet-Serret frame.[3] Thus modulo the boosts between local rest spaces, the spacetime Frenet-Serret frame is intimately connected with the various relative Frenet-Serret frames which shape our picture of what is happening in the gravitational field of the black hole.

To describe accelerated photon circular orbits, one can choose the limits $\nu \to \pm 1$ of $\gamma^{-1}U$ and $\gamma^{-2}a$ (corresponding to a unit energy affine parameter[16]), yielding respectively the 4-momenta $P_{\pm} = n \pm \partial_{\hat{\phi}}$ and the corresponding acceleration vectors

$$\vec{a}_{\pm} = \vec{k} + \vec{A} \pm 2\vec{\theta}_{\hat{\phi}} \ , \tag{22}$$

where the $+/-$ sign refers to the corevolving/counter-revolving photon orbits. These two vectors give the directions of the asymptotes of the timelike circular orbit acceleration hyperbola, which are the the diagonals of the two parallelograms formed by the vectors $\pm\vec{\theta}_{\hat{\phi}}$ and $(\vec{k} + \vec{A})/2$ (modulo a factor of 2) thought of as tangent to the hyperbola center at $(-\vec{k} + \vec{A})/2$, as the third line of equation (20) reveals.

It is worth noting that these vectors and their average can be expressed in the form

$$\vec{a}_{\pm} = \mp(\tilde{g}_{\phi\phi})^{1/2}\vec{\nabla}\zeta_{\pm} \ , \qquad (\vec{a}_{+} + \vec{a}_{-})/2 = \vec{k} + \vec{A} = -\vec{\nabla}\ln(\tilde{g}_{\phi\phi})^{1/2} \tag{23}$$

where $\zeta_{\pm} = -N^{\phi} \pm (\tilde{g}_{\phi\phi})^{-1/2}$ are the photon angular velocities (see equation (33)), $\tilde{g}_{\phi\phi} = N^{-2}g_{\phi\phi}$ is the corresponding optical metric component, and $\vec{\nabla} = (\partial_{\hat{r}}, \partial_{\hat{\theta}})$. Thus their lines of force can be represented as orthogonal trajectories to these three potential functions. The surfaces associated with the last potential function are the generalizations to the rotating black hole case of the von Zeipel cylinders discussed for Schwarzschild by Abramowicz.[19] Along the lines of force of the vector field $\vec{k} + \vec{A}$ in the ρ-z coordinate plane, r decreases in region A (r component

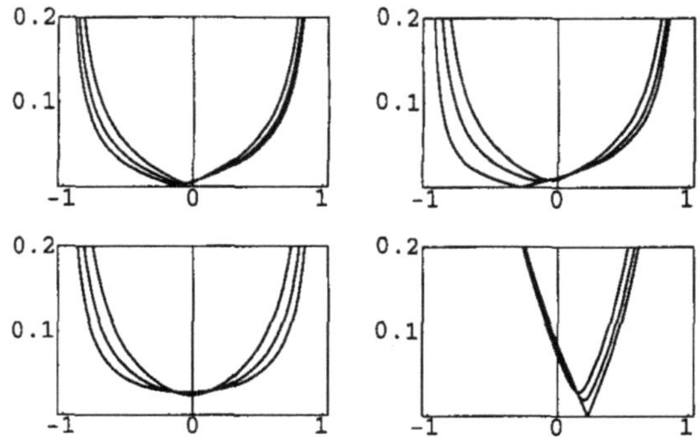

Figure 8. The magnitude $||\omega_{(\mathrm{FW})}||$ of the Fermi-Walker angular velocity versus velocity shown at angles $\theta = \pi/3$, $\pi/2.5$, $\pi/2.01$ at the same radii as in Fig. 3 for $\bar{a} = 0.5$: (a) $\bar{r} = 5$ (region A, upper left), (b) $\bar{r} = 4$ (region A, upper right), (c) $\bar{r} = 3$ (region B, lower left), (d) $\bar{r} = 2.25$ (region C, lower right).

negative, points radially inward) and increases in region C (r component positive, points radially outward), with the transition between these inward and outward behaviors occurring in region B. The lines of force of the photon accelerations $\vec{a}_{\pm}$ are respectively more inward (+) and more outward (−) compared to this vector field whose direction lies nearly midway between the two.

By extremizing the distance between the acceleration hyperbola and its center, one easily finds its vertex rapidity $\alpha_{(\mathrm{vert})}$ to satisfy the following condition (which can be represented equivalently in the second form)

$$\begin{aligned}\tanh 4\alpha_{(\mathrm{vert})} &= -2\left(\frac{\vec{k}+\vec{A}}{2}\right)\cdot\vec{\theta}_{\hat{\phi}}/(||\frac{\vec{k}+\vec{A}}{2}||^2+||\vec{\theta}_{\hat{\phi}}||^2)\\ \alpha_{(\mathrm{vert})} &= \frac{1}{4}\ln(||\vec{a}_-||/||\vec{a}_+||\ ,\end{aligned} \tag{24}$$

which also determines the vertex of the closely related Frenet-Serret angular velocity hyperbola at which $||\omega_{(\mathrm{FS})}||$ is minimized. The corresponding relative velocity $\nu_{(\mathrm{vert})} = \tanh\alpha_{(\mathrm{vert})}$ is shown in Fig. 4 of Ref. 8, counter-revolving until very near the horizon where it becomes corevolving.

Figure 8 shows a plot of the magnitude $||\omega_{(\mathrm{FW})}||$ of the Fermi-Walker angular velocity versus velocity for selected radii and angles. The minima occur at the vertex observer velocity $\nu_{(\mathrm{vert})}$. Approaching the equatorial plane a kink develops corresponding to the limiting minimum value of zero assumed by $||\omega_{(\mathrm{FW})}||$ at the extremely accelerated observers there in region A (counter-revolving) and in region C (corevolving), while a smooth minimum occurs in region B in the equatorial limit where the velocity makes a transition between the null limiting velocities at the AB (counter-revolving) and BC (corevolving) interfaces. The latter observers are referred to as the spin critical observers in Ref. 9, but Fig. 7(b) there incorrectly shows their velocity going beyond $|\nu| = 1$ rather than terminating on the circular photon geodesic points being approached in the acceleration vertex velocity curve in Fig. 4(c) of Ref. 8. The gravitomagnetic effects associated with the expansion

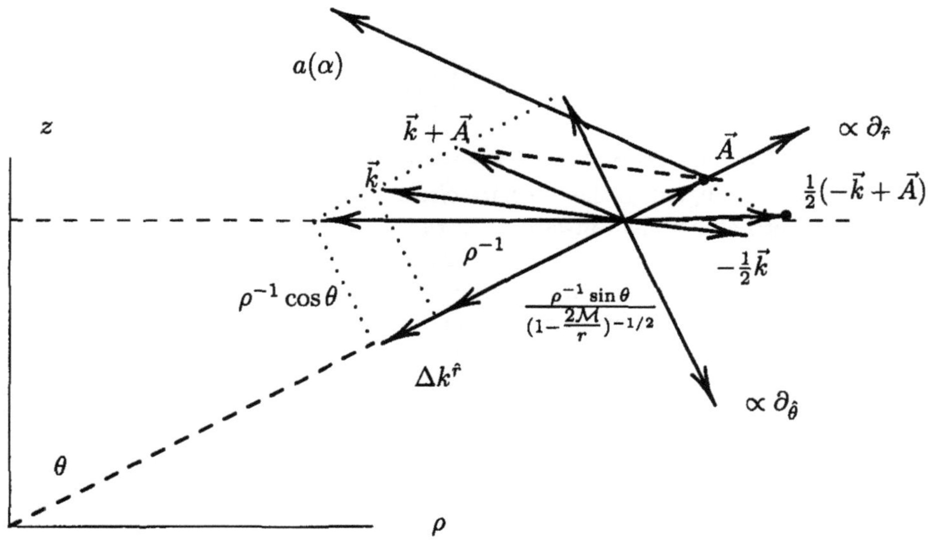

Figure 9. In the Schwarzschild case, the geometry of the ZAMO Lie relative curvature $\vec{k}$ tilted by the radial contraction $\Delta k^{\hat{r}}$, the ZAMO acceleration $\vec{A}$, the photon acceleration $\vec{k}+\vec{A}$, the timelike circular orbit degenerate acceleration hyperbola half line $a(\alpha)$ and its center at $\frac{1}{2}(-\vec{k}+\vec{A})$ in region A for the typical angle $\theta = \pi/3$, shown with the correct orientation of the acceleration plane in the ρ-z coordinate plane. The additional tilt of the half line $a(\alpha)$ above the dashed line is due to the proper time dilation factor multiplying $\vec{A}$.

tensor due to the rotation of the black hole are only appreciable near the hole. The nonrotating Schwarzschild case offers an opportunity to examine the behavior of the remaining gravitoelectric and space curvature effects in the absence of the former effect. Turning on the rotation one can then see how things change.

Consider then the Schwarzschild case. The expansion tensor is zero and the nonzero orthonormal components of the acceleration vector are

$$\vec{a} = (\vec{k}+\vec{A})\sinh^2\alpha + \vec{A} = \frac{1}{2}(\vec{k}+\vec{A})\cosh 2\alpha + \frac{1}{2}(-\vec{k}+\vec{A}) \,, \tag{25}$$

where

$$\begin{aligned} \vec{k} &= -(r\sin\theta)^{-1}((1-2\mathcal{M}/r)^{1/2}\sin\theta, \cos\theta) \,, \\ \vec{A} &= ((1-2\mathcal{M}/r)^{-1/2}\mathcal{M}/r^2, 0) \,. \end{aligned} \tag{26}$$

The relative normal $\vec{k}$ is tilted away from the equatorial plane compared to the horizontal inward flat spacetime relative normal $-\rho^{-1}(\sin\theta, \cos\theta)$ by the change $\rho^{-1} \to (\rho/\sqrt{1-2\mathcal{M}/r})^{-1}$ reflecting the dilation of radial arclength compared to coordinate radius near the hole. This contracts the radial component of the relative normal, tilting the normal away from its horizontal direction as shown in Fig. 9. As one approaches the horizon $\bar{r} = 2$, the radial component of $\vec{k}$ goes to zero and $\vec{k}$ becomes tangent to the horizon, pointing away from the equatorial plane.

The circular photon acceleration is the vector $\vec{k}+\vec{A}$ shown in Fig. 9 parallel to the acceleration half line. This vector has the same direction as the optical Lie curvature vector[16]

$$\tilde{\vec{k}} = N(\vec{k} + \vec{A}) = -\rho^{-1}\left(\left(1 - \frac{3\mathcal{M}}{r}\right)\sin\theta, \left(1 - \frac{2\mathcal{M}}{r}\right)^{1/2}\cos\theta\right) . \tag{27}$$

Its radial component goes to zero at the interface $\bar{r} = 3$ between regions A and C where $\tilde{\vec{k}}$ becomes tangent to the interface circle, inside of which it the vector points radially outward rather than inward as in region A, becoming purely radial at the horizon. This is the famous "reversal of the optical centrifugal force" effect which occurs passing from region A into region C.[20]

At zero ZAMO relative velocity $\nu = 0 = \alpha$, the acceleration starts out at $\vec{A}$ on the radial axis, but the resultant of the centripetal acceleration term $\vec{k}\sinh^2\alpha$ and $\vec{A}$ scaled by the factor $\gamma^2 = 1 + \sinh^2\alpha$ (reflecting the change in the proper time parametrization of the second time derivative) leads to an additional tilting of the acceleration half line (due to the term $\sinh^2\alpha$) compared to the resultant of the centripetal acceleration and $\vec{A}$ alone. Far from the hole the decrease in the radial component of $\vec{k}$

$$\Delta k^{\hat{r}} = [1 - (1 - 2\mathcal{M}/r)^{1/2}]\rho^{-1}\sin\theta \qquad \begin{cases} \overset{r\to\infty}{\longrightarrow} \frac{\mathcal{M}}{r^2}\left(1 + \frac{\mathcal{M}}{2r}\right) \\ \overset{2\mathcal{M}}{\longrightarrow} \frac{1}{2\mathcal{M}} \end{cases} \tag{28}$$

is exactly compensated for by the observer acceleration

$$A^{\hat{r}} = \frac{\mathcal{M}}{r^2}\left(1 - \frac{2\mathcal{M}}{r}\right)^{-1/2} \qquad \begin{cases} \overset{r\to\infty}{\longrightarrow} \frac{\mathcal{M}}{r^2}\left(1 + \frac{\mathcal{M}}{r}\right) \\ \overset{2\mathcal{M}}{\longrightarrow} \infty \end{cases} \tag{29}$$

in the vector sum $(-\vec{k} + \vec{A})/2$ defining the center of the degenerate acceleration hyperbola, putting the center on the horizontal axis $\chi = \frac{\pi}{2} - \theta$ (see Fig. 9), but moving it away from the equatorial plane on the opposite side of this axis as the observer acceleration grows without bound approaching the horizon while the radial shortening factor is bounded. Both $\Delta k^{\hat{r}}$ and $A^{\hat{r}}$ are decreasing functions of r satisfying $\Delta k^{\hat{r}} < A^{\hat{r}}$.

Since $k^{\hat{r}} + A^{\hat{r}} = -r^{-1}(1 - 2\mathcal{M}/r)^{-1/2}(1 - 3\mathcal{M}/r)$ changes sign (negative in region A, positive in region C) while $A^{\hat{r}} > 0$, the component $a^{\hat{r}}$ can change sign (so that the acceleration curve crosses the $\partial_{\hat{\theta}}$-axis) only in region A. As shown in Fig. 6, for $\bar{r} > 3$ (region A) the degenerate acceleration hyperbola half line always intersects the $\partial_{\hat{\theta}}$-axis, but at $\bar{r} = 3$ it is parallel to this axis with a constant radial acceleration component, and for $2 < \bar{r} < 3$ (region C) this half line tilts away from the $\partial_{\hat{\theta}}$-axis, corresponding to the fact that the acceleration must always be radially outward to resist the strong attraction to the hole no matter what value the velocity has.

The acceleration is symmetric under the transformation $\nu \to -\nu$ (or $\alpha \to -\alpha$) which interchanges the corevolving and counter-revolving orbits in the Schwarzschild case. To leading order in the black hole angular momentum parameter a, the nonzero observer expansion tensor components and additional acceleration component are

$$(\theta^{\hat{r}}{}_{\hat{\phi}}, \theta^{\hat{\theta}}{}_{\hat{\phi}}) \to \frac{\mathcal{M}a}{r^3}\sin\theta\left(-3, \frac{a^2}{r^2}\sin 2\theta\right), \; A^{\hat{\theta}} \to -\frac{\mathcal{M}a^2}{r^3}\sin 2\theta . \tag{30}$$

As one turns on the angular momentum parameter a from 0 to a small positive value, the leading order effect is the introduction of a small radially inward symmetry breaking term $-(3\mathcal{M}a/r^2)\sin\theta\sinh 2\alpha$ which separates the corevolving and counter-revolving acceleration vectors in the radial direction by equal amounts about the $\bar{a}=0$ acceleration half line, with the counter-rotating/counter-rotating orbits experiencing an increased/decreased radial acceleration due to the gravitomagnetic field. The small nonzero component $A^{\hat{\theta}}$ shifts the vertex of the acceleration hyperbola slightly off the $\partial_{\hat{r}}$-axis, tilting $\vec{A}$ away from the equatorial plane. These effects qualitatively describe the situation for all physical values $0<\bar{a}\le 1$ since the vector $\vec{\theta}_{\hat{\phi}}$ is always radial to within less than about 1% at its most extreme value not far from the horizon ($|\theta^{\hat{\theta}}{}_{\hat{\phi}}/\theta^{\hat{r}}{}_{\hat{\phi}}|\lesssim 0.01$), while $\vec{A}$ is always radial to within less than about 5% at its most extreme value not far from the horizon. Thus the acceleration half line parallel to $(\vec{k}+\vec{A})\cosh 2\alpha$ terminating at $\vec{A}$ (combined spatial geometry and gravitoelectric terms) is split into the two halves of the one branch of the acceleration hyperbola by the addition of the nearly radial vector fields $\pm 2\vec{\theta}_{\hat{\phi}}|\sinh 2\alpha|$ (the gravitomagnetic term) which pulls the counter-rotating acceleration farther away from the symmetry axis $\rho=0$.

7 Concluding remarks

The Frenet-Serret properties of circular orbits analyzed from the point of view of the ZAMOs help bridge the gap between the Newtonian and general relativistic pictures of the gravitational field of a central source. The past decade has seen a large number of articles written about circular orbits and their interpretation in terms of space-plus-time concepts, but the situation has remained somewhat cloudy. By combining the relative observer machinery with the Frenet-Serret approach in various ways, one begins to see how our nonrelativistic ideas stretch to fit into the relativistic setting and how they relate to the Fermat's principle (conformal) approach.

Appendix: The Kerr computational quantities

To study the geometry of the acceleration and Frenet-Serret angular velocity hyperbolae in the Kerr spacetime, one needs the ZAMO computational quantities for the Kerr metric.[21] With some convenient abbreviations associated with the Boyer-Lindquist coordinate system, including formulas for the radius of the horizon ($N=0$) and ergosphere, and the angular velocities

$$\zeta_\pm = -N^\phi \pm N(g_{\phi\phi})^{-1/2} \tag{31}$$

of the circular photon orbits, which follow from the general angular-physical velocity component relations

$$\zeta = -N^\phi + N(g_{\phi\phi})^{-1/2}\nu^{\hat{\phi}}_{(n)} \tag{32}$$

obtained by solving for the roots of the expression (2.3) of Ref. 9 for the coordinate gamma factor of U, one needs the lapse, shift, and spatial metric coordinate components for the ZAMO observers, the ZAMO Lie curvature vector components, and the ZAMO kinematical quantities, namely the nonzero components of the acceleration vector A^α (minus the gravitoelectric field g) and expansion tensor $\theta^\alpha{}_\beta$ (minus half the Lie gravitomagnetic tensor field $H^\alpha{}_\beta$).[16] The optical Lie curvature vector components $\tilde{\vec{k}} = N(\vec{k} + \vec{A})$ are also given, together with the angular velocity variable roots of the acceleration components $a^{\hat{r}}$ and $a^{\hat{\theta}}$. Finally the curves separating the regions A, B, and C in the ρ-z coordinate plane are derived.

Convenient abbreviations and formulas

$$\begin{aligned}
&(c, s) = (\cos\theta, \sin\theta) \\
&\rho^2 = r^2 + a^2c^2 \ , \ \epsilon = r^2 - a^2s^2 \\
&\Delta = r^2 + a^2 - 2\mathcal{M}r \\
&\Lambda = (r^2 + a^2)\rho^2 + 2\mathcal{M}ra^2s^2 \\
&r_{(\mathrm{erg})} = \mathcal{M} + \sqrt{\mathcal{M}^2 - a^2c^2} \\
&r_{(\mathrm{h})} = \mathcal{M} + \sqrt{\mathcal{M}^2 - a^2} \\
&\zeta_\pm = \frac{2a\mathcal{M}rs \pm \rho^2\sqrt{\Delta}}{\Lambda s} = \frac{\rho^2 - 2\mathcal{M}r}{s(-2a\mathcal{M}rs \pm \rho^2\sqrt{\Delta})}
\end{aligned} \tag{33}$$

Metric (ZAMO decomposition)

$$\begin{aligned}
&(N, N^\phi) = \left(\sqrt{\frac{\Delta\rho^2}{\Lambda}}, -\frac{2a\mathcal{M}r}{\Lambda}\right) \\
&(g_{rr}, g_{\theta\theta}, g_{\phi\phi}, \sqrt{g}) = \left(\frac{\rho^2}{\Delta}, \rho^2, \frac{\Lambda s^2}{\rho^2}, \frac{\rho^2 s}{N}\right)
\end{aligned} \tag{34}$$

ZAMO relative curvature vector and optical counterpart

$$\begin{aligned}
(k^{\hat{r}}, k^{\hat{\theta}}) &= \left(-\frac{\sqrt{\Delta}}{\rho^3\Lambda}[r\rho^4 - \mathcal{M}a^2s^2\epsilon], -\frac{c}{\rho^3 s\Lambda}[\Lambda(r^2 + a^2) - \Delta a^2s^2\rho^2]\right) \\
\tilde{k}^{\hat{r}} &= \frac{1}{\rho^2\Lambda^{3/2}}\left[m\rho^2(r^4 - a^4) + \Delta[ma^2s^2(3r^2 - a^2c^2) - r\rho^4]\right] \\
\tilde{k}^{\hat{\theta}} &= -\frac{\sqrt{\Delta}}{\rho^2\Lambda^{3/2}}\left[\rho^2\Lambda\cot\theta + 4a^2mrsc(r^2 + a^2)\right]
\end{aligned} \tag{35}$$

ZAMO kinematical quantities

$$(A^{\hat{r}}, A^{\hat{\theta}}) = \left(\frac{\mathcal{M}}{\rho^3\Lambda\sqrt{\Delta}}[\rho^2(r^4 - a^4) + 2\Delta r^2a^2s^2], -\frac{2a^2\mathcal{M}r}{\rho^3\Lambda}(r^2 + a^2)sc\right) \tag{36}$$

$$(\theta^{\hat{r}}{}_{\hat{\phi}}, \theta^{\hat{\theta}}{}_{\hat{\phi}}) = \left(\frac{a\mathcal{M}}{\Lambda\rho^3}[(r^2 + a^2)(a^2 - 3r^2) - a^2s^2(a^2 - r^2)]s, \frac{2a^3\mathcal{M}r\sqrt{\Delta}}{\Lambda\rho^3}s^2c\right)$$

Roots of acceleration components (angular velocity)

$$\zeta_{\pm}^{(r)} = s^{-1}[as \mp \rho^2 \sqrt{r/(\mathcal{M}\epsilon)}]^{-1} \tag{37}$$

$$\zeta_{\pm}^{(\theta)} = \frac{2\mathcal{M}ra(r^2+a^2 \pm ia\rho^2\sqrt{2\mathcal{M}r\Delta})}{(r^2+a^2)^3 - \Delta a^2 s^2 (r^2+a^2+\rho^2)} \tag{38}$$

The regions A, B and C

Timelike circular orbits with vanishing radial acceleration have 4-velocity $U_{\pm}^{(r)} = \Gamma_{\pm}^{(r)}(\partial_t + \zeta_{\pm}^{(r)}\partial_\phi)$ and coordinate angular velocity $\zeta_{\pm}^{(r)}$ given above. They become null (i.e., coincide with photon orbits) at the interface radii $r_{AB}(\theta)$ (counter-rotating photon: $-$) and $r_{BC}(\theta)$ (corotating photon: $+$) characterized by the condition $\zeta_{\pm}^{(r)} = \zeta_{\pm}$. A straightforward calculation shows that this relation is equivalent to

$$\sqrt{\Delta} \mp as = \sqrt{\mathcal{M}\epsilon/r}, \tag{39}$$

or

$$a^2s^2(r-\mathcal{M}) \mp 2ars\sqrt{\Delta} + [\Delta r - \mathcal{M}(r^2-a^2)] = 0\ , \tag{40}$$

which are implicit equations for r as a function of θ that are quadratic in s and so easily solved for θ as a function of r for the two curves $\theta = \theta_{\rm AB}(r)$ and $\theta = \theta_{\rm BC}(r)$

$$\begin{aligned} \theta_{\rm AB}(r) &= \arcsin S\ ,\ 0 \le S \le 1\ , \\ \theta_{\rm BC}(r) &= \arcsin S\ ,\ -1 \le S \le 0\ , \\ S &= \frac{r\sqrt{\Delta} - \sqrt{m}\sqrt{2r^3 - 3mr^2 + ma^2}}{a(r-m)}\ . \end{aligned} \tag{41}$$

Figures 4(a),(b),(c) are plots of these curves for the cases $\bar{a} = 0, 0.5, 1$ respectively. In the extreme case $\bar{a} = 1$, equation (40) can be factored into the form

$$(r-\mathcal{M})[s^2 \pm 2rs + (r^2 - \mathcal{M}^2 - 2\mathcal{M}r)] = 0 \tag{42}$$

which exhibits a bifurcating behaviour at the horizon for the curve $r = r_{\rm BC}(\theta)$. The $r = r_{\rm BC}(\theta)$ solution of (42) consists of two branches: a) $r = r_{\rm BC1}(\theta) = \mathcal{M} = r_{\rm (h)}$, namely the horizon itself, and b) $r = r_{\rm BC2}(\theta)$, the solution of the quadratic equation (in either r or in s) $s^2 \pm 2rs + (r^2 - \mathcal{M}^2 - 2\mathcal{M}r) = 0$. The limiting angle at which $r = r_{\rm BC2}(\theta)$ intersects the horizon is $\theta_* = \arcsin(\sqrt{3}-1) \approx 47.1°$. This angle coincides with the maximum angle for which the stationary observers with the same fixed angular velocity as the hole $\zeta_{\rm (h))} = a^2/(a^2+r_+^2) = 1/2$ are timelike near the horizon, and with the minimum angle of the single corotating horizon skimming[7] photon spherical geodesic orbit at the horizon. The latter follows by expressing the coefficients of equation (29) of Ref. 7 for $z = \cos^2\theta$ entirely in terms of the energy constant E using equation (22), leading to $E^2 = (z^2-1)/(z^2+6z-3)$; the limit $E \to \infty$ for a photon corresponds to $z = 2\sqrt{3}-3$, equivalent to $\theta = \theta_*$.

References

1. Bini D, Carini P and Jantzen R T 1999 *Proceedings of the Eighth Marcel Grossmann Meeting on General Relativity (1997)* ed T Piran (Singapore: World Scientific)
2. Bini D, de Felice F and Jantzen R T 1999 *Proceedings of the Third William Fairbank Meeting on the Lense-Thirring Effect (1998)* ed C Sigismondi (Singapore: World Scientific)
3. Bini D, de Felice F and Jantzen R T 1999 *Class. Quantum Grav.* **16** 2105
4. Iyer B R and Vishveshwara C V 1993 *Phys. Rev.* **D48** 5706
5. Hönig E, Schücking E L and Vishveshwara C W 1974 *J. Math. Phys.* **15** 744
6. Synge J L 1960 *Relativity: The General Theory* (Amsterdam: North Holland)
7. Wilkins D 1972 *Phys. Rev.* **5** 814
8. Bini D, Jantzen R T and Merloni A 1999 *Class. Quantum Grav.* **16** 1333
9. Bini D, Carini P and Jantzen R T 1997 *Int. J. Mod. Phys.* **D6** 143
10. Page D 1998 *Class. Quant. Grav.* **15** 1669
11. Semerák O 1996 *Gen. Relativ. Grav.* **28** 1151
12. Semerák O 1998 *Gen. Relativ. Grav.* **30** 1203
13. Semerák O 1994 *Astron. Astrophys.* **291** 679
14. Bini D, Carini P and Jantzen R T 1992 *Proceedings of the Third Italian-Korean Astrophysics Meeting* ed S. Kim, H. Lee and K.T. Kim *J. Korean Phys. Soc.* **25** S190
15. de Felice F 1995 *Class. Quantum Grav.* **12** 1119; de Felice F and Usseglio-Tomasset S 1991 *Class. Quantum Grav.* **8**, 1871
16. Bini D, Carini P and Jantzen R T 1997 *Int. J. Mod. Phys.* **D6** 1
17. Jantzen R T, Carini P and Bini D 1992 *Ann. Phys. (N.Y.)* **215** 1
18. Abramowicz M A, Nurowski P Wex N 1995 *Class. Quantum Grav.* **12** 1467
19. Abramowicz M A 1990 *Mon. Not. R. Ast. Soc.* **245** 733
20. Abramowicz M A and Prasanna A R 1990 *Mon. Not. R. Ast. Soc.* **245** 720
21. Misner C W, Thorne K S and Wheeler J A 1973 *Gravitation* (San Francisco: Freeman)

CENTRIPETAL ACCELERATION AND CENTRIFUGAL FORCE IN GENERAL RELATIVITY

D. BINI
Istituto per Applicazioni della Matematica, C.N.R., I-80131 Napoli, Italy and I.C.R.A., Università di Roma, I-00185 Roma, Italy
e-mail: binid@icra.it

F. DE FELICE
Dipartimento di Fisica, Università di Padova, I-35131 Padova, Italy and INFN, Sezione di Padova, Italy
e-mail: fernando.defelice@pd.infn.it

R. T. JANTZEN
Dept. of Math. Sciences, Villanova University, Villanova, PA 19085, USA and I.C.R.A., Università di Roma, I-00185 Roma, Italy
e-mail: jantzen@ucis.vill.edu

In nonrelativistic mechanics noninertial observers studying accelerated test particle motion experience a centripetal acceleration which, once interpreted as a centrifugal force acting on the particle, allows writing the particle's equation of motion in a Newtonian form, simply by adding the inertial force contribution to that of the external forces in the acceleration-equals-force equation. In general relativity centripetal and centrifugal acceleration generalizing the classical concepts must be properly (geometrically) defined. This requires a relative Frenet-Serret frame approach based on a family of test observers.

1 Introduction

A standard approach for studying the intrinsic properties of spacetime curves is via the introduction of an "absolute" or spacetime Frenet-Serret frame.[1,2] The obvious space-plus-time generalization of this analysis with respect to a preferred family of observers in a spacetime with symmetry (i.e. a "relative" Frenet-Serret frame) has only recently been introduced[3] and provides another tool for understanding the properties of test particle motion in the same framework in which we ourselves experience spacetime.[4]

Given a family of test observers, a relative Frenet-Serret 3-frame may be introduced independent of any spacetime symmetries in an obvious way which only depends on the choice of a geometrically defined relative temporal derivatives along the particle world line (for example a projected Lie or Fermi-Walker derivative). This relative Frenet-Serret 3-frame lives in the local rest space of the family of test observers and reflects the way in which those observers "see" the motion of test particles.

Analogously, for a given family of test observers, one can introduce another complementary relative Frenet-Serret frame which lives in the test particle local rest space and this is determined by the unique Fermi-Walker derivative along its world line. Such a "comoving relative Frenet-Serret 3-frame" is corresponds to a reconstruction by the test particle of its motion relative to the given family of preferred observers and leads naturally to a geometrical way of re-introducing the elusive notion of centrifugal force within general relativity.

After a brief introduction to the spacetime splitting associated with a family of test observers and their analysis of a test particle motion (see section 2), the relative Frenet-Serret frame along a test particle world line is introduced in the local rest space of the observer family. *Based on the direction of the relative motion of* the test particle with respect to the observer, this splitting geometrizes the usual decomposition of acceleration into longitudinal and transverse parts with respect to the relative velocity unit vector of the particle with respect to the observer and gives rise to the definition of a relative centripetal acceleration.

A "relative comoving Frenet-Serret frame" in the test particle local rest space, may be unambiguously defined by using the natural Fermi-Walker derivative along the test particle world line. Based on the the direction of the relative motion of the observer with respect to the particle (i.e. in the local rest space of the particle itself) this splitting geometrizes instead that of the acceleration into longitudinal and transverse parts, but with respect to the relative velocity unit vector of the observer with respect to the particle, and gives rise to the definition of a relative centrifugal acceleration. This is discussed in section 4.

Thus the concepts of centripetal and certifugal accelerations in general relativity have different geometric origin whereas they coincide (up to the sign) in the classical context. Section 5 discusses finally how they both enter the splitting of the acceleration-equals-force equation.

2 Spacetime splitting language

A spacetime manifold $\mathcal{M}$ which admits a congruence $\mathcal{C}_u$ of timelike lines with unit tangent vector field u can be locally split into space plus time through the orthogonal decomposition of the tangent spaces on $\mathcal{C}_u$ into the local rest space LRS_u and the local time direction u. The congruence $\mathcal{C}_u$ and the associated (local) 1+3 splitting of the spacetime identifies a family of observers on $\mathcal{M}$ whose world lines are the congruence curves and whose 4-velocity is u. Let τ_u be a proper time parameter defined on each such curve.

Tensors or tensor fields which have no component along u are called spatial (with respect to u). The spatial projection (mixed) tensor $P_{(u)}$

$$P_{(u)}{}^{\alpha}{}_{\beta} = \delta^{\alpha}{}_{\beta} + u^{\alpha}u_{\beta} \tag{1}$$

projects out the spatial part of any tensor field by the usual process. Any spacetime tensor field may be represented by a collection of spatial tensor fields which result from their orthogonal decomposition. Let $\eta_{(u)\alpha\beta\delta} = u^{\sigma}\eta_{\sigma\alpha\beta\delta}$ denote the spatial alternating tensor associated with the spatial metric $P_{(u)\alpha\beta}$ (the inner product on LRS_u), which may be used in an obvious way to define a spatial cross product $\times_u$ of spatial vector fields and a spatial duality operation $^{*(u)}$ for antisymmetric spatial tensor fields.

As described in Refs. 4–6, one may also spatially project various derivative operators so that the result of the derivative of any tensor field is always spatial. Two useful spatial derivatives are the spatial Lie derivative $\pounds_{(u)X} = P_{(u)}\pounds_X$ for any vector field X, and the spatial covariant derivative $\nabla_{(u)\alpha} = P_{(u)}P_{(u)}{}^{\beta}{}_{\alpha}\nabla_{\beta}$, where the projection on all free indices is implied after the application of the derivative. Similarly three useful temporal derivatives (i.e., along u) are the temporal Lie derivative $\nabla_{(\mathrm{lie},u)} = P_{(u)}\pounds_{u}$, the Fermi-Walker temporal derivative $\nabla_{(\mathrm{fw},u)} = P_{(u)}u^{\alpha}\nabla_{\alpha}$, and the corotating Fermi-Walker temporal derivative $\nabla_{(\mathrm{cfw},u)}$ related to the first two by the kinematical quantities (acceleration $a_{(u)}{}^{\alpha}$, vorticity $\omega_{(u)}{}^{\alpha}{}_{\beta} = \eta_{(u)}{}^{\alpha}{}_{\beta\gamma}\omega_{(u)}{}^{\gamma}$, expansion $\theta_{(u)}{}^{\alpha}{}_{\beta}$ defined for instance in Ref. 5) of the observer congruence through the relations

$$\nabla_{(\mathrm{cfw},u)}X^{\alpha} = \nabla_{(\mathrm{fw},u)}X^{\alpha} + \omega_{(u)}{}^{\alpha}{}_{\beta}X^{\beta} = \nabla_{(\mathrm{lie},u)}X^{\alpha} + \theta_{(u)}{}^{\alpha}{}_{\beta}X^{\beta} \tag{2}$$

valid only for a spatial vector field X^{α}, and easily extended to any spatial tensor field in the usual way.

Moreover, when dealing with more than one family of test observers, say u and U, then the mixed projection map $P_{(U,u)} = P_{(u)}P_{(U)}$ from LRS_U to LRS_u arises naturally. In this case the orthogonal decomposition of U made by the family of observers u identifies the spatial relative velocity $\nu_{(U,u)} = ||\nu_{(U,u)}||\hat{\nu}_{(U,u)}$ of U relative to u, where $\hat{\nu}_{(U,u)}$ is a unit vector (the unit relative velocity), by the relation

$$U = \gamma_{(U,u)}[u + ||\nu_{(U,u)}||\hat{\nu}_{(U,u)}] \,, \tag{3}$$

and vice versa

$$u = \gamma_{(u,U)}[U + ||\nu_{(u,U)}||\hat{\nu}_{(u,U)}] \,, \tag{4}$$

where $\gamma_{(U,u)} = \gamma_{(u,U)} \equiv \gamma = (1-\nu^2)^{-1/2}$ with $\nu \equiv ||\nu_{(U,u)}|| = ||\nu_{(u,U)}||$. Note that in this general discussion, the speed ν is nonnegative. Of course in a specific case it can be useful ν to be relaxed to signed quantity.

This decomposition of U, upon differentiation along U and splitting with respect to u, leads to a corresponding splitting of the 4-acceleration $a_{(U)}$ of U

$$a_{(U)} = \frac{\mathrm{D}}{\mathrm{d}\tau_U}U = \gamma_{(U,u)}[A_{(U,u)} \cdot \nu_{(U,u)}\, u + A_{(U,u)}] \,, \tag{5}$$

where

$$A_{(U,u)} = \gamma^{-1}P_{(U,u)}a(U) = P_{(u)}\frac{\mathrm{D}}{\mathrm{d}\tau_{(U,u)}}[\gamma_{(U,u)}\nu_{(U,u)}] - F^{(G)}_{(\mathrm{fw},U,u)} \,, \tag{6}$$

and $\mathrm{D}/\mathrm{D}\tau_U = \nabla_U$ is the intrinsic covariant derivative along the world lines of U, while $\tau_{(U,u)} = \int \gamma d\tau_U$ is a Cattaneo relative standard time parameter along those world lines. The spatial gravitational force field $F^{(G)}_{(\mathrm{fw},U,u)}$ depends of the kinematical quantities of u in the following way

$$F^{(G)}_{(\mathrm{fw},U,u)} = -P_{(u)}\frac{Du}{d\tau_U} = -P_{(u)}\nabla_U u \tag{7}$$

$$= \gamma\left\{g_{(u)} + \nu\left[\frac{1}{2}\hat{\nu}_{(U,u)} \times_u H_{(u)} - \theta_{(u)} \llcorner \hat{\nu}_{(U,u)}\right]\right\} . \tag{8}$$

$g_{(u)} = -a_{(u)}$ is the gravitoelectric field, while $H_{(u)} = 2\omega_{(u)}$ is the gravitomagnetic field. In the spatial projection of the acceleration-equals-force equation

$$P_{(U,u)}a_{(U)} = \gamma F_{(U,u)} \,, \tag{9}$$

where $F_{(U,u)} = \gamma^{-1}P_{(U,u)}f_{(U)}$ is the relative force, the gravitational force term may be passed to the righthand side of the equation where it has the correct sign for a force-like term driving the observed rate of change of the spatial momentum $p_{(U,u)} = \gamma\nu_{(U,u)}$

$$\frac{\mathrm{D}_{(\mathrm{fw},U,u)}}{\mathrm{d}\tau_{(U,u)}}p_{(U,u)} = F_{(U,u)} + F^{(G)}_{(\mathrm{fw},U,u)}\,, \tag{10}$$

where the abbreviated notation for the projected covariant derivative along U with respect to any parameter λ along its world lines

$$\frac{\mathrm{D}_{(\mathrm{fw},U,u)}}{\mathrm{d}\lambda} = P_{(u)}\frac{\mathrm{D}}{\mathrm{d}\lambda}\,, \tag{11}$$

is called the spatial Fermi-Walker derivative along U. It is also convenient to introduce the Cattaneo standard length parameter along the world lines of U

$$\ell_{(U,u)} = \int \gamma\nu d\tau_U = \int \nu d\tau_{(U,u)}\,. \tag{12}$$

$\ell_{(U,u)}$ corresponds to the spatial arc length along the world line of U as observed by u, as long as $\nu \neq 0$. If $\nu = 0$, the test particle remains fixed on a single observer world line and there is no relative motion and no curve in the quotient space of observer world lines, so the parametrization is not valid.

All of the above continues to hold for a single test particle world line with 4-velocity U, except that when differentiating any tensor field defined only along that world line, the symbol $\mathrm{D}/\mathrm{D}\lambda$ for the intrinsic covariant derivative along the world line must be taken seriously and not just thought of as an abbreviation for $d\tau_U/\mathrm{d}\lambda\,\nabla_U$.

3 The relative Frenet-Serret frame

One can further decompose the spatial parts of tensor fields using the decomposition of the local rest space of the observer u along the unit relative velocity $\hat{\nu}_{(U,u)}$ of U (longitudinal directions) and orthogonal to it (transverse directions) in an orthogonal $1+(1+2)$ refinement of the $1+3$ relative observer decomposition.[5,7] It is exactly this decomposition of the local rest space LRS_u which corresponds to the longitudinal-transverse decomposition of acceleration in the ordinary nonrelativistic classical mechanics of a point particle. It is most naturally accomplished by introducing the spatial orthonormal 3-frame defined by the differential properties of the unit relative velocity itself in complete analogy with the Frenet-Serret frame approach in an ordinary Riemannian 3-manifold.

The first vector of the orthonormal relative Frenet-Serret spatial frame is the unit relative velocity vector $\hat{\nu}_{(U,u)}$. The remaining two vectors are the relative normal $\hat{\eta}_{(\mathrm{fw},U,u)}$, which apart from the sign is defined as the normalized $P_{(u)}$-projected covariant derivative of $\hat{\nu}_{(U,u)}$ along U, and its cross product with the unit relative velocity vector, namely the relative binormal $\hat{\beta}_{(\mathrm{fw},U,u)} = \hat{\nu}_{(U,u)} \times_u \hat{\eta}_{(\mathrm{fw},U,u)}$. Let $\{E_{(\mathrm{fw},U,u)\hat{a}}\} = \{\hat{\nu}_{(U,u)}, \hat{\eta}_{(\mathrm{fw},U,u)}, \hat{\beta}_{(\mathrm{fw},U,u)}\}$ denote this righthanded spatial frame. The relative normal and binormal span the transverse subspace $LRS_u \cap LRS_U$ of the local rest space.

The covariant derivatives along U of these spatial frame vectors $E_{(\mathrm{fw},U,u)\hat{a}}$, after projection in the rest frame of u and reparametrizing the derivative by the spatial arc length, are the relative Frenet-Serret equations which define the Fermi-Walker relative curvature, torsion, normal, and binormal

$$\frac{\mathrm{D}_{(\mathrm{fw},U,u)}}{\mathrm{d}\ell_{(U,u)}}\hat{\nu}_{(U,u)} = k_{(\mathrm{fw},U,u)}\hat{\eta}_{(\mathrm{fw},U,u)} \,, \tag{13}$$

$$\frac{\mathrm{D}_{(\mathrm{fw},U,u)}}{\mathrm{d}\ell_{(U,u)}}\hat{\eta}_{(\mathrm{fw},U,u)} = -k_{(\mathrm{fw},U,u)}\hat{\nu}_{(U,u)} + \tau_{(\mathrm{fw},U,u)}\hat{\beta}_{(\mathrm{fw},U,u)} \,, \tag{14}$$

$$\frac{\mathrm{D}_{(\mathrm{fw},U,u)}}{\mathrm{d}\ell_{(U,u)}}\hat{\beta}_{(\mathrm{fw},U,u)} = -\tau_{(\mathrm{fw},U,u)}\hat{\eta}_{(\mathrm{fw},U,u)} \,. \tag{15}$$

The Frenet-Serret relation (13) only fixes the product of the curvature $k_{(\mathrm{fw},U,u)}$ and the second frame vector $\hat{\eta}_{(\mathrm{fw},U,u)}$ and not their individual signs. Once these signs are chosen, the third vector $\hat{\beta}_{(\mathrm{fw},U,u)} = \hat{\nu}_{(U,u)} \times_u \hat{\eta}_{(\mathrm{fw},U,u)}$ is then chosen to make the frame righthanded, after which the sign of the torsion is then determined. The sign of the curvature may be chosen conveniently.

These relations may be written in a more compact form as

$$\frac{\mathrm{D}_{(\mathrm{fw},U,u)}}{\mathrm{d}\ell_{(U,u)}}E_{(\mathrm{fw},U,u)\hat{a}} = \omega_{(\mathrm{fw},U,u)} \times_u E_{(\mathrm{fw},U,u)\hat{a}} \,, \tag{16}$$

where

$$\omega_{(\mathrm{fw},U,u)} = \tau_{(\mathrm{fw},U,u)}\hat{\nu}_{(U,u)} + k_{(\mathrm{fw},U,u)}\hat{\beta}_{(\mathrm{fw},U,u)} \tag{17}$$

defines the Fermi-Walker relative angular velocity of the frame. The test particle world line with 4-velocity U will be said *Fermi-Walker u-relatively straight* if the Fermi-Walker relative curvature vanishes $k_{(\mathrm{fw},U,u)} = 0$ and *Fermi-Walker u-relatively flat* if the Fermi-Walker relative torsion vanishes $\tau_{(\mathrm{fw},U,u)} = 0$. In the first case the normal $\hat{\eta}_{(\mathrm{fw},U,u)}$ is not defined, but one can introduce any spatial unit vector in its place which is orthogonal to $\hat{\nu}_{(U,u)}$ and transported along the world line so that its projected covariant derivative is zero in order to obtain a frame satisfying the above equations with vanishing curvature and torsion. For cases when the relative curvature vanishes at an isolated point, one must allow it to have either sign when nonzero in order to extend the relative normal continuously through this point.

4 The comoving relative Frenet-Serret frame

While centripetal acceleration is an acceleration term measured by the observer in its own rest space, a centrifugal force is an "inertial force" felt by a test particle in its rest space. In nonrelativistic mechanics, these rest spaces share the same Newtonian time, and one can simply move the centripetal acceleration term to the force side of the acceleration-equals-force equation to obtain the corresponding centrifugal force term. However, in general relativity, the observer and test particle in relative motion have different local rest spaces, so one must look for a geometric description of centrifugal force in the local rest space of the test particle if one wishes to generalize the *nonrelativistic concept*. Classically it is also a relative effect, involving the relative motion with respect to the implicit Newtonian family of inertial observers, so the relativistic generalization must also involve the relative motion of the test particle with respect to some family of observers. The generalization developed here will involve the comparison of the rotation (with respect to gyro-fixed axes) of the relative velocity of the test particle with respect to the observer family as seen by the test particle itself.

The two vectors u and U define a unique boost $B_{(u,U)}$ of each tangent space into itself which maps $(u, \nu_{(U,u)})$ onto $(U, -\nu_{(u,U)})$ in the relative observer plane of u and U and leaves the orthogonal 2-space $LRS_u \cap LRS_U$ (spanned by $\hat{\eta}_{(\text{fw},U,u)}$ and $\hat{\beta}_{(\text{fw},U,u)}$) invariant. Let

$$\hat{\mathcal{V}}_{(u,U)} = -\hat{\nu}_{(u,U)} = \gamma(\nu u + \hat{\nu}_{(U,u)}) \tag{18}$$

be the relative velocity of U with respect to u as seen by U.

It is only natural to differentiate vectors in LRS_U along U by projecting the covariant derivative along U into LRS_U. This leads to the usual Fermi-Walker derivative along U measuring the rate of change of vectors in LRS_U with respect to a Fermi-Walker transported orthonormal 3-frame in LRS_U, interpreted as test gyro fixed axes carried by the test particle

$$\frac{\mathrm{D}_{(\text{fw},U)}}{\mathrm{d}\tau_U} X = \gamma\nu \frac{\mathrm{D}_{(\text{fw},U)}}{\mathrm{d}\ell_{(U,u)}} X = P_{(U)} \frac{\mathrm{D}}{\mathrm{d}\tau_U} X \,, \qquad X \in LRS_U \,. \tag{19}$$

One can construct a Frenet-Serret-like frame starting from the first frame vector $\hat{\mathcal{V}}_{(u,U)}$ by adding two new vectors in the usual way, this time unambiguously since there is only one natural derivative to use

$$\{\mathcal{E}_{(\text{fw},u,U)\hat{a}}\} = \{\hat{\mathcal{V}}_{(u,U)}, \hat{\mathcal{N}}_{(\text{fw},u,U)}, \hat{\mathcal{B}}_{(\text{fw},u,U)} = \hat{\mathcal{V}}_{(u,U)} \times_U \hat{\mathcal{N}}_{(\text{fw},u,U)}\} \,, \tag{20}$$

with Frenet-Serret relations

$$\begin{aligned}
\frac{\mathrm{D}_{(\text{fw},U)}}{\mathrm{d}\ell_{(U,u)}} \hat{\mathcal{V}}_{(u,U)} &= \mathcal{K}_{(\text{fw},u,U)} \hat{\mathcal{N}}_{(\text{fw},u,U)} \,, \\
\frac{\mathrm{D}_{(\text{fw},U)}}{\mathrm{d}\ell_{(U,u)}} \hat{\mathcal{N}}_{(\text{fw},u,U)} &= -\mathcal{K}_{(\text{fw},u,U)} \hat{\mathcal{V}}_{(u,U)} + \mathcal{T}_{(\text{fw},u,U)} \hat{\mathcal{B}}_{(\text{fw},u,U)} \,, \\
\frac{\mathrm{D}_{(\text{fw},U)}}{\mathrm{d}\ell_{(U,u)}} \hat{\mathcal{B}}_{(\text{fw},u,U)} &= -\mathcal{T}_{(\text{fw},u,U)} \hat{\mathcal{N}}_{(\text{fw},u,U)} \,,
\end{aligned} \tag{21}$$

where

$$\mathcal{K}_{(\text{fw},u,U)} \hat{\mathcal{N}}_{(\text{fw},u,U)} = \gamma k_{(\text{fw},U,u)} \hat{\eta}_{(\text{fw},U,u)} + \hat{\nu}_{(U,u)} \times_u [\hat{\nu}_{(U,u)} \times_u F^{(G)}_{(\text{fw},U,u)}] \,. \tag{22}$$

This is derived by differentiating (18) and manipulating the result. Thus the two vector curvature quantities differ by a term which also belongs to $LRS_u \cap LRS_U$. Since the test particle is not moving in its own rest space, there is no natural "rest arclength" parameter to introduce. In fact it makes sense to continue to use the arclength as seen by the observers u in the parametrization of the derivatives because the test particle is essentially reconstructing in its own rest space what the observers u are seeing it do.

In compact notation the comoving relative Frenet-Serret relations become

$$\frac{\mathrm{D}_{(\mathrm{fw},U)}}{\mathrm{d}\ell_{(U,u)}}\mathcal{E}_{(\mathrm{fw},u,U)\hat{a}} = \Omega_{(\mathrm{fw},u,U)} \times_U \mathcal{E}_{(\mathrm{fw},u,U)\hat{a}} \ , \tag{23}$$

where $\Omega_{(\mathrm{fw},u,U)} = \mathcal{T}_{(\mathrm{fw},u,U)}\hat{\mathcal{V}}_{(u,U)} + \mathcal{K}_{(\mathrm{fw},u,U)}\hat{\mathcal{B}}_{(\mathrm{fw},u,U)}$ defines the Fermi-Walker angular velocity of this new frame, which describes how the test particle sees the direction of its velocity relative to the observer family u change with respect to gyro-fixed axes. Again no assumption is made about the choice of sign for the curvature and torsion in this general discussion.

5 The longitudinal-tranverse splitting of the force equation

The longitudinal-transverse splitting of the relative acceleration defines the relative centripetal acceleration, which is rigidly tied to the idea of centrifugal force in non-relativistic classical mechanics. Applying this further splitting to the spatial part of the acceleration-equals-force equation reveals how the centripetal term enters the observed equation of motion. This splitting is accomplished by projecting onto the relative Frenet-Serret frame.

Let $||p_{(U,u)}|| = \gamma||\nu_{(U,u)}||$ be the magnitude of the specific (i.e., divided by the test mass) spatial momentum of U as seen by u, so that $p_{(U,u)} = ||p_{(U,u)}||\hat{\nu}_{(U,u)}$, while the gamma factor is the corresponding specific energy $E_{(U,u)} = \gamma$. They satisfy the identity

$$E^2_{(U,u)} - ||p_{(U,u)}||^2 = 1 \ . \tag{24}$$

Then the spatial projection onto LRS_u of the acceleration-equals-force equation $a(U) = f(U)$ takes the form[4]

$$\frac{\mathrm{d}||p_{(U,u)}||}{\mathrm{d}\tau_{(U,u)}}\hat{\nu}_{(U,u)} + \gamma{\nu_{(U,u)}}^2 k_{(\mathrm{fw},U,u)}\hat{\eta}_{(\mathrm{fw},U,u)} = F_{(U,u)} + F^{(G)}_{(\mathrm{fw},U,u)} \ . \tag{25}$$

The relative Frenet-Serret components of this are

$$\begin{aligned}
\frac{\mathrm{d}E_{(U,u)}}{\mathrm{d}\ell_{(U,u)}} &= [F_{(U,u)} + F^{(G)}_{(\mathrm{fw},U,u)}]\cdot\hat{\nu}_{(U,u)} \\
\gamma{\nu_{(U,u)}}^2 k_{(\mathrm{fw},U,u)} &= [F_{(U,u)} + F^{(G)}_{(\mathrm{fw},U,u)}]\cdot\hat{\eta}_{(\mathrm{fw},U,u)} \\
0 &= [F_{(U,u)} + F^{(G)}_{(\mathrm{fw},U,u)}]\cdot\hat{\beta}_{(\mathrm{fw},U,u)} \ .
\end{aligned} \tag{26}$$

On the other hand, the acceleration-equals-force equation may also be directly viewed in the test particle rest frame in a reconstruction by the particle itself of the description of its relative motion with respect to the family of observers. In this case the Frenet-Serret frame of the unit relative velocity $\hat{\mathcal{V}}_{(u,U)}$ naturally makes its appearance.

By applying the derivative $\mathrm{D}/\mathrm{D}\tau_U$ to the test particle decomposition (4) of u, solving for $a_{(U)}$ and using the relation (7), one obtains

$$\gamma^{-1}\frac{\mathrm{d}||p_{(u,U)}||}{\mathrm{d}\tau_U}\hat{\mathcal{V}}_{(u,U)} = f_{(U)} + \gamma^{-1}P_{(u,U)}F^{(G)}_{(\mathrm{fw},U,u)} - \nu^2\mathcal{K}_{(\mathrm{fw},u,U)}\hat{\mathcal{N}}_{(\mathrm{fw},u,U)} \ . \tag{27}$$

If there is any geometrical way of introducing a force term that generalizes the familiar centrifugal force in nonrelativistic classical mechanics, it can only be through the last term in the right-hand side of this equation. This term will be called the "generalized centrifugal force" keeping in mind that there is no clean separation of "gravitational forces" and "inertial forces" in the general relativistic domain. Of course it depends on the choice of observer family, and the relative gravitational force term itself reflects the "inertial forces" due to the noninertial motion of the family of observers. For constant speed orbits such as circular orbits in stationary axisymmetric spacetimes, the left hand side vanishes so the resulting equation (27) may be interpreted as describing a force balance in the local rest space of the test particle. Its nonrelativistic limit in flat spacetime leads to the familiar centrifugal force relative to the usual family of inertial observers.

The comoving relative Frenet-Serret decomposition of equation (27) is

$$\begin{aligned} \frac{dE_{(U,u)}}{d\ell_{(U,u)}} &= [f_{(U)} + \mathcal{F}^{(G)}_{(\mathrm{fw},u,U)}] \cdot \hat{\mathcal{V}}_{(u,U)} \\ \nu^2 \mathcal{K}_{(\mathrm{fw},u,U)} &= [f_{(U)} + \mathcal{F}^{(G)}_{(\mathrm{fw},u,U)}] \cdot \hat{\mathcal{N}}_{(\mathrm{fw},u,U)} \\ 0 &= [f_{(U)} + \mathcal{F}^{(G)}_{(\mathrm{fw},u,U)}] \cdot \hat{\mathcal{B}}_{(\mathrm{fw},u,U)} \ , \end{aligned} \tag{28}$$

where

$$\mathcal{F}^{(G)}_{(\mathrm{fw},u,U)} = \gamma^{-1} P_{(u,U)} F^{(G)}_{(\mathrm{fw},U,u)} = -\gamma^{-1} \frac{D_{(\mathrm{fw},U)} u}{d\tau_U} = -\frac{D_{(\mathrm{fw},U)} u}{d\tau_{(U,u)}} \ . \tag{29}$$

These equations are valid as long as ν belongs to the open interval $(0,1)$ (and with a slight modification when $\nu = 1$).[5] When $\nu = 0$, the spatial arclength parametrization is singular since one can no longer use the arclength in the observer quotient space as a parameter along the world line because the test particle path in that quotient space collapses to a point. However, since they are just projections of the spacetime acceleration-equals-force equation, these equations continue to hold in the limit $\nu \to 0$ including the scalar term $\nu^2 \mathcal{K}_{(\mathrm{fw},u,U)}$ which goes to zero and can be interpreted as a generalized centrifugal force term, but the factor $\mathcal{K}_{(\mathrm{fw},u,U)}$ by itself goes to infinity.

One may use the terminology "comoving relatively straight" and "comoving relatively flat" for those world lines for which the comoving relative curvature and torsion respectively vanish. By solving equation (25) for the gravitational force term and inserting the result into equation (22), one finds the relation

$$\mathcal{K}_{(\mathrm{fw},u,U)} \hat{\mathcal{N}}_{(\mathrm{fw},u,U)} = \gamma^{-1} k_{(\mathrm{fw},U,u)} \hat{\eta}_{(\mathrm{fw},U,u)} - \hat{\nu}_{(U,u)} \times_u [\hat{\nu}_{(U,u)} \times_u F_{(U,u)}] \ . \tag{30}$$

For a geodesic where $F_{(U,u)} = 0$, the two curvatures $\mathcal{K}_{(\mathrm{fw},u,U)}$ and $k_{(\mathrm{fw},U,u)}$ only differ by a gamma factor and the two directions $\hat{\mathcal{N}}_{(\mathrm{fw},u,U)}$ and $\hat{\eta}_{(\mathrm{fw},U,u)}$ coincide. Thus for geodesics the notions of relative Fermi-Walker straight and comoving relatively straight world lines agree. A relationship between the relative and comoving torsion may be established too[8] which, in the special case of geodesics, reduces to

$$\mathcal{T}_{(\mathrm{fw},u,U)} = \gamma^4 \tau_{(\mathrm{fw},U,u)} \ , \tag{31}$$

so the notions of relative Fermi-Walker flatness and comoving relative Fermi-Walker flatness also agree. Finally let us observe that for general test particle world lines and observers in a general spacetime, these relative Frenet-Serret frames are not closely related to the spacetime Frenet-Serret frame.

6 Concluding remarks

The natural Frenet-Serret machinery from the point of view of an observer family has been presented and linked to the natural geometrization of the usual notion of a centrifugal force in the rest space of a test particle world line (as opposed to a centripetal acceleration in the rest space of the observers).

For the most interesting case in which these concepts may be illustrated, namely circular orbits in stationary axisymmetric spacetimes, it has been shown in a related paper[8] that this comoving relative generalized centrifugal force is nothing other than the effect of the orbital rotation relative to gyro-fixed axes of the relative velocity, itself rigidly attached to stationary axisymmetric axes. For equatorial plane orbits in the Kerr spacetime, it agrees with our nonrelativistic intuition about centrifugal force near the circular geodesic orbits. For equatorial plane orbits at $r = 3\mathcal{M}$ of a Schwarzschild black hole, this generalized centrifugal force then vanishes due to the phase-locking[9] of gyro-fixed axes with stationary axisymmetric axes.

For more general noncircular orbits in stationary spacetimes, or completely general trajectories in any spacetime, this comoving relative generalized centrifugal force is the first such inertial force introduced by decomposing the test particle acceleration in its own rest space, which is exactly how the usual nonrelativistic centrifugal force is experienced.

Further applications of this relative Frenet-Serret approach to the case of gravitational wave spacetime solution are still in progress.

References

1. Honig E, Schücking E L, and Vishveshwara C W 1974 *J. Math. Phys.* **15** 744
2. Iyer B R and Vishveshwara C V 1993 *Phys. Rev.* **D48** 5706
3. Bini D, Merloni A. and Jantzen R T 1998 *Nuovo Cimento* **113B** 611
4. Jantzen R T, Carini P and Bini D 1992 *Ann. Phys. (N.Y.)* **215** 1
5. Bini D, Carini P and Jantzen R T 1997 *Int. J. Mod. Phys.* **D6** 1
6. Bini D, Carini P and Jantzen R T 1997 *Int. J. Mod. Phys.* **D6** 143
7. Bini D, Carini P and Jantzen R T 1995 *Class. Quantum Grav.* **12** 2549
8. Bini D, de Felice F and Jantzen R T 1999 *Class. Quantum Grav.* **16** 2105
9. de Felice F 1991 *Class. Quantum Grav.* **12** 1119; de Felice F and Usseglio-Tomasset S 1991 *Class. Quantum Grav.* **8** 1871

ON THE (NON) EXISTENCE OF A GRAVITOMAGNETIC DYNAMO

J.-F. PASCUAL-SÁNCHEZ

Dept. de Matemática Aplicada Fundamental, Sección Facultad de Ciencias, Universidad de Valladolid, 47011, Valladolid, Spain
E-mail: jfpascual@wamba.cpd.uva.es

Due to the resemblance between Maxwell and the gravitomagnetic equations obtained in the weak field and slow motion limit of General Relativity, one can ask if it is possible to amplify a seed intrinsic rotation or spin motion by a gravitomagetic dynamo, in analogy with the well-known dynamo effect. Using the Galilean limits of the gravitomagnetic equations, the answer to this question is negative,due to the fact that a "magnetic" Galilean limit for the gravitomagnetic equations is physically inconsistent.

1 Gravitomagnetic and Maxwell equations

I am sure that this subject is well-known to the reader, so I will try to go quickly in this introductory section.

This work is devoted to the possibility of a gravitomagnetic dynamo, which would amplify and maintain a initial seed spin motion. The starting point is the resemblance between Maxwell and the linear and slow motion approximation of the Einstein's equations

Hence, I do not start from the full non-linear Einstein's equations, to develop, after the projection into the local rest spaces of a congruence of observers, the Maxwell analogy in General Relativity, based on the correspondence between the Faraday tensor of the electromagnetic field and the Weyl tensor of the gravitational tidal field. This analogy has been developed in several recent papers, however, it was put forward and clearly exposed in the Refs. 1–5

This approach was done without any aproximation and, in this framework, the Bianchi identities are dynamical and the Einstein equations can be interpreted as constitutive relations of a 4-dim non-linear elastic continuum, (this can be seen in Ref. 6).

As I said, I will follow a more simple path in this work and I will start from the linearized Einstein's equations. On this line, to begin with, I will basically use the corresponding chapters of the following books,[7,8] with some changes in the notation and new symbols.

To center the subject, we begin with several historical remarks, perhaps not always sufficiently remembered. In the Newtonian theory of gravity, no fundamental gravitational force is associated with the rotation of a mass. In this theory, if a body rotates, the gravitational force it exerts on other masses changes only to the extent that the matter distribution within the body is affected by the rotation. The Newtonian gravitational force is only associated to the distribution of mass at a time, but not with the state of intrinsic rotation of this mass.

The development of electrodynamics in the last century and the close analogy between Coulomb's law of electrostatics and Newton's law of gravity, led to Holzmüller (1870) and Tisserand (1872), to suggest that a gravitational "magnetic" (or better rotational) force is associated with a rotating mass due to the "mass current" generated by the rotation, in close analogy with electrodynamics.

Later Heaviside (1893), in an appendix of his book *On Electromagnetic Theory* and assuming wave propagation for the gravitational field as for the electromagnetic one, wrote Maxwell's equations for gravitating bodies, changing in this case, either the sign of the sources or the sign of the fields, to account for the attractive nature of the gravitostatic Newtonian field between masses.

Heaviside, in a successive paper, applied these Maxwell-like equations for gravity to the motion of the Sun-Earth system through the cosmic ether, in order to explain the excess perihelion precession of Mercury by the new gravitational "magnetic" force.

The Mercury perihelion precession was explained several years later by the Einstein non-linear theory of gravity. Thus, the *ad hoc* introduction of a gravitational "magnetic" field by Heaviside became moot until Lense and Thirring (1918) and Thirring (1921) showed that, a certain gravitomagnetic field is indeed associated with the rotation of a mass, in the framework of the weak field approximation to General Relativity.

Beginning with the Einstein equations

$$R_{ab} - \frac{1}{2} g_{ab} R = -8\pi T_{ab}, \tag{1}$$

where $a, b = 0, 1, 2, 3$, are space-time indices and we use geometrized units, $G = c = 1$. If the gravity field is weak the metric (gravitational potential) g_{ab}, can be descomposed as

$$g_{ab} = \eta_{ab} + h_{ab}, \tag{2}$$

where $\eta_{ab} = \text{diag}\,(1, -1, -1, -1)$ is the Minkowski metric and h_{ab} is the perturbation due to the weak gravity field. Usually, it is more convenient to use a "new" perturbation $\bar{h}_{ab}$, for calculational purposes and physical reasons.

$$g_{ab} = \eta_{ab} + \bar{h}_{ab} - \frac{1}{2}\,\eta_{ab}\,\bar{h}, \tag{3}$$

where $\bar{h} = \bar{h}^a_a$ is the trace. Introduce now the object

$$G^{abc} = \frac{1}{4}(\bar{h}^{ab,c} - \bar{h}^{ac,b} + \eta^{ab}\,\bar{h}^{cd}{}_{,d} - \eta^{ac}\,\bar{h}^{bd}{}_{,d}), \tag{4}$$

where "," represents the partial derivative. Impose now a Lorentz-like gauge condition in analogy with electrodynamics, which represents the choice of harmonic coordinates:

$$\bar{h}^{ab}{}_{,b} = 0. \tag{5}$$

From (4) and (5)

$$G^{a[bc]} = G^{abc}, \tag{6}$$

$$G^{[abc]} = 0, \tag{7}$$

$$G^{d[ab,c]} = 0, \tag{8}$$

where $[\cdot\cdot]$ is the antisymmetry symbol.

Putting (3) and (4) into (1) and keeping only linear terms, one obtains the weak field equations:

$$\frac{\partial G^{abc}}{\partial x^c} = -4\pi T^{ab}. \tag{9}$$

Introduce, now, new symbols

$$\begin{aligned}
&\boldsymbol{g} = (g^1, g^2, g^3) \qquad g^i = G^{00i} \qquad\qquad i = 1,2,3 \\
&\boldsymbol{a} = (a^1, a^2, a^3) \qquad a^i = \tfrac{1}{4}\bar{h}^{0i} \\
&\boldsymbol{b} = (b^1, b^2, b^3) \qquad b^1 = G^{023} \qquad b^2 = G^{031} \qquad b^3 = G^{012} \\
&\boldsymbol{b} = \nabla \wedge \boldsymbol{a} \qquad\qquad G^{0ij} = a^{i,j} - a^{j,i}.
\end{aligned} \tag{10}$$

Put (10) into (8) and (9), one obtains, when the first order effects of the motion of the sources are taken into account, the following Maxwell-like (*gravitomagnetic*) equations, invariant under the Poincaré (or even conformal) group.

$$\nabla \boldsymbol{g} = -4\pi T^{00} = -4\pi\,\rho, \tag{11}$$

$$\nabla \wedge \boldsymbol{b} = -4\pi\,\rho\,\boldsymbol{u} + \frac{\partial \boldsymbol{g}}{\partial t}, \tag{12}$$

$$\nabla \wedge \boldsymbol{g} = -\frac{\partial \boldsymbol{b}}{\partial t}, \tag{13}$$

$$\nabla \boldsymbol{b} = 0. \tag{14}$$

Also, from the geodesic equation (equations of motion)

$$\frac{d^2x^a}{d\tau^2} + \Gamma^a_{bc}\,\frac{dx^b}{d\tau}\frac{dx^c}{d\tau} = 0, \tag{15}$$

for a weak stationary field, one obtains the Lorentz-like force law

$$\frac{d\boldsymbol{u}}{dt} = \boldsymbol{g} + 4\,\boldsymbol{u} \wedge \boldsymbol{b}, \tag{16}$$

where $\boldsymbol{u}$ is the velocity of the test particle. Note the factor 4 in the gravitomagnetic force term. Moreover, for a weak stationary field one obtains the gravitomagnetic potential

$$\boldsymbol{a} = -\frac{1}{2}\frac{\boldsymbol{S} \wedge \boldsymbol{r}}{r^3}, \tag{17}$$

and the gravitomagnetic field

$$\boldsymbol{b} = \nabla \wedge \boldsymbol{a} = -\frac{1}{2}\frac{3\boldsymbol{n}(\boldsymbol{S}\cdot\boldsymbol{n}) - \boldsymbol{S}}{r^3}, \tag{18}$$

where $\boldsymbol{S}$ is the intrinsic angular momentum of the source and $\boldsymbol{n}$ is the unit position vector. These equations are analogous to the electromagnetic ones, changing the magnetic dipole moment by minus twice the angular momentum. They are used in the GP-B gyroscope and LAGEOS III experiments to obtain the precession of a test body due to $\boldsymbol{b}$ field. Finally, note that the electromagnetic analog further iluminates the interdependence of the mass-energy currents effects (Lense-Thirring precession) and mass-energy effects (DeSitter-Fokker or geodetic precession), which in the electric-Galilean limit (will be discussed), are related as

$$\boldsymbol{b}' = \boldsymbol{b} - \boldsymbol{v} \wedge \boldsymbol{g}, \tag{19}$$

where $\boldsymbol{b}'$ would be the gravitomagnetic field measured, for instance, in the proper frame of the gyroscope, $\boldsymbol{b}$ would be measured in a "fixed" laboratory (at a terrestial pole, for example) and $\boldsymbol{v} \wedge \boldsymbol{g}$, contributes to the geodetic precession.

2 Magnetic dynamo theory

The term "dynamo action or effect" in magnetohydrodynamics (hereafter MHD) is generically used to describe the systematic and sustained generation of magnetic energy as a result of the stretching action of a velocity field $\boldsymbol{u}$, on a magnetic field $\boldsymbol{B}$. In other words, if a conducting fluid moves in a magnetic field $\boldsymbol{B}$, the flow will be affected by the force due to the interaction between $\boldsymbol{B}$ and the currents of the fluid. Also, $\boldsymbol{B}$ will be modified by the currents of the fluid and this is just the dynamo effect.

We do not write down the whole forbidding set of PDE of MHD, whose complexity precludes any hope of a systematic analytic or even numerical treatment. Instead, one studies particular aspects or makes simplifyng hypotheses. One of them is the so-called kinematic approach to the dynamo effect. In this approach, the velocity field $\boldsymbol{u}$ of the fluids regarded as known for any time and fixed and the back reaction of the magnetic field $\boldsymbol{B}$ on $\boldsymbol{u}$, via the Lorentz force, is assumed negligible. The kinematic dynamo is the most simple case of self-excited one and considers the evolution (amplification) of the magnetic field according to the induction equation:

$$\frac{\partial \boldsymbol{B}}{\partial t} = \nabla \wedge (\boldsymbol{u} \wedge \boldsymbol{B}) + \frac{1}{4\pi} \eta_e \Delta \boldsymbol{B}, \tag{20}$$

being η_e, the resistivity or difussivity (for insulators is infinite, for plasmas is cero), the reciprocal of the electric conductivity σ.

$$\eta_e = \frac{1}{\sigma} \tag{21}$$

The induction equation is obtained by taking the "macroscopic" magnetic Galilean limit (will be discussed) of Maxwell's equations, when the displacement current is neglected. In this limit the Ampère equation is:

$$\nabla \wedge \boldsymbol{B} = 4\pi \boldsymbol{J}. \tag{22}$$

From this and from the Ohm's law:

$$\boldsymbol{J} = \sigma(\boldsymbol{E} + \boldsymbol{u} \wedge \boldsymbol{B}), \tag{23}$$

one can obtain a expression for the electric field $\boldsymbol{E}$ and substituting it into the Faraday's equation

$$\nabla \wedge \boldsymbol{E} = -\frac{\partial \boldsymbol{B}}{\partial t}, \tag{24}$$

one obtains the MHD induction equation (20). Of course, we will not study in this work the usual magnetic dynamo coupled to the gravitomagnetic field of a Kerr metric. This very interesting study has been recently initiated and developed by Khanna and Camenzind in several papers. They build its formalism upon the membrane paradigm for black hole horizons of Damour and Thorne & Macdonald. See for instance.[9,10]

Instead, I will try to propose a similar mechanism to the magnetic dynamo effect in gravitomagnetism, to amplify $\boldsymbol{b}$ and hence the intrinsic angular momentum $\boldsymbol{S}$, due to the fact that we have Maxwell-like equations for gravity at our disposal. However, the key equation of the kinematic magnetic dynamo, which stablish the loop to amplify $\boldsymbol{B}$ is the Ohm's law. Do we have a similar equation in gravity?

3 An analog for gravitomagnetism of the Ohm's law

Our main radical and new idea, (perhaps senseless), is that in order to have a gravitomagnetic dynamo, the source fluid can not be a perfect fluid. The fluid must be "not dry", wet, and hence must have viscosity. But, as viscosity is a tensorial object and as we need a scalar, we only consider its trace, the viscous pressure, neglecting the shear viscosity. Viscosity (viscous pressure) η, will be the analog in gravitomagnetism of the resistivity η_e, for a conducting electrical medium.

Our Ohm's-like law for the moving viscous fluid, in a moving frame, will be:

$$\boldsymbol{j} = \rho \boldsymbol{u} = \delta\,(\boldsymbol{g} + \boldsymbol{u} \wedge \boldsymbol{b})\,, \tag{25}$$

where $\boldsymbol{j}$ is the mass current that appears in the first term of the r.h.s. of (13) and being $\delta = 1/\eta$, the "dryness" of the viscous fluid.

The reason of this analogy is based on the following physical picture of the Lense-Thirring effect. Consider, for instance, the GP-B gyroscope which will be Fermi transported along its trajectory. Its precession will be measured by a Frenet-Serret transported frame (or fixed with respect to distant stars). This precession could be pictured by considering a spinning sphere immersed in a viscous fluid and a rod (the axis of the spinning gyroscope). If the rod is placed, at the poles, into the viscous fluid, orthogonal to the axis of rotation of the sphere, it would precess in the same direction as the sphere rotates. However, if the rod is placed at the equator, also orthogonal to the axis of rotation of the sphere, then, it would precess on the opposite direction to the sense of rotation of the sphere.

4 Galilean limits of the gravitomagnetic equations

It is well-known that Maxwell's equations and the Lorentz force law have two different kinds of Galilean limits: electric and magnetic. This is due, from the mathematical point of view, to the existence of two different kinds of Galilean four-vectors. Starting from a Lorentz four-vector, for instance $(\rho_e, \boldsymbol{J})$, this can be more timelike, i.e., $|\rho_e| \gg |\boldsymbol{J}|$, and in this electric Galilean limit, its transformation under the Galilean inertial one is

$$\rho'_e = \rho_e, \qquad \boldsymbol{J}' = \boldsymbol{J} - \boldsymbol{v}\,\rho_e, \tag{26}$$

and for the electric $\boldsymbol{E}$ and magnetic field $\boldsymbol{B}$ one obtains:

$$\boldsymbol{E}' = \boldsymbol{E}, \qquad \boldsymbol{B}' = \boldsymbol{B} - \boldsymbol{v} \wedge \boldsymbol{E}. \tag{27}$$

Alternatively, the current four-vector can be more spacelike, i.e., $|\rho_e| \ll |\boldsymbol{J}|$. This case corresponds to the magnetic Galilean limit and the corresponding transformation formulas are:

$$\rho'_e = \rho_e - \boldsymbol{v}\,\boldsymbol{J}, \qquad \boldsymbol{J}' = \boldsymbol{J} \tag{28}$$

$$\boldsymbol{E}' = \boldsymbol{E} + \boldsymbol{v} \wedge \boldsymbol{B}, \qquad \boldsymbol{B}' = \boldsymbol{B}. \tag{29}$$

Physically, in the electric limit one describes situations where isolated electrical charges move at low velocities. On the other hand, the magnetic Galilean limit is the usual situation at the macroscopic level where magnetic effects are dominant, due to the balance between negative and positive electric charges.

This magnetic Galilean limit is the proper one that is used in magnetic dynamo theory but it is not possible in gravitomagnetism, where we do not have negative masses at our disposal. Thus, in gravitomagnetism, if we take a Galilean limit, this must necessarily be of the electric (almost Newtonian) kind and describe situations where isolated masses move at low velocities. In this electric (almost Newtonian) limit, the gravitomagnetic equations (12)–(14) have the same expressions, but there is an important difference, in this limit the Faraday-like equation (14) has not induction term, this equation now read

$$\nabla \wedge \boldsymbol{g} = 0, \tag{30}$$

And the Lorentz-like force law (16), is reduced in this electric (almost Newtonian) limit, by imposing Galilean invariance, to:

$$\frac{d\boldsymbol{u}}{dt} = \boldsymbol{g}, \tag{31}$$

And finally, in this electric (almost Newtonian)-Galilean limit, the transformations laws of the fields, under Galilean inertial transformations, read as:

$$\boldsymbol{g}' = \boldsymbol{g}, \tag{32}$$

$$\boldsymbol{b}' = \boldsymbol{b} - \boldsymbol{v} \wedge \boldsymbol{g}. \tag{33}$$

Hence the gravitomagnetic field $\boldsymbol{b}$ exists, but has no observable effects at the level of forces and torques. Moreover, *in this limit, the proper one for gravity, it is impossible to build a gravitomagnetic dynamo even if we have an Ohm's like law for gravity as (25)*, because we do not have, at our disposal, an induction term in the Faraday's equation.

The only possibility that remains, in my opinion, to construct a gravitomagnetic dynamo, would be to consider a non-relativistic generalized Newtonian theory of gravity of the kind introduced by Bel.[11] This possibility will be explored in a future work.

Acknowledgments

I am grateful to A. San Miguel and F. Vicente for discussions and TeX help and, after the completion of this paper, to Ll. Bel for drawing my attention to one of his publications. This work has been partially supported by the project VA61/98 of Junta de Castilla y León (Spain).

References

1. A. Matte, *Canadian J. Math.*, **5**, 1, (1953).
2. Ll. Bel, *C.R. Acad Sci.*, **247**, 1094, (1958).
3. R. Penrose, *Ann. of Phys.*, **10**, 161, (1960).
4. G.F.R. Ellis, in *General Relativity and Cosmology*, ed. R.K. Sachs (N.Y.: Academic Press, 1971).
5. R.T. Jantzen, P. Carini, D. Bini, *Ann. of Phys.*, **215**, 1, (1992).
6. J.-F. Pascual-Sánchez, *Gen. Rel. Grav.*, **20**, 869, (1988).
7. H.C. Ohanian & R. Ruffini, *Gravitation and Spacetime*, W.W.Norton & Company, 1994.
8. R.M. Wald, *General Relativity*, The University of Chicago Press, 1984.
9. R. Khanna, & M. Camenzind, *Astron.& Astrophys.*, **313**, 1028, (1996).
10. M. Nuñez, *Phys. Rev. D*, **54**, 7506, (1996).
11. Ll. Bel, in *Recent Developments in Gravitation*, ed. E. Verdaguer, J. Garriga, J. Céspedes, (World Scientific, 1990).

ROTATION OF THE POLARIZATION PLANE IN THE GRAVITATIONAL FIELD OF ROTATING OBJECTS

VOLKER PERLICK
TU Berlin, Sekr. PN 7-1, 10623 Berlin, Germany
E-mail: vper0433@w421zrz.physik.tu-berlin.de

For observers in a rotating reference system the polarization plane of linearly polarized light undergoes an apparent rotation. This effect, which is of relevance in the gravitational field of rotating bodies, is sometimes called the "gravitational Faraday effect". Here I investigate whether the gravitational Faraday effect would cause a rotation of the polarization plane *relative to the image of the light source.*

1 Introduction

Many celestial bodies emit electromagnetic radiation that is linearly polarized to a degree of 20 % or more. The underlying mechanism is well understood in principle. As a simple example, let us consider a rotating body with the shape of a rotational ellipsoid and a magnetic field parallel to the rotation axis, see Fig. 1. Then charged particles moving in the equatorial plane on circular paths around the rotation axis would emit *synchrotron radiation* which, according to standard electrodynamics, is linearly polarized, with the electric field vector oscillating in the direction perpendicular to the exterior magnetic field and to the particle velocity. Hence, we expect the polarization direction as given by the thick line in Fig. 1.

If the observed polarization direction differs from the theoretically expected one by an angle ψ as indicated in Fig. 1, then it is reasonable to assume that some mechanism has rotated the polarization plane on its way from the light source to the observer. An obvious candidate for such a mechanism is the familiar *Faraday effect* according to which the polarization direction of electromagnetic radiation undergoes a rotation if it passes through a magnetized medium, e.g., through a magnetized plasma.

It is interesting to inquire whether the spacetime geometry (i.e., the gravitational field) could produce an additional rotation of the polarization direction. In view of the *gravito-magnetic analogy*, according to which magnetic fields are formally analogous to certain inertial forces in a rotating reference frame, one would expect a gravitational analogue of the Faraday effect to take place in the gravitational field of rotating objects. This question can be discussed in a local and in a cosmological context. In the local context one would ask whether the gravitational field of a rotating body such as a neutron star or a galaxy has an effect on the polarization plane of electromagnetic radiation emitted by this body. In the cosmological context one would ask whether a hypothetical rotation of our universe would have an effect on the polarization plane of electromagnetic radiation coming to us from distant galaxies. The latter question was investigated by Korotkiĭ and Obukhov[6] who discussed observations by Birch[3] on the distribution of the polarization directions over the sky. (As a matter of fact, Birch's data refer to double radio galaxies; so the geometry is different from Fig. 1. The common feature is that there is a preferred axis relative to which the polarization direction can be measured.) The

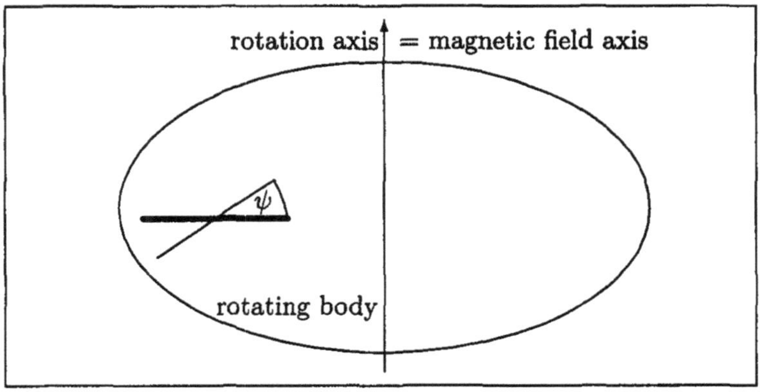

Figure 1: This picture illustrates the situation that we receive linearly polarized electromagnetic radiation from a celestial body with the shape of a rotational ellipsoid. The angle ψ indicates the deviation of the observed polarization direction (thin line) from the theoretically expected polarization direction (thick line).

significance of Birch's data, and of material more recently published by Nodland and Ralston,[7] is still under debate.

In all discussions of this kind it is crucial to realize that the polarization direction is to be measured *with respect to the image*, e.g., with respect to the major axis of the elliptical disk in Fig. 1. In the following I investigate whether a gravitational analogue of the Faraday effect could really produce such a rotation of the polarization direction relative to the image.

2 The kinematical origin of the gravitational Faraday effect

The gravitational Faraday effect was investigated for the first time in 1957 by Skrotskiĭ.[1] Later references can be found in Perlick and Hasse.[5] Like all other gravitomagnetic effects the gravitational Faraday effect can be studied not only on curved spacetime but also in a rotating reference frame on flat spacetime. This is an instructive preparation for more general studies since it clearly illustrates the simple kinematical origin of this effect. The following discussion should be seen in connection with the *equivalence principle* according to which an accelerated reference frame can be viewed as mimicking a gravitational field.

Figure 2 shows a light source which is at rest in an inertial system on Minkowski space. We assume that this light source emits linearly polarized light which is observed by a family of rotating observers. By elementary optics, the plane spanned by the propagation vector **k** and the polarization direction **e** remains parallel to itself. Hence, an observer in the rotating reference frame will see an apparent rotation of the polarization direction *relative to neighboring observers in the rotating reference frame.* This very simple kinematical consideration explains the origin of the "gravitational Faraday effect". In this Minkowski space situation it is obvious that there is no rotation of the polarization direction *relative to the image of the light source* because the latter undergoes exactly the same apparent rotation as

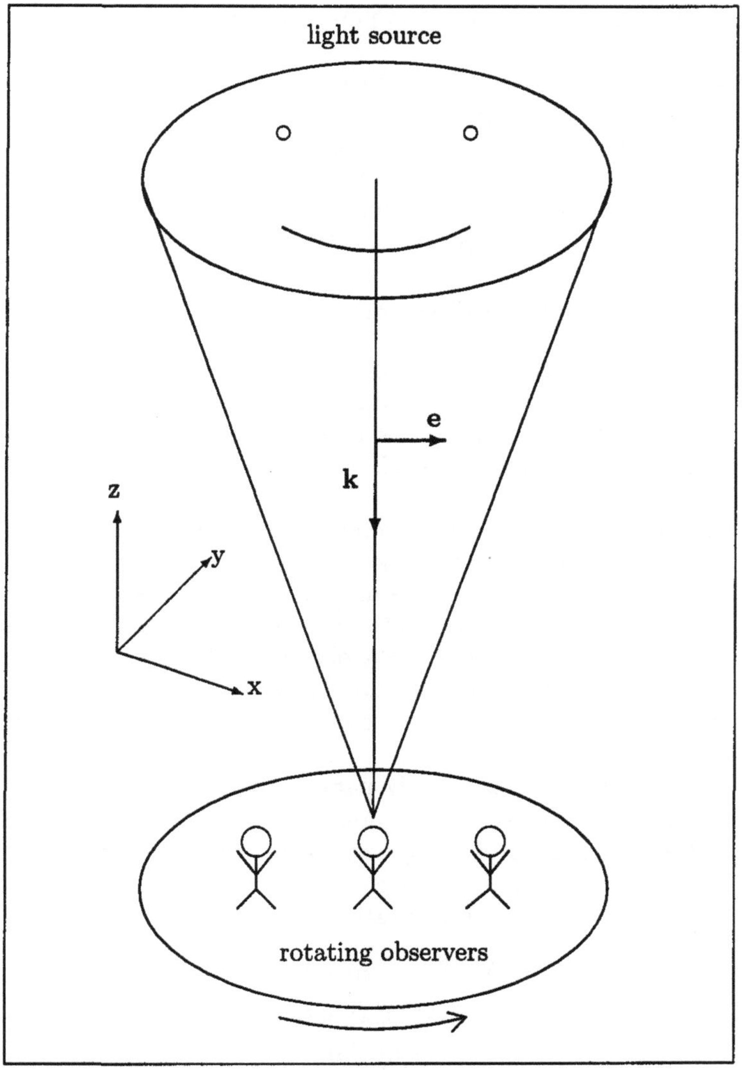

Figure 2: This picture illustrates the situation that polarized light is emitted from a light source which is at rest in an inertial system on Minkowski space. In this situation, the polarization plane, spanned by the propagation vector **k** and by the direction vector **e** of the electric field, remains fixed in the inertial system. For observers in a rotating reference frame, this will cause an apparent rotation of the polarization direction *relative to neighboring observers in the rotating reference frame*. However, it is evident that there is no rotation of the polarization direction *relative to the image of the light source*. (The smiling face is meant to indicate a structure of the light source relative to which the polarization direction is to be measured.)

the former. In the next section we inquire whether this is still true on curved spacetimes.

Incidentally, our simple kinematical consideration indicates that the gravitational Faraday rotation is independent of the wavelength λ. This is different from the ordinary Faraday rotation which is proportional to λ^2. There is another difference insofar as the ordinary Faraday effect takes place in a magnetized *medium* whereas there is no analogue of a medium in the gravitational case. This shows that the gravito-magnetic analogy has its limits.

3 Rotation of polarization direction in an arbitrary spacetime

It is the goal of this section to determine, for arbitrary observers on arbitrary Lorentzian manifolds, the apparent rotation of the polarization direction *relative to the image of the light source.* The same question was investigated by Panov and Sbytov[4] whose result was criticized by Korotkiĭ and Obukhov.[6] I shall comment on their work later.

As a spacetime model I consider a 4-dimensional manifold M with a Lorentzian metric g of class C^2, and I denote the Levi-Cività connection of g by ∇. I restrict to geometric optics in vacuo, i.e., I assume that light rays are lightlike geodesics and that the polarization plane of polarized light is ∇-parallely propagated along each ray. (The "polarization plane" is the observer-independent plane spanned by the tangent to the ray and by the direction of the electric field.) It is well known that the formalism of geometric optics can be derived from Maxwell's equations by a limit procedure, see Ehlers.[2]

Let $\lambda\colon [0, s_o] \longrightarrow M$ be an affinely parametrized lightlike geodesic, with the parametrization interpreted as past-oriented. We want to consider a *homocentric bundle of light rays*, consisting of all rays near λ that reach the same point $\lambda(0)$. We assume that the bundle is "infinitesimally thin", i.e., we model its rays as vector fields $J\colon [0, s_o] \longrightarrow TM$ along λ which satisfy

$$\nabla_{\dot\lambda}\nabla_{\dot\lambda}J = R(J, \dot\lambda, \dot\lambda)\,, \tag{1}$$

$$g\big(\nabla_{\dot\lambda}J, \dot\lambda\big) = 0\,, \tag{2}$$

$$J(0) = 0\,. \tag{3}$$

(1) is the *Jacobi equation*, with R denoting the curvature tensor of ∇, which makes sure that "the tip of the arrow J" traces a *geodesic* near λ in linear approximation. By (2), this neighboring geodesic is again lightlike. By (3), it meets λ at the point $\lambda(0)$. It is easy to verify that (1), (2) and (3) imply that

$$g\big(J, \dot\lambda\big) = 0\,. \tag{4}$$

Conversely, it is obvious that (4) implies (2). We can thus restrict to vector fields along λ that satisfy (4) and our bundle is determined by (1) and (3).

For the evaluation of (1) and (3) it is convenient to introduce two spacelike vector fields $E_1, E_2\colon [0, s_o] \longrightarrow TM$ along λ with the properties

$$g\big(E_A, \dot\lambda\big) = 0\,, \quad g\big(E_A, E_B\big) = \delta_{AB}\,, \quad \nabla_{\dot\lambda}E_A = 0\,, \tag{5}$$

for $A, B = 1, 2$. Clearly, such a basis exists and is unique up to transformations

$$\begin{aligned} E_1(s) &\longmapsto \tilde{E}_1(s) = \cos\varphi\, E_1(s) - \sin\varphi\, E_2(s) + f_1(s)\,\dot{\lambda}(s)\,, \\ E_2(s) &\longmapsto \tilde{E}_2(s) = \sin\varphi\, E_1(s) + \cos\varphi\, E_2(s) + f_2(s)\,\dot{\lambda}(s)\,, \end{aligned} \tag{6}$$

where f_1, f_2 are arbitrary functions on $[0, s_o]$ and φ is a constant angle. If an observer field has been chosen, i.e., if there is a timelike vector $V(s)$ distinguished at each point $\lambda(s)$, then it is natural to choose E_1 and E_2 perpendicular to V. This fixes f_1 and f_2. We can then interpret E_1 and E_2 as vectors in the local rest space of V, spanning the two-space perpendicular to the direction of the ray λ. With E_1, E_2 chosen along λ, (4) requires J to be of the form

$$J(s) = J^1(s)\,E_1(s) + J^2(s)\,E_2(s) + h(s)\,\dot{\lambda}(s) \tag{7}$$

with some functions J^1, J^2, and h. Here h can be set equal to zero without loss of generality, thereby fixing the parametrization for each "neighboring ray". If we introduce the matrix notation

$$\begin{pmatrix} J^1(s) \\ J^2(s) \end{pmatrix} = \begin{pmatrix} A_{11}(s) & A_{12}(s) \\ A_{21}(s) & A_{22}(s) \end{pmatrix} \begin{pmatrix} J^1(s_o) \\ J^2(s_o) \end{pmatrix}, \tag{8}$$

(1), (3) and (8) require that the matrix valued function $s \longmapsto \mathbf{A}(s) = (\!(\, A_{BC}(s)\,)\!)$ has to satisfy the equations

$$\ddot{A}_{BC} = \sum_{D=1}^{2} g\big(E_B, R(E_D, \dot{\lambda}, \dot{\lambda})\big) A_{DC}\,, \tag{9}$$

$$A_{BC}(0) = 0\,, \quad A_{BC}(s_o) = \delta_{BC}\,. \tag{10}$$

As g is assumed to be of class C^2, $g\big(E_B, R(E_D, \dot{\lambda}, \dot{\lambda})\big)$ is a continuous function from $[0, s_o]$ into the real numbers for any choice of λ, E_B, E_D. Thus, the standard existence and uniqueness theorem for linear differential equations guarantees that there is a unique C^2 solution $s \longmapsto \mathbf{A}(s) = (\!(\, A_{BC}(s)\,)\!)$ of (9) and (10) on the interval $[0, s_0]$. This matrix valued function $\mathbf{A}$ determines the totality of all solutions J of (1), (2), (3) via (8), where $J^1(s_o), J^2(s_o)$ ranges over all pairs of real numbers.

The solution $\mathbf{A}$ of (9), (10) determines the shape of a bundle of light rays that reaches the eye of an observer at the point $\lambda(0)$, measured against a ∇-parallel basis. The bundle undergoes a pure deformation, without a rotation, with respect to the ∇-parallel basis if and only if the matrix $\mathbf{A}(s)$ is symmetric for all s. This follows from the familiar *polar decomposition* of matrices. Since, according to our geometric optics assumption, the polarization plane of linearly polarized light is ∇-parallel along the ray, this implies the following result.

There is no rotation of the polarization plane with respect to the image if and only if the solution $\mathbf{A}$ *of* (9), (10) *is symmetric.*

We shall see that the question of whether $\mathbf{A}$ is symmetric depends on the conformal curvature tensor along λ. To work this out, we first use the familiar decomposition of the curvature tensor R into *Ricci tensor* Ric, *conformal curvature tensor* C and *scalar curvature*. This puts (9) into the form

$$\ddot{\mathbf{A}} = -\begin{pmatrix} \operatorname{Re}\Psi_0 & \operatorname{Im}\Psi_0 \\ \operatorname{Im}\Psi_0 & -\operatorname{Re}\Psi_0 \end{pmatrix} \mathbf{A} - \Phi_{00}\,\mathbf{A}\,. \tag{11}$$

Here we have used the abbreviations

$$\Phi_{00} = \tfrac{1}{2}\operatorname{Ric}(\dot\lambda,\dot\lambda)\,, \quad \Psi_0 = g\big(E_2, C(E_2,\dot\lambda,\dot\lambda)\big) - i\,g\big(E_2, C(E_1,\dot\lambda,\dot\lambda)\big)\,, \tag{12}$$

which are in agreement with the standard notation of the *Newman-Penrose formalism*. Please note that Φ_{00} is real whereas Ψ_0 is complex. If we change the basis E_1, E_2 according to (6), these quantities transform as

$$\Phi_{00}(s) \longmapsto \tilde{\Phi}_{00}(s) = \Phi_{00}(s)\,, \qquad \Psi_0(s) \longmapsto \tilde{\Psi}_0(s) = \Psi_0(s)\,e^{2i\varphi}\,. \tag{13}$$

Thus, Φ_{00} is unique and Ψ_0 is unique up to a *constant* phase factor. As a further preparation we need the following proposition.

Proposition 1. The solution $\mathbf{A} = ((A_{BC}))$ of (9) and (10) satisfies
(a) $\det \mathbf{A}(s) \neq 0$ for almost all $s \in [0, s_o]$;
(b) $\dot{\mathbf{A}}\mathbf{A}^T = \dot{\mathbf{A}}^T\mathbf{A}$, where $(\ )^T$ denotes the transpose, $(A^T)_{BC} = A_{CB}$.
Proof. (a) is a standard result from the theory of linear ordinary differential equations. To prove (b), one first verifies from (11) that $\mathbf{Q} = \dot{\mathbf{A}}\mathbf{A}^T - \dot{\mathbf{A}}^T\mathbf{A}$ is a first integral, $\dot{\mathbf{Q}} = \mathbf{0}$. Since $\mathbf{A}(0) = \mathbf{0}$ this implies $\mathbf{Q} = \mathbf{0}$. *Q.E.D.*

We are now ready to prove the following criterion for symmetry of $\mathbf{A}$.

Proposition 2. The solution $\mathbf{A}$ of (9) and (10) is symmetric if and only Ψ_0 is of the form $\Psi_0(s) = q(s)e^{i\varphi}$ with a real function q and a constant phase φ.
Proof. To prove the "only if" part, we recall the transformation behavior (13) of Ψ_0 under a change of bases (6). Since, by assumption, Ψ_0 has constant phase, we can switch to a new basis such that $\tilde{\Psi}_0$ is real (viz., equal to q). Owing to (11), together with the boundary conditions (10), this implies that the transformed matrix $\tilde{\mathbf{A}}$ is diagonal. Thus, the original matrix $\mathbf{A}$ must be symmetric. – To prove the "if" part, we decompose $\mathbf{A}$ into a symmetric traceless part and a multiple of the unit matrix, i.e., we express $\mathbf{A}$ in terms of a complex scalar a and a real scalar k according to

$$\mathbf{A} = \begin{pmatrix} \operatorname{Re}a & \operatorname{Im}a \\ \operatorname{Im}a & -\operatorname{Re}a \end{pmatrix} + \begin{pmatrix} k & 0 \\ 0 & k \end{pmatrix}. \tag{14}$$

This puts (11) into the form

$$\ddot{a} + \Phi_{00}\,a = -k\operatorname{Re}\Psi_0\,, \tag{15}$$

$$\ddot{k} + \Phi_{00}\,k = -\Psi_0\,\bar{a}\,. \tag{16}$$

Moreover, from Proposition 1 (b) we find

$$\operatorname{Re}\dot{a}\operatorname{Im}a = \operatorname{Re}a\operatorname{Im}\dot{a}\,. \tag{17}$$

Now we distinguish two cases. *Case A*: a is identically zero on the whole interval $[0, s_0]$. Then, by Proposition 1 (a), k is different from zero almost everywhere, so

(15) implies that $\operatorname{Re}\Psi_0$ vanishes almost everywhere. Since our C^2 assumption on the metric guarantees that Ψ_0 is continuous, we have thus proven that Ψ_0 has the desired form with $\varphi = \frac{\pi}{2}$. *Case B*: a is not identically zero on $[0, s_0]$. Then the existence and uniqueness theorem for the linear differential equation (15) implies that at all points with $a(s) = 0$ we must have $\dot{a}(s) \neq 0$. By (17), on each interval on which a has no zeros the phase of a must be constant. Since a is a C^2 function (and, in particular, a C^1 function) these two facts together guarantee that a is of the form $a(s) = u(s)e^{i\varphi}$ with a real function u and a constant phase φ. From (16) we read that $\Psi_0\bar{a}$ is real. At all points where a is different from zero (i.e., almost everywhere) we can divide by $\bar{a}$ and find that Ψ_0 is a real multiple of $e^{i\varphi}$. Since Ψ_0 is continuous, this must be true everywhere. *Q.E.D*

This result has the following consequences.

There is no rotation of the polarization plane relative to the image if and only if Ψ_0 is real up to a constant phase factor. This shows that the rotation of the polarization plane relative to the image is a conformal curvature effect. In particular, there is no such rotation in conformally flat spacetimes since then $\Psi_0 = 0$ along any ray.

The question of whether the polarization plane is rotated with respect to the image is observer-independent. The choice of an observer field V enters into the whole discussion only insofar as the basis vectors E_1 and E_2 are to be chosen perpendicular to V if the shape of the bundle is to be interpreted in the rest space of the observers chosen. This can be achieved by a transformation of the form (6) with some functions f_1 and f_2. By (13), this has no influence on the question of whether Ψ_0 has constant phase.

These observations are in perfect agreement with the results of Panov and Sbytov[4] but not with those of Korotkiĭ and Obukhov.[6] (However, some arguments in the Panov-Sbytov paper are not convincing in my view.) The origin of this discrepancy is easily detected. Korotkiĭ and Obukhov consider a non-homocentric bundle, i.e., their light rays do not meet in the eye of an observer. Therefore their resulting formula does not give the rotation of the polarization plane relative to the image *as it is seen by an observer*.

Acknowledgments

I wish to thank Wolfgang Hasse and Yuri Obukhov for stimulating discussions.

References

1. G. V. Skrotskiĭ, Sov. Phys. Dokl. **2**, 226 (1957).
2. J. Ehlers, Z. Naturforsch. **22a**, 1328 (1967).
3. P. Birch, Nature **298**, 451 (1982).
4. V. F. Panov and Yu. G. Sbytov, Sov. Phys. JETP **74**, 411 (1992).
5. V. Perlick and W. Hasse, Class. Quantum Grav. **10**, 147 (1993); Erratum **10**, 1935 (1993).
6. V. A. Korotkiĭ and Yu. N. Obukhov, JETP **81**, 1031 (1995).
7. B. Nodland and J. P. Ralston, Phys. Rev. Lett. **78**, 3043 (1997).

Section C

The GP-B Mission: The Orbiting Gyroscope Experiment around the Earth

GENERAL TREATMENT OF GEODETIC AND LENSE–THIRRING EFFECTS ON AN ORBITING GYROSCOPE

R.J.ADLER, A.S.SILBERGLEIT
Gravity Probe B, W.W.Hansen Experimental Physics Laboratory, Stanford University, Stanford, CA 94305-4085, USA
E-mail: adler@relgyro.stanford.edu gleit@relgyro.stanford.edu

We give a general derivation of the metric of a spinning body of any shape and composition using linearized general relativity theory (LGRT), and also obtain the same metric using a simple transformation argument. The latter derivation makes it clear that the linearized metric contains only the Eddington γ and $\alpha\,(\equiv 1)$ parameters, so no new parameter is involved in any frame–dragging or Lense–Thirring (LT) effects. We then calculate the precession of an orbiting gyroscope in a general gravitational field, described by a Newtonian potential (gravito-electric field) and a vector potential (gravito-magnetic field). Finally we do a multipole analysis and give the general spherical harmonics expansion of the precession in terms of multipoles of the scalar and vector potentials, i.e., moments of the density distribution. In particular, in regard to the Gravity Probe B (GP-B) experiment, we find that the effect of the Earth's quadrupole moment J_2 on the geodetic precession is large enough to be measured by GP-B (a previously known result), but the effect on the LT precession is somewhat beyond the expected GP-B accuracy.

1 Introduction

The Gravity Probe B satellite is scheduled to fly in the year 2000.[1] It contains a set of gyroscopes intended to test the predictions of general relativity (GR) that a gyroscope in a low (altitude$\approx 650\,km$) circular polar orbit will precess about 6.6 *arcsec/year* in the orbital plane (geodetic precession) and about 42 *milliarcsec/year* perpendicular to the orbital plane (LT precession, see Ref. 2; Ref. 3: Secs. 4.7 and 7.8; Ref. 4: Sec. 9.1). In this paper we review the theoretical derivation of these effects and in particular consider the Earth's quadrupole and higher multipole fields' contribution to them.

We first review the derivation of the metric for a rotating body using the standard LGRT approach (Refs. 2 and 5; Ref. 3: Secs. 4.7 and 7.8). The metric is characterized by a Newtonian scalar potential (called the gravito-electric field) and a vector potential (called the gravito-magnetic field) (Ref. 3: Sec. 3.5). We then obtain the same result with a simple transformation argument which clarifies the physical meaning of the metric (Ref. 4: Sec. 4.3). Specifically it makes clear that if the metric of a point mass contains fundamental parameters such as the Eddington parameters α and γ, then to lowest order the metric of a rotating body contains no **new** fundamental parameters.[6] Thus there is no new "LT parameter" to be measured by GP-B – or any other experiment.

We then derive the precession equations for a gyroscope in a general way, that is for any scalar and vector potential fields.[2] The calculation is valid to first order in the fields and velocities of the source body and the satellite. The gravitational field of the earth is described by the scalar and vector potentials which depend on the shape of the body and the mass distribution inside it; in addition, the gravito-magnetic potential depends also on the rotation velocity (Ref. 3: Sec. 3.5). We treat both of these fields by a multipole expansion and express the precessions in series of spherical harmonics whose coefficients are combinations of scalar and vector potential multipoles, in other words, of spherical harmonics moments of the density distribution. In particular we show that up to the order $l \leq 2$, both precessions depend only on the tensor of inertia of the earth. The major contributions to the GP-B precessions are from the Earth's quadrupole moment and both have a magnitude of about 1 part in 10^3. The contribution to the geodetic precession is detectable by GP-B and quite important for the determination of the parameter γ which is to be measured to about 1 part in 10^5, the most accurate measurement envisioned (Ref. 3: Sec. 3.5 and in particular Table 14.2; Ref. 7); the contribution to the LT precession is somewhat beyond the GP-B accuracy.

2 The Lense–Thirring Metric and Eddington Parameters

2.1 Derivation by Linearized General Relativity

This is a standard derivation, so we review it briefly.[2]

The metric of a rotating body such as the Earth in LGRT is obtained by introducing a small perturbation $h_{\mu\nu}$ of the Lorentz metric $\eta_{\mu\nu}$, that is $g_{\mu\nu} = \eta_{\mu\nu} + h_{\mu\nu}$. The perturbation is assumed to be independent of time and isotropic in space, $h_{11} = h_{22} = h_{33} = h_s$. The energy–momentum tensor of slow-moving and low density matter with negligible pressure is $T^{\mu\nu} = \rho u^\mu u^\nu$, where u^μ is the 4–velocity and ρ is the matter density, the field equations are then ($R_{\mu\nu}$ is the Ricci tensor):

$$R_{\mu\nu} = -8\pi G(T_{\mu\nu} - (1/2)g_{\mu\nu}T), \qquad T = T^\sigma_\sigma\,. \tag{1}$$

The calculation of $R_{\mu\nu}$ and $T_{\mu\nu}$ to the lowest order in the perturbation is straightforward and results in the following form of equations (1):

$$\begin{aligned} \Delta(h_{00}/2) = \Delta(h_s/2) &= 4\pi G\rho\,, \\ \Delta h_{0i} - h_{0l|li} &= 16\pi G\rho\, v^i\,. \end{aligned} \tag{2}$$

In the last expression we have kept only first order terms in the velocity in the off-diagonal metric elements. The first of equations (2) is the Poisson equation for the metric perturbation, so by correspondence with the classical theory the latter should be related to the classical gravitational potential by the well-known equality:[2]

$$h_{00} = h_s = 2\Phi\,. \tag{3}$$

Equations (2) may be solved using the Green function as ($\vec{h} = \{h_{01}, h_{02}, h_{03}\}$)

$$\begin{aligned} \Phi(\vec{r}) &= -G\int \frac{\rho(\vec{r}\,')\,d^3\vec{r}\,'}{|\vec{r} - \vec{r}\,'|}, \\ \vec{h}(\vec{r}) &= 4G\int \frac{\rho(\vec{r}\,')\vec{v}(\vec{r}\,')\,d^3\vec{r}\,'}{|\vec{r} - \vec{r}\,'|}\,. \end{aligned} \tag{4}$$

Note that these expressions are analogs of the equations of electrostatics and magnetostatics, which is why one may speak about gravitoelectric and gravitomagnetic effects in LGRT described by gravitoelectric potential (Φ) and gravitomagnetic vector potential[2] $\vec{h}$. In summary, we may write the Lense–Thirring line element as

$$ds^2 = (1+2\Phi)dt^2 - (1-2\Phi)d\vec{r}^2 + 2\vec{h}\cdot d\vec{r}\,dt \tag{5}$$

2.2 Derivation by Transformation; Eddington Parameters

It is possible to obtain the above result from a different and physically interesting perspective, and moreover introduce parameters convenient for discussing experimental measurements. Following Eddington, consider the metric of a massive point with a geometric mass m at a large distance r, $m/r \ll 1$. Expand the Schwarzschild solution in isotropic coordinates for this situation as

$$\begin{aligned} ds^2 &= \frac{(1-m/2r)^2}{(1+m/2r)^2}dt^2 - (1+m/2r)^4 d\vec{r}^2 \\ &= \left(1 - \frac{2m}{r} + \frac{2m^2}{r^2} + \ldots\right)dt^2 - \left(1 + \frac{2m}{r} + \frac{3m^2}{r^2} + \ldots\right)d\vec{r}^2 . \end{aligned}$$

Eddington suggested that this be written in terms of parameters as[8,9]

$$ds^2 = \left(1 - \alpha\frac{2m}{r} + \beta\frac{2m^2}{r^2} + \ldots\right)dt^2 - \left(1 + \gamma\frac{2m}{r} + \ldots\right)d\vec{r}^2 . \tag{6}$$

The Eddington parameters α, β and γ are equal to 1 for GR. The power series (5) is clearly a rather general form for the metric far from a spherical body. Since the m which appears in the metric (6) is merely a constant of integration representing the mass of the central body (specifically $m = GM/c^2$), we may absorb parameter α into it, which is equivalent to taking $\alpha \equiv 1$. This is consistent as long as no independnt nongravitational determination of the mass of the central body (the earth) is considered. We shll nevertheless display the α in our calculations as a book-keeping device.

The parameters may be used as a tool for tracking the terms in the metric which contribute to some gravitational effect; alternatively they may be viewed as numbers which may **not** be equal to 1 if a metric theory other than GR is actually valid; in either case they provide a convenient way to express the results of experimental tests of gravity as giving value to parameters. This parametrized approach has been extended to include many other parameters and has been highly developed under the name parametrized post–Newtonian theory, or PPN.[4] In this paper we take the viewpoint that GR is to be tested and emphasize that we are **not** using the more general PPN approach.

We consider only phenomena in which the term $\sim m^2/r^2$ in g_{00} of (6) is unimportant, that is in which we may ignore β and thus have an underlying linear theory. Then, for a *stationary* mass,

$$ds^2 = \left(1 - \alpha\frac{2m}{r}\right)dt^2 - \left(1 + \gamma\frac{2m}{r}\right)d\vec{r}^2 . \tag{7}$$

Since this is nearly the Lorentz metric, we may generalize it to a *moving* mass point by simply using the frame transformation that is Lorentzian to the first order in velocity: $t_r = t - vx, \quad x_r = x - vt$; here the subscript r labels the system in which the mass is at rest, and which moves at v in the positive x direction. This gives the metric for the *moving* point mass as

$$ds^2 = \left(1 - \alpha\frac{2m}{r}\right) dt^2 - \left(1 + \gamma\frac{2m}{r}\right) d\vec{r}^{\,2} + (\alpha + \gamma)\frac{4m}{r}\, v dx\, dt,$$

which obviously generalizes for motion in any direction to

$$ds^2 = \left(1 - \alpha\frac{2m}{r}\right) dt^2 - \left(1 + \gamma\frac{2m}{r}\right) d\vec{r}^{\,2} + (\alpha + \gamma)\frac{4m}{r}\,(\vec{v} \cdot d\vec{r})\, dt\,. \tag{8}$$

Since we assume that our theory is linear to this order, we can superpose the fields of any distribution of such point masses and write $\Phi(\vec{r})$ of (4) in place of m/r and $\vec{h}(\vec{r})$ of (4) in place of $4mG\vec{v}/r$, resulting in

$$ds^2 = (1 + 2\alpha\Phi)\, dt^2 - (1 - 2\gamma\Phi)\, d\vec{r}^{\,2} + (\alpha + \gamma)(\vec{h} \cdot d\vec{r})dt\,. \tag{9}$$

This agrees with the LGRT formula (5) when $\alpha = \gamma = 1$ but now contains appropriate combinations of Eddington parameters. We emphasize that line element (9) has been obtained for any slowly moving mass distribution from the parametrized metric for a stationary mass point by transformation and superposition, and thus no new parameter appears in its expression. Therefore a measurement of a phenomenon which depends on the cross term in (9) provides a value for $\alpha + \gamma$ and **not** for some new parameter, i.e., **does not** provide an independent test of gravitational theory.[6]

3 Precession Equations

An orbiting gyroscope has its spin axis parallel displaced in accord with metric (9), that is (Ref. 2; Ref. 3: Secs. 4.7 and 7.8; Ref. 4: Sec. 9.1):

$$\frac{dS^\mu}{ds} + \Gamma^\mu_{\nu\sigma} S^\nu \frac{dx^\sigma}{ds} = 0\,. \tag{10}$$

Assuming that the gyro spin 4-vector is perpendicular to the velocity 4-vector (that is the spin has no zero component it its rest frame), we work to the first order in the potential and the velocities, $\vec{v}$ and $\vec{V}$, of the central body and orbiting gyro, respectively. We calculate the Christoffel symbols from metric (9) and use the corresponding Euler–Lagrange equations of motion to convert (10) into the final 3D equation of spin evolution with antisymmetric and symmetric parts separated:

$$\frac{d\vec{S}}{dt} = \vec{\Omega} \times \vec{S} - \frac{1}{2}\{2\gamma(\vec{V} \cdot \nabla\Phi)\vec{S} - \alpha[(\vec{S} \cdot \vec{V})\nabla\Phi + (\vec{S} \cdot \nabla\Phi)\vec{V}]\}\,. \tag{11}$$

Here $\vec{\Omega} = \vec{\Omega}_G + \vec{\Omega}_{LT}$ is the sum of the geodetic and Lense–Thirring precessions given by

$$\begin{aligned} \vec{\Omega}_G &= \left(\frac{\alpha + 2\gamma}{2}\right) \nabla\Phi \times \vec{V}, \\ \vec{\Omega}_{LT} &= \left(\frac{\alpha + \gamma}{2}\right) \frac{1}{2} \nabla \times \vec{h}, \end{aligned} \tag{12}$$

whereas the second term is responsible for stretching of $\vec{S}$, which effect is extremely small under the conditions similar to GP-B. When averaged over the closed orbit, this term gives a zero contribution, as found from the motion equation $\nabla\Phi = -\dot{\vec{V}}$ (c. f. $\nabla\Phi \cdot \dot{\vec{V}} = 0.5\, dV^2/dt$, etc.); in reality, a small orbit variation might produce some very small average which is inversely proportional to the time and may be neglected, anyway. Therefore the average drift rate of the gyro spin is

$$\langle \dot{\vec{S}} \rangle = \dot{\vec{S}}_G + \dot{\vec{S}}_{LT}, \qquad \dot{\vec{S}}_A = \langle \vec{\Omega}_A \rangle \times \vec{S}, \quad A = G,\ LT, \tag{13}$$

with the averaged values of precessions found from (12).

4 Effect of Distant Masses

When some distant masses M_n, such as the moon, move with velocities v_n relative to the central body, the density distribution $\rho(\vec{r})$ may be expressed as

$$\rho_N(\vec{r}) = \rho(\vec{r}) + \sum_{n=1}^{N} M_n \delta(\vec{r} - \vec{r}_n)\,.$$

In their turn, the scalar and vector potentials (4) are now replaced by

$$\Phi_N(\vec{r}) = \Phi(\vec{r}) - G \sum_{n=1}^{N} \frac{M_n}{|\vec{r} - \vec{r}_n|},$$
$$\vec{h}_N(\vec{r}) = \vec{h}(\vec{r}) + 4G \sum_{n=1}^{N} \frac{M_n \vec{v}_n}{|\vec{r} - \vec{r}_n|}\,.$$

From (12) we find, in particular, the LT precession including the effect of distant masses ($\vec{L}_n = M_n(\vec{r} - \vec{r}_n) \times \vec{v}_n$ is the angular momentum):

$$\vec{\Omega}_{LT}^N = \vec{\Omega}_{LT} - (\alpha + \gamma) G \sum_{n=1}^{N} \frac{\vec{L}_n}{|\vec{r} - \vec{r}_n|^3},$$

5 Solid Body Rotation: Multipole Expansions

5.1 Solid Body Rotation: Vector Potential $\Pi(\vec{r})$

From now on, we assume, for brevity, the GR values $\alpha = \gamma = 1$, and study the case of rigid rotation: $\vec{v} = \vec{\omega} \times \vec{r}$. We see from (4) that it is natural to introduce a new vector potential $\vec{\Pi}(\vec{r})$ by setting

$$\vec{h}(\vec{r}) = 4\omega \times \vec{\Pi}(\vec{r}), \qquad \vec{\Pi}(\vec{r}) = \int \frac{\rho(\vec{r}\,')\vec{r}\,'\, d^3\vec{r}\,'}{|\vec{r} - \vec{r}\,'|}\,. \tag{14}$$

From (4) and (14) we derive the important relation $\nabla \cdot \vec{\Pi} = -(\Phi + \vec{r} \cdot \nabla\Phi)$ which allows us to rewrite expression (4) for $\vec{\Omega}_{LT}$ as

$$\vec{\Omega}_{LT} = 2[-(\vec{\omega} \cdot \nabla)\vec{\Pi} + \vec{\omega}(\nabla \cdot \vec{\Pi})] = -2[(\vec{\omega} \cdot \nabla)\vec{\Pi} + \vec{\omega}(\Phi + \vec{r} \cdot \nabla\Phi)], \tag{15}$$

or, choosing the z axis along $\vec{\omega}$,

$$\vec{\Omega}_{LT} = -2\omega[\vec{\Pi}_{,z} + (\Phi + r\Phi_{,r})]\hat{e}_3, \qquad \vec{\omega} = \omega\,\hat{z}\,. \tag{16}$$

Note also that the metric (5) in terms of $\Pi(\vec{r})$ is:

$$ds^2 = (1+2\Phi)dt^2 - (1-2\Phi)d\vec{r}^{\,2} + 8(\vec{\omega}\times\vec{\Pi})\cdot d\vec{r}\,dt\,. \tag{17}$$

5.2 *Multipole Expansions: General*

We choose the origin of spherical coordinates $\{r,\theta,\varphi\}$ at the center of mass of the earth, introduce spherical harmonics

$$Y^{\nu}_{lm}(\theta,\varphi) = P^m_l(\cos\theta)\begin{cases}\cos m\varphi, & \nu = c\\ \sin m\varphi, & \nu = s\end{cases},$$

and expand the potentials Φ and $\vec{\Pi}$ in corresponding series:

$$\begin{aligned}\Phi(\vec{r}) &= -\frac{GM}{r}\left[1 + \sum_{l\geq 2,\,m,\,\nu} a^{\nu}_{lm}\left(\frac{R}{r}\right)^l Y^{\nu}_{lm}\right],\\ \Pi_i(\vec{r}) &= \frac{GMR}{r}\sum_{l\geq 1,\,m,\,\nu} p^{i\nu}_{lm}\left(\frac{R}{r}\right)^l Y^{\nu}_{lm}\end{aligned} \tag{18}$$

Here R is the characteristic size of the body (we will use the equatorial radius for the Earth), and the coefficients are related to the mass distribution by

$$a^{\nu}_{lm} = \frac{(2-\delta_{m0})}{M}\frac{(l-m)!}{(l+m)!}\int \rho(\vec{r})\left(\frac{r}{R}\right)^l Y^{\nu}_{lm}\,d^3\vec{r}, \quad m = 0,1,\ldots,l; \tag{19}$$

$$p^{i\nu}_{lm} = \frac{(2-\delta_{m0})}{M}\frac{(l-m)!}{(l+m)!}\int \rho(\vec{r})\left(\frac{r}{R}\right)^l \frac{x_i}{R} Y^{\nu}_{lm}\,d^3\vec{r}, \quad \nu = c,s;\ i = 1,2,3\,. \tag{20}$$

(For convenience, we shall write $a^c_{l0} \equiv a_{l0}$, $a^s_{l0} \equiv 0$ $p^{ic}_{l0} \equiv p^i_{l0}$, $p^{is}_{l0} \equiv 0$). Among many reasons for doing the multipole expansions, one is that the gravity coefficients a^{ν}_{lm} are very well measured for the Earth at least up to $l = 18$ (Ref. 10).

From (12) and (16) the multipole expansions for $\vec{\Omega}_G$ and $\vec{\Omega}_{LT}$ may be computed; the first calculation is straightforward since it reduces to finding $\nabla\Phi$; the second one would be rather difficult but for expression (16) which simplifies it significantly. Thus, with the z axis along $\vec{\omega}$, we have a surprisingly simple expansion for $\vec{\Omega}_{LT}$:

$$\Omega^i_{LT} = \frac{GM\omega}{r}\sum_{l\geq 2,\,m,\,\nu}\left[(l-m)\,p^{i\nu}_{l-1\,m} - l\,a^{\nu}_{lm}\delta_{i3}\right]\left(\frac{R}{r}\right)^l Y^{\nu}_{lm}, \quad \vec{\omega} = \omega\,\hat{z}\,. \tag{21}$$

For any frame rotated from this one the corresponding expansion is found in the standard way by means of rotation matrices (see Ref. 11).

5.3 Multipole Expansions for $l \leq 2$ and the Inertia Tensor

If the shape of the central body and the mass distribution inside it are known, then all the pertinent quantities, including multipole expansion coefficients a_{lm}^{ν} and $p_{lm}^{i\nu}$, may be found by integration, but this is rarely the case. Even when a_{lm}^{ν} are measured, as for the Earth, all the $p_{lm}^{i\nu}$, and the LT effect with them, remain entirely undetermined. However, for a body of any shape and composition, a_{2m}^{ν}, $(l = 2)$ and $p_{1m}^{i\nu}$, $(l = 1)$ can be expressed in terms of elements $I_{ij} = \int \rho(\vec{r})\,(r^2\delta_{ij} - x_i x_j)\, d^3\vec{r}$ of the tensor of inertia $\mathbf{I}$ (we write $I_{ii} \equiv I_i$):

$$
\begin{aligned}
a_{20} = -(2I_3 - I_2 - I_1)/(2MR^2), \qquad a_{22}^{c} &= (I_2 - I_1)/(4MR^2) \\
p_{10}^{3} &= -(I_3 - I_2 - I_1)/(2MR^2), \\
p_{11}^{1c} &= -(I_3 + I_2 - I_1)/(2MR^2), \\
p_{11}^{2s} &= -(I_3 - I_2 + I_1)/(2MR^2), \\
a_{21}^{c} = -p_{10}^{1} = p_{11}^{3c} &= I_{13}/(MR^2), \\
a_{21}^{s} = -p_{10}^{2} = p_{11}^{3s} &= I_{23}/(MR^2), \\
-a_{22}^{s} = p_{11}^{1s}/2 = p_{11}^{2c}/2 &= I_{12}/(MR^2).
\end{aligned}
\tag{22}
$$

This is done by comparing the integrals (18) (with $l = 2$) and (19) (with $l = 1$) to I_{ij} using explicit expressions of Legendre functions with $l = 1, 2$; formulas for a_{20} and a_{22}^{c} are known and used in geodesy for the determination of Earth's moments of inertia.[12]

Introducing (22) into (17) and dropping the terms with $l > 2$ for Φ and $l > 1$ for $\vec{\Pi}$, we first obtain the $l \leq 2$ formulas for the potentials:

$$
\begin{aligned}
\Phi(\vec{r}) &= -\frac{G}{r}\left\{M + \frac{1}{2r^2}\left[\operatorname{tr}\mathbf{I} - \frac{3}{r^2}(\mathbf{I}\vec{r}\cdot\vec{r})\right]\right\}, \\
\vec{\Pi}(\vec{r}) &= -\frac{G}{r^3}\left[\mathbf{I}\vec{r} - \frac{1}{2}(\operatorname{tr}\mathbf{I})\vec{r}\right]
\end{aligned}
\tag{23}
$$

and then from those, by doing differentiation in (12), we obtain for the precessions:

$$
\begin{aligned}
\vec{\Omega}_G &= \frac{3G}{2r^3}\left\{\left[M + \frac{3}{2r^2}\left(\operatorname{tr}\mathbf{I} - \frac{5}{r^2}(\mathbf{I}\vec{r}\cdot\vec{r})\right)\right](\vec{r}\times\vec{V}) + \frac{3}{r^2}(\mathbf{I}\vec{r}\times\vec{V})\right\}, \\
\vec{\Omega}_{LT} &= \frac{2G}{r^3}\left\{\mathbf{I}\vec{\omega} - 3\left[\frac{1}{2}(\operatorname{tr}\mathbf{I}) - \frac{1}{r^2}(\mathbf{I}\vec{r}\cdot\vec{r})\right]\vec{\omega} + 3\,\frac{\vec{\omega}\cdot\vec{r}}{r^2}\left[\frac{1}{2}(\operatorname{tr}\mathbf{I})\vec{r} - \mathbf{I}\vec{r}\right]\right\}.
\end{aligned}
\tag{24}
$$

(Of course, the same expression for $\vec{\Omega}_{LT}$ is also obtained from (22) and (21) with $l = 2$). These expressions are valid under either of two conditions:

1) *far field*, $R/r \ll 1$;
2) *high symmetry* (all higher order moments are small).

They alter our notion of the geodetic and Lense–Thirring effects: the first one is proportional not only to the orbital momentum, but to $\mathbf{I}\vec{r}\times\vec{V}$ as well, and the LT precession points not in the direction of the angular momentum $\vec{L} = \mathbf{I}\vec{\omega}$, but has also components parallel to $\vec{\omega}$, $\vec{r}$, and $\mathbf{I}\vec{r}$. Two particular cases of the inertia tensor are of special interest:

a) Spherical symmetry, $\mathbf{I} = \mathrm{diag}\{I,\, I,\, I\}$. In this case, the classical Lense–Thirring formula follows immediately from (24):

$$\vec{\Omega}_{LT} = \frac{2GI}{r^3}\left[-\vec{\omega} + \frac{3}{r^2}(\vec{\omega}\cdot\vec{r})\vec{r}\right]. \tag{25}$$

Note that we have thus shown this to be the exact result for a spherical earth with any *radial* density distribution $\rho = \rho(r)$.
b) Symmetric top, $\mathbf{I} = \mathrm{diag}\{I_1,\, I_1,\, I\}$, $I_1 \neq I$. For $\vec{\omega} = \omega\hat{z}$, the previous expression remains true; to the main order in the oblateness, this proves to be the exact result for a slightly oblate uniform ellipsoid of revolution rotating about its semiminor axis.

6 Earth Models and Results for GP-B

6.1 Earth Models

To go beyond the $l \leq 2$ approximation, one must make some assumptions about the shape of the earth and density distribution inside it, and use the data of gravitational potential measurements, if available. In the case of the Earth and GP-B conditions ($R/r \approx 0.9$), one needs to check the validity of the $l \leq 2$ approximation to ensure that theoretical predictions match the expected experimental accuracy (10^{-5} for geodetic effect, $\sim 10^{-2}$ for LT effect). Here is our set of reasonable assumptions.

1. Gravitational potential For the required accuracy, it is enough to include only the quadrupole moment into Φ, i. e., to set

$$-a_{20} \equiv J_2 \approx 1.083 \times 10^{-3}, \qquad a^{\nu}_{lm} = 0 \quad \text{for} \quad l = 2,\, m > 0; \quad l > 2, \tag{26}$$

because all gravitational coefficients other than the Earth's oblateness J_2 are at least 2 orders of magnitude smaller.[10] Then, by (18) and (22), the $l \leq 2$ expression (23) for Φ is valid with

$$\mathbf{I} = \mathrm{diag}\{I_1,\, I_1,\, I\}, \quad I_1 = I - J_2 M R^2 .$$

Also, by (12), (23) and (26), the geodetic precession is completely determined including the J_2 correction calculated for the first time by Breakwell.[7] Note that small and/or slow motion of the Earth rotation axis relative to the Earth centered inertially fixed frame may be neglected.
2. Shape It is sufficient to assume that the Earth is a slightly oblate ellipsoid of revolution (the Clairaut formula, see Roy[13]), so that to the first order in eccentricity $\epsilon\,(\approx 3.353 \times 10^{-3})$ the surface equation is $r = r_s(\theta, \varphi) = R(1 - \epsilon\cos^2\theta)$.
3. Mass distribution We examine two different models:
a) With ρ_0, $\Delta\rho$ being arbitrary functions of their arguments, set

$$\rho(\vec{r}) = \rho_0(r) + \Delta\rho(\theta, \varphi), \qquad \int_{\text{unit sphere}} \Delta\rho(\theta, \varphi)\sin\theta\, d\theta d\varphi = 0 \tag{27}$$

The first term here describes any depth variation of the *average* density, and the only assumption is that the angular variations are depth–independent.

b) For arbitrary ρ_0 and ρ_s, set

$$\rho(\vec{r}) = \rho_0(r) + \rho_s(\theta,\varphi)\delta(r - r_s(\theta,\varphi)) \tag{28}$$

Contrary to the previous model, here all angular variations of the density are concentrated at the Earth's surface, which is rather realistic, since the estimated thickness of the layer where the mass distribution varies significantly in the angular directions is about 30 *km*.

Note that instead of 3a) and 3b) an entirely different assumption is used in geodesy, namely, that the sum of gravitational and centrifugal potentials is constant at the Earth's ellipsoid surface.[14] It allows one to relate a_{l0} to the eccentricity and the Earth angular velocity only (in particular, to obtain J_2 with a surprisingly good accuracy), but gives zero values to a^{ν}_{lm}, $m \neq 0$ and leaves $p^{i\nu}_{lm}$ undetermined.

Evidently, our two models should give the bounds for the corrections to the $l \geq 2$ values of the LT effect. Moreover, for both of them it is possible to find the corrections explicitly by means of the following procedure. Working all the way to the first order in the eccentricity, we: 1) using (19), fit (27), model a), (or (28), model b),) to the values of a^{ν}_{lm} assumed in (26) ; this permits one to determine completely the function $\Delta\rho(\theta,\varphi)$ (respectively, $\rho_s(\theta,\varphi)$); 2) using the latter, calculate $p^{i\nu}_{lm}$ for $l > 1$ by (20), determining thus the LT precession by (21). It turns out that, *for both models*, the only additional non-zero coefficients are those with $l = 3,\ m = 0, 1$; their values for model a) are

$$p^{1c}_{31} = p^{2s}_{31} = (5/49)\, J_2; \qquad p^{3}_{l0} = -(15/49)\, J_2; \tag{29}$$

for model b) they are larger by just the factor 7/5.

6.2 Results for GP-B

Using the above results, let us finally give the values of geodetic and LT precessions averaged over the circular polar orbit of 650 *km* altitude perturbed by the quadrupole moment. To the main order in the oblateness J_2,

$$\langle \Omega_G \rangle = \frac{3GMV}{2r^2}\left[1 - \frac{9}{8} J_2 \left(\frac{R}{r_0}\right)^2\right]$$

$$\langle \Omega_{LT} \rangle = \frac{GI\omega}{2r^3}\left[1 + \frac{9}{8} J_2 \left(\frac{R}{r_0}\right)^2 \left(1 - \frac{88}{147}\frac{MR^2}{I}\right)\right]$$

where the values (29) for model a) are used. The corrections in both cases are close to 0.1%, which is beyond GP-B accuracy for the LT effect but two orders of magnitude larger than the expected measurement error for geodetic effect. Since GP-B is intended to measure the geodetic precession and parameter γ to about a part in 10^5, this is a critically important correction.

Acknowledgments

This work was supported by NASA grant NAS 8-39225 to Gravity Probe B. We are grateful to the members of GP-B Theory Group, especially, to C.W.F.Everitt,

G.M.Keiser and R.V.Wagoner, for many fruitful discussions and enlightening comments.

References

1. G.M. Keiser in *These Proceedings.*
2. C.W. Misner, K.S. Thorne, J.A. Wheeler, *Gravitation,* (W.H.Freeman and Company, New York, 1973, sec. 40.7).
3. H.C. Ohanian and R. Ruffini, *Gravitation and Spacetime,* (W.W. Norton and Co., New York, 1994).
4. C.M. Will, *Theory and Experiment in Gravitational Physics,* (Cambridge University Press, Cambridge, 1993).
5. J. Lense and H.Thirring, Phys. Z. **19**, 156 (1918).
6. K. Nordtvedt in *These Proceedings.*
7. J.V. Breakwell in *Near Zero,* eds. J.D. Fairbank, B.S. Deaver, Jr., C.W.F. Everitt, P.F. Michelson (W.H. Freeman and Company, New York, 1988, pp. 685–690).
8. A.S. Eddington, *The Mathematical Theory of Relativity, p. 105,* (Cambridge University Press, New York, 1988).
9. H.P. Robertson in *Space Age Astronomy,* eds. A.J. Deutsch and W.B. Klemperer, (Academic Press, New York, 1962, p. 128).
10. *World Geodetic System 1984,* (DMA Technical Report, 1987).
11. E.M. Rose, *Elementary Theory of Angular Momentum,* (John Wiley and Sons, New York, 1957).
12. M. Bursa, Studia geoph. and geod. **36**, 109 (1992).
13. A.E. Roy, *Orbital Motion,* (Adam Hilger, Bristol and Philadelphia, 1988).
14. W.M. Kaula, *Theory of Satellite Geodesy,* (Blasidell Publ. Co., Waltham (MA), Toronto, London, 1966).

CLASSICAL TORQUE ERRORS IN GRAVITY PROBE B EXPERIMENT

A.S.SILBERGLEIT, M.I.HEIFETZ, G.M.KEISER
Gravity Probe B, W.W.Hansen Experimental Physics Laboratory,
Stanford University, Stanford, CA 94305-4085, USA
E-mail: gleit@relgyro.stanford.edu

Classical torques are the first to come into one's mind when thinking about sources of systematic errors in Gravity Probe B (GP-B) Relativity Science Mission, which is why they have been most thoroughly investigated. In this paper we give a brief survey of these studies and their results. First, the classification of all known non-relativistic torques that is used in the GP-B Error Tree is given, and the Error Tree spreadsheet is described. Second, the theory of electrostatic support dependent torques is outlined, including the universal expression for the housing-fixed components of the torques, the definition of the fifteen torque coefficients and their relation to the rotor shape. Examples of the final formulas for the drift rates of several particular support dependent and support independent torques are presented. Finally, contributions of different groups of torques to the total expected experimental error are compared, and the top contributing torques are discussed.

1 Introduction

The Gravity Probe B satellite, which is planned to fly in the year 2003, contains a set of high-precision superconducting gyroscopes (the outline of the GP-B experiment is given in Ref. 1). They are to test the predictions of general relativity that a gyroscope in a low (altitude$\approx$ 650 *km*) circular polar orbit will precess about 6.6 *arcsec/year* in the orbital plane (geodetic precession, East-West (EW) direction) and about 42 *milliarcsec/year* perpendicular to the orbital plane (frame-dragging precession, North-South (NS) direction; for the theory of both effects, see c.f. Ref. 2). The required total error of the gyro drift rate determination is 0.5 *milliarcsec/year*; we expect even a better accuracy, about few parts in 10^5 for the measurement of the geodetic effect.

Non-relativistic torques on the GP-B gyroscopes had been primarily considered as the most important source of experimental errors; that is why the experimental hardware is required to provide the performance allowing not more than 0.3 *milliarcsec/year* contribution of all the classical torques. To prove that this is possible, as well as to develop proper verification procedures, the classical torques have therefore been extensively investigated for a long time. Earlier GP-B classical torque studies are reported in Ref. 3; this paper is a brief survey of state of the art results from Refs. 4 and 5. A large part of it is devoted to the theory of electrostatic torques as developed in Ref. 4; unlike the previous versions, it is based mostly on symmetry considerations and thus avoids lengthy technical complications characteristic of earlier approaches (see Sec. 3).

2 GP-B Error Tree and Classification of Non-Relativistic Torques

There are two different types of torques: support dependent (SD) torques, which are caused by control voltages of the electrostatic gyro suspension system, and support

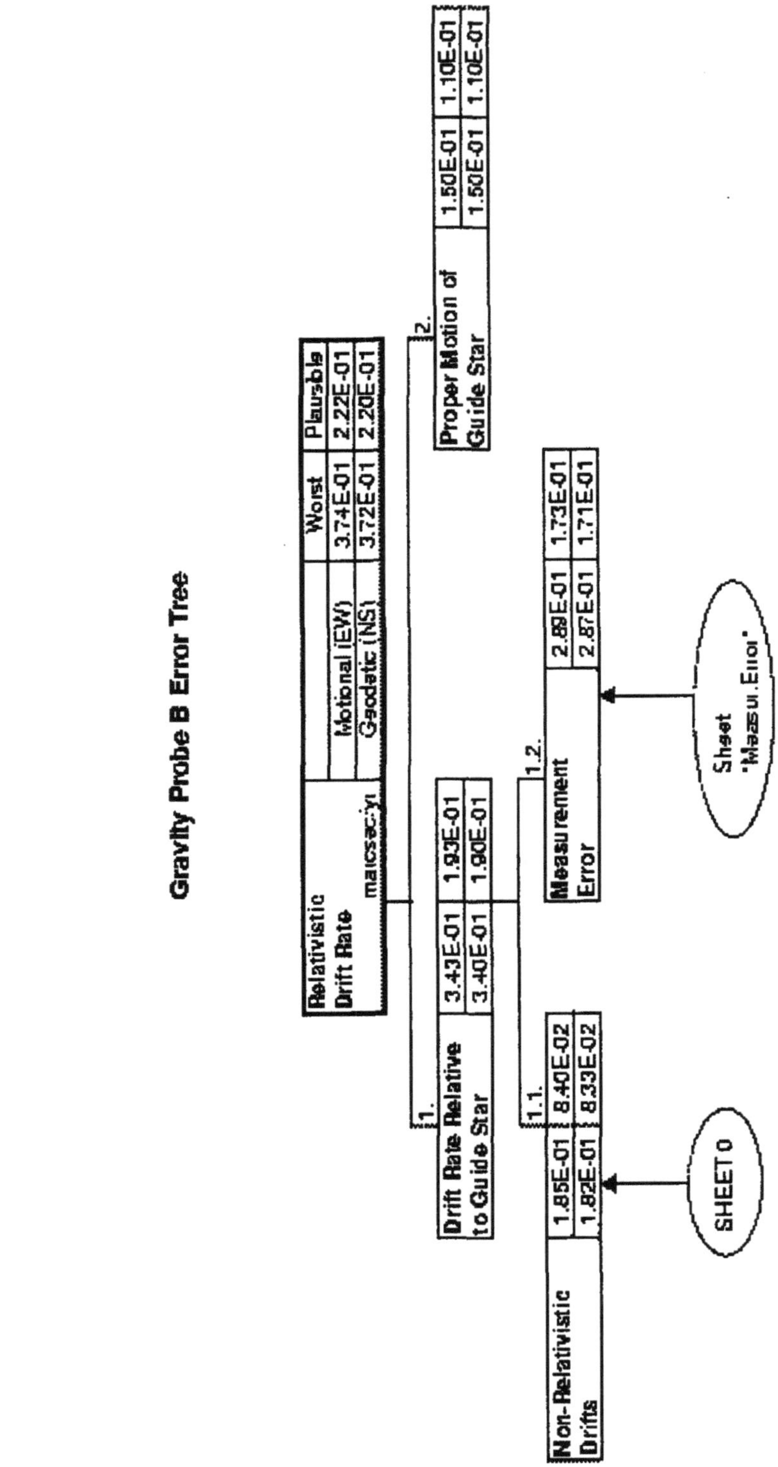

Figure 1. 'Top' sheet of the NET

independent ones (SI), which act on the gyro even if it is not supported (the torques due to position bridge voltages belong here). Extensive results on torques of either group are given in Refs. 4 and 5, respectively.

A complete classification of all non-relativistic torques and evaluation of the gyro drift caused by them is done in the GP-B New Error Tree (NET) (see Refs. 6 and 7) implemented as an Excel spreadsheet workbook. The NET consists of two main parts describing non-relativistic drifts and measurement (readout) errors, respectively. The NET is a 'live' one, i.e., one only has to specify (type in) numerical values of all necessary parameters in the 'input' sheets (*"Constants"*, *"Parameters"*, *"Coefficients"*, *"Preloads"*, and, partly, *"Accelerations"* and *"Accelerations Squared"*), and the values of all classical drift rates are immediately computed by the spreadsheet. After combining the result with the measurement error and the uncertainty in the proper motion of the Guide Star, the total experimental error is also calculated. The three resulting errors and their total (in *milliarcsec/year*) are seen in the 'top' sheet of the NET, Fig. 1, for the drift rate in the NS and EW directions (the way those totals are computed is explained below).

Each error there and in all boxes of all the 'lower level' sheets is represented by two numbers corresponding to the so called worst and plausible case, respectively. The numbers are calculated from a different sets of parameters in the input sheets mentioned above (only the table *"Constants"* contains one set of values, because those are either fundamental constants, like the speed of light, or have a well specified value in the GP–B project, such as the rotor radius). The total number of parameters is around one hundred. The worst case values are either taken directly from the requirements, or derived from them, or else are the result of our estimate of what the worst case value may be for a particular parameter in reality. The plausible case values either are already achieved experimentally, or are expected to be achieved 'with the 99%' certainty. So, the plausible case shows what we strongly expect to have in the real experiment, while the worst case represents the upper bound of errors: they definitely are not going to be worse than that if all the hardware works as required. Naturally, the NET book is arranged in such way that a change of a worst case value automatically implies variations of dependent worst case values *only*, not affecting the plausible case at all, and vice versa.

The principal classification of the torques is found in the next (after the 'top' one) sheet of the NET shown in Figs. 2 and 3, along with the drift rate contribution of each particular group of torques. SD torques are first subdivided in two major groups. The first group contains the torques which depend on accelerations, i.e., which stem from how the suspension system reacts to the forces applied to the gyro when trying to keep it centered. The second group consists of torques which depend on the preloads, in other words, on the voltages provided by the suspension system when it is just turned on, even if no force at all is acting on the gyro. The next subdivision in both major groups is related to the rotor's *position* in which the torques are produced, namely, the *nominal*, *misaligned*, and *miscentered* position. (We call the rotor's position nominal when the rotor is perfectly centered in the housing and its spin axis is aligned with the roll axis of the spacecraft; accordingly, in the misaligned position there is an angle between the spin and roll axes, while

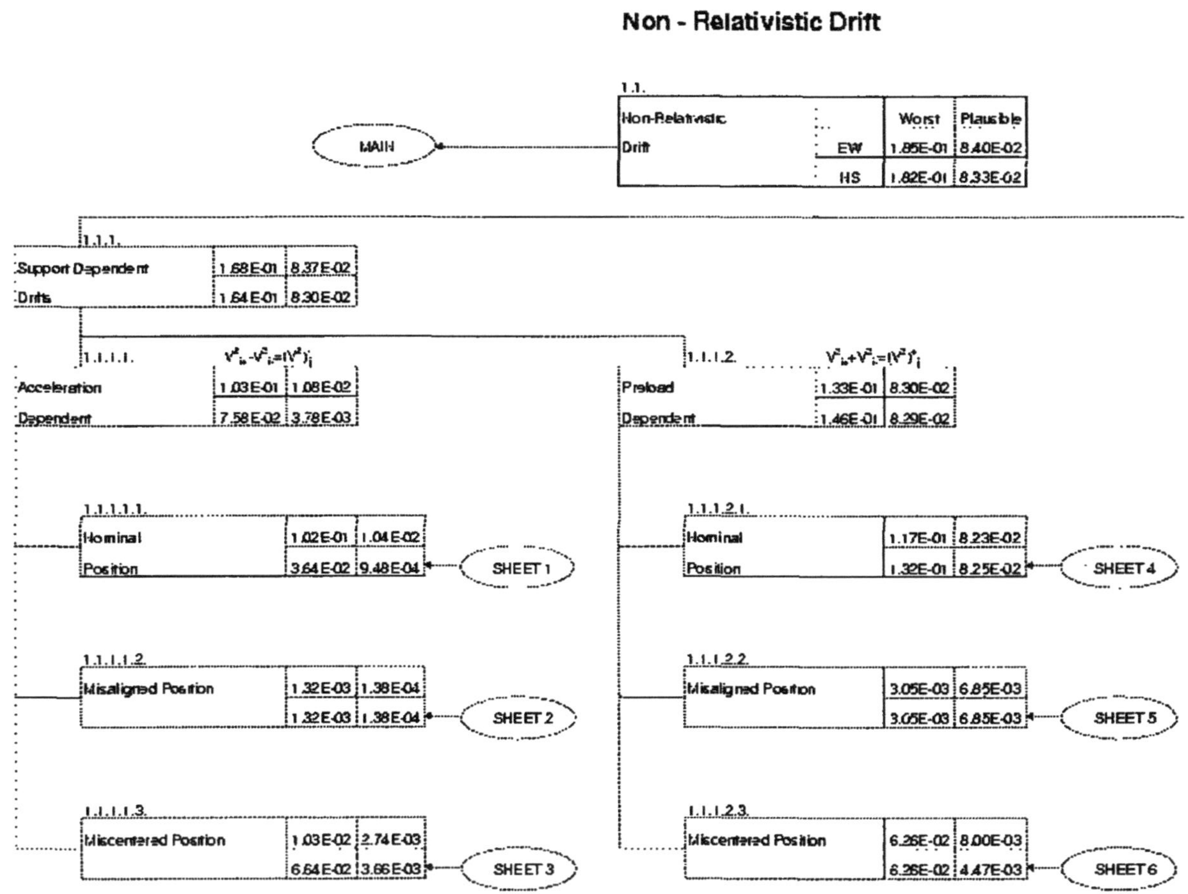

Figure 2. Non-Relativistic Drift. Left-hand side of the scheme

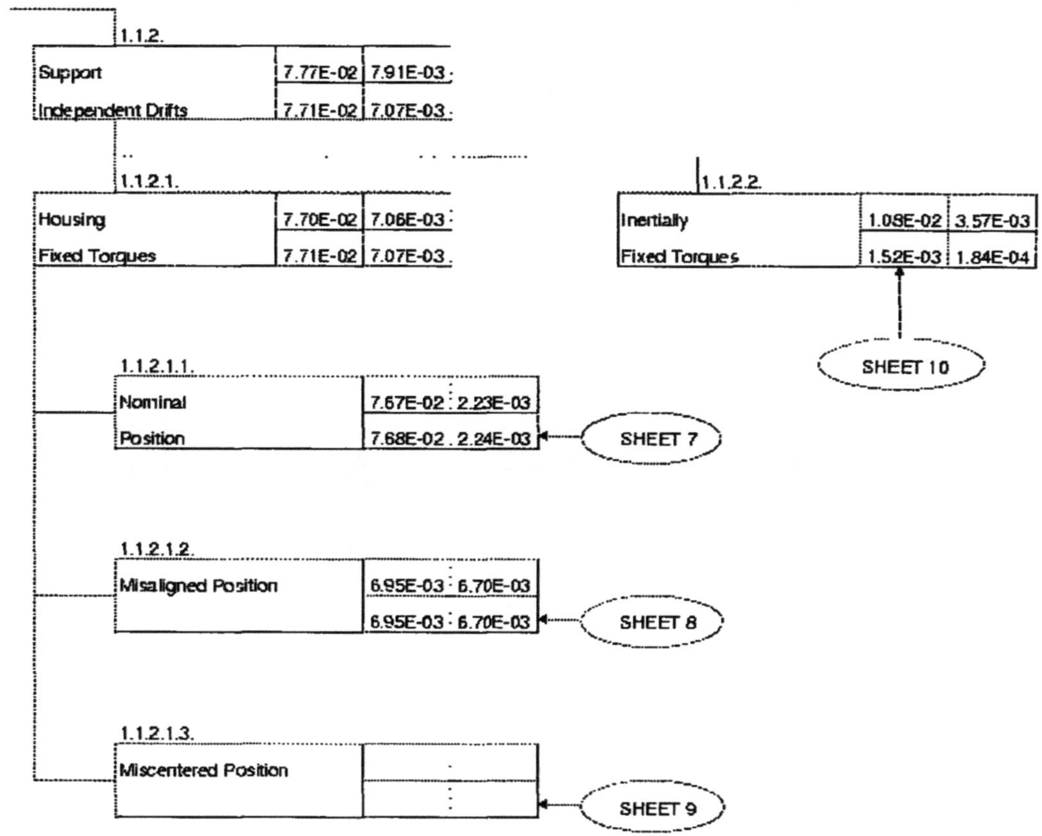

Figure 3. Non-Relativistic Drift. Right-hand side of the scheme

in the miscentered position the center of mass of the rotor is displaced from the center of the housing.)

Therefore there are exactly $6 = 2 \times 3$ sheets with SD torques from *"Sheet 1"* (acceleration dependent, nominal position torques) to *"Sheet 6"* (preload dependent, miscentered position torques). In each of the sheets, torques are again subdivided into several larger groups, such as those caused by accelerations either parallel or perpendicular to the spacecraft roll axis, or by preloads or preload differences, etc. Finally, the torques are sorted out by their physical source, and every such particular torque is represented by its own 'lowest level' box. There are over a hundred of different torques in the NET now; even more are known, but quite a few of them are not included in the NET because their contribution turns out to be entirely negligible.

The boxes are connected by lines in an ascending way, so that the structure of every sheet with torques is that of a tree graph: each box of a lower level is connected to a single upper level box. An example of *"Sheet 2"* with acceleration dependent, misaligned position SD torques is shown in Fig. 4; this sheet has been chosen because its structure is rather simple, other sheets include more branches and boxes than that one.

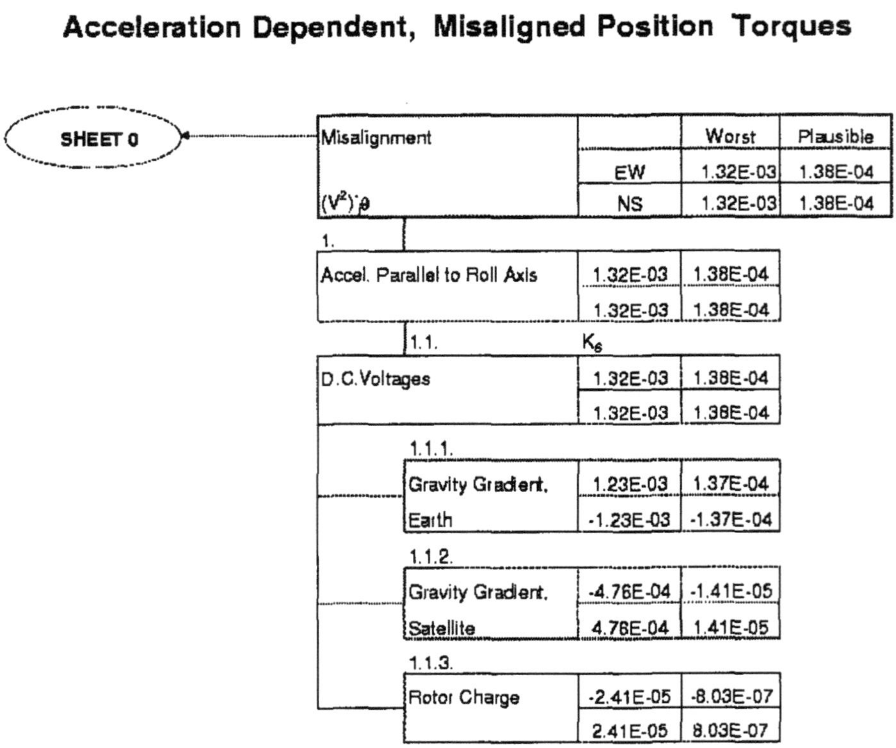

Figure 4. Acceleration Dependent, Misaligned Position Torques

SI torques are first classified as housing fixed and inertially fixed, Figs. 2–3. The former (sheets 7–9) are again sorted relative to the nominal, misaligned and miscentered rotor's position; all inertially fixed torques are listed in sheet 10. Particular SI torques are dealt with in each of the sheets 7–10 in a similar manner as SD torques in sheets 1–6, except that the grouping is done now by the physical origin of the torques (electrostatic, magnetis, etc.). As an example, *"Sheet 7"* with SI, housing-fixed, nominal position torques is shown in Fig. 5.

All the ten sheets with the torques are filled in in the following way. The lowest level boxes of every sheet show drift rates due to a torque which belongs to some particular group and is caused by a particular physical source. The four values of drift rates given in the box (NS worst, NS plausible, EW worst, EW plausible) are calculated by the proper formula from Refs. 4 or 5 in which either worst or plausible values of the pertinent parameters from the input sheets are used. As soon as the values of all the input quantities are specified, the numbers appear in the lowest level boxes, as well as in all upper level ones, all the way up to the top sheet shown in Fig. 1. The numbers in the upper level boxes are automatically computed by means of a built-in procedure operating by the rule based on the assumption that different torques act independently. So, say, the squares of EW, worst case drift rates from all the boxes of a group at any given level are summed up, and the square root of this sum is inserted as a value of EW, worst case drift rate into the next level box to which the group is attached. This operation is repeated level after level from the bottom to the top of a given sheet with torques, which results in the numbers given in the uppermost box of the sheet representing the combined contribution of all the torques included in it (for instance, SD, acceleration dependent, misaligned position torques from *"Sheet 2"* in Fig. 3). Those uppermost boxes of sheets 1–10 serve as the lowest level boxes of the sheet in Figs. 2–3, whose uppermost box, containing the total drift rate from all classical torques, serves, in its turn, as one of the three lowest level boxes of the top sheet (Fig. 1), along with the box containing the total measurement error and the other one with the contribution of proper motion of the Guide Star.

So, one needs some formulas to calculate every separate 'lowest level' torque; in the next two section we outline their derivation as given in Ref. 4, 5 and discuss various related questions.

3 Theory of Support Dependent Torques

3.1 Derivation of Housing-Fixed SD Torques

GP-B gyro suspension system consists of three pairs of electrodes (labeled $a_\pm$, $b_\pm$, $c_\pm$) positioned symmetrically as shown in Fig. 6. Cartesian coordinates $\{x, y, z\}$ are fixed in the housing, with the $\hat{\mathbf{z}}$ unit vector pointed in the direction of the roll axis of the spacecraft. Thus the unit vectors to the centers of the electrodes are given by

$$\hat{\mathbf{e}}_{a\pm} = \mp\frac{\sqrt{2}}{2}(\hat{\mathbf{y}} - \hat{\mathbf{z}}); \qquad \hat{\mathbf{e}}_{b\pm} = \pm\frac{\sqrt{2}}{2}(\hat{\mathbf{y}} + \hat{\mathbf{z}}); \qquad \hat{\mathbf{e}}_{c\pm} = \pm\hat{\mathbf{x}} \tag{1}$$

The first complete and consistent derivation of GP-B SD torques had been carried out by G.M. Keiser in Ref. 9, whose approach was independently checked

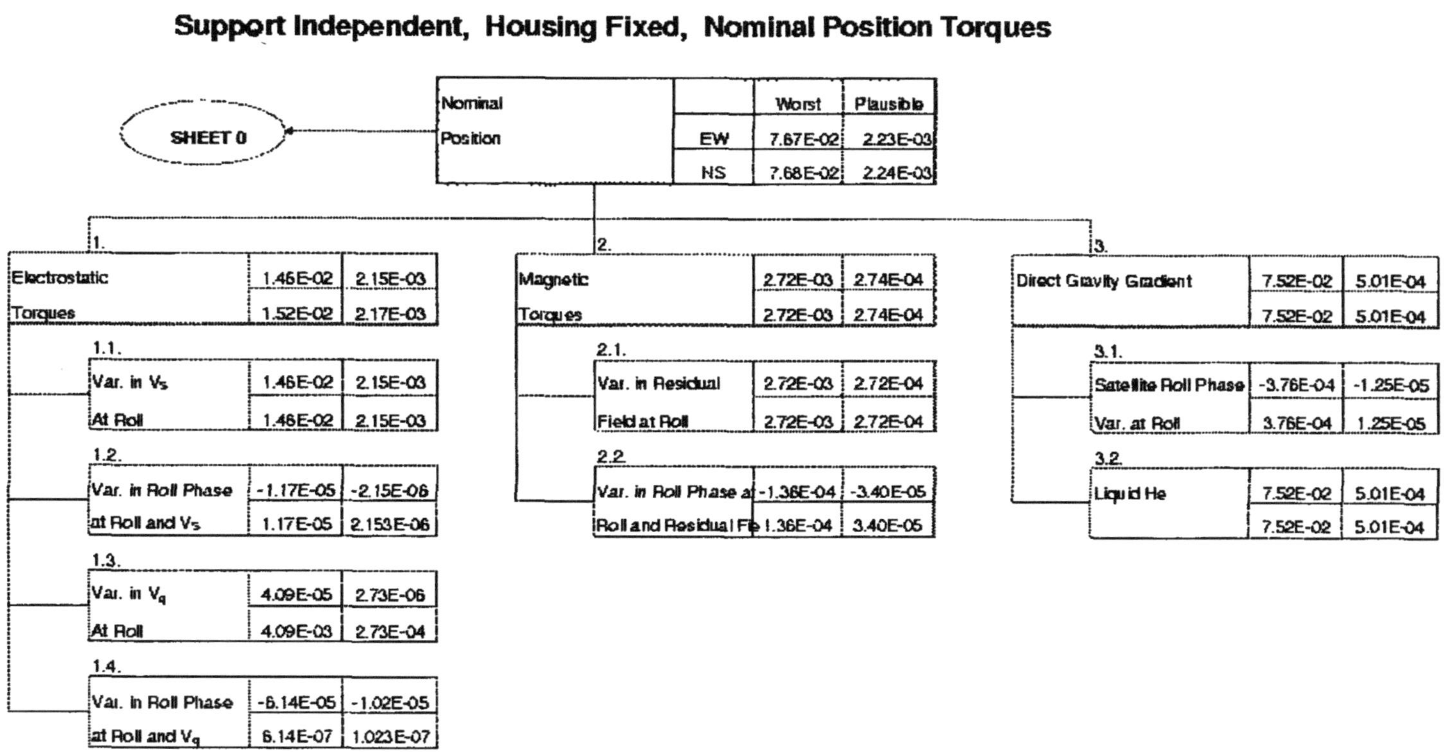

Figure 5. Support Independent, Housing Fixed, Nominal Position Torques

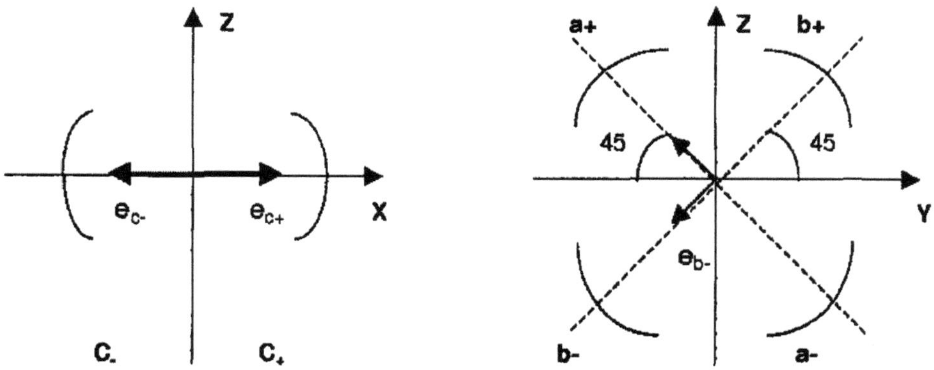

Figure 6. GP-B gyro suspension system consisting of three pairs of electrodes (labeled $a_\pm$, $b_\pm$, $c_\pm$) positioned symmetrically

later in Ref. 10. An approximate explicit expression for the capacitance and a rather complicated technique of spherical harmonics and rotation matrices was used there. Here we follow another version of the theory developed in Ref. 4 which is based on the symmetry arguments and thus avoids explicit expression of the capacitance (until the very end, when the torque coefficients are calculated, see Sec. 2.4), allowing one to establish the structure of SD torques without it.

The calculation of SD torques starts with the general formula for an electrostatic torque acting on a conducting body (c.f. Ref. 8, Sec. 3.14, (6)):

$$\mathbf{T} = \frac{1}{2}V^2 \frac{\partial C}{\partial \eta}\hat{\eta}, \tag{2}$$

which is a direct implication of the energy conservation. Here V is the electrode to rotor potential, C is the electrode to rotor capacitance, and η is the rotation angle about the direction $\hat{\eta}$.

The orientation of the rotor may be described by any three Euler angles α, β, γ. It is convenient to choose them as shown in Fig. 7 , so that α and β specify the position of the gyro spin axis, $\vec{\omega}_s = \omega_s \hat{\mathbf{s}}$, relative to the housing, while γ represents rotation about $\vec{\omega}_s$. Since $\hat{\mathbf{z}}$ points along the roll axis, the angle β is the spin–roll axes misalignment. The spin axis unit vector is given, evidently, by

$$\hat{\mathbf{s}} = \hat{\mathbf{s}}(\alpha, \beta) = (\hat{\mathbf{x}}\cos\alpha + \hat{\mathbf{y}}\sin\alpha)\sin\beta + \hat{\mathbf{z}}\cos\beta$$

By applying formula (2) to each of the Euler angles for any electrode and then projecting the obtained equalities on the $\{x, y, z\}$ axes using

$$\hat{\alpha} = \hat{\mathbf{z}}, \quad \hat{\beta} = -\hat{\mathbf{x}}\sin\alpha + \hat{\mathbf{y}}\cos\alpha, \quad \hat{\gamma} = (\hat{\mathbf{x}}\cos\alpha + \hat{\mathbf{y}}\sin\alpha)\sin\beta + \hat{\mathbf{z}}\cos\beta,$$

for the housing-fixed components of the torque we obtain:

$$T_x = \frac{1}{2}V^2\left(-\sin\alpha\frac{\partial C}{\partial\beta} - \cos\alpha\cot\beta\frac{\partial C}{\partial\alpha} + \frac{\cos\alpha}{\sin\beta}\frac{\partial C}{\partial\gamma}\right)$$

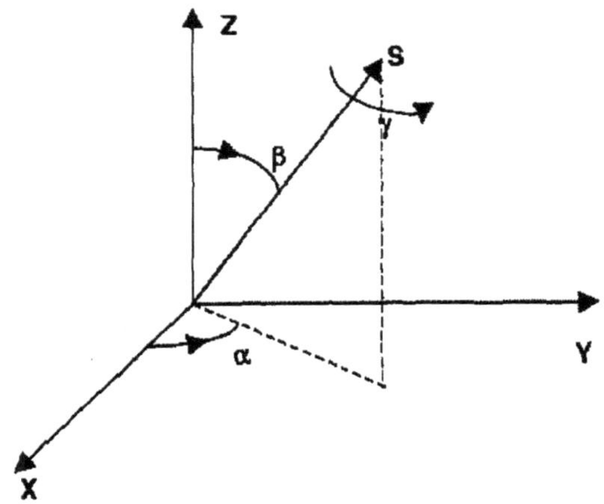

Figure 7. The orientation of the rotor may be described by any three Euler angles α, β, γ.α and β specify the position of the gyro spin axis, $\vec{\omega}_s = \omega_s \hat{\mathbf{s}}$, relative to the housing, while γ represents rotation about $\vec{\omega}_s$

$$T_y = \frac{1}{2}V^2\left(\cos\alpha\,\frac{\partial C}{\partial\beta} - \sin\alpha\cot\beta\,\frac{\partial C}{\partial\alpha} + \frac{\sin\alpha}{\sin\beta}\frac{\partial C}{\partial\gamma}\right) \tag{3}$$

(we are not interested here in the torque T_z which changes the rotor spin speed, so we do not give its expression). The primary simplification of these formulas comes from the first of our three basic assumptions. All the assumptions are good approximations under the GP-B conditions, and they were used in all the previous versions of the theory, sometimes implicitly.

Assumption 1 (spin averaging) *Instantaneous spin axis direction* $\hat{\mathbf{s}}$ *coincides with the (inertially fixed) direction of the angular momentum; the voltage does not vary at the spin frequency.*

Under Assumption 1, the torques (3) averaged over the rotation of the rotor become

$$T_x^i = -\frac{1}{2}V_i^2\left(\sin\alpha\,\frac{\partial C^i}{\partial\beta} + \cos\alpha\cot\beta\,\frac{\partial C^i}{\partial\alpha}\right),$$

$$T_y^i = \frac{1}{2}V_i^2\left(\cos\alpha\,\frac{\partial C^i}{\partial\beta} - \sin\alpha\cot\beta\,\frac{\partial C^i}{\partial\alpha}\right). \tag{4}$$

The index i enumerates the electrodes, $i = a_\pm$, $b_\pm$, $c_\pm$; C^i is the spin averaged value of the capacitance which, given a real rotor shape, depends on the rotor's position in the housing. The latter can be described by the misalignments relative to the housing-fixed axes,

$$\theta_x = \sin\beta\cos\alpha \simeq \beta\cos\alpha \qquad \theta_y = \sin\beta\sin\alpha \simeq \beta\sin\alpha$$

which are small because $\beta \sim 10^{-5}$ in GP-B, and another small parameter, the relative miscentering, $\mathbf{r}_\circ = (x_\circ\,\hat{\mathbf{x}} + y_\circ\,\hat{\mathbf{y}} + z_\circ\,\hat{\mathbf{z}})\,/d$. Here d is the nominal gap between the rotor and the housing, and vector $\mathbf{r}_\circ$ goes from the geometrical center of the housing to the rotor's center of mass.

So, generally, $C^i = C^i(\alpha, \beta, \mathbf{r}_\circ)$, and we need its derivatives in the angles involved in (4) to the first order in β (or θ_x, θ_y) and $r_\circ = |\mathbf{r}_\circ|$. According to our classification of torques, the derivatives are needed for the three positions of the rotor: *1) nominal position*, ($\beta = 0$, $r_\circ = 0$); *2) misaligned position*, ($\beta \neq 0$, small; $r_\circ = 0$); *3) miscentered position*, ($\beta = 0$; $r_\circ \neq 0$, small). Of course, no progress can be made unless we somehow restrict the functional dependence $C^i(\alpha, \beta, \mathbf{r}_\circ)$. The suitable restriction can be formulated as

Assumption 2 *The spin-averaged electrode-to-rotor capacitance depends only on the angle between the spin and electrode axes, and on two projections of miscentering on those directions:*

$$C^i = C^i(\xi_i,\, \lambda,\, \mu_i), \qquad i = a_\pm,\, b_\pm,\, c_\pm$$

where $\xi_i = \hat{e}_i \cdot \hat{\mathbf{s}}$, $\quad \lambda = \hat{\mathbf{s}} \cdot \mathbf{r}_\circ$, $\quad \mu_i = \hat{e}_i \cdot \mathbf{r}_\circ$.

This is a natural conjecture confirmed by direct calculations in Ref. 9, 10; it reduces the number of independent arguments of function C^i from five to three.

Assumption 2 allows us to essentially simplify $\partial C^i/\partial\alpha$, $\partial C^i/\partial\beta$ and to write the housing–fixed torques for the three positions as follows:

$$\mathbf{T}^i_{nom} = \frac{V_i^2}{2} \tilde{K}^i \left[-(\hat{e}_i \cdot \hat{\mathbf{y}})\hat{\mathbf{x}} + (\hat{e}_i \cdot \hat{\mathbf{x}})\hat{\mathbf{y}}\right] \tag{5}$$

$$\mathbf{T}^i_{misal} = \frac{V_i^2}{2}\left\{\tilde{K}^i (\hat{e}_i \cdot \hat{\mathbf{z}})(\theta_y\hat{\mathbf{x}} - \theta_x\hat{\mathbf{y}}) + \tilde{L}^i \left[-(\hat{e}_i \cdot \hat{\mathbf{y}})^2\theta_y\hat{\mathbf{x}} + (\hat{e}_i \cdot \hat{\mathbf{x}})^2\theta_x\hat{\mathbf{y}}\right]\right\} \tag{6}$$

$$\mathbf{T}^i_{misc} = \frac{V_i^2}{2}\left\{\tilde{D}^i_1 \left(-\frac{y_\circ}{d}\hat{\mathbf{x}} + \frac{x_\circ}{d}\hat{\mathbf{y}}\right) + \left[\tilde{D}^i_2 \frac{z_\circ}{d} + \tilde{D}^i_3 (\hat{e}_i \cdot \mathbf{r}_\circ)\right]\left[-(\hat{e}_i \cdot \hat{\mathbf{y}})\hat{\mathbf{x}} + (\hat{e}_i \cdot \hat{\mathbf{x}})\hat{\mathbf{y}}\right]\right\} \tag{7}$$

$$\tilde{K}^i = \frac{\partial C^i}{\partial \xi_i}, \quad \tilde{L}^i = \frac{\partial^2 C^i}{\partial \xi_i^2}, \quad \tilde{D}^i_1 = \frac{\partial C^i}{\partial \lambda}, \quad \tilde{D}^i_2 = \frac{\partial^2 C^i}{\partial \lambda \partial \xi_i}, \quad \tilde{D}^i_3 = \frac{\partial^2 C^i}{\partial \mu_i \partial \xi_i}; \tag{8}$$

the derivatives here are taken at $\beta = 0$, $\mathbf{r}_\circ = 0$.

This is an intermediate answer, and it looks rather simple. However, there are still 30 coefficients involved in it; our last assumption, which has always been used in GP-B, slashes their number twice.

Assumption 3 (electrode symmetry) *Capacitances between electrodes and rotor averaged over the spin period, as functions of their arguments, coincide pairwise:*

$$C^{c+}(\xi,\, \lambda,\, \mu) \equiv C^{c-}(\xi,\, \lambda,\, \mu) \equiv C(\xi,\, \lambda,\, \mu)$$

$$C^{a+}(\xi,\, \lambda,\, \mu) \equiv C^{b+}(\xi,\, \lambda,\, \mu) \equiv C^{+}(\xi,\, \lambda,\, \mu)$$

$$C^{a-}(\xi,\, \lambda,\, \mu) \equiv C^{b-}(\xi,\, \lambda,\, \mu) \equiv C^{-}(\xi,\, \lambda,\, \mu)$$

According to this assumption and the definitions (8), the number of coefficients is reduced by way of

$$K^{c+} = K^{c-} \equiv K, \quad K^{a+} = K^{b+} \equiv K^+, \quad K^{a-} = K^{b-} \equiv K^-$$

$$L^{c+} = L^{c-} \equiv L, \quad L^{a+} = L^{b+} \equiv L^{+}, \quad L^{a-} = L^{b-} \equiv L^{-},$$

etc.

Using the new and consistently denoted coefficients, the housing-fixed torques (5)–(7) in all the three positions can be written as sums of terms of the standard form:

$$K_\nu \times \left(V_{i+}^2 \pm V_{i-}^2\right) \times \left(\text{position factor}\right), \qquad i = a, b, c, \tag{9}$$

where the position factor equals to 1 for the nominal position, to one of the misalignments θ_x, θ_y for the misaligned position, and to some projection of the miscentering $\mathbf{r}_\circ$ for the miscentered position; K_ν, $\nu = 1, 2, \ldots, 15$ are the fifteen torque coefficients studied in Sec. 2.4.

3.2 SD Torques in the Inertial Frame. Relation to Accelerations and Preloads

To obtain the final formulas for the drift rates used in the NET from the expressions for the housing-fixed torques based on (9), one needs to 1) transform the torques to the inertial frame; 2) relate the squares of voltages to either accelerations or preloads (and specify the latter); 3) average the result over the roll period.

The first step, though cumbersome, is otherwise straightforward, since, with enough accuracy, we can consider the inertial frame just rotating with the roll frequency about the z axis of the housing-fixed frame. At the second step we make use of the expression for the net force on the gyro exerted by a pair of electrodes[8] which relates thus the differences of squares of opposite electrode voltages to the acceleration g (see Ref. 9):

$$V_{i+}^2 - V_{i-}^2 = -A\,g, \quad A = m_g \, \frac{8d^2}{\pi\epsilon_0 r_g^2} \simeq 5.09 \times 10^4 \; volts^2/m\,sec^{-2}$$

(m_g and r_g are the mass and radius of the rotor). The corresponding sums, on the other hand, define the preload acceleration h_i according to

$$V_{i+}^2 + V_{i-}^2 = \frac{1}{2} A \left(h_i + \frac{g^2}{h_i}\right)$$

(acceleration is also involved here, but typically the second term on the R.H.S. is much smaller than the first one. Nevertheless, such terms are also taken care of in the NET, and the inclusion of the input sheet *"Accelerations squared"* is caused by them). The values of preloads are specified by GP-B requirements; the exhaustive list of forces acting on the GP-B gyro (and the expressions for the corresponding accelerations used in the NET) has been compiled in Ref. 11; its shorter version is also found in Ref. 4.

The roll averaging is significantly complicated by the fact that, along with the D.C. components, accelerations, preloads *and* the miscentering may also vary at roll, twice roll, and some other frequencies, which variations generally do contribute to the average SD torques. That is why even the list of expressions for the averaged inertial SD torques in terms of voltages (and not yet accelerations and preloads!) occupies almost five pages in Ref. 4; it is thus not included here. Instead, we give just two examples of the final expressions for the drift rates which are typical in their structure and, we think, give one a good idea of all the formulas exploited in the lowest level boxes of sheets 1–6 of the NET.

3.3 Two Examples of Formulas for SD Torques

For brevity, we give the expressions only for the East–West (geodetic) component, $\dot{EW}$, of the drift rate; the North–South component looks absolutely similar.

1. Preload dependent, nominal position torque due to variation of preload difference $h_a - h_b$ at roll (NET sheet 4, box 1.2.1; see Fig. 5) Here

$$\dot{EW} = \frac{K_3}{I\omega_s}\langle (V^2)_a^+ - (V^2)_b^+ \rangle_{co} = \frac{K_3}{I\omega_s}\frac{A}{4}\left(h_{ac} - h_{bc}\right),$$

where I is the moment of inertia, h_{ic} is the amplitude at cosine of roll of the i-th preload, we denote $\left(V^2\right)_i^{\pm} \equiv V_{i+}^2 \pm V_{i-}^2$, and the angle brackets mean the averaging over the roll period with the factor of either cosine (*'co'*) or sine (*'si'*) of the roll phase.

2. Acceleration dependent, nominal position torque due to Earth gravity gradient (NET sheet 1, box 1.2.1) In this case

$$\dot{EW} = -\frac{K_1}{I\omega_s}\langle (V^2)_c^- \rangle_{co} + \frac{K_4}{I\omega_s}\langle (V^2)_a^- - (V^2)_b^- \rangle_{si} = \frac{K_1 + K_4\sqrt{2}}{I\omega_s}\frac{A}{2}g_e$$

where the acceleration g_e is given by[11,4]

$$g_e = \frac{3\mu_e D}{16R^3} J_2\left(\frac{R_e}{R}\right)^2 \sin 2\delta$$

Here μ_e is GM value for Earth, D is the distance of the gyro in question to the unsupported gyro, J_2 is the Earth oblateness, R_e is the Earth radius, R is the orbital radius, and δ is the declination of the Guide Star.

Same as these two, all the final expressions of the drift rates are sums of ratios of products of various input parameters.

3.4 Torque Coefficients and the Rotor Shape

To finish with the SD torques, it remains only to describe how the torque coefficients are evaluated. Suppose that the exact shape of the rotor is known in terms of the spherical harmonics expansion,

$$r = R(\theta', \varphi') = \sum_{l=0}^{\infty}\sum_{m=-l}^{l} r'_{lm}Y'_{lm} \equiv \sum_{l,m} r'_{lm}Y'_{lm} \qquad (10)$$

Here the primed frame is fixed in the body of the rotor, r'_{lm} are called the rotor shape coefficients, and

$$Y'_{lm} = Y_{lm}(\theta', \varphi'); \quad Y_{lm}(\theta, \varphi) = \sqrt{\frac{2l+1}{4\pi}\frac{(l-m)!}{(l+m)!}}\, P_l^m(\cos\theta)e^{im\varphi};$$

$P_l^m(\cos\theta)$ are the Legendre functions. If the spin axis is oriented along the z' axis, which is, we emphasize, randomly fixed in the body when measuring the shape, i.e., if $\hat{\mathbf{s}} = \hat{\mathbf{z}}'$, then, as shown in Ref. 4,

$$K_i' = \kappa \sum_{l=even/odd,\, l>0} \hat{K}_i(l) r_{l0}' \quad i = 1, 2, \ldots, 15; \qquad \kappa = \frac{\varepsilon_0 r_g^2}{2d}, \tag{11}$$

where $\hat{K}_i(l)$ are known functions of l, such as

$$\hat{K}_1(l) = \sqrt{\frac{\pi}{2l+1}}\, P_l^1(0) \left[P_{l-1}\left(\sqrt{3}/2\right) - P_{l+1}\left(\sqrt{3}/2\right)\right], \qquad l > 0,$$

etc. (the complete list of expressions for the multipliers $\hat{K}_i(l)$ is found in Ref. 4). Note that no electrostatic torque acts on a perfect sphere, and, accordingly, no $l = 0$ term with $r_{00}' = \sqrt{4\pi} r_g$ is present in (11).

However, we need to know the torque coefficients for an *arbitrary* orientation of the spin axis in the body of the rotor, that is, when the unit spin vector $\hat{\mathbf{s}}$ is at any Euler angles α, β to the body fixed axis z'. The problem of their determination was solved in Ref. 12 with the result (* means complex conjugate):

$$K_i(\alpha, \beta) = \kappa \sum_{l=even/odd,\, l>0} \hat{K}_i(l) \left(\frac{4\pi}{2l+1}\right)^{1/2} \sum_{m=-l}^{l} r_{lm}'^{*} Y_{lm}(\alpha, \beta) \tag{12}$$

Analytical formulas for the torque coefficients including polhode motion of the spin axis are also obtained in Ref. 12; they prove that polhoding reduces torque coefficients and thus the total drift error.

The only difficulty with using the formula (12) in practice is that it is impossible to precisely measure the shape of the coated GP-B flight rotors without a serious damage to the rotor's surface. To circumvent that, three coated GP-B flight class gyros have been anodized, to strengthen their surface layer, and then their shape was measured, giving the set of the rotor shape coefficients r_{lm}', $l \leq 16$ for each ball. Then the torque coefficients as functions of the spin axis position in the rotor's body have been tabulated by means of formula (12). Thus, the ranges of their variation with the spin axis direction have been determined. The details of this study, as well as the theoretical analysis of the torque coefficients, are given in Ref. 12. Here we emphasize the following results: 1) the mass unbalance ($l = 1$) due to nonuniformity of the rotor coating extends the ranges of odd harmonics torque coefficients; 2) the centrifugal asphericity ($l = 2$) due to the equatorial bulge of the spinning rotor just shifts the ranges of even harmonics torque coefficients; 3) with the measured mass unbalance and centrifugal asphericity included, all the three rotors meet the GP-B requirements on the classical drift with a good margin (almost an order of magnitude for the plausible case). The last result is obtained, naturally, with the help of the NET.

4 Remarks on Support Independent Torques

In contrast with SD torques, it is impossible to develop a uniform theory of SI torques on the GP-B gyro, because of the difference in their physical origin. With this lack of a unifying structure behind, we face a large variety of SI torque derivations collected in Ref. 5. One can though detect some similarities in calculation of torques of the same physical nature, such as electrostatic, magnetic, differential damping torques, etc. Our first point here is that several electrostatic SI torques are expressed in terms of the same torque coefficients as the SD torques. For example, let us consider

1. *Housing fixed, misaligned position electrostatic torque due to rotor charging.* The drift rates caused by it are shown to be

$$\dot{EW} = -2NS_0 \frac{K_2 + 2K_5}{I\omega_s} V_r^2, \qquad \dot{NS} = 2EW_0 \frac{K_2 + 2K_5}{I\omega_s} V_r^2,$$

where V_r is the rotor to ground potential, and NS_0, EW_0 are the NS and EW misalignments, respectively.

As another example of the SI torques we present the largest of them, namely

2. *Housing fixed, misaligned position magnetic torque due to coupling of the London moment (M_L) with the local shield.* In this case the drift rates have been originally calculated in Ref. 13 as

$$\{\dot{EW}, \dot{NS}\}^T = 7.6 \times 10^6 \frac{\mu_0 M_L^2}{I\omega_s R_0^3} \{NS_0, EW_0\}^T,$$

where R_0 is the pick-up loop radius, the misalignments are in *marcsec*, and the numerical coefficient produces the answer in *marcsec/yr* as soon as all other quantities in the R.H.S. are in the MKS units.

An interesting example of calculation of the torque caused by eddy currents in normal metals surrounding the rotor due to the magnetic field of fluxons on the rotor's surface may be found in Ref. 14.

5 Discussion. Top Contributing Torques

The 'top' sheet of the NET in Fig. 1 shows that our estimates of the drift caused by the classical torques in both the worst and plausible case meets the GP-B requirement of 0.3 *marcsec/yr* with a good margin. It also demonstrates that the largest expected GP-B error is the measurement error (see Ref. 15 for details), and that the classical torques contribution is on the level of the uncertainty in the Guide Star proper motion.

From the sheet with the non-relativistic drift in Figs. 2–3 it is clear that the contribution of the SD torques dominates that of the SI torques, and that the nominal position, preload dependent torques lead among the former.

Finally, the top torques contributing to the drift in either direction are found in the table in Fig. 8, also taken form the NET. The selection includes all torques with the drift rate larger than 0.001 *marcsec/yr.* It is seen that only two of eight (EW) or nine (NS) top torques are support independent ones, and only one of them (given in the previous section) makes the third place in the case of the NS drift.

The torques listed in Fig. 8 provide well over 99% of the classical drift rate, and only seven of the fifteen torque coefficients are involved in them. Moreover, the contribution of only three largest torques is slightly below 99%, and just three torque coefficients and few other parameters are present in their expressions. On

Top Torques (plausible case)					
	Category	Sht.	Box	Name / Position	Dr. R. marcs/yr
				I. EW	
1	SD	4	1.2.1.	h_a-h_b at roll, Nominal	8.23E-02
2	SD	8	2.1.	Coupling with local shield, Misaligned, Housing fixed	6.70E-03
3	SD	5	1.1.1.	h_a+h_b , h_c D.C., Misaligned	6.85E-03
4	SD	1	1.2.1.	Gravity gradient, Earth, perp. to roll axis, Nominal	1.04E-02
5	SD	6	2.1.1.	h_a+h_b , h_c D.C. + Miscentering at roll	7.91E-03
6	S I	10	1.1.	Direct gravity gradient, Earth, Inertially fixed	3.56E-03
7	S I	7	1.1.	Var. in sensing voltage at roll, Nominal	2.15E-03
8	SD	6	2.2.1.	h_a+h_b , h_c at roll + D.C. Miscentering	1.23E-03
				II. NS	
1	SD	4	1.2.1.	h_a-h_b at roll, Nominal	8.23E-02
2	S I	8	2.1.	Coupling with local shield, Misaligned	6.70E-03
3	SD	5	1.1.1.	h_a+h_b , h_c D.C., Misaligned	6.85E-03
4	SD	4	1.2.2.3.	Gravity gradient, Earth, $g^2{}_a$-$g^2{}_b$, Nominal	6.31E-03
5	SD	3	2.2.2.	Difference in Miscenter. Amplitudes at ω_r +/- $2\omega_0$	2.99E-03
6	S I	7	1.1.	Variation in sensing voltage at roll, Nominal	2.15E-03
7	SD	6	2.1.1.	h_a+h_b , h_c D.C. + Miscentering at roll	4.29E-03
8	SD	4	2.2.1.	h_a+h_b , h_c at roll, Nominal	1.67E-03
9	SD	6	2.1.1.	h_a+h_b , h_c at roll + D.C. Miscentering	1.23E-03

Figure 8. Top Torques (plausible case)

the one hand, this fact makes our estimates of the torques 'robust', that is, less sensitive to parameter changes; on the other hand, it indicates priorities of the post-flight verification regarding non-relativistic torque errors of the experiment.

Acknowledgments

This work was supported by NASA grant NAS 8-39225 to Gravity Probe B. We would like to appreciate the efforts of many people who have contributed, over two decades, to the study of classical torques on the GP-B gyros: Penny Axelrad, Brian DiDonna, Francis Everitt, Dale Gill, Christian Gothier, Jeremy Kasdin, Jim Lockhart, Frane Marcelja, John Turneaure, Richard Vassar, Todd Walter.

References

1. G.M.Keiser in *These Proceedings.*
2. R.J.Adler, A.S.Silbergleit in *These Proceedings.*
3. C.W.F.Everitt, *Report on a Program to Develop a Gyro Test of General Relativity in a Satellite and Associated Control Technology*, Stanford University, Stanford, 1980.
4. G.M.Keiser, A.S.Silbergleit, *New Error Tree: Support Dependent Torques*, (GP-B doc. S0263, Stanford University, 1998).
5. G.M.Keiser, A.S.Silbergleit, *New Error Tree: Support Independent Torques*, (GP-B doc. S0264, Stanford University, 1996).
6. G.M.Keiser, M.I.Heifetz, A.S.Silbergleit, *New Live Error Tree*, (GP-B doc. S0292, Stanford University, 1998).
7. M.I.Heifetz, C.W.F.Everitt, G.M.Keiser, A.S.Silbergleit, in *Proc. of the 8th Marcel Grossman Meeting on General Relativity*, (in press).
8. W.R.Smythe, *Static and Dynamic Electricity*, 3rd edition, (Hemisphere Publ. Corp., New York, Washington, Philadelphia and London, 1989).
9. G.M.Keiser, *Suspension Torques on a Gimballed Electrostatically Supported Gyroscope and Requirements on the Gyroscope and Spacecraft for the Relativity Gyroscope Experiment*, (GP-B doc. S0019, Stanford University, 1985).
10. B.DiDonna, *Suspension Torque Prediction for the Relativity Gyroscope Experiment*, (Undergraduate Honors Thesis, Dept. of Physics, Stanford University, 1994).
11. G.M.Keiser, *Non-Relativistic Forces Acting on GP-B Gyroscopes*, (GP-B doc. S0269, Stanford University, 1996).
12. A.S.Silbergleit, *GP-B Rotor Shape and Torque Coefficients*, (GP-B doc. S0289, Stanford University, 1998).
13. L.B.Holdeman, J.T.Holdeman, Jr., J. Appl. Phys. **57**, 584 (1980).
14. G.M.Keiser, A.S.Silbergleit, Techn. Phys. **43**, 131 (1998).
15. M.I.Heifetz, G.M.Keiser, A.S.Silbergleit in *These Proceedings.*

DATA REDUCTION IN THE GRAVITY PROBE B EXPERIMENT: OPTIMAL ESTIMATION AND FILTERING

M.I.HEIFETZ, G.M.KEISER, A.S.SILBERGLEIT
Gravity Probe B, W.W.Hansen Experimental Physics Laboratory, Stanford University, Stanford, CA 94305-4085, USA
E-mail: misha@relgyro.stanford.edu

Data analysis is one of the most essential components of the Gravity Probe B (GP-B) experiment. We discuss three main problems that need to be resolve in the data reduction: 1)direct observation of the GP-B gyroscope's precession angle; 2)precise estimation of the gyroscope's geodetic and frame-draggig average drift rates; 3)gyroscope-telescope matching: elimination of the reference direction pointing error. We formulate these problems from the point of view of the modern control theory and describe our approach to the GP-B data analysis as the "bank" of filters that estimate the model-dependent system state vectors and calculate corresponding covariance matrices. For the problems 1),2),and 3) estimation recursive algorithms are obtained based on the two-step nonlinear filtering approach and these filters will be used in the GP-B data reduction.

1 Introduction

Gravity Probe B (GP-B) is an experiment designed to measure the geodetic and Lense-Thirring (frame-dragging) precessions relative to the fixed guide star, of a gyroscope placed in a polar orbit about the Earth. For a polar orbit these two precessions are orthogonal. The geodetic effect occurs in the plane of the orbit and has a magnitude of 6.6 arcseconds/year for a 650-km altitude orbit, while the orbit-averaged frame-dragging effect is perpendicular to the orbit plane and has a magnitude of 0.042 arcsec/year. The goal of the Gravity Probe B experiment is to measure these precessions to better than 0.001% and 0.2%, respectively.[1,2]

An on-board telescope keeps the spacecraft pointed toward a fixed guide star, providing the distant inertial reference from which changes in the gyroscope's spin axis direction are measured. Measurements are made by the SQUID-based readout system and sent to the ground via telemetry, where a data reduction scheme is used to estimate the relativistic drifts. The readout system scale factor is precisely calibrated during the experiment using the optical aberration of starlight due to the spacecraft motion around the Earth and the Sun. The GP-B spacecraft also rotates at a constant roll rate about the line of sight to the guide star. Rolling the satellite allows to move the science signal to the roll frequency, where the readout measurement noise (which has 1/f spectrum) is lower. Rolling also allows a single gyroscope pick-up loop and its readout to measure both the geodetic and frame-dragging precessions.

The GP-B gyroscope has been designed as a near perfect inertial instrument: Newtonian precession due to the classical torques is supposed to be less than 0.3 milliarcsec/year.[1] This means that the gyroscope's measured precession angle is assumed to be caused mainly by the relativistic effects. A detailed analysis of classical torques acting on the GP-B gyroscope is presented in Ref. 3.

In this paper we describe our approach to the GP-B data analysis as the 'bank' of filters that estimate the model-dependent system state vectors and calculate covariance matrices that represent statistical errors of the relativistic drift measurements due to the gyroscope and telescope readout noise.

2 GP-B Science Signal Models

The GP-B readout system is based on the effect of magnetic field generated by a spinning superconductor (London moment).[4] Precession of the gyroscope angular momentum results in the variations of the London moment, that is aligned with the instantaneous gyroscope spin axis. The science signal measured by the readout system represents the London moment magnetic flux through the gyroscope pick-up loop converted to the output voltage by the SQUID magnetometer.

The model of the gyroscope science signal can be represented as

$$z_G(t) = C_g\Big[\big(NS(t) - \varepsilon_1\big)\cos(\omega_r t + \delta\phi) - \big(EW(t) - \varepsilon_2\big)\sin(\omega_r t + \delta\phi)\Big]$$

$$+b(t) + C_g\Theta(t) + \nu(t), \tag{1}$$

where C_g is the readout system scale factor, $NS(t)$ and $EW(t)$ are North-South (in the orbit plane) and East-West (normal to the orbit plane) components of the gyroscope drift angle, ω_r is the spacecraft roll rate, $\delta\phi$ is the roll phase offset, ε_1 and ε_2 are optical aberration components, b is the readout system bias, and $\Theta(t)$ is the telescope body-fixed pointing error. Measurement noise $\nu(t)$ is assumed to be a white noise with zero mean and known spectral density.

The expressions for the aberration signals in the case of a circular orbit have the following form:[5]

$$\varepsilon_1 = \Lambda_o \sin(\omega_o t + \phi_o) + \Lambda_{a1}^{NS}\sin(\omega_a t + \phi_a) + \Lambda_{a2}^{NS}\cos(\omega_a t + \phi_a)$$

$$\varepsilon_2 = \Lambda_{a1}^{EW}\sin(\omega_a t + \phi_a) + \Lambda_{a2}^{EW}\cos(\omega_a t + \phi_a) \tag{2}$$

Here $\Lambda_o = v_s/c$, v_s is the spacecraft orbital velocity, ω_o is the orbital angular velocity, ϕ_o is the initial orbital phase, ω_a and ϕ_a are the angular velocity and initial phase of the Earth's motion around the Sun. Also

$$\Lambda_{a1}^{NS} = \frac{v_E}{c}\sin\delta\cos\lambda, \quad \Lambda_{a2}^{NS} = \frac{v_E}{c}\big(-\sin\delta\sin\lambda\cos\alpha + \cos\delta\sin\alpha\big)$$

$$\Lambda_{a1}^{EW} = -\frac{v_E}{c}\sin\lambda, \qquad \Lambda_{a2}^{EW} = -\frac{v_E}{c}\cos\lambda\cos\alpha, \tag{3}$$

where v_E is the orbital velocity of the Earth, δ and λ are the declination and right ascension of the Guide Star, and α is the angle between the ecliptic and equatorial plane of the Earth. Orbital and annual aberrations have different time signatures and therefore can be used as the major sources to provide observability of the scale factor and roll phase offset, in other words, to calibrate these science instrument parameters.

Gyroscope precession angles in the North-South and East-West directions can be modeled as

$$NS(t) = NS_0 + R_g t, \quad EW(t) = EW_0 + R_f t, \tag{4}$$

where NS_0 and EW_0 are North-South and East-West initial misalignments, R_g and R_f are the average drift rates over the observation time T_{obs}:

$$[R_g, R_f] = \frac{1}{T_{obs}} \int_0^{T_{obs}} [NS(t), EW(t)]dt \,. \tag{5}$$

The GP-B spacecraft and its attitude control system are designed to point the telescope continuously towards the Guide Star (or towards its apparent position) and to minimize the body-fixed pointing error $\Theta(t)$. The GP-B science telescope is used as a sensor for this servosystem and therefore the telescope output signal provides information about the pointing error Θ. This additional information from the science telescope can be expressed as

$$z_T(t) = C_T\Theta(t) + b_T + \mu(t) \tag{6}$$

where C_T, b_T, and μ are the telescope scale factor, bias and noise, respectively.

The accuracy required in the GP-B experiment demands resolving numerous problems of the 'optimal' data processing of the GP-B science signals. Here we discuss only four of them, namely:

- Estimation (measurement) of the relativistic geodetic and frame-dragging average drift rates of the GP-B gyroscope
- Direct observation (estimation) of the GP-B gyroscope's precession angle;
- Elimination of the reference direction pointing error by using the GP-B science telescope as a source of additional information (Gyroscope/Telescope scale factor 'matching');

3 Relativistic drift rate estimation

It is natural and convenient to consider the GP-B data reduction problem as an estimation problem in the time domain.[6] In this section we assume that there is no systematic pointing error Θ and the random pointing error has been already included in the readout system measurement noise.

Introduce the state vector:

$$\mathbf{X} = \Big[R_g, \; R_f, \; C_g, \; \delta\phi, \; NS_0, \; EW_0, \; b\Big]^T, \tag{7}$$

and assume that all components of this vector are either constant parameters or can be represented by the random walk stochastic process:

$$x_{k+1} = x_k + w_k, \tag{8}$$

where the process noise w_k is the Gaussian one with a zero mean and covariance matrix Q. The measurement equation (1) can then be written in a general form as

$$z_k = F(x_k, t_k, \varepsilon_1(t_k), \varepsilon_2(t_k), \omega_r) + \nu_k, \qquad k = 1, 2, \ldots, N \tag{9}$$

We can formulate now the following optimal estimation problem:

Problem 1

• Based on the measurements $\{z(t_1), z(t_2), \ldots\ldots, z(t_N),\}$, the known (measured or calculated) signals $\varepsilon_1(t)$, $\varepsilon_2(t)$, and the model (8)—(9), find the estimate $\hat{\mathbf{x}}$ of the state vector $\mathbf{x}$ that minimizes the least–square cost function

$$J = \frac{1}{2}\left[\sum_{k=1}^{N}(z_k - F(x_k,t_k))^T R_k^{-1}(z_k - F(x_k,t_k)) + \sum_{k=1}^{N-1} w_k^T Q_k^{-1} w_k\right] \tag{10}$$

It is important to emphasize that the problem (8)—(10) is a *nonlinear* least-square estimation problem and as it is well-known,[6,7] there is no exact solution for the optimal estimators in the nonlinear case.

As was shown by simulations, the standard nonlinear estimators, such as the extended Kalman filter (EKF) and the iterated extended Kalman filter (IEKF), give, for the above problem, a *biased* estimates of relativistic drifts R_g and R_g.[4] The reason is that both EKF and IEKF linearize the measurement equation (9) and the cost function (10). To overcome this difficulty, a new nonlinear recursive two–step estimator has been developed.[8]

Instead of linearizing the cost function, it breaks the minimization procedure into two steps. A new set of states is defined for the first–step filter using nonlinear combinations of the unknowns, so that the measurement equation becomes linear with respect to the new ones. The choice of first step states is dependent on the particular problem being addressed. The first–step linear problem can be solved optimally by exploiting a linear Kalman filter.[6] The second-step states are then calculated by treating the first step state estimates as 'new' measurements and by using an iterative Gauss–Newton algorithm.

The two–step estimation approach was applied to the problem (8)—(10). By choosing the first step states as

$$\begin{aligned} y = f(x) = \Big[&C_g\big(NS_0\cos\delta\phi - EW_0\sin\delta\phi\big),\ -C_g\big(NS_0\sin\delta\phi + EW_0\cos\delta\phi\big),\\ &C_g\big(R_g\cos\delta\phi - R_f\sin\delta\phi\big),\ -C_g\big(R_g\sin\delta\phi + R_f\cos\delta\phi\big),\\ &-C_g\cos\delta\phi,\ C_g\sin\delta\phi,\ b\Big]^T \end{aligned} \tag{11}$$

we convert the nonlinear measurement equation (1)—(4) into a linear one:

$$z = H(t)y + \nu, \tag{12}$$

where

$$\begin{aligned} H = \Big[&\cos\omega_r t,\ \sin\omega_r t,\ t\cos\omega_r t,\ t\sin\omega_r t,\\ &\varepsilon_1\cos\omega_r t + \varepsilon_2\sin\omega_r t,\ \varepsilon_1\sin\omega_r t - \varepsilon_2\cos\omega_r t,\ 1\Big]^T \end{aligned} \tag{13}$$

Applying now the two–step estimator[7] we obtain the following recursive estimation procedure.

First-Step Optimization

Measurement Update ($\hat{y}_k$ and $P_{y,k}$ are the optimal first-step estimate and corresponding covariance matrix after processing k data points):

$$\hat{y}_k = \bar{y}_k + P_{y,k} H_k^T R_k^{-1} (z_k - H_k \bar{y}_k) \tag{14}$$

$$P_{y,k} = \left(M_{y,k}^{-1} + H_k^T R_k^{-1} H_k \right)^{-1} \qquad k = 1, 2, \ldots, N \tag{15}$$

Time Update:

$$\bar{y}_k = \hat{y}_{k-1}$$

$$M_{y,k} = P_{y,k-1} + \left(\frac{\partial f}{\partial x} \right) Q \left(\frac{\partial f}{\partial x} \right)^T \Bigg|_{x=\hat{x}_{k-1}}; \tag{16}$$

the Jacobian $\partial f / \partial x$ of the nonlinear transformation $y = f(x)$ is calculated at the current point $\hat{x}_{k-1}$.

Second-Step Optimization

Measurement Update: ($\hat{x}_k$ and $P_{x,k}$ are the optimal first-step estimate and corresponding covariance matrix after processing k measurement data points; i is the iteration index)

$$\hat{x}_{k,i+1} = \hat{x}_{k,i} - P_{x,k,i}\, q_{k,i}^T; \tag{17}$$

$$P_{k,i+1}^{-1} = \left(\frac{\partial f}{\partial x} \right)^T P_{y,k}^{-1} \left(\frac{\partial f}{\partial x} \right) \Bigg|_{x=\hat{x}_{k,i}} \tag{18}$$

$$q_{k,i} = -\left(\hat{y}_k - f_k(\hat{x}_{k,i})\right)^T P_{y,k}^{-1} \left(\frac{\partial f_k}{\partial x_k} \right) \Bigg|_{x_k=\hat{x}_{k,i}} \tag{19}$$

Time Update:

$$\bar{x}_k = \hat{x}_{k-1}; \qquad M_{x,k} = P_{x,k-1} + Q_k \tag{20}$$

Matrices H, Q and R, as well as nonlinear transformation $y = f(x)$, are defined above.

The baseline data reduction two-step filter (estimator) has been used intensively for the general error analysis of the GP-B experiment. Measurement errors of the geodetic (R_g) and frame-dragging (R_f) drifts depend on several experimental operational parameters. Table 1 gives the list of the key parameters and provides the expected overall measurement error calculated for the "worst" and "plausible" cases that are defined and discussed in Ref. 3.

Parameters	Worst Case Value	Plausible Case Value
Experiment Duration (month)	12	13
Gyroscope Spin frequency (Hz)	100	120
Roll Period (min)	3	2
Readout noise (marcsec/($\sqrt{Hz}$)	190	190
Start Day	January 1	April 10
Fraction of orbit Guide Star is Visible	0.4	0.5
Total R_g Error (marcsec/yr)	0.379	0.187
Total R_g Error (marcsec/yr)	0.380	0.189

Table 1. Main Operational Parameters and Measurement Errors

4 Direct observation of the GP-B gyroscope's precession

One of the main technical goals of the GP-B Experiment and its Science Instrument is to observe the precession of a near perfect (that practically does not have a drift due to the classical Newtonian torques) earth-orbiting and free-falling gyroscope. By processing data it would be very important to observe the drift of the GP-B gyroscope "as is" making no assumptions about models of its precession. This problem differs from the problem discussed above where the matter of interest was a gyroscope's average drift rate. From the estimation point of view, direct observation of the gyroscope precession angle is a more difficult problem than the average drift rate estimation because now one has to determine time-varying functions $NS(t)$ and $EW(t)$ without making any assumption about their dynamic behaviour. Let us introduce two separate state vectors - gyroscope orientation vector $\mathbf{X_D}$ and the Science Instrument vector $\mathbf{X_I}$:

$$\mathbf{X_D} = [NS(t),\ EW(t)]^T; \qquad \mathbf{X_I} = [C_g,\quad \delta\phi,\quad b]^T. \tag{21}$$

The only assumption we are making here is that the state vector $\mathbf{X_I}$ is a constant vector and there is no systematic pointing error Θ. Precession angles $NS(t)$ and $EW(t)$ are expected to be slowly varying functions of time and without loss of generality could be considered as step-functions; i.e. constants within some interval (e.g. 12 hours) and varying between intervals.

It is convenient also to separate the orbital and annual components of the optical aberrations ε_1 and ε_2 and to rewrite the measurement equation (1) as

$$z(t) = C_g\Big[\big(NS(t) + l_1\cos(\omega_a t) + l_2\sin(\omega_a t) + A_0\cos(\omega_o t)\big)\cos(\omega_r t + \delta\phi) -$$

$$\big(EW(t) + l_3\cos(\omega_a t) + l_4\sin(\omega_a t)\big)\sin(\omega_r t + \delta\phi)\Big] + b + \nu(t), \tag{22}$$

where l_1–l_4 and A_0 are known constants. The problem of the gyroscope drift angle determination now can be formulated as follows:

Problem 2

• From the available data $Z_G = z(t_1), z(t_2), ..., z(t_N)$, and the measurement model (22) find the best estimate of the time-varying step-function $\hat{\mathbf{X}}_D(t)$ and the estimation error covariance matrix P_D.

The qualitative approach to Problem 2 implies that in order to determine $\hat{NS}(t)$ and $\hat{EW}(t)$ one has to calculate $\hat{C}_g$ and $\hat{\delta\phi}$. That can be done based on the known information about the optical aberration. By sequentially processing packets corresponding to 12 hours of data one can improve knowledge of the Science Instrument vector $\mathbf{X}_\mathbf{I}$, This process can be treated as the natural "calibration" of the Science Instrument. Then based on the "best" current knowledge of the $C_g, \delta\phi, b$ one can obtain the "best" estimates $\hat{NS}$ and $\hat{E}W(t)$ of the within 12 hour time intervals.

The quantitative solution is based on the two-step estimation approach. Introduce new first-step variables:

$$y_1(t) = C_g \cos\delta\phi NS(t) - C_g \sin\delta\phi EW(t), \tag{23.a}$$

$$y_2(t) = -C_g \sin\delta\phi NS(t) - C_g \cos\delta\phi EW(t), \tag{23.b}$$

$$y_3 = A_0 C_g \cos\delta\phi; \qquad y_4 = -A_0 C_g \sin\delta\phi, \tag{23.c}$$

$$y_5 = l_1 C_g \cos\delta\phi - l_3 C_g \sin\delta\phi, \qquad y_6 = l_2 C_g \cos\delta\phi - l_4 C_g \sin\delta\phi, \tag{23.d}$$

$$y_7 = -l_1 C_g \sin\delta\phi - l_3 C_g \cos\delta\phi, \qquad y_8 = -l_2 C_g \sin\delta\phi - l_4 C_g \cos\delta\phi, \tag{23.e}$$

$$y_9 = b,$$

For these variables the measurement equation becomes a linear one:

$$z(t) = H(t)y + \nu(t),$$

where $y = [y_1, y_2, ..., y_9]^T$, and

$$H(t) = \Big[\cos\omega_r t,\ \sin\omega_r t,\ \cos\omega_r t\cos\omega_o t,\ \sin\omega_r t\cos\omega_o t,$$

$$\cos\omega_r t\cos\omega_a t,\ \cos\omega_r t\sin\omega_a t,\ \sin\omega_r t\cos\omega_a t,\ \sin\omega_r t\sin\omega_a t,\ 1\Big]^T \tag{24}$$

Let $\hat{y}^L$ and ${P_y}^L$ be the estimate of the first step state vector y and covariance matrix P_y obtained after processing L 12-hour packets of science data $\mathbf{Z}$. The next, $L+1$ data packet, that has N_{L+1} data points, will be processed by using the linear recursive Kalman Filter.[6,7]

Starting from the initial condition (the estimate and covariance matrix after processing L data-packets)

$$\hat{y}_0 = y^L; \qquad S_0 = {P_y}^L/\sigma_\nu^2;$$

the optimal estimate $\hat{y}^{L+1}$ can be calculated recursively:

$$\hat{y}^{L+1} = \hat{y}_{k=N_{L+1}}.$$

$$\hat{y}_k = \hat{y}_{k-1} + S_k H_k^T \Big(z_k^{L+1} - H_k \hat{y}_{k-1} \Big); \qquad k = 1, 2, \ldots, N_{L+1}; \tag{25}$$

The first-step covariance matrix ${P_y}^{L+1}$ is given by

$$P_y^{L+1} = \sigma_\nu^2 S_{k=N_{L+1}},$$

$$S_k = S_{k-1} + S_{k-1} H_k^T \Big(I - H_k S_{k-1} {H_k}^T \Big)^{-1} H_k S_{k-1} \tag{26}$$

The next step is the calculation of the estimates $\hat{C}_g^{L+1}$, $\hat{\delta\phi}^{L+1}$, and the corresponding covariance matrix $P_{C,\delta\phi}^{L+1}$. To do that we consider estimates $[\hat{y}_3, \hat{y}_4, \ldots, \hat{y}_8]^{L+1}$ as the new "measurements" with the model equations (23.c,d,e) and a "measurement noise" defined by the matrix $R_{3,8}$ that consists of the corresponding elements of P_y^{L+1}. Quantities $\hat{C}_g^{L+1}$, $\hat{\delta\phi}^{L+1}$, and $P_{C,\delta\phi}^{L+1}$ can be obtained now by using the iterative Gauss-Newton search method described in the section 3 as the second step optimization. From the structure of the representation (23.c,d,e) one can see that the estimates $\hat{C}_g$ and $\hat{\delta\phi}$, are determined from the orbital and/or annual aberrations and can be treated as the "natural" calibration of the GP-B gyroscope readout system.

Going on to the estimation of the precesson angles $\hat{NS}^{L+1}$ and $\hat{EW}^{L+1}$ we again treat transformation (23.a, 23.b) as the measurement equations:

$$\hat{y}_1^{L+1} = [\hat{C}_g \cos\hat{\delta\phi}]^{L+1} NS - [\hat{C}_g \sin\hat{\delta\phi}]^{L+1} EW + e_1,$$

$$\hat{y}_2^{L+1} = -[\hat{C}_g \sin\hat{\delta\phi}]^{L+1} NS - [\hat{C}_g \cos\hat{\delta\phi}]^{L+1} EW + e_2, \tag{27}$$

where terms e_1 and e_2 represent the "measurement noise". Finally, the solution of the linear equations (27) concludes the estimation $\hat{NS}$ and $\hat{EW}$ and the covariance matrix $P_{NS,EW}$ of the gyroscope precession angle during the $(L+1)$'s 12-hour interval.

It is important to note here that each new 12-hour data packet provides additional information for 'better' estimates of C_g and $\delta\phi$. Therefore after determining the optimal estimates $\hat{C}_g$ and $\hat{\delta}\phi$ we can use the above described procedure to recalculate all previous estimates $(\hat{NS}, \hat{EW})^l, l = 1, 2, \ldots L+1$. This backward update should be considered as a specific case of the optimal smoothing technique.[6]

5 Gyroscope/Telescope scale factors 'matching' problem

Let us go back to the original structure of the gyroscope science signal (1) and assume now that there is a systematic pointing error $\Theta(t)$ that should be compensated in the Data Reduction. The science telescope provides an indirect information about this error as it has its own scale factor, bias and noise, as one can see from the equation (6).

We can formulate here the following estimation problem: How to optimally combine measurements $[Z_G]$ and $[Z_T]$, match the scale factors C_g and C_T, separate the pointing error $\Theta(t)$ from the gyroscope drift and to obtain the best estimate and covariance matrix of the state vector (7).

We make no assumptions about the dynamic behaiviour of $\Theta(t)$ and cannot solve the formulated problem simply by augmenting the state vector (7) with the telescope scale factor C_T and bias b_T and by using signals (1) and (6): the pointing error $\Theta(t)$ is an unobservable variable. To make it observable the GP-B spacecraft control system would be able to provide a slow harmonic dithering of the spacecraft around the roll axis. In other words, during the dithering the pointing error has the following stucture:

$$\Theta^{dither}(t) = \Theta(t) + A_d \sin(\omega_d t + \phi_d), \qquad (29)$$

where A_d and ϕ_d are unknown amplitude and phase of dithering, ω_d is a known dithering frequency that is naturally different from the roll, orbital and annual frequencies. Now the structure of measurement equations (1),(6) (where term $\Theta(t)$ should be replaced with the representation (28)) allows to implement the following estimation procedure:

1) By applying a linear Kalman Filter to the data $[Z_G]$ and $[Z_T]$, one can get the estimates of the variables $(C_g A_d)$ and $(C_T A_d)$;

2) The scale factors matching coefficient can be determined:

$$m = \frac{(C_g\hat{} A_d)}{(C_T\hat{} A_d)} \qquad (29)$$

3) Signals $Z_G(t)$ and $Z_T(t)$ will be combined together as a new signal Z_{GT}:

$$Z_{GT}(t) = Z_G(t) - mZ_T(t) \qquad (30)$$

4) The signal (30), that now does not have a pointing error component, will be used as an input signal for the filters that estimate the gyroscope precession motion and the relativistic drift rate as it has been discussed in the sections 3 and 4.

Acknowledgements

This work was supported by NASA grant NAS 8-39225 to Gravity Probe B. We are indebted to Prof. D.DeBra for his insightful discussions of a broad range of concepts related to GP-B Data Analysis.

References

1. C.W.F.Everitt, *The Stanford Relativity Gyroscope Experiment: History and Overview*, in *Near Zero,pp.587-639, W.H.Freeman and Co., New York, NY, 1980.*
2. Relativity Mission (GP–B). *12 Fundamental Science Requirements.* GP-B document T002, update July, 1998 .
3. A.S.Silbergleit *et al Classical Torque Errors in Gravity Probe B Experiment* in *These Proceedings.*

4. F.London, *Superfluids Vol 1: Macroscopic Theory of Superconductivity*,Wiley, New York, 1953.
5. G.T.Haupt, *Development and Experimental Verification of a Nonlinear Data Reduction Algorithm for the Gravity Probe B Rlativity Mission*,Ph.D. Thesis, Dept. Aero/Astro, Stanford University, 1996.
6. A.E.Bryson,Y.Ho, *Applied Optimal Control*,Hemisphere, New York, 1975.
7. T.F.Elbert , *Estimation and Control of Systems*,Van Nostrand Reinhold, New York, 1984.
8. G.Haupt,N.Kasdin,G.Keiser,B.Parkinson *Optimal Recursive Iterative Algorithm for Discrete Nonlinear Least-Squares Estimation*, Journ. of Guidance, Control and Dynamics, **19**, 3, 643–649, 1996.

Section D

Probing the Lense-Thirring Effect around the Earth with Twin Satellites

THE GRAVITOMAGNETIC FIELD AND ITS MEASUREMENT WITH THE LAGEOS SATELLITES

IGNAZIO CIUFOLINI
Dip. Aerospaziale, Univ. Roma "La Sapienza", Rome and IFSI-CNR, Rome, Italy;
E-mail: ciufoli@nero.ing.uniroma1.it

ERRICOS PAVLIS
Joint Center for Earth System Technology, Univ. of Maryland-Baltimore County, Baltimore, MD, USA

This paper is structured in four sections. In the first one we briefly describe the gravitomagnetic field and the Lense-Thirring effect in general relativity. In the second and third sections we review the method proposed to measure the Lense-Thirring effect by analyzing the orbits of the two laser-ranged satellites LAGEOS and LAGEOS II. We then report on our measurement of the Lense-Thirring effect by analyzing the data of LAGEOS and LAGEOS II with the orbital programs GEODYN-SOLVE, using the Earth's Gravitational Models JGM-3 and EGM-96, and the method explained in section 2. The first detection was obtained in 1995, the most accurate measurements were obtained in 1998 using EGM-96. Finally, in the fourth section, we briefly describe the proposed LARES experiment to measure the Lense-Thirring effect with accuracy of about 2%–3%.

1 Spinning Bodies, Gravitomagnetic Field and Lense-Thirring Effect

The contention of Ernst Mach that the inertial forces arise from rotations and accelerations with respect to the distant masses in the universe[1,2] was one of the inspiring ideas for Einstein to develop the general theory of relativity. General relativity satisfies only in part this "Mach principle", nevertheless, Einstein's theory predicts the occurrence of peculiar phenomena in the vicinity of a spinning body, caused by its rotation, which may be considered as weak manifestations of Mach's ideas on the origin of inertia and which have not yet been measured.[2–4]

When a clock that co-rotates very slowly around a spinning body returns to its starting point, it finds itself advanced relative to a clock kept there at "rest" (with respect to "distant stars"). Indeed, synchronization of clocks all around a closed path near a spinning body is not possible, and light co-rotating around a spinning body would take less time to return to a fixed point than light rotating in the opposite direction.[2] Similarly, the orbital period of a particle co-rotating around a spinning body would be longer than the orbital period of a particle counter-rotating on the same orbit. Furthermore, an orbiting particle around a spinning body will have its orbital plane "dragged" around the spinning body in the same sense as the rotation of the body, and small gyroscopes that determine the axes of a local, freely falling, inertial frame, where "locally" the gravitational field is "unobservable," will rotate with respect to "distant stars" because of the rotation of the body. This phenomenon—called "dragging of inertial frames" or, more simply, "frame dragging," as Einstein named it—is also known as the Lense-Thirring effect.[1–3] In Einstein's general theory of relativity, all of these phenomena are the result of the rotation of the central mass.

Rotation, inertia, and the "fictitious" inertial forces arising in a rotating system have been central issues and problems of mechanics since the time of Galileo and Newton.[4] Mach thought that the centrifugal forces were the result of rotation with respect to the masses in the universe, and Einstein's development of the general theory of relativity was influenced by Mach's ideas on the origin of inertia and inertial forces. Today, the level at which general relativity satisfies Mach's ideas on inertia is still debated and discussed. However general relativity satisfies at least a kind of "weak manifestation" of Mach's ideas: the dragging of inertial frames.[2] Indeed, in Einstein's gravitational theory, the concept of an inertial frame has only a local meaning, and a local inertial frame is "rotationally dragged" by mass-energy currents because moving masses influence and change the orientation of the axes of a local inertial frame (that is, the gyroscopes); thus, a current of mass such as the spinning Earth "drags" and changes the orientation of the gyroscopes with respect to the distant stars.

To understand these phenomena of general relativity associated with the rotation of a mass, one may use a formal analogy with the classical theory of electromagnetism. Newton's law of gravitation has a formal counterpart in Coulomb's law of electrostatics; however, Newton's theory has no phenomenon formally analogous to magnetism. On the other hand, Einstein's theory of gravitation predicts that the force generated by a current of electrical charge, described by Ampere's law, should also have a formal counterpart "force" generated by a current of mass. The detection and measurement of this "gravitomagnetic" force is the subject of this paper.

In general relativity, for a localized, stationary, mass-energy distribution, in the weak-field and slow-motion limit, the three "vector" components of the metric tensor are given by: $\mathbf{h} \cong -2(\mathbf{J} \times \mathbf{x})/\mathbf{r^3}$, where $\mathbf{h}$ is known as the gravitomagnetic potential. The gravitomagnetic field $\mathbf{H}$ is given by $\mathbf{H} = \nabla \times \mathbf{h} \cong \mathbf{2}\, [\frac{(\mathbf{J} - 3(\mathbf{J} \cdot \hat{\mathbf{x}})\, \hat{\mathbf{x}})}{\mathbf{r^3}}]$. By using the geodesic equation, in the weak-field and slow-motion limit, one has then for a test particle: $m\, \frac{d^2\mathbf{x}}{dt^2} \cong m\, (\mathbf{G} + \frac{d\mathbf{x}}{dt} \times \mathbf{H})$, where $\mathbf{G} \cong -\frac{M}{\mathbf{r}^2}\, \hat{\mathbf{x}}$ is the standard Newtonian acceleration. Thus, a central object with angular momentum $\mathbf{J}$, via the second term in the "force", $\dot{\mathbf{x}} \times \mathbf{H}$, drags the orbital plane (and the orbital angular momentum) of a test particle in the sense of rotation of the central body. This dragging of the whole orbital plane is described by the formula (2), below, for the rate of change of the longitude of the nodes. Similarly, by integrating the above equation of motion of the test particle, one finds the formula (3), below, for the secular rate of change of the longitude of the pericenter. On a gyroscope with angular momentum $\mathbf{S}$ there is a torque $\tau \cong (1/2)\ \mathbf{S} \times \mathbf{H} = d\mathbf{S}/dt \equiv \dot{\mathbf{\Omega}} \times \mathbf{S}$, therefore, the gyroscope precesses with respect to an asymptotic inertial frame with angular velocity $\dot{\mathbf{\Omega}}$ given by formula (1) below; on the gyroscope there is also a force exerted by the gravitomagnetic field.

Thus, an external current of mass, such as the spinning Earth, "drags" and changes the orientation of gyroscopes. Indeed, a test gyroscope precesses with respect to "an asymptotic inertial frame" with angular velocity:

$$\dot{\mathbf{\Omega}} = -\frac{1}{2}\mathbf{H} = \frac{[-\mathbf{J} + 3\,(\mathbf{J} \cdot \hat{\mathbf{x}})\,\hat{\mathbf{x}}]}{|\mathbf{x}|^3}, \tag{1}$$

where **J** is the angular momentum of the central object, **H** its gravitomagnetic field generated by **J**, and **x** is the gyroscope's position vector. This formula quantifies the Lense–Thirring effect for a gyroscope.[1,2]

The whole orbital plane of a test particle is itself a kind of enormous gyroscope (for a motion under a central force) dragged by the gravitomagnetic field. Indeed, the orbit of a test particle around a central body with angular momentum **J** has a secular rate of change of the longitude of the line of the nodes (intersection between the orbital plane of the test particle and the equatorial plane of the central object), discovered by Lense-Thirring:[3]

$$\dot{\Omega}^{Lense-Thirring} = \frac{2\mathrm{J}}{[a^3\,(1-e^2)^{3/2}]}, \tag{2}$$

where a is the semimajor axis of the test particle, and e its orbital eccentricity. The pericenter of an orbiting test particle is also a kind of enormous gyroscope (for a motion under a central force $\sim 1/r^2$). Indeed, the argument of the pericenter, ω, that is the angle from the equatorial plane to the pericenter, and thus the longitude of the pericenter $\tilde{\omega} = \Omega + \omega$ (determining the Runge-Lenz vector), have a secular rate of change:

$$\dot{\tilde{\omega}}^{Lense-Thirring} = \frac{2J\,(\hat{\mathbf{J}} - 3\,cos\,I\,\hat{\mathbf{l}})}{[a^3(1-e^2)^{3/2}]}, \tag{3}$$

where $\hat{\mathbf{l}}$ is the orbital angular momentum, unit vector, of the test particle, and I its orbital inclination (angle between the orbital plane and the equatorial plane of the central object): $cos\, I \equiv \hat{\mathbf{J}} \cdot \hat{\mathbf{l}}$.

In astrophysics, a supermassive spinning black hole and its gravitomagnetic field might explain the constant orientation of the long jets of matter ejected from the cores of quasars and active galactic nuclei.[5] These jets have emission times that may reach millions of years.

To characterize the gravitomagnetic field generated by the intrinsic angular momentum of a body and the Lense-Thirring effect, and to distinguish it from other relativistic phenomena such as the de Sitter effect–due to the motion of a local inertial frame through a static gravitational field–, one may give a description of this gravitomagnetic field in terms of spacetime-curvature invariants.[2,6] The pseudoinvariant $^*\mathbf{R}\cdot\mathbf{R} = \frac{1}{2}\,\varepsilon^{\alpha\beta\sigma\rho}\,R_{\sigma\rho}{}^{\mu\nu}\,R_{\alpha\beta\mu\nu}$, built from the Riemann tensor **R** and its dual $^*\mathbf{R}$, gives an invariant characterization of this gravitomagnetic phenomenon since it is non-zero in the field of a central body if and only if the body is rotating. Indeed the pseudoinvariant $^*\mathbf{R}\cdot\mathbf{R} \sim \frac{JM}{r^7} cos\theta$ is proportional to the angular momentum of the central body. Therefore, using this invariant one can determine whether or not there is a contribution of angular momentum, and mass currents, to the spacetime geometry and curvature. Thus, one may describe this gravitomagnetic effect, as that phenomenon of nature, that is an "absolute" (coordinate-independent) feature of general relativity, such that *spacetime curvature is generated by the spin of a body.*[2,6]

Thus, the direct measurement of the Lense-Thirring effect described in sections 2 and 3 confirms one of the remaining fundamental predictions of general relativity, that a current of mass-energy, such as a spinning mass, as a result of its mass-energy-motion, changes the geometry of the universe[2] by generating spacetime curvature.

2 Measurement of the Lense-Thirring Effect with LAGEOS and LAGEOS II

Our detection and measurement of the Lense-Thirring effect, one of the phenomena due to the gravitomagnetic field, was obtained by using the satellite laser-ranging data of the satellites LAGEOS (LAser GEOdynamics Satellite, NASA) and LAGEOS II (NASA and ASI, the Italian Space Agency) and the Earth gravitational field models, JGM-3 and EGM-96.

The measurement of distances has always been a fundamental issue in astronomy, engineering, and science in general. So far, laser ranging has been the most accurate technique for measuring the distances to the moon and artificial satellites.[7,8] Short-duration laser pulses are emitted from lasers on Earth, aimed at the target through a telescope, and then reflected back by optical cube-corner retroreflectors on the moon[9] or the artificial satellite,[10] such as LAGEOS. By measuring the total round-trip travel time, one can determine the distance to a retroreflector on the moon with an accuracy of about 2 cm and to the LAGEOS satellites with a few millimeters accuracy.

The LAGEOS satellites are heavy brass and aluminum satellites, of about 406 kg weight, completely passive and covered with retroreflectors, orbiting at an altitude of about 6,000 km above the surface of Earth. LAGEOS, launched in 1976 by NASA, and LAGEOS II, launched by NASA and ASI in 1992, have an essentially identical structure but they have different orbits. The semimajor axis of LAGEOS is $a \cong 12,270$ km, the period $P \cong 3.758$ hr, the eccentricity $e \cong 0.004$, and the inclination $I \cong 109.9°$. The semimajor axis of LAGEOS II is $a_{II} \cong 12,163$ km, the eccentricity $e_{II} \cong 0.014$, and the inclination $I_{II} \cong 52.65°$.

We analyzed the laser-ranging data using the principles described in Ref. 11 and adopted the IERS conventions[12] in our modeling, except that, in the 1998 analysis, we used the recently released static and tidal EGM-96 model.[13] Error analysis of the LAGEOS orbits indicated that the EGM-96 errors can only contribute periodic root-sum-square errors of 2 to 4 mm radially, and in all three directions they do not exceed 10 to 17 mm. The initial positions and velocities of the LAGEOS satellites were adjusted for each 15-day batch of data, along with a constant and a variable acceleration in the along-track direction (to account for the residual drag at the satellites' altitude) and very small variations in their reflectivities. Solar radiation pressure, Earth albedo, and anisotropic thermal effects were also modeled.[14–17] In modeling the thermal effects, the orientation of the satellite spin axis was obtained from Ref. 18. Lunar, solar, and planetary perturbations were also included in the equations of motion, formulated according to Einstein's general theory of relativity with the exception of the Lense-Thirring effect, which was purposely set to zero. All of the tracking station coordinates were adjusted (accounting for tectonic motions) except for those defining the TRF terrestrial reference frame. Polar motion was also adjusted, and Earth's rotation was modeled from the very long baseline interferometry-based series SPACE96.[19] We analyzed the orbits of the LAGEOS satellites using the orbital analysis and data reduction software GEODYN II.[20]

The node and perigee of LAGEOS and LAGEOS II are dragged by the Earth's angular momentum. From the Lense-Thirring formula (2), we get $\dot{\Omega}_I^{Lense-Thirring} \cong 31$ mas/yr and $\dot{\Omega}_{II}^{Lense-Thirring} \cong 31.5$, where mas is a millisecond of arc. The argument of pericenter (perigee in our analysis), ω, also has a Lense-Thirring drag. From formula (3) we get for LAGEOS: $\dot{\omega}_I^{Lense-Thirring} \cong 32\, mas/yr$, and for LAGEOS II: $\dot{\omega}_{II}^{Lense-Thirring} \cong -57\, mas/yr$. The nodal precessions of LAGEOS and LAGEOS II can be determined with an accuracy of the order of 1 mas/year. Over our total observational period of about 4 years, we obtained a root mean square of the node residuals of about 4 mas for LAGEOS and about 7 mas for LAGEOS II. For the perigee, the observable quantity is the product $e \cdot a \cdot \dot{\omega}$, where e is the orbital eccentricity of the satellite. Thus, the perigee precession $\dot{\omega}$ for LAGEOS is difficult to measure because its orbital eccentricity e is $\sim 4 \times 10^{-3}$. The orbit of LAGEOS II is more eccentric, with $e \sim 0.014$, and the Lense-Thirring drag of the perigee of LAGEOS II is almost twice as large in magnitude as that of LAGEOS. Over about 4 years, we obtained a root mean square of the residuals of the LAGEOS II perigee of about 25 mas, whereas the total Lense-Thirring effect on the perigee is, over 4 years, $\cong -228$ mas.

To precisely quantify and measure the gravitomagnetic effects we have introduced the parameter μ that is by definition 1 in general relativity[2] and zero in Newtonian theory.

The main error in this measurement is due to the uncertainties in the Earth's even zonal harmonics and their time variations. The unmodeled orbital effects due to the harmonics of lower order are comparable to, or larger than, the Lense-Thirring effect. However, by analyzing both the JGM-3 and the EGM-96 models with their uncertainties in the even zonal harmonic coefficients, and by calculating the secular effects of these uncertainties on the orbital elements of LAGEOS and LAGEOS II, we find[21] that the main sources of error in the determination of the Lense-Thirring effect are concentrated in the first two even zonal harmonics, J_2 and J_4. We can, however, use the three observable quantities $\dot{\Omega}_I$, $\dot{\Omega}_{II}$ and $\dot{\omega}_{II}$ to determine μ,[21,22] thereby avoiding the two largest sources of error, those arising from the uncertainties in J_2 and J_4. We do this by solving the system of the three equations for $\delta\dot{\Omega}_I$, $\delta\dot{\Omega}_{II}$ and $\delta\dot{\omega}_{II}$ in the three unknowns μ, J_2 and J_4, obtaining:

$$\delta\,\dot{\Omega}^{Exp}_{LAGEOS I} + c_1\,\delta\,\dot{\Omega}^{Exp}_{LAGEOS II} + c_2\,\delta\,\dot{\omega}^{Exp}_{LAGEOS II} = \mu\,(31 + 31.5\,c_1 - 57\,c_2)\,mas/yr{+}other\ errors \cong \mu\,(60.2\,mas/yr)\,, \quad (4)$$

where $c_1 = 0.295$ and $c_2 = -0.35$. Equation (4) for μ does not depend on J_2 and J_4 nor on their uncertainties; thus, the value of μ that we obtain is unaffected by the largest errors, which are due to δJ_2 and δJ_4, and is sensitive only to the smaller errors due to δJ_{2n}, with $2n \geq 6$.

Similarly, regarding tidal, secular, and seasonal changes in the geopotential coefficients, the main effects on the nodes and perigee of LAGEOS and LAGEOS II, caused by tidal and other time variations in Earth's gravitational field,[23,24] are due to changes in J_2 and J_4. However, the tidal errors in J_2 and J_4 and the

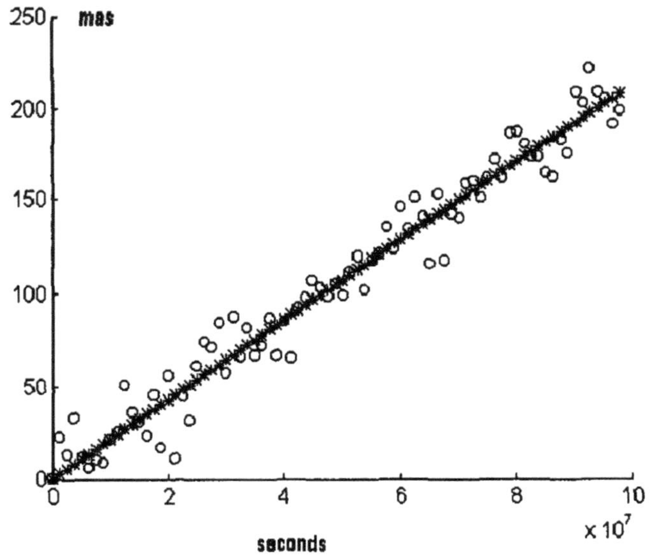

Figure 1. Sum of the residuals (o) of the nodes of LAGEOS and LAGEOS II and perigee of LAGEOS II from November 1992 to December 1995, using the Earth's gravitational model JGM-3 and with the method described in the text. On the vertical axis, we plotted: (node residuals of LAGEOS)+0.295(node residuals of LAGEOS II)- 0.35(perigee residuals of LAGEOS II). The best-fit line (*) shown through these combined residuals has a slope of about $\mu_{Measured} \cong 1.1$ (compared with $\mu_{GR} \equiv 1$ in general relativity). Together with the slope we fitted and removed ten periodic signals from the combined residuals. The units on the vertical axis are milliseconds of arc (mas), and on the horizontal axis seconds of time. See Fig. 2 to 5 in the next pages.

errors resulting from other unmodeled, medium and long period, time variations in J_2 and J_4, including their secular and seasonal variations, are eliminated by our combination of residuals of nodes and perigee. In particular, most of the errors resulting from the 18.6- and 9.3-year tides, associated with the lunar node, are eliminated in our measurement. An extensive discussion of the various error sources that can affect the result is given in Ref. 24; only a brief discussion of the error sources is given here.

In Fig. 1, we show the linear combination of the residuals of the nodes of LAGEOS and LAGEOS II and perigee of LAGEOS II according to Eq.(4) to eliminate the δJ_2 and δJ_4 errors, using the Earth gravitational model JGM-3, over about 3.1-year period, and after removing ten small periodic residual signals and the small observed inclination residuals.[24]

In Fig. 2, we display the improved analysis[25] (obtained with the linear combination of the residuals of the nodes of LAGEOS and LAGEOS II and perigee of LAGEOS II according to Eq. (4)) using the more recent static and tidal Earth gravitational model EGM-96, we also refined the non-gravitational perturbations model, the total period of observations was of 4 years, longer of about 1 year with respect to the observational period corresponding to Fig. 1. We only removed four small periodic residual signals and the small observed inclination residuals. The removal of the periodic terms was achieved by a least squares fit of the residuals

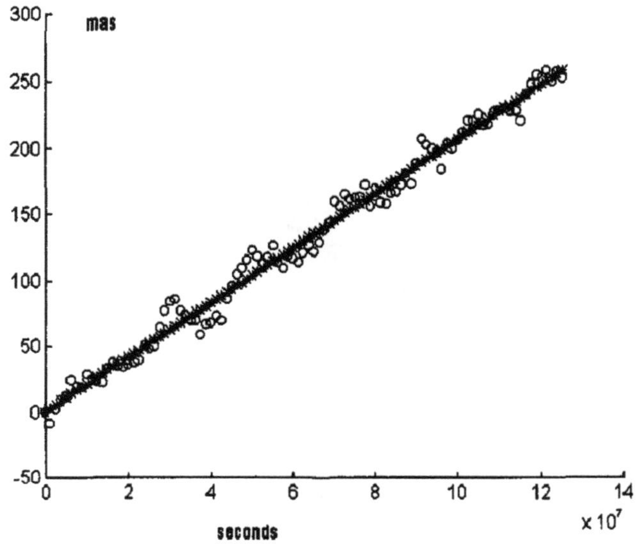

Figure 2. Sum of the residuals (o) of the nodes of LAGEOS and LAGEOS II and perigee of LAGEOS II from January 1993 to January 1997. On the vertical axis, we plotted: (node residuals of LAGEOS)+0.295(node residuals of LAGEOS II)- 0.35(perigee residuals of LAGEOS II). The best-fit line (*) shown through these combined residuals has a slope of about 66 mas/year (the total integrated effect corresponds to about 16 m at the LAGEOS altitude); that is, $\mu_{measured} \cong 1.1$ (compared with $\mu_{GR} \equiv 1$ in general relativity). The standard deviation of the slope is about 2 mas/year, and the root mean square of the post-fit residuals is about 9 mas. Because of modeling errors (secular and periodic) and random errors, we estimate the total error in our measurement of μ to be $\lesssim$20%–30% of μ. Together with the slope we fitted and removed from the combined residuals four small periodic signals, with 1044-, 820-, 569-, and 365.25-day periods, corresponding to, respectively, the nodal period of LAGEOS, the perigee and nodal periods of LAGEOS II, and 1 year.

using a secular trend and four periodic signals with 1044-, 820-, 569-, and 365.25-day periods, corresponding to, respectively, the nodal period of LAGEOS, the perigee and nodal periods of LAGEOS II, and 1 year. The 820-day period is the period of the main odd zonal harmonics perturbations of the LAGEOS II perigee; the 1044- and 569-day periods are the periods of the main tidal orbital perturbations, with l=2 and m=1, which were not eliminated using Eq. (4). Some combinations of these frequencies correspond to the main non-gravitational perturbations of the LAGEOS II perigee. We stress that the present analysis using EGM-96 and its accurate tidal model is substantially independent of the removed signals, whereas the previous analysis,[24] corresponding to Fig. 1, depended on the periodic terms included in the fit. In othe r words, our value (Fig. 2) of the secular trend does not significantly change by fitting additional periodic perturbations, and indeed, even the fit of the residuals with a secular trend only, with no periodic terms, just increases the slope by less than 10 %. Nevertheless, in this case, the root mean square of the post-fit residuals increases by about four times with respect to Fig. 2.

Figure 3 shows the fit of the residuals obtained as in Fig. 2 but with removal of only three periodic signals with 1044-, 820-, and 569-day periods.

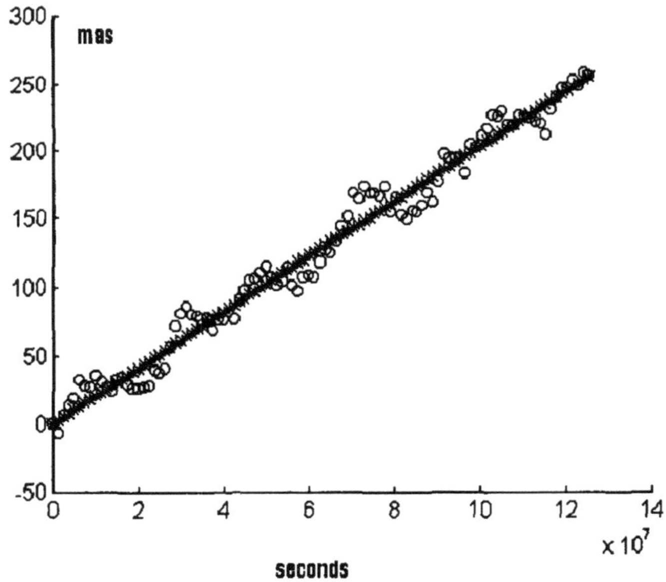

Figure 3. Same as in Fig. 2 but by just removing three small periodic signals, with 1044-, 820-, and 569-day periods, corresponding to, respectively, the nodal period of LAGEOS, and the perigee and nodal periods of LAGEOS II. The best-fit line shown through these combined residuals has a slope of about $\mu_{Measured} \cong 1.1$.

Our best-fit straight lines of Figs. 1–3, through the combined residuals of nodes and perigee, have respectively a slope $\mu_1{}^{Measured} \cong 1.1$, $\mu_2{}^{Measured} \cong 1.1 \pm 0.03$, and $\mu_3{}^{Measured} \cong 1.1 \pm 0.03$, where 0.03 is the standard deviation of the fits corresponding to Figs. 2 and 3 using the EGM-96 gravitational model. This combined, measured, gravitomagnetic perturbation of the satellites' orbits corresponds, in a 1-year period, to about 16 m at the LAGEOS altitude, that is, about 265 mas. The root mean square of the post-fit combined residuals corresponding to Figs. 2 and 3 is about 9 mas. Our total error is estimated to be of the order of 30%–40% of μ_{GR} corresponding to Fig. 1, and of the order of 20%–30% of μ_{GR} corresponding to Figs. 2 and 3. Using the JGM-3 covariance matrix, we found the errors due to the uncertainties in the even zonal harmonics J_{2n}, with $2n \geq 6$, to be: $\delta\,\mu^{even\ zonals: J_{2n} \geq J_6} \lesssim 17\,\%$ of μ_{GR}, and using the EGM-96 covariance matrix: $\delta\,\mu^{even\ zonals: J_{2n} \geq J_6} \lesssim 13\,\%$ of μ_{GR}. The errors in the modeling of the perigee rate of LAGEOS II due to the uncertainties in the odd zonal harmonics J_{2n+1}, are, with EGM-96: $\delta\,\mu^{odd\,zonals} \lesssim 2\,\%$ of μ_{GR}. Using the EGM-96 tidal model, we estimated the effect of tidal perturbations and other variations of Earth gravitational field to be $\delta\,\mu^{tides+} \lesssim 4\,\%$ of μ_{GR}. On the basis of analyses[24] of the non-gravitational perturbations—in particular, those on the perigee of LAGEOS II—we found $\delta\,\mu^{non-gravitational} \lesssim 13\%$–$20\%$ of μ_{GR}, including uncertainties in the modeling of the satellites' reflectivities, and the error due to uncertainties in the orbital inclinations of LAGEOS and LAGEOS II was estimated as $\delta\,\mu^{inclination} \lesssim 5\,\%$ of μ_{GR}. Taking into account all of these error sources, we arrived at a total root-sum-square error $\lesssim 20\%$–30% of μ_{GR}. Therefore, over an observational period of 4 years and using EGM-96, we determined $\mu^{Measured} = 1.1 \pm 0.3$, where 0.3 is the estimated total uncertainty due to all the error sources.

Based on the analysis of the orbits of the laser-ranged satellites LAGEOS and LAGEOS II, we conclude that the gravitomagnetic or Lense-Thirring effect exists, and its value is within 10 %, plus or minus $\sim$ 20%–30%, of what is predicted by Einstein's theory of general relativity, that is, μ_{GR}=1.

3 Testing Our Method to Measure the Lense-Thirring Effect and Its Error Budget

A basic concern in our analyses was to estimate the total error in our measurement of the Lense-Thirring effect and, in particular, to quantify the potential absorption, by adjusting the reflectivity of LAGEOS II, of part of the Lense-Thirring effect and of the secular effects due to J_2 and J_4 on the perigee of LAGEOS II.

It is important to observe that we have analyzed the laser-ranging data of the LAGEOS satellites using a method different to that used by CSR with UTOPIA. Indeed, in our analysis we have allowed GEODYN-SOLVE to adjust the reflectivity of LAGEOS II by a very small amount: the average of the differences between the post-fit values of the reflectivity, C_R^{II}, of LAGEOS II and the a-priori values of C_R^{II} are of the order of $10^{-6}\,C_R^{II}$. An average variation of this size in C_R^{II} corresponds to an acceleration on the LAGEOS II satellite of the order of only $4 \times 10^{-15}\,m/s^2$, and thus is by far too small to significantly absorb, or increase, the value of the Lense-Thirring effect on the perigee of LAGEOS II. However, in order to obtain an accurate fit of the orbits of LAGEOS II (and LAGEOS), we have allowed GEODYN-SOLVE to adjust the along-track acceleration (along the velocity vector of the satellites), a constant part and a once-per-revolution variable one. The small changes in the once-per-revolution along-track acceleration have the potentiality to absorbe, or increase, the Lense-Thirring effect on the perigee, however: (a) a fundamental consideration, (b) a basic test, and (c) a blind-test simulation show that this potential effect was *not* significantly affecting our analysis and our measurement of the Lense-Thirring effect.

(a) The Lense-Thirring shift of node and perigee of an Earth satellite is due to the gravitomagnetic field **H** and to the corresponding "force" $m\,\frac{d^2\mathbf{x}}{dt^2} \cong m\,(\frac{d\mathbf{x}}{dt} \times \mathbf{H})$, as explained in section 1, valid in the weak-field and slow-motion approximation. This gravitomagnetic force is formally identical to the Lorentz force on a moving charge, q, due to a magnetic induction **B** in electrodynamics: $m\,\frac{d^2\mathbf{x}}{dt^2} = (\frac{q}{c})\,(\frac{d\mathbf{x}}{dt} \times \mathbf{B})$. Thus, as it is well known, the Lorentz force is always orthogonal to the velocity of the moving charge; similarly the gravitomagnetic force on LAGEOS and LAGEOS II is always orthogonal to the velocity of the moving satellite. Therefore, by adjusting the component of the acceleration of LAGEOS II along the velocity vector of the satellite (along-track acceleration) one cannot absorb, or generate, any acceleration, such as the gravitomagnetic acceleration (that is always orthogonal to the along-track acceleration) without producing anomalous effects on the orbit of LAGEOS II. Indeed, a ficticious along-track acceleration, that GEODYN would generate to absorb an acceleration orthogonal to the along-track one, would produce anomalous effects on the Keplerian orbital elements that are affected by the along-track accelerations, that is, eccentricity, mean anomaly and semi-major axis. This basic consideration applies also to the secular effects of J_2 and J_4 on the LAGEOS orbits. Thus, by trying to absorb the gravitomagnetic force, orthogonal to the satellite velocity, by an along-track acceleration, this would generate anomalous changes in the mean motion, eccentricity and semi-major axis, and thus would generate anomalous orbital residuals.

In order to support this basic consideration we have performed a test and a preliminary simulation, explained in the next paragraphs (b) and (c).

(b) Testing our solution for μ using node, perigee, mean anomaly, eccentricity and semimajor axis of LAGEOS II, and node of LAGEOS.

To test our measurement of the Lense-Thirring effect with the residuals of node and perigee of LAGEOS II and node of LAGEOS, on the basis of our previous consideration (a), we have also analyzed the residuals of eccentricity, mean anomaly and semimajor axis of LAGEOS II. Thus, we have produced a solution for μ by using all these orbital elements. Indeed, using the residuals of these orbital elements, we have a system of equations (for $\delta\Omega_I$, $\delta\Omega_{II}$, $\delta\omega_{II}$, δe_{II}, δM_{II}, and δa_{II}) in the unknowns: Lense-Thirring effect, δJ_2 and δJ_4 errors, and a constant and a variable, once-per-revolution, along-track acceleration. Therefore, the solution of this system for μ and the corresponding fit (shown in Fig. 4) are *completelyindependent* from

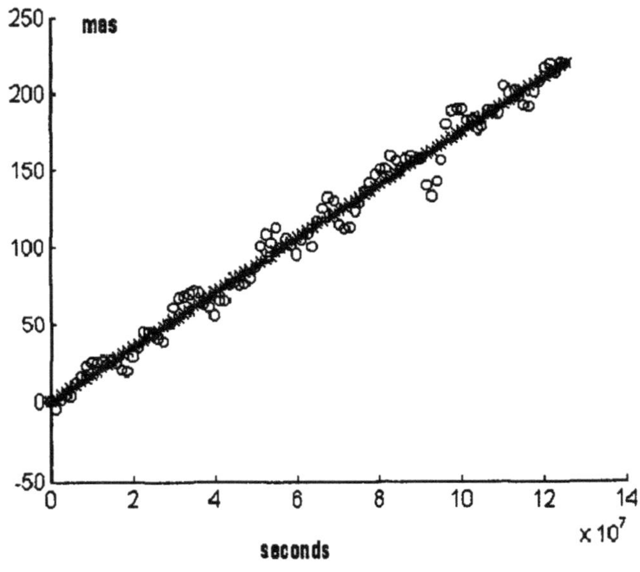

Figure 4. The fit of the parameter μ using the combined residuals of six orbital elements $\delta\Omega_I$, $\delta\Omega_{II}$, $\delta\omega_{II}$, δe_{II}, δM_{II}, and δa_{II}. This fit and the corresponding value of μ are completely independent from the adjusted along-track accelerations of LAGEOS II and from the errors δJ_2 and δJ_4 (see section 3). Together with the secular trend shown in this figure we have contemporarily fitted four periodic signals with the periods of perigee and node of LAGEOS II, node of LAGEOS and 365-day. The measured value of the Lense-Thirring effect obtained with this method is $\mu_{Measured} \cong 1$.

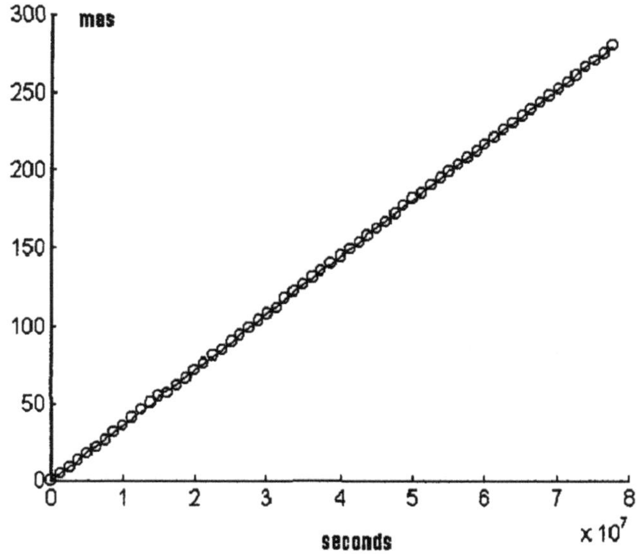

Figure 5. Result of a *preliminary* simulation and blind-test analysis. It is shown the fit (solid line) of the parameter μ using about two and a half years of simulated data (o) of the LAGEOS satellites; their simulated orbits were obtained by a model of perturbations in which the a-priori value of the Lense-Thirring effect was *twice* its general relativistic value. The shown secular trend of the combined post-fit residuals is more than 90 % of the a-priori Lense-Thirring value (twice the general relativistic effect) set up in the simulation of the orbits.

the adjusted accelerations and from the errors δJ_2 and δJ_4. The result of our solution with the residuals of these orbital elements, and the corresponding fit for μ, is shown in Fig. 4: the measured value of the Lense-Thirring effect is $\mu_{Measured} \cong 1$. Thus, this further, new, analysis and the corresponding fit test and confirm our previous result.[25]

(c) Preliminary Blind-Test-Simulation.

Finally, following a suggestion by the group at CSR of UT-Austin,[26] we have performed a *preliminary* simulation and blind-test analysis. Using GEODYN we have simulated about two and a half years of data of the LAGEOS satellites with a model of perturbations in which the a-priori value of the Lense-Thirring effect was *twice* its general relativistic value. Then, we have performed our standard analysis explained in section (2), using the same model of perturbations but without the Lense-Thirring effect. In figure (5) is shown the result of our fit of the residuals of node and perigee of LAGEOS II and node of LAGEOS, obtained with the analysis of the simulated data of the orbits of these two satellites. By our combination (4) of the residuals of their orbital elements we found a value of μ larger than 1.8 times the Lense-Thirring effect, that is, more than 90 % of the value a-priori set up in our simulation of their orbits and of the corresponding laser-ranging data. We observe that this test was just a preliminary blind-test-simulation; indeed the simulated orbits and the corresponding laser-ranging data have to be simulated by using a model of perturbations including variations in all the relevant parameters (such as J_2 and J_4) within the known uncertainties. Nevertheless, this preliminary simulation was important in that it showed the consistency of our method and of the preliminary error budget.

4 LARES, LAser RElativity Satellite

4.1 Scientific Objectives of the LARES experiment

The main scientific objectives of the LARES mission[27] are:

•• (1) High precision tests of Einstein's theory of general relativity, in particular:
• (1a) A 2%–3% measurement of the Lense-Thirring effect due to the Earth's angular momentum and a high precision test of the Earth's gravitomagnetic field.
• (1b) An improved, high precision, test of the inverse square law for very weak-field gravity and test of the equivalence principle (2).
• (1c) A 10^{-6} measurement of the PPN (Parametrized-Post-Newtonian) parameter α_1 testing the existence of preferred frames in some alternative metric theories of gravitation.
• (1d) A 10^{-3} measurement of the general relativistic perigee precession of LARES and a high precision measurement of the corresponding combination of the PPN (Parametrized-Post-Newtonian) parameters β and γ. The PPN parameters β and γ test Einstein's theory of gravitation versus other metric theories of gravitation.
• (1e) Other tests of general relativity and gravitation.
•• (2) Measurements and improved determinations in geodesy and geodynamics.

4.2 The Concept of the LARES experiment

The fundamental idea[21] (see also Ref. 23) of the LARES (LAser RElativity Satellite) experiment can be divided into two parts:

- Position measurements of laser ranged satellites, of the LAGEOS type, are accurate enough to detect the very tiny effect due to the gravitomagnetic field: the Lense–Thirring precession (Equations 2 and 3).

- For 2%–3% measurement of the Lense-Thirring effect we need to "cancel out" the enormous perturbations due to the non-sphericity of the Earth gravity field, and we thus need a new satellite: LARES, with inclination supplementary to that of LAGEOS, and with the semimajor axis, a, equal to that of LAGEOS.

The new LARES satellite has been designed to be about 4 times lighter than LAGEOS, with a total weight of about 100 kg, and to be smaller than LAGEOS, with a total radius of about 16 cm.[27]

In Refs. 21, 23 and 27 is shown how, by combining the the measured nodal precessions of LAGEOS, $\dot{\Omega}_{LAGEOS}$, and LARES, $\dot{\Omega}_{LARES}$, we can get a very accurate measurement of the Lense-Thirring effect, $\dot{\Omega}^{Lense-Thirring}$.

In 1988, NASA and the Italian Space Agency[23] formed study groups for the purpose of performing a comprehensive analysis and a comprehensive numerical simulation of an experiment using two supplementary inclination satellites (called LAGEOS III).

The NASA-ASI study[23] proved the following results regarding the total error in the Lense-Thirring measurement:

- A covariance analysis confirmed a 1-σ error of 8%, or less, (*in* 1989) in the measurement of the Lense-Thirring effect, with no improvements in the knowledge of the various parameters and in the measurement errors *(however see the present–1998–covariance analysis of the LARES mission, in the LARES phase A study, Ref. 27, and below, showing a 1-σ error of 3%, or less).*
- Blind simulations and tests gave a root-mean square difference between values arbitrarily assumed for the Lense-Thirring effect and the values recovered from the simulated data, of 8%.
- Therefore, both the covariance analysis and the blind tests confirmed the previous comprehensive error analyses,[23] with a total statistical error of 8% or less *(however the present–1998–budget for the total error in the measurement of the Lense-Thirring effect with the LARES mission is 3%, or less: see the LARES phase A study, Ref. 27, and below).*

4.3 The Present (1998) error budget of the LARES experiment

The 1998 error budget of the LARES experiment is dramatically reduced with respect to the previous estimates: present analyses show a *total statistical error in the LARES experiment of about 3%* $\dot{\Omega}^{Lense-Thirring}$*, or less,* over a 3 year period.[27]

The main improvements for this substantial reduction of the total statistical error are briefly outlined.

- Improvements in the modeling of geopotential and Earth tides over the previous GEM-T1 solution, with the new JGM-3, and EGM-96, gravity field solutions,[13] and improvements in the modeling of seasonal variations in low degree geopotential harmonics, also due to the data from the new LAGEOS II satellite, launched in October 1992 by the Italian Space Agency and NASA. The total statistical error due to the uncertainties in geopotential and Earth tides is now estimated to be of about 1% (see the LARES phase A study[27]).
- The improvements in the knowledge of the geopotential plus the orbital data available from LAGEOS II imply a much smaller error (with respect to LAGEOS III) in the LARES experiment due to orbital injection errors.
- Accurate measurements (and theoretical studies) of the spin axis orientation and of the spin rate of LAGEOS, carried out with various techniques (microwave Doppler, infrared Doppler, and optical glints from front reflectors surfaces) by the University of Maryland and other groups, together with accurate measurements of thermal and optical properties of LAGEOS type satellites, will allow a modeling and a reduction of the error due to anisotropic thermal radiation uncertainties to about 1%, or 2%.
- The effect of the Earth albedo on the orbit of the LAGEOS satellite has been accurately calculated using data from the meteorological satellites, ERBE (Earth

Radiation Budget Experiment) and NOAA9 and NOAA10 by C. Martin and D. Rubincam. The effect of albedo on the LAGEOS nodal longitude turns out to be negligible with respect to the Lense-Thirring drag. Furthermore, a study of the effect of Earth albedo on the bisector of the nodal lines of two LAGEOS satellites with supplementary inclinations (that is, the albedo effect in the LARES experiment) has shown that the albedo gives a substantially negligible contribution to the rate of change of the bisector and therefore gives a negligible contribution to the uncertainty in the LARES gravitomagnetic measurement.

We then observe that:

- The 3% error budget in the LARES experiment to measure the Lense-Thirring effect corresponds to a 3-year period of data analysis, however, since the lifetime of LAGEOS type satellites has been estimated to be of the order of half a million years, and the accuracy in the modeling of the various perturbations (including the Earth gravity field: see the GRACE mission) and in the measurements with laser ranging and VLBI are improving steadily, the total error will be substantially reduced with longer periods of observations.

- For the LARES experiment is just needed a satellite of the type of the LAGEOS *passive* satellite (the satellites LAGEOS and LAGEOS II are completely passive); therefore, the LARES satellite may be built in about 1 year, thus it will be a very low risk and very low cost space experiment.

In addition to the high precision measurement of the Lense-Thirring effect due to the rotation of the Earth mass, the LARES experiment will provide other important measurements, among which:

- Improved bounds on very weak, long-range, new gravitational forces, and tests of the equivalence principle,[27] corresponding to ranges of the order of few thousands km.

  In regard to the use of the pericenter of LARES to test gravitational physics, Ken Nordtvedt has clearly pointed out[27] that: "precise periastron precession measurements are among the most sensitive means for detecting such new physics, ..., the launching of another LAGEOS-type satellite into an orbit of the largest possible eccentricity, and the subsequent ranging to that satellite, would allow for an unprecedentedly precise search for any very weakly coupled, Yukawa-like interaction having a range in the vicinity of one or two Earth radii and which acts between Earth and the satellite. *Because the ability to measure perigee precession scales as the inverse of the orbital eccentricity, ranging to a new, more eccentric LAGEOS-type satellite could be an even more sensitive probe for presence of a finite range force acting between Earth and satellite.* So measuring α [coupling constant of a new interaction] to precision of 10^{-11} or better, using the precision observation of a new LAGEOS satellite's slightly eccentric orbit (specifically, its perigee precession rate) can extend our search for a new interaction of this range by at least *four orders of magnitude* beyond present knowledge. It should also be noted that the perigee precession test using the LAGEOS satellite will also detect or constrain a finite range interaction which is *metric*, and to comparable precision, while this possibility would be essentially undetectable by universality of free fall experiments."

In the LARES phase-A study[27] is then shown that by measuring the LARES perigee rate we could also improve our present tests of the equivalence principle by about two orders of magnitude.[27]

- A very accurate measurement, of the order of 10^{-6} (improved by about two orders of magnitude) of the PPN parameter α_1 testing metric theories of gravitation with preferred frames.
- A very accurate, of the order of 10^{-3} over a period of observation of several years, measurement of the general relativistic pericenter precession of a test particle (LARES) *in the field of Earth*, and a high precision measurement of the corresponding combination of the PPN (Parametrized-Post-Newtonian) parameters β and γ. Improved determinations of these PPN parameters put tighter limits on gravitational theories alternative to general relativity.
- Other tests of general relativity and gravitation, such as improvements of the current limits on hypothetical spatial anisotropies of the gravitational interaction.
- Important measurements and improved determinations in geodesy and geophysics, in areas such as global plate tectonic, crustal deformation and determinations of variability of Earth's rotation.[23,27]

Acknowledgements

We thank and appreciate unconditional help from Despina E. Pavlis and D. D. Rowlands. We also thank Antonio Paolozzi and Davide Lucchesi for significant contributions.

References

1. C.W. Misner, K.S. Thorne, J.A. Wheeler, *Gravitation* (Freeman, San Francisco, 1973).
2. I. Ciufolini and J.A. Wheeler, *Gravitation and Inertia* (Princeton University Press, Princeton, New Jersey, 1995).
3. J. Lense and H. Thirring, *Phys. Z.* **19**, 156 (1918); see also English translation by B. Mashhoon, F.W. Hehl, D.S. Theiss, *Gen. Relativ. Gravit.*, **16**, 711 (1984).
4. In the famous Newton's gedanken experiment of a rotating bucket filled with water, the question was "which is the origin of the inertial forces that curve the surface of the water in the bucket?" "Are these forces produced by the relative rotation of the bucket with respect to absolute space or with respect to the other masses of the universe?"

5. K.S. Thorne, R.H. Price, D.A. MacDonald, eds., *Black Holes, the Membrane Paradigm* (Yale University Press, New Haven and London, 1986).
6. I. Ciufolini, *Class. and Quantum Grav.*, **11**, A73 (1994).
7. S.C. Cohen, et al., ed., *LAGEOS Scientific Results, J. Geophys. Res.*, **90** (B11), 9215 (1985).
8. P. Bender and C.C. Goad, in *Proceedings of the Second International Symposium on the Use of Artificial Satellites for Geodesy and Geodynamics*, Vol. II, G. Veis, E. Livieratos, eds. (National Technical University of Athens, 1979), p. 145.
9. K. Nordtvedt, *Physics Today*, **49**, 26 (1996).
10. J.J. Degnan and E.C. Pavlis, *GPS World*, **5**, 62 (1994).
11. *International Earth Rotation Service (IERS) Annual Report, 1996* (Observatoire de Paris, Paris, July 1997).
12. D. McCarthy, *The 1996 IERS Conventions* (Observatoire de Paris, Paris, 1996).
13. F.G. Lemoine, et al., to appear in *Proceedings of the 1997 IAG Scientific Assembly* (Rio de Janeiro, Brazil, September 5-9, 1997).
14. D.P. Rubincam, *J. Geophys. Res.*, **93** (B11), 13803 (1988).
15. D.P. Rubincam, *J. Geophys. Res.*, **95** (B11), 4881 (1990).
16. D.P. Rubincam and A. Mallama, *J. Geophys. Res.*, **100** (B10), 20285 (1995).
17. C.F. Martin and D.P. Rubincam, *J. Geophys. Res.*, **101** (B2), 3215 (1996).
18. P. Farinella, D.V.Y. Vokrouhlicky, F. Barlier, *J. Geophys. Res.*, **101** (B8), 17861 (1996).
19. R.S. Gross, *J. Geophys. Res.*, **101** (B4), 8729 (1996).
20. D.E. Pavlis, et al. *GEODYN II, Operations Manual*, 3, (1997).
21. I. Ciufolini, *Phys. Rev. Lett.* **56**, 278 (1986).
22. I. Ciufolini, *Nuovo Cimento A*, **109**, 1709 (1996).
23. I. Ciufolini, *Int Journ. Mod. Phys. A*, **4**, 3083 (1989). See also: B. Tapley, I. Ciufolini, J.C. Ries, R.J. Eanes, M.M. Watkins, *NASA-ASI Study on LAGEOS III*, (1989).
24. I. Ciufolini, et al. *Class. and Quantum Grav.*, **14**, 2701 (1997).
25. I. Ciufolini, E. C. Pavlis, et al. *Science*, **279**, 2100 (1998).
26. R. Eanes and J. Ries, private communication (1998).
27. I. Ciufolini, A. Paolozzi, et al. *LARES phase A study for ASI*, (1998).

LENSE-THIRRING PRECESSION DETERMINATION FROM LASER RANGING TO ARTIFICIAL SATELLITES

J. C. RIES, R. J. EANES, B. D. TAPLEY

Center for Space Research, The University of Texas at Austin
Austin, TX 78712, USA
E-mail: ries@csr.utexas.edu

Detection of the Lense-Thirring precession of the orbit of an Earth satellite has been a challenge for decades, but as yet, no dedicated mission has been flown to accomplish this goal. The difficulty is not the ability to observe the precession of an orbit with sufficient precision, but rather, it is the uncertainty in the knowledge of the even zonal harmonics of the Earth's gravity field which limits the experiment using existing satellites. As the gravity models for the Earth have improved, there has been a re-examination of whether these errors still limit this fundamental test of general relativity. While this goal is tantalizingly close, it is argued here that the uncertainties are still too large for a confident and robust test of general relativity. However, the prospects for a successful test will improve significantly with the launch of the GRACE gravity mission.

1 Introduction

Earth satellites cannot generally compete with Very Long Baseline Interferometry (VLBI), Lunar Laser Ranging (LLR) and interplanetary ranging in the area of tests of general relativity. Many of the relativistic effects are large enough to be observable, but the contribution of the orbit perturbations from a variety of gravitational and nongravitational sources limits the quality of the experiment. As a consequence, while the relativistic effects are usually modeled in satellite orbit determination, using Earth satellites to test the general relativity theory has not been particularly successful.

In the case of the Lense-Thirring (LT) precession of a satellite's orbital plane [6], the errors in the estimates for the even zonal spherical harmonic coefficients of the Earth's gravity field model are the principal limitation. The even zonals cause a precession of the orbital plane (i.e., in the longitude of the node Ω) in the same manner as the LT precession. Because the Lense-Thirring precession is so small, the model for the even zonals would need to be extremely accurate. However, an alternative to requiring such extreme precision was proposed by Ciufolini [1] in which a second satellite (LAGEOS-3), physically identical to the LAGEOS satellite, would be launched into an orbit with an inclination supplementary to LAGEOS. The possibility for an observation of the LT precession exists because of the fact that for the even zonals, the magnitude and sign of the precession of the node is dependent on the sine of the orbit inclination i, whereas the LT precession is always in the same direction as the Earth's rotation. Choosing two orbits with equal and opposite precession due to the even zonals would allow a determination of the common precession due to LT. A study conducted in 1989 indicated that such an experiment could measure the LT precession with an accuracy of approximately 8% [10,13]. With the progress in the gravity models since then, the error estimate has improved to the 4% level [9].

As the gravity models for the Earth have continued to improve, there has been a re-examination of whether these errors still limit this fundamental test of general relativity. In this paper, we examine the recent analyses in which the LAGEOS-1 and LAGEOS-2 satellites are used to observe the LT effect using two of the most accurate gravity models,

JGM-3 [12] and EGM96 [5]. The analysis by Ciufolini et al. [3] uses a simplified but innovative method to isolate the trends due to the even zonals and LT to arrive at a confirmation of the general relativity prediction at the 20% level. There are, however, some concerns with this analysis, and the assumptions in the error estimation appear to be overly optimistic. The simulations performed at the University of Texas Center for Space Research (CSR) indicate that, even assuming that the gravity model covariances are not underestimating the errors, the uncertainty in observing the LT precession is at best in the 50-100% range. This is tantalizingly close but not yet a robust detection of this fundamental relativity signal.

2 Lense-Thirring precession from LAGEOS-1 and LAGEOS-2

2.1 Using orbit nodes only

Past analyses have tended to concentrate on the longitude of the node as the best place to observe the LT precession in a satellite orbit, for good reason. In the case of a high inclination orbit, the effect is predominantly in the node. For orbits with lower inclinations, the LT precession effects both the node (Ω) and the argument of perigee (ω), but the perigee has typically been a poorer orbit element to examine. The motion of the orbit plane is generally far easier to observe than the precession of the elliptical orbit within the orbital plane. The magnitude of the effect of a change in perigee on a laser range measurement, for example, is diminished by a factor proportional to the orbital eccentricity (e), and there is strong interaction between the determination of the eccentricity and perigee. For low eccentricity orbits, the perigee can be extremely difficult to determine accurately, and typically a set of nonsingular elements are used rather than work with perigee and eccentricity [4,8]. An additional problem with analyses of the perigee is that the variety of surface forces tend to have much more dramatic effects on the in-plane orbit elements. The LAGEOS satellites are plagued with thermal forces which cause dramatic variations in the in-plane orbit elements [8,11]. These variations are sufficiently large as to limit the usefulness of the LAGEOS satellites for observing the time variations in the odd zonal gravity coefficients [4].

If only the nodal precessions of LAGEOS-1 and 2 are used to determine the LT precession, one has essentially two observations but many unknowns. For the purpose of a simplified analysis, however, it is reasonable to consider only the unknowns that have the most significant contributions to the nodal precessions, which are the even zonals up to degree eight (i.e., J2, J4, J6 and J8) and the LT precession itself. From this, it is possible to form, in the manner described in Ciufolini et al. [3] (hereafter referred to as C98), a "J2-free" observation of the LT precession, but the contribution of the other zonals is about 48% of the signal predicted by Generally Relativity, based on the EGM96 gravity model solution covariance. Since this considers no other contributions to the error, the true error can be expected to be larger than 48%, and even this must assume that the EGM96 covariance accurately represents the errors in a very small number of coefficients in the gravity model.

This "J2-free" result can be compared with the analysis by Peterson [9], which indicates that the error in the determination of the LT precession from LAGEOS-1 and 2, using the JGM-3 gravity model, is in the range of 100-200%. This was based on a full simulation of the entire LAGEOS-1 and 2 processing and modeling, including the effects of all known aspects of the orbit dynamics (such as the thermal forces) and measurement models. The complete effects on all the orbit elements are implicitly taken into account through the orbit dynamics and the fitting of the simulated tracking data. If the signal from an orbit element other than the node was useful, the simulation would automatically reflect this. This was the same method used to arrive at the 4% error assessment for the LAGEOS-3 experiment. The EGM96 covariance predicts almost a factor of two improvement over JGM-3 in the lower degree coefficients, which would lead to an assessment of an error of 50-100% using EGM96. Whether EGM96 actually is a factor of two better than JGM-3 is uncertain. Orbit fits do not tend to support this much of an increase in accuracy (orbits tend to test the longer wavelengths of the gravity model), while geoid comparisons do suggest a significant improvement (which tends to be a test of the shorter wavelengths).

2.2 *Using both nodes and perigee*

To remove the limitation in the LT determination due to having only the two orbit node precessions as observations, C98 attempts to provide a third observation using the LT effect on the perigee. Since the eccentricity of LAGEOS-1 is small ($e = 0.004$), only the perigee of LAGEOS-2 ($e = 0.014$) is used. Thus, in theory, the three observations allows the determination of a "J2/J4-free" observation of a LT precession parameter μ defined by

$$\mu = (\delta\dot{\Omega}_1 + 0.295\delta\dot{\Omega}_2 - 0.35\delta\dot{\omega}_2) \,/\, 60.2 \text{ mas/year}$$

where μ has been normalized to be dimensionless and equals 1 for general relativity.

The method used to determine the node and perigee rates in C98 is based on the comparison of the endpoints of a sequence of 15-day arcs. Any secular signal in the node or perigee not accounted for in the modeling should be observable in these endpoint differences. When accumulated over time, any secular mismatch between the model and the true orbit reflects secular errors in the model. While relatively simple compared to a full estimation of all parameters simultaneously, it appears to work well for recovering the secular signals being sought here. However, the issue is not the observability of the trends in the orbit elements. As noted earlier, these can generally be observed with relatively high precision. That fact that perigee can be observed with a precision much smaller than the LT signal does not imply that the LT signal can be reliably determined. The real issue is the differences between the model and the observed phenomenon, since this is what is reflected in the accumulated endpoint differences. If the model has systematic errors, or the parameters estimated while fitting the observations can absorb part of the desired signal, or if part of the signal simply remains in the observation residuals due to an imperfect fit, then the interpretation of the residual trends is not reliable.

Based on four years of laser tracking to LAGEOS-1 and 2, C98 determines a value of μ=1.1 (i.e., within 10% of the general relativity prediction). It is risky, however, to put too much weight on the fact that the estimate is close to the general relativity prediction, because the current models for the gravity field and the thermal forces have been developed using the same data and models that already have assumed that general relativity is correct. Recovering a time series that gives back the *a priori* assumption is not necessarily a robust test. The analysis should be conducted in a manner which removes dependence on the *a priori* models. A simultaneous recovery of the gravity field, the LT precession and all the other adjusted parameters would be a more reliable experiment, and, as discussed below, would provide a much more rigorous error estimate.

2.3 *Error analysis of Ciufolini et al. (1998)*

The analysis of the 15-day orbit endpoint differences in C98 does not provide an explicit covariance for the solution. Because the correlations between the LT parameter and the other model parameters are not available, an error analysis to account for these dependencies has been improvised. The resulting error estimate of 20% was dominated by an estimate of 13% from the nongravitational effects and the same for the even zonals other than J2 and J4. These are the main areas of the error analysis that must be examined more closely.

2.3.1 Nongravitational effects on perigee

Examining the contributions from the surface forces first, the models adopted for the surface forces, as well as the adjustment of parameters in these models, can have profound effects on the residual perigee signals. The adjustment of a solar radiation pressure scale factor every 15 days can accommodate some of the variability in the LAGEOS-2 perigee. The other parameters being adjusted may also be having a substantial effect. In Figs. 1 and 2, the "J2/J4-free" time series from LAGEOS-1 and 2 using the orbit endpoint difference technique is shown, where the only difference between the two figures is whether the reflectivity coefficient was estimated for each 15-day arc. Figure 2 is intended to be the same quantity as plotted in Fig.2 of C98, but there are some differences. The details of the station coordinate and polar motion estimation were not specified in C98, so these quantities were not estimated in this analysis. In Fig. 1, the solar radiation pressure reflectivity coefficient is not estimated. The resulting variations are very large and make the determination of a slope problematic. The large quasiperiodic variations beginning after year 3 are indications of the influence of the thermal forces acting on LAGEOS 2. In

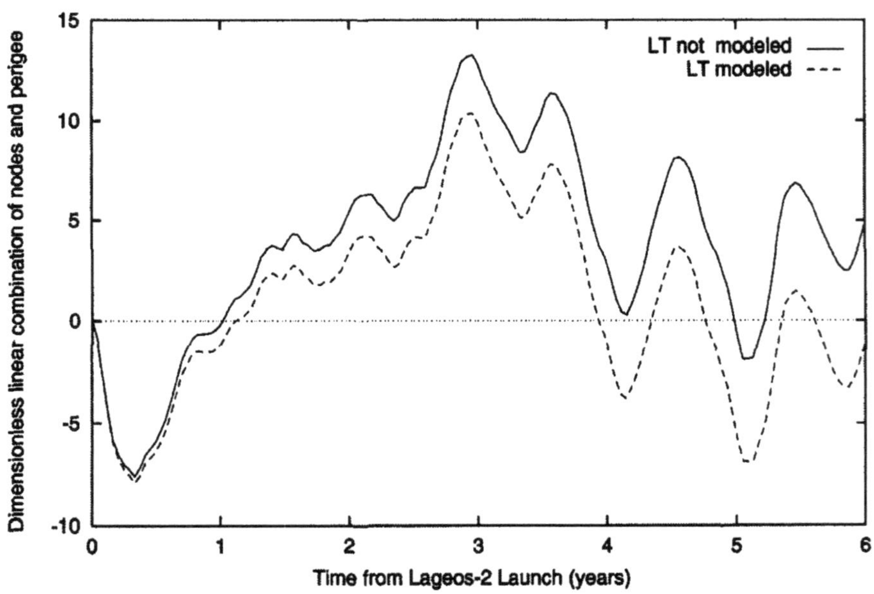

Figure 1. The "J2/J4-free" time series from LAGEOS-1 and 2 using the orbit endpoint difference technique. The solar radiation pressure reflectivity coefficient was not estimated in each 15-day arc.

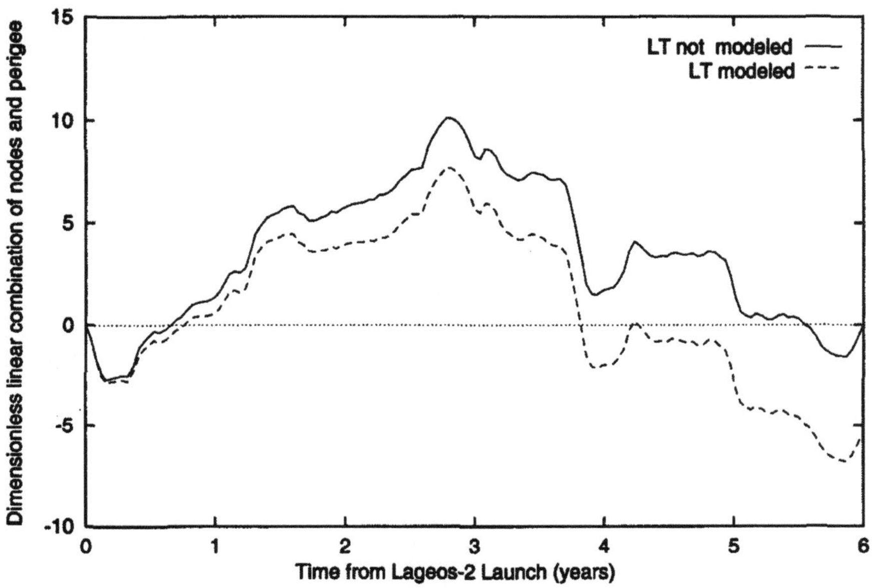

Figure 2. The "J2/J4-free" time series from LAGEOS-1 and 2 using the orbit endpoint difference technique. The solar radiation pressure reflectivity coefficient was estimated for each 15-day arc. In comparison to Fig. 1, the variations have been significantly reduced.

Fig. 2, these variations are significantly reduced, although there is still considerably more variation in this plot than in C98. The reason for this is not clear, but it suggests that the combination of parameters being estimated in the analysis of C98 is removing a great deal more power from the time series, and it is difficult to judge how much this influences the slope or the error estimate. The difference between including or not including the LT precession in the model is clearly visible, indicating that the technique is able to distinguish the slope difference, but the absolute slope (which determines the estimate of μ) will vary considerably depending on which span of data is used. The variations seen in the 15-day reflectivity estimates cannot represent actual changes in the satellite reflectivity; hence, they must be absorbing other signals, presumably the errors in the thermal force models. The effect of the interaction between the solar reflectivity and the thermal forces on the perigee and eccentricity of LAGEOS-type satellites is well documented [8].

Figure 3 illustrates the value of the LT parameter μ recovered depending on the span of data in Fig. 2. Even after 3 years, the estimate for μ substantially disagrees with general relativity and also with the results in Ciufolini et al. [2]. Only after 6 years of data does the estimate approach the prediction of general relativity. When the LT precession is included in the modeling, the value of μ approaches zero as expected, but more data is required to determine if the estimates will converge near the expected values or continue to decrease. The formal error in the slope determination (not considering any systematic contributions from the gravity field) can be expected to decrease with the data span. The contributions of the surface forces will continue to average down, as long as the signal is not being distorted by the estimation of the reflectivity or other parameters. Simulations indicate that the estimation of the satellite reflectivity every 15 days does not bias the LT

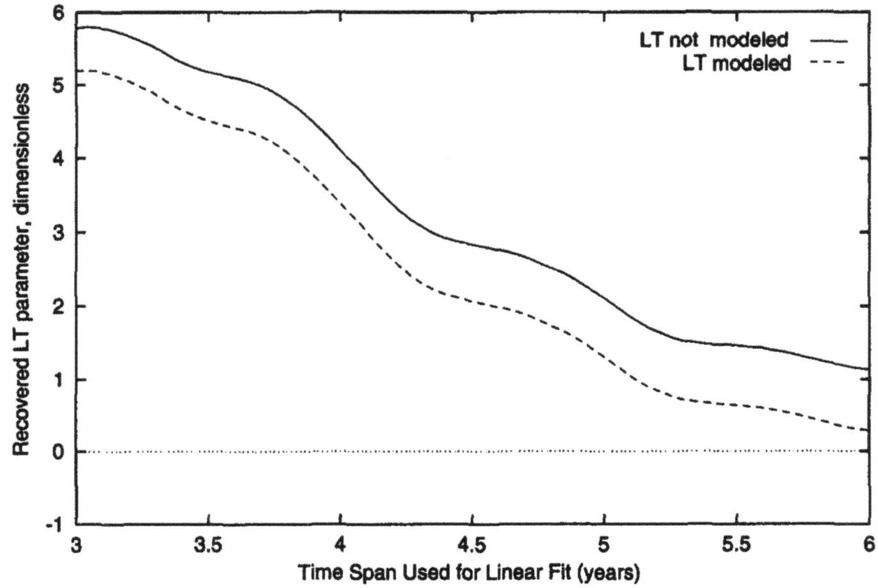

Figure 3. The CSR estimate for the LT parameter μ as a function of the data span used. This is the slope of the time series in Fig. 2. as the time series is extended. The two curves should be differ by exactly 1, since the only difference is whether the LT precession was included in the modeling or not, but the actual difference ranges from 0.6 to 0.9.

parameter recovery by more than about 10% if a sufficiently long time span of data is used. This is consistent with the results in Fig. 3, where the difference between the two cases with 6 years of data is not exactly 1, but rather approximately 0.9. The estimation of the satellite reflectivity appears to have a 'dampening' effect on the LT signal.

The contribution from the error in the models for the thermal forces, however, is much more difficult to quantify. The current models have been successful in explaining the dominant part of the observed orbit perturbations, but they are still far from exact representations. The fact that it is necessary to estimate the reflectivity every 15 days, and the unrealistic variations that result, indicates that the reflectivity parameter is compensating for deficiencies in the thermal force model. However, it cannot be assumed that this cures all of the modeling ills. Consequently, the influence of these unknown forces on the perigee cannot be confidently quantified. It can be shown analytically that very small one-cycle-per-revolution (1-cpr) accelerations (less than 1 picometer/sec/sec) can erase the entire LT signal in the LAGEOS-2 perigee if they are at the right frequency. This frequency corresponds to a modulation of 1-cpr by exp(-i ω) because that term resonates with the eccentricity vector (a non-singular combination of eccentricity and perigee) excitations and causes a secular trend in perigee [4]. When the CSR eccentricity vector excitations residuals for LAGEOS-2 are examined, the amplitude of the variations in the region of the retrograde 1 cycle/perigee frequency is more than 10 times the size of the excitations that cause the LT perigee advance. It appears to be questionable to expect that the LT perigee signal can be extracted from the nongravitational noise with any reliability. This is a concern for any signal, relativistic or classical, which is hoped to be seen in the perigee of a satellite orbit.

In addition to the satellite reflectivity, virtually the entire SLR network and polar motion was adjusted every 15 days in C98. However, there would seem to be little justification for this, since these quantities are known sufficiently well for this problem, and including this large sink for error signals appears to be, at the very least, unnecessary and only increases the uncertainty of the error analysis.

2.3.2 Gravity model error analysis

Because there is no covariance provided by the method of analysis used in C98, the errors due to the other even zonals was estimated indirectly. The assumption was that only J6 and J8 needed to be considered. Using the EGM96 covariance (including the correlations), an estimate of roughly 13% was arrived at. There are, however, several concerns regarding this conclusion.

Considerable effort has been made to assure that the covariance for the JGM-3 and EGM96 gravity solutions are as accurate as possible, but there are a number of caveats to this process. The calibration is performed by various tests in which the data weight is adjusted until the output covariance provides reasonable results [7]. This test is imperfect since it attempts to find a weighting that affects the entire covariance uniformly, although it is often clear that the calibration should probably be dependent on the wavelengths that a coefficient is associated with. Thus some coefficient errors are likely to be calibrated to be too small or too large, even though the covariance for the gravity field as a whole provides reasonable error predictions. The calibration is accomplished by simply increasing the noise-only standard deviation of the input data. In reality, the full covariance of the data should be used, but this is unknown and thus not a realistic option. Furthermore, the covariance (and thus the weighting) can only be affected by the model parameters being estimated. The covariance cannot reflect uncertainty in parameters which are not estimated or systematic errors in the models which are not known.

The usual method of increasing the noise-only standard deviation of the data is a poor substitute for all this missing information, but it has so far been the only workable technique. Used carefully, and with judicious interpretation of the results, the solution covariance is able to provide a reliable indication of the overall errors. However, interpretation of specific details of the covariance, such as the error for a few specific coefficients, is problematic. In particular, the coefficients most likely to be affected by the systematic errors in the modeling are the zonals, since these tend to be determined largely by their long-period or secular contributions to the orbit. This is precisely the same frequency range that the surface forces act, and the errors predicted for the zonals are among the most likely to be optimistic. Dependence on the knowledge of the errors in just a few gravity coefficients cannot provide a robust test of general relativity. This is the special advantage of the supplementary satellite proposed by Ciufolini [1]; such exact *a priori* knowledge is not critical to the success of the LAGEOS-3 experiment.

In addition to the basic uncertainty in the magnitude of the error estimate assigned to the low degree zonals in any gravity solution, the "J2/J4-free" LT estimate benefits from a remarkably favorable correlation coefficient between J6 and J8. Where, individually, J6 and J8 would contribute 40% or more each, which is much more consistent with the analysis of Peterson [9], the strong negative correlation largely cancels their contributions in the expression for μ, resulting in only a 13% uncertainty. This strong negative correlation is a common characteristic of most current gravity models, as demonstrated in Fig. 4, and is the result of a poor separation of the various even zonals. The correlation

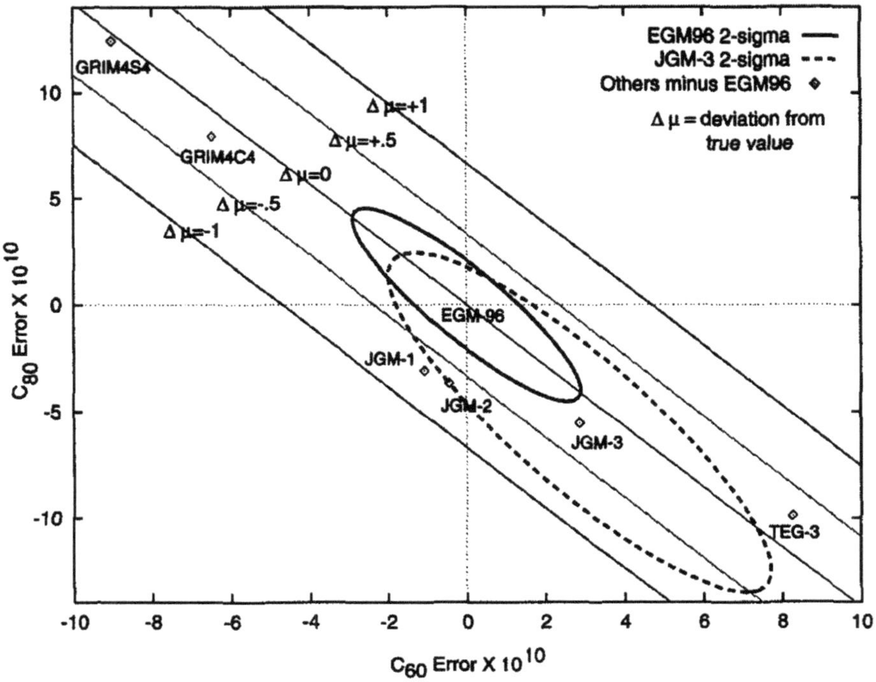

Figure 4. The change in the LT parameter due to the correlated C60/C80 errors. The lack of good separation between the estimates of the even zonals in the current gravity models, leading to a significantly negative correlation coefficient (-0.89 in the case of EGM96), produces an remarkably favorable error estimate for the LT determination.

between J2 and J4 for EGM96 is –0.93; for J4 and J6, it is –0.80; between J6 and J8, it is –0.89. JGM-3 has similar values. As noted earlier, the dominant signal for the even zonals is their secular effect on the orbit, and without a sufficient number of unique satellite orbital inclinations, the separation between the zonals is relatively poor. However, caution would suggest that a test of general relativity should not rest on the favorable correlation of two important sources of error, where the correlation is simply the result of an inadequate separation of the individual geopotential coefficients. It should also be noted that the various gravity solution covariances only represent the errors in the static gravity field (i.e. the multiyear average). The time variations in the even zonals are not accounted for by the covariance, and the errors in any given year in time may be quite different from those predicted by the covariance. This again illustrates the advantage of the LAGEOS-3 mission in which all the even zonal errors, static or variable, cancel out.

It is unlikely that the error estimates would be the same if the LT precession were estimated simultaneously with the even zonals, which highlights what may be one of the most problematic aspect of the error analysis. Most of the useful satellites have already been incorporated into the JGM-3 and EGM96 gravity solutions, and LAGEOS-1 and 2 generally have considerable influence. The analysis described above, however, holds all the errors fixed while using LAGEOS-1 and 2 again to determine a new parameter, a parameter which is known to be highly correlated with the even zonals. It is very likely that the correlations and errors would change significantly if the LT precession parameter were included with the estimation of the even zonals.

In addition to the above considerations, significant sensitivity to J3 was found, both from the mean value and its time variations, in the analysis of the proposed LAGEOS-3 mission [10]. This has been ignored in the C98 analysis and warrants closer attention. This effect is included in the simulations at CSR and would automatically be accounted for if the LT parameter were adjusted simultaneously with the geopotential coefficients.

The simulations performed at CSR included a host of time variable gravity model errors (secular, seasonal and tidal) [9,10,13]. When the same simulation process that was applied successfully to the analysis of LAGEOS-1/LAGEOS-3 (to obtain the current error estimate of ~4%) was applied to LAGEOS-1/LAGEOS-2, the results were errors that ranged from 100 to 200 percent [9]. These results would improve if the EGM96 covariance was adopted, but that would improve the results by less than a factor of two. Consequently, it is estimated that the error for the current 'detection' of Lense-Thirring is in the range of 50 to 100%, even using the EGM-96 covariance, and there is evidence suggesting that the EGM96 covariance may be even more optimistic than most gravity solutions for the low degree geopotential coefficients. Thus the error estimate is much more likely to be closer to 100%. Most of the important errors are systematic and cannot be reduced by simply accumulating a longer time series. Only a significant reduction in the gravity model errors (and not simply the static field as represented by the JGM-3 or EGM-96 covariances) will improve the results from LAGEOS-1/LAGEOS-2.

3 Future Prospects

The Gravity Recovery and Climate Experiment (GRACE) mission, a joint project between the National Aeronautics and Space Administration (NASA) and the Deutsches Zentrum für Luft und Raumfahrt (DLR), has recently been approved for launch in mid-2001 and has the potential for a dramatic improvement in the determination of the Lense-Thirring precession from the LAGEOS satellites. The primary objective of the GRACE mission is to provide, with unprecedented accuracy, estimates of the global high-resolution models of the Earth's gravity field for a period of up to five years. The temporal sequence of gravity field estimates will yield the mean Earth gravity field, as well as a time history of its variability. An additional science objective of the GRACE mission is to provide several hundred globally distributed profiles each day of the excess delay of the GPS measurements due to the ionosphere and the atmosphere, using GPS limb-sounding.

The mission objective will be achieved by making measurements of the inter-satellite range between two co-planar, low altitude (485-350 km) near-polar orbiting satellites, using a microwave tracking system. In addition, each satellite will carry geodetic quality Global Positioning System (GPS) receivers and high accuracy accelerometers. The path of a satellite in Earth orbit is dependent on the integrated effect of the mass distribution within the Earth system, and the orbits of the two GRACE satellites, sensing these effects at slightly different phases, will be perturbed differentially. This difference in the perturbations is manifested in the intersatellite range, and the microwave ranging instrument will provide very accurate measurements of these range changes.

The expected results from the GRACE mission are illustrated in Fig. 5, where the errors in the gravity field (in terms of geoid height) are compared to the current gravity field knowledge as represented by the EGM96 gravity solution covariance, as well as to other gravity signals of interest. As much as two orders of magnitude improvement are

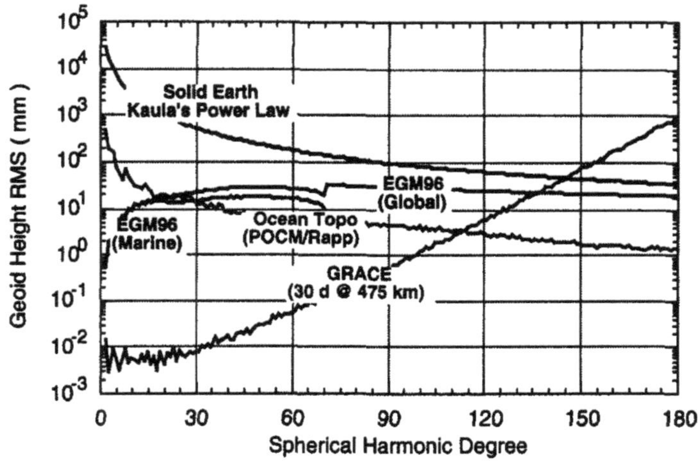

Figure 5. Expected performance of the joint NASA/DLR GRACE gravity recovery mission

anticipated at some of the longer wavelengths. This suggests the potential for a measurement of the Lense-Thirring precession that might approach the accuracy of the proposed LAGEOS-1/LAGEOS-3 experiment. However, considerably more analysis of the errors will be required after the actual GRACE data is processed to determine if this ambitious level of improvement has been realized. In addition, the quality of the Lense-Thirring experiment would still rest on the estimated errors in the low degree even zonal harmonics and their temporal variations. Error predictions at this extremely small level will be very difficult to verify, although efforts to design methods for such a validation are in progress.

4 Conclusions

At this time, a robust detection of the Lense-Thirring precession does not appear to be obtainable from the constellation of Earth satellites currently available. The technique used by Ciufolini et al. [3] is innovative but requires a number of optimistic assumptions to arrive at an error estimate at the 20-30% level. A more rigorous analysis in which the Lense-Thirring effect is included as part of the gravity model solution is required to see the true error covariance, and even then, considerable care would be needed to make sure the effects of the nongravitational perturbations are properly considered. CSR simulations and analysis of the actual LAGEOS-1 and 2 data indicate that the error is more likely to be at least 50–100%, and this requires an assumption that the small errors predicted by the current gravity solutions for the low degree zonals are correct, an error estimate that is as yet impossible to confidently verify.

A significant improvement in the knowledge of the Earth's gravity field will be provided by the upcoming GRACE mission. This will remove the dependence on the perigee signal and dramatically reduce the errors due to uncertainties in the even degree zonals, but an accurate verification of the residual gravity model errors from the GRACE mission at the longest wavelengths will still be a concern. The proposed LAGEOS-3 satellite [1,10] may still be the most likely experiment to provide a confident and robust observation of the Lense-Thirring precession of satellite orbits to the few percent level.

References

1. Ciufolini I., Measurement of the Lense-Thirring drag on high-altitude laser-ranged artificial satellites. *Phys. Rev. Lettz.*, **56** (1986) pp. 278–281.
2. Ciufolini I., Lucchesi D., Vespe F., and Chieppa F., Measurement of gravitomagnetism. *Europhys. Lett.* **39** (1997) pp. 359–364.
3. Ciufolini I., Pavlis E., Chieppa F., Fernandes-Vieira E., and Pérez-Mercader P., Test of general relativity and measurement of the Lense-Thirring effect with two Earth satellites. *Science* **279** (1998) pp. 2100–2103.
4. Eanes R. J., A study of temporal variations in Earth's gravitational field using LAGEOS-1 laser range observations. CSR Report 95-08 (University of Texas, Center for Space Research, Austin, Texas, 1995).
5. Lemoine F. G. *et al.*, The development of the joint NASA GSFC and the National Imagery and Mapping Agency (NIMA) geopotential model EGM96, in *Gravity, Geoid, and Marine Geodesy*, ed. by J. Segawa, H. Fujimoto, and S. Okubo, (International Association of Geodesy Symposium Vol. 117, Springer-Verlag, Berlin, 1997), p. 461.
6. Lense J. and Thirring H., Uber die Einfluss der Eigenrotation der Zentralkorper auf die Bewegung der Planeten und Monde nach der Einsteinschen Gravitationstheorie. *Phys. Zeitschr.* **19**, 156 (1918) [English translation by Mashoon B. *et al.*, *Gen. Relativ. Gravitation* **16**, 711 (1984)].
7. Lerch F. J. *et al.*, An improved error assessment for the GEM-T1 gravitational model. *J. Geophys. Res.* **96** (1991) pp. 20023-20040.
8. Métris G., Vokrouhlicky D., Ries J. C., and Eanes R. J., Non-gravitational phenomena and the LAGEOS eccentricity excitations. *J. Geophys. Res.* **102** (1997) pp. 2711–2729.
9. Peterson G. E., Estimation of the Lense-Thirring precession using laser-ranged satellites. CSR Report 97-01 (Univ. of Texas, Center for Space Research, Austin, Texas, 1997)
10. Tapley B. D., Ciufolini I. *et al.*, Results of a joint NASA/ASI study on LAGEOS III. CSR Report 89-03 (Univ. of Texas, Center for Space Research, Austin, Texas, 1989)
11. Tapley B. D., Schutz B. E., Eanes R. J., Ries J. C., and Watkins M. M., LAGEOS laser ranging contributions to geodynamics, geodesy, and orbital dynamics, in *Contributions of Space Geodesy to Geodynamics: Earth Dynamics, Geodynamics* Series, **24**, ed. by Smith D. E. and Turcotte D. L., (American Geophysical Union, Washington, D. C., 1993) pp. 147–173.
12. Tapley B. D. *et al.*, The JGM-3 geopotential model. *J. Geophy. Res.* **101** (1996) pp. 28029–28049.
13. Watkins M. M., Tapley B. D., Eanes R. J., and Ries J. C., Detailed simulation of the LAGEOS-3 measurement of the Lense-Thirring precession, *The First William Fairbank Conference on Relativistic Gravitational Experiments in Space*, ed. by M. Demlanksi and C. W. F. Everitt, Advanced Series in Astrophysics and Cosmology – Vol 7 (World Scientific Publishing Company, Singapore, 1993), pp. 189-194.

ON GRAVITOMAGNETIC AND ROTATIONAL ANALOGUE OF THE HALL EFFECT: A POSSIBILITY TO DETECT LENSE-THIRRING FIELD OF EARTH

B.J. AHMEDOV

Institute of Nuclear Physics, Ulughbek, Tashkent 702132, Uzbekistan
E-mail: ahmedov@suninp.tashkent.su

A possible new galvanogyroscopic effect which is the rotational (gravitomagnetic) analogue of Hall effect in electromagnetism has been proposed. As a consequence of a generalized Ohm's law, the effect of the Coriolis force on the conduction current is predicted to give rise to an azimuthal potential difference V_{gg} in a spinning rotor carrying radial electric current i_r. The potential difference developed by the galvanogyroscopic effect is proportional to angular velocity Ω and to the electric current both. An experiment is proposed for measuring the earth's gravitomagnetic field by using galvanogyroscopic effect. Due to the Lense-Thirring effect potential difference about $10^{-19}V$ will be developed in a current carrying conductor on a platform which is nonrotating relative to the distant stars.

1 Introduction

It is generally known that the equations of motion corresponding to the rotating frame of reference in Minkowski space-time contain Coriolis terms proportional to angular velocity Ω and centrifugal terms proportional to Ω^2. Moreover, according to general relativity, rotating gravitational mass produces a field proportional to Ω and called, by analogy with electromagnetism, as gravitomagnetic field.

Recently it has been accepted (see, for instance Ref. 1) that there exists exact dynamical similarity between centrifugal forces (gravitational acceleration) and electric field, on one hand, and between Coriolis force (gravitomagnetic field) and magnetic field on another hand. It has been experimentally established that centrifugal forces (and gravitational acceleration) produce potential drop in a spinning conductor[2,3] (in a conductor embedded in the earth's field of gravity, see Ref. 4 for review) and consequently act on the electrons within accelerating conductors as true electric field applied to the conductor. Our obligation is to mention here the significant contribution of Fairbank[5] (to the memory of whom this meeting is devoted) on the experimental study of electric field induced by gravity in metals.

In the year 1879 Hall experimentally observed[6] influence of magnetic field on conduction current. In the Hall effect, a voltage $V = R_H \mathbf{j} \times \mathbf{B}$ appears in a conductor to which an electric current $\mathbf{j}$ and a magnetic field $\mathbf{B}$ are applied in perpendicular directions, where R_H is the Hall constant. Thus the question arises: Does the Coriolis (gravitomagnetic) force acts on the electric current, that is, moving charged particles, in the rotating conductor as does the magnetic field, i.e. is there rotational (gravitational) analogue of the Hall effect? To get an answer to this question is exteremely interesting, not only from a purely theoretical point of view, but also because it can suggest new electrodynamic tests of relativistic effects.

In the next section we extend Ohm's law to include the effects of Coriolis force. Then the effect of Coriolis force on conduction current in the spinning rotor is considered by using extended Ohm's law. This effect, called as a galvanogyroscopic one, can be summarized by saying that an inner electric field perpendicular to the current and angular velocity is induced by rotation in a spinning crosslike rotor with a radial electric current i_r. Then the voltmeter applied in azimuthal direction will show the appearance of galvanogyroscopic voltage V_{gg}. This effect has the following properties: i) galvanogyroscopic effect is odd, that is galvanogyroscopic voltage changes the sign in dependence of polarity of either angular velocity or electric current; ii) in the first approximation voltage V_{gg} is proportional to Ω and described by formula

$$V_{gg} = (R_{gg}/d) i_r \Omega \tag{1}$$

where R_{gg} is the coefficient of proportionality and d is the thickness of rotor's plate.

In the third section we shall present possible electrodynamic test of general relativistic gravitomagnetic force of earth using galvanogyroscopic effect.

2 On the galvanogyroscopic voltage developed by Coriolis force in spinning rotor

We first consider the effects of Coriolis force on the normal current. The expression for Ohm's law in the rotating conducting disc can be obtained by extending the usual derivation of constitutive equations to include noninertial acceleration. The equation of motion for conduction electrons with mass m in a rotating frame can be written as

$$m\frac{d\mathbf{v}}{dt} - m\mathbf{\Omega} \times (\mathbf{r} \times \mathbf{\Omega}) - 2m\mathbf{v} \times \mathbf{\Omega} = -e\left(\mathbf{E} + \frac{1}{c}\mathbf{v} \times \mathbf{B}\right) + \frac{ne^2}{\lambda}\mathbf{v} \tag{2}$$

where the last term is due to the resistance force acting on conduction electrons from continuous medium with conductivity λ, $\mathbf{v}$ and n are the relative velocity and density of conduction electrons, $\mathbf{E}$ and $\mathbf{B}$ are the usual electric and magnetic fields.

Hereafter we are doing calculations in the linear in $\Omega r \ll c$ approximation and therefore neglect terms arising from centrifugal forces. Taking into account the fact that the steady state $\frac{d\mathbf{v}}{dt} = 0$ establishes within the conductor in extremely short time ($\tau \approx 10^{-13}s$), one arrives at the Ohm's law for the current density $\mathbf{j} = ne\mathbf{v}$ which flows in the rotating conductor

$$\mathbf{E} = \frac{1}{\lambda}\mathbf{j} + R_H \mathbf{j} \times \mathbf{B} - R_{gg}\mathbf{j} \times \mathbf{\Omega}, \tag{3}$$

where coefficient of proportionality $R_{gg} = \frac{2m}{ne^2}$. Except for ordinary classical terms, the current flowing in the rotating conductor has Coriolis effect contribution which is represented by the term $R_{gg}\mathbf{j} \times \mathbf{\Omega}$.

The Eq. (3) can be also obtained directly from the general relativistic constitutive equation

$$F_{\alpha\beta}u^\beta = \frac{1}{\lambda}\jmath_\alpha + R_H(F_{\nu\alpha} + u_\alpha u^\sigma F_{\nu\sigma})\jmath^\nu$$
$$+ \Lambda^{-1/2} \overset{\perp}{\nabla}_\alpha \tilde{\mu} - \beta\Lambda^{-1/2} \overset{\perp}{\nabla}_\alpha \tilde{T} - b\jmath^\beta A_{\alpha\beta} \tag{4}$$

obtained in Ref. 7 for conductors embedded in the fields of gravity and inertia. Here $\tilde{T} = \Lambda^{1/2}T, \tilde{\mu} = \Lambda^{1/2}\mu$, μ is the chemical potential per unit charge, T is the temperature, $\Lambda = -\xi^\alpha \xi_\alpha$, ξ_α is a timelike Killing vector being parallel to the four velocity of the conductor u^α, β is normal differential thermopower coefficient, $F_{\alpha\beta} = A_{\beta,\alpha} - A_{\alpha,\beta}$ is the electromagnetic field tensor, A_α is the vector potential, $A_{\beta\alpha} = u_{[\alpha,\beta]} + u_{[\beta} w_{\alpha]}$ is the relativistic rate of rotation, $w_\alpha = u_{\alpha;\beta} u^\beta$ is the absolute acceleration and b is the parameter for the conductor, $\overset{\perp}{\nabla}_\alpha$ and $[\cdots]$ denote the spatial part of covariant derivative and antisymmetrization. The last term in the right hand side of Eq. (4) is responsible for gravitomagnetic and Coriolis force effects on conduction current.

Ohm's law for conduction current has been generalized to include effects of gravity and inertia in the recent papers,[3,7–10] but the effect of Coriolis (gravitomagnetic) force upon an electric current has been taken into account only in Ref. 7. Recently Khanna[11] derived general relativistic Ohm's law for plasma, which includes gravitomagnetic terms.

Indeed if the conductor is at rest in the rotating frame of reference

$$ds^2 = -(c^2 - \Omega^2 r^2)dt^2 + 2\Omega r^2 d\varphi dt + dr^2 + r^2 d\varphi^2 + dz^2 \tag{5}$$

then the absolute acceleration of conductor has the following components

$$w_\alpha \left\{0, -\frac{\Omega^2 r}{c^2}(1 - \Omega^2 r^2/c^2)^{-1}, 0, 0\right\}$$

and only one component

$$A_{21} = -\frac{\Omega r}{c}(1 - \Omega^2 r^2/c^2)^{-3/2}$$

of the relativistic rate of rotation is nonvanishing. Thus the general-relativistic Ohm's law given by Eq. (4) in the rotating frame of reference in Eq. (5) takes a form coinciding with Eq. (3), if we suppose $\tilde{\mu}$ ($grad\tilde{\mu}$ produced by gravity and inertia is proportional to absolute acceleration for homogeneous conductors,[8,10] and in our case, to Ω^2) and $\tilde{T}$ are constant and neglect terms quadratic in Ω.

Consider now a crosslike rotor carrying radial current and spinning with a high angular velocity as in Figure. Suppose that the external magnetic field is absent. Charges moving radially inward or outward will experience Coriolis force, which will cause some motion of charges relative to the lattice in azimuthal direction. This motion will be compensated by an azimuthal electric field arizing from charge separation on opposite sides of rotor. As we can see from Ohm's law Eq. (3), effect of the Coriolis force on current gives rise to an electric field perpendicular to the current, whose magnitude is defined by Eq. (1). Here we have ignored the influence of magnetic field arising from effect of rotation on radial electric field because it depends from i^2 and can be eliminated by the special design of the experiment.

The potential difference produced should be minute in metals, but appreciable in semiconductors because the concentration of electric carriers $n \approx 10^{-14} cm^3$ much smaller in semiconductors with compare that in conductors. The sign (+ or -) of the coefficient $R_{gg} \approx 0.8 \times 10^{-22} s$ depends on the electric charge of the majority carriers, hence measurement of the R_{gg} coefficient should determine both the sign and the concentration of the majority carriers.

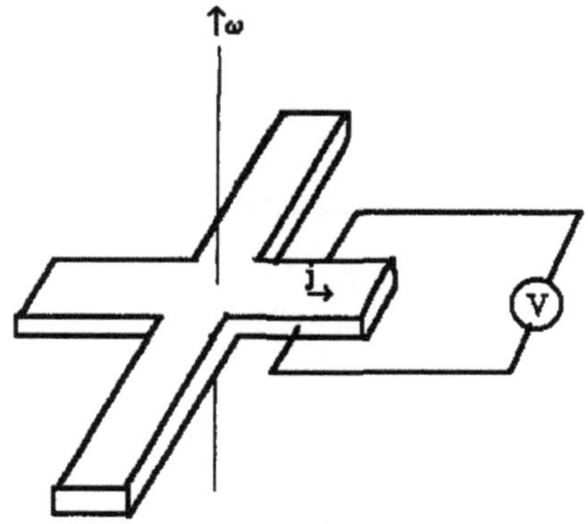

For typical values of parameters $\Omega = 10^3 s^{-1}$, $i_r = 10^3 A$ and $d = 10^{-2} cm$ in the absence of any additional effects one would expect that $V_{gg} = 0.7 \times 10^{-2} V$. An experiment to observe this voltage is to rotate a rotor carrying radial current and measure the potential difference through it in azimuthal direction (radial current in the rotor can be supported by thermoelectric effects as in experiment[12]). The most dangerous sources of experimental error, arising from earth's magnetic field, ionization in surrounding air etc should be circumvented during the experiment with the required accuracy. Measuring the galvanogyroscopic potential difference as a function of conduction current thus yields a direct measure of rotation rate which, in principle, can be used for construction of new type devices as high sensitivity speedometers of angular velocity.

If the semiconductor carrying radial current is at rest with respect to the earth, then there would be a voltage due to the earth's diurnal rotation with angular velocity $\Omega_\oplus \approx 7.3 \times 10^{-5} s^{-1}$. From the real estimations, for parameters $i_r = 10^3 A$ and $d = 10^{-2} cm$ one can find that azimuthal voltage $V_\oplus \approx 0.5 \times 10^{-10} V$ will be produced by galvanogyroscopic effect across the semiconductor rigidly rotating together with earth. The measurement of this voltage should reprove the diurnal rotation of earth, and in this sence, is analogous to mechanical measurements made with Foucault pendulum[13] for observation of earth's rotation.

Note that Eq. (1) is similar to the Hall voltage, produced by magnetic field $B = -2mc\Omega/e$. So we can say that $-2mc\Omega/e$ plays a role analogous to the magnetic field B. This analogy has also been pointed out by Moorhead and Opat in the recent paper[3] during derivation of EMF induced by centrifugal (not Coriolis) forces in rotating coils, and by Mashhoon[14] during formulation of the gravitational analogue of Larmor's theorem.

If the rod is bent into a closed ring galvanogyroscopic effect gives rise to a current in the ring. The magnetic field created by this current around rotating conductor with radial current has been detected in Ref. 12 and correctly explained in Ref. 7.

3 Possibility to measure earth's Lense-Thirring field using galvanogyroscopic effect

The spacetime metric describing the exterior field of the slow rotating gravitational object with gravitational radius α and angular momentum a is[15]

$$ds^2 = -\left(1 - \frac{\alpha}{r}\right)(dx^0)^2 + \left(1 - \frac{\alpha}{r}\right)^{-1} dr^2 + r^2 d\theta^2 + r^2 \sin^2\theta d\varphi^2$$

$$- \frac{2a\alpha \sin^2\theta}{cr} dx^0 d\varphi. \quad (6)$$

Thus according to the general relativity rotating body drags inertal frames around it with angular velocity $\Omega_{LT} = \frac{a\alpha}{r^3}$ which is known as Lense-Thirring one.

If a platform is nonrotating relative to the distant stars as determined by telescopes, then it would rotate with respect to the local inertial frames if a nearby rotating body was present because of the general relativistic Lense-Thirring field.[16] The Lense-Thirring field of the earth, for example, causes a small precession Ω_{LT} of the local inertial frames with respect to the distant stars whose direction is the same as the angular velocity $\Omega_\oplus$ of the earth but whose magnitude is about $10^{-9}\Omega_\oplus$. Really, if the rotating body is the earth (which is approximately spherically symmetric sphere of uniform density with radius $R_\oplus$) then $a_\oplus = \frac{2}{5}\Omega_\oplus R_\oplus^2$. For the earth with $\alpha_\oplus = 2 \times 0.44cm, R_\oplus \approx 6.37 \times 10^8 cm$ we have

$$\Omega_{LT} = \frac{2}{5}\frac{\alpha_\oplus}{R_\oplus}\Omega_\oplus \approx 0.55 \times 10^{-9}\Omega_\oplus .$$

Suppose a semiconductor is placed on the platform at rest with respect to the distant stars, then the semiconductor has angular velocity $-\Omega_{LT}$ with respect to the local inertial frames. Therefore, there would be small potential difference which is proportional to the Lense-Thirring field. This is the general relativistic generalization of galvanogyroscopic effect. Hence the general relativistic correction $V_{LT} = (R_{gg}/d)i_r\Omega_{LT}$ to the potential difference along the semiconductor, at rest on the platform, can be computed from Eq. (4) written in the metric defined by Eq. (6). If $i_r = 10^3 A$ and $d = 10^{-2}$ cm then the developed voltage is $0.3 \times 10^{-19} V$. Measuring this voltage would confirm the existence of the Lense-Thirring field. Such voltages have been measured previously; however, in our case, there is serious problem arising from magnetic field of the earth, which should be absolutely shielded. If it proves to be technically feasible to measure a voltage drop of $10^{-22}V$,[17] say by using a superconducting quantum interference device (SQUID) magnetometer, the sensitivity of device with $i_r = 10^3 A$ and $d = 10^{-3} cm$ is of the order $10^{-11}\Omega_\oplus$, which could be improved by increasing the value of electric current and decreasing the semiconductor's thickness. In proposed experiment superconducting loop of SQUID can be connected accross carrying radial current semiconductor which will form normal layer of Josephson junction with source of constant voltage V_{LT}. The nonvanishing potential difference V_{LT} would lead to a time varying current and magnetic flux through the loop which can be measured.

Two formidable difficulties in performing this experiment are (i) ensuring that the semiconductor is non-rotating with respect to the distant stars to the required accuracy, and (ii) shielding the earth's magnetic fields which can induce much bigger voltage due to the Hall effect. The first difficulty can be overcome by doing the experiment on a satellite.

Conclusion

In this paper, we have shown that the effect of the Coriolis force on the conduction current is to induce a galvanogyroscopic potential difference. Thus electrodynamic analogue of the well known mechanical Beer's law arising from the effect of the Coriolis force on river's stream has been found. Each element of water's flow in river experiences Coriolis force perpendicular to the flow and earth's axis of rotation. Thus in Northern hemisphere Coriolis force is directed to the right bank of river, and in Southern hemisphere on the left one, and consequently, the corresponding bank of river is more steep than opposite one. Similarly, according to galvanogyroscopic effect one side of carrying current conductor in the rotating frame of reference has more electric carriers relative to the opposite one. The Beer's and galvanogyroscopic effects arise from linear order of angular velocity of rotation and appear in the presence of relative velocity.

The galvanogyroscopic effect illustrates the fact when carrying current semiconductor non rotating relative to the distant stars is subject to general-relativistic gravitational field of earth, the potential difference $V_{LT} \approx 10^{-19}V$ will be developed across it. The semiconductor may be on a satellite or on the earth and moving so that its angular velocity relative to the distant stars has no component normal to the plane of semiconductor's plate as determined by telescope.

Thus the general relativistic generalization of galvanogyroscopic effect seems to be experimentally verifiable at present within capacity of modern technical measurements. Because of the great importance of gravitomagnetic fields for astrophysics and fundamental physics, such laboratory experiment can be done which would constitute the direct measurement of the general relativistic Lense-Thirring field.

Acknowledgments

The part of work was done when the author was visiting the Abdus Salam International Center for Theoretical Physics, Trieste and he is grateful to the ICTP, for the financial support which made this visit possible. The author greately acknowledges a grant from Remo Ruffini for his participation at the III W. Fairbank Meeting and First ICRA Network Workshop on "Lense-Thirring Effect".

References

1. I. Ciufolini and J.A. Wheeler, *Gravitation and Inertia* (Princeton Univ. Press, New Jersey, 1995).
2. J.W. Beams, *Phys. Rev. Lett.* **21**, 1093 (1968).
3. G.F. Moorhead and G.I. Opat, *Class. Quantum Grav.* **13**, 3129 (1996).
4. T.W. Darling, F. Rossi, G.I. Opat and G.F. Moorhead,*Rev. Mod. Phys.* **64**, 237 (1992).

5. F.C. Witterbonk and W.M. Fairbank, *Nature* **220**, 436 (1968); *Phys. Rev. Lett.* **19**, 1049 (1967); *Rev. Sci. Instrum.* **48**, 1 (1977); J.M. Lockart, F.C. Witterbonk and W.M. Fairbank, *Phys. Rev. Lett.* **38**, 1220 (1977).
6. E.H. Hall, *Am. J. Math.* **2**, 287 (1951).
7. B.J. Ahmedov, *Gravitation & Cosmology* **4**, 139 (1998).
8. J. Anandan, *Phys. Lett.* A **105**, 280 (1984); *Gen. Rel. Grav.* **16**, 33 (1984); in *New Techniques and Ideas in Quantum Measurement Theory*, ed. D. Greenberger (Ann. New York Acad. Sci., 1986) **480** p. 224.
9. T.J. Davis and G.I. Opat, *Class. Quantum Grav.* **5**, 1011 (1988).
10. B.J. Ahmedov and L.Ya. Arifov, *Gen. Rel. Grav.* **26**, 1187 (1994).
11. R. Khanna, *Mon. Not. R. Astron. Soc.* **294**, 673 (1998).
12. B.V. Vasil'ev, *JETP Lett.* **60**, 47 (1994).
13. L. Foucault, *Travaux scientifiques de Foucault* (Paris, 1878).
14. B. Mashhoon, *Phys. Lett.* A **173**, 347 (1993).
15. B. Mashhoon, F.W. Hehl and D.S. Theiss, *Gen. Rel. Grav.* **16**, 711 (1984).
16. H. Thirring, *Phys. Z.* **19**, 33 (1918); *Phys. Z.* **22**, 29 (1921); J. Lense and H. Thirring, *Phys. Z.* **19**, 156 (1918).
17. A.K. Jain, J. Lukens and J.S. Tsai, *Phys. Rev. Lett.* **58**, 1165 (1987).

DETERMINATION OF THE PPN PARAMETER γ THROUGH GEODETIC VLBI

M. CATALDO, L. GUERRIERO
Physics Department, University of Bari,
E-mail: cataldo@ba.infn.it, guerriero@ba.infn.it

G. BIANCO, R. LANOTTE
Space Geodesy Centre "G. Colombo", Matera,
E-mail: bianco@mt.asi.it, lanotte@mt.asi.it

F. DILETTUSO
Astronomy Department, University of Bologna,
E-mail: dilett@mt.asi.it

The analysis of a large data set (from 1990 to 1995) of VLBI group delays has provided a new γ estimate ($\gamma = 1.0008 \pm 0.0012$), highlighting some critical issues in treating data which have undergone processing in many different steps and places.

1 State of the art

The PPN parameter γ has been measured with different experimental methods. The traditional approach based on determination of light deflection has obtained a major improvement from the astrometric satellite Hipparcos and from direct time delay measurement from Viking probes, however the VLBI localization of radio emitters allows continuos and more precise measurements. Figure 1 shows the present status in γ estimate.

2 This work

Following the way pioneered by D.H. Robertson a new γ estimate has been obtained using the geodetic VLBI database from 1990 to 1995 available at Matera's Space

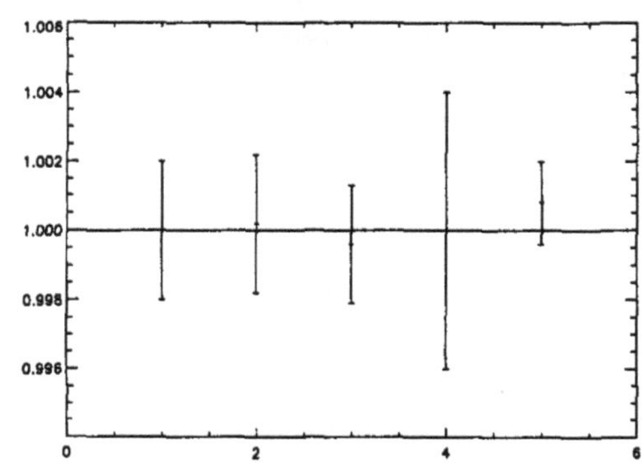

Figure 1. Most significant γ estimates. From left to right: Viking probes (Reasenberg, Shapiro et al. [1]), geodetic VLBI (Robertson [2]), astrometric VLBI (Lebach [3]), Hipparcos [4], geodetic VLBI (this work)

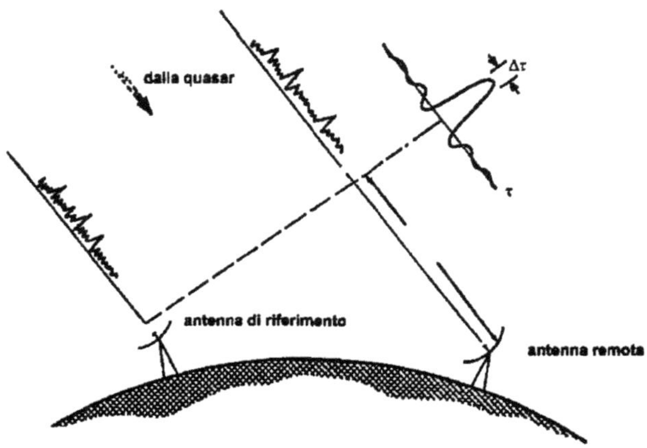

Figure 2. A schematic view of a VLBI observation

Geodesy Center. The total database includes 1300 VLBI experiments. Only the "best" ones (about 100) have been selected for the present analysis.

3 The VLBI observable: group delay

A VLBI experiment (or session) consists of several radio sources observations (each one lasting a few minutes) carried out by a network of antennas over a period of about 24 hours. Each experiment gives a large group delay data set corresponding to the varying orientation of the baseline relative to the radio sources. Group delay is modeled as

$$\tau = \tau_{gm} + \tau_n + \tau_d, \tag{1}$$

where τ_{gm} is the difference in propagation times along the vacuum signal path (geometric delay), τ_n includes other non dispersive terms (clock behavior, troposphere, etc.), and τ_d is a frequency dependent term related to the dispersive elements in the path (ionosphere, solar corona). Important information of geodetic interest (tectonic displacements, tides, polar wobbles, etc.) are obtained from the delay, sensitive to the baseline length variation in time (Dickey, Eubanks [5]).

4 Analysis software

Standard VLBI data analysis software (CALC-SOLVE) from Goddard Space Flight Center has been used in data reduction. It essentially performs a multi parameter least squares fit, comparing theoretical and experimental delay values to estimate first order corrections to a priori parameters (Himwich [6]). Each fit includes group delay data from a single experiment: an independent γ estimate has been obtained as solution for each observational session. Experimental delay values are given coupled to an estimated a priori instrumental error which has proved itself to be not a valid representation of the actual data quality. This value should be modified

τ_{rel}	angular distance from the body's center				
body	Grazing	1°	30°	90°	175°
Sun	169 *ns*	45 *ns*	1.5 *ns*	0.4 *ns*	17 *ps*
Jupiter	1.5	11 *ps*	0.4 *ps*	0.1 *ps*	
Saturn	0.5	2	0.05		
Uran	0.18	0.1			
Neptune	0.23	0.1			
Mercury	4 *ps*	0.04			
Venus	33	0.4	0.02		
Mars	6.5	0.04			
Moon	1.1	0.47	0.02		
Earth	21				

Figure 3. Relativistic (Shapiro) term in group delay (Klioner [8])

by a scale factor plus a noise floor (of the order of 20 ps) in order to agree with the experimental error (Ray, Corey [7]). For this reason a constant Formal Error (FEC) is added to the instrumental one for all measures from the same observational session during the least squares fit. The FEC value is chosen in order to obtain a distribution of group delay residuals consistent with the total error, that is

$$\chi^2_\nu = \frac{\Sigma_\kappa \frac{r^2_\kappa}{\sigma'^2_{\tau\kappa}}}{\nu} = 1, \quad (2)$$

where $\sigma'_\tau = \sqrt{\sigma^2_\tau + \sigma^2_{FEC}}$, r_κ and $\sigma'_{\tau\kappa}$ denote, respectively, the post-fit residual and standard error for the κ-th observation and n the degrees of freedom.

5 Floor error on γ

The geometric delay includes the effect of signal path deflection introduced by the Solar System bodies' gravitational field. Within the Parametrized Post Newtonian framework this contribution depends only on the parameter γ. We can approximately write the delay in the form

$$\tau_{gm} = \tau_{cl} + (\gamma + 1)\tau_{rel}\,. \quad (3)$$

For $\gamma = 1$ (General Relativity) the values of the relativistic contributions to the geometric delay for a baseline perpendicular to the source direction are given in Figure 3.

The presence of the "noise floor" on measures sets a limit on the maximum reachable precision on γ estimate through VLBI. A rough value for it can be derived as follows

$$\gamma = \frac{\tau - \tau_{cl}}{\tau_{rel}} - 1\,. \quad (4)$$

Assuming a 20 ps error floor and a geometry maximizing the relativistic component of the geometric delay we obtain

$$\delta\gamma = \frac{\delta\tau}{\tau_{rel}} = 6x10^{-5}\,. \quad (5)$$

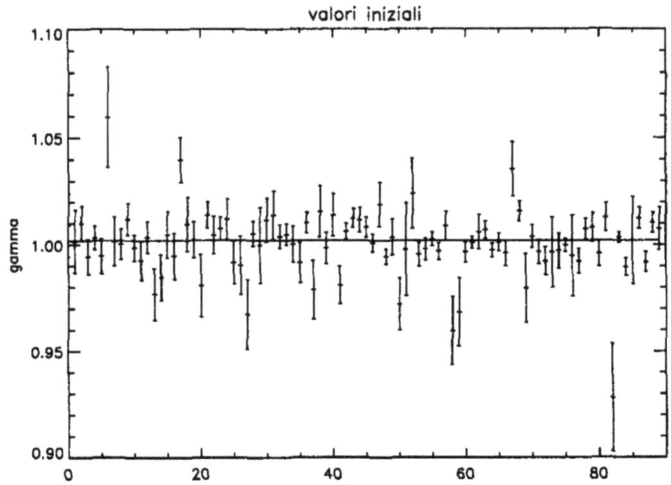

Figure 4. Plot of γ estimates from the selected experiments displayed with formal error bars.

6 Experiments' selection from database

Two criteria were used to extract the most suitable experiments from the entire database. The geometry should minimize:

(i) the formal error (from covariance matrix) on γ best fit correction

(ii) the correlation between γ and the other estimated parameters

Preliminary results for the best selection (90 experiments) are shown in Figure 4.

The value obtained for the average γ is

$$\gamma = 1.0020 \pm 0.0006\,. \tag{6}$$

However there is evidence that other error sources are not considered in the adopted model. In fact

$$\chi^2_\nu = 2.0044, \tag{7}$$

is too high even though we introduced unmodeled errors through FECs. Several global parameters (not strictly dependent on the single experiment conditions) are used in the model assuming for them the values reported in the official a priori files established by IERS. The standard deviations of all these a prioris are not considered during the least squares fit. In particular, the errors on sources a priori positions can have a strong effect on γ estimate and explain the experimental residuals distribution. To correctly evaluate the effects of these errors in the knowledge of sources positions one should introduce these uncertainties, known from the catalogue (we used IAU95), in SOLVE least squares procedure. In order not to modify the CALC - SOLVE software we evaluated the effect of these uncertainties via a MonteCarlo-like method.

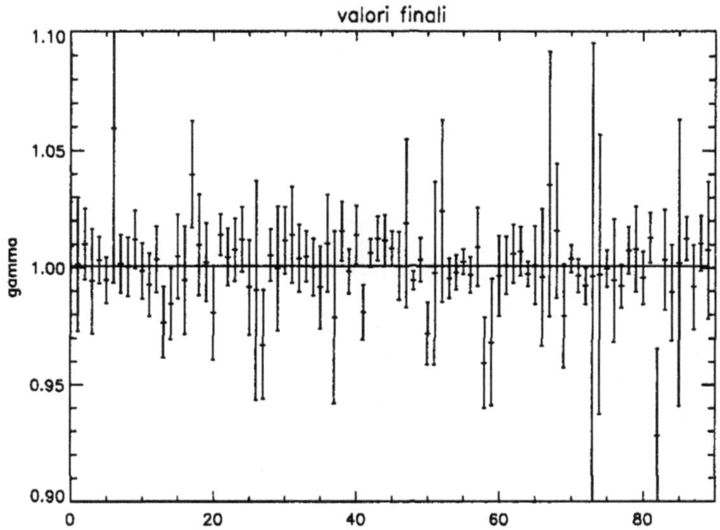

Figure 5. Final results with standard errors inflated through MonteCarlo simulation of new sources positions.

7 Procedure

To find out the contribution to the γ values dispersion due to sources positions uncertainties

(i) we repeated the least squares fit for each experiment twenty times with modified sources coordinates. These were obtained randomizing the nominal position values with gaussian probability according to the known associated uncertainties

(ii) we evaluated the dispersion of γ values for a single experiment

(iii) for each experiment we maintained the gs from the most probable a priori values but the total associated error was taken as the root sum square of the fit output error (formal error) and the evaluated dispersion.

8 Final results

$$\gamma = 1.0008 \pm 0.0012, \tag{8}$$

$$\chi^2_\nu = 0.6525 \,. \tag{9}$$

9 Conclusions

The evidence of unmodeled errors in VLBI data appears linked to a procedure lack: the imperfect knowledge of a prioris was not considered during the multiparameter fit. Introduction of errors on sources positions gives agreement between residuals and standard errors for the γ values and supports good confidence for the standard deviation estimated on the mean γ value. The results encourage further analysis in order to determine whether our total error has been overestimated since the effect due to the sources positions could be already partially corrected by the FEC calculation. In addition it seems proper to introduce these uncertainties in the least squares fit model since the source position error can influence the fitted γ value.

References

1. R.D. Reasenberg, I.I. Shapiro et al., ApJ **234**, L219 (1979).
2. D.S.Robertson, W.E.Carter, W.H. Dillinger, Nature **349**, 768 (1991).
3. D.E.Lebach at al.,Phys.Rev.Lett. **75**, 1434 (1995).
4. http://astro.estec.esa.nl/SA-general/Projects/Hipparcos/news-scires.html.
5. J.O.Dickey, T.M.Eubanks, Space Geodesy and Geodynamics, 221, Academic Press Inc., London (1986).
6. E.Himwich, Estimation according to SOLVE, GSFC internal memo (1990).
7. J.R.Ray, B.E.Corey, Current precision of VLBI multi-band delay observables, Proc. AGU Chapman Conference on Geodetic VLBI: Monitoring Global Change (1991).
8. S.A.Klioner, General relativistic models of VLBI observables, Proc. AGU Chapman Conference on Geodetic VLBI: Monitoring Global Change (1991).

QUANTUM TESTS OF LENSE–THIRRING TYPE EFFECTS

C. LÄMMERZAHL
Laboratoire des Physique des Lasers, Institut Galilée, Université Paris 13, F - 93430 Villetaneuse, France
and
Department of Physics, University of Konstanz, Fach M 674, D - 78457 Konstanz, Germany
e-mail: claus.laemmerzahl@uni-konstanz.de

The possibility of quantum tests of the Lense–Thirring effect is analyzed. These quantum tests are (i) atomic interferometry with spin flip, (ii) atomic interferometric measurement of the acceleration, and (iii) Hughes–Drever type experiments. With the present day techniques it is not possible to observe the Lense–Thirring effect quantum mechanically.

1 Introduction

There is no need to emphasize the importance of the experimental verification of the Lense–Thirring effect.[1] This effect is a prediction of Einstein's General Relativity which is not present in Newtonian gravity. It has been claimed that this effect has been observed using the trajectories of satellites.[2] With the Gravity Probe B project it is planned to test this effect within the next two years to an accuracy of 1%.[3]

In this short contribution we want to analyze possible quantum tests of this effect. We embed the description of these quantum tests in a general theory considering quantum tests of gravitational theories. Our main results are: (i) it is not possible with the present day technology to compete with the tests on the classical level. (ii) many effects in generalized theories of gravity are of Lense–Thirring type and it needs more considerations to distinguish the various contributions.

Measuring the Lense–Thirring effect amounts to the measurement of a rotation. The typical rotations which are present on the earth are (see e.g. Ref. 3) the rotation of the earth $\omega_{\text{earth}} \sim 10^{-5}$ sec^{-1}, $\Omega_{\text{geod.prec.}} = \frac{3}{2}\boldsymbol{v} \times \boldsymbol{\nabla} U \sim 10^{-13}$ sec^{-1}, and $\Omega_{\text{LT}} = \frac{1}{2}\frac{G}{c^3}\nabla \times \frac{\boldsymbol{J}\times\boldsymbol{r}}{r^3} \sim 10^{-14}\,\text{sec}^{-1} \sim 10^{-9}\omega_{\text{earth}}$. The question we want to answer is whether it is possible with quantum devices to measure these rotations.

2 The generalized Pauli equation

The equation we use for the description of gravitational effects in quantum systems is the generalized Pauli equation (GPE) in a rotating and accelerating frame. The first step in deriving this equation is a generalized Dirac equation which can be derived from the requirements of (i) deterministic evolution, (ii) superposition principle, (iii) finite propagation speed, and (iv) a conservation law. It is now a mathematical result[4] that fields obeying these principles must obey a first order system of partial differential equations,

$$i\widetilde{\gamma}^{\mu}\partial_{\mu}\varphi - \widetilde{M}\varphi = 0 \qquad (1)$$

which we call a generalized Dirac equation (GDE). Here, the matrices $\tilde{\gamma}^\mu$ are not assumed to fulfill a Clifford algebra. The GDE is a hyperbolic 1st order system. Since we want to restrict ourselves to 2 spin states we assume $\varphi(x) \in \mathbf{C}^4$. The GDE defines a maximal velocity which we call c. For further discussion, see Ref. 5.

After a specification of the possible gravitational fields and performing a non-relativistic limit we arrive at the generalized Pauli equation (GPE)[5]

$$
\begin{aligned}
i\frac{\partial}{\partial t}\varphi = -\frac{1}{2m}\Big(\delta^{ij} \underbrace{-\frac{\delta m_{\mathrm{i}}^{ij}}{m} - \frac{\delta \bar{m}_{\mathrm{i}k}^{ij}\sigma^k}{m}}_{\substack{\text{spin–dependent tensor}\\ \text{of anom. inertial mass}}}\Big)\nabla_i\nabla_j\varphi + \underbrace{\Big(cA_j^i + \frac{1}{m}a_j^i\Big)\sigma^j i\nabla_i\varphi}_{\text{spin–momentum–coupling}} \\
+\Big[\boldsymbol{\omega}\cdot\boldsymbol{J} + \boldsymbol{\Omega}_{\mathrm{LT}}\cdot\boldsymbol{J} + \boldsymbol{\Omega}_{\text{geod. prec.}}\cdot\boldsymbol{J} + mU(\boldsymbol{x}) \qquad (2) \\
+\underbrace{\boldsymbol{C}\cdot\boldsymbol{\sigma}\, mU(\boldsymbol{x}) + \delta m_{\mathrm{g}ij}U^{ij}(\boldsymbol{x})}_{\substack{\text{spin–dependent tensor}\\ \text{of anom. gravitational mass}}} + \underbrace{c\boldsymbol{T}\cdot\boldsymbol{\sigma}}_{\substack{\text{torsion–like}\\ \text{coupling}}} + \underbrace{mc^2\boldsymbol{B}\cdot\boldsymbol{\sigma}}_{\substack{\text{spin–dep.}\\ \text{‘rest–mass’}}}\Big]\varphi,
\end{aligned}
$$

where we introduced the coupling to various rotations and neglected products of rotation and anomalous terms. In this equation the following quantities appear: $\delta m_{\mathrm{i}}^{ij}$ and $\delta\bar{m}_{\mathrm{i}k}^{ij}$ are anomalous inertial mass tensors (the second one depends on the spin of the particle), $cA_j^i + \frac{1}{m}a_j^i$ gives rise to a spin–momentum coupling, $\boldsymbol{\omega}$ is the rotation of the frame, $\boldsymbol{\Omega}_{\mathrm{LT}}$ and $\boldsymbol{\Omega}_{\text{geod. prec.}}$ are the rotation of the frame due to the Lense–Thirring effect and to the geodesic precession, respectively, $\boldsymbol{J}$ is the total angular momentum, $\delta m_{\mathrm{g}ij}$ and $\boldsymbol{C}$ are anomalous gravitational mass contributions (the second one again depends on the spin), $\boldsymbol{T}$ is a torsion–like coupling, and $\boldsymbol{B}$ gives a spin–dependent 'rest mass" contribution. The quantities $\delta m_{\mathrm{i}}^{ij}$, $\delta\bar{m}_{\mathrm{i}k}^{ij}$, A_j^i, a_j^i, $\boldsymbol{T}$, and $\boldsymbol{B}$ violate Local Lorentz Invariance, $\delta m_{\mathrm{g}ij}$ and $\boldsymbol{C}$ violate Local Position Invariance, and together they violate the Einstein Equivalence Principle, see e.g. Ref. 6.

The term $\boldsymbol{\omega}\cdot\boldsymbol{J}$ gives the Sagnac effect as well as the spin–rotation coupling[7], and $\boldsymbol{\Omega}_{\mathrm{LT}}\cdot\boldsymbol{J}$ and $\boldsymbol{\Omega}_{\text{geod. prec.}}\cdot\boldsymbol{J}$ lead to analogous effects. $mU(\boldsymbol{x})$ yields the usual gravitational acceleration effects leading e.g. in neutron interferometry the COW experiment.

From the GPE we can calculate the classical acceleration

$$
\begin{aligned}
a^i = -\left(\delta^{ij} + \frac{\delta m_{\mathrm{i}}^{ij}}{m} + \left(\frac{\delta\bar{m}_{\mathrm{i}k}^{ij}}{m} + \delta^{ij}C_k\right)S^k\right)\partial_j U - \delta^{ij}\frac{\delta m_{\mathrm{g}kl}}{m}\partial_j U^{kl} \\
+2\boldsymbol{\Omega}^{\mathrm{total}}\times\boldsymbol{v} - \boldsymbol{\Omega}^{\mathrm{total}}\times(\boldsymbol{\Omega}^{\mathrm{total}}\times\boldsymbol{r}). \qquad (3)
\end{aligned}
$$

The first line is the gravitational acceleration including anomalous terms characterized by the quantities $\delta m_{\mathrm{i}}^{ij}$, $\delta m_{\mathrm{g}}^{ij}$, and C_k. The first term in the second line is the Coriolis force connected with the total angular velocity $\boldsymbol{\Omega}^{\mathrm{total}} = \boldsymbol{\omega}_{\mathrm{earth}} + \boldsymbol{\Omega}_{\mathrm{LT}} + \boldsymbol{\Omega}_{\text{geod. prec.}}$.

Inspection of Eqn. (3) shows that on the quantum level, that is in Eqn. (2), there are more possibilities to violate EEP than on classical level, even if one uses spin–polarized matter[8,9] (see also the review[10]). We get the complete set of EEP–violating parameters by including the dynamical behavior of spin expectation value in classical approximation:

$$\frac{d}{dt}\boldsymbol{S} = \boldsymbol{\Omega} \times \boldsymbol{S} \tag{4}$$

with

$$\Omega_i := \frac{1}{2m}\frac{\delta\bar{m}^{kl}_{ii}}{m}p_k p_l + \left(\frac{1}{m}a^j_i + cA^j_i\right)p_i + mc^2 B_i + cT_i + C_i m U(x) + \Omega^{\text{total}}_i \tag{5}$$

That means that for testing the complete set of EEP parameters one needs the path *and* the spin of quantum particles. It is not enough to measure the acceleration of quantum particles. However, measuring the precession of the spin expectation value is not very accurate. Therefore one uses spectroscopy in order to measure these spin effects. Consequently, we are going to describe spectroscopic tests of the Lense–Thirring effect instead of spin–precession tests.

3 Quantum Tests

According to the Hamiltonian (2) we have the possibility to measure effects on the spin and on the wave analogy of the path motion. Especially for rotation effects, both the spin and the motion of wave packets couple to rotation. Therefore, especially for the detection of rotation, it is worth to consider effects on the path of wave packets as well as effects on the spin of quantum objects. In addition, we have the possibility to consider the dynamics of elementary particles and the dynamics of composed systems. For the latter we have in addition the possibility to observe the energy of the bound states giving additional information about the couplings in question. It can be shown[11] that if the single particle Hamiltonian has the form (2) then the Hamiltonian for the center–of–mass motion as well has the Hamiltonian for the energy levels have the same structure except that the coefficients are different combinations of the coefficients of the single particles. Therefore we carry on using Hamiltonians of the form (2).

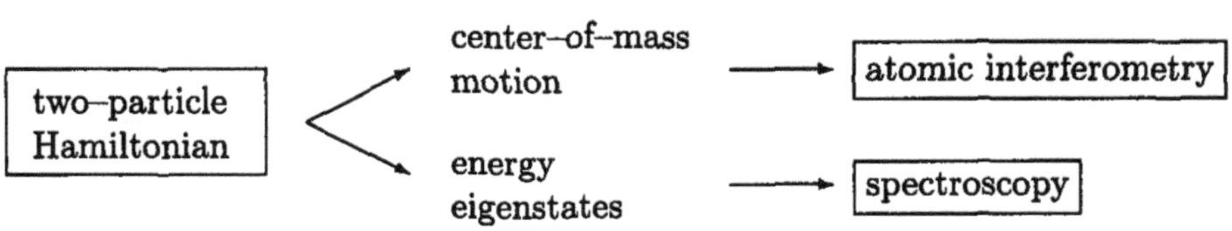

3.1 *Atomic interferometry*

We propose two different kinds of interference experiments: The first is an interference experiment in spin space using a spin–flip technique, the second one is an interference experiment of the Kasevich and Chu type[12] measuring the gravitational acceleration of atoms. The used atoms are Cesium atoms with a mass $m_{\text{Cs}} = 2.21 \cdot 10^{-25}\,\text{kg}$. In both cases we take an interaction time of $\Delta t = 0.1\,\text{sec}$, an accuracy $\Delta\phi/\phi \leq 10^{-10}$ for the relative phase shift ϕ,[13] and an absolute accuracy of the phase $\Delta\phi \leq 10^{-4}$.

Spin–flip–experiment

In spin–flip experiments a beam of polarized atoms is split coherently in spin space using a spin flip mechanism where half of the atoms stay in the original spin state and the other half have a reversed spin state. After a time Δt again a spin–flip mechanism is applied leading to a superposition of the two spin states. If there is a spin–dependent interaction, the two different spin states accumulate different amounts of energy leading to an observable phase shift

$$\delta\phi = \frac{1}{\hbar}\left[H(p,S) - H(p,-S)\right]\Delta t \tag{6}$$

Using the above Hamiltonian for vanishing electromagnetic field leads to the phase shift

$$\delta\phi = \frac{1}{\hbar}\left(\frac{\delta\bar{m}^{ij}_{\text{I}k}}{m}p_i\delta_{jl}l^l - 2\left(\hbar\lambda^i_k + mc\Lambda^i_k\right)\delta_{ij}l^j \right.$$
$$\left. + \left(2mc^2B_k + 2cT_k + 2C_k mU + \Omega^{\text{tot}}_k\right)\Delta t\right)S^k\,. \tag{7}$$

If we assume that the experimental outcome will give a null result, then the accuracy of the Kasevich–Chu atomic interferometer[12] will lead to the estimates given in Table 1 which generalizes a previous work.[14] For some coefficients one gets better results with Hughes–Drever experiments.

The accuracy of experiment allows the detection of rotations of about 0.9 sec^{-1}. This is far too less for measuring ω_{earth}, $\Omega_{\text{geod.prec.}}$, or Ω_{LT}.

Measurement of acceleration

Atomic interferometry in gravitational fields leads to the acceleration induced phase shift[12]

$$\phi^{\text{atom}} = \boldsymbol{k}\cdot\boldsymbol{a}\,T^2\,, \tag{8}$$

where $\boldsymbol{a}$ is the acceleration and T is the flight of time. Using the acceleration (3) this gives

$$\delta\phi = -k_i\left(\delta^{ij} + \frac{\delta m^{ij}_{\text{i}}}{m} + \left(\frac{\delta\bar{m}^{ij}_{\text{i}k}}{m} + \delta^{ij}C_k\right)S^k\right)\partial_j U - k^j\frac{\delta m_{gkl}}{m}\partial_j U^{kl}T^2$$
$$+2\boldsymbol{k}\cdot(\boldsymbol{\omega}\times\boldsymbol{v})T^2\,, \tag{9}$$

where we neglected squares of rotation.

The angular momentum part gives the Sagnac effect and effects due to geodesic precession and the Lense–Thirring effect. Since the rotation in our cases is very small, we neglect the last term. The present–day sensitivity of atomic interferometers measuring rotation[15] $2\cdot 10^{-8}$ (rad/sec)/$\sqrt{\text{Hz}}$ allows one to detect rotations of about $\omega \sim 10^{-7}\,\text{sec}^{-1}$. The rotation of the earth has been observed this way[15] within an accuracy of $\Delta\omega \leq 0.5\cdot 10^{-6}\,\text{Hz}$. The present accuracy of this apparatus is of the order $\Delta\omega \sim 10^{-8}\,\text{Hz}$. It is clear that this accuracy is not enough to measure the geodesic precession or the Lense–Thirring effect. However, there is a big effort in order to improve the capabilities of the interferometers for example by enlarging the time of flight of the atoms (or the length of the interferometer, or by transferring more than one photon of the laser beam (which gives the amount of the beam splitting) to the atoms.

The quantities appearing in the acceleration parts (first line of Eqn. (9)) can be estimated to

$$\left|\Delta\frac{\delta m_{gij}}{m}\right|, \left|\Delta\frac{\delta m_{izx}}{m}\right|, |\Delta C_k| \leq 10^{-10} \tag{10}$$

if one uses two different kinds of atoms and assumes a null result.

3.2 Spectroscopy

Doing spectroscopy of atoms, that is, observing energy shifts and especially the splitting of energy singlet states, with an adiabatically time varying quantization axis amounts to a Hughes–Drever–experiment, see e.g. Ref. 6. Usually one takes an atomic nucleus consisting in a $J = 0$ core and a valence proton with $L = 1$. Therefore we do spectroscopy of states $|J, M_J\rangle$ with $J = 3/2$. A standard calculation with the Hamiltonian (2) gives

$$\begin{aligned} E(\tfrac{3}{2},\tfrac{3}{2}) - E(\tfrac{3}{2},\tfrac{1}{2}) &= \delta + \bar{\delta}_1 + \bar{\delta}_2 \\ E(\tfrac{3}{2},\tfrac{1}{2}) - E(\tfrac{3}{2},-\tfrac{1}{2}) &= \bar{\delta}_2 \\ E(\tfrac{3}{2},-\tfrac{1}{2}) - E(\tfrac{3}{2},-\tfrac{3}{2}) &= -\delta + \bar{\delta}_1 + \bar{\delta}_2 \, , \end{aligned} \tag{11}$$

with

$$\begin{aligned} \delta &:= -\frac{\hbar^2}{3a^2}\left(\frac{\delta m_{\rm i}^{xx}}{m^2} + \frac{\delta m_{\rm i}^{yy}}{m^2} - 2\frac{\delta m_{\rm i}^{zz}}{m^2}\right) \\ \bar{\delta}_1 &:= -\frac{\hbar^2}{a^2}\left(\frac{\delta \bar{m}_{\rm iz}^{xx}}{m^2} + \frac{\delta \bar{m}_{\rm iz}^{yy}}{m^2} - 2\frac{\delta \bar{m}_{\rm iz}^{zz}}{m^2}\right) \\ \bar{\delta}_2 &:= -\frac{5\hbar^2}{3a^2}\frac{\delta \bar{m}_{\rm iz}^{zz}}{m^2} + \frac{2}{3}(mc^2 B_z + mC_z U) + \frac{\hbar}{c}T_z + \omega_z + \Omega_z + \Omega_z \, . \end{aligned} \tag{12}$$

Here we used the harmonic oscillator potential as nuclear binding potential, and a is the radius of nucleus.

The accuracy of present day apparatus[16,17,18] is given by $\delta E < 0.45 \cdot 10^{-6}$ Hz, so that ($a = 1.5$ fm and the gravitational potential as that of our galaxy $U/c^2 \approx 10^{-7}$) and the experimental result that no shift of the type (11) was observed leads to the following estimates

$$\left|\frac{\delta \bar{m}_{\rm iz}^{xx}}{m}\right| \leq 5 \cdot 10^{-30}, \quad \left|\frac{\delta m_{\rm i}^{xx}}{m}\right| \leq 5 \cdot 10^{-30}, \quad |B_z| \leq 3 \cdot 10^{-31},$$
$$|C_z| \leq 3 \cdot 10^{-24}, \quad |T_z| \leq 1.5 \cdot 10^{-15}\,\mathrm{m}^{-1},$$

provided that no fortuitous cancelations occurred. Due to the result $\Delta E(\omega_{\rm earth}) \sim 10^{-4}\ \mathrm{sec}^{-1}$ the rotation of the earth has an observable effect on the spectrum of atoms. Indeed, in order to make a consistent interpretation of the observed energy levels, one has to include corrections coming from the rotation of the earth[19] which

Table 1. List of the parameters appearing in the GPE which violate LLI or LPI, together with estimates from proposed and already done experiments. In this list we do not consider the combinations which are actually measured but apply the experimental result to each term assuming no cancelation. Entries without numbers indicate that these quantities cannot be measured by the corresponding class of experiments.

Class	Parameter	interference experiments		Hughes–Drever
		pure spin–flip	acceleration	
earth's rot.	$\boldsymbol{\omega}$	not measurable	measurable	measurable
geod. prec.	$\boldsymbol{\Omega}_{\text{geod.prec.}}$	not measurable	not measurable	not measurable
Lense–Th.	$\boldsymbol{\Omega}_{\text{LT}}$	not measurable	not measurable	not measurable
LLI$_0$	$\lvert\delta m_{\text{i}}^{ij}/m\rvert$	—	$\leq 10^{-10}$	$\leq 1.6 \cdot 10^{-28}$
	$\lvert\delta\bar{m}_{\text{i}k}^{ij}/m\rvert$	$\leq 5 \cdot 10^{-15}$	$\leq 10^{-10}$	$\leq 5 \cdot 10^{-29}$
	Λ_j^i	$\leq 8 \cdot 10^{-24}$	—	—
	B_i	$\leq 3 \cdot 10^{-30}$	—	$\leq 6 \cdot 10^{-30}$
	λ_j^i	$\leq 5 \cdot 10^{-6}\,\text{m}^{-1}$	—	—
	T_i	$\leq 3 \cdot 10^{-9}\,\text{m}^{-1}$	—	$\leq 1.5 \cdot 10^{-15}\,\text{m}^{-1}$
LPI	$\lvert\delta m_{\text{g}}^{ij}/m\rvert$	—	$\leq 5 \cdot 10^{-10}$	—
	C_i	$\leq 3 \cdot 10^{-23}$	$\leq 10^{-10}$	$\leq 3 \cdot 10^{-23}$

also is a confirmation of the spin–rotation coupling.[20] Such type of tests can also be used for an estimate on the strength of torsion.[21]

The corresponding angular velocities of the geodesic precession and of the Lense–Thirring effect yield an energy shift $\Delta E(\omega_{\text{geod.prec.}}) = 10^{-13}\ \text{sec}^{-1}$ and $\Delta E(\omega_{\text{LT}}) = 10^{-14}\ \text{sec}^{-1}$, respectively. These angular velocities are too small to be detected spectroscopically. However, since the performance of these type of tests on the earth requires a long–term stability of the apparatus (the quantization axis – the magnetic field – is attached to the surface of the earth), the corresponding experiment in space can be more accurate because there the quantization axis can be changed rather quickly. This means that the accuracy of the experiment is determined by the short term stability of the apparatus which usually is much more better than the long–term stability.

4 Conclusion

Staring from our generalized Pauli equation (2), we calculated the phase shifts due to the three rotational effects, namely the earth's rotation, geodesic precession, and Lense–Thirring effect, for three types of experiments: a spin–flip interferometric experiment, an interferometric acceleration measurement, and a spectroscopic experiment. The accuracy of the present–day apparatus tell us that up tp now only the earth's rotation is measurable with atomic interferometry by measuring the Coriolis force on the atoms.

These results are compared with general quantum tests of non–Riemannian space–time structures, see Table 1. It is interesting that the same class of experiments give extremely accurate estimates on hypothetical anomalous inertial mass tensors and other anomalous terms, but is not capable to measure the Lense–Thirring effect or the geodesic precession. One reason for this may be that according to (12) the search for an anomalous mass tensor is a high–energy test while any coupling to rotation does not involve the high (kinetic) energies of the valence proton.

Acknowledgments

I thank Ch. Bordé, H. Dehnen, F.W. Hehl, A. Peters, T. Pfau, and St. Schiller for discussions, Ch. Bordé and Ch. Chardonnet for their hospitality at the Laboratoire de Physique des Lasers, and the CNRS (France) and the "Optik–Zentrum" of the University of Konstanz for financial support.

References

1. C W Misner, K S Thorne, and J A Wheeler: *Gravitation*, Freeman, San Francisco 1973.
2. I Ciufolini *et al.*, *Class. Quantum Grav.* **14**, 2701 (1997).
3. S Buchmann and F Everitt *et al.*, *Class. Quantum Grav.* **13**, A 185 (1996).
4. J Audretsch and C Lämmerzahl **in:** U Majer and H-J Schmidt (eds.): *Semantical Aspects of Space-Time Geometry*, BI Verlag, Mannheim 1993.
5. C Lämmerzahl, *Class. Quantum Grav.* **15**, 13 (1998).
6. C M Will: *Theory and Experiment in Gravitational Physics* (Revised Edition), Cambridge University Press, Cambridge, 1993.
7. B Mashhoon B., *Phys. Rev. Lett.* **61** 2639 (1988).
8. C–H Hsieh *et al.*: *Mod. Phys. Lett.* **4**, 1597 (1989).
9. R C Ritter, L I Winkler, and G T Gillies: *Phys. Rev. Lett.* **70**, 701 (1993).
10. C Lämmerzahl **in:** P G Bergmann *et al.* (eds.): *Spin in Gravity*, WorldScientific, Singapore 1998.
11. C Lämmerzahl and U Bleyer: A Quantum Test Theory for Basic Principles of Special and General Relativity with applications to atomic interferometry and Hughes–Drever type experiments, *preprint*, University of Konstanz 1998.
12. M Kasevich and S Chu: *Phys. Rev. Lett.* **67**, 181 (1991); *Appl. Phys.* **B 54**, 321 (1992).
13. A Peters: *High Precision Gravity Measurements Using Atom Interferometry*, Dissertation, Stanford University (1998).
14. J Audretsch, U Bleyer, and C Lämmerzahl, *Phys. Rev.* **A 47**, 4632 (1992).
15. T L Gustavson, P Bouyer, and M A Kasevich: *Phys. Rev. Lett.* **78**, 2046 (1997).
16. T E Chupp *et al.*: *Phys. Rev. Lett.* **63**, 1541 (1989).
17. J D Prestage *et al.*: *Phys. Rev. Lett.* **54**, 2387 (1985).
18. S K Lamoreaux *et al.*: *Phys. Rev. Lett.* **57**, 3125 (1986); *Phys. Rev.* **A 39**, 1082 (1989), E **42**, 5763 (1990).
19. B J Venema *et al.*: *Phys. Rev. Lett.* **68**, 135 (1992).
20. B Mashhoon: *Phys. Lett.* **A 198**, 9 (1995).
21. C Lämmerzahl: *Phys. Lett.* **A 228**, 223 (1997).

Section E

Probing the Gravitomagnetic Lense-Thirring effect with Neutron Stars and Black Holes

RELATIVISTIC PERIASTRON AND NODAL PRECESSION AND QUASI-PERIODIC OSCILLATIONS FROM LOW MASS X-RAY BINARIES

LUIGI STELLA

Osservatorio Astronomico di Roma, Via Frascati 33, I-00040 Monteporzio Catone (Roma), Italy, e-mail stella@coma.mporzio.astro.it; affiliated to I.C.R.A.

The relativistic precession model for quasi periodic oscillations, QPOs, in low mass X-ray binaries is reviewed. The behaviour of three simultaneous types of QPOs is well matched in terms of the fundamental frequencies for geodesic motion in the gravitational field of the accreting compact object, for reasonable star masses and spin frequencies. The model ascribes the higher frequency kHz QPOs, the lower frequency kHz QPOs and the horizontal branch oscillations to the Keplerian, periastron precession and nodal precession frequencies of matter inhomogeneities orbiting close to the inner edge of the accretion disk. The remarkable correlation between the centroid frequency of QPOs in both neutron star and black hole candidate low mass X-ray binaries is very well fit by the model. Some testable predictions are described. QPOs from low mass X-ray binaries might provide an unprecedented laboratory to test general relativity in the strong field regime.

1 Introduction

Old accreting neutron stars, NSs, in low mass X-ray binaries, LMXRBs, display a complex variety of quasi-periodic oscillation, QPO, modes in their X-ray flux. The *low frequency* QPOs ($\sim 1 - 100$ Hz) that were discovered and studied from high luminosity Z-sources in the eighties are further classified into horizontal, normal and flaring branch oscillations (HBOs, NBOs and FBOs, respectively), depending on the simultaneous position occupied by a source in the X-ray colour-colour diagram (for a review see van der Klis 1995). The kHz QPOs (range of ~ 0.2 to ~ 1.3 kHz) that were revealed and investigated with RXTE in a number of NS LMXRBs (see van der Klis 1998, 1999, 2000 and references therein) involve timescales similar to the dynamical timescales close to the NS. A common phenomenon is the presence of a pair of kHz QPOs (centroid frequencies of ν_1 and ν_2) which drift in frequency while mantaining their frequency difference $\Delta\nu \equiv \nu_2 - \nu_1 \approx 250 - 360$ Hz approximately constant. kHz QPOs show remarkably similar properties across NS LMXRBs of the Z and Atoll groups, the luminosity of which differs by a factor of ~ 10 on average. During type I bursts from six Atoll sources, a nearly coherent signal at a frequency of $\nu_{burst} \sim 290 - 580$ Hz has also been detected. In a few cases ν_{burst} is consistent, to within the errors, with the frequency separation of the kHz QPO pair, $\Delta\nu$, or twice its value, $2\Delta\nu$.

The presence of HBOs has been firmly established in both Z-type and Atoll LMXRBs; their frequency, ν_{HBO}, ranges from ~ 15 to ~ 60 Hz. The frequency changes of the kHz QPOs and HBOs are positively correlated with the instantaneous accretion rate. In a few Atoll sources some evidence has been found for an equivalent of the NBOs and FBOs of Z-sources (Wijnands et al. 1999; Psaltis, Belloni & van der Klis 1999).

The origin of these QPOs is still debated. According to beat frequency models, BFMs, disk inhomogeneities at the magnetospheric boundary and the sonic point radius are accreted at the beat frequency between the local Keplerian frequency, ν_ϕ, and the NS spin frequency, ν_s, giving rise to the HBOs and the lower frequency kHz QPOs, at ν_1, respectively (Alpar & Shaham 1985; Lamb et al. 1985; Miller et al. 1998). The higher frequency kHz QPOs are attributed to the Keplerian motion ($\nu_2 = \nu_\phi$) at the sonic point radius. The frequency separation of the QPO pair therefore, yields the NS spin frequency ($\nu_s = \Delta\nu$). The inferred ν_s possess a narrow distribution ($\sim 250 - 360$ Hz), far from the mass shedding limit of any NS model and considerably less extended than that of fast millisecond radio pulsars (~ 600 Hz). The NSs of LMXRBs should therefore be equilibrium rotators with a narrow frequency range despite their different average mass transfer rates, magnetic field strenghts and evolutionary histories. This implies that angular momentum is transferred outward at the same rate at which is received through accretion. Possibilities include gravitational radiation losses (Bildsten 1998; Andersson et al. 1999) or magnetic torques (White & Zhang 1997; Psaltis et al. 1999). Moreover, if LMXRBs are the progenitors of millisecond radio pulsars, then the BFM would require a different evolutionary path leading to the formation of radio pulsars with $\nu_s > 400$ Hz. Recent findings undermine the applicability of simple BFMs to a growing number of LMXRBs with kHz QPOs. First, the separation $\Delta\nu$ decreases by up to ~ 100 Hz as ν_2 increases in Sco X-1 (van der klis et al. 1997), 4U1608-52 (Mendez et al. 1998a,b), 4U1735-44 (Ford et al. 1998) and 4U1728-34 (Mendez & van der Klis 1999). Owing to poor statistics, a similar variation of $\Delta\nu$ in other sources would have remained undetected (Psaltis et al. 1998). Second, there are currently two sources (4U1636-53, Mendez et al. 1999, and 4U1728-34, Mendez & van der Klis 1999) for which ν_{burst} is significantly different from $\Delta\nu$ and its harmonics. BFMs are not applicable to black hole candadates, BHCs, since the *no hair theorem* excludes the possibility that an offset magnetic field or radiation beam can be stably anchored to the black hole, as required to produce the beating with the disk Keplerian frequency.

A variety of alternative QPO models has been proposed. The basic features of the relativistic precession model, RPM, are reviewed here (Stella & Vietri 1998a, 1999; Stella, Vietri & Morsink 1999). In the RPM the QPO signals arise from the fundamental frequencies of the motion of blobs in the vicinity of the NS. The orbits of these blobs are supposed to be slightly eccentric and tilted. As in BFMs, the higher frequency kHz QPOs at ν_2 are produced by the ϕ-motion (*i.e.* the Keplerian motion) of inhomogeneities orbiting the inner disk boundary, while the lower frequency QPO signal at ν_1 originates from their periastron precession, which is primarily determined by strong-field effects. The HBOs are due to the nodal precession of the same inhomogeneities, an effect which is dominated by frame-dragging around fast-rotating collapsed stars. The RPM can be applied to BHCs as well. Within this model both the quadratic dependence of the HBO frequency on ν_2 and the decreasing separation $\Delta\nu$ for increasing ν_2, observed in several Z and Atoll sources are naturally explained. Moreover the RPM fits very well the remarkable correlation between the centroid frequencies of QPOs (or peaked noise components) from low mass X-ray binaries,that was discovered by Psaltis, Belloni & van der Klis (1999). This correlation extends over nearly 3 decades in frequency and encompasses both NS and BHC systems. If this interpretation were confirmed, kHz QPOs and HBOs would provide an unprecedented test for strong–field general relativity, GR.

2 Periastron Precession and kHz QPOs

We consider here only infinitesimally eccentric and tilted blob orbits, under the assumption that the blobs' motion is dictated by the star's gravity alone. In the case of a circular geodesic in the equatorial plane ($\theta = \pi/2$) of a Kerr black hole of mass M and specific angular momentum a, the coordinate frequency measured by a static observer at infinity is (Bardeen et al. 1972)

$$\nu_\phi = \pm M^{1/2} r^{-3/2} [2\pi(1 \pm a M^{1/2} r^{-3/2})]^{-1} \quad (1)$$

(we use units such that $G = c = 1$). ν_ϕ deviates from its Keplerian value at small radii. The upper sign refers to prograde orbits. If we slightly perturb a circular orbit introducing velocity components in the r and θ directions, the coordinate frequencies of the small amplitude oscillations within the plane (the epicyclic frequency ν_r) and in the perpendicular direction (the vertical frequency ν_θ) are given by (Okazaki, Kato & Fukue 1987; Kato 1990; de Felice & Usseglio-Tomasset 1996; Perez et al. 1997)

$$\nu_r^2 = \nu_\phi^2 (1 - 6Mr^{-1} \pm 8aM^{1/2} r^{-3/2} - 3a^2 r^{-2}) \ , \quad (2)$$

$$\nu_\theta^2 = \nu_\phi^2 (1 \mp 4aM^{1/2} r^{-3/2} + 3a^2 r^{-2}) \ . \quad (3)$$

In the Schwarzschild limit ($a = 0$) ν_θ coincides with ν_ϕ, such that the nodal precession frequency $\nu_{nod} \equiv \nu_\phi - \nu_\theta$ is identically zero. ν_r, instead, is always lower than the other two frequencies, reaching a maximum for $r = 8M$ and going to zero at $r_{ms} = 6M$. This qualitative behaviour of ν_r is preserved in the Kerr field ($a \neq 0$). Therefore the periastron precession frequency $\nu_{per} \equiv \nu_\phi - \nu_r$ is dominated by a "Schwarzschild" term over a wide range of parameters (Stella & Vietri 1999).

In the RPM the higher and lower frequency kHz QPOs are identified with $\nu_2 = \nu_\phi$ and $\nu_1 = \nu_{per}$, respectively. Therefore $\Delta\nu \equiv \nu_2 - \nu_1 = \nu_\phi - (\nu_\phi - \nu_r) = \nu_r$. For $a = 0$, Eqs. (1)–(2) give

$$\nu_r = \nu_\phi (1 - 6M/r)^{1/2} = \nu_\phi [1 - 6(2\pi\nu_\phi M)^{2/3}]^{1/2} \ . \quad (4)$$

The curves in Fig. 1A show ν_r vs. ν_ϕ for $a = 0$ and selected values of M, the only free parameter in Eq. (4). It is easy to see that ν_r decreases after a maximum of $\nu_r(max) = 358(2M_\odot/M)$ Hz for a ϕ-frequency of $\nu_\phi = 2\nu_r(max) \simeq 0.65\ \nu_\phi(r_{ms})$. The measured $\Delta\nu$ vs. ν_2 for eleven NS LMXRBs is also plotted. It is apparent that for NS masses in the 2 $M_\odot$ range, the simple model outlined above is in qualitative agreement with the measured values, including the decrease of $\Delta\nu$ for increasing ν_2 seen in Sco X-1, 4U1608-52, 4U1735-44 and 4U1728-34. The model above is only an approximation: first, the spacetime around a fast rotating NS is different from a Kerr spacetime (due to the star's oblateness induced by rotation); second, the blobs' orbits might possess a finite (though small) eccentricity. Analytical formulae to partly correct for these effects were derived by Stella & Vietri (1999); Fig. 1B shows the fit to the observed $\Delta\nu$ vs. ν_2 relationship in Sco X-1 that was obtained

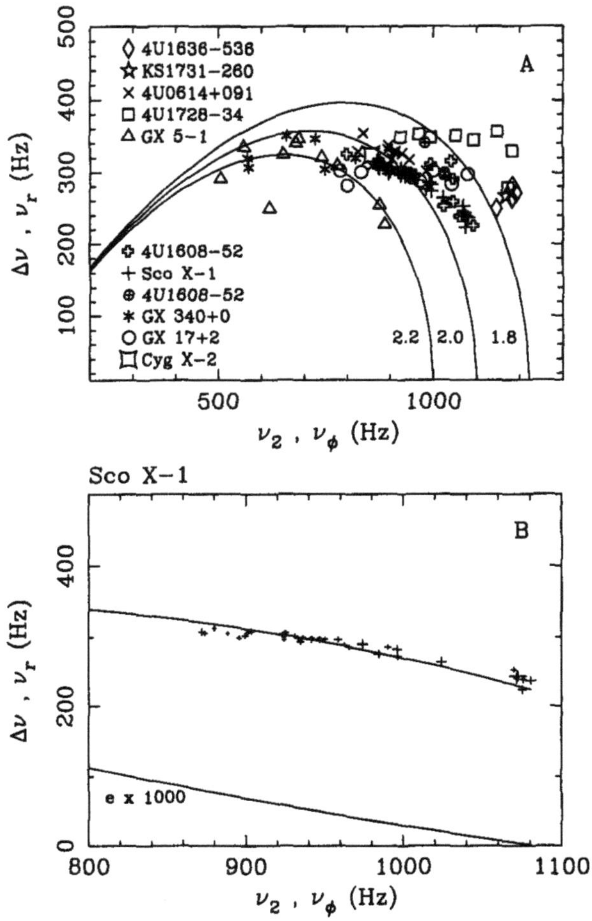

Figure 1. (A) kHz QPO frequency difference $\Delta\nu$ versus higher QPO frequency ν_2 for eleven LMXRBs. Error bars are not plotted for the sake of clarity. The curves give the r- and ϕ-frequencies of matter in nearly circular orbit around a non-rotating neutron star, of mass 2.2, 2.0 and 1.8 $M_\odot$. (B) $\Delta\nu$ versus ν_2 in Sco X-1. The best fit model corresponds to the r- and ϕ-frequencies of matter orbiting a non-rotating 1.90 $M_\odot$ neutron star at a periastron distance of 6.25 M (17.5 km). The line marked with e gives the orbital eccentricity ($\times 1000$) as a function of the ϕ-frequency.

through them. The orbital eccentricity, in particular, was varied in order to obtain different frequencies, while keeping the periastron distance $r_p = a(1-e)$ fixed. The best model for a non-rotating NS is shown in Fig. 1B. The model reproduces fairly accurately the data with a minimum number of free parameters, the NS mass ($M \sim 1.9$ M$\odot$) and periastron distance ($r_p \simeq 6.2\ M$). The latter value is close to the marginally stable orbit radius. When ν_s is allowed a finite value (say $\nu_s \sim 300-600$ Hz), fits of very similar quality are obtained, the parameters of which differ only slightly from those given above. In essence, the effects induced by the NS rotation on ν_r are small, though non-negligible.

The behaviour of the curves in Fig. 1A and 1B and therefore the ability of the model to match the observations, reflects the properties of the strong field Schwarzschild metric, since lower order expansions fail to reproduce the observed frequencies (see Stella & Vietri 1999). This can be seen through a simple argument. In the RPM interpretation, the epicyclic frequency ν_r is $\ll \nu_\phi$ in kHz QPO sources. This is unusual, the epicyclic frequency in most cases being close to ν_ϕ. However, if one remembers that $4\pi^2\nu_r^2 \equiv d^2V_{eff}/dr^2$, where V_{eff} is the effective potential, one sees that the inequality above is natural for orbits close to the marginally stable one, for which the effective potential notoriously has an inflection point, $d^2V_{eff}/dr^2 = 0$. Thus the relevant orbits must be close to the innermost stable orbit.

Within the RPM, the maximum value of $\nu_r = \Delta\nu$ depends mainly on the mass of the compact object. The NS masses deduced from the simple modelling in Fig. 1A,B are in the $\sim 1.8-2.0$ M$_\odot$ range, in agreement with the only relatively accurate mass measurement from optical spectro-photometry in any of these systems (Cyg X-2; $M = 1.8 \pm 0.2$ M$_\odot$; Orosz & Kuulkers 1999). In general within the RPM, $\Delta\nu$ should hold no relation to the NS spin frequency ν_s. Therefore, it is natural to identify ν_s with ν_{burst}, *i.e.* the stable frequency seen during type I X-ray bursts (for a review see Strohmayer 2000). Infact in a given source, the frequency during each burst, after some transient variation, has been observed to remain stable up to a part in $\sim 10^4$ across time intervals of a few years (e.g. in 4U1728-34; Strohmayer et al. 1998). The distribution of NS spins inferred in this way is fairly wide ($\sim 290-580$ Hz); it compares well with that of millisecond radio pulsars and the 401 Hz spin of SAX J1808.4-3658, the only bursting LMXRB displaying coherent pulsations in its persistent emission (Wijnands & van der Klis 1998; Chakrabarty & Morgan 1998). None of the LMXRBs of the (high luminosity) Z-group has yet displayed burst oscillations; therefore their spin period is still to be measured. Cyg X-2 and GX 17+2, the only type I X-ray bursters in the group, might provide this important piece of information.

3 Nodal Precession and HBOs

If the blobs' orbits giving rise to the signals at ν_ϕ and ν_{per} are slightly tilted relative to the equatorial plane, nodal precession will take place around the spin axis. In the RPM the HBO frequency is related to the nodal precession frequency. From Eqs. (1) and (3) this can be written in the slow rotation limit $(a/M \ll 1)$

$$\nu_{nod} \simeq 4\pi a\nu_\phi^2 \simeq 6.2 \times 10^{-5}(a/M)m\nu_\phi^2 \text{ Hz} \simeq 4.4 \times 10^{-8}\, I_{45}m^{-1}\nu_\phi^2\nu_s \text{ Hz}, \quad (5)$$

where $M = m$ M$_\odot$. This is the well known Lense-Thirring nodal precession formula. The latter equality refers to a rotating NS, where $aM = 2\pi\nu_s I$ with $I = 10^{45} I_{45}$ g cm^2 its moment of inertia.

If ν_ϕ and ν_s are measured, the only parameter in Eq. (5) that is not identified from observations is $I_{45}m^{-1}$; this can vary over a limited range, $0.5 < I_{45}m^{-1} < 2$, for virtually any mass and EOS (see the rotating NS models of Friedman, Ipser & Parker 1986 and Cook, Shapiro & Teukolsky 1994). The stellar oblateness induced by the star's rotation (and thus the quadrupole and higher multipole terms in the gravitational potential), gives rise to correction terms in the nodal precession frequency also (see Morsink 1999 and Morsink & Stella 1999 for a post-Newtonian formula). Their relative importance increases for high values of ν_s and ν_ϕ. Yet the Lense-Thirring term dominates over a wide range of parameters, such that a $\sim \nu_\phi^2$ dependence is expected.

Since the relativistic nodal precession model was proposed, an approximately quadratic dependence of ν_{HBO} on the higher frequency kHz QPOs has been measured in a number of LMXRBs. This dependence was originally suggested on the basis of a few power spectra of the Atoll source 4U1728-34 (Stella & Vietri 1998a). Ford & van der Klis (1998) analysed a large set of power spectra from the same source and determined that the frequency ν_{low} of the $\sim 10-50$ Hz QPOs scales as $\nu_2^{2.11\pm0.06}$. Note that the low frequency QPO vs. ν_2 relation of this source appears to be double-branched, with the centroid frequency shifting by a factor of ~ 2 across different observations. This suggests that on occasion the 2nd harmonic of ν_{HBO} is excited instead of the fundamental. Stella & Vietri (1998b) first noticed that the HBO frequency of the Z-source GX 17+2 displays a nearly quadratic dependence on ν_2. Psaltis et al. (1999) carried out a systematic study of Z-sources and determined that the HBO frequency is consistent with a ν_2^2 scaling (Cyg X-2 and Sco X-1 show evidence for a somewhat flatter dependence). for frequencies up to $\nu_2 \sim 900-1000$ Hz. A somewhat flatter dependence or even a saturation is observed sometimes at higher frequencies. In essence these results confirmed one of the basic features of the RPM, namely the nearly quadratic dependence of the nodal precession frequency on the ϕ-frequency.

If the NS spin frequency is measured, then for any value of ν_ϕ the model yields a predicted nodal precession frequency which is uncertain only by a factor of a few, mainly due to the allowed range of $I_{45}m^{-1}$ (Stella & Vietri 1998a). Only in the Atoll source 4U1728-34 burst oscillations and simultaneous kHz QPOs and HBOs have been detected unambiguously (Strohmayer et al. 1996; Ford & van der Klis 1998; Mendez & van der Klis 1999).[a] Therefore the QPO frequencies of 4U1728-34 can be used to test both the ν_{HBO} and the $\Delta\nu$ versus ν_2 relationships predicted by the RPM, when the NS spin derived from burst oscillations is used ($\nu_{burst} \simeq 364$ Hz). In order to take fully into account of all the effects that contribute determining the blobs' geodetic motion in the vicinity of the NS, we adopted a numerical approach and computed the spacetime metric of the star using Stergioulas' (1995) code, an equivalent of that of Cook et al.(1992) (see also Stergioulas and Friedman 1995). ν_{nod} was calculated according to the prescription of Morsink & Stella (1999), whereas ν_r was obtained from the second radial derivative of the effective potential (see Stella, Vietri & Morsink 1999). ν_r and ν_{nod} were derived as a function of ν_ϕ for infinitesimally small tilt angles and eccentricities (Morsink & Stella 1999; Stella, Vietri & Morsink 1999).

Figure 2 shows the measured values of $\Delta\nu$ and ν_{HBO} versus ν_2 in 4U1728-34. Relatively high NS masses (see also Sect. 2) and stiff EOSs such as AU and UU (Wiringa et al. 1988) are required in this application of the RPM. The solid

[a]Note that in the case of KS1731-260, for which $\nu_{burst} \simeq 524$ Hz, it is still unclear whether the ~ 26 Hz QPOs, that where were observed together with kHz QPOs at ~ 910 and ~ 1170 Hz, represent HBOs or the equivalent of NBOs in an Atoll source see Smith et al. 1997; Wijnands & van der Klis 1997; Psaltis, Belloni & van der Klis 1999.

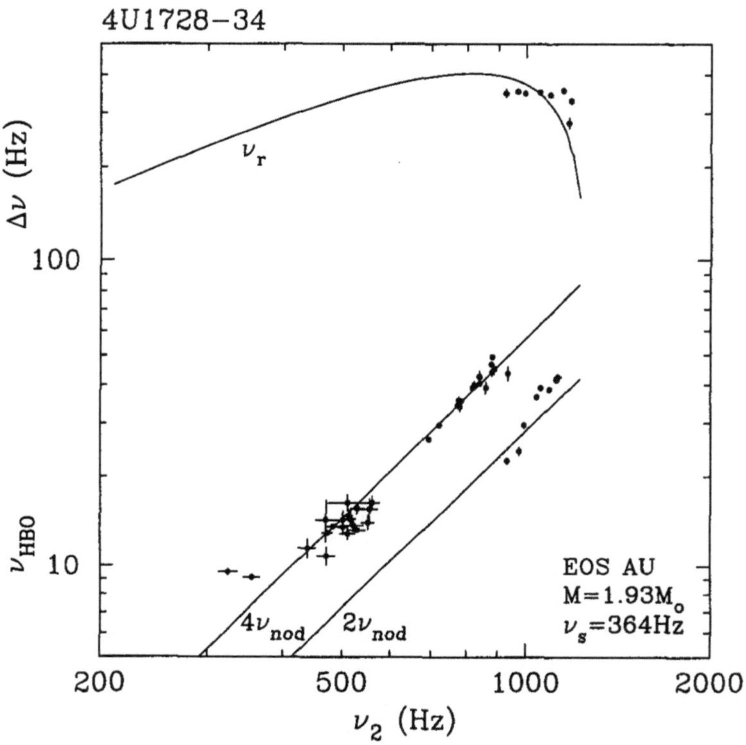

Figure 2. kHz QPO frequency difference $\Delta\nu$ and (double-branched) HBO frequency versus higher QPO frequency ν_2 in 4U1728-34 (Strohmayer et al. 1996; Ford & van der Klis 1998; Mendez& van der Klis 1999). The solid lines gives the r-frequency and the 2nd and 4th harmonics of the nodal precession frequency ν_{nod} as a function of the ϕ-frequency, for infinitesimally eccentric and tilted orbits in the spacetime of a 1.93 $M_\odot$ neutron star spinning at 364 Hz (EOS AU).

lines in Fig. 2 are for a 1.93 $M_\odot$ NS with EOS AU and ν_s = 364 Hz. A good agreement is obtained if the HBO frequency, the lower of the two branches seen in 4U1728-34, is identified with the 2nd harmonics of ν_{nod} (*i.e.* $2\nu_{nod}$) (see also Morsink & Stella 1999; Stella & Vietri 1999a). Correspondingly the upper HBO branch is well fit by $4\nu_{nod}$. The geometry of tilted orbits in the innermost disk regions might be such that a stronger signal is produced at the even harmonics of the nodal precession frequency (e.g. Psaltis et al. 1999). The frequency range and trend of the epicyclic frequency ν_r in this model are also in reasonable agreement with the $\Delta\nu$ measurements; a more complex model is clearly required in order to fit these data more accurately. The analogy with the modelling of the Sco X-1 kHz QPO frequencies described in Sect. 2, suggests that introducing a finite (though small) eccentricity in the blobs' orbits would greatly improve the quality of the fit. Detailed modelling based on finite eccentricity geodesics in the spacetime of fast rotating NSs are in progress and will be reported elsewhere.

In summary, the model presented here is capable of reproducing the salient features of both the $\Delta\nu$ versus ν_2 and ν_{HBO} versus ν_2 relationships, with just two free parameters (M and the EOS), the allowed range of which is fairly limited (moreover the EOS cannot even be varied continuously!). Concerning Z-sources and all other Atoll sources for which burst oscillations have not been detected yet, the NS spin can still be regarded as a free parameter. The application of the RPM to the HBOs of these sources can therefore be used to constrain the spin of their NSs. Conversely, in those Atoll source in which ν_{burst} is measured, but HBOs have not been detected yet, the RPM can be used to predict ν_{HBO}. These issues are briefly addressed in the next Section.

4 The RPM and The PBV Correlation

Psaltis, Belloni & van der Klis (1999) recently identified two QPOs and peaked noise components the frequency of which follows a tight correlation over nearly three decades. This correlation (hereafter PBV correlation) involves both NS and BHC LMXRBs spanning different classes and a wide range of luminosities (see the points in Fig. 3). In kHz QPO NS systems, these components are the lower frequency kHz QPOs, ν_1, and the low frequency, HBO or HBO-like QPOs, ν_{HBO}. For BHC systems and lower luminosity NS LMXRBs the correlation involves either two QPOs, or a QPO and a peaked noise component. In all cases the frequency separation is about a decade and an approximate linear relationship ($\nu_{HBO} \sim \nu_1^{0.95}$) holds. The QPO frequencies from the peculiar NS system Cir X-1 varies over nearly a decade while closely following the PBV correlation and bridging its low and high frequency ends. Psaltis, Belloni & van der Klis (1999) noted also the that ν_2 vs. ν_1 relations of different Atoll and Z-sources line-up with good accuracy (see also Psaltis et al. 1998).

We show here that the frequencies predicted by the RPM match precisely the PBV correlation, without resorting to any additional assumption (see Stella, Vietri & Morsink 1999). We assume that in all QPO sources, including BHCs, $\nu_{HBO} \simeq 2\nu_{nod}$ as in 4U1728-34 (see Sect.3). Figure 3A shows $2\nu_{nod}$ and ν_ϕ obtained from Eqs. (1)–(3) as a function of ν_{per} for corotating orbits and selected values of M and a/M. The high frequency end of each line is dictated by the orbital radius reaching the marginally stable orbit at r_{ms} (where $\nu_\phi = \nu_{per}$, i.e. $\nu_r = 0$, one of the defining characteristics of the marginally stable orbit). It is apparent that r_{ms} depends mainly on M and to a lesser extent on a/M.

The separation of the lines in Fig. 3A testifies that while ν_{nod} depends weakly on the mass and more strongly on a/M, the opposite is true for ν_ϕ. By taking the weak field ($M/r \ll 1$) and slow rotation ($a/M \ll 1$) limit of Eqs. (1)–(3) the relevant first order dependence is made explicit,

$$\nu_\phi \simeq (2\pi)^{-2/5} 3^{-3/5} M^{-2/5} \nu_{per}^{3/5} \simeq 33\ m^{-2/5} \nu_{per}^{3/5}\ \text{Hz}\ , \tag{6}$$

$$\nu_{nod} \simeq (2/3)^{6/5} \pi^{1/5} (a/M) M^{1/5} \nu_{per}^{6/5} \simeq 6.7 \times 10^{-2}\ (a/M) m^{1/5} \nu_{per}^{6/5}\ \text{Hz}\ . \tag{7}$$

For the case of rotating NSs where the effects of stellar oblateness introduces modifications relative to a Kerr spacetime, we adopt the numerical approach outlined in Sect. 3. Results are shown in Fig. 3B for a NS mass of 1.95 $M_\odot$, EOS AU and $\nu_s = 300$, 600, 900 and 1200 Hz (corresponding to $a/M = 0.11$, 0.22, 0.34 and 0.47, respectively). While also in this case ν_ϕ depends only very weakly on ν_s, the approximately linear dependence of ν_{nod} on ν_s is apparent. Note that the approximate scalings in Eqs. (6)–(7) remain valid over a wide range of frequencies. Only for the largest values of ν_{per} and ν_s, ν_{nod} departs substantially from the $\sim \nu_{per}^{6/5}$ dependence.

The measured QPO and peaked noise frequencies giving rise to the PBV correlation are also plotted in Fig. 3B. Higher kHz QPO frequencies from NS systems (ν_2) are included (for the sake of clarity NBOs and FBOs were excluded). The agreement over the range of frequencies spanned by each kHz QPO NS system should not be surprising: together with the accurate matching of the corresponding $\nu_1 - \nu_2$ relationship in Z-sources, this is indeed part of the evidence on which the RPM model was proposed. However the fact that the dependence of ν_{nod} on ν_{per} matches the observed $\nu_{HBO} - \nu_1$ correlation to a good accuracy over ~ 3 decades in frequency (down to ν_1 of a few Hz), encompassing both NS and BHC systems, provides additional independent evidence in favor of the RPM. The observed variation of ν_{HBO} and ν_1 in individual sources (Cir X-1 is the most striking example, see Fig. 3B) further supports the scaling predicted by the RPM. The matching of the observed ν_2 vs. ν_1 relation in terms of ν_ϕ vs. ν_{per} is also quite accurate.

In the case of NS LMXRBs the best matching is obtained for relatively stiff EsOS (*e.g.* AU, as in Fig. 3B, and UU), which can also achieve the relatively high masses ($m \sim 1.8 - 2.0$) that are required to match the observed ν_1 and ν_2 relationship. For EOS AU and $m = 1.95$, the ν_{HBO} vs. ν_1 values of most NS LMXRBs are best matched for ν_s in the ~ 600 to 900 Hz range. It is apparent from Fig. 3B that ν_s as low as ~ 300 Hz are required for the Atoll sources with ν_{HBO} somewhat below the main PBV correlation: these are 4U1728-34 (see Sect. 3) and 4U1608-52 (from which burst oscillations have not been detected yet). The values above are close to the range of ν_s inferred from ν_{burst} in a number of other Atoll sources (van der Klis 1999). Z-type LMXRBs, appear to require ν_s in the ~ 600 to 900 Hz range, a possibility that is still open since for none of these sources there exists yet a ν_s measurement based on burst oscillations or coherent pulsations (see also Stella, Vietri & Morsink 1999b). Note that the upper HBO branch of 4U1728-34 matches well the main PBV correlation. In the interpretation of Sect. 3 the lower branch corresponds to $2\nu_{nod}$ and the upper branch to $4\nu_{nod}$. One could further speculate that sources following the main PBV correlation, Z-sources in particular, are also in the upper HBO branch at $4\nu_{nod}$; in this case their ν_s might be expected in the $\sim 300 - 400$ Hz range.

For BHC LMXRBs the scatter around the PBV correlation implies values of a/M of $\sim 0.1 - 0.3$ (see Fig. 3B). Note that so far no evidence has been found for a BHC QPO (or peaked noise) signal that is the equivalent of ν_2 in NS systems. The points from XTE J1550-564, while inconsistent with any single value of a/M, might lie along two distinct branches separated by a factor of ~ 2 in ν_{HBO}, similar to the case of 4U1728-34. Interpreting the PBV correlation within the RPM implies that both NS and BH systems share relatively high values of their specific angular momentum ($a/M \sim 0.1 - 0.3$ for BHCs or $\nu_s = 300 - 900$ Hz for relatively stiff EOS NSs). Population synthesis calculations show that a significant fraction of low magnetic field, relatively stiff EOS NSs in LMXRBs can be spunup by accretion torques to $\nu_s \sim 1$ kHz, provided that $> 0.3 - 0.4$ $M_\odot$ are accreted (Burderi et al.

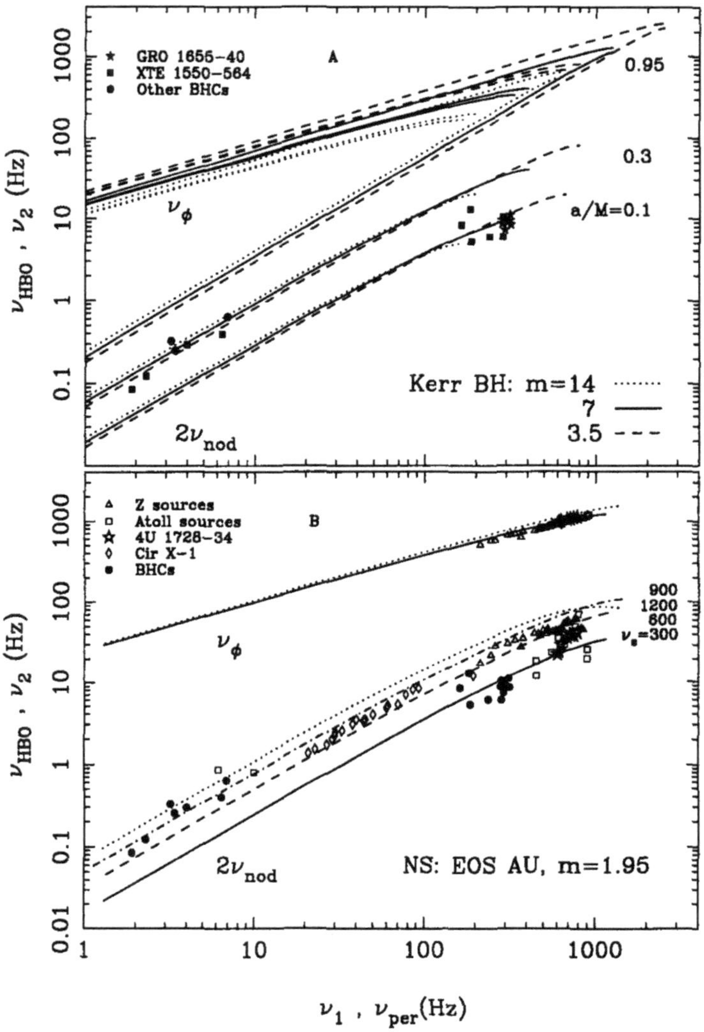

Figure 3. Twice the nodal precession frequency, $2\nu_{nod}$, and ϕ-frequency, ν_ϕ, vs. periastron precession frequency, ν_{per}, for black hole candidates of various masses and angular momenta (panel A) and rotating neutron star models (EOS AU, $m = 1.95$) with selected spin frequencies (panel B). The measured QPO (or peaked noise) frequencies ν_1, ν_2 and ν_{HBO} giving rise to the PBV correlation are also shown in panel B for both BHC and NS LMXRBs and in panel A for BHC LMXRBs only; errors bars are not plotted (see Psaltis, Belloni & van der Klis 1999 for a complete list of references). We included only those cases in which QPOs at ν_1 were unambiguously detected. NBO and FBO frequencies are not plotted.

1999). The combination of the relatively high NS mass spin frequency required by the RPM is at least qualitatively in agreement with evolutionary scenarios for LMXRBs. In the Kerr metric, the increase in ν_{nod} for increasing values of a/M shows no signs of saturation (Fig. 3A). BHCs can in principle achieve extreme values of $a/M \simeq 1$ (NSs are instead limited to $a/M < 0.6 - 0.7$, cf. Salgado et al. 1994).

In practice spinup to high values of a/M is unlikely to occur in BHC LMXRBs, because of the limited mass that can be accreted from the mass donor star (King & Kolb 1999). In the RPM the high frequency BHC QPOs (*e.g.* the ~ 300 Hz QPOs of GRO1655-40) are interpreted terms of $\nu_1 = \nu_{per}$ (while the lower frequency QPOs $\nu_{HBO} = 2\nu_{nod}$) implying moderate values of $a/M \sim 0.1 - 0.3$. This is at variance with the $\nu_1 = \nu_{nod}$ interpretation of Cui et al. (1998), which requires high values of a/M (~ 0.95 in GRO1655-40), in contrast with BHC accretion-driven spinup scenarios. From Fig. 3A it is apparent that the ~ 300 Hz QPOs from GRO 1655-40 lie close to the high frequency end of the $a/M = 0.1$, $m = 7$ line. Since the mass of the BHC in GRO 1655-40 determined through optical observations is ~ 7 $M_\odot$ (Shahbaz et al. 1999), we conclude that, according to the RPM, $\nu_{per} \simeq 300$ Hz close to the marginally stable orbit, where by definition $\nu_\phi = \nu_{per}$. This suggests that any additional QPO signal at $\nu_2 = \nu_\phi$ would be very close to or even blended with the QPO peak at ν_1. The detection of two closeby or even partially overlapping QPO peaks close to ~ 300 Hz in GRO 1655-40 would therefore provide further evidence in favor of the RPM interpretation. Similar considerations might apply to the ~ 284 Hz QPOs of XTE J1550-564 (Homan et al. 1999), provided its BHC mass is in the ~ 7 $M_\odot$ range.

5 Discussion

The RPM makes several predictions that can be tested through future analyses and/or observations:

(a) The frequency difference $\Delta\nu = \nu_r$ is expected to decrease also for low values of $\nu_2 = \nu_\phi$ (see Eq. (4) and Fig. 1). Moreover if the highest $\nu_2 = \nu_\phi$ frequencies do originate from nearly circular orbits (see Fig. 1A) then $\Delta\nu$ should quickly decrease as ν_2 increases further and the orbital radius approaches the marginally stable orbit.

(b) $\nu_2 = \nu_\phi$ is expected to scale as $\nu_1^{3/5} = \nu_{per}^{3/5}$. Extending the ν_2 vs. ν_1 correlation in NS systems toward lower frequencies and detecting the signal at ν_2 in BHC systems would provide important new tests.

In the RPM the QPO signals are produced at $r/M \sim 100\ m^{-2/5}\nu_{per}^{-2/5}$, which, for individual sources, must decrease for increasing mass accretion rates (as $\dot{M}$ is positively correlated with *e.g.* ν_2). The inferred radii range from close to the marginally stable ($r/M \sim 6$) to $r/M \sim 30$ over the frequency span covered by the PBV correlation. Many NS and BHC LMXRBs display two component X-ray spectra consisting of a soft thermal component, usually interpreted in terms of emission from an optically thick accretion disk, and a harder, often power-law like, component, likely due to a hot inner disk region. The QPOs might originate at the transition radius between the optically thick disk and the hot inner region as a result of occultation by orbiting blobs or modulated emission by self-luminous blobs (Stella, Vietri & Morsink 1999; see also Di Matteo & Psaltis 1999). For kHz QPO NS systems the inferred radii are remarkably close to the marginally stable orbit. A non-negligible eccentricity would also alter these frequencies slightly (see Sect. 2).

Mechanisms that can induce a finite (though small) eccentricity and tilt in the blobs' orbits, are currently under investigation. In the case of NSs, some kind of resonance between the star spin and the motion of the blobs (see *e.g.* Vietri & Stella 1998) might be responsible for the close commensurability of ν_{burst} and $\Delta\nu$ (or $2\Delta\nu$) observed in a few sources. These issues will be addressed elsewhere. Note that in the approach explored by Psaltis & Norman (1999) (see also Psaltis 2000), the same frequencies as in the RPM (plus an additional frequency at $\nu_\phi + \nu_r$) are selected by the response of the innermost annulus of the disk, when this is subject to a wide-band input noise; in this case no blobs would be required.

If confirmed, the RPM will provide an unprecedented opportunity to test GR effects in the strong field regime, such as the periastron precession in the vicinity of the marginally stable orbit and the radial dependence of the Lense-Thirring nodal precession frequency. In principle, accurately measured kHz QPO and HBO frequencies would yield crucial information on the compact object such as its mass and angular momentum (*e.g.* by solving Eqs. (1)–(3) for m, a/M and r). Should suitable, additional observables be found, it might become possible to obtain a self-consistency check of the RPM, together with tests of GR vs. alternative theories of gravity in the strong field regime.

Acknowledgements

I am grateful to my collaborators in this project, S. Morsink and M. Vietri.

References

1. Alpar, A. & Shaham, J. 1985, Nature, 316, 239
2. Andersson, N., Kokkotas, K.D. & Stergioulas, N. 1999, ApJ, 516, 307
3. Bardeen J.M., Press W.H. & Teukolski S.A. 1972, ApJ, 178, 347
4. Bildsten, L. 1998, ApJ, 501, L89
5. Burderi, L., Possenti, A., Colpi, M., Di Salvo, T. & D'Amico, N. 1999, ApJ, 519, 285
6. Chakrabarty D. & Morgan E.H. 1998, Nature, 394, 346
7. Cook, G.B., Shapiro, S.L., & Teukolsky, S.A. 1992, ApJ, 398, 203
8. Cui, W., Zhang, S.N., Chen, W. 1998, ApJ, 492, L53
9. de Felice F. & Usseglio-Tomasset S. 1996, Gen. Rel. Grav., 28, 179
10. Di Matteo, T. & Psaltis, D. 1999, ApJ, 526, L101
11. Ford, E.C. & van der Klis, M. 1998, ApJ, 506, L39
12. Ford, E.C. et al. 1998, ApJ, 508, L155
13. Friedman, J.L, Ipser, J.R. & Parker, L. 1986, ApJ, 304, 115
14. Homan, J., van der Klis, M., Wijnands, R., Vaughan, B, & Kuulkers, E. 1998, ApJ, 499, L41
15. Homan, J., Wijnands, R. & van der Klis, M. 1999, IAU Circ. 7121
16. Kato S. 1990, PASJ, 42, 99
17. King, A.R. & Kolb, U. 1999, MNRAS, 305, 654
18. Lamb, F.K., Shibazaki, N., Alpar, A. & Shaham, J. 1985, Nature 317, 681
19. Mendez, M. et al. 1998a ApJ, 494, L65
20. ——— 1998b ApJ, 505, L23
21. Mendez, M. & van der Klis, M. 1999, ApJ, 517, L51

22. Mendez, M., van der Klis, M., Ford, E.C., Wijnands, R. & van Paradijs, J. 1999, ApJ, 511, L49
23. Miller, M.C., Lamb, F.K. & Psaltis, D. 1998, ApJ, 508, 791
24. Morsink, S. 2000, this volume
25. Morsink, S. & Stella, L. 1999, ApJ, 513, 827
26. Perez C.A., Silbergleit A.S., Wagoner P.V. & Lehr D.E. 1997, ApJ, 476, 589
27. Orosz, J.A. & Kuulkers, E. 1999, MNRAS, 305, 132
28. Psaltis, D. 2000, in "X-ray Astronomy '99: Stellar Endpoints, AGNs and Diffuse Background", Ap. Lett. & Comm., in press
29. Psaltis, D. & Norman, C. 2000, ApJ, in press.
30. Psaltis, D., et al. 1998, ApJ, 501, L95
31. ——— 1999, ApJ, 520, 763
32. Psaltis, D., Belloni, T., & van der Klis, M. 1999, ApJ, 520, 262
33. Salgado, M., Bonazzola, S., Gourgoulhon, E. & Haensel, P. 1994, A&A, 291, 155.
34. Shahbaz, T. et al. 1999, MNRAS, 306, 89
35. Smith, D.A., Morgan, E.H. & Bradt, H. 1997, ApJ, 479, L137
36. Stella, L. & Vietri, M. 1998a, ApJ, 492, L59
37. ——— 1998b, in "The active X-ray sky", Nucl Phys B (Proc. Suppl.), 69, 135
38. ——— 1999, Phys. Rev. Lett, 82, 17
39. Stella, L., Vietri, M. & Morsink, S.M. 1999, ApJ, 524, L63
40. Stergioulas, N. 1995, code available via ftp from `pauli.phys.uwm.edu` in directory `/pub/rns`
41. Stergioulas, N., & Friedman, J.L. 1995, ApJ 444, 306
42. Strohmayer, T.E. et al. 1996, ApJ, 469, L9
43. Strohmayer, T.E et al. 1998, ApJ, 503, L147
44. Strohmayer, T.E. 2000, in "X-ray Astronomy '99: Stellar Endpoints, AGNs and Diffuse Background", Ap. Lett. & Comm., in press
45. van der Klis, M. 1995,in "X-ray Binaries", Eds. W. H. G. Lewin, J. van Paradijs & E. P. J. van den Heuvel (Cambridge University Press), p. 252
46. ——— 1998, in: Proc. NATO ASI "The many faces of neutron stars", Series C, Vol. 515, p. 337
47. ——— 2000, this volume.
48. ——— 2000, in "X-ray Astronomy '99: Stellar Endpoints, AGNs and Diffuse Background", Ap. Lett. & Comm., in press
49. van der Klis, M. et al. 1997 ApJ, 481, L97
50. Vietri, M. & Stella, L. 1998b, ApJ, 503, 350
51. White, N.E. & Zhang, W. 1997, ApJ, 490, L87
52. Wijnands R.A.D. & van der Klis M. 1997, ApJ, 482, L65
53. ——— 1998, Nature, 394, 344
54. Wijnands, R., van der Klis, M. & Rijkhorst, E.J. 1999, ApJ, 512, L39
55. Wiringa, R.B., Fiks, V. & Fabrocini, A. 1988, Phys. Rev. C, 38, 1010

ON GRAVITOMAGNETIC PRECESSION AND QPO IN BLACK HOLE CANDIDATES

WEI CUI
Center for Space Research, Massachusetts Institute of Technology, Cambridge, MA 02139, USA
E-mail: cui@space.mit.edu

WAN CHEN *
NASA/Goddard Space Flight Center, Code 661, Greenbelt, MD 20771, USA
E-mail: chen@milkyway.gsfc.nasa.gov

S.N. ZHANG
Department of Physics, University of Alabama-Huntsville, Huntsville, AL 35812, USA
E-mail: Shuang.Zhang@msfc.nasa.gov

The X-ray observation of black hole candidates provides a valuable tool to probe regions very close to the central black hole, where strong-field relativistic effects become important. Recent studies have shown that these effects seem to manifest themselves in the observed X-ray spectrum, in terms of the shape of X-ray continuum and the profile of emission lines, and in the X-ray light curves, in terms of certain quasi-periodic oscillations (QPOs). The latter will be the focus of this paper. We will review the proposed observational evidence for gravitomagnetic precession in black hole candidates, in light of new observational data and alternative models. Quantitative comparison will be made between the data and the models. We will comment on recent theoretical and observational efforts to address whether the gravitomagnetic precession of accreted, orbiting matter around a rotating black hole is a viable process for producing the observed QPOs. The results are inconclusive. We will mention a few areas where further progress can be made to possibly shed more light on the issue.

1 Background

1.1 X-ray Binaries

X-ray binaries are among the brightest X-ray sources in the sky. An X-ray binary consists of a compact object (neutron star or black hole) and a normal companion star, orbiting around each other. At certain stage of the binary evolution, the separation of the two stars becomes so close that material from the upper atmosphere of the companion star begins to flow toward the compact object under the influence of the latter's gravity. This process is known as mass accretion due to "Roche-lobe overflow". The accreted matter, carrying large amount of angular momentum from the orbital motion, circulates around the compact object and forms an accretion disk. As the matter spirals in toward the compact object, due to angular momentum losses from viscous processes, its gravitational energy is converted into heat. Depending upon the mass accretion rate, the temperature of the inner accretion

*Also Department of Astronomy, University of Maryland, College Park, MD 20742, USA

disk can reach more than one million degrees so that X-rays are produced. Therefore, the X-ray observation of such sources provides a valuable tool to probe regions very close to the compact object, where relativistic effects are strong thus important. It should be noted, however, that Roche-lobe overflow does not always occur. For some sources (especially those with a massive companion star), the accretion process may simply involve the capture of stellar wind from the companion star by the compact object. A small accretion disk can still form in such "wind-fed" systems, due to any residual angular momentum of the captured matter.

1.2 Black Hole Binaries

In some cases, the companion star is visible optically. By carefully monitoring its orbital motion, we can derive the so-called mass function, which is defined as

$$f_c(M) \equiv \frac{M_x^3 \sin^3 i}{(M_x + M_c)^2} = \frac{P_{orb} K_c^3}{2\pi G}, \tag{1}$$

where M_x and M_c are the mass of the compact object and its companion star respectively, i is the inclination angle of the binary orbit with respect to the line-of-sight (with $i = 0$ for face-on systems), P_{orb} is the binary orbital period, and K_c is the semi-amplitude of the optical radial velocity curve. Both M_x/M_c and i can be further constrained by modeling ellipsoidal variation in the optical light curves,[1] and M_c by the spectral type of the companion star. For roughly a dozen X-ray binaries, M_x is estimated to be greater than $\sim 3M_\odot$, which is considered to be a reliable upper limit to the mass of a neutron star,[2] for any values of M_c and i within their respective allowed parameter spaces. These sources are, therefore, good candidates for black hole binaries (BHBs). Table 1 lists all such systems known to date. It is clear from Eq. (1) that without making any assumptions the mass function sets a firm lower limit to the mass of the compact object. Therefore, those with the measured mass function greater than $3M_\odot$ are the best BHB candidates (BHBCs).

1.3 Spin of Black Hole

BHBs are thought to be formed by the core collapse of very massive stars which are usually fast rotators. The conservation of stellar angular momentum would lead to extremely rotating black holes, if the progenitor stars behave like rigid bodies. In reality, however, the stellar core and envelop are likely decoupled, with the ejected envelop carrying away significant amount of angular momentum. There may also be other processes that can cause additional loss of angular momentum during the formation. Therefore, while we expect that all black holes rotate to some extent, it is not clear whether any can be formed rapidly spinning.

In a binary configuration, the process of mass accretion can, in principle, result in the accretion of angular momentum by the black hole and, thus, the hole can be subsequently spun up.[15,16] However, most BHBs are transient sources (see Table 1, except for the top three sources) with a very low duty cycle of X-ray outbursts (when the mass accretion rate surges by orders of magnitude), so the lifetime-

Table 1. Stellar-Mass Black Hole Candidates†

Sources	M_x ($M_\odot$)	M_c ($M_\odot$)	i (degrees)	$f_c(M)$ ($M_\odot$)
Cyg X-1[3]	4.7–14.7	11.7–19.2	27–67	0.25
LMC X-3[4]	7–14	4–8	50–70	2.3
LMC X-1[5]	4–8	17-24	40–63	0.14
GS 2023+338[6]	10–15	0.5–1.0	52–60	6.26
A 0620-00[7]	5.1–17.1	0.2–0.7	31–54	3.18
GS 1124-68[8]	4–11	0.5–0.8	39–74	3.1
GS 2000+25[9]	4.8–14.4	< 0.7	43–85	4.97
GRO J1655-40[10]	6.8–7.2	2.2–2.5	69.5	3.24
H 1705-250[11,12]	4.9–7.9	0.07–0.42	48–51	4.65
GRO J0422+32[13]	> 9	0.2–0.4	13–31	1.21
4U 1543-47[14]	2.7–7.5	2.3–2.6	24–36	0.22

†In case of multiple measurements, the numbers are taken from the most recent papers. See references in these papers for proper credits on individual results.

averaged accretion rate is extremely low. Consequently, the accretion-induced spin-up of black holes is probably not effective in these sources (unlike in active galactic nuclei[15]). Other theoretical considerations, such as the presence of sub-Keplerian accretion flows in the vicinity of black holes,[17,18] would only make the mechanism less effective.

Recently, evidence has been found for the presence of rapidly spinning black holes in GRO J1655-40 and GRS 1915+105,[19] the only two Galactic sources that occasionally display spectacular radio jets with superluminal motion, analogous to radio-loud quasars, which is why they are sometimes referred to as "microquasars". Also discovered are slowly or moderately spinning black holes in several "normal" BHBCs. Not only has the study raised questions about the formation of rapidly rotating black holes, it has also brought into new focus possible observational consequences of black hole rotation in BHBs. Subsequently, a lot of excitement has been generated about the prospect of using BHBs to probe relativistic effects in strong gravity regime.[19,20,21]

2 GRO J1655-40: A Test Case

2.1 Existence of A Spinning Black Hole?

The binary parameters are known to such a high accuracy for GRO J1655-40 (with uncertainties typically only a few percent; see Table 1) that makes the interpretation of the X-ray data for this source much more reliable than for any other sources. Despite being a microquasar, GRO J1655-40 appears as a normal BHBC in terms of the observed spectral properties. The X-ray spectrum can be well described by an ultra-soft component at low energies and a power law at high energies,[22] which is canonical for BHBCs.[23] The ultra-soft component is generally attributed to the emission from the hot inner region of an accretion disk, while the power-law tail is thought to be the product of soft photons being Compton upscattered by energetic electrons in the region.

In practice, the ultra-soft component is often modeled by a multi-color blackbody.[24] In this model, the accretion disk is assumed to be geometrically thin, optically thick (i.e., the standard α disk[25,26]). The emission from such a disk is a blackbody, but the temperature of the blackbody is a function of the distance to the black hole. There are only two free parameters in the model: the radial distance of the inner disk edge to the black hole and the effective temperature at the inner edge, once the distance of the source and the inclination angle of the disk (which is usually assumed to be equal to the inclination of the binary orbit) are known. The model has worked remarkably well for a number of BHBCs (see a review by Tanaka and Lewin[23]). One of the most interesting results is that the inferred radius of the inner disk edge remains fairly stable and constant as the observed fluxes vary greatly,[23] strongly implying that the last (marginally) stable orbit is reached.

Since the model assumes Newtonian gravity, appropriate corrections need to be made to include relativistic effects.[27,28] Also, gravitational shift and gravitational focusing cause both the observed color temperature and integrated flux to deviate from the local values, depending on the inclination angle of the disk and the spin of the black hole.[29] Furthermore, in the hot inner region of an accretion disk, electron scattering is likely to dominate over free-free absorption, so the inner disk actually radiates approximately as a "diluted" blackbody with the peak effective temperature lower than that derived from fitting the spectrum.[30] Unfortunately, this "color correction factor" is still poorly determined, although it appears to depend only weakly on the properties of the black hole, mass accretion rate, or the location on the disk.[31]

Taking into account these factors, we found that for several BHBCs the inferred radius of the inner disk edge is consistent with that of the last stable orbit around non-rotating black holes.[19] When we applied the same model to the spectrum of GRO J1655-40 during an outburst, however, we found that the inner edge of the disk is only about 1.2 Schwarzschild radii away from the black hole,[19] which is, of course, impossible if the black hole does not rotate. Therefore, we were forced to conclude that GRO J1655-40 contains a *rapidly* spinning black hole. The spin of the black hole was determined to be $\sim$93% maximal rotation for this source.

The "ordinary" nature of GRO J1655-40 as a BHBC (in terms of its spectral properties) is strongly supported by recent observations with the *Rossi X-ray Timing Explorer* (RXTE). The source was monitored by RXTE extensively throughout its recent (1996–1997) X-ray outburst. From these observations, Sobczak et al.[32] found that the observed X-ray spectrum can be well characterized by the canonical spectral shape of BHBCs (soft multi-color blackbody plus hard power law) throughout the entire period, which confirms the previous findings[22] (although the actual values of the parameters differ slightly). In addition, they found that the inferred radius of the inner disk edge remained roughly constant (with occasional low points) as the X-ray flux varied by a factor of 3–4, similar to normal BHBCs.[23]

However, these authors mistakenly concluded that the corrections for relativistic effects are negligible for this source (actually they mis-quoted Zhang et al.[19] on this issue). As a matter of fact, the general relativistic effects are very significant in this case.[19] In particular, the inner boundary condition (i.e., torque free at the last stable orbit) can drastically change the temperature profile of the inner portion of the accretion disk under Newtonian gravity.[27,28] For GRO J1655-40, the correction amounts to a factor of ~2, after taking into account all the effects. Consequently, the radius of the inner disk edge in Sobczak et al. was over-estimated by the same amount.[32]

Our interpretation of the same data would, therefore, put the last stable orbit roughly at 1.5 Schwarzschild radii, which confirms the need for the presence of a rapidly rotating black hole in this system (although the spin of the black hole is ~78% maximal rotation in this case). We take note of occasional low values of the inferred radius when the measured X-ray luminosity is relatively high.[32] We interpret them as the times when the adopted spectral model breaks down. It is likely that when accretion rate is high the inner portion of a standard α disk may become unstable to radiation pressure and fragmentation.[33,34] During the outburst, GRO J1655-40 might be situated close to the critical point for such instability, as in the case of Cyg X-1 (A. Zdziarski, private communication).

In contrast, Sobczak et al. proposed that the last stable orbit was reached by the accretion disk *only* at the *smallest* radius (i.e., the lowest point) obtained from their spectral analysis. To explain why the inner edge of the disk stays far away from the last stable orbit during nearly the entire outburst, they invoked strong radiation drag. Intuitively, however, the radiation drag should be stronger in the higher luminosity state (when the "low points" occur), which would be opposite to their line of arguments. Moreover, it seems unlikely that the radiation drag is capable of significantly affecting gas dynamics in an *optically thick* disk (which is the most fundamental assumption in the model), since any external radiation fields (if present) can hardly penetrate the surface of the disk and the internal emission percolates through the surface preferentially in the direction perpendicular to the thin disk. Finally, from a practical point of view, it seems risky to base the entire argument on a single data point which is barely allowed physically,[32] especially considering the fact that not all the "low points" reach the same level.

In summary, while further studies are required to reveal important details about accretion processes in BHBCs, it seems highly plausible, as suggested by observations,[23] that during an X-ray outburst the accretion disk is of the standard form[25,26] in transient BHBCs and the inner disk edge reaches and remains at the last stable orbit. GRO J1655-40 is no exception. Quantitative modeling of its X-ray spectrum has led to the conclusion that a rapidly rotating black hole is likely to be present in this source.

2.2 Gravitomagnetic Precession

A natural consequence of curved spacetime in the vicinity of a spinning black hole is the gravitomagnetic precession of orbits that are tilted with respect to the equatorial plane perpendicular to the spin axis.[35] For circular orbits, which are probably most relevant to accretion processes in X-ray binaries, the precession frequency can be expressed as $\nu_{FD} = \nu_\theta \Delta\Omega/2\pi$, where ν_θ is the orbital frequency of the polar (or θ) motion, and $\Delta\Omega$ is the angle by which the line of nodes of a circular orbit are dragged during each orbital period. Note that this definition is slightly different form the one adopted by Cui et al.[21] As correctly pointed out by Merloni et al.,[36]

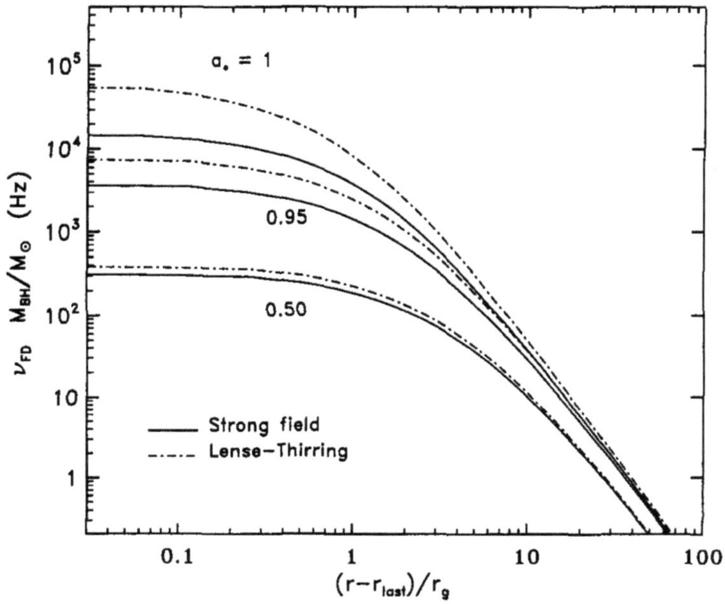

Figure 1. Gravitomagnetic precession frequency (multiplied by black hole mass), as a function of the distance from the last stable orbit, for different black hole spin. The weak-field limits are also shown in dash-dotted line for comparison.

with the new definition, ν_{FD} approaches the rotation frequency of the black hole at the event horizon, which is physically expected. Figure 1 shows ν_{FD} as a function of black hole spin. Comparing it with Fig. 1 of Cui et al.,[21] we find that, as expected, *the difference becomes significant only in the case of extremely spinning black holes* when the last stable orbit approaches the event horizon. Therefore, the results of Cui et al.[21] hardly change, even for microquasars, neither do any of the conclusions that they reached (see also discussion in the following section).

2.3 "Stable" QPOs

Quasi-periodic oscillations (QPOs) have been observed in the X-ray light curves of BHBCs over a wide frequency range, from mHz to roughly a few hundred Hz (reviews by van der Klis[37] and Cui[38]). While it is clear that no single process can account for all QPOs, the origins of QPOs are not known.

One type of QPOs in BHBCs is of particular interest here. Such QPO was first reliably established in GRS 1915+105 with the detection of a 67 Hz QPO.[39] One unique characteristic of this QPO is the stability of its frequency against any variation in the X-ray flux. The QPO appears to represent a transient event, and seems to be present only in certain spectral states. Shortly after this discovery, a similar signal was detected at ~300 Hz for GRO J1655-40,[40] *only* when the X-ray luminosity was relatively high[32] (i.e., at the "low points", as discussed in § 2.1).

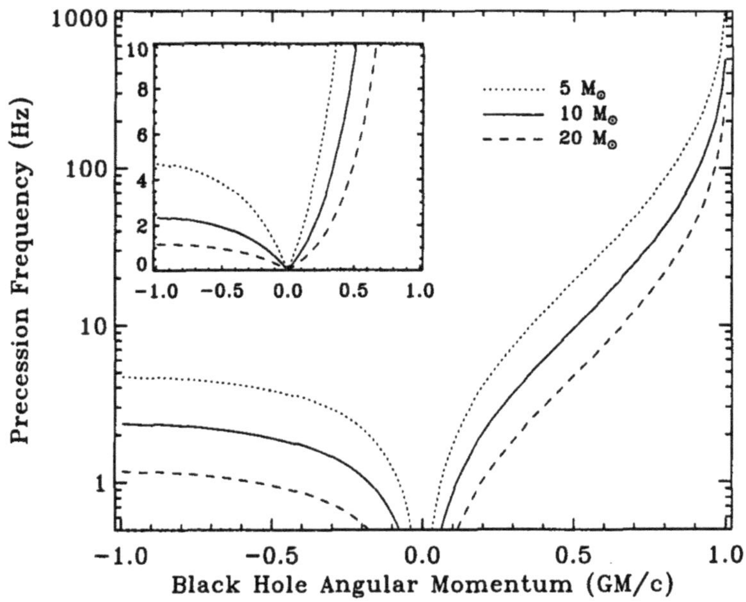

Figure 2. Gravitomagnetic precession frequency of a disk annulus, at which the effective disk temperature peaks, as a function of black hole spin (assuming $Q = 1$).

We subsequently proposed that these "stable" QPOs are perhaps the observational manifestation of gravitomagnetic precession of accreted, orbiting matter around spinning black holes.[21]

To facilitate comparison with observations, we have computed the expected gravitomagnetic precession frequency of a disk annulus at which the effective disk temperature peaks. The results are plotted in Fig. 2, as a function of black hole spin for three values of black hole mass. Comparing this figure to Fig. 2 of Cui et al.,[21] we find, again as expected, significant difference exists only for large (and positive) black hole spin. Note that the results depend little on the value of Q,[21,36] which is a constant of motion that specifies the tilt of precessing orbits.[41]

For GRO J1655-40 (where $M_{bh} = 7M_{\odot}$), the spin of the black hole would be about 97% maximal rotation were the observed 300 Hz QPO to be attributed to the gravitomagnetic precession (as shown in Fig. 2). This result assumes that the QPO originates in the modulation of the X-ray emission from the disk, which does not necessarily have to be the case, as will be discussed in § 4.3. To derive a lower limit to the black hole spin, we computed the frequency of gravitomagnetic precession right at the last stable orbit, as a function of the spin, and set the result equal to the QPO frequency. We found that in this case the spin would be ~87% maximal rotation. The results are consistent with those obtained spectroscopically (see also Fig. 3).

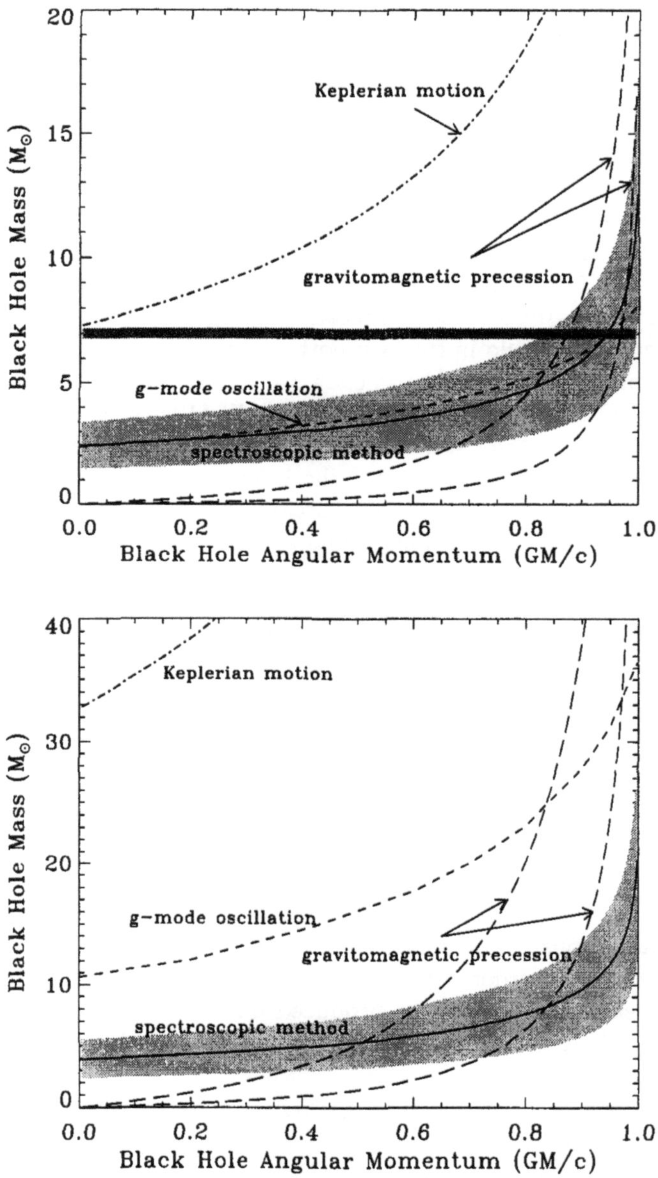

Figure 3. (top) Allowed parameter space for the mass and spin of the black hole in GRO J1655-40. The solid line shows the spectroscopic results of Zhang et al. (see § 2.1), with the lightly-shaded region indicating the estimated uncertainty. The dot-dashed, short-dashed, and long-dashed lines show, respectively, the predictions of three proposed models: Keplerian motion, g-mode oscillation, and gravitomagnetic precession. For the last model, two limiting cases are shown for frequencies of gravitomagnetic precession at the inner edge of the disk (upper curve) and at where the effective disk temperature peaks (lower curve). The heavily-shaded area represents the confidence region of the measured black hole mass. (bottom) Similar to the top panel, but for GRS 1915+105. In this case, the mass of the black hole is not known.

3 Application to Other Sources

Attempts have been made to apply the model to QPOs observed in "normal" BHBCs.[21] The closest examples are perhaps the so-called "very-high-state" QPOs that were observed only twice, in GX 339-4[42] (~6 Hz) and GS 1124-68[43] (5–8 Hz), in a high luminosity state (i.e., the very-high state). The frequency of such QPOs is fairly stable against flux variation, and any change in the QPO frequency can be explain by the variation in radiation pressure.[21] The low values of the QPO frequency would imply the presence of only slowly rotating black holes in these sources were these QPOs to be produced by the gravitomagnetic precession of accreted matter. This would in turn support the speculation that the relativistic jets observed in microquasars might somehow related to the spin of black holes.[19]

Similar QPOs have also been observed in GS 1124-68[43] and Cyg X-1[44] during the transition between the low and high state. However, in these cases, the situation is much more complicated and, thus, the interpretation is much more uncertain, because of the variable nature of the sources and the transient nature of the QPO phenomenon in general.[38] For a particular source, a different set of QPOs are often present at different times (or fluxes). It is, therefore, difficult to associate one QPO (if any) with the gravitomagnetic precession of accreted matter. For instance, a lower-frequency QPO (6–8 Hz) was often present in GRO J1655-40 during the same outburst when the 300 Hz QPO was discovered.[45] To make further progress, an systematic effort is imperative to reliably characterize and classify QPOs in BHBCs.

4 Unsettled Issues

4.1 Origin of Tilted Orbits

Accreted matter must be in a non-equatorial orbit, with respect to the spin axis of the black hole, in order for the orbit to undergo gravitomagnetic precession. In principle, such orbits could be produced during the formation of a black hole binary, due to the misalignment of the spin axis of the black hole with the axis of the binary motion.[46]

If mass accretion proceeds through a flat viscous disk, the disk can still be warped, due to irradiation by a central X-ray source.[47] However, in this case, differential gravitomagnetic precession generates twists in the disk and the inner region of the disk tends to gradually realign with the equatorial plane due to the viscous torque.[48] Once the realignment takes place, gravitomagnetic precession modes are damped out and, thus, the QPO should simply disappear. On the other hand, this process can help explain why the QPOs seem to be present *only* during transitional or unstable periods, when the inner region of the disk experiences significant changes.

As mentioned in § 2.1, the inner portion of the disk may become unstable,[33,34] when the accretion rate is relatively high. Once this occurs, the disk might become fragmented. The gravitomagnetic precession of the fragments is probably only weakly damped, so it might be capable of producing the observed QPOs.

4.2 *Excitation of Gravitomagnetic Modes*

A recent theoretical development involves the discovery of a class of high-frequency gravitomagnetic modes that are only weakly damped, in addition to the strongly-damped low-frequency ones.[49] These high-frequency modes are very localized spiral corrugations of the inner portion of the disk. They are perhaps most relevant to the observed QPOs, because X-ray emission from BHBCs is likely to originate very close to the central black hole. However, it remains to be seen whether the modes can be excited easily and whether they are capable of modulating X-ray emission.[49]

It should be pointed out that two critical assumptions are made in the study: small tilt angle and (pseudo) Newtonian potential. The former simplifies the equation by ignoring the non-linear aspect of the problem, which might affect the damping time scales of gravitomagnetic modes in a significant way; non-linear calculations are required to shed light on this issue. The latter might still represent a good approximation for systems that contain a relatively slowly rotating neutron star, but is certainly invalid for rapidly rotating black holes in which the stable QPOs are observed. These assumptions, therefore, tend to limit the scope of the applicability of the results. At present, it would seem premature to draw any definitive conclusions about confirming or rejecting the gravitomagnetic origin of the stable QPOs in microquasars.

4.3 *X-ray Modulation Mechanisms*

The gravitomagnetic precession of accreted matter only provides a natural frequency for the QPOs. To actually see them, certain physical processes are required to produce X-ray modulations. The processes are entirely unknown at present. Possibilities include (by no means exclusively): (1) variation in the X-ray emitting area of the disk, due to precession; (2) Doppler or gravitational shift of photon energy, due to the Keplerian motion of self-illuminating clumps or "hot spots" in the accretion disk; (3) oscillation that modulates the emission from the disk; and (4) occultation of a highly compact X-ray emitting region by disk fragments or the inner part of the disk itself.

The transient nature of the QPOs seems to imply that the first possibility, if viable at all, must be only a small effect. Any processes that involve discrete clumps would unavoidably also produce X-ray modulation at the orbital (both polar and azimuthal) frequencies.[36,53] These QPOs have not been seen in BHBCs, despite the high quality of RXTE data (the QPOs of comparable strength have been detected at around kHz, using the RXTE data, in more than a dozen neutron star systems[54]). Therefore, such processes also seem unlikely to be responsible for the observed QPOs. Hot spots or oscillations that modulate the emission from the accretion disk can also modulate Comptonized hard X-ray emission, if the seed photons originate in the disk emission. However, it seems difficult for these processes to amplify the fractional amplitude of the QPOs toward high photon energies, which is observed.[39,40] Such energy dependence of the QPO amplitude might, in our view, hold the key to our understanding of the origin of the QPOs.

The only remaining possibility in the above list is the occultation of a highly compact hard X-ray emitting region by detached disk annuli or the inner disk itself.

Interestingly, both observations[50] and theories[51,52] support the presence of such X-ray emitting regions in transient BHBCs during an X-ray outburst. As an example, Laurent & Titarchuk[52] recently showed that the observed power-law type hard X-ray spectrum of BHBCs may be the result of soft photons from the accretion disk being inverse-Compton scattered by the bulk motion of relativistic electrons in the vicinity of central black holes. The Comptonizing electrons are spatially confined almost entirely within the inner edge of the disk. In the context of this model, harder X-rays should, on average, originate closer to the black hole. Therefore, a precessing ring of matter occults a larger fraction of the harder X-ray emitting region, which might explain the observed energy dependence of the QPO amplitude, although a definitive answer still awaits detailed calculations based on this model.

4.4 Alternative Models

The stable QPOs in GRS 1915+105 and GRO J1655-40 can conceivably be associated with the Keplarian motion of self-illuminating clumps right at the last stable orbit around the black hole.[39,40] In the case of GRO J1655-40 whose black hole mass is accurately known, the model would be able to explain the observed QPO frequency (300 Hz) only if the black hole does not rotate.[40] This is clearly incompatible with the requirement for the presence of a rapidly rotating black hole in this source from modeling the observed X-ray spectrum. A more serious (and model independent) problem with this interpretation is due to the fact that general relativistic effects cause the temperature of the accretion disk to vanish at the last stable orbit.[27,28] Consequently, no X-ray emission is expected from the inner edge of the disk.

Another proposal has been made to associate the stable QPOs to epicyclic oscillation modes in the accretion disk.[58] Not only can such modes be supported by the disk, they also become trapped in the innermost portion of the disk, purely due to relativistic effects.[55,56,57,59,60] It is also realized that g-mode oscillations affect the largest area of a region in the disk where most X-rays are emitted and are, therefore, perhaps most relevant to observations.[58] This model can explain the measured frequencies of the stable QPOs in microquasars. However, like other disk-based processes, it may have difficulty in explaining the observed energy dependence of the QPO amplitude. Furthermore, it cannot account for the very-high-state QPOs or other relatively low-frequency QPOs in BHBCs (see § 3), although arguments can be made that these are fundamentally different from the stable QPOs in microquasars.

To quantify the comparison among the three possible models (gravitomagnetic precession, Keplerian motion, and g-mode oscillation), we have applied the models to both GRO J1655-40 and GRS 1915+105. For each model, a relationship between the mass and spin of the black hole is derived from the measured frequency of the QPO. The results are shown in Fig. 3, for both sources. In the gravitomagnetic precession model, two limiting cases are shown for frequencies at the inner edge of the disk and at where the effective disk temperature peaks, since we do not know the actual X-ray modulation mechanism (see § 4.3). Also shown in the figure is the same relationship derived from modeling the observed X-ray spectrum for each source.[19] The intersection between the latter and each of the three model curves, therefore, yields a solution to the mass and spin of the black hole, for a particular source, as predicted by that model. For GRO J1655-40, the mass of the black hole is known quite accurately, so it can be used to determine which models are compatible with the observed properties.

The uncertainty of the spectroscopic results is likely dominated by that of the color correction factor (up to $\sim$ 12%[31]), which can contribute as much a uncertainty as $\sim$24% in the derived radius of the last stable orbit.[19] The next major source of uncertainty comes from determining the source distance, to which the radius of the last stable orbit is directly proportional. It amounts to $\sim$6% for GRO J1655-40[61] and $\sim$12% for GRS 1915+105.[62] Likely minor contributions include uncertainties in the binary inclination angle, in the derived X-ray flux and the peak effective temperature of the disk, and in various relativistic correction factors.[19] To be relatively conservative, we have adopted an overall uncertainty of 40% on the radius of the last stable orbit for both sources. The lightly-shaded area in Fig. 3 reflects this measurement uncertainty.

Clearly, the Keplerian motion model is ruled out, since it is far from being consistent with the spectroscopic results. While the g-mode oscillation model seems to work for GRO J1655-40, it fails to yield any physical solutions for GRS 1915+105. The difference is so large that any reconciliation between the model and the data seems unlikely, unless the spectroscopic method has severely underestimated the radius of the last stable orbit (or the method does not apply in this case). On the other hand, the gravitomagnetic precession model seems to explain the observations well for both sources. For GRS 1915+105, the model constrains the mass ($\sim$3–14 $M_\odot$) and spin ($\sim$37–90% maximal rotation) of the black hole. Future infrared observations of the source might be able to test the mass constraint.

Acknowledgments

We would like to thank Andrea Merloni for sharing with us the results from his work prior to publication. W. Cui is also grateful to Ron Remillard for keeping him updated of the results on GRO J1655-40 and for many helpful discussions, despite some difference in their interpretation of the data. This work is supported in part by NASA through Contract NAS5-30612.

References

1. Y. Avni, in *Physics and Astrophysics of Neutron Stars and Black Holes*, eds. R. Giacconi and R. Ruffini (Amsterdam: North-Holland), 42 (1978).
2. C.E. Rhoades and R. Ruffini, *Phys. Rev. Lett.* **32**, 324 (1974).
3. A. Herrero *et al*, *A&A* **297**, 556 (1995).
4. A. Cowley *et al*, *ApJ* **272**, 118 (1983).
5. J.B. Hutchings *et al*, *AJ* **94**, 340 (1987).
6. T. Shahbaz *et al*, *MNRAS* **271**, L10 (1994).
7. T. Shahbaz *et al*, *MNRAS* **268**, 756 (1994).
8. T. Shahbaz *et al*, *MNRAS* **285**, 607 (1997).
9. G. Beekman *et al*, *MNRAS* **281**, L1 (1996).

10. J.A. Orosz and C.D. Bailyn, *ApJ* **477**, 860 (1997).
11. A.C. Martin *et al*, *MNRAS* **274**, 46 (1995).
12. E.T. Harlaftis *et al*, *AJ* **114**, 1170 (1997).
13. G. Beekman *et al*, *MNRAS* **390**, 303 (1997).
14. J.A. Orosz *et al*, *ApJ* **499**, 375 (1998).
15. J.M. Bardeen, *Nature* **226**, 64 (1970).
16. K.S. Thorne, *ApJ* **191**, 507 (1974).
17. R. Narayan and I. Yi, *ApJ* **444**, 231 (1995).
18. S.K. Chakrabarti and L.G. Titarchuk, *ApJ* **455**, 623 (1995).
19. S.N. Zhang *et al*, *ApJ* **482**, L155 (1997).
20. M.A. Nowak *et al*, *ApJ* **477**, L91 (1997).
21. W. Cui *et al*, *ApJ* **492**, L53 (1998).
22. S.N. Zhang *et al*, *ApJ* **479**, 381 (1997).
23. Y. Tanaka and W.H.G. Lewin, in *X-ray Binaries*, eds. W.H.G. Lewin *et al* (Cambridge: Cambridge U. Press), 126 (1995).
24. K. Mitsuda *et al*, *PASJ* **36**, 377 (1984).
25. N.I. Shakura and R.A. Sunyaev, *A&A* **24**, 337 (1973).
26. I.D. Novikov and K.S. Thorne, in *Black Holes*, eds. C. DeWitt & B.S. DeWitt (New York: Gordon & Breach) (1973).
27. D.N. Page and K.S. Thorne, *ApJ* **191**, 499 (1974).
28. T. Hanawa, *ApJ* **341**, 948 (1989).
29. C.T. Cunningham, *ApJ* **202**, 788 (1975).
30. T. Ebisuzaki *et al*, *PASJ* **36**, 551 (1984).
31. T. Shimura and F. Takahara, *ApJ* **445**, 780 (1995).
32. G.J. Sobczak *et al*, *ApJ*, submitted (1998) (astro-ph/9809195).
33. A.P. Lightman and D.M. Eardley, *ApJ* **187**, L1 (1974).
34. J.H. Krolik, *ApJ* **498**, L13 (1998).
35. J. Lense and H. Thirring, *Physik* Z. **19**, 156 (1918).
36. A. Merloni *et al*, *MNRAS*, submitted (1998).
37. M. van der Klis, in *X-ray Binaries*, eds. W.H.G. Lewin *et al* (Cambridge: Cambridge U. Press), 252 (1995).
38. W. Cui, in *Proc. High-Energy Processes in Accreting Black Holes*, eds. J. Poutanen & R. Svensson (San Francisco: PASP) (1998) (astro-ph/9809408).
39. E. Morgan *et al*, *ApJ* **482**, 993 (1997).
40. R. Remillard *et al*, *ApJ*, in press (1998) (astro-ph/9806049).
41. D.C. Wilkins, *Phys. Rev.* D **5**, 814 (1972).
42. S. Miyamoto *et al*, *ApJ* **383**, 784 (1991).
43. T. Belloni *et al*, *A&A* **322**, 587 (1997).
44. W. Cui *et al*, *ApJ* **484**, 383 (1997).
45. M. Méndez *et al*, *ApJ* **499**, L187 (1998).
46. J.G. Hills, *ApJ* **267**, 322 (1983).
47. J.E. Pringle, *MNRAS* **281**, 357 (1996).
48. J.M. Bardeen and J.A. Petterson, *ApJ* **195**, L65 (1975).
49. D. Marković and F.K. Lamb, *ApJ* **507**, in press (1998) (astro-ph/9801075).
50. J.E. Grove *et al*, *ApJ* **502**, 45 (1998).

51. R. Narayan, in *Accretion Phenomena and Related Outflows*, eds D. T. Wickramasinghe, G. V. Bicknell, & L. Ferrario, ASP Conf. Series Vol. 121 (1997) (astro-ph/9611113).
52. P. Laurent and L. Titarchuk, *ApJ*, in press (1998) (astro-ph/9808015).
53. S.M. Morsink and L. Stella, *ApJ*, in press (1998) (astro-ph/9808227).
54. M. van der Klis, *Proc. NATO Advanced Institue "The Many Faces of Neutron Stars"* (1997) (astro-ph/9710016).
55. S. Kato and J. Fukue, *PASJ* **32**, 377 (1980).
56. M.A. Nowak and R.V. Wagoner, *ApJ* **393**, 697 (1992).
57. M.A. Nowak and R.V. Wagoner, *ApJ* **418**, 187 (1993).
58. M.A. Nowak *et al*, *ApJ* **477**, L91 (1997).
59. X. Chen and R.E. Taam, *ApJ* **441**, 354 (1995).
60. C.A. Perez *et al*, *ApJ* **476**, 589 (1997).
61. R.M. Hjellming and M.P. Rupen, *Nature* **375**, 464 (1995).
62. I.F. Mirabel and L.F. Rodríguez, *Nature* **371**, 46 (1994).

MILLISECOND PHENOMENA IN ACCRETING NEUTRON STARS – AN UPDATE

M. VAN DER KLIS

Astronomical Institute "Anton Pannekoek", University of Amsterdam
Kruislaan 403, 1098 SJ Amsterdam, The Netherlands
E-mail: michiel@astro.uva.nl

Since the initial discoveries with the Rossi X-ray Timing Explorer (RXTE) in 1996 of kilohertz quasi-periodic oscillations (kHz QPOs) and burst oscillations in a number of low-mass X-ray binaries (LMXBs) containing low-magnetic-field neutron stars, a very active field has developed. I briefly summarize some of the developments since those early days, which include the discovery of the first accreting millisecond pulsar, further claims of the detection of strong-field general-relativistic effects, intriguing correlations of the kHz QPO properties with QPOs at lower frequencies, strong challenges for the beat-frequency interpretation of the twin kHz QPO peaks and a number of new theoretical ideas, some of which involving frame dragging. Twenty LMXBs have now been seen to exhibit periodic or quasi-periodic phenomena with frequencies exceeding $10^{2.5}$ Hz. Although the commensurabilities between the observed frequencies that suggest a beat-frequency interpretation have conclusively been shown to be not precise, the preponderance of the evidence still is in favour of the idea that there exists some kind of beat-frequency relation between the twin kHz peaks. However, that evidence is only coming from 4 of the 17 sources showing twin kHz QPOs.

1 Introduction

In this review, I attempt to summarize what happened in the field of millisecond X-ray variability from accreting low-magnetic field neutron stars in the period of slightly over one year since I submitted my review for the proceedings of the Lipari NATO ASI (van der Klis 1998). So, for anything that happened before October 1997 and for much of the introductory material, I refer to that article. As in that article, I refer to the tables summarizing the phenomenology for most of the references. A brief account of what went before:

Twin kilohertz quasi-periodic oscillation (QPO) peaks began to be discovered with the Rossi X-ray Timing Explorer (RXTE) early 1996, briefly after its launch. The peaks had associated rms amplitudes of $\lesssim$1% up to several 10% of the total source flux, peak separations of ~250–360 Hz, coherences between Q ~ 1–10^2, and their frequencies (of the higher-frequency peak of the two) varied between 500 and 1200 Hz with mass accretion rate $\dot{M}$. Only a short time later another series of RXTE discoveries began, of a different type of rapid oscillations. These came from some of the same sources, but only during type 1 X-ray bursts, and had frequencies between 330 and 590 Hz, depending on the source. These burst oscillations were interpreted in terms of the spin frequency of a layer in the neutron star atmosphere that is presumably close to the neutron star spin frequency itself. Finally, quite recently, the first true spin frequency was detected (at 401 Hz) in an accreting low-magnetic field neutron star. In several sources it was found that the twin kHz peak separation was equal to the frequency of the burst oscillations, or half that, to within a few percent. This strongly suggested a beat-frequency interpretation, where the higher-frequency peak (hereafter the upper peak) is identified with orbital motion at some preferred radius in the disk and the lower-frequency kHz QPO peak (hereafter the lower peak) is at the beat frequency between this orbital frequency and the spin frequency of the neutron star:

$$\nu_{lo} = \nu_{upp} - \nu_s \, .$$

It was immediately realized that if we are indeed seeing the signature of orbital motion around neutron stars with orbital frequencies of up to 1200 Hz, then we are probing a region of space-time where strong-field general relativity is required to describe orbital motion, and constraining the mass-radius relation of neutron stars and thereby the equation of state (EOS) of supranuclear-density matter. This spurred a large amount of both observational and theoretical work on these phenomena, some of which I will summarize here.

A few ballpark numbers are useful to keep in mind. Orbital motion around a M=1.4M$_\odot$ neutron star with a frequency ν_K of 1200 Hz takes place at an orbital radius of r_K=15 km; the expressions are

$$\nu_K = \left(\frac{GM}{4\pi^2 r_K^3}\right)^{1/2} \approx 1200\,\mathrm{Hz}\left(\frac{r_K}{15\,\mathrm{km}}\right)^{-3/2}\left(\frac{M}{1.4\mathrm{M}_\odot}\right)^{1/2},$$

or

$$r_K = \left(\frac{GM}{4\pi^2 \nu_K^2}\right)^{1/3} \approx 15\,\mathrm{km}\left(\frac{\nu_K}{1200\,\mathrm{Hz}}\right)^{-2/3}\left(\frac{M}{1.4\mathrm{M}_\odot}\right)^{1/3}$$

(this is exact for a Schwarzschild geometry and as measured at infinity; for realistic neutron stars the frame dragging corrections are moderate). The radius of the innermost stable circular orbit (ISCO) calculated in the same geometry is $R_{ISCO} = 6GM/c^2$, which for 1.4M$_\odot$ works out to 12.5 km.

2 Phenomenology

2.1 Millisecond pulsar

The first accreting millisecond pulsar was finally discovered, more than two years after RXTE's launch, by Rudy Wijnands using RXTE on April 13, 1998 in the soft X-ray transient SAX J1808.4−3658 (Fig. 1; Wijnands and van der Klis 1998a,b). The pulse frequency is 401 Hz.

This discovery, a long sought after "Holy Graal" of X-ray astronomy, had been anticipated for nearly 20 years, because magnetospheric disk accretion theory as well as the example of the non-accreting millisecond *radio* pulsars indicated that such rapid spin frequencies had to exist among accreting low-magnetic field neutron stars – yet despite numerous searches such rapid pulsars had not turned up in observations of LMXBs. Now that one such accreting millisecond pulsar is known, of course, the question becomes "why only one?". The pulsations in SAX J1808.4−3658 are not particularly weak and we know for sure that pulsars of similar amplitude and observed (as opposed to intrinsic) coherence are not present in many other LMXBs observed in a similar (rather standard) way as SAX J1808.4−3658 was with RXTE. The answer to this question may lie in part in the orbital characteristics of the source. The 2.0-hr orbital period was discovered, and the orbit measured, with

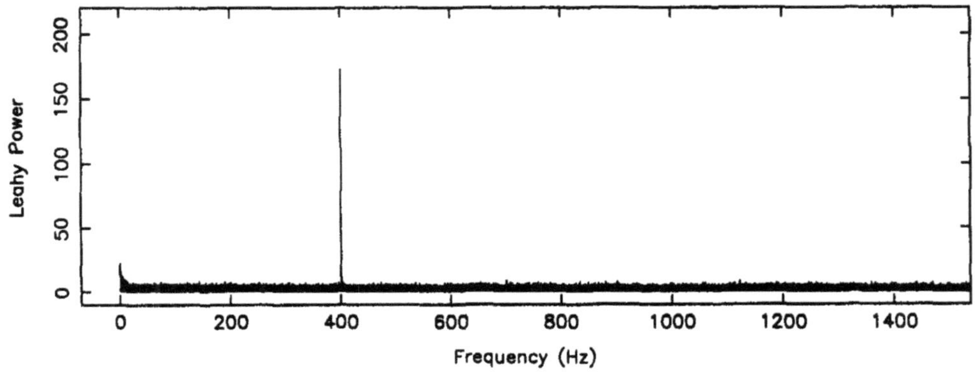

Figure 1. The discovery power spectrum of the first accreting millisecond X-ray pulsar. (Wijnands and van der Klis 1998b)

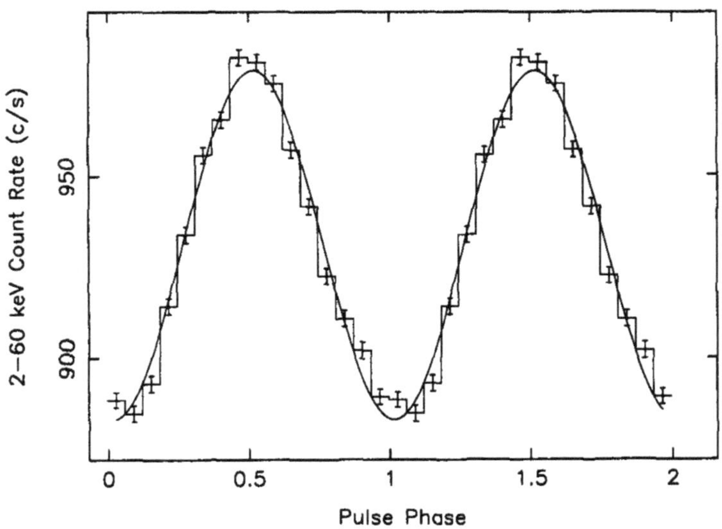

Figure 2. The pulse profile of SAX J1808.4−3658. (Wijnands and van der Klis 1998b)

admirable speed by Chakrabarty and Morgan (1998a,b). With a projected orbital radius $a \sin i$ of only 63 light *milli*seconds and a mass function of $3.8\,10^{-5}M_\odot$, the companion is either very low mass, or we are seeing the orbit nearly pole-on.

If we see the orbit pole-on, then this may be what allows us to see the pulsations in this system, and a different orientation of the orbit what hides them in many other LMXBs. The fact that the pulse profile of SAX J1808.4−3658 (Fig. 2) is nearly sinusoidal might be related to this pole-on orientation. In any case, of course, the low radial-velocity amplitude of the orbit reduces the Doppler shifts and keeps the pulsar more nearly coherent, facilitating its discovery without the use of special deacceleration techniques (which are, however, being used to try and discover other pulsars; e.g., Vaughan et al. 1994; van der Klis et al. 1999, in prep.).

Another possibility is that SAX J1808.4−3658 is a different type of system from the other well-studied LMXBs. Perhaps the fact that it is an unspectacular little transient is related to an accretion history that favours the preservation of a magnetic field configuration (strength, orientation) that produces a pulsar, and some *other* unspectacular little transients (with, perhaps, low-mass companions) will turn out to be millisecond pulsars as well.

It would be of enormous interest to find burst oscillations (§2.2) or kHz QPOs (§2.3) in SAX J1808.4−3658, as this would allow right away to confirm or reject various models for these phenomena involving the neutron star spin, which in SAX J1808.4−3658, uniquely among accreting low-magnetic field neutron stars, is known accurately and with certainty. However, the RXTE observations during the first few days of the April 1998 outburst, when the source was brightest, were relatively limited. Perhaps for this reason, although SAX J1808.4−3658 is known from its September 1996 SAX discovery observation (in 't Zand et al. 1998) to be a type 1 X-ray burster, no bursts were seen in the RXTE observations of the April 1998 outburst (the source was unobservable for SAX during that time), nor were kHz QPOs detected. Early in the outburst, when from comparing to other LMXBs the chances of seeing kHz QPOs were best, due to the limited observing time RXTE was not very sensitive to them (Wijnands and van der Klis 1998c). Plans are in place to spend a large amount of observing time with RXTE on SAX J1808.4−3658 during its next outburst, which should allow us to do a good job with respect to this and various other issues related to this interesting source.

2.2 Burst oscillations

Six sources have now shown burst oscillations. The frequencies are listed in Table 1. These oscillations are not detected in each burst. In four of these sources (the top four in Table 1), twin kHz peaks in the persistent emission have also been observed (see §2.3). It turns out that in each case, the burst QPO frequencies are close to the frequency *differences* between the twin peaks, or twice that. These frequencies have been listed in Table 1 in the columns labeled Frequency 1 and Frequency 2, respectively.

Table 1. Burst oscillations

Source	Frequency 1 (Hz)	Frequency 2 (Hz)
4U 1636−53	290	581
4U 1702−43	330	—
4U 1728−34	363	—
KS 1731−260	—	524
MXB 1743−29	589	
Aql X-1	549	

References see Tables 2, 3 and 4. Source identification of MXB 1743−29 is uncertain. Frequency 1 is close to the kHz QPO peak separation frequency, frequency 2 to twice that. For MXB 1743−29 and Aql X-1 this separation frequency is unknown.

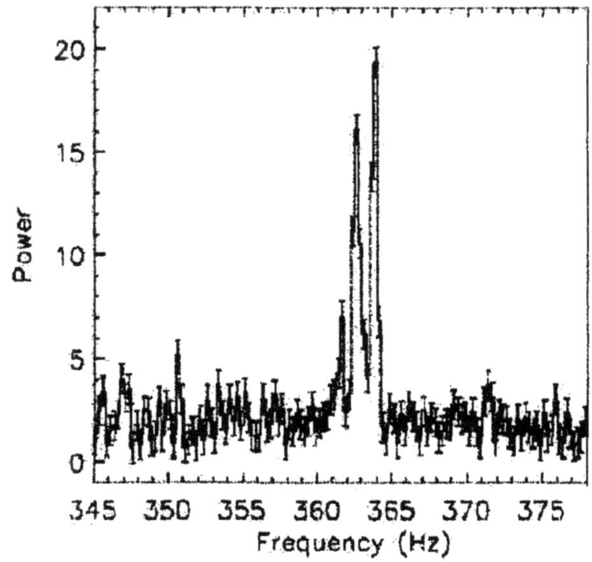

Figure 3. A drifting burst oscillation in 4U 1728−34. (Strohmayer et al. 1996c)

The burst QPOs have a relatively large coherence. In KS 1731−260 a Q of 900 was reported (Smith, Morgan and Bradt 1997). However, usually drifts by a few Hz are observed in the QPO frequency (Strohmayer et al. 1998c; Fig. 3). These drifts are suggestive of the bursting layer slightly expanding and then recontracting, changing its rotation rate to conserve angular momentum and thus modulating the QPO frequency. In 4U 1728−34 the QPO amplitude decreases and the fitted black-body radius increases during the burst rise in a way that is very suggestive of a modulation caused by an expanding hot spot (Strohmayer, Zhang and Swank 1997b). The high Q value, the small frequency drifts and the amplitude variations of the burst QPOs therefore support the view that their origin lies in the neutron star spin, and that their frequency is close to the spin frequency.

In addition to this evidence, it has been shown by Strohmayer et al. (1998b) that the frequency to which the oscillation approaches near the end of the burst (Fig. 4) is constant to within about 0.01% between bursts far apart in time. In the above interpretation this asymptotic frequency is the rotation frequency of the bursting layer when it approaches its quiescent rotation state, one that may be more nearly in corotation with the underlying star. This is another argument in favour of the near-constant neutron star spin as the process underlying the observed, slightly drifting frequencies.

Miller (1998), from an analysis of five bursts in 4U 1636−53, reports the presence of a 290 Hz burst oscillation at half the frequency of, and simultaneous with, the stronger 580 Hz oscillation, and interprets this in terms of the presence of two antipodal hot spots on the surface. If correct, this would indicate a neutron star spin frequency of 290 Hz (more nearly equal to the kHz peak separation, see §2.3),

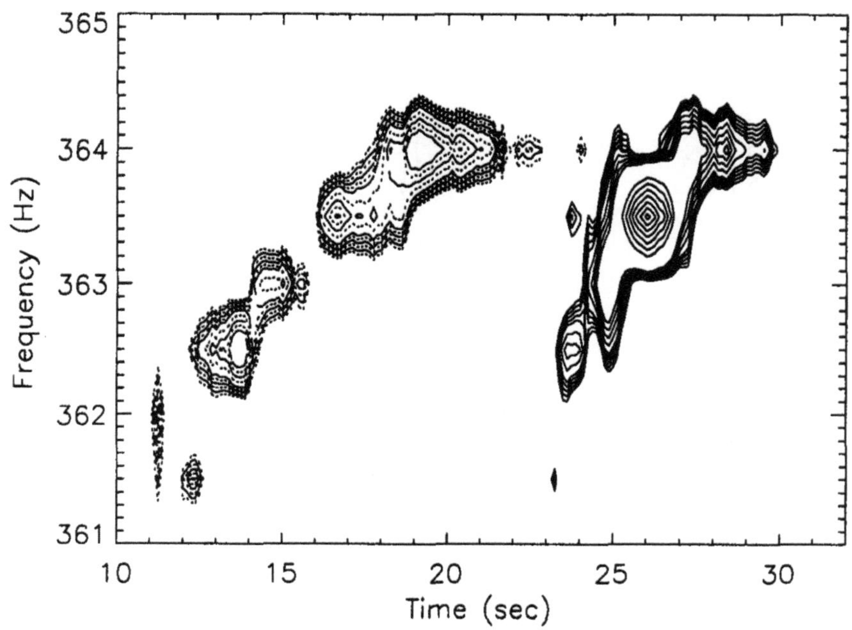

Figure 4. Dynamic power spectrum of burst oscillations in two bursts separated by 1.6 yr in 4U 1728−34 showing near-identical asymptotic frequencies. (Strohmayer et al. 1998b)

and strongly suggests we are seeing the signature of the presence of two magnetic poles. As the oscillation frequency drifts, we can *not* be looking at the magnetic poles themselves, which are of course fixed in the frame corotating with the spin of the star. Perhaps we are seeing the effect of the accumulation of extra nuclear fuel at the poles, which becomes decoupled from the magnetic field lines by convection during the bursts (Bildsten 1996, priv. comm.). The required near-simultaneous ignition of both poles at the burst onset may be a problem for this interpretation.

2.3 Kilohertz quasi-periodic oscillations

The number of sources that have shown kilohertz quasi-periodic oscillations (kHz QPOs) is now[a] eighteen. Together with the one source that showed burst oscillations (§2.2) but no kHz QPOs, and the millisecond pulsar (§2.1), which so far showed neither, this brings the total number of sources that has shown kilohertz periodic or quasi-periodic phenomena (here defined as phenomena at frequencies exceeding $10^{2.5}$ Hz) to twenty. All but one of the kHz QPO sources have shown twin kHz peaks in their power spectrum (Fig. 5); the exception with only a single peak so far is Aql X-1. Tables 2, 3 and 4 summarize the results on these sources.

[a]December 21, 1998

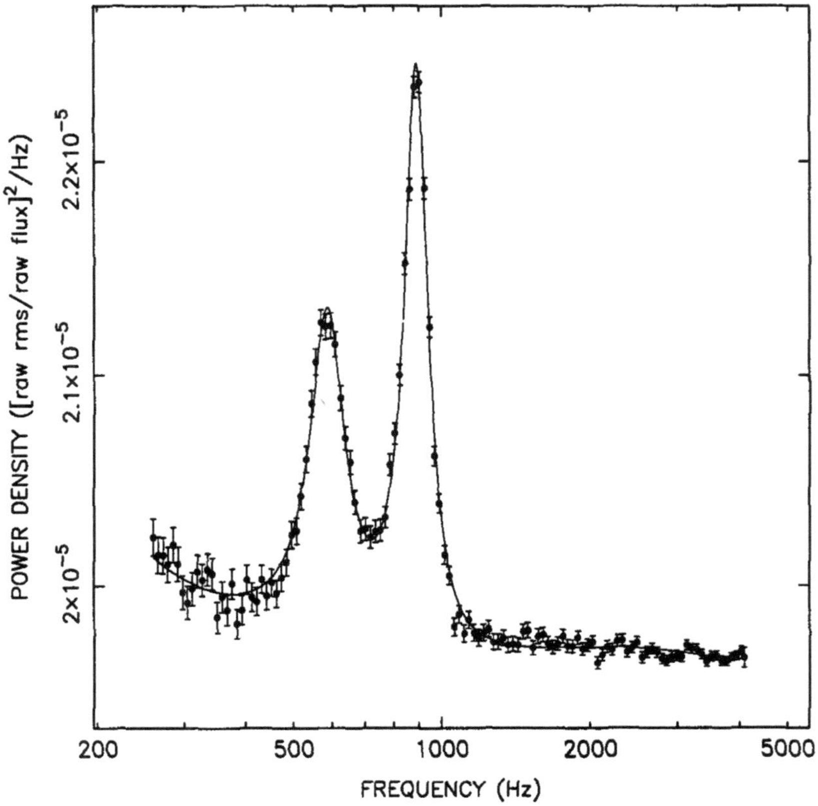

Figure 5. Twin kHz peaks in Sco X-1. (van der Klis et al. 1997b)

The two kHz QPO peaks increase in frequency with inferred mass accretion rate $\dot{M}$ both in Z and in atoll sources (see Hasinger and van der Klis 1989 or the review by van der Klis 1995 for the introduction of these subtypes of LMXBs). In Z sources, the QPOs are nearly always seen down to the lowest inferred $\dot{M}$ levels these sources reach, in atoll sources the QPOs tend to occur in the middle of the $\dot{M}$ range of each source (Fig. 6). Kilohertz QPOs are seen in a similar frequency range (500–1200 Hz for the upper peak) in sources that differ in average X-ray luminosity L_x by 2.5 orders of magnitude (where L_x is here defined simply as $4\pi d^2 f_x$ with f_x the X-ray flux and d the distance), and the kHz QPO frequency seems to be determined more by the difference between average and instantaneous L_x of a source than by L_x itself. This is unexplained and must mean that another parameter than $\dot{M}$, related in some way to the average L_x, affects the QPO frequency (van der Klis 1997). Perhaps this parameter is the neutron star magnetic field strength.

In 4 sources, both twin kHz peaks and burst oscillations are seen, and the fact that the burst oscillation frequency is near 1 or 2 times the kHz peak separation frequency is the main argument for the beat frequency interpretation of kHz QPOs. The evidence for this is summarized in Table 5. In 4U 1636−53, Méndez

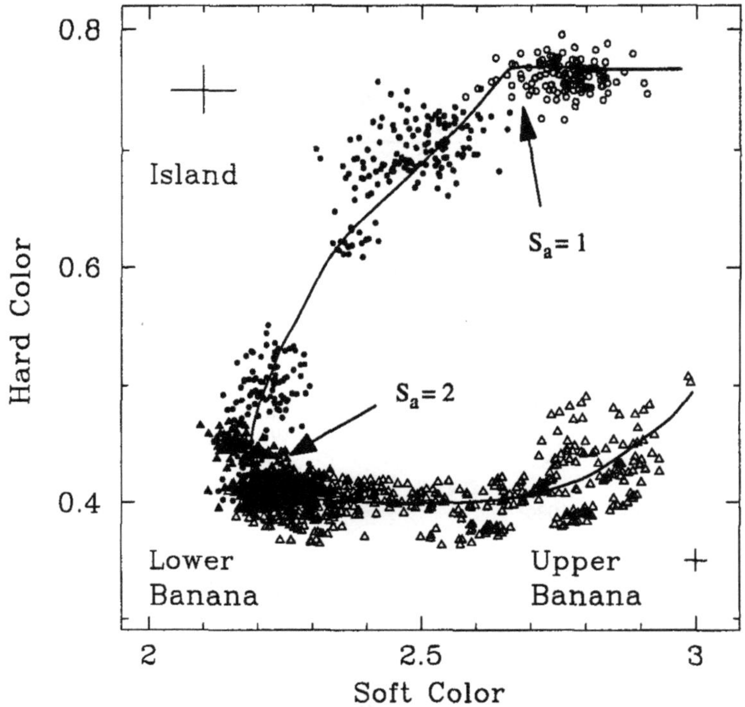

Figure 6. X-ray color-color diagram of 4U 1608−52. Mass accretion rate is inferred to increase counterclockwise, approximately along the drawn curve. kHz QPO detections are indicated with filled symbols. (Méndez et al. 1998d)

et al. (1998c) have shown that the correspondence is not exact. Yet, although we would dearly like to have a few more examples, the preponderance of the evidence still seems to indicate that at least an approximate beat-frequency relation exists between the three frequencies in those sources where they are all observed.

In an interpretation where the burst oscillations and the kHz peak separation are both close to the neutron star spin, this peak separation is predicted to be approximately constant. It is clear that this is not always the case: both in Sco X-1 and in 4U 1608−52 this separation decreases systematically when the kHz QPO frequencies increase (Fig. 7), and indications for this exist in 4U 1735−44 and 4U 1702−43 as well (see Tables 3 and 4).

The possibility to detect evidence in kHz QPOs for the existence of the innermost stable circular orbit (ISCO) predicted by general relativity, which would constitute the first direct detection of a strong-field general-relativistic effect, has fascinated since the beginning (Kaaret, Ford and Chen 1997; Zhang, Strohmayer and Swank 1997). It has been conjectured that when the inner edge of the accretion disk reaches the ISCO, the QPO frequency might level off and remain constant while $\dot{M}$ continues rising. For this reason, the recent measurement of an apparent leveling off of the increase of QPO frequency with X-ray count rate in 4U 1820−30 (Zhang et al. 1998b; Fig. 8) attracted considerable attention.

Table 2. Observed frequencies of kilohertz QPOs in Z sources

Source	Lower peak freq. (Hz)	Upper peak freq. (Hz)	Peak separation (Hz)	Burst osc. freq. (Hz)	References
Sco X-1	565	870	307±5		Van der Klis et al. 1996a,b,c,1997b
	↓	↓	↓		
	845	1080	237±5		
		↓			
		1130			
GX 5−1	215	505			Van der Klis et al. 1996e
	↓	↓	298±11		Wijnands et al. 1998c
	660	890			
	↓				
	700				
GX 17+2		645			Van der Klis et al. 1997a
		↓			Wijnands et al. 1997b
	480	783			
	↓	↓	294±8		
	780	1080			
		↓			
		1085			
Cyg X-2		730			Wijnands et al. 1998a
		↓			
	530	855			
	↓	↓	346±29		
	660	1005			
GX 340+0	250	570			Jonker et al. 1998
	↓	↓	325±10		
	500	820			
	↓				
	625				
GX 349+2	710	980	266±13		Zhang et al. 1998a
	1020[a]				Kuulkers and van der Klis 1997

These notes refer also to Tables 3 and 4. Frequencies with no quoted errors were rounded to the nearest 5, or, for burst QPOs, 1 Hz. Arrows indicate ranges over which the frequency was observed or inferred to vary; these can be made up of several overlapping ranges from different observations. Frequencies not connected by arrows are measurements at different epochs. Frequencies in the same row were observed simultaneously (except for burst QPOs). Entries straddling the upper and lower peak columns are of single, unidentified peaks. [a] Marginal detection. [b] Special detection method; Méndez et al. (1998a,b,c). [c] Source identification uncertain. [d] Pulsar.

The difficulty in measurements of this kind is not in measuring the frequency, but in measuring $\dot{M}$. We have known since the days of EXOSAT that variations in X-ray count rate or even X-ray flux in the band accessible to our instruments do not necessarily track $\dot{M}$ (Hasinger and van der Klis 1989, van der Klis et al. 1990, Hasinger et al. 1990), and RXTE observations have confirmed this beautifully. A clear example of this is seen in 4U 1608−52 (Méndez et al. 1998d; Fig. 9 left). This figure, containing many more measurements, looks much less nice than Fig. 8. With sparser coverage it could easily have happened to give a similar impression of leveling off. When plotting frequency vs. X-ray color, which in this case may be

Table 3. Observed frequencies of kilohertz QPOs in atoll sources - Part I

Source	Lower peak freq. (Hz)	Upper peak freq. (Hz)	Peak separation (Hz)	Burst osc. freq. (Hz)	References
4U 0614+09		540			Ford et al. 1996, 1997a,b
		↓			Van der Klis et al. 1996d
	485	840			Méndez et al. 1997
	↓	↓	323±4		Vaughan et al. 1997,1998
	800	1145			Kaaret et al. 1998
4U 1608−52	270				Van Paradijs et al. 1996
	440	765	325±7		Berger et al. 1996
	↓	↓	↓		Vaughan et al. 1997, 1998
	475	800[b]	326±3		Yu et al. 1997
	↓	↓	↓		Méndez et al. 1998a,b,d
	865	1090[b]	225±12		Kaaret et al. 1998
	↓				
	895				
4U 1636−53	830				Zhang et al. 1996, 1997a
	↓				Zhang 1997
	900	1150	257±20		Van der Klis et al. 1996d
	↓	↓			Wijnands et al. 1997a
	950	1190	276±10		Vaughan et al. 1997
		↓			Miller 1998
		1228			Strohmayer et al. 1998b
	830				Méndez et al. 1998c
	↓	1190[b]	251±4[b]	290,581	
	1050				
4U 1702−43	625				Markwardt et al. 1998
	↓	1058[b]	333±5[b]	330	
	825				
	655	1000[b]	344±7[b]		
	↓	↓	↓		
	700	1040[b]	337±7[b]		
	↓	↓	↓		
	770	1085[b]	315±11[b]		
	902				
4U 1705−44	775	1075[a]	298±11		Ford et al. 1998a
	↓				
	870				

Notes: see Table 2.

a better measure of $\dot{M}$ (though not a perfect one either), there is no sign of any saturation (Fig. 9).

A number of intriguing correlations has been found between the kHz QPOs and the phenomena at lower frequencies. Wijnands and van der Klis (1998d) point out, that the low $\dot{M}$ power-spectral similarity between BHCs and low-luminosity low-magnetic field neutron stars found with EXOSAT and Ginga (van der Klis 1994) also holds for the millisecond pulsar SAX J1808.4-3658 (§ 2.1), and may even extend to Z sources in *their* lowest $\dot{M}$ states (the left end of the so-called horizontal

Table 4. Observed frequencies of kilohertz QPOs in atoll sources - Part II

Source	Lower peak freq. (Hz)	Upper peak freq. (Hz)	Peak separation (Hz)	Burst osc. freq. (Hz)	References
4U 1728–34		325[f]			Strohmayer et al. 1996a,b,c,
		↓			1997b,1998b
		500			Ford and van der Klis 1998
		↓			
	640	990			
	↓	↓	355±5	364	
	790	1120			
KS 1731–260	900	1160	260±10	524	Morgan and Smith 1996
		↓			Smith et al. 1997,
		1205			Wijnands and Van der Klis 1997
4U 1735–44	640	980	341±7		Wijnands et al. 1996, 1998b
	↓	↓	↓		Ford et al. 1998b
	730	1025	296±12		
	900[a]	1150	249±15		
		↓			
		1160			
MXB 1743–29[c]				589	Strohmayer et al.
					1996d,1997a
SAX J1808.4-3658				401[d]	Wijnands and van der Klis
					1998a,b,c
4U 1820–30		660			Smale et al. 1996, 1997
		↓			Zhang et al. 1998b
	500	860	358±43		
	↓	↓	↓		
	796	1065	275±8		
Aql X-1		677			Zhang et al. 1998c
		↓		549	Cui et al. 1998
		871			Yu et al. 1998
4U 1915-05		820			Barret et al. 1997,1998
		↓			
	555	935			
	↓	↓	355±?		
	655	1005			
XTE J2123–058	850	1110[a]	261±10[a]		Homan et al. 1998a,b
	↓	↓	↓		
	855	1123	268±9		
	↓	↓	↓		
	870[a]	1140	270±5[a]		

Notes: see Table 2.

branch), where by the way these sources are still quite luminous. The power spectra look very similar (Fig. 10), and with the better RXTE data it is now evident that nearly always, in addition to the break at low frequency, a QPO-like feature above the break is present. The correlation between the break frequency and the QPO frequency is excellent (Fig. 11), and encompasses both neutron stars and black-hole candidates. This suggests that both the band-limited noise component with the 0.03–30 Hz break frequency, and the 0.2–70 Hz QPO (with the possible exception

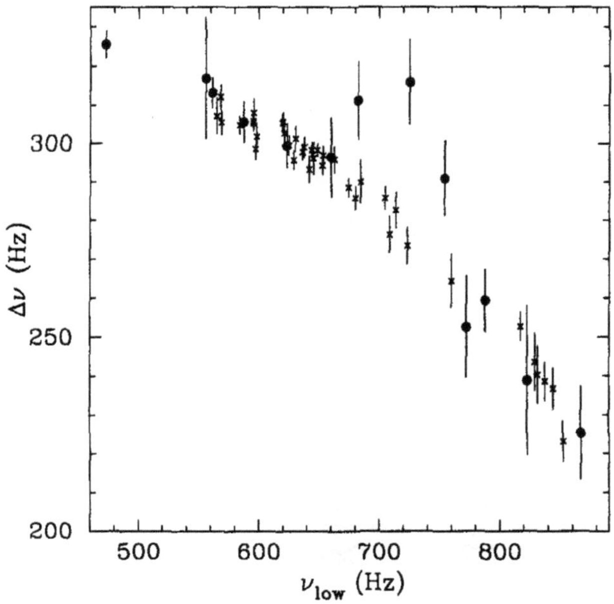

Figure 7. The variations in kHz QPO peak separation in Sco X-1 and 4U 1608−52 as a function of the lower peak frequency. (Méndez et al. 1998b)

Table 5. Commensurability of kHz QPOs and burst oscillation frequencies

Source	Burst oscillation frequency (Hz)	kHz QPO separation frequency (Hz)	Ratio (burst/separation frequency)
4U 1636−53	290	251±4	1.155±0.018
"	581	" "	2.315±0.037
4U 1702−43	330	333±5	0.991±0.015
4U 1728−34	364	355±5	1.025±0.014
KS 1731−260	524	260±10	2.015±0.078

References see Tables 3 and 4.

of the Z sources, which are slightly off the main relation) are found in both neutron stars and black holes. This would exclude spin-orbit beat-frequency models and any other models requiring a material surface, an event horizon, a magnetic field, or their absence, for their explanation, essentially implying these phenomena are generated in the accretion flow towards *any* low-magnetic field compact object and are most likely disk variability features. Ford and van der Klis (1998) studied the relation between both of these features (the band-limited noise and the low-frequency QPO-like feature) and the kHz QPOs in 4U 1728−34, and find generally good correlations (there is one deviant data set), suggesting that kHz QPOs also fit into schemes of this kind in some way. This of course leaves one with the

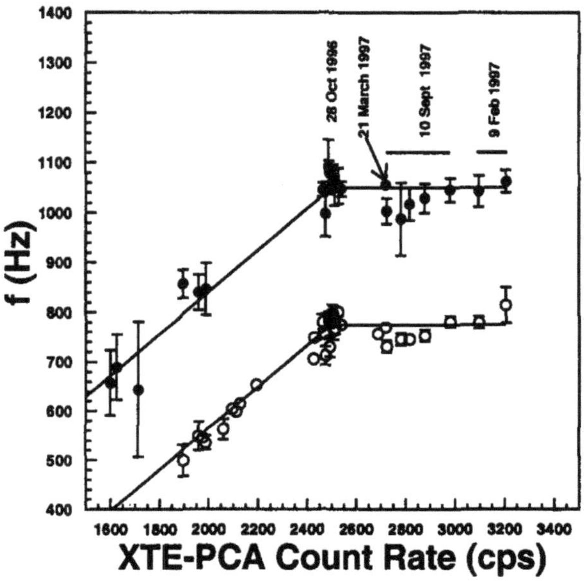

Figure 8. Evidence for a leveling off of the kHz QPO frequency with count rate in 4U 1820−30. (Zhang et al. 1998b)

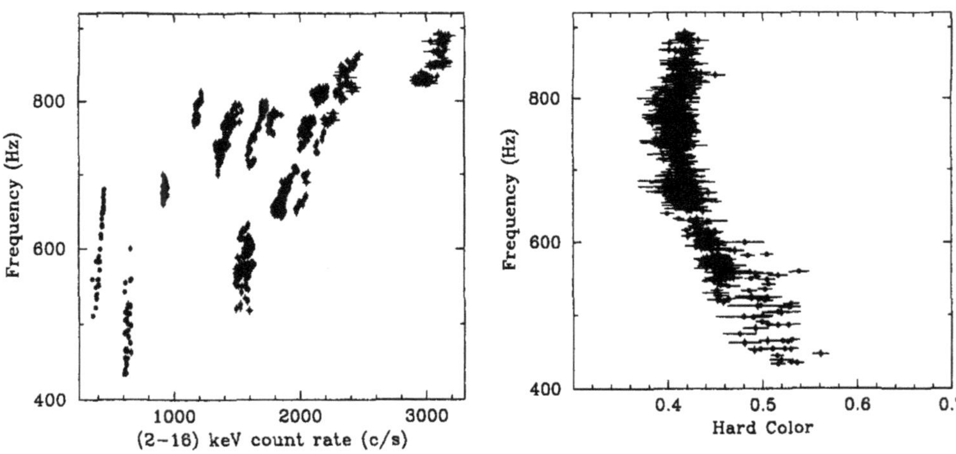

Figure 9. In 4U 1608−52 a QPO frequency vs. count rate plot shows no clear correlation. When plotted vs. X-ray color it is clear there is no saturation of QPO frequency. (Méndez et al. 1998d)

question how the absence of kHz QPOs of the type discussed in this article from black hole candidates fits in with this (the highest-frequency QPOs seen in black hole candidates have frequencies up a few 100 Hz, but although the amount of information on these QPOs is relatively limited, they appear to have properties that are different from kHz QPOs in neutron stars).

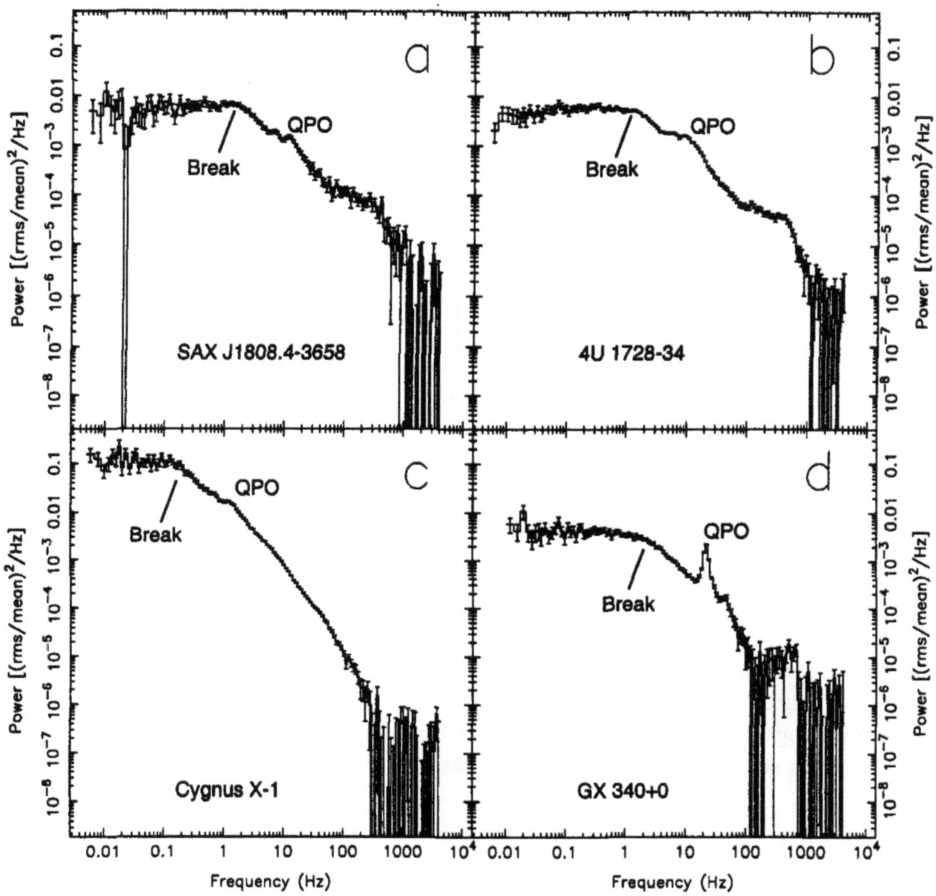

Figure 10. Broad-band power spectra of, respectively, the millisecond pulsar, an atoll source, a black-hole candidate and a Z source. (Wijnands and van der Klis 1998d)

A coincidence of QPO properties pointed out by Psaltis, Belloni and van der Klis (1998) may provide an answer to this question. Power spectra of many Z and atoll sources, of Cir X-1 and of a few low luminosity neutron stars and BHCs show two QPO or broad noise phenomena whose centroid frequencies, when plotted vs. each other, seem to line up (Fig. 12). This would not only identify the low-frequency (few 10 Hz) QPOs in Z and atoll sources with even lower-frequency QPOs in Cir X-1 (3–10 Hz) and BHCs (0.3–1 Hz) as also suggested by the Wijnands and van der Klis (1998d) results, but would *also* indicate that the *lower* kHz peak in Z and atoll sources (the same argument does not hold for the upper kHz peak), the broad 20–100 Hz bumps found in Cir X-1 (Shirey et al. 1996, 1998, see also Tennant 1987), and even lower-frequency (0.2–1 Hz) bumps in some low luminosity neutron stars and BHCs are due to the same physical phenomenon. While this conjecture remains to be confirmed, the currently available data are suggestive. The interpretation would be similar to that mentioned above, particularly, it would

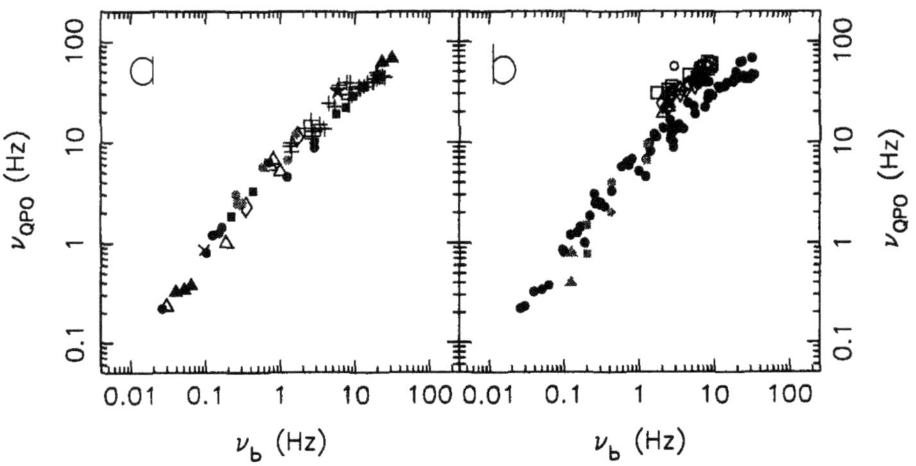

Figure 11. Relation between noise break frequency and QPO frequency for sources of the types shown in the previous figure. (Wijnands and van der Klis 1998d)

imply that the lower kHz QPO is not unique to neutron star systems either, and therefore presumably a general feature of the accretion flow onto a compact object. Of course, *orbital motion* in the disk is such an accretion flow phenomenon and remains an attractive interpretation for some of the observed frequencies. The phenomenology is quite complex; further detailed work will show to what extent these various suggestions will stand the test of additional data.

3 Some final remarks

A number of new theoretical ideas has been proposed for explaining various aspects of the phenomenology described in the previous sections. Examples are the papers by Alpar and Yılmaz (1997), Ghosh (1998), Kluźniak (1998), Titarchuk et al. (1998). An interesting discussion has also begun about the possibility, suggested by the beat-frequency interpretation of the kHz QPOs, that all these neutron stars spin at approximately the same rate (Bildsten 1998, Anderson et al. 1998, Levin 1998).

For the purpose of these final remarks let us concentrate on one aspect of the recent discussions and postulate that any successful model should be able to expain at least:

(i) The apparent beat-frequency relation between the twin kHz peaks and the burst oscillations (Table 5).
(ii) The relations between kHz QPOs and lower-frequency QPOs in Z and atoll sources described above.
(iii) The systematic variations in kHz QPO peak separation with $\dot{M}$ (Fig. 7).

Here I concentrate on what may be shaping up to be two rivals: the spin-orbit beat-frequency interpretation, and the general-relativistic apsidal motion/precession interpretation. In versions of the beat-frequency model (BFM) considered by Miller

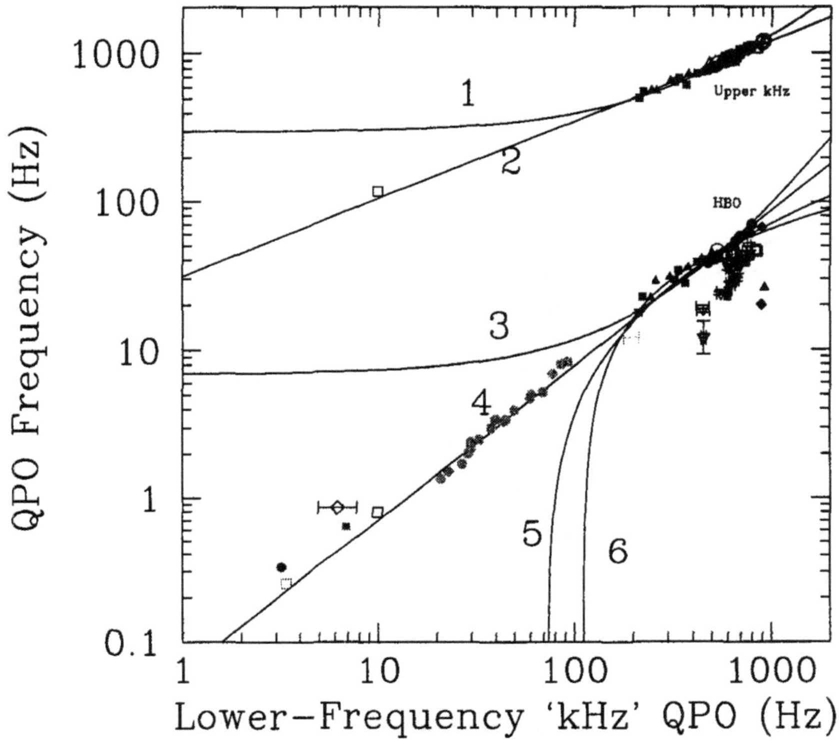

Figure 12. When plotting the frequencies of various QPO and broad noise components seen in accreting neutron stars and black holes vs. each other, a diagram emerges that suggests that similar components occur in all these sources and span a very wide range in frequency. Power-law relations describe the correlations reasonably well. The curved lines illustrate various attempts to fit these relations with spin-orbit beat-frequency models at the same spin frequency for all sources. Such models always converge to a constant frequency and therefore do not fit. (Psaltis, Belloni and van der Klis 1998)

et al. (1998) and Psaltis et al. (1998), the kHz QPOs are identified with the orbital frequency at the sonic radius and its beat with the neutron star spin, and the lower-frequency HBO in Z sources with the beat of the orbital motion at the *magnetospheric* radius with the same neutron star spin. In general-relativistic (GR) precession/apsidal motion models discussed by Stella and Vietri (1998a,b) the kHz peaks are due to orbital motion and GR apsidal motion of a slightly eccentric orbit, and the low-frequency QPO to Lense-Thirring precession of this same orbit. Both types of model may be able, with some difficulty, to explain (ii) (Psaltis et al. 1998). While of course BFMs are built around, and so have no difficulty to explain, (i), in these models the question is: how to explain, in a spin-orbit beat model, the varying peak separation (iii). The GR models may be able to explain (iii) (Stella and Vietri 1998b), but seem to have no obvious way to predict a beat-frequency relation (i). Surprisingly, ways out may exist from both dilemma's. In the sonic-point beat-frequency model, the beat interaction between spin and orbital frequencies takes place at the sonic radius by an interaction between an orbiting clump and an X-ray beam of the central, unseen, pulsar. However, the signal associated with this interaction is produced only when the matter released from this clump arrives at the neutron star surface. As pointed out by Lamb (1998, priv. comm.), if the clump is in an orbit that is gradually spiraling down, then the observed frequency will be higher than the actual beat frequency at which beam and clump interact, because the propagation time for the matter from the clump to the surface will be gradually diminishing. This will put the lower kHz peak closer to the upper one, and thus decrease the kHz peak separation, more so when due to stronger radiation drag at higher L_x the spiralling-down is faster, as observed.

In the GR apsidal motion model, the upper kHz peak is at the orbital frequency and the lower one at the apsidal motion frequency of a slightly eccentric orbit. Interestingly, the difference between these two frequencies is the frequency at which the orbiting clump goes through periastron. So, perhaps in a model of this type one could explain the fact that the difference frequency between the kHz peaks is sometimes seen back during X-ray bursts by some kind of interaction at periastron with, e.g., an expanding bursting layer. Whether this would lead to a viable model remains to be seen. In particular it might be a problem to account for the apparent constancy of the burst oscillation frequency in each source.

At the conclusion of my 1997 Lipari review to which the current text is intended to be an update I said that "eventually, most LMXBs will likely exhibit the new phenomenon", and that this would help enormously in constraining the models. Indeed, this is turning out to be the case. A further synthesis among what is now a considerable and fortunately fast-growing body of data on the properties of kHz QPOs, low-frequency QPOs and X-ray spectra observed simultaneously is likely to bring much progress in the year to come. Further long RXTE observations of these sources will be indispensible in reaching a full understanding of the systematics of the phenomenology.

References

1. Alpar, M.A., Yılmaz, A., 1997, New Astronomy 2, 225
2. Anderson, N., Kokkotas, K.D., Stergioulas, N., 1998, astro-ph/9806089
3. Barret, D., Olive, J.F., Boirin, L., Grindlay, J.E., Bloser, P.F., Chou, Y., Swank, J.H., Smale, A.P., 1997, IAU Circ. 6793
4. Barret, D., Boirin, L., Olive, J.F., Grindlay, J.E., Bloser, P.F., Swank, J.H., Smale, A.P., 1998, paper presented at the Third INTEGRAL Workshop (The Extreme Universe, September 14-18th, 1998, Taormina, Italy); Astrophysical Letters and Communications, in press
5. Berger, M., van der Klis, M., van Paradijs, J., Lewin, W.H.G., Lamb, F., Vaughan, B., Kuulkers, E., Augusteijn, T., Zhang, W., Marshall, F.E., Lapidus, I., Lochner, J.C., Strohmayer, T.E., 1996, ApJ 469, L13
6. Bildsten, L., 1998, ApJ 501, L89
7. Cui, W., Barret, D., Zhang, S.N., Chen, W., Boirin, L., Swank, J., 1998, ApJ 502, L49
8. Chakrabarty, D., Morgan, E.H., 1998a, IAU Circ. 6877
9. Chakrabarty, D., Morgan, E.H., 1998b, Nature, 394, 346

10. Ford, E., Kaaret, P., Tavani, M., Harmon, B.A., Zhang, S.N.,Barret, D., Bloser, P., Grindlay, J., 1996, IAU Circ. 6426
11. Ford, E., Kaaret, P., Tavani, M., Barret, D., Bloser, P., Grindlay, J., Harmon, B.A., Paciesas, W.S., Zhang, S.N., 1997a, ApJ 475, L123
12. Ford, E.C., Kaaret, P., Chen, K., Tavani, M., Barret, D., Bloser, P., Grindlay, J., Harmon, B.A., Paciesas, W.S., Zhang, S.N., 1997b, ApJ 486, L47
13. Ford, E.C., van der Klis, M., Kaaret, P., 1998a, ApJ 498, L41
14. Ford, E.C., van der Klis, M., van Paradijs, J., Méndez, M., Wijands, R., Kaaret, P. 1998b, ApJ 508, L155
15. Ford, E.C., van der Klis, M., 1998, ApJ 506, L39
16. Ghosh, P., 1998, ApJ 506, L109
17. Gilfanov, M., 1998, A&A 338, L83
18. Hasinger, G. & Van der Klis, M. 1989, A&A 225, 79
19. Hasinger, G. et al., 1990, A&A 235, 131
20. Homan, J., van der Klis, M., van Paradijs, J., Méndez, M., 1998a, IAU Circ. 6971
21. Homan, J., et al. 1998b, ApJ submitted
22. Jonker, P.G., Wijnands, R., van der Klis, M., Psaltis, D., Kuulkers, E., Lamb, F.K., 1998, ApJ 499, L191
23. Kluźniak, W., 1998, ApJ 509, L37
24. Kuulkers, E., van der Klis, M., 1998, A&A 332, 845
25. Kaaret, P., Ford, E. & Chen, K., 1997, ApJ 480, L27
26. Kaaret, P., Yu, W., Ford, E.C., Zhang, S.N., 1998, ApJ 497, L93
27. Levin, Y., 1998, astro-ph/9810471
28. Markwardt, C.B., Strohmayer, T.E., Swank, J.H., 1998, ApJ submitted
29. Méndez, M., van der Klis, M., van Paradijs, J., Lewin, W.H.G., Lamb, F.K., Vaughan, B.A., Kuulkers, E., Psaltis, D., 1997, ApJ 485, L37
30. Méndez, M., van der Klis, M., van Paradijs, J., Lewin, W.H.G., Vaughan, B.A., Kuulkers, E., Zhang, W., Lamb, F.K., Psaltis, D., 1998a, ApJ 494, L69
31. Méndez, M., van der Klis, M., Wijnands, R., Ford, E.C., van Paradijs, J., Vaughan, B.A., 1998b, ApJ 505, L23
32. Méndez, M., van der Klis, M., van Paradijs, J., 1998c, ApJ 506, L117
33. Méndez, M., van der Klis, Ford, E.C., M., Wijnands, R., van Paradijs, J., 1998d, ApJ 511 in press; `astro-ph/9811261`
34. Miller, M.C., 1998, ApJ submitted
35. Miller, M.C., Lamb, F.K. & Psaltis, D. 1998, ApJ 508, 791
36. Morgan, E.H. & Smith, D.A. 1996, IAU Circ. 6437
37. Psaltis, D. et al., 1998, ApJ in prep.
38. Psaltis, D., Belloni, T., van der Klis, M., 1998, ApJ submitted
39. Shirey, R.E., 1996, ApJ 469, L21
40. Shirey, R.E., 1998, in prep.
41. Smale, A.P., Zhang, W. & White, N.E. 1996, IAU Circ. 6507
42. Smale, A.P., Zhang, W. & White, N.E. 1997, ApJ 483, L119
43. Smith, D.A., Morgan, E.H. & Bradt, H. 1997, ApJ 479, L137
44. Stella, L. & Vietri, M., 1998a, ApJ 492, L59
45. Stella, L. & Vietri, M., 1998b, Phys. Rev. Letters in press, astro-ph/9812124

46. Strohmayer, T., Zhang, W. & Swank, J. 1996a, IAU Circ. 6320
47. Strohmayer, T., Smale, A., Day, C., Swank, J., Titarchuk, L., Lee, U., 1996b, IAU Circ. 6387
48. Strohmayer, T.E., Zhang, W. , Swank, J.H., Smale, A., Titarchuk, L., Day, C., 1996c, ApJ 469, L9
49. Strohmayer, T., Lee, U. & Jahoda, K. 1996d, IAU Circ. 6484
50. Strohmayer, T.E., Jahoda, K. Giles, A.B., & Lee, U. 1997a, ApJ 486, 355
51. Strohmayer, T., Zhang, W., Swank, J.H., 1997b, ApJ 487, L77
52. Strohmayer, T.E., Zhang, W., Swank, J.H., White N.E., 1998a, ApJ 498, L135
53. Strohmayer, T.E., Zhang, W., Swank, J.H., Lapidus, I., 1998b, ApJ 503, L147
54. Tennant, A.F., 1987, MNRAS226, 971
55. Titarchuk, L., Lapidus, I., Muslimov, A., 1998, ApJ 499, 315
56. Van der Klis, M., 1994, ApJS 92, 511
57. Van der Klis, M. 1995, in *X-Ray Binaries*, W.H.G. Lewin, J. van Paradijs & E.P.J. van den Heuvel (Eds), Cambridge University Press, p. 252
58. Van der Klis, M. 1997, in: Proc. Wise Obs. 25th Anniv. Symp., Dec 1996 "Astronomical Time Series", Kluwer Astroph. and Space Science Library 218, 121
59. van der Klis, M., 1998, in: Proc. NATO ASI "The many faces of neutron stars", Series C, Vol. 515, p. 337; astro-ph/9710016
60. Van der Klis, M. et al. 1990, ApJ 360, L19
61. Van der Klis, M., Swank, J., Zhang, W., Jahoda, K., Morgan, E., Lewin, W., Vaughan, B., van Paradijs, J., 1996a, IAU Circ. 6319
62. Van der Klis, M., Wijnands, R., Chen, W., Lamb, F.K., Psaltis, D., Kuulkers, E., Lewin, W.H.G., Vaughan, B., van Paradijs, J., Dieters, S., Horne, K., 1996b, IAU Circ. 6424
63. Van der Klis, M., Swank, J.H., Zhang, W., Jahoda, K., Morgan, E.H., Lewin, W.H.G., Vaughan, B., van Paradijs, J., 1996c, ApJ 469, L1
64. Van der Klis, M., van Paradijs, J., Lewin, W.H.G., Lamb, F.K., Vaughan, B., Kuulkers, E., Augusteijn, T., 1996d, IAU Circ. 6428
65. Van der Klis, M., Wijnands, R., Kuulkers, E., Lamb, F.K., Psaltis, D., Dieters, S., van Paradijs, J., Lewin, W.H.G., Vaughan, B., 1996e, IAU Circ. 6511
66. Van der Klis, M., Homan, J., Wijnands, R., Kuulkers, E., Lamb, F.K., Psaltis, D., Dieters, S., van Paradijs, J., Lewin W.H.G., Vaughan, B.A., 1997a, IAU Circ. 6565
67. Van der Klis, M., Wijnands, R.A.D., Horne, K., Chen, W., 1997b, ApJ 481, L97
68. Van Paradijs, J., Zhang, W., Marshall, F., Swank, J.H., Augusteijn, T., Kuulkers, E., Lewin, W.H.G., Lamb, F., Lapidus, I., Lochner, J., Strohmayer, T., van der Klis, M., Vaughan, B., 1996, IAU Circ. 6336
69. Vaughan, B.A., et al., 1994, ApJ 435, 362
70. Vaughan, B.A., van der Klis, M., Méndez, M., van Paradijs, J., Wijnands, R.A.D. Lewin, W.H.G., Lamb, F.K., Psaltis, D., Kuulkers, E., Oosterbroek, T., 1997, ApJ 483, L115
71. Vaughan, B.A., van der Klis, M., Méndez, M., van Paradijs, J., Wijnands, R.A.D. Lewin, W.H.G., Lamb, F.K., Psaltis, D., Kuulkers, E., Oosterbroek,

T., 1998, ApJ 509, L145
72. Wijnands, R.A.D. and Van der Klis, M., 1997, ApJ 482, L65
73. Wijnands, R. and van der Klis, M., 1998a, IAU Circ.6876
74. Wijnands, R. and van der Klis, M., 1998b, Nature, 394, 344
75. Wijnands, R. and van der Klis, M., 1998c, ApJ 507, L63
76. Wijnands, R. and van der Klis, M., 1998d, ApJ in press; astro-ph/9810342
77. Wijnands, R.A.D., van der Klis, M., van Paradijs, J., Lewin, W.H.G., Lamb, F.K., Vaughan, B., Kuulkers, E., Augusteijn, T., 1996, IAU Circ. 6447
78. Wijnands, R.A.D. et al., 1997a, ApJ 479, L141
79. Wijnands, R., Homan, J., van der Klis, M., Méndez, M., Kuulkers, E., van Paradijs, J., Lewin, W.H.G., Lamb, F.K., Psaltis, D., Vaughan, B., 1997b, ApJ 490, L157
80. Wijnands, R., Homan, J., van der Klis, M., Kuulkers, E., van Paradijs, J., Lewin, W.H.G., Lamb, F.K., Psaltis, D., Vaughan, B., 1998a, ApJ 493, L87
81. Wijnands, R.A.D. et al. 1998b, ApJ 495, L39
82. Wijnands, R., Méndez, M., van der Klis, M., Psaltis, D., Kuulkers, E., Lamb, F.K., 1998c, ApJ 504, L35
83. Yu, W., Zhang, S.N., Harmon, B.A., Paciesas, W.S., Robinson, C.R., Grindlay, J.E., Bloser, P., Barret, D., Ford, E.C., Tavani, M., Kaaret, P., 1997, ApJ 490, L153
84. Yu, W., Li, T.P., Zhang, W., Zhang, S.N., 1998, ApJ submitted
85. Zhang, W. 1997, Paper presented at the 190th AAS Meeting, Winston-Salem, USA and priv. comm.
86. Zhang, W., Strohmayer, T.E. & Swank, J.H., 1997, ApJ 482, L167
87. Zhang, W., Lapidus, I., White, N.E. & Titarchuk, L. 1996, ApJ 469, L17
88. Zhang, W., Lapidus, I., Swank, J.H., White, N.E., Titarchuk, L., 1997a, IAU Circ. 6541
89. Zhang, W., Strohmayer, T.E., Swank, J.H., 1998, ApJ 500, L167
90. Zhang, W., Smale, A.P., Strohmayer, T.E., Swank, J.H., 1998b, ApJ 500, L171
91. Zhang, W., et al., 1998c, ApJ 495, L9

VARIABILITY OF BLACK-HOLE BINARY SOURCES, AND LENSE-THIRRING ORBITAL PRECESSION

V. KARAS AND O. SEMERÁK

Astronomical Institute, Department of Theoretical Physics, Faculty of Mathematics and Physics, Charles University, V Holešovičkách 2, CZ-180 00 Praha, Czech Republic;
E-mail: karas@mbox.troja.mff.cuni.cz, semerak@mbox.troja.mff.cuni.cz

F. DE FELICE

Department of Physics "G. Galilei", University of Padova, Via Marzolo 8, I-35131 Padova, Italy;
E-mail: defelice@padova.infn.it

We discuss a possibility that observed radiation from active galactic nuclei (AGN) can be visibly modulated by collisions between stars (or secondary black holes) and an accretion disc around a central black hole. It is proposed that material swept out of the disc partly obscures the innermost region of the disc, while contributing by its own radiation. Under suitable circumstances, an individual orbiter can produce periodic modulation. If the central black hole is rotating and the companion star is at a low orbit, the Lense-Thirring orbital precession will show up in the signal and parameters of the central black hole can be thus determined.

1 Introduction

In this contribution we argue that the orbital Lense-Thirring precession could be detected in AGN with periodic modulation of variability if this is due to passages of a secondary object through an accretion disc around a rotating supermassive black hole. The observed orbital and Lense-Thirring frequencies can be used to estimate black hole's parameters.

An extensive amount of data on variability properties of AGN in the range of wavelengths from radio to hard X-rays and γ has been collected during recent decades (Duschl, Wagner & Camenzind 1991). It is widely presumed, although not yet proven, that massive black holes with accretion discs reside in the cores of AGN (Kato, Fukue & Mineshige 1998). Under assumptions of the model with a massive central black hole surrounded by an accretion disc, most of luminosity originates inside the innermost regions, mainly in the form of reprocessed X-rays. The observed signal shows irregular, featureless fluctuations with very complex behaviour at frequency about $\lesssim 10^{-2}$ Hz. Available data do not currently allow to discriminate between theoretical models of fluctuations which have been proposed in the literature, but, most probably, more than one single mechanism is responsible for this phenomenon:

(i) Various instabilities and wave-modes propagating in the disc material (Krolik et al. 1991; Mineshige, Ouchi & Nishimori 1994; Ipser 1994; Kanetake, Takeuti & Fukue 1995). They may result in quasi-periodic oscillations similar to those discussed in connection with Lense-Thirring precession around stellar-mass objects (Stella & Vietri 1998).

(ii) Orbital motion of irregularities (regions of enhanced emissivity, or, bright spots) which persist on the surface over many orbital periods (Abramowicz, Bao, Lanza & Zhang 1991; Wiita, Miller, Carini & Rosen 1991; see however Karas 1997 for a critical discussion of restrictions which can be imposed on this model, and for further references).

(iii) Variable obscuration of the disc surface by material which has been pulled out of the disc by collisions with orbiting stars and compact objects (Syer, Clarke & Rees 1991; Zurek, Siemiginowska & Colgate 1994; Karas & Vokrouhlický 1994; Pineault & Landry 1994; Rauch 1995). This gas is first shocked and heated up to very high temperetures, then it cools down rapidly and obscures the underlying surface of the disc (Ivanov, Igumenshchev & Novikov 1998).

In our present discussion we will concentrate on the last mechanism listed above. The three mechanisms can however be combined; e.g., collisions with the disc may be a source of waves in the disc and spots on its surface. An order-of-magnitude estimate shows that $\approx 10^7$ stars concentrated in a dense star-cluster, within the innermost 1 parsec around the core, should pull out of the disc enough material to visibly modulate observed radiation even if the disc itself were steady. One can however speculate further: if an orbiting object happens to be on an (almost) spherical orbit at a distance of $\approx$ 10–100 gravitational radii, it will modulate the source radiation periodically on the time-scale of the order of hours or days.[a] Both the orbital and associated Lense-Thirring periods will appear in the observed signal, and we will discuss below how this can be used to determine parameters of the central black hole (see Semerák, de Felice & Karas 1998 for further details). Observational detection of suitable sources requires high temporal and frequency resolution which is planned for future X-ray satellites.[b]

2 Set of equations for determining black-hole parameters

2.1 Assumptions

The following is assumed about the system of a central black hole, an accretion disc, and an orbiting secondary:

(i) Periodic component of variability is due to a secondary object orbiting within a few tens of gravitational radii around the central black hole and passing periodically through the disc. Three distinct variability time-scales show up in the signal as a result of orbital motion, Lense-Thirring precession of the orbital plane, and coupled radial-azimuthal oscillation of the disc material which can be excited by the orbiter and confined in a restricted range of radii, in the inner disc region.

[a]Processes leading to gradual circularization of eccentric orbits have been discussed by many authors; in the context of AGN, see Syer, Clarke & Rees 1991; Rauch 1995; Vokrouhlický & Karas 1998. An almost circular orbit not much inclined to the disc plane is a natural outcome of successive collisions with gaseous material of the accretion disc.

[b]Some interest has recently been raised in connection with the source IRAS 18325-5926 because of the claims, based on ASCA observations (Fabian et al. 1998), that the source shows periodic variability.

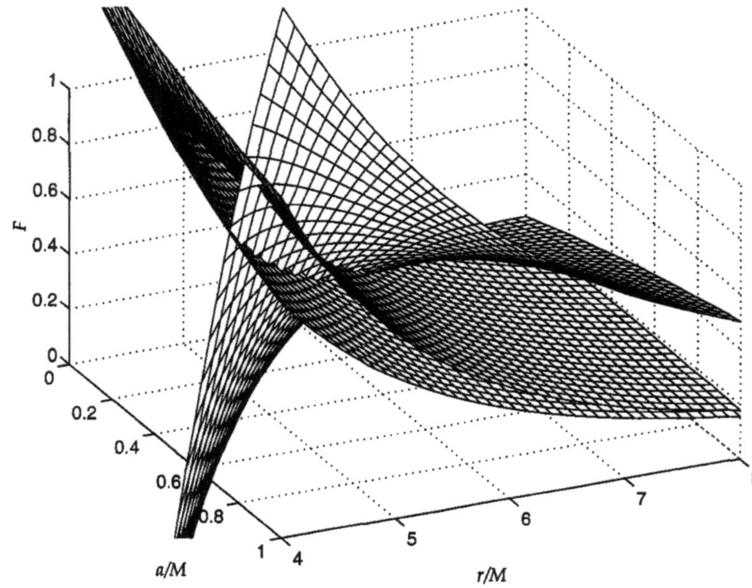

Figure 1. Graphs of functions $\kappa^2/0.002$, $\omega_+^2/0.01$, and $\Omega_\infty^2/0.01$ for corotating orbits (arranged from the top to bottom at large r, F stands for each of the three functions in convenient normalization). The way to determine black-hole parameters described in this contribution employs the fact that these directly measurable functions depend on parameters of the model (a, M, and r). See the text for details.

(ii) The secondary orbits almost freely along a quasi-circular trajectory with small non-zero inclination. Perturbation due to passages through the disc is very weak (secular). Effect on the disc is however strong enough to result in observable photometric and spectroscopic features.

(iii) The gravitational field within the nucleus is that of the (Kerr) black hole, the secondary and the disc contribute negligibly. In particular, the mass of the secondary is much less than that of the central object.

We note that, apart from older works mentioned in Introduction, motion outside the disc was studied in rather detailed manner in the recent work by Vokrouhlický & Karas (1998), while hypersonic transition of a secondary black hole across the disc was treated by Ivanov et al. (1998). These works show how the long-term evolution of orbital parameters of the secondary can be determined. Predicted line profiles have been discussed by numerous authors but what we really need here is temporal information (Karas & Kraus 1996) which has not yet been investigated in much detail because X-ray spectra resolved in time have become available only recently and still with rather limited resolution (e.g. Lee et al. 1998). Low-frequency oscillations of relativistic discs have been described, e.g., in Kato et al. (1998; this type of waves appears relevant for our present discussion because they are trapped near the central hole, approximately in the same region where the orbiter crosses the disc).

2.2 Equations

We summarize, in standard notation and in geometrized units, relations from which one can determine basic parameters of the system, i.e. central mass M, specific angular momentum a, and radius r of the orbit of the secondary. For further details and derivation see Semerák et al. (1998). We assume that the orbit is almost circular with small inclination relative to the disc. Then the azimuthal angular velocity $\omega = \mathrm{d}\phi/\mathrm{d}t$ (measured by a distant observer at rest) is

$$\omega_{\pm} = \left(a + 1/y_{\pm}\right)^{-1}, \tag{1}$$

where $y_{\pm} = \pm\sqrt{M/r^3}$, with upper and lower signs corresponding to co-rotating and counter-rotating orbits, respectively. For the angular frequency Ω_{∞} of small harmonic latitudinal oscillations of the secondary about the equatorial plane we obtain

$$\Omega_{\infty}^2 = \omega_{\pm}^2 \left(1 - 4ay_{\pm} + 3a^2/r^2\right). \tag{2}$$

Even if one succeeds in measuring both frequencies, $\omega_{\pm}$ and Ω_{∞}, an additional observable is needed to determine all three unknowns of the model, a, M, and r.

Here we briefly suggest two possibilities. First, the orbiter can excite low-frequency relativistic oscillations in the disc. This type of oscillations has typical modes which are governed by epicyclic frequency (Kato et al. 1998)

$$\kappa^2 = \omega_{\pm}^2 \left(1 - 6M/r + 8ay_{\pm} - 3a^2/r^2\right). \tag{3}$$

It is only within a restricted range of radii near the centre where these modes can be trapped and modulate total luminosity of the source. An alternative way to get required information comes from measuring rotational broadening of a spectral feature (obviously, we mainly have the iron line at $\approx 6.4\,\mathrm{keV}$ in mind). Let us note that the line formation is also affected by the star-disc collisions, making the observed profile variable in time, but one assumes to be able to resolve that contribution in the signal (still rather optimistic assumption in view of currently available data). Width of the line is determined by variation δg of the redshift factor $g = \nu_{\mathrm{obs}}/\nu_{\mathrm{em}}$ during the orbital period. Ratio of δg and of the corresponding variation $\delta e_{\hat{\phi}}$ in the locally measured direction cosine $e_{\hat{\phi}}$ of photons reads:[c]

$$\delta^2 \equiv \frac{(\delta g)^2}{(\delta e_{\hat{\phi}})^2} = \frac{M}{r}\,\frac{1 - 2M/r + a^2/r^2}{1 - 3M/r + 2ay_{\pm}}. \tag{4}$$

The four algebraic equations (1)–(4) express measurable quantities $\omega_{\pm}$, Ω_{∞}, κ, and δ in terms of the unknown parameters M, a, and r. (Graphs of relevant functions are illustrated in Fig. 1 for corotating orbits.) Knowledge of three of these quantities is sufficient to determine the desired parameters. Note that they are

[c]Semerák et al. (1998) discuss how this quantity could be measured. One should however bear in mind that a more sophisticated approach is necessary to estimate the observed width of the reflection line observed in AGN; cf. Martocchia, Karas & Matt (1998).

obtained in physical units rather than scaled with gravitational radius (as usual). Naturally, the above-described method can be combined with independent estimates of the parameters; for example, other proposed ways to determine M can be involved.

3 Conclusions

We illustrated how the black-hole parameters (a, M) and radius (r) of the secondary orbit can be *determined in physical units*. Our basic assumption is that three time-scales (related to orbital motion, to Lense-Thirring precession of the orbit of the secondary, and to global disc oscillations) are discovered in the signal. Well-resolved line profile is needed which originates in the region near the black hole where the secondary collides with the disc. Although such data are not currently at our disposal, the prospect of determining the three parameters is very tempting, and foreseeable.

Acknowledgments

We acknowledge kind invitation from the Conference organizers and support from the grants GAUK 63/98, GAUK 230/96, and GACR 202/96/0206.

References

1. Abramowicz M. A., Bao G., Lanza A., Zhang X.-H., AA **245**, 454 (1991).
2. Duschl W. J., Wagner S. J., Camenzind M., eds, Variability of Active Galaxies (Springer-Verlag, Berlin, 1991).
3. Fabian A. C., Lee J. C., Brandt W. N., Iwasawa K., Jahoda K., Reynolds C. S., MNRAS, 1998, in press; astro-ph/9803075.
4. Ipser J. R., ApJ **435**, 767 (1994).
5. Ivanov P. B., Igumenshchev I. V., Novikov I. D., 1998, ApJ, in press.
6. Kanetake R., Takeuti M., Fukue J., MNRAS **276**, 971 (1995).
7. Karas V., MNRAS **218**, 12 (1997).
8. Karas V., Kraus P., PASJ **48**, 771 (1996).
9. Karas V., Vokrouhlický D., ApJ **422**, 218 (1994).
10. Kato S., Fukue J., Mineshige S., Black-Hole Accretion Disks (Kyoto University Press, Kyoto, 1998).
11. Krolik J. H., Horne K., Kallman T. R., Malkan M. A., Edelson R. A., Kriss G., ApJ **371**, 541 (1991).
12. Lee J. C., Fabian A. C., Reynolds C. S., Iwasawa K., Brandt W. N., MNRAS, 1998, in press; astro-ph/98050198.
13. Martocchia A., Karas V., Matt G., Proc. of the the Third National Congress on Active Galactic Nuclei, in press (Rome, 1998).
14. Mineshige S., Ouchi N. B., Nishimori H., PASJ **46**, 97 (1994).
15. Pineault S., Landry S., MNRAS **267**, 557 (1994).
16. Rauch K. P., MNRAS **275**, 628 (1995).
17. Semerák O., de Felice F., Karas V., submitted, 1998; astro-ph/9802025.

18. Stella L., Vietri M., ApJ Lett. **492**, L59 (1998).
19. Syer D., Clarke C. J., Rees M. J., MNRAS **250**, 505 (1991).
20. Vokrouhlický D., Karas V., MNRAS **293**, 53 (1998).
21. Wiita P. J., Miller H. R., Carini M. T., Rosen A., in IAU Colloqium 129, C. Bertout et al., eds, (Editions Frontiéres, Gif-sur-Yvette, 1991) p. 557.
22. Zurek W. H., Siemiginowska A., Colgate S. A., ApJ **434**, 46 (1994); Addendum ApJ **470**, 652 (1996).

MAGNETOSPHERES AROUND ROTATING BLACK HOLES

M. DOVČIAK, V. KARAS
Astronomical Institute, Charles University,
V Holešovičkách 2, CZ-180 00 Praha, Czech Republic
E-mail: dovciak@mbox.troja.mff.cuni.cz;
karas@mbox.troja.mff.cuni.cz

Influential and recent selected works concerning mutual interaction between electromagnetic fields and a rotating (Kerr) black hole are summarized. Vacuum solutions (exact and perturbative), electrodynamic approximation of test particles in electrovacuum fields, and force-free magnetohydrodynamic (MHD) flows are mentioned. Equations and corresponding boundary conditions which govern waves in the background magnetosphere are formulated, keeping properly all the dragging terms which become important near the black-hole horizon.

1 Introduction

Historically, interaction between rotating black holes, electromagnetic fields, and surrounding material has been studied in three steps — with increasing complexity and relevance to realistic astrophysical situations: (i) electrovacuum fields with test particles moving under combined influence of electromagnetic forces and gravity; (ii) stationary force-free magnetospheres (axially symmetric solutions); and (iii) more complicated solutions (non-axisymmetric perturbations including MHD waves, in particular). We will briefly record the above-mentioned items and then we will look in more detail into equations relevant for description of waves in a magnetosphere around a rotating black hole.

In the 1970s, a general framework was developed for constructing stationary electrovacuum Maxwell equations in the fixed background metric of the Kerr black hole. Pre-1973 works were summarized by Ruffini,[1] more recently by e.g. Karas[2] with particular emphasis on the effects of rotation. Wald's[3] construction of axisymmetric test fields in the Kerr background metric has been used by many authors as a 'seed' for describing aligned test fields. It was found that aligned stationary fields are expelled out of the black-hole horizon when its angular-momentum parameter increases, and the magnetic flux across the horizon of the maximally rotating black hole vanishes completely in the extremely rotating case.[4,5] It was speculated[6] that this behaviour of magnetic fluxes threading black holes might constrain the power which can be electromagnetically extracted from a black hole, and the situation is even more complex if the magnetic field is to be supported by an astrophysically realistic accretion disk.[7]

As an alternative to the approach of test fields in a pre-determined metric, exact solutions of coupled Einstein-Maxwell equations were found for certain simplified configurations: a magnetized black hole immersed in an aligned field[8,9,10] is a particularly important example. Karas & Vokrouhlický[11] then examined a non-trivial relation between Wald's solution and the corresponding exact solution in the limit of a weak electromagnetic field (but not necessarily weak gravitational field near the black-hole horizon).

In this contribution we briefly summarize older works which declared original motivations. We formulate equations for MHD waves in force-free magnetospheres and write them explicitly under specific assumptions.

2 Force-free magnetospheres around rotating black holes

Analysis of electrovacuum stationary test fields resulted in beautiful analytical solutions, astrophysical relevance of which remains, however, rather limited. As the first approximation, motion of cold plasma (consisting of electrically charged particles with negligible inertia) [a] was treated by Damour *et al.*[13] The main motivation to adopt the cold-plasma approximation was to study possible effects of collimation in very rarefied cosmic matter but hydrodynamic description appears more relevant in many situations. It was the well-known work of Blandford & Znajek[14] which greatly revived interest in these topics by studying electrodynamic effects in a force-free magnetosphere of the Kerr black hole. The main idea of their paper (which is linked to analogous problems of the pulsar physics) articulated the hope that rotational energy of a black hole could be transferred electromagnetically to accelerated particles. Distant particles might become collimated and form outflowing jets. The original formulation required major improvements which are the subject of numerous follow-up papers during 1980s. Phinney[15] wrote the energy-momentum tensor of the material as a sum of contributions from ideal fluid plus a contribution from electromagnetic field (including in this way also the effects of inertia of the fluid) but he still assumed the force-free condition. Difficulties in treating black-hole force-free magnetospheres resulted in critical reconsiderations[16] and consistent MHD treatment had to be developed. An alternative approach has been described by Camenzind[17] and collaborators who examined, in a series of articles, acceleration and collimation of relativistic hydromagnetic winds (see also Kennel *et al*;[18] Okamoto[19]). Linear perturbation theory of the force-free magnetospheres has been developed by Uchida[20] in geometrical optics approximation. Very recently, Khanna[21] applied MHD equations describing two-component plasma and concluded that gravitomagnetic terms do not significantly modify Ohm's law unless either the plasma is charged (in its local rest frame) or MHD approximation breaks down. Quite naturally, effects of general relativity may become important only close to the black-hole horizon where gravity is strong and the terms proportional to $g_{t\phi}$ cannot be neglected.

2.1 Basic assumptions and equations for a stationary and axisymmetric force-free magnetosphere

In this section we will summarize basic equations describing a stationary axisymmetric force-free magnetosphere around a rotating black hole. We assume that the gravity of plasma and electromagnetic field are weak compared with the gravity of the black hole. Therefore the background geometry of the magnetosphere is

[a] See Karas & Vokrouhlický[12] for a critical discussion of the notion of plasma horizons and for a detailed construction of flow lines in the guiding-centre approximation.

described by the Kerr metric

$$ds^2 = -\left(1 - \frac{2Mr}{\Sigma}\right) dt^2 - \frac{4Mr}{\Sigma} a \sin^2\theta \, dt \, d\varphi + \frac{\Sigma}{\Delta} dr^2 + \Sigma \, d\theta^2 + \frac{\mathcal{A}}{\Sigma} \sin^2\theta \, d\varphi^2 \, ,$$

where $\Sigma = r^2 + a^2 \cos^2\theta$, $\Delta = r^2 - 2Mr + a^2$ and $\mathcal{A} = (r^2 + a^2)^2 - \Delta a^2 \sin^2\theta$. Throughout this paper we use Boyer-Lindquist's coordinates and geometrized units with $c = G = 1$ and $\mu_0 = 4\pi$ in standard notation.

The electromagnetic field is described by the tensor $F_{\mu\nu}$ which satisfies Maxwell's equations

$$F_{[\mu\nu,\sigma]\text{cycl.}} = 0 \ , \ F^{\mu\nu}{}_{;\nu} = 4\pi J^{\mu} \ , \tag{1}$$

where J^{μ} is a four-vector of the electric current induced by the fluid.

Here we describe plasma as a perfect fluid with four-velocity U^{μ}, mass density ρ, pressure p, macroscopic electric charge density ρ_e, and the corresponding current $J^{\mu} = \rho_e U^{\mu}$. (The last assumption is actually a strong condition on the character of plasma but detailed discussion of its validity remains beyond the scope of the present paper.) The fluid is assumed to be barotropic and its relative permitivity and permeability are as in vacuum, i.e. $p = p(\rho)$ and $\varepsilon_r = \mu_r = 1$. Velocity of the fluid is normalized to unity, $g_{\mu\nu}U^{\mu}U^{\nu} = -1$. The electric charge is conserved: [b]

$$(\rho_e U^{\mu})_{;\mu} = 0, \tag{2}$$

and, by definition of the force-free magnetosphere, the electric field vanishes in the fluid rest frame, that is

$$F_{\mu\nu}U^{\nu} = 0. \tag{3}$$

The motion of the fluid is governed by the following equations of motion:

$$0 = T_{\text{m} \ ;\nu}^{\mu\nu} = [(\rho + p)U^{\mu}U^{\nu} + pg^{\mu\nu}]_{;\nu} \ . \tag{4}$$

There should be also an electromagnetic part present in the above equations, but the force-free condition ensures that this part is zero. This means that the fluid and the electromagnetic field do not interact directly with each other, which is not much surprising since the force-free condition sets the Lorenz force on the elements of the fluid equal to zero.

The last assumption we make about the background magnetosphere concerns stationarity and axial symmetry around the rotation axis of the the black hole. This means that every quantity is independent of t and φ, in other words, time and angular derivatives of each quantity vanish everywhere.

Notice that there exist four constants that are conserved along each flow line: (*i*) the angular velocity of magnetic field lines $\Omega^{\text{F}} \equiv F_{tr}/F_{r\varphi} = F_{t\theta}/F_{\theta\varphi}$; (*ii*) the toroidal magnetic field $B_{\text{T}} \equiv {}^{*}F_{t\varphi} = (\Delta \sin\theta/\Sigma)F_{r\theta}$; (*iii*) energy $e \equiv -\mu U_t$, and (*iv*) the angular momentum $l \equiv \mu U_{\varphi}$ of the fluid per unit charge. Here, $\mu \equiv (p + \rho)/\rho_e$. The above-given constants of motion are functions of A_{φ} only ($A^{\mu}(r, \theta)$ is the vector potential of the electromagnetic field, $F_{\mu\nu} = A_{\nu,\mu} - A_{\mu,\nu}$) and their conservation along the flow lines follows from the

[b] Phinney[15] uses the conservation law for the rest mass ρ_0 instead.

fact that the four-velocity of the fluid is perpendicular to the gradient of the φ-component of the vector potential.

The fluid velocity and the tensor of the electromagnetic field can be expressed in terms of the charge density ρ_e, the quantity μ, the vector potential A_φ, the metric and the four constants of motion in the following way:

$$\begin{aligned} U^r &= \frac{1}{4\pi\sqrt{-g}\rho_e}\frac{\mathrm{d}B_\mathrm{T}}{\mathrm{d}A_\varphi}A_{\varphi,\theta} \,, \quad U^t = \frac{1}{\mu R^2}(g_{\varphi\varphi}e + g_{t\varphi}l) \,, \\ U^\theta &= -\frac{1}{4\pi\sqrt{-g}\rho_e}\frac{\mathrm{d}B_\mathrm{T}}{\mathrm{d}A_\varphi}A_{\varphi,r} \,, \quad U^\varphi = -\frac{1}{\mu R^2}(g_{t\varphi}e + g_{tt}l) \,, \end{aligned} \tag{5}$$

$$\begin{aligned} F_{t\varphi} &= 0 \,, \quad F_{r\theta} = \frac{\Sigma}{\Delta\sin\theta}B_\mathrm{T} \,, \\ F_{t\mu} &= \Omega^\mathrm{F} F_{\mu\varphi} = \Omega^\mathrm{F} A_{\varphi,\mu} \,, \quad \mu = r,\theta \end{aligned} \tag{6}$$

where $R^2 = g_{t\varphi}^2 - g_{tt}g_{\varphi\varphi} = \Delta\sin^2\theta$. These expressions guarantee that all basic equations (1)–(4) but for six of them are satisfied. The six remaining equations represent the r-component (or, equivalently, the θ-component) of the force-free condition (3), the normalization of the four-velocity, t- and φ-components of inhomogeneous Maxwell's equations (1), contraction of the equations of motion (4) with four-vector of the fluid velocity, and contraction of the equations of motion with a vector linearly independent of the four-velocity and time and axial Killing vectors [c] $k^\mu = (\partial_t)^\mu$, $m^\mu = (\partial_\varphi)^\mu$. These relations have the form of nonlinear differential equations that, on the condition that the functions Ω^F, B_T, e and l are specified, determine the unknown functions ρ, ρ_e and the vector potential A^μ.

Boundary conditions for the electromagnetic field and for the fluid's velocity on the horizon ($r = r_\mathrm{H} = M + \sqrt{M^2 - a^2}$, $\Omega^\mathrm{H} \equiv a/(r^2 + a^2)$) are

$$U^r = \frac{r^2 + a^2}{\Sigma}(U_t + \Omega^\mathrm{H} U_\varphi) \,, \quad B_\mathrm{T} = \frac{r^2 + a^2}{\Sigma}\sin\theta(\Omega^\mathrm{F} - \Omega^\mathrm{H})F_{\theta\varphi} \,. \tag{7}$$

3 MHD waves in black-hole magnetospheres

In this section we consider a small-amplitude non-axisymmetric perturbation superposed on the unperturbed state from the previous paragraphs. We assume that the perturbations have the form of an axially propagating wave, i.e. $\xi(t,r,\theta,\varphi) = \bar{\xi}(r,\theta)\,\mathrm{e}^{\mathrm{i}(\omega t - m\varphi)}$ where ξ, $\bar{\xi}$, ω/m are the perturbation, the amplitude of the perturbation and the axial velocity of the wave (m is an integer), respectively. Arbitrary nonstationary and non-axisymetric perturbation can be resolved into (or, vice-versa, constructed from) these particular perturbations. Functions $\bar{\xi}(r,\theta)$ represent the Fourier-transformed perturbations $\xi(t,r,\theta,\varphi)$.[d]

[c] The contraction of the equations of motion with these Killing vectors implies conservation of energy e and angular momentum l per unit charge along the flow lines.

[d] Here one should write $\bar{\xi}(r,\theta;\omega,m)$, to be exact, but we will prefer the brief notation.

Let us introduce the new functions e_r, e_θ, e_φ, b_r, b_θ and b_φ:

$$\overline{f}_{\mu\nu} = \begin{pmatrix} 0 & -\frac{g_{rr}}{\sqrt{-g}} e_r & -\frac{g_{\theta\theta}}{\sqrt{-g}} e_\theta & -\frac{1}{\sqrt{-g}} e_\varphi \\ -\overline{f}_{tr} & 0 & \frac{g_{rr} g_{\theta\theta}}{\sqrt{-g}} b_\varphi & -\frac{g_{rr}}{\sqrt{-g}} b_\theta \\ -\overline{f}_{t\theta} & -\overline{f}_{r\theta} & 0 & \frac{g_{\theta\theta}}{\sqrt{-g}} b_r \\ -\overline{f}_{t\varphi} & -\overline{f}_{r\varphi} & -\overline{f}_{\theta\varphi} & 0 \end{pmatrix} , \tag{8}$$

where $\overline{f}_{\mu\nu}$ is the amplitude of the perturbation of the electromagnetic field. Because $\overline{f}_{\mu\nu}$ is expressed in Boyer-Lindquist coordinates, it is divergent on the horizon. Therefore we introduced functions e_μ and b_μ in such a way that they don't diverge. Now we will write down all the basic equations which will be linearized in perturbations. Because of antisymmetry of the electromagnetic tensor there are only three independent equations out of the four force-free conditions. They can be rewritten in the following form (we emphasize the first-order perturbations by underlining the corresponding quantities twice):

$$\begin{aligned} & A_{\varphi,\theta} \left(l_{\mathrm{S}} \underline{\underline{\delta \overline{e}}} - e_{\mathrm{S}} \underline{\underline{\delta \overline{l}}} \right) + \Sigma \sin\theta \mu^2 B_{\mathrm{T}} U_{\mathrm{S}}^t \underline{\underline{\overline{u}^r}} \\ & \quad = \frac{\mu U_{\mathrm{S}}^t}{\sin\theta} \left[\left(g_{\varphi\varphi} e + g_{t\varphi} l \right) \underline{\underline{e_\theta}} - \left(g_{t\varphi} e + g_{tt} l \right) \underline{\underline{b_r}} \right] - \Sigma \sin\theta \mu^2 U_{\mathrm{S}}^t U^r \underline{\underline{b_\varphi}} , \end{aligned} \tag{9}$$

$$\begin{aligned} & \sin\theta \left(A_{\varphi,r} \underline{\underline{\overline{u}^r}} + A_{\varphi,\theta} \underline{\underline{\overline{u}^\theta}} \right) \\ & \quad = -U^\theta \underline{\underline{b_r}} + \frac{1}{\Delta} \left[-\frac{1}{\Sigma \sin^2\theta} \frac{1}{\mu} \left(g_{\varphi\varphi} e + g_{t\varphi} l \right) \underline{\underline{e_\varphi}} + U^r \underline{\underline{b_\theta}} \right] , \end{aligned} \tag{10}$$

$$\frac{1}{\sin\theta} B_{\mathrm{T}} \underline{\underline{e_\varphi}} = -A_{\varphi,\theta} \left(\underline{\underline{e_r}} - \Omega^{\mathrm{F}} \underline{\underline{b_\theta}} \right) + \Delta A_{\varphi,r} \left(\underline{\underline{e_\theta}} + \Omega^{\mathrm{F}} \underline{\underline{b_r}} \right) , \tag{11}$$

where $\delta \overline{e} = -\mu \overline{u}_t$, $\delta \overline{l} = \mu \overline{u}_\varphi$ and $\overline{u}^\mu$ are the amplitudes of the perturbations of the fluid velocity. Quantities e_{S}, l_{S} and U_{S}^t are the corresponding quantities for fluid which would move purely axially with the angular velocity $\mathrm{d}\varphi/\mathrm{d}t = U_{\mathrm{S}}^\varphi / U_{\mathrm{S}}^t = \Omega^{\mathrm{F}}$ (with the same pressure, mass and charge density). One can easily verify that this purely axially moving fluid satisfies the force-free condition, however it does not necessarily have to satisfy other basic equations.

As we express the electromagnetic tensor perturbations in terms of the functions e_μ and b_μ (rather than vector potential), homogeneous Maxwell's equations must be also treated:

$$\frac{\Delta \sin\theta}{\Sigma} \left[-\frac{1}{\Delta} \left(\frac{1}{\sin\theta} \underline{\underline{e_r}} \right)_{,\theta} + \frac{1}{\sin\theta} \underline{\underline{e_{\theta,r}}} \right] + \mathrm{i}\omega \underline{\underline{b_\varphi}} = 0 , \tag{12}$$

$$\Delta \left(\frac{1}{\Sigma} \underline{\underline{e_\varphi}} \right)_{,r} + \mathrm{i} (m \underline{\underline{e_r}} - \omega \underline{\underline{b_\theta}}) = 0 , \tag{13}$$

$$\sin\theta \left(\frac{1}{\Sigma \sin\theta} \underline{\underline{e_\varphi}} \right)_{,\theta} + \mathrm{i} (m \underline{\underline{e_\theta}} + \omega \underline{\underline{b_r}}) = 0 . \tag{14}$$

Inhomogeneous Maxwell's equations are:

$$\Delta\left[\frac{1}{\Delta}(g_{\varphi\varphi}\underline{\underline{e_r}}+g_{t\varphi}\underline{\underline{b_\theta}})\right]_{,r}+\sin^2\theta\left[\frac{1}{\sin^2\theta}(g_{\varphi\varphi}\underline{\underline{e_\theta}}-g_{t\varphi}\underline{\underline{b_r}})\right]_{,\theta}-\mathrm{i}m\underline{\underline{e_\varphi}}$$
$$=\frac{4\pi\Sigma\sin\theta}{\mu}\left[\underline{\underline{\delta\bar\rho_\mathrm{e}}}(g_{\varphi\varphi}e+g_{t\varphi}l)+\rho_\mathrm{e}(g_{\varphi\varphi}\underline{\underline{\delta\bar e}}+g_{t\varphi}\underline{\underline{\delta\bar l}})\right]\,, \tag{15}$$

$$\underline{\underline{b_{\varphi,\theta}}}-\frac{\mathrm{i}}{\Delta\sin^2\theta}\left[(g_{\varphi\varphi}\omega+g_{t\varphi}m)\underline{\underline{e_r}}+(g_{t\varphi}\omega+g_{tt}m)\underline{\underline{b_\theta}}\right]$$
$$=4\pi\Sigma\sin\theta\,(\underline{\underline{\delta\bar\rho_\mathrm{e}}}U^r+\rho_\mathrm{e}\underline{\underline{\bar u^r}})\,, \tag{16}$$

$$\underline{\underline{b_{\varphi,r}}}+\frac{\mathrm{i}}{\Delta\sin^2\theta}\left[(g_{\varphi\varphi}\omega+g_{t\varphi}m)\underline{\underline{e_\theta}}-(g_{t\varphi}\omega+g_{tt}m)\underline{\underline{b_r}}\right]$$
$$=-4\pi\Sigma\sin\theta\,(\underline{\underline{\delta\bar\rho_\mathrm{e}}}U^\theta+\rho_\mathrm{e}\underline{\underline{\bar u^\theta}})\,, \tag{17}$$

$$\Delta\left[\frac{1}{\Delta}(g_{t\varphi}\underline{\underline{e_r}}+g_{tt}\underline{\underline{b_\theta}})\right]_{,r}+\sin^2\theta\left[\frac{1}{\sin^2\theta}(g_{t\varphi}\underline{\underline{e_\theta}}-g_{tt}\underline{\underline{b_r}})\right]_{,\theta}+\mathrm{i}\omega\underline{\underline{e_\varphi}}$$
$$=\frac{4\pi\Sigma\sin\theta}{\mu}\left[\underline{\underline{\delta\bar\rho_\mathrm{e}}}(g_{t\varphi}e+g_{tt}l)+\rho_\mathrm{e}(g_{t\varphi}\underline{\underline{\delta\bar e}}+g_{tt}\underline{\underline{\delta\bar l}})\right]\,, \tag{18}$$

where $\delta\bar\rho_\mathrm{e}$ is the amplitude of the perturbation of the fluid charge density. Explicit form of the equations of motion reads

$$-U^r\underline{\underline{\delta\bar\rho_{,r}}}-U^\theta\underline{\underline{\delta\bar\rho_{,\theta}}}+\left[(\rho_{,r}U^r+\rho_{,\theta}U^\theta)\frac{1}{p+\rho}\left(1+\frac{\mathrm{d}p}{\mathrm{d}\rho}\right)-\mathrm{i}(\omega U^t-mU^\varphi)\right]\underline{\underline{\delta\bar\rho}}$$
$$+\mu\left(U^r\underline{\underline{\delta\bar\rho_{\mathrm{e},r}}}+U^\theta\underline{\underline{\delta\bar\rho_{\mathrm{e},\theta}}}\right)+\left[-\frac{1}{\rho_\mathrm{e}}(\rho_{,r}U^r+\rho_{,\theta}U^\theta)+\mathrm{i}\mu(\omega U^t-mU^\varphi)\right]\underline{\underline{\delta\bar\rho_\mathrm{e}}}$$
$$+(-\rho_{,r}+\mu\rho_{\mathrm{e},r})\underline{\underline{\bar u^r}}+(-\rho_{,\theta}+\mu\rho_{\mathrm{e},\theta})\underline{\underline{\bar u^\theta}}=0\,, \tag{19}$$

$$\rho_\mathrm{e}e_{,r}\underline{\underline{\bar u^r}}+\rho_\mathrm{e}e_{,\theta}\underline{\underline{\bar u^\theta}}+\rho_\mathrm{e}U^r\underline{\underline{\delta\bar e_{,r}}}+\rho_\mathrm{e}U^\theta\underline{\underline{\delta\bar e_{,\theta}}}+\mathrm{i}\rho_\mathrm{e}(\omega U^t-mU^\varphi)\underline{\underline{\delta\bar e}}$$
$$+\frac{1}{\mu}\left(1+\frac{\mathrm{d}p}{\mathrm{d}\rho}\right)e\left(U^r\underline{\underline{\delta\bar\rho_{,r}}}+U^\theta\underline{\underline{\delta\bar\rho_{,\theta}}}\right)+\left[\frac{e}{\mu}K_1(r,\theta)-\mathrm{i}\omega\frac{\mathrm{d}p}{\mathrm{d}\rho}\right]\underline{\underline{\delta\bar\rho}}$$
$$-e\left(U^r\underline{\underline{\delta\bar\rho_{\mathrm{e},r}}}+U^\theta\underline{\underline{\delta\bar\rho_{\mathrm{e},\theta}}}\right)+eK_2(r,\theta)\underline{\underline{\delta\bar\rho_\mathrm{e}}}=0\,, \tag{20}$$

$$\rho_\mathrm{e}l_{,r}\underline{\underline{\bar u^r}}+\rho_\mathrm{e}l_{,\theta}\underline{\underline{\bar u^\theta}}+\rho_\mathrm{e}U^r\underline{\underline{\delta\bar l_{,r}}}+\rho_\mathrm{e}U^\theta\underline{\underline{\delta\bar l_{,\theta}}}+\mathrm{i}\rho_\mathrm{e}(\omega U^t-mU^\varphi)\underline{\underline{\delta\bar l}}$$
$$+\frac{1}{\mu}\left(1+\frac{\mathrm{d}p}{\mathrm{d}\rho}\right)l\left(U^r\underline{\underline{\delta\bar\rho_{,r}}}+U^\theta\underline{\underline{\delta\bar\rho_{,\theta}}}\right)+\left[\frac{l}{\mu}K_1(r,\theta)-\mathrm{i}m\frac{\mathrm{d}p}{\mathrm{d}\rho}\right]\underline{\underline{\delta\bar\rho}}$$
$$-l\left(U^r\underline{\underline{\delta\bar\rho_{\mathrm{e},r}}}+U^\theta\underline{\underline{\delta\bar\rho_{\mathrm{e},\theta}}}\right)+lK_2(r,\theta)\underline{\underline{\delta\bar\rho_\mathrm{e}}}=0\,, \tag{21}$$

where the complex functions $K_1(r,\theta)$ a $K_2(r,\theta)$ are

$$K_1(r,\theta) = \left[\frac{\mathrm{d}^2p}{\mathrm{d}\rho^2} - \frac{1}{p+\rho}\left(1+\frac{\mathrm{d}p}{\mathrm{d}\rho}\right)^2\right](\rho_{,r}U^r + \rho_{,\theta}U^\theta) + \mathrm{i}\left(1+\frac{\mathrm{d}p}{\mathrm{d}\rho}\right)(\omega U^t - mU^\varphi)\,, \tag{22}$$

$$K_2(r,\theta) = \frac{1}{p+\rho}(\rho_{,r}U^r + \rho_{,\theta}U^\theta) - \mathrm{i}(\omega U^t - mU^\varphi)\,. \tag{23}$$

These equations are contractions of the equation of motion with the vector of the perturbed fluid's velocity and with the time and axial Killing vectors. The fourth equation of motion is the contraction of the equation of motion with the vector linearly independent of these vectors.
The normalization condition for the perturbed velocity of the fluid is

$$U^r\underline{\underline{\overline{u}^r}} + \Delta U^\theta \underline{\underline{\overline{u}^\theta}} - \frac{1}{\Sigma\sin^2\theta}\frac{1}{\mu^2}\left[(g_{\varphi\varphi}e + g_{t\varphi}l)\underline{\underline{\delta\overline{e}}} + (g_{t\varphi}e + g_{tt}l)\underline{\underline{\delta\overline{l}}}\right] = 0\,. \tag{24}$$

The conservation law for the electric charge for perturbed state is

$$U^r\underline{\underline{\delta\overline{\rho}_{\mathrm{e},r}}} + U^\theta\underline{\underline{\delta\overline{\rho}_{\mathrm{e},\theta}}} - \left[\frac{1}{p+\rho}(\rho_{,r}U^r + \rho_{,\theta}U^\theta) - \mathrm{i}(\omega U^t - mU^\varphi)\right]\underline{\underline{\delta\overline{\rho}_{\mathrm{e}}}}$$

$$+\rho_{\mathrm{e}}\underline{\underline{\overline{u}^r{}_{,r}}} + \rho_{\mathrm{e}}\underline{\underline{\overline{u}^\theta{}_{,\theta}}} + \frac{1}{\Sigma}(\Sigma\rho_{\mathrm{e}})_{,r}\underline{\underline{\overline{u}^r}} + \frac{1}{\Sigma\sin\theta}(\Sigma\sin\theta\rho_{\mathrm{e}})_{,\theta}\underline{\underline{\overline{u}^\theta}}$$

$$+\,\mathrm{i}\frac{1}{\Delta\sin^2\theta}\frac{\rho_{\mathrm{e}}}{\mu}\left[(g_{\varphi\varphi}\omega + g_{t\varphi}m)\underline{\underline{\delta\overline{e}}} + (g_{t\varphi}\omega + g_{tt}m)\underline{\underline{\delta\overline{l}}}\right] = 0\,. \tag{25}$$

Similarly as in the unperturbed state, we can derive the boundary conditions on the horizon of the black hole for the perturbations:

$$\underline{\underline{\overline{u}^r}} = -\frac{r^2+a^2}{\mu\Sigma}\left[\underline{\underline{\delta\overline{e}}} - \Omega^{\mathrm{H}}\underline{\underline{\delta\overline{l}}}\right]\,,\quad \underline{\underline{e_r}} = a\Sigma^{-1}\underline{\underline{e_\varphi}}\,, \tag{26}$$

$$\underline{\underline{b_\varphi}} = -\frac{r^2+a^2}{\Sigma}\left(\underline{\underline{e_\theta}} + \Omega^{\mathrm{H}}\underline{\underline{b_r}}\right),\quad \underline{\underline{b_\theta}} = \frac{r^2+a^2}{\Sigma}\underline{\underline{e_\varphi}}\,, \tag{27}$$

where all quantities are evaluated on the horizon. Another boundary condition can be found by assuming that the perturbed quantities (in particular $F_{\hat{\mu}\hat{\nu}}$) in Kerr coordinates have a nondivergent derivative (on the horizon) with respect to the Kerr coordinate $\hat{r}$:

$$\underline{\underline{\xi_{,r}}} = \mathrm{i}\frac{r^2+a^2}{\Delta}(\omega - \Omega^{\mathrm{H}}m)\underline{\underline{\xi}}\,, \tag{28}$$

where ξ denotes one of the functions $\delta\overline{p}$, $\delta\overline{\rho}_{\mathrm{e}}$, $\overline{u}^r$, $\overline{u}^\theta$, $\delta\overline{e}$, $\delta\overline{l}$, e_μ, b_μ and again all quantities are evaluated on the horizon.

4 Conclusions

We derived equations (9)–(25) that must be satisfied by nonstationary and non-axial perturbations of the stationary and axisymmetric magnetosphere. We also expressed boundary conditions (26)–(28) for perturbations and their r-derivatives. All equations except the fourth equation of motion are given in an explicit form and they are well-suited for numerical calculations (work in progress).

Acknowledgments

We acknowledge kind invitation and support from the Conference organizers and the grants GAUK 63/98 and GACR 205/97/1165.

References

1. R. Ruffini in *Black Holes*, ed. C. DeWitt & B. S. DeWitt (Gordon and Breach, New York 1973), p. 451.
2. V. Karas, *Phys. Rev.* D **40**, 2121 (1989).
3. R. M. Wald, *Phys. Rev.* D **10**, 1680 (1974).
4. J. Bičák and L. Dvořák, *Gen. Rel. Gravit.*, 7, 959 (1976).
5. A. R. King, *Math. Proc. Camb. Philos. Soc.*, 81, 149 (1977).
6. J. Bičák and V. Janiš, *MNRAS*, 212, 899 (1985).
7. P. Ghosh and M. A. Abramowicz, *MNRAS*, 292, 887 (1997).
8. D.V. Gal'tsov and V.I. Petukhov, *Zh. eksper. teor. fiz.*, 74, 801 (1978).
9. F.J. Ernst and W.J. Wild, *J. Math. Phys.*, 17, 182 (1976).
10. A.G. Díaz, *J. Math. Phys.*, 26, 155 (1985).
11. V. Karas and D. Vokrouhlický, *J. Math. Phys.*, 32, 714 (1991).
12. V. Karas and D. Vokrouhlický, *Journal de Physique*, 1, 1005 (1991).
13. T. Damour, R.S. Hanni, R. Ruffini and J.R. Wilson, *Phys. Rev.* D **17**, 1518 (1978).
14. R.D. Blandford and R.L. Znajek, *MNRAS*, 179, 433 (1977).
15. E. S. Phinney, *Thesis* (Institute of Astronomy, Cambridge 1983).
16. B. Punsly and F.V. Coroniti, *Phys. Rev.* D **40**, 3834 (1989).
17. M. Camenzind, *A&A*, 156, 137 (1986).
18. C.F. Kennel, F.S. Fujimura and I. Okamoto, *J. Astrophys. Geophys. Fluid Dyn.*, 26, 147 (1983).
19. I. Okamoto, *MNRAS*, 185, 69 (1978).
20. T. Uchida, *MNRAS*, 291, 125 (1997).
21. R. Khanna, *MNRAS*, 294, 673 (1998).

LENSE–THIRRING EFFECT IN THE SUPERFLUID INSIDE NEUTRON STARS

H. CASINI, R. MONTEMAYOR

Instituto Balseiro and Centro Atómico Bariloche Universidad Nacional de Cuyo and CNEA 8400 S.C. de Bariloche, Río Negro, Argentina

In this work we analyze the most significant manifestations of the Lense-Thirring effect in the superfluid component of a neutron star, which consist in the line vortex diffraction, the lowering of the vortex density and the covariant Magnus force. These provide a very interesting framework for studying the Lense-Thirring effect in quantum coherent systems.

1 Introduction

In astrophysical systems the manifestations of the Lense-Thirring effect are usually small. A noteworthy exception is provided by neutron stars, extremely massive and highly magnetized objects with a very large angular velocity. The interior of these stars is essentially constituted by superfluid matter in presence of gravitational fields strong enough to produce non-trivial relevant effects. For these stars we can assume that there is an approximate spheric symmetry broken by the angular momentum of the star, and thus we can write the metric up to linear terms in the angular velocity

$$ds^2 = -e^{-2\Phi}dt^2 + e^{-2\Lambda}dr^2 + r^2(d\theta^2 + \sin^2\theta d\varphi^2) - 2r^2\sin^2\theta\varpi d\varphi dt\,, \qquad (1)$$

where $\varpi(\theta, \varphi)$ is the angular velocity at the point (θ, φ) of a free falling object from infinity.

The neutron stars are detected as pulsars, and their observable dynamical variable is the angular velocity in the form of the frequency of the electromagnetic radiation pulses. The medium and long time dynamics is mainly governed by the electromagnetic torques acting on the magnetic dipolar moment of the star and the dynamics of the component superfluid. Details of these dynamics are discussed in several papers.[1] In fact, the Lense-Thirring effect on the electromagnetic torques is negligible, as it has been already discussed,[2] but this is not the case for the behavior of the superfluid matter, where it generates noticeable contributions. This paper is devoted to the analysis of two interesting effects of the frame dragging on a superfluid in a gravitational background given by the metric (1), which are the gravitational distortion of the vortex lines and the dragging correction to the covariant Magnus force.

2 The covariant superfluid

The superfluid phase in the neutron stars originates in the formation of a condensate which spontaneously brakes the baryonic symmetry. The neutron field Ψ has a global $U(1)$ symmetry, $\Psi \to e^{i\alpha}\Psi$, related with the conservation of the baryonic number, which is broken by the formation of neutron pairs.[3] These pairs have a non-null expectation value and the $U(1)$ baryonic symmetry is broken into Z_2, the subgroup with $\alpha = 0$ and $\alpha = \pi$. This spontaneous symmetry breaking produces Goldstone bosons with zero energy in the null momentum limit. The transformation group that acts on the Goldstone bosons corresponds to $\phi \to \phi + \alpha$. Given that ϕ parametrizes $U(1)/Z_2$, ϕ and $\phi + \pi$ are equivalents.

The $U(1)$ invariant Lagrangian density $\mathcal{L}$ must be a function of the derivatives of ϕ and the neutron field Ψ, and the baryonic Noether current will be given by $J^\mu = \frac{\delta L}{\delta(\phi_{,\mu})}$. The most general Lagrangian that we could propose has non-local terms, but the non-locality has a characteristic length of the order of the coherence length of the superfluid. We are interested in the macroscopic dynamics, so we will consider only local terms that give an effective description of the large scale behavior. These terms can be written in term of the gradient of the fermionic condensate ϕ, the metric and non-dynamical tensors which characterize the interaction with the Ψ field. Thus we can write the Lagrangian:

$$\mathcal{L} = \frac{1}{2}\lambda_{\mu\nu}(x)\phi^{,\mu}\phi^{,\nu} + \lambda_\mu(x)\phi^{,\mu} + \lambda(x) \,. \tag{2}$$

The symmetric tensor $\lambda_{\mu\nu} = g_{\mu\nu} + \tilde{\lambda}_{\mu\nu}$ contains a contribution due to the free propagation of the boson, plus a second one which is part of the effective interaction with the medium. The second term in $\mathcal{L}$ describes the interchange of energy-momentum with the medium, while the third term is generated by the background.[4] The energy-momentum tensor is:

$$T^\mu_\nu = \lambda^\mu_\sigma\phi^{,\sigma}\phi_{,\nu} + (\lambda^\mu(x)\phi_{,\nu} + \lambda_\nu(x)\phi^{,\mu}) - g^\mu_\nu\left(\frac{1}{2}\lambda_{\sigma\tau}\phi^{,\sigma}\phi^{,\tau} + \lambda_\sigma\phi^{,\sigma} + \lambda\right) . \tag{3}$$

For a superfluid at rest we have the solution $\phi = \mu t$ for the wave equation. This represents the fundamental state of the superfluid. The spherical symmetry and the stationary state of the star implies that $\lambda \equiv \lambda(r)$ and that the non-null components of $\lambda^{\mu\nu}$ are:

$$\lambda^{00} = a(r) \ , \ \lambda^{rr} = b(r) \ , \ \lambda^{\theta\theta} = r^2 c(r) \ , \ \lambda^{\varphi\varphi} = \sin^2\theta\, r^2 c(r) \,. \tag{4}$$

If we also assume that there is no dissipation, i.e. $\lambda^\mu = 0$, the non-null components of T^μ_ν are:

$$\rho = T^0_0 = \frac{1}{2}a(r)\mu^2 - \lambda(r) \,, \tag{5}$$

$$-p = T^r_r = T^\theta_\theta = T^\varphi_\varphi = -\frac{1}{2}a(r)\mu^2 - \lambda(r) \,, \tag{6}$$

such that we can identify

$$p - \rho = 2\lambda \ , \ p + \rho = a\mu^2 \,. \tag{7}$$

Besides this, the Lagrangian baryonic current is $j^\alpha = \lambda^{\alpha\beta}\phi_{,\beta}$, which in this case reduces to $j^0 = n = a\mu$. This last relation together with Eq. (7) allows us to state the physical meaning of the μ parameter:

$$\mu = \frac{\rho + p}{n} \,, \tag{8}$$

which is the chemical potential of the condensate.

The current component j_φ can also be written as $n_0(r)\, u_\varphi$, where $n_0(r) = \frac{1}{\sqrt{\gamma}} \frac{dn}{dV}$ is the baryon number density in the fluid rest frame and γ is the determinant of the spacial metric tensor.[5] From here we obtain

$$u_\varphi = \frac{n(r)}{n_0(r)} \partial_\varphi \phi = \frac{2}{m^*(r)} \partial_\varphi \phi \ , \tag{9}$$

where $m^* = n_0/n$ is a scalar quantity of dimension one. This parameter can be interpreted as the effective mass of the quasiparticles, and for a neutron star it takes a value of order of the neutron mass.[6] The Nambu-Goldstone field ϕ should be compared with the double of the phase of the Ginzburg-Landau wave function, which is the origin of the factor two in the preceding equation.[3]

Inside the neutron star the gravitational potentials $\Phi(r)$ and $\Lambda(r)$ are determined by the Einstein equations

$$\frac{d\Phi}{dr} = \frac{m + 4\pi r^3 p}{r(r-2m)} \quad , \quad e^{-2\Lambda} = 1 - \frac{2m}{r} \ , \tag{10}$$

and the state equation for the matter, the Oppenheimer-Volkoff equation,

$$\frac{dp}{dr} = -\frac{(\rho + p) + (m + 4\pi r^3 p)}{r(r-2m)} \ , \tag{11}$$

where $m(r) = 4\pi \int_0^r dr\, r^2 \rho(r)$.

The remaining component of the metric, $\varpi(\theta, \varphi)$, is given by the Einstein equation corresponding to the component $R_{\varphi t}$ of the Ricci tensor. We can write this equation in terms of $\Omega = u^\varphi / u^t$, where u^μ is the quadrivelocity of the fluid. The magnitude Ω is the angular velocity of the system measured in a frame at rest with respect to the fluid. The stable configuration, which corresponds to minimum energy, has Ω constant. It is reasonable to consider the fluid very close to a perfect one, and thus the corresponding component of the energy-momentum tensor is

$$T_{\varphi t} = -r^2 \sin^2\theta ((\rho + p)\Omega - \rho\varpi) \ . \tag{12}$$

With this, and expanding $\varpi(\theta, \varphi) = -\sum_{l=1}^{\infty} \varpi^l(r) \sin^{-1}\theta\, \frac{dP_l}{d\theta}$, where P_l is a Legendre polynomial of degree l, the Einstein equation can be written:

$$\varpi^l_{,rr} + \left(\frac{4}{r} - \Lambda' - \Phi'\right) \varpi^l_{,r} + \frac{2}{r}\left(\frac{1}{r} + \Phi' - \Lambda' - \frac{l(l+1)}{2r} e^{2\Lambda}\right) \varpi^l = 16\pi e^{2\Lambda} \left(\frac{1}{2}(\rho + 3p)\varpi^l - (\rho + p)\Omega \delta^l_1\right) \ . \tag{13}$$

The angular velocity of the star, $\Omega = u^\varphi/u^t = \left(u_\varphi g^{\varphi\varphi} + u_t g^{\varphi t}\right)/u^t$, with Eq.(9) becomes $\Omega = \frac{2\, \partial_\varphi \phi}{m^*\, r^2 \sin^2\theta} e^\Phi + \omega$ when we keep up to the first order terms in ω. The factor e^Φ gives the redshift of the effective mass. From the axial symmetry, the periodicity and the continuity of the ϕ field outside singularities, we obtain $2\phi = \sigma(r, \theta)\, \varphi$, where $\sigma(r, \theta)$ is the sum of the indexes of the field singularities held by a circular closed path in a plane orthogonal to the rotation axis, centered on this axis and containing the point (r, θ). The $\sigma(r, \theta)$ function is also equal to the number of vortices surrounded by the closed path, because each vortex has a topological number one to minimize the energy. In consequence

$$\Omega = \frac{\sigma(r,\theta)}{m^* \, r^2 \sin^2\theta} \; e^{\Phi} + \omega \; . \tag{14}$$

This equation generalizes the expression describing the angular velocity and vortex density relation in flat space.[7] This relation shows that the phase introduced by the vortices in the superfluid is equal to the sum of the kinetic phase plus a gravitational phase. If there are no vortices it is $\sigma(r,\theta) = 0$ and hence $\Omega = \omega$. Therefore, in this case the Einstein equation for ϖ for a superfluid star is:

$$\varpi^l_{,rr} + \left(\frac{4}{r} - \Lambda' - \Phi'\right)\varpi^l_{,r} + \frac{2 - l(l+1)}{r^2} e^{2\Lambda}\varpi^l = 0 \, . \tag{15}$$

For Λ and Φ regular in the origin the only regular solution is $\varpi = 0$. Therefore, there is no solution for rotating stars without vortices. This conclusion is also valid for bosonic stars.

3 Gravitational diffraction of vortex lines

Given the angular velocity Ω and the star structure at rest, we can state the vortex distribution in a superfluid star. First we compute ω from Eq.(13), with the reasonable assumption that the vortex interaction energy is negligible with respect to the rotation energy. Once ω is obtained we can compute the vortex distribution and shape from Eq.(14). The resulting vortex distribution minimizes the total energy of the fluid, leading to a macroscopic motion equal to the corresponding one without vortices. The formula (14) shows that the number of vortices will decrease compared with the flat space situation. Since ω is a decreasing function of the radius, the decrease is greater at the axis of the star. Outside the star we have $\omega(r) = \frac{2GJ}{r^3}$, where J is the total angular momentum. From here and applying Eq.(14) with $\theta = \frac{\pi}{2}$ and $r = R$, we get the total number of vortices $N = m^*(R) \; (\Omega R^2 - \frac{2GJ}{R^3}) \; e^{-\Phi(R)}$. Even if we do not know the star mass distribution, which is model dependent, it is possible to give an upper bound for the average number of vortices ν_g per invariant area unit, $dA = 2\pi r e^{\Lambda} dr$. Given that $e^{\Lambda(r)} > e^{-\Phi(R)}$ for $r < R$ this upper bound is:

$$\nu_g = \frac{N}{A} = m^* \left(\Omega - \frac{2GJ}{R^3}\right) \frac{R^2 e^{-\Phi}}{A} \leq \frac{m^*}{\pi}\left(\Omega - \frac{2GJ}{R^3}\right). \tag{16}$$

Thus the relative decrease of the average vortex number density with respect to the one at a flat space, $\nu_0 = \frac{m^*\Omega}{\pi}$, is:

$$\Delta = \frac{\nu_0 - \nu_g}{\nu_0} \geq \frac{2GI}{c^2 R^3} \, , \tag{17}$$

where I is the star moment of inertia. Thus, we get the decrease rate bound as $(\frac{R_s}{R}).(\frac{\bar{r}^2}{R^2})$, where $R_s = \frac{2GM}{c^2}$ is the Schwarzschild radius of the star and $\bar{r}$ is the radius of gyration. For different neutron star models[6] this lower bound varies from 11% to 15%.[7] The vortex density decrease is greater as the star is more relativistic.

Besides this, the gravitational field produces a line vortex diffraction in such a way that these lines are not parallel to the rotation axis except on the equatorial plane. This shows a clear difference with the flat space case, where these lines are parallel to the rotation axis at all points. The angle β between the rotation axis of the star and the direction of the vortex lines at the point (r, θ) is given by

$$\sin\beta = \frac{1}{2}\frac{\kappa(r)\ \sin(2\theta)}{\left(1+\kappa(r)(\kappa(r)-2)\ \sin^2(\theta)\right)^{1/2}}\ , \tag{18}$$

where

$$\kappa(r) = \frac{r}{2}\left(\frac{m^*\,r^2\sin^2\theta}{\sigma(r,\theta)}e^{-\Phi}\frac{d\omega}{dr} - \frac{1}{m*}\frac{dm*}{dr} + \frac{d\Phi}{dr}\right).$$

In particular, the Lense-Thirring diffraction is given by the first contribution in the brakets, characterized by $\kappa_{L-T}(r) = \frac{m^*\,r^3\sin^2\theta}{2\sigma(r,\theta)}e^{-\Phi}\frac{d\omega}{dr}$.

4 The generalized Magnus force

Another interesting consequence of the gravitational background is contained in the expression of the force acting on a vortex immersed in a fluid. To simplify the discussion we will consider a free massless boson field

$$S_\phi = \frac{\kappa}{2}\int d^4x\sqrt{-g}\phi_{,\mu}\phi^{,\mu}\ , \tag{19}$$

where κ is a constant given by $\kappa = n_0 e^{-2\Phi}/\mu$. The interaction between the vortex and the fluid is rather difficult to analyze in these terms, but it becomes easier using an equivalent Kalb-Ramond action:

$$S_B = \frac{\kappa}{12}\int d^4x\sqrt{-g}\ H^{\mu\nu\sigma}H_{\mu\nu\sigma}\ , \tag{20}$$

with $H^{\mu\nu\sigma} = \partial^\mu B^{\nu\sigma} + \partial^\nu B^{\sigma\mu} + \partial^\sigma B^{\mu\nu}$. Both descriptions are related by a duality transformation at the level of the Lagrangian currents given by[8] $\partial_\mu\phi = -\sqrt{-g}\epsilon_{\mu\nu\sigma\tau}D^\nu B^{\sigma\tau}$. Using the Kalb-Ramond action we can describe the interaction between the vortex and the fluid by an interaction term in the Lagrangian. Introducing the antisymmetric tensor $J^{\mu\nu} = \eta\int d\sigma^{\mu\nu}\delta^4\left(x - y(\sigma,\tau)\right)$, with a (σ,τ) parametrization of the world sheet of the vortex, we can write for a fluid interacting with a vortex the Lagrangian

$$\mathcal{L} = \frac{\kappa}{12}\ H^{\mu\nu\sigma}H_{\mu\nu\sigma} + B_{\mu\nu}J^{\mu\nu}\ , \tag{21}$$

which leads to the equation of motion

$$D_\mu H^{\mu\nu\sigma} = 2J^{\nu\sigma}\ , \tag{22}$$

and to the energy-momentum tensor

$$T^\mu_\nu = \frac{\kappa}{2}\left(H^{\mu\rho\sigma}H_{\nu\rho\sigma} - \frac{1}{6}g^\mu_\nu H^2\right).$$

For a free field this is a conserved tensor, but the presence of a vortex generates a density of force on the fluid given by

$$f^\nu = D_\mu T^{\mu\nu} = J_{\mu\sigma} H^{\mu\sigma\nu} \; . \tag{23}$$

The solution of Eq. (22) can be written as the solution of the homogeneous equation $B_o^{\mu\nu}$ plus a particular solution with the vortex as a source, $B_v^{\mu\nu}$, i.e. $H^{\mu\sigma\nu} = H_o^{\mu\sigma\nu} + H_v^{\mu\sigma\nu}$. If the radius of curvature of the vortex in a given point is much larger than the penetration length of the fluid the contribution of the particular solution is negligible. Hence, the force density exerted by the fluid on the vortex is :

$$\tilde{f}^\nu = -f^\nu \simeq -J_{\mu\sigma} H_o^{\mu\sigma\nu} \; . \tag{24}$$

The vortex is characterized at each point by the tangent vector $m^\nu$of the vortex line, which satisfies $m^\nu m_\nu = 1$, and the quadrivelocity u^ν of this point, with $u^\nu u_\nu = -1$. Thus we have

$$d\sigma^{\mu\nu} = \left(u^\mu m^\nu - m^\mu u^\nu\right) d\sigma d\tau \; , \tag{25}$$

and considering that $j^\mu = \kappa \partial^\mu \varphi = n_0 v^\mu$, where v^μ is the quadrivelocity of the fluid, it finally becomes

$$H_o^{\mu\nu\sigma} = \frac{1}{\sqrt{-g}} \epsilon^{\mu\nu\sigma\tau} \partial_\tau \varphi = \frac{1}{\sqrt{-g}} \frac{n_0}{\kappa} \epsilon^{\mu\nu\sigma\tau} g_{\tau\kappa} v^\kappa \; , \tag{26}$$

and the force per unit of length acting on the vortex results:

$$\tilde{f}^\mu = -\frac{2 n_0 \eta}{\kappa} \sqrt{-g} g^{\mu\rho} \epsilon_{\rho\nu\sigma\tau} v^\nu u^\sigma m^\tau \; . \tag{27}$$

Introducing $\delta u^\mu = u^\mu - v^\mu$ and for a vortex with $m^\mu = (0, \mathbf{m})$, we have:

$$\tilde{\mathbf{f}} = \frac{2 n_0 \eta}{\kappa} \sqrt{-g} \left[G \cdot \left(v^0 \delta \mathbf{u} + \delta u^0 \mathbf{v} \right) \times \mathbf{m} + \mathbf{g} \left(\mathbf{m} \cdot \delta \mathbf{u} \times \mathbf{v} \right) \right] \equiv \tilde{\mathbf{f}}_M + \tilde{\mathbf{f}}_{L-T} \; , \tag{28}$$

where $g^{lj} = G$ and $g^{0j} = \mathbf{g}$. The first term is directly related with the usual Magnus force, whereas the second one is a manifestation of the Lense-Thirring effect. For a flat space-time and at first order on v/c this force reduces to the well-known Magnus force $\tilde{\mathbf{f}} = \frac{2 n_0 \eta}{\kappa} (\delta \mathbf{u} \times \mathbf{m})$.

5 Final remarks

In this paper we have analyzed two effects on the vortices in the superfluid matter in neutron stars, due to the dragging of the reference frames originated by the star angular momentum. One of them consists in a diffraction of the vortex lines, and the other in the appearance of a new force acting on a vortex moving respect to the background fluid, oriented according to $\mathbf{r} \times \mathbf{J}$, where $\mathbf{r}$ is the position vector of the considered point in the vortex line and $\mathbf{J}$ the angular momentum of the star.

These results are interesting at least for two reasons. On the one hand, they can be a significant ingredient for analyzing the dynamics of neutron stars, provided the development of the high pressure nuclear matter theory and pulsar models are accurate enough for describing observations. On the other hand, they provide a framework for the verification of the equivalence principle at the level of quantum systems by checking the effect of frame dragging, which at the moment has only been observed as an inertial effect using gyrometric devices.[9]

Acknowledgments

This work was realized with partial support from the Consejo Nacional de Investigaciones Científicas y Técnicas (CONICET), Argentina.

References

1. S.L. Shapiro and S.A. Teukolsky, *Black Holes, White Dwarfs and Neutron Stars: The Physics of Compact Objects* (J. Wiley and Sons, New York, 1983); H.Casini and R.Montemayor, Ap.J. **503**, 374 (1998).
2. A. Muslinov and A.K. Harding, Ap.J. 485, 735 (1997).
3. S.Weinberg, Prog.Theor.Phys.Sup. **86**, 43 (1986).
4. A. Bressan, *Relativistic Theory of Materials* (Springer Verlag, New York, 1978).
5. L.D.Landau and E.Lifchitz, *The Classical Theory of Fields* (Pergamon Press, Oxford, 1980).
6. R.B. Wiringa, V. Fiks and A. Fabrocini, Phys. Rev. **C38** , 1010 (1988).
7. F. Rothen, Astron. Astrophys. **98**, 36 (1981); H.Casini and R. Montemayor, Gen. Rel. and Grav. **30**, 849 (1998); *Covariant description for superfluids in gravitational fields,* gr-qc/9910115.
8. R.L. Davies and E.P.S. Shellard, Phys. Lett. **B214**, 219 (1988); Phys Rev. **D63**, 2021 (1989).
9. O. Avenel, P. Hakonen and E. Varoquaux, Phys. Rev. Lett. **78**, 3602 (1997).

GENERAL RELATIVISTIC CALCULATIONS OF PRECESSION AROUND RAPIDLY ROTATING NEUTRON STARS

SHARON M. MORSINK
Department of Physics, University of Wisconsin-Milwaukee,
P.O. Box 413, Milwaukee, WI, 53201, USA
E-mail: morsink@pauli.phys.uwm.edu

Neutron stars offer a unique opportunity to explore the interplay of strong gravitational fields and matter at super-nuclear density via relativistic effects such as frame-dragging. I will explain how a detailed description of the Lense-Thirring precession of orbits can be obtained using the equations of general relativity. The calculations presented go beyond the usual weak field approximations for the frame-dragging and quadrupole precessions and provide an approximate expression for the precession of orbits in the region *close* to a neutron star. The results in this report show that the exact strong field effects are very important for neutron stars which are described by a stiff equation of state: the precession frequencies for these stars differ greatly from the usual weak field approximations.

1 Introduction

The dragging of inertial frames is usually discussed in the context of the weak-field limit of general relativity. The original calculations by Lense and Thirring,[1] for example, concerned the precession of the planes of orbit of Jupiter's moons due to Jupiter's frame-dragging effect. This precession is so small that it is still not possible to measure. The satellite experiment Gravity Probe B[2] is designed to measure the precession of gyroscopes in orbit around the Earth to an accuracy in the percent range, while a lower accuracy measurement of frame-dragging precession of the LAGEOS satellites has been reported.[3] (See reports in these proceedings for more details on these experiments.) These experiments are impressive, but will only be capable of testing the weak-field limit of general relativity since the gravitational field of the Earth is rather weak. If we wish to observe the full non-linear effects of general relativity, we need to look at objects with very strong gravitational fields. The strongest gravitational fields in the universe are caused by neutron stars and black holes, so it is natural to study these astrophysical objects in the hope of observing frame-dragging.

It is useful to briefly consider the simplest manifestation of the frame-dragging effect. Suppose that an apple is held at rest with respect to the fixed stars so that it has no angular momentum. The apple is then released and allowed to fall freely. According to Newton's theory of gravity, the apple falls along a radial path connecting the centre of the apple to the centre of the Earth. The fact that the Earth rotates does not matter to the apple. (Note that this discussion refers to measurements in the *inertial* frame. If we were to consider the motion of the apple in the *rotating* frame, where the Earth appears not to rotate, the discussion would be complicated by the Coriolis force.) In general relativity, which is Einstein's description of strong gravitational fields, the picture is different. General relativity states that the motion of test particles near a body is affected by both the body's mass and its motion. In the weak field limit, the dependence on motion contributes a magnetic type force (often called gravito-magnetism) which acts perpendicularly to the test particle's direction of motion (see the book by Ciufolini and Wheeler[4] for a nice introduction to the analogy with magnetism). In the case of the apple, the force causes the apple to be deflected by a tiny amount in an east-ward direction from its straight Newtonian course. This has the bizarre result that the zero-angular momentum apple (ZAMA) aquires a non-zero angular velocity! In other words, the apple is dragged along by the Earth's rotation. This gravito-magnetic force also causes a precession of a test particle's orbital plane if it is in an orbit tilted with respect to the equator. This is what is known as the Lense-Thirring effect. In the weak field limit the frequency of precession for a test particle orbiting a body with mass M, radius R and angular momentum J is given by the standard formuala

$$\nu_{LT} = \frac{1}{\pi} \times \frac{Jc}{M^2G} \times \left(\frac{GM}{Rc^2}\right)^2 \times \frac{c}{R} \times \left(\frac{R}{r}\right)^3 \tag{1}$$

where r is the distance from the centre of the body to the test particle. The precession depends quadratically on the dimensionless measure of the body's compactness GM/Rc^2, which is only 7×10^{-10} for the Earth. The value of compactness for a black hole is 1/2, and for neutron stars the compactness ranges from about 1/10 to 4/9. Clearly the values of ν_{LT} will be much larger for compact objects than for solar system objects. However, the displayed formula for the Lense-Thirring precession is only strictly valid for compact objects when the test particle is far away. In this article, I will discuss the problem of finding the correct precession frequencies in the region close to a rapidly rotating neutron star. It is useful to have a simple formula for the precession, so I will also present a more accurate approximate formula which is based on a post-Newtonian, slow rotation expansion of the equations of motion.

The motivation for this work is the recent discussion by Stella and Vietri[5] of the possibility that Lense-Thirring precession of accretion disks orbiting neutron stars may have been observed by the Rossi X-ray Timing Explorer satellite (RXTE). The X-ray observations and astrophysics will be discussed in detail in the contributions to this volume by Luigi Stella and Michiel van der Klis. A more detailed exposition of precessing orbits around neutron stars can be found elsewhere.[6]

2 Models of Rapidly Rotating Relativistic Neutron Stars

A neutron star is the densest configuration of matter possible. In other words, if one takes the densest stable neutron star and tries to make denser, it will collapse in on itself and form a black hole. (I am using the term "neutron star" in a very liberal sense to include stars made of more exotic matter such as strange quark matter or skyrmions.) The density near the centre of the neutron star can be up to an order of magnitude larger than nuclear density. This makes it difficult to describe a neutron star, since we don't know how matter behaves when it is this dense. Nuclear theorists have come up with a number of theories for how neutron star matter behaves. Each theory predicts a specific equation of state (EOS), which is a relation between the star's pressure and density. It isn't known which EOS is the correct description of nature, and observations of neutron stars and determinations of their properties are the best way to rule out certain EOSs.

The neutron star models which we compute are assumed to be stationary, axisymmetric, uniformly rotating perfect fluid solutions of the Einstein field equations. The assumptions of stationarity and axisymmetry allow the introduction of two coordinates ϕ and t on which the space-time metric doesn't depend. The metric, $g_{\alpha\beta}$ can be written as

$$ds^2 = -e^{\gamma+\rho}dt^2 + e^{\gamma-\rho}\bar{r}^2 \sin^2\theta\,(d\phi - \omega dt)^2 + e^{2\alpha}\left(d\bar{r}^2 + \bar{r}^2 d\theta^2\right), \qquad (2)$$

where the metric potentials ρ, γ, α and ω depend only on the coordinates $\bar{r}$ and θ. The function $\frac{1}{2}(\gamma+\rho)$ is the relativistic generalization of the Newtonian gravitational potential; the time dilation factor between an observer moving with angular velocity ω and an observer at infinity is $e^{\frac{1}{2}(\gamma+\rho)}$. The coordinate $\bar{r}$ is not the same as the Schwarzschild coordinate r. In the limit of spherical symmetry, $\bar{r}$ corresponds to the isotropic Schwarzschild coordinate. Circles centred about the axis of symmetry have circumference $2\pi r$ where r is related to our coordinates $\bar{r}, \theta$ by

$$r = e^{\frac{1}{2}(\gamma-\rho)}\bar{r}\sin\theta. \qquad (3)$$

The metric potential ω is the angular velocity about the symmetry axis of zero angular momentum observers (ZAMOs) and is responsible for the Lense-Thirring effect. The fourth metric potential, α specifies the geometry of the two-surfaces of constant t and ϕ. When the star is non-rotating, the exterior geometry is that of the isotropic Schwarzschild metric, with

$$e^{\frac{1}{2}(\gamma+\rho)} = \frac{1 - M/2\bar{r}}{1 + M/2\bar{r}},\; e^{\frac{1}{2}(\gamma-\rho)} = e^{\alpha} = (1 + M/2\bar{r})^2\,,\; \omega = 0. \qquad (4)$$

Notice that the metric of a spinning neutron star (2) is different from the metric for a rotating black hole described by the Kerr solution. It is well known that a result of Birkoff's theorem (see e.g. Wald[7]) is that the exterior of a non-rotating neutron star is described by the the static Schwarzschild black hole solution. However, Birkoff's theorem only holds in spherical symmetry, so there isn't a similar correspondence between rotating stars and black holes as is commonly misconstrued.

We shall investigate uniformly rotating perfect fluid stars. We will numerically integrate the field equations by assuming various zero-temperature, barotropic equations of state of the form $\epsilon = \epsilon(p)$. The numerical method used to integrate the Einstein field equations for the metric potentials is identical to the method described by Cook, Shapiro & Teukolsky[8] which is based on the self-consistent field method presented by Komatsu, Eriguchi & Hachisu.[9] The code which we used to integrate the background gravitational field is a public domain code written by Nikolaos Stergioulas.[10] More details about the numerical method can be found in these papers.

Once the geometry of the star is determined, the total mass, baryonic mass and angular momentum of the star, M, M_B and J respectively can be computed. This involves integration of the metric potentials and the stress tensor over all space.[9]

2.1 Geodesic Motion for Equatorial, Circular Orbits

The assumptions that the neutron star is stationary and axisymmetric has the consequence that two constants of motion are associated with any orbit, corresponding to the energy per unit mass, E, and orbital angular momentum per unit mass, L. As a result, the set of all circular orbits confined to the star's equatorial plane is a two-parameter family depending only on E and L, once the background metric is determined. The theory of these orbits and their perturbations is presented by Bardeen.[11] Bardeen introduces a one-parameter family of orbits corresponding to the motion of zero-angular momentum observers (ZAMOs) which have $L = 0$. Although these observers have zero angular momentum, the dragging of inertial frames by the rotating star causes the ZAMOs to orbit the star with angular velocity ω, as seen by an observer at infinity.

It is useful to consider the motion of other observers with reference to the ZAMOs. Consider a particle confined to a circular orbit in the star's equatorial plane with four-velocity

$$U^\alpha := \frac{dx^\alpha}{d\tau} = \left[\frac{dt}{d\tau}\right]_{\pi/2} \left(t^\alpha + \left[\frac{v_\infty}{r}\right]_{\pi/2} \phi^\alpha\right) \tag{5}$$

where τ is the particle's proper time and v_∞ is the particle's physical three-velocity as measured by an observer at infinity. Square brackets with subscript $\pi/2$ denotes that the quantity within the brackets is evaluated in the star's equatorial plane, $\theta = \pi/2$. The four-momentum per unit particle mass is

$$p_\alpha = g_{\alpha\beta} U^\beta = -E\nabla_\alpha t + L\nabla_\alpha \phi\,. \tag{6}$$

From the definition of four-momentum, the particle's four-velocity is determined by E and L through the relations

$$\left[\frac{dt}{d\tau}\right]_{\pi/2} = \left[(E - \omega L)e^{-(\gamma+\rho)}\right]_{\pi/2}, \tag{7}$$

$$[v_\infty]_{\pi/2} = \left[\omega r + \frac{L}{E - \omega L}\frac{e^{\gamma+\rho}}{r}\right]_{\pi/2}. \tag{8}$$

A ZAMO will measure the particle to move with a velocity, v defined by

$$[v]_{\pi/2} = \left[e^{-(\gamma+\rho)/2}(v_\infty - \omega r)\right]_{\pi/2}. \tag{9}$$

The constants of motion E and L can be rewritten explicitly in terms of v:

$$L = \left[\frac{vr}{\sqrt{1 - v^2}}\right]_{\pi/2}, \tag{10}$$

$$E - [\omega]_{\pi/2} L = \left[\frac{1}{\sqrt{1 - v^2}} e^{\frac{1}{2}(\gamma+\rho)}\right]_{\pi/2}, \tag{11}$$

where $(E - \omega L)\exp - \frac{1}{2}(\gamma + \rho)$ is the energy of the particle measured by a ZAMO. The solution of the geodesic equation yields two possible values for the three-velocity:[11]

$$[v_\pm]_{\pi/2} = \left[\frac{e^{-\rho}\bar{r}^2\omega_{,\bar{r}} \pm \left(e^{-2\rho}\bar{r}^4\omega_{,\bar{r}}^2 + 2\bar{r}(\gamma_{,\bar{r}} + \rho_{,\bar{r}}) + \bar{r}^2(\gamma_{,\bar{r}}^2 - \rho_{,\bar{r}}^2)\right)^{1/2}}{2 + \bar{r}(\gamma_{,\bar{r}} - \rho_{,\bar{r}})}\right]_{\pi/2}, \quad (12)$$

where v_+ and v_- correspond to co-rotating and counter-rotating orbits respectively. The Kepler frequencies, $\nu_{K\pm}$ are defined as the orbital frequency of the prograde and retrograde circular orbits as measured by asymptotic observers

$$\nu_{K\pm} := \left[\frac{v_{\infty\pm}}{2\pi r}\right]_{\pi/2} = \left[\frac{1}{2\pi}\left(v_\pm \frac{e^\rho}{\bar{r}} + \omega\right)\right]_{\pi/2}. \quad (13)$$

2.2 Precession of Tilted Orbits

We now turn to circular geodesics which are not confined to the star's equatorial plane. It is simplest to consider motion where the angle of inclination between the orbital and equatorial planes is infinitesimally small.

The four-velocity of a particle moving on a circular orbit inclined to the star's equatorial plane is

$$U^\alpha = \frac{dt}{d\tau}\left(t^\alpha + \frac{v_\infty}{r}\phi^\alpha\right) + \frac{d\theta}{d\tau}\delta_\theta^\alpha . \quad (14)$$

The values of $\frac{dt}{d\tau}$ and v_∞ on the star's equatorial plane are given by Eqs. (7) and (8) respectively. The normalization condition $U^\alpha U_\alpha = -1$ leads to an equation for $\theta(\tau)$ which takes the form of particle motion in a one-dimensional potential,

$$0 = \frac{1}{2}\left(\frac{d\theta}{d\tau}\right)^2 + V(\theta) , \quad (15)$$

where the potential is[11]

$$V(\theta) = \frac{1}{2}\frac{e^{-2\alpha}}{\bar{r}^2}\left(-1 + e^{-(\gamma+\rho)}(E - \omega L)^2 + \frac{L^2}{\bar{r}^2\sin^2\theta}e^{\rho-\gamma}\right). \quad (16)$$

Consider an orbit which only makes small oscillations out of the star's equatorial plane. This orbit is defined by the conditions $[V(\theta)]_{\pi/2} = [\partial_\theta V]_{\pi/2} = 0$. The first condition is automatically met by the assumed form of the four-velocity. The symmetry of the spacetime guarantees that the second condition is met. The frequency of the oscillation is given by

$$(d\theta/d\tau)^2 = [\partial_\theta\partial_\theta V]_{\pi/2} . \quad (17)$$

The precession frequency ν_p of disk's orbital plane about the star's axis of symmetry is the difference between the frequency of oscillations of the particle in the longitudinal and latitudinal directions, $2\pi\nu_p := d\phi/dt - d\theta/dt$, as observed by observers at infinity. The explicit formula is

$$2\pi\nu_{p\pm} = \left[\omega + \frac{v_\pm}{\bar{r}}e^\rho\left(1 - e^{\alpha+\frac{1}{2}(\gamma-\rho)}X(r,\theta)\right)\right]_{\pi/2} \quad (18)$$

where ν_{p+} and ν_{p-} are the precession frequencies of prograde and retrograde orbits respectively, and the function $X(r,\theta)$ is defined by

$$X(r,\theta) := \left(1 + \frac{1}{2}\partial_\theta\partial_\theta(\rho-\gamma) + \frac{1}{2v_\pm^2}\partial_\theta\partial_\theta(\rho+\gamma) + \frac{\bar{r}}{v_\pm}e^{-\rho}\partial_\theta\partial_\theta\omega\right)^{1/2}. \tag{19}$$

Once the star's geometry is known, it is fairly easy to compute the precession frequency using equations (18) and (19). All that is required is the numerical evaluation of the derivatives in (19). More details about how these equations are evaluated are given elsewhere.[6]

For slowly rotating stars with dimensionless rotation parameter $j = J/M^2 \ll 1$, it is meaningful to split Eq. (18) into two terms, a precession frequency due to the Lense-Thirring effect,

$$\nu_{LT} = (\nu_{p+} + \nu_{p-})/2 = \omega/2\pi + O(j^3) \tag{20}$$

and a precession due to the star's oblateness,

$$\nu_{oblate} = (\nu_{p+} - \nu_{p-})/2 \tag{21}$$

which is the relativistic generalization of the Newtonian quadrupole precession. Thus, to leading order, the precession of orbits due to the L-T effect is equal to the orbital frequency of a ZAMO ($\omega/2\pi$) in an equatorial orbit around the star. It should be remembered, however, that the frame-dragging potential ω makes contributions to the velocity, $v_\pm$, and the frequency which we have named ν_{oblate}. If the star is rotating rapidly, such a split may not be meaningful.

3 Asymptotic Expansions

Our main interest in this paper is in the motion of test particles near a neutron star rotating arbitrarily quickly. The motion of a particle making small oscillations out of the star's equatorial plane is determined exactly by the formulae presented in section 2, once the spacetime's geometry has been determined numerically. Although, this method provides a sufficient description of the motion, it is useful to examine the weak field, slow rotation limit to elucidate the nature of the orbital precession. This corresponds to a double expansion in two small parameters, M/r and $j := J/M^2$, which isn't generally a good description of the strong gravitational field near a rapidly rotating neutron star. However, from this double expansion, the Newtonian quadrupole formula will emerge, which will allow us to discuss the validity of the Newtonian formula.

A weak field, slow rotation limit of the precession frequencies can be derived in the limit $j\,(M/r)^{3/2} \ll 1$. A more detailed description of this expansion can be found elsewhere.[6] In this limit, occurences of $v_\pm$ and $\bar{r}$ can be replaced by ν_K and the star's mass, through the relation

$$\frac{ve^\rho}{\bar{r}} = 2\pi\nu_K + O\left[j\left(\frac{M}{r}\right)^{3/2}\right] = \sqrt{\frac{M}{r^3}} + O\left[j\left(\frac{M}{r}\right)^{3/2}\right]. \tag{22}$$

In this limit, the Lense-Thirring precession frequency, Eq. (20) is

$$\nu_{LT} = \frac{8\pi^2 I \nu_s \nu_K^2}{c^2 M}, \tag{23}$$

where factors of G and c have been restored and $\nu_s := \Omega/2\pi$ is the spin frequency of the star.

The precession frequency due to the star's oblateness can be split into terms

$$\nu_{oblate} = \nu_{quad}\left(1 + c_{PN}\frac{M}{r}\right) + \nu_{cent}. \tag{24}$$

In equation (24), ν_{quad} and ν_{cent} correspond to the the Newtonian quadrupole precession and a centrifugal precession due to other relativistic effects which enter at the same order as the Newtonian quadrupole term. The constant c_{PN} is the coefficient of the first post-Newtonian correction to the Newtonian quadrupole precession. The Newtonian quadrupole precession is

$$\nu_{quad} = -\frac{3\nu_K}{2}\frac{\Phi_2}{Mr^2} \tag{25}$$

$$= -3\Phi_2\left(\frac{2\pi^2}{G^2M^5}\right)^{1/3}\nu_K^{7/3}. \tag{26}$$

The post-Newtonian correction parameter has the value

$$c_{PN} = 3 - 2j^2\frac{M^3}{\Phi_2}. \tag{27}$$

For black holes $\Phi_2 = \frac{G^2}{c^4}j^2M^3$, so that $c_{PN} = 1$. However, for neutron stars this is not the case. It has been shown by Laarakkers and Poisson[12] that $\Phi_2 \simeq a(M, \mathrm{EOS})\frac{G^2}{c^4}j^2M^3$ with $a(M, \mathrm{EOS})$ a constant which depends on the star's mass and equation of state. For EOS from the Arnett and Bowers catalogue,[13] $a \sim 2-12$. The constants $a(M, \mathrm{EOS})$ increase with decreasing mass for a fixed EOS, while they increase for increasing stiffness for fixed mass.[12] It follows that the first post-Newtonian correction term is most important for the stiffest equations of state and the smallest mass stars. At distances close to $6M$, the post-Newtonian correction term can correspond to as much as a 50% correction to the quadrupole precession, and can't be neglected.

The centrifugal precession frequency is defined by

$$\nu_{cent} := \nu_K\left[\left(1 - e^{-\alpha + \frac{1}{2}(\gamma-\rho)}\right)\right]_{\pi/2}. \tag{28}$$

The function $\left[1 - \exp(-\alpha + \frac{1}{2}(\gamma - \rho))\right]_{\pi/2}$ measures the fractional difference in the proper length elements of two infinitesimal curves, both centred about the same point on the star's equatorial plane. The first curve is defined by constant values of the coordinates $\phi, \bar{r}$ and t, and has proper length $dl_1 = \bar{r}\left[\exp\alpha\right]_{\pi/2} d\theta$. The second curve is restricted to the star's equatorial plane and has proper length $dl_2 = \bar{r}\left[\exp\frac{1}{2}(\gamma - \rho)\right]_{\pi/2} d\phi$. If the spacetime is spherically symmetric (or flat), the proper length of the two curves must agree and the frequency ν_{cent} vanishes. When the star is rotating, the proper lengths will not agree, but the condition of asymptotic flatness requires that their difference falls off at least as fast as $1/r^2$. The percent magnitude of the centrifugal precession relative to the quadrupole precession is

$$\zeta := \frac{\nu_{cent}}{\nu_{quad}} \times 100 \tag{29}$$

and is displayed in the table. The size of this term relative to the quadrupole precession depends on the star's mass and equation of state, but typically corresponds to a correction of $1 - 15\%$. It should be noted that the split between centrifugal and quadrupole precession made here is *not* coordinate invariant.

Since it turns out that the centrifugal precession frequency is rather small compared to the quadrupole precession, it can generally be neglected. Similarly in the limit of large quadrupole moments, the post-Newtonian coefficient is roughly $c_{PN} \sim 3$. As a result, we can write a simple approximate formula for the precession of orbits near a neutron star in the form

$$\nu_p \simeq \nu_{LT} - \nu_{quad}(1 + 3\frac{GM}{rc^2}), \tag{30}$$

where the Lense-Thirring precession is

$$\nu_{LT} \simeq 13.2\, \frac{I}{\bar{M}}\, \nu_{s,2.5}\, (\nu_{K,3})^2 \;\; Hz, \tag{31}$$

the Newtonian quadrupole precession is

$$\nu_{quad} \simeq 3.5\; \Phi_{2,43}\; \left(\bar{M}\right)^{-5/3} (\nu_{K,3})^{7/3} \;\; Hz, \tag{32}$$

and the post-Newtonian correction term is

$$3\frac{GM}{rc^2} \simeq 0.3\, (\bar{M})^{2/3}\, (\nu_{K,3})^{2/3}. \tag{33}$$

In Eqs. (31), (32) and (33) we have defined a number of scaled variables defined by: $\bar{M} = M/M_{\odot}$, $I_{45} = I/10^{45}$, $\Phi_{2,43} = \Phi_2/10^{43}$, $\nu_{s,2.5} = \nu_s/300$ and $\nu_{K,3} = \nu_K/10^3$.

In terms of the form factors $a(M, \mathrm{EOS})$ given by Laarakkers and Poisson,[12] the Newtonian quadrupole precession is

$$\nu_{quad} \simeq 6.5 \frac{a(M,\mathrm{EOS})}{10} \left(\frac{I}{\bar{M}}\, \nu_{s,2.5}\right)^2 (\bar{M})^{-2/3} (\nu_{K,3})^{7/3} \;\; Hz. \tag{34}$$

The formula (30) gives a good approximation to the precession of orbits near neutron stars which rotate with spin frequency $\nu_s \sim 300$ Hz. If the star is rotating much faster than this, the formula will not be accurate at radii close to the star.

4 Description of neutron star models selected

We have chosen a set of four equations of state: A, WFF3, C, L (in order of increasing stiffness) which cover a range of possible neutron star properties. Equations of state A (Ref. 14), C (Ref. 15) and L (Ref. 16) are labeled as in the Arnett and Bowers catalogue.[13] EOS A is the softest equation of state which allows a stable $1.4 M_{\odot}$ neutron star with zero angular velocity. EOS L is one of the stiffest EOS in the Arnett and Bowers catalogue. EOS WFF3 is a more modern EOS[17] computed using a variational framework. This EOS uses data from nucleon-nucleon scattering and the known properties of nuclear matter to constrain the two- and three-nucleon potentials in the Hamiltonians to derive a causal "best fit" equation of state for high densities. The equation of state WFF3 is a match of the EOS denoted UV14+TNI[17] at high densities to the FPS[18] EOS low densities and corresponds to a conservative estimate of the correct neutron star EOS.

The neutron stars observed by RXTE have spin frequencies very close to 300 Hz. In particular, one source, 4U 1728-34[19] spins with a frequency of 360 Hz. In this contribution, only results for stars spinning with a frequency of 360 Hz will be presented. A more complete sampling of spin frequencies are presented elsewhere.[6]

The masses of the neutron stars observed by RXTE have not been determined. We will examine a wide range of masses for each equation of state.

5 Numerical Results

The results of the numerical computation of the precession frequencies for corotating orbits inclined an infinitesimal angle to the star's equatorial plane are now presented. For each stellar model selected by the criteria laid out in the preceding section, the spacetime geometry and the worldlines of geodesics were computed using the formalism described in section 2. The equilibrium properties of these models are listed in the table. For each model we have also listed the values of the orbital frequency and the Lense-Thirring, and total relativistic precession frequencies computed at the innermost stable circular orbit (ISCO). If the ISCO radius is smaller than the star's radius, its value is reported with the symbol "—", and the frequencies reported are for orbits at the star's surface. All values of radius reported in these tables refer to the Schwarzschild-like coordinate r defined by Eq. (3).

The key to Tables 1–4 is as follows:

ν_s	Star's spin frequency $\nu_s = \Omega/2\pi$ [Hz]
$\bar{M}$	Total mass, $\bar{M} = M/M_\odot$
$\bar{M}_B$	Baryon (or rest) mass, $\bar{M}_B = M_B/M_\odot$
R	Radius of star, measured at the equator [km]
$I/\bar{M}$	Ratio of moment of inertia to mass [10^{45} g cm^2]
Φ_2	Quadrupole moment[10^{43} g cm^2]
j	Rotation parameter cJ/GM^2
ζ	Centrifugal parameter (cf. Eq. (29))
r_+	Radius of innermost stable circular orbit [km]
ν_K	Kepler frequency at $r = r_+$ [kHz]
$\omega/2\pi$	Lense-Thirring precession frequency at $r = r_+$ [Hz]
ν_p	Total precession frequency at $r = r_+$ [Hz]

These tables illustrate a number of features typical of the orbits of particles at the minimum allowed orbital radius for a variety of masses and EOS. Two competing effects affect the precession frequency of a particle in a prograde orbit: the Lense-Thirring effect which scales as j and is positive, and the precession due to oblateness, which scales as j^2 and is negative. With this knowledge of the scaling, it can be understood how the the total prograde precession frequency varies with $\bar{M}$ and ν_s.

If ν_s is held fixed and $\bar{M}$ allowed to vary, $j = J/M^2$ will increase quadratically, causing the precession due to oblateness to grow rapidly. The overall effect will be to reduce the total precession. If $\bar{M}$ is held fixed while ν_s increases, j increases linearly. As long as j is small, ($j \leq 0.2$) a small linear increase in j will tend to increase the Lense-Thirring precession while the oblateness precession will not change very much. Hence, we expect that the total precession will increase. Once j is larger than ~ 0.2, the quadratic dependence of the oblateness precession will cause ν_p to decrease with increasing j. This general behaviour is observed in the models presented in the tables which have radius larger than the ISCO radius ($R > r_+$).

Within beat frequency models,[20] the observed variations in the centroid frequency of the kHz QPOs in a given source, correspond to changes in radial position of the inner edge of the accretion disk around the neutron star. As the radius changes, the orbital and precession frequencies will change in a correlated fashion. If the precession is dominated by the L-T effect, then the precession should vary as the square of orbital frequency.[5] Close to the star relativistic effects may be important and large deviations from the quadratic behaviour may occur.

In Figures 1 and 2 graphs of precession frequency versus Keplerian (or orbital) frequency are plotted for a subset of the stars featured in the table. Each graph

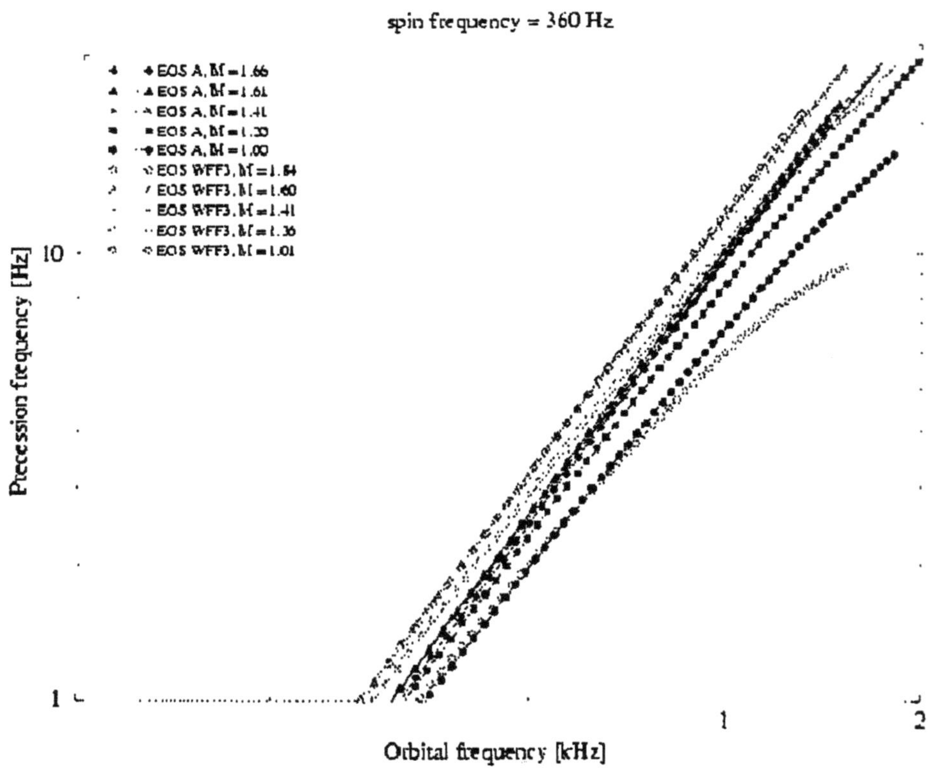

Figure 1. Precession frequency versus Keplerian (or orbital) frequency.

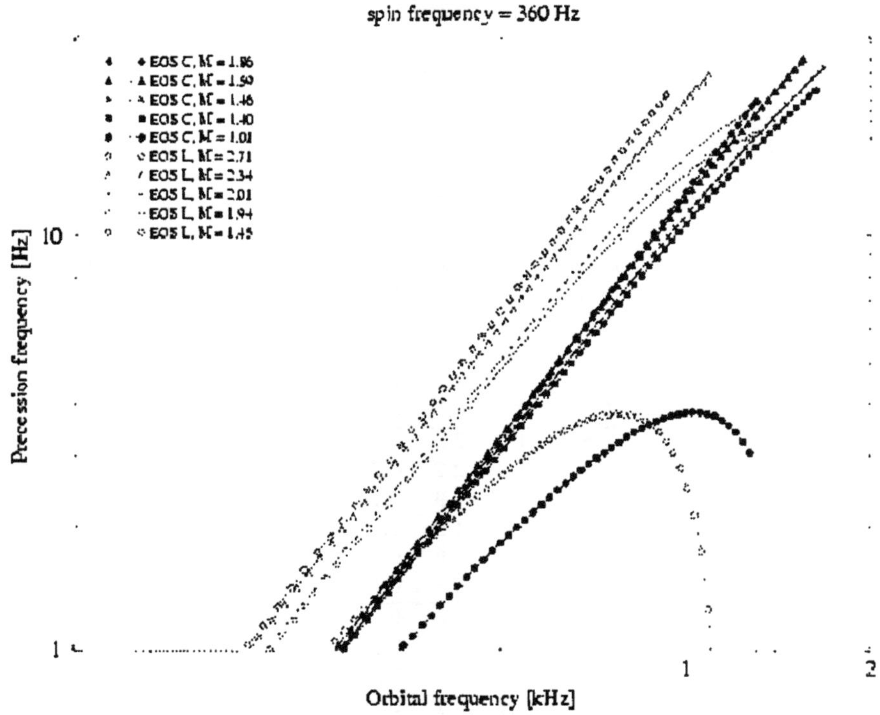

Figure 2. Precession frequency versus Keplerian (or orbital) frequency.

features a number of stars with angular frequency $\nu_s = 360$ Hz. The frequencies are plotted log-log to make it easy to discern departures from a power-law behaviour.

Each curve corresponds to a different stellar model. For each equation of state, the curve with the largest mass corresponds to the maximum mass allowed for a non-rotating star. (Although rotation will allow heavier stars, this increase in mass is negligible for the rotation frequencies considered in this article.) Each point on the curve corresponds to a different radius at which the precession and orbital frequencies are evaluated. It is clear from the curves presented that the precession around stars with masses close to the maximum allowed value is dominated by the Lense-Thirring effect; in these cases the precession versus orbital frequency curves are almost straight lines. For lower mass stars the curves are bent downwards as the orbital frequency increases. The effect is most striking for the stiffest equation of state, L. For these models, the strong field effects are very important and the usual weak-field formula is insufficient to describe the curve.

The termination of each curve corresponds to the smallest radius (or highest orbital frequency) at which a particle can orbit the star. This limiting radius is either the surface of the star or the innermost stable circular orbit.

As an example of the use of these graphs, consider the power spectra displayed in Figure 3 of Strohmayer et al.[19] for the Atoll source 4U 1728-34, the neutron star of which rotates at $\nu_s \sim 360$ Hz. Broad peaks with centroids near 20, 26 and 35

Table 1. Precession frequencies for stars with $\nu_s = 360$ Hz

EOS	M	M_B	R	I/M	Φ_2	j	ζ	r_+	ν_K	$\omega/2\pi$	ν_p
A	1.00	1.08	10.1	0.7	0.9	0.17	2	—	1.81	35.2	16.9
	1.03	1.12	10.0	0.7	0.9	0.17	2	—	1.85	37.0	18.7
	1.20	1.33	9.9	0.7	0.8	0.15	3	9.9	2.01	46.2	28.4
	1.41	1.60	9.6	0.7	0.7	0.13	5	11.7	1.71	34.1	26.7
	1.61	1.87	9.1	0.7	0.5	0.11	10	13.4	1.49	24.9	22.1
	1.66	1.95	8.5	0.6	0.3	0.10	15	13.9	1.43	21.3	19.6
WFF3	1.01	1.08	11.3	0.8	1.6	0.21	1	—	1.53	30.3	9.3
	1.12	1.22	11.2	0.8	1.5	0.19	2	—	1.62	35.5	15.1
	1.36	1.50	11.1	0.9	1.4	0.17	3	11.2	1.80	46.5	26.6
	1.41	1.57	11.1	0.9	1.4	0.16	3	11.6	1.73	43.4	27.3
	1.60	1.82	10.8	0.9	1.2	0.15	6	13.2	1.52	34.3	26.6
	1.84	2.16	9.9	0.9	0.7	0.12	12	15.3	1.30	23.6	21.0
C	1.01	1.08	12.8	1.0	2.5	0.24	1	—	1.27	24.9	3.0
	1.25	1.36	12.4	1.0	2.2	0.20	2	—	1.48	34.9	14.1
	1.40	1.54	12.1	1.0	1.9	0.18	3	—	1.62	42.3	22.4
	1.46	1.61	12.0	1.0	1.9	0.17	3	12.0	1.67	45.3	25.4
	1.59	1.77	11.7	1.0	1.6	0.16	5	13.0	1.54	38.4	26.6
	1.86	2.14	10.3	0.9	0.8	0.12	12	15.4	1.30	23.8	21.1
L	1.45	1.57	15.5	1.6	8.3	0.29	1	—	1.14	34.5	0.1
	1.94	2.16	15.5	1.8	8.2	0.24	3	15.6	1.29	49.8	17.5
	2.01	2.25	15.5	1.8	8.1	0.24	4	16.2	1.25	47.3	19.8
	2.34	2.69	15.3	1.9	7.1	0.21	6	18.7	1.08	36.9	24.3
	2.71	3.22	14.3	1.9	4.8	0.18	13	21.8	0.93	26.1	21.9

Hz are clearly seen, which Stella and Vietri[5] propose to interpret as due to L-T precession. At the same time, the higher kHz QPOs, interpreted as the orbital frequency, are observed at $\sim$ 900, 980 and 1100 Hz respectively. These data points do not lie along any of the curves on in either of the figures.

6 Conclusion

Although the observational data for 4U 1728-34 do not coincide with any of the theoretical curves, it can be seen that stellar models exist for which the observed QPOs are close to twice the theoretical precession frequencies. It is fairly natural for a disk precession model to produce peaks at twice the precession frequency. For example, if our line of view is directed exactly at the star's equator, the disk will block our view of half the star two times every precession period.

The data from Strohmayer et al.[19] is not enough to conclusively decide whether frame-dragging has been observed. New data on the source 4U 1728-34 will be presented in Michiel van der Klis' contribution to these proceedings. Comparisons of the new data with the curves presented in this report should help resolve the matter.

Acknowledgments

It is a pleasure to thank Luigi Stella and Mario Vietri for many helpful discussions. This work was funded in part by NSERC of Canada and NSF grant PHY 95-07740.

References

1. Lense, J., & Thirring, H. 1918, Physik. Zeitschr. 19, 156.
2. Everitt, C.W.F. et al. 1993, in *Relativistic Gravitational Experiments in Space*, Eds. Demiansky, M. & Everitt, C.W.F. (World Scientific: Singapore), P. 309.
3. Ciufolini, I., Pavlis, E., Chieppa, F., Fernandes-Vieira, E. & Perez-Mercader, J. 1998, Science, 279, 2100.
4. Ciufolini, I., & Wheeler, J.A. 1995, *Gravitation and inertia*, (Princeton University Press, Princeton, NJ).
5. Stella, L., & Vietri, M. 1998, ApJ 492, L59.
6. Morsink, S.M., & Stella, L. 1999, ApJ, in press, (astro-ph/9808227).
7. Wald, R.M. 1984, *General Relativity*, (University of Chicago Press, Chicago, IL).
8. Cook, G.B., Shapiro, S.L., & Teukolsky, S.A. 1992, ApJ 398, 203; 1994, ApJ 422, 227; 1994, ApJ 424, 823.
9. Komatsu, H., Eriguchi, Y., & Hachisu, I. 1989, Mon. Not. R. astr. Soc. 237, 355.
10. Stergioulas, N. 1996, public domain code available by anonymous ftp from pauli.phys.uwm.edu in directory /pub/rns or from the web-site http://www.gravity.phys.uwm.edu/Relativity/Code/rns/
11. Bardeen, J.M. 1970, ApJ 161, 103; Bardeen, J.M. 1973, in Black Holes, Ed. C. DeWitt and B.S. DeWitt, (Gordon and Breach, New York);
12. Laarakkers, W.G., & Poisson, E. 1997, ApJ, in press, (gr-qc/9709033).
13. Arnett, W.D., & Bowers, R. L. 1977, ApJ Sup., 33, 415.
14. Pandharipande, V.R. 1971, Nucl. Phys. A 174, 641.
15. Bethe, H.A. & Johnson, M.B. 1974, Nucl. Phys. A 230, 1.
16. Pandharipande, V.R. & Smith, R.A. 1975, Phys. Lett. B 59, 15.
17. Wiringa, R.B., Fiks, V. & Fabrocini, A. 1988, Phys. Rev. C 38, 1010.
18. Lorenz, C.P., Ravenhall, D.G., & Pethick, C.J. 1993, Phys. Rev. Lett., 70, 379.
19. Strohmayer, T.E., Jahoda, K., Giles, A.B., & Lee, U. 1997, ApJ 486, 355.
20. Miller, M.C., Lamb, F.K. & Psaltis, D. 1998, ApJ, in press (astro-ph/9609157).

RELATIVISTIC DISKOSEISMOLOGY: C–MODES AND THE LENSE–THIRRING EFFECT

A.S. SILBERGLEIT
Gravity Probe B, W.W.Hansen Experimental Physics Laboratory, Stanford University, Stanford, CA 94305-4085, USA
E-mail: gleit@relgyro.stanford.edu

R.V. WAGONER
Department of Physics, Stanford University, Stanford, CA 94305-4060, USA
E-mail: wagoner@leland.stanford.edu

We will briefly review how we investigate the modes of oscillation trapped within the inner region of accretion disks by the strong-field gravitational properties of a black hole (or a compact, weakly-magnetized neutron star). The focus here will be on the 'corrugation'(c)–modes, nearly incompressible perturbations of the inner disk. The lowest c–mode has an eigenfrequency which corresponds to the Lense–Thirring frequency in the weak-field or slow-rotation limit. Its radial extent is a decreasing function of the angular momentum, so a major part of the disk is excited for slowly rotating black holes. We compare the c and g('gravity')–mode frequencies with stable frequency features detected in the X-ray power spectra of two galactic 'microquasars'.

1 Introduction

For eighteen years (beginning with Kato & Fukue[1]), it has been known that general relativity can trap normal modes of oscillation near the inner edge of accretion disks around black holes. The strong gravitational fields that are required can also be produced by neutron stars that are sufficiently compact (requiring a soft equation of state) and weakly magnetized to produce a gap between the surface of the star and the innermost stable orbit of the accretion disk. Although we shall not explicitly consider such neutron stars here, the results obtained will also apply to them to first order in the dimensionless angular momentum parameter $a = cJ/GM^2$, since their exterior metric is identical to that of a black hole to that order.

These modes of oscillation provide a potentially powerful probe of both strong gravitational fields and the physics of accretion disks, since they do not exist in Newtonian gravity. In addition, their frequencies depend upon the angular momentum as well as the mass of the black hole. The fractional frequency spread of each mode depends upon the elusive viscosity parameter of the accretion disk.

This subject of 'relativistic diskoseismology' has recently been reviewed by Wagoner.[2] In this paper we shall focus on the 'corrugation'(c) modes, previously studied (less generally) by Kato,[3,4] Ipser,[5,6] and Perez.[7] We shall also compare these modes with the 'gravity'(g) modes, studied most extensively by Perez *et al.*[8]

2 Basic Assumptions and Equations

2.1 *Structure of the Unperturbed Accretion Disc*

We employ the Kerr metric to study a thin accretion disk, neglecting its self-gravity. The stationary ($\partial/\partial t = 0$) and axisymmetric ($\partial/\partial\varphi=0$) equilibrium disk is taken to be described by the standard relativistic α-disk model (Novikov & Thorne,[9] Page & Thorne[10]). The velocitiy components $v^r = v^z = 0$, and the disk semi-thicknness $h(r) \sim c_s/\Omega \ll r$, where $c_s(r,z)$ is the speed of sound. Its key frequencies are

$$\begin{aligned} \Omega(r) &= (r^{3/2}+a)^{-1}\,, \\ \Omega_\perp(r) &= \Omega(r)\left(1-\frac{4a}{r^{3/2}}+\frac{3a^2}{r^2}\right)^{1/2}\,, \\ \kappa(r) &= \Omega(r)\left(1-\frac{6}{r}+\frac{8a}{r^{3/2}}-\frac{3a^2}{r^2}\right)^{1/2}\,; \end{aligned} \tag{1}$$

the rotational, vertical epicyclic, and radial epicyclic angular velocities, respectively. The angular momentum parameter a is less than unity in absolute value. We take $c = 1$, and express all distances in units of GM/c^2 and all frequencies in units of c^3/GM (where M is the mass of the central body).

To simplify the analysis, we here consider only barotropic disks [$p = p(\rho)$, vanishing buoyancy frequency], in which case vertical hydrostatic equilibrium provides the vertical density and pressure profiles

$$\rho = \rho_c(r)(1-y^2)^{1/(\Gamma-1)}\,, \quad p = p_c(r)(1-y^2)^{\Gamma/(\Gamma-1)} \qquad (\Gamma > 1)\,. \tag{2}$$

The disk surfaces are at $y = \pm 1$, and y is related to the vertical coordinate z by

$$y = \frac{z}{h(r)}\sqrt{\frac{\Gamma-1}{2\Gamma}}\,.$$

The adiabatic index $\Gamma = 4/3$ within any radiation pressure dominated region of the disk, and $\Gamma = 5/3$ within any gas pressure dominated region. The disk is symmetric about its midplane $z = 0$. Its inner edge is at the radius of the last stable Keplerian orbit $r = r_i(a)$, where the epicyclic frequency $\kappa(r_i) = 0$. This radius $r_i(a)$ is a decreasing function of a, from $r_i(-1) = 9$ through $r_i(0) = 6$ to $r_i(1) = 1$.

2.2 Equations for the Disk Perturbations

To investigate the eigenmodes of the disk oscillations, we apply the general relativistic formalism that Ipser and Lindblom[11] developed for perturbations of purely rotating perfect fluids. It allows one to express Eulerian perturbations of all physical quantities through a single function $\delta V \propto \delta p/\rho$ which satisfies a second-order PDE. Due to the stationary and axisymmetric background, the angular and time dependences are factorized out as $\delta V = V(r,z)\exp[i(m\phi+\sigma t)]$, where σ is the eigenfrequency. Then the assumption of *slow variation* of modes *in the radial direction* (characteristic radial wavelength $\lambda_r \ll r$) ensures the WKB-separability of variables in the PDE for the functional amplitude, $V(r,z) = V_r(r)V_y(r,y)$. Function V_y varies slowly with r. The resulting ODEs for the vertical (V_y) and radial (V_r) eigenfunctions are (see Nowak & Wagoner[12] and Perez *et al.*[8] for details):

$$(1-y^2)\frac{d^2V_y}{dy^2} - \frac{2y}{\Gamma-1}\frac{dV_y}{dy} + \frac{2\omega_*^2}{\Gamma-1}\left[1+\frac{\Psi-\omega_*^2}{\omega_*^2}(1-y^2)\right]V_y = 0\,, \tag{3}$$

$$\frac{d^2V_r}{dr^2} - \frac{1}{(\omega^2-\kappa^2)}\frac{d}{dr}(\omega^2-\kappa^2)\frac{dV_r}{dr} + \alpha^2(\omega^2-\kappa^2)\left(1-\frac{\Psi}{\omega_*^2}\right)V_r = 0\,. \tag{4}$$

Together with the appropriate homogeneous boundary conditions (discussed in Ref. 8), these equations generate the vertical and radial eigenvalue problems, respectively. The radial conditions depend on the type of mode and its capture zone, as discussed in the next section. Like V_y, the vertical eigenvalue (separation function) $\Psi = \Psi(r)$ varies slowly with radius, as does the dimensionless parameter ω_*:

$$\omega_*(r) = \omega(r)/\Omega_\perp(r)\,, \qquad \omega(r) = \sigma + m\Omega(r) \tag{5}$$

(note that V_y, Ψ and ω_* depend also on σ). We have introduced the corotation frequency $\omega(r)$. Up to a slowly varying relativistic factor of order unity, the coefficient $\alpha(r)$ coincides with $1/c_s(r,0)$. From these two eigenvalue problems, one can obtain the spectrum of eigenfrequences σ and the corresponding eigenvalues $\Psi(r)$.

3 Classification of Modes

One can see from the radial equation (4) that a mode oscillates in a domain of $r > r_i$ where the quantity $(\omega^2-\kappa^2)(1-\Psi/\omega_*^2)$ is positive. The first factor here (for σ in the allowed range) changes sign twice at points

$$r = r_\pm(m, a, \sigma), \;\; r_i < r_- \le r_+$$

determined by equations (1) and (5). This factor is negative between r_- and r_+ and positive otherwise [in Figure 1 $r_\pm$ are plotted as functions of σ for a few values of m and a]. Thus the classification of modes is based on the behavior of the second factor, containing the vertical eigenvalue.

There are essentially three possibilities, corresponding to the following types of modes:

- *g–modes:* $\Psi >$(typically $\gg$) ω_*^2; capture zone: $r_- < r < r_+$.
- *p–modes:* $\Psi <$(typically $\ll$) ω_*^2; capture zone: $r_i < r < r_-$, or $r > r_+$.
- *c–modes:* $\Psi \approx \omega_*^2$; capture zone: $r_i < r < r_c < r_-$, or this interval together with $r_- < r < r_+$.

The capture, or trapping, zone is where the mode oscillations occur.

The g (inertial-gravity)–modes are centered on the radius r_m where $r_- = r_+$. The eigenfrequencies $|\omega|$ corresponding to the lowest mode numbers are close to $\kappa(r_m)$. The p (inertial-pressure)–modes involve the uncertain physics at either the inner or outer radius of the disk.

The c–modes are typically non-radial ($m = \pm 1$) vertically incompressible waves near the inner edge of the disk that precess around the angular momentum of the black hole. Their fundamental frequency is shown below to coincide with the Lense-Thirring (LT) frequency (produced by the dragging of inertial frames by the angular momentum of the black hole) at their capture zone boundary $r_c(m,a)$ in the appropriate weak-field or slow-rotation limit. This is one reason we are concentrating on the c–modes in this paper. The other reason is that g– and p–modes have been rather extensively studied in Ref. 8 and 12, respectively (and in earlier papers therein and in a brief summary of results in Ref. 2).

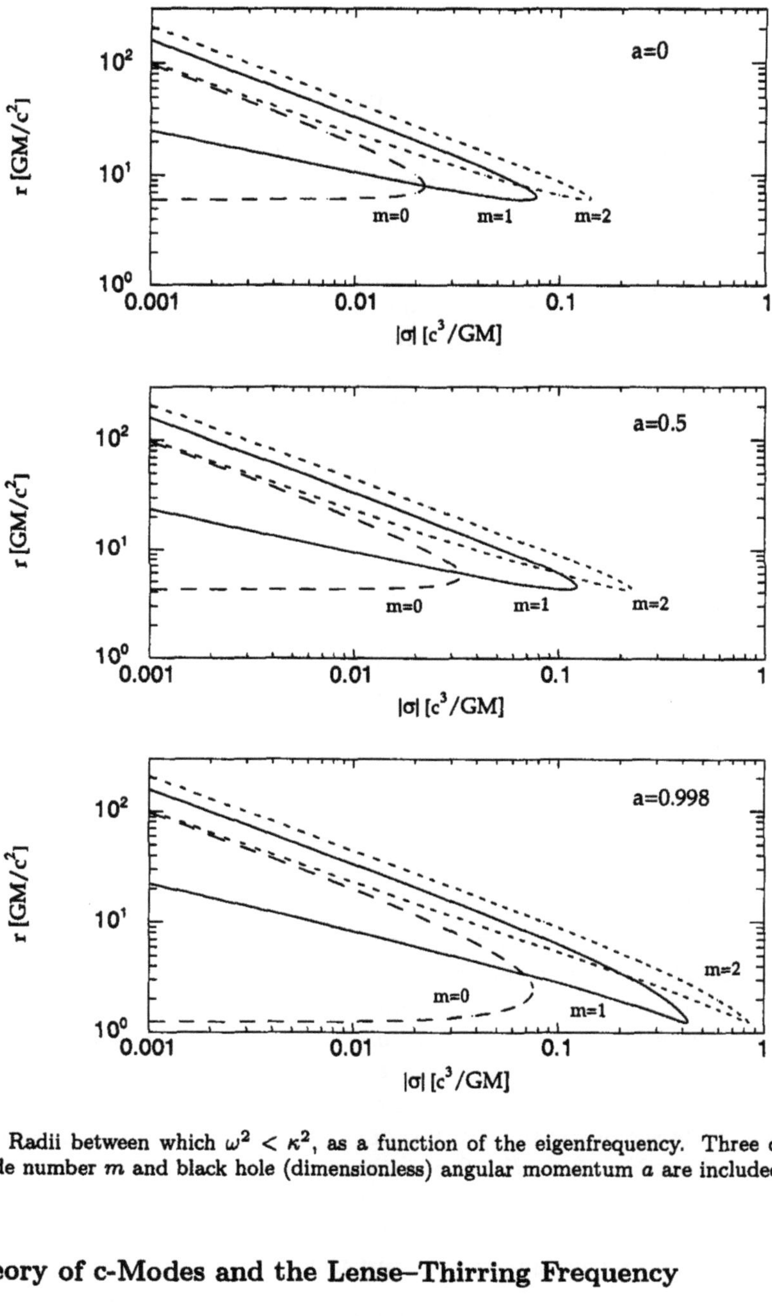

Figure 1. Radii between which $\omega^2 < \kappa^2$, as a function of the eigenfrequency. Three choices of axial mode number m and black hole (dimensionless) angular momentum a are included.

4 Theory of c-Modes and the Lense–Thirring Frequency

4.1 Vertical Eigenvalue Problem: A Selection Rule for the Angular and Vertical Mode Numbers

According to the definition of c–modes, we set

$$\Psi(r)/\omega_*^2(r) = 1 - \chi(r)\,, \qquad |\chi(r)| \ll 1\,, \tag{6}$$

and write the vertical equation (3) as

$$(1-y^2)\frac{d^2V_y}{dy^2} - \frac{2y}{\Gamma-1}\frac{dV_y}{dy} + \frac{2\omega_*^2}{\Gamma-1}\left[1-\chi(1-y^2)\right]V_y = 0\,, \tag{7}$$

with the boundary condition $|V_y(\pm 1)| < \infty$. The last term containing χ is considered a small perturbation. We then immediately see that the unperturbed equation coincides with that for the Gegenbauer polynomials (c.f. Ref. 13)

$$C_j^\lambda(y),\ \lambda = (3-\Gamma)/2(\Gamma-1) > -1/2,\ j = 0,1,2,\dots\,,$$

which are its only solutions regular at $y = \pm 1$. Therefore the coefficient of V_y must coincide with the corresponding eigenvalues: $2\omega_*^2/(\Gamma-1) = j(j+2\lambda)$. However, the left hand side of this equality is a (slowly varying) function of r, so it can be at best valid only 'to the main order'. To clarify the meaning of that, we make use of the definition (5) of ω_* and the fact that the difference between Ω and $\Omega_\perp$ is typically very small, except for values of a close to 1 and simultaneously r close to $r_i \approx 1$. So (anticipating that $|\sigma| \ll \Omega_\perp$ in the same regime) we introduce a small parameter ϵ by setting

$$\omega_* = \frac{\sigma + m\Omega}{\Omega_\perp} \equiv m + \epsilon, \quad \epsilon = \epsilon(r,\sigma,a) = \frac{\sigma + m(\Omega - \Omega_\perp)}{\Omega_\perp}, \quad |\epsilon(r)| \ll 1, \tag{8}$$

and expand the eigenvalue and eigenfunction in it:

$$\chi = \epsilon\chi_1 + O(\epsilon^2)\,, \qquad V_y = V^0 + \epsilon V^1 + O(\epsilon^2)\,. \tag{9}$$

From equations (7) and (9), to zero order we have

$$V^0(y) \equiv V_j^0(y) = C_j^\lambda(y) \tag{10}$$

and $2m^2/(\Gamma-1) = j(j+2\lambda)$, or

$$2m^2 = j^2(\Gamma-1) + j(3-\Gamma)\,, \qquad m = 0,\pm 1,\pm 2,\dots; \quad j = 0,1,2,\dots\,. \tag{11}$$

Note that j is the number of vertical eigenfunction nodes through the disk.

Equation (11) is the criterion for existence of the studied c-modes. It is also a *selection rule* for the angular mode numbers m and j because, depending on the value of Γ, the modes may not exist for *any* m and j. On the other hand, given a rational Γ, there might exist many integer values of m^2 and j satisfying equation (11). (Using some results from number theory,[14] it is possible to show that equation (11) has a *finite* set of solutions if and only if $2pq = k^2$ for some integer k, with coprime p, q from $\Gamma = 1 + p/q$. For instance, we have an infinite sequence of c–modes for both $\Gamma = 4/3$ and $\Gamma = 5/3$, but only a finite number of them if, say, $\Gamma = 3/2$.) However, the only solutions *independent* of Γ are $j = m = 0$ (which does not provide radial mode trapping and must thus be discarded) and

$$j = 1 = m^2\,, \tag{12}$$

which specifies the fundamental c-mode. Its vertical eigenfunction to the main order is independent of r and linear in the vertical coordinate, $V_1^0 \propto y$. Some other possible modes are $m^2 = 4$, $j = 3$ and $m^2 = 25$, $j = 10$ for $\Gamma = 4/3$; and $m^2 = 16$, $j = 6$ for $\Gamma = 5/3$. To be specific, from now on we assume $m > 0$, $a > 0$, unless indicated otherwise.

When equation (11) is satisfied, the first correction χ_1 to the vertical eigenvalue is found in a standard way for the discrete spectrum perturbation, from the solvability condition of the problem for V^1:

$$\chi_1 = \chi_1(m, j, \Gamma) = \frac{2}{m} \frac{\langle V^0, V^0 \rangle}{\langle (1-y^2)\, V^0, V^0 \rangle} > 0, \tag{13}$$

where $\langle \cdot, \cdot \rangle$ is the proper scalar product with the Gegenbauer polynomial weight $(1-y^2)^{\lambda - 1/2}$. From equations (6) and (9), the eigenvalue is then obtained from

$$\Psi(r)/\omega_*^2(r) = 1 - \epsilon\chi_1 + \mathcal{O}(\epsilon^2)\,. \tag{14}$$

The integrals in equation (13) allow for a closed form in the general case and are elementary to calculate for $j = m = 1$, giving

$$\chi_1 = 3\Gamma - 1\,. \tag{15}$$

The vertical eigenvalue and eigenfunction for all c-modes are completely determined to the needed order by equations (9), (10), (11), (13), and (14).

4.2 Radial Behavior of the Vertical Eigenvalue: the LT Frequency at the Capture Zone Boundary is the Eigenfrequency

We now need to study the behavior of the vertical eigenvalue as a function of the radius. From our definitions of the quanitites involved, it follows that for $a > 0$, $m > 0$ the product $\Omega_\perp(r)\epsilon(r)$ is a decreasing function of the radius. Thus in this case $\epsilon(r)$ has *at most* one root between r_i and r_{max}(the outer radius of the disk), since $\Omega_\perp(r) > 0$. If $\epsilon(r_i) < 0$, then $\epsilon(r)$ remains negative in the whole range $r_i < r < r_{max}$, or, equivalently, $\Psi(r)/\omega_*^2(r) > 1$ there; but in such a case we are dealing, by definition, with a g–mode, rather than with a c–mode.

Hence we have to assume that $\epsilon(r_i) > 0$, which converts into a restriction on the eigenfrequency, $\sigma + m\,[\Omega(r_i) - \Omega_\perp(r_i)] > 0$. Then at any value of the root r_c, equation (8) gives

$$\sigma = -m\,[\Omega(r_c) - \Omega_\perp(r_c)]\,, \qquad r_i < r_c < r_{max}\,. \tag{16}$$

Thus $\Psi(r)/\omega_*^2(r) < 1$ for $r_i < r < r_c$, and $\Psi(r)/\omega_*^2(r) > 1$ for $r_c < r < r_{max}$; so the c-mode is trapped in $r_i < r < r_c$, and r_c is the outer boundary of the capture zone.

Moreover, for $r > r_i(a) \geq 1$, $|a| < 1$ is smaller than any positive power of r. In particular, $a/r^{1/2} \ll 1$, except for a very close to 1 (where some of the above considerations do not apply, anyway; e.g., $\epsilon(r_i) \to \infty$ as $a \to 1$). Using this fact, from equations (1) and (16) we obtain

$$\sigma = -m\frac{2a}{r_c^3}\left[1 + \mathcal{O}\left(\frac{a}{r_c^{1/2}}\right)\right]\,. \tag{17}$$

In physical units, the quantity $2a/r_c^3$ is the LT frequency $2GJ/(c^2 r_c^3)$. Thus, to the main order in $(a/r_c)^{1/2}$, *the c-mode eigenfrequency $|\sigma|$ is an integer multiple of the Lense–Thirring frequency at $r = r_c$. In particular, the eigenfrequency of the fundamental c–mode ($m^2 = j = 1$) coincides with the Lense–Thirring frequency* *at its outer trapping zone boundary.* For $a/r_c^{3/2} < a/r_c^{1/2} \ll 1$, the fundamental c–mode is a low–frequency oscillation.

We can now replace the eigenfrequency σ by its expression via r_c [equation (17), or in general equation (16)], and use the latter as a spectral parameter in the radial eigenvalue problem. Its eigenvalue will fix unambigously the value of r_c and thus σ. Unlike other recent suggestions regarding the manifestation of the LT effect in accretion disks around black holes,[16] the value of the radius corresponding to a Lense–Thirring frequency is not a (loosely constrained) parameter, but is predicted by the theory.

Note that in all the remaining cases $a > 0$, $m < 0$ and $a < 0$ the fundamental result (16) holds as well. However, for a counterrotating disk ($a < 0$) the mode capture zone proves to be $r_c < r < r_-$, and the radius r_c is its *left* boundary.

Returning to the case $a > 0$, we remark that in principle a c–mode may have a second oscillation domain $r_- < r < r_+$, since the coefficient of V_r in the radial equation (4) is also positive there. However, in the slow rotation limit $a \to 0$, the radii (r_c, r_-, r_+) and the distances separating them ($r_- - r_c$, $r_+ - r_-$) tend to infinity, so this second domain could be entirely outside the disk. Also (independent of the value of a), for our theory to be consistent the parameter $\epsilon(r)$ must be small *at all the pertinent radii.* This is not true already at $r = r_-$, where (as shown above)

$$\epsilon(r_-) = -m \pm \kappa(r_-)/\Omega_\perp(r_-) = \mathcal{O}(1) \ (\text{unless } m = 1, |a| \ll 1)\,.$$

Therefore the existence of c-modes with two trapping zones, as well as the modes relation to the Lense–Thirring effect, are unclear; and our present approach is not suited to such an investigation, which therefore will be carried out elsewhere.

4.3 Radial Eigenvalue Problem: Determining the Capture Zone Boundary

Here we treat the c–modes trapped near the inner edge of the disk, $r_i < r < r_c$; to be particular, we consider the case $a > 0$, $m > 0$. We require the modes to decay within $r_c < r < r_-$. which requirement provides the boundary condition at $r = r_c$ for the radial equation (4). The boundary condition at the inner edge is not easy to specify, so we write the general linear combination,

$$\cos\theta\,\frac{dV_r}{dr} - \sin\theta\,V_r\bigg|_{r=r_i} = 0\,, \qquad 0 \le \theta \le \frac{\pi}{2}\,; \tag{18}$$

keeping the lack of knowledge parametrized by θ. By introducing a new independent variable·

$$\tau(r) = \int_{r_i}^{r} \left[\omega^2(r') - \kappa^2(r')\right] dr'\,, \qquad \tau_c \equiv \tau(r_c)\,; \tag{19}$$

we rewrite equation (4) in the WKB form, along with the boundary condition (18), in terms of τ:

$$\frac{d^2V_r}{d\tau^2} + \frac{\alpha^2\chi_1\epsilon}{\omega^2-\kappa^2}V_r = 0\,, \qquad \cos\theta\,\omega^2(r_i)\frac{dV_r}{d\tau} - \sin\theta\,V_r\bigg|_{\tau=0} = 0\,. \tag{20}$$

The coefficient in the first equation here is obtained from equation (14) to lowest order in ϵ.

For brevity, the following calculations are carried out for $m^2 = 1$. In this case from equations (1) and (8) one finds

$$\omega^2(r) - \kappa^2(r) = 6/r^4$$

to lowest order in a; so to this accuracy,

$$\tau(r) = 2\left(1/r_i^3 - 1/r^3\right)\,, \qquad \tau_c = 2\left(1/r_i^3 - 1/r_c^3\right)\,. \tag{21}$$

In particular, this means that τ_c is small enough in all cases we are interested in (c. f., $\tau_c \to 1/108 \ll 1$ though, as shown below, $r_c \to \infty$ when $a \to 0$).

This is another reason which allows us to expect that a good approximation to the eigenvalues can be obtained by linearizing the coefficient in equation (20) near the turning point $\tau = \tau_c$ ($r = r_c$). Since, from equations (8), (17), and (21),

$$\epsilon = 2ar^{3/2}\left(1/r^3 - 1/r_c^3\right) = ar^{3/2}\left(\tau_c - \tau\right)$$

to lowest order in a, equation (20) after the linearization of its coefficient becomes

$$\frac{d^2V_r}{d\tau^2} - q\left(\tau - \tau_c\right)V_r = 0\,, \tag{22}$$

where

$$q(r_c, a, \Gamma) = (1/6)a\chi_1\alpha^2(r_c)r_c^{11/2}\,. \tag{23}$$

The solution to equation (22) which decays for $r > r_c$ is the Airy function of the first kind:

$$V_r = \mathrm{Ai}\left(q^{1/3}\left(\tau - \tau_c\right)\right)\,. \tag{24}$$

The nth eigenvalue τ_c is determined from the boundary condition (20) as

$$q^{1/3}\tau_c = \gamma_n,$$

with

$$\gamma_n = \gamma_n(\theta, r_c, a, \Gamma)\ (\gamma_1 < \gamma_2 < \gamma_3 \ldots)$$

being the nth root of

$$\cos\theta\,\omega^2(r_i)q^{1/3}\mathrm{Ai}'(-\gamma_n) - \sin\theta\,\mathrm{Ai}(-\gamma_n) = 0; \tag{25}$$

n is the number of nodes of the radial eigenfunction (24). The (eigen)value of the radius r_c is now found, from equation (21), by means of the resulting equation

$$2q^{1/3}(r_c, a, \Gamma)\left(1/r_i^3 - 1/r_c^3\right) = \gamma_n(\theta, r_c, a, \Gamma)\,. \tag{26}$$

Generally, equations (25) and (26) form a system of transcendental equations for γ_n and r_c which is not easy to analyze.

However, for boundary conditions of the first or second kind ($\theta = \pi/2$ or $\theta = 0$), γ_n is just a number independent of any parameters (*including* r_c). Namely, it is the negative of the nth root of either the Airy function or its derivative, respectively. For example, $\gamma_1 \approx 2.338$ for $\theta = \pi/2$, and $\gamma_1 \approx 1.019$ for $\theta = 0$. In this case an explicit asymptotic expression for r_c in the limit of small $a/r_c^{1/2}$ can be obtained. Indeed, as seen from equation (23), the only dependence of q on r_c remaining unspecified is $\alpha(r_c) \propto 1/c_s(r_c, 0)$. Since the speed of sound drops as some power of the radius at large distances in the standard model of accretion disks,[15] one can write

$$\alpha(r) \approx \alpha_* r^{\nu} \quad (r \to \infty)\, ; \qquad \nu > 0\, , \quad \alpha_* = \text{constant}\, . \tag{27}$$

Together with equations (23) and (26), this determines the outer boundary of the trapping zone (for boundary conditions of the first or second kind at the inner edge) to be

$$r_c = r_* \, a^{-\mu}\, , \qquad \mu = 1/(2\nu + 11/2)\, , \quad r_*(n, \Gamma) = \left(\frac{3\gamma_n^3 r_i^9}{4\chi_1 \alpha_*^2}\right)^{\mu} \tag{28}$$

in the limit of small a ($r_c \gg r_i \cong 6$). Our main small parameter in this case $a/r_c^{1/2} = r_*^{-1/2} a^{1+\mu/2}$ might turn out to be numerically small even if a is not very close to zero, and the whole approximation might still work fine. For example, for $a = 0.5$, $\Gamma = 4/3$, $n = j = 1$, one finds $r_i \approx 4.23$, and $r_c \approx 4.30$ (Perez[7]), so $a/r_c^{1/2} \approx 0.24$ is still small enough.

The slower the rotation of the central mass, the larger the radial extension of the c–mode and the lower its frequency. The size of the trapping zone also increases with the radial mode number n. Note that for the common case of a gas-pressure dominated, optically thick region of the accretion disk, the exponents $\nu = 9/20$ and $\mu = 5/32$. Also in this case, $\alpha_* \propto L^{-2/5}$, where L is the luminosity radiated from the disk.

For $m > 1$ the low–frequency power law (28) for r_c remains valid with

$$\mu = 1/(2\nu + 7/2), \qquad r_*(m, j, n, \Gamma) = \left[\frac{4\gamma_n^3 r_i^6}{3m(m^2 - 1)\chi_1 \alpha_*^2}\right]^{\mu}$$

Let us finally give explicit results for the lowest fundamental c–mode, $m = n = j = 1$. From equations (15), (17), and (28) we obtain:

$$r_* = \left[\frac{A}{(3\Gamma - 1)\, \alpha_*^2}\right]^{\mu}\, , \qquad |\sigma| = \frac{2a}{r_c^3} = \frac{2}{r_*^3} a^{1+3\mu}\, , \tag{29}$$

where $A = 9.659 \times 10^7$ if $\theta = \pi/2$ and $A = 7.997 \times 10^6$ if $\theta = 0$.

5 Microquasars GRS 1915+105 and GRO J1655-40

In principle, any of these (p, g, or c) modes of oscillation could modulate the radiation from accretion disks, in the ultraviolet for supermassive black holes and at soft x-ray energies for stellar mass black holes. We now discuss two black-hole candidate binary X-ray sources whose power spectra (obtained with the RXTE satellite) each contained a high-frequency feature which could be produced by one of these modes. (It is intriguing that these two sources are the strongest 'microquasars' in our galaxy.) It is intriguing that (unlike lower frequency features that have been seen in various black hole candidates) the frequency did not change as the source luminosity (and presumably mass accretion rate) did.

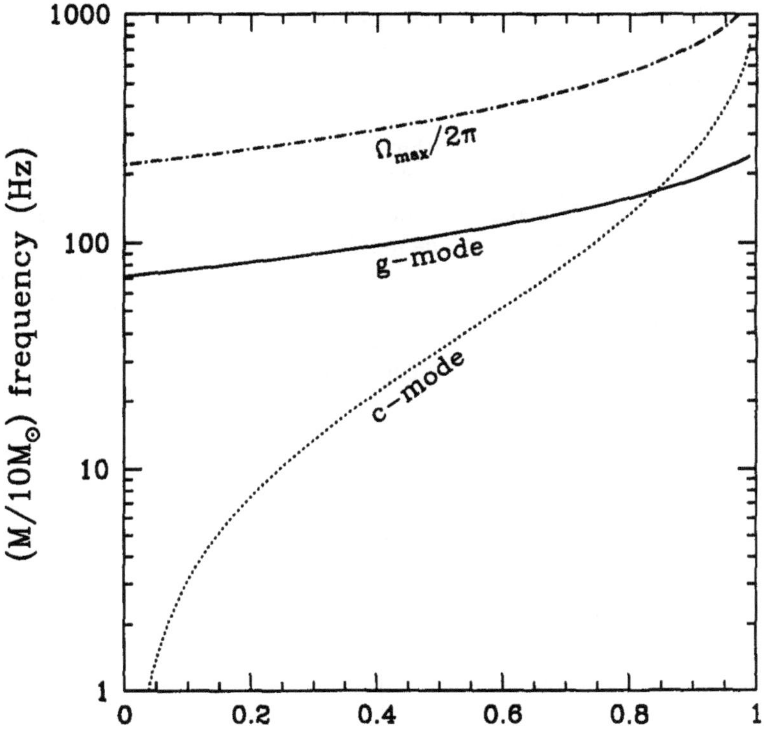

Figure 2. The dependence of three characteristic disk frequencies on the angular momentum of the black hole. The top one is that of a 'blob' rotating at the inner edge of the accretion disk.

This lack of dependence of the frequency on the properties of the accretion disk is clearly predicted by the relation[8]

$$f = 714(1 - \epsilon_{nj})(M_\odot/M)F(a)\ \text{Hz}\,, \qquad \epsilon_{nj} \approx \left(\frac{n+1/2}{j+\delta}\right)\frac{h}{r_m} \tag{30}$$

for the frequency of the lowest radial ($m = 0$) g-modes. It is plotted in Figure 2. ($M_\odot$ is the mass of the Sun.) The properties of the accretion disk enter only through the small mode-dependent term ϵ_{nj}, which involves the disk thickness $2h(r_m)$ and the radial (n) and vertical (j) mode numbers, and $\delta = \delta(\Gamma) \sim 1$. Typically $h(r_m)/r_m \sim 0.1L/L_{Edd}$ for a radiation-pressure dominated optically thick disk region. (L_{Edd} is the effective maximum luminosity.) The dimensionless function $F(a)$ increases monotonically from $F(-1) = 0.60$ through $F(0) = 1$ to $F(1) = 3.5$. Although the approximate center of the radial g-mode eigenfunction is close to the temperature maximum of the disk, the mode's relatively small width appears to limit the luminosity modulation to at most a few percent. The modes with higher m have a somewhat different dependence[8] upon the angular momentum of the black hole.

Also plotted in Figure 2 is the frequency of the fundamental c–mode. For values of a approaching 1, the numerical calculations of Perez[7] were used. Although the mode is almost incompressible, the changing projected area of the mode could modulate the luminosity.

Morgan, Remillard, and Greiner[17] have detected a feature at $f = 67$ Hz with $\Delta f/f \approx 0.05$ in the microquasar GRS 1915+105. During the two periods that the feature was detectable, the source luminosity varied by a factor of ~ 2 while the frequency changed by less than 3%. The amplitude of the peak varied from 1.5% at 3 keV to 6% at 15 keV. If this feature is produced by a g–mode oscillation, equation (30) predicts a black hole mass of $10.6 M_\odot$ if it is nonrotating to $36.3 M_\odot$ if it is maximally rotating. Other aspects of this identification are explored by Nowak et al.[18] On the other hand, a c-mode identification places much weaker limits on the mass.

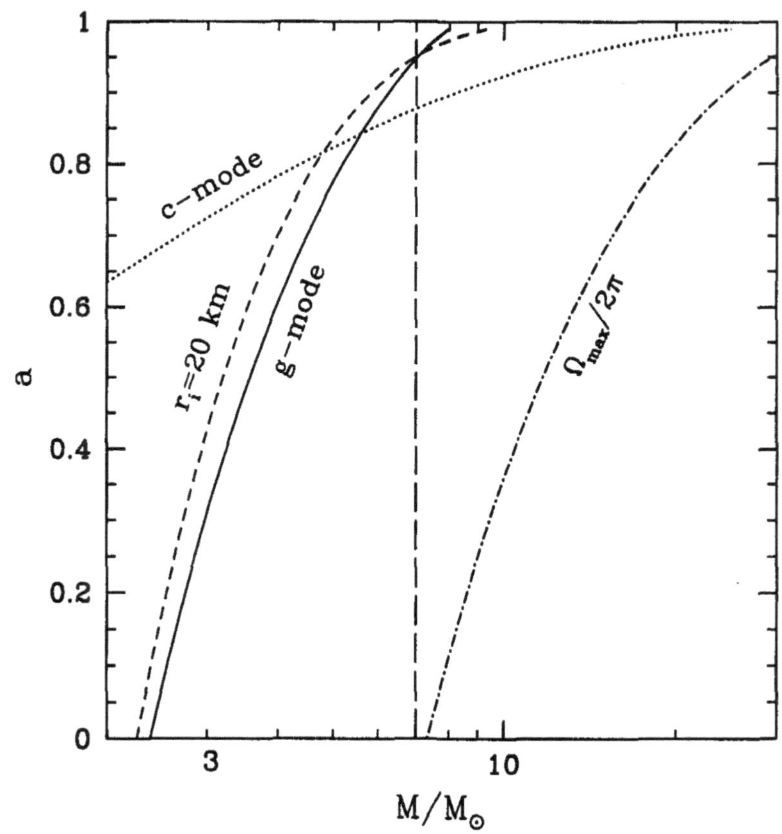

Figure 3. Relations between black hole angular momentum and mass for the microquasar GRO J1655–40. The three frequencies indicated are identified with the 300 Hz feature in its power spectrum. The value of inner radius was obtained by the spectroscopic (Stefan-Boltzmann) method.[21] The black hole mass, as determined by Ref. 20, is also indicated.

Remillard et al.[19] have detected a feature at $f = 300$ Hz with $\Delta f/f \approx 0.4$ in the microquasar GRO J1655-40. It was only detected during 7 observations that showed the hardest X-ray spectra. Although the feature was clearly identified (with an amplitide of 0.8%) only by summing all observations, its frequency appeared to be stable. From extensive optical observations of the companion star, Orosz and Bailyn[20] have determined the black hole candidate mass to be $7.02 \pm 0.22 M_{\odot}$ and the inclination of the binary to be 70°. This agrees with the estimated direction of the jets (if they are perpendicular to the orbital plane and the accretion disk) in both microquasars.

In Figure 3, we plot the relations between black hole angular momentum and mass if we identify the feature as being produced by the fundamental g–mode, c–mode, or a (commonly invoked) 'blob' rotating at the inner disk radius (r_i). Using a distance of 3.2 kpc and the observed energy spectrum of this source, Zhang, Cui, & Chen[21] obtained an inner disk radius of approximately 20 km (using a relativistic Stefan-Boltzmann equation). This yields the other relation between a and M shown. The value of the black hole mass is also indicated.

Note that only by identifying the frequency with that of the g–mode is in agreement with the spectroscopic determination attained, leading to a black hole angular momentum approximately 93% of its maximum possible value. However, the uncertainties in the spectroscopic method might allow a concordance with the c-mode frequency, at a value $a = 0.88$. A major problem with this identification is the very small area occupied by the c-mode at such a large vale of a. (The same would apply to an inner p-mode.) However, the dependence of the c-mode frequency [equation (16)] on the luminosity [through $r_c(\alpha_*)$] should not be appreciable since $r_c \approx r_i$ for such a value of a.

Acknowledgements

This work was supported by NASA grant NAS 8-39225 to Gravity Probe B, and NASA grant NAG 5-3102 to RVW. We are grateful to Lev Kapitanski for finding the number theory results for the Diophantine equation (11), and to Dana Lehr for her remarks and providing the figures.

References

1. S.Kato, J.Fukue, PASJ **32**, 377 (1980).
2. R.V.Wagoner, Phys. Reports, in press (1998) (astro-ph/9805028).
3. S.Kato, PASJ **42**, 99 (1990).
4. S.Kato, PASJ **45**, 219 (1993).
5. J.R.Ipser, Ap. J. **435**, 767 (1994).
6. J.R.Ipser, Ap. J. **458**, 508 (1996).
7. C.A.Perez, *Ph. D. thesis*, Stanford University, 1993.
8. C.A.Perez, A.S.Silbergleit, R.V.Wagoner, D.E.Lehr, Ap. J. **476**, 589 (1997).
9. I.D.Novikov, K.S.Thorne, in *Black Holes*, ed. C.DeWitt & B.S.DeWitt (Gordon and Breach, New York, 1973).
10. D.N.Page, K.S.Thorne, Ap. J. **191**, 499 (1974).

11. J.R.Ipser, L.Lindblom, Ap. J. **389**, 392 (1992).
12. M.A.Nowak, R.V.Wagoner, Ap. J. **378**, 656 (1991).
13. H.Bateman, A.Erdély, *Higher Transcendental Functions, vol. 2*, (McGraw-Hill Book Co., New York–Toronto–London, 1953).
14. Hua Loo Keng. *Introduction to Number Theory*, (Springer Verlag, New York, 1982).
15. N.I.Shakura, R.A.Sunyaev, Astron.Ap. **24**, 337 (1973).
16. W.Cui, S.N.Zhang, W.Chen, Ap. J. Lett. **492**, L53 (1998).
17. E.H.Morgan, R.A.Remillard, J.Greiner, Ap. J. Lett. **473**, L107 (1996).
18. M.A.Nowak, R.V.Wagoner, M.C.Begelman, D.E.Lehr, Ap. J. Lett. **477**, L91 (1997).
19. R.A.Remillard, E.H.Morgan, J.E.McClintock, C.D.Bailyn, J.A.Orosz, Ap. J., submitted (1998) (astro-ph/9806049).
20. J.A.Orosz, C.D.Bailyn, Ap. J. **477**, 876 (1997).
21. S.N.Zhang, W.Cui, W.Chen, Ap. J. Lett. **482**, L155 (1997).

GENERATION AND EVOLUTION OF MAGNETIC FIELDS IN THE GRAVITOMAGNETIC FIELD OF A KERR BLACK HOLE

RAMON KHANNA

Landessternwarte Königstuhl,
D-69117 Heidelberg, Germany
E-mail: Ramon.Khanna@lsw.uni-heidelberg.de

I study the generation and evolution of magnetic fields in the plasma surrounding a rotating black hole. Attention is focused on effects of the gravitomagnetic potential. The gravitomagnetic force appears as battery term in the generalized Ohm's law. The generated magnetic field should be stronger than fields generated by the classical Biermann battery. The coupling of the gravitomagnetic potential with electric fields appears as gravitomagnetic current in Maxwell's equations. In the magnetohydrodynamic induction equation, this current re-appears as source term for the poloidal magnetic field, which can produce closed magnetic structures around an accreting black hole. In principle, even self-excited axisymmetric dynamo action is possible, which means that Cowling's anti dynamo theorem does not hold in the Kerr metric. Finally, the structure of a black hole driven current is studied.

1 Introduction

The influence of the gravitomagnetic field of a rotating compact object on electromagnetic fields has been studied for some 25 years (Wald 1974, Bicak & Dvorak 1976). It was shown by Blandford & Znajek (1977) that the coupling of the gravitomagnetic potential with a magnetic field results in an electromotive force. Currents driven by this electromotive force may extract rotational energy from a black hole. Cast in the language of the 3+1 split of the Kerr metric, Maxwell's equations, together with the ingoing wave boundary condition for electromagnetic fields at the horizon, led to the *Membrane Paradigm* (Thorne et al. 1986), enforcing the analogy between a rotating black hole, immersed in an external magnetic field, with pulsars. There are, however, some difficulties with this pulsar analogy, namely that (i) it is not easy to transport Poynting flux from close to the black hole's horizon out to infinity (Punsly & Coroniti 1989, Punsly 1996), and (ii) a black hole does not carry its own magnetic field, not to mention kGauss fields. Magnetic fields have to be accreted onto the black hole either from the outer accretion disk, or must be generated in the plasma surrounding the black hole.

In this talk I address the question of the generation of magnetic fields by a battery operating in the plasma close to a rotating black hole and the evolution and the structure of magnetic fields that are brought in by an accretion disk. I show that the gravitomagnetic force may play a crucial role in the battery and that the coupling between the gravitomagnetic potential and an electric field is a source for the poloidal magnetic field, which may produce closed magnetic loops in the accreting plasma around the hole. In principle, even an axisymmetric gravitomagnetic dynamo is possible, i.e. Cowling's anti-dynamo theorem is not valid in the Kerr-metric.

2 The MHD description of an electron-ion plasma

Here I summarize the relativistic definition of a plasma as center-of-mass fluid of its components, and the derivation of the generalized Ohm's law (Khanna 1998a). The plasma is assumed to be a perfect fluid and is defined by the sum of the ion and electron stress-energy tensors, which contain a collisional coupling term:

$$(\rho_{\rm m}^{'} + p^{'})W^{\alpha}W^{\beta} + p^{'}g^{\alpha\beta} \equiv T^{\alpha\beta} = \sum_{x=i,e} (\rho_{\rm mx}^{\rm x} + p_{\rm x}^{\rm x})W_{\rm x}^{\alpha}W_{\rm x}^{\beta} + p_{\rm x}^{\rm x}g^{\alpha\beta} + T_{\rm x\,coll}^{\alpha\beta} \,. \quad (1)$$

Subscripts i, e refer to ion and electron quantities, respectively. Superscripts denote the rest-frame in which the quantity is defined, where $'$ refers to the plasma rest-frame. In the 3+1 split (into hypersurfaces of constant Boyer-Lindquist time t, filled with stationary zero angular momentum *fiducial observers*) $T^{\alpha\beta}$ splits into the total density of mass-energy ϵ and momentum density $\vec{S}$

$$\epsilon \equiv (\rho_{\rm m}^{'} + p^{'}v^2)\gamma^2 \approx \rho_{\rm m}^{'}\gamma^2 \qquad \vec{S} \equiv (\rho_{\rm m}^{'} + p^{'})\gamma^2\vec{v} \approx \rho_{\rm m}^{'}\gamma^2\vec{v} \quad (2)$$

and the stress-energy tensor of 3-space with metric $\overleftrightarrow{h}$

$$\overleftrightarrow{T} \equiv (\rho_{\rm m}^{'} + p^{'})\gamma^2\vec{v}\otimes\vec{v} + p^{'}\,\overleftrightarrow{h} \approx \rho_{\rm m}^{'}\gamma^2\vec{v}\otimes\vec{v} + p^{'}\,\overleftrightarrow{h}\;. \quad (3)$$

The approximate expressions hold for a 'cold' plasma. Adding the current density 4-vectors of each plasma component and splitting into charge density and current density yields

$$\rho_{\rm c} \equiv \rho_{\rm ci} + \rho_{\rm ce} = Zen_{\rm i}\gamma_{\rm i} - en_{\rm e}\gamma_{\rm e} \qquad \vec{j} \equiv \vec{j}_{\rm i} + \vec{j}_{\rm e} = Zen_{\rm i}\gamma_{\rm i}\vec{v}_{\rm i} - en_{\rm e}\gamma_{\rm e}\vec{v}_{\rm e}\,. \quad (4)$$

All quantities resulting from the split are measured locally by FIDOs.

2.1 The generalized Ohm's law in the 3+1 split of the Kerr metric

In the 'cold' plasma limit, the local laws of momentum conservation for each species can be re-written as equations of motion, which can then be combined to yield the generalized Ohm's law for an electron-ion plasma

$$\frac{\vec{j}}{\sigma\gamma_{\rm e}} \approx \vec{E} + \frac{Zn_{\rm i}\gamma_{\rm i}}{n_{\rm e}\gamma_{\rm e}}\vec{v}\times\vec{B} - \frac{\vec{j}\times\vec{B}}{en_{\rm e}\gamma_{\rm e}} + \frac{\vec{\nabla}(\alpha_{\rm g}p_{\rm e}^{\rm e})}{en_{\rm e}\gamma_{\rm e}\alpha_{\rm g}} + \frac{4\pi\gamma_{\rm e}}{\omega_{\rm pe}^2}\rho_{\rm c}^{'}\vec{g} + \frac{\rho_{\rm c}^{'}\gamma\vec{v}}{\sigma\gamma_{\rm e}}$$
$$- \frac{4\pi e(Zn_{\rm i}\gamma_{\rm e}^2 - n_{\rm e}\gamma_{\rm i}^2)}{\omega_{\rm pe}^2\gamma_{\rm e}\gamma_{\rm i}^2}\left(\frac{d(\gamma\vec{v})}{d\tau_{\rm p}} - \overleftrightarrow{H}\cdot(\gamma^2\vec{v})\right)\,, \quad (5)$$

with the conductivity $\sigma = e^2n_{\rm e}/m_{\rm e}\nu_{\rm c} \equiv \omega_{\rm pe}^2/4\pi\nu_{\rm c}$, the electron plasma frequency $\omega_{\rm pe}$, the factor of gravitational redshift $\alpha_{\rm g}$ (with $\vec{g} = -\vec{\nabla}\ln\alpha_{\rm g}$) and the gravitomagnetic tensor field $\overleftrightarrow{H} \equiv \alpha_{\rm g}^{-1}\vec{\nabla}\vec{\beta}$. $\vec{\beta} = \beta^{\phi}\vec{e}_{\phi} \equiv -\omega\vec{e}_{\phi}$ is the gravitomagnetic potential, which drags space into differential rotation with angular velocity ω. Note that, in the single fluid description, the gravitomagnetic force drives currents only, if the plasma is charged in its rest frame. $\tau_{\rm p}$ is the proper time in the plasma rest frame. The derivation was made with the assumption that the species are coupled sufficiently strong that their bulk accelerations

$$\frac{d(\gamma_{\rm x}\vec{v}_{\rm x})}{d\tau_{\rm x}} \equiv \left[\frac{\gamma_{\rm x}}{\alpha_{\rm g}}\frac{\partial}{\partial t} + \gamma_{\rm x}\left(\vec{v}_{\rm x} - \frac{\vec{\beta}}{\alpha_{\rm g}}\right)\cdot\vec{\nabla}\right](\gamma_{\rm x}\vec{v}_{\rm x}) \tag{6}$$

are synchronized. The same is required for the gravitomagnetic accelerations, i.e. $|\overleftrightarrow{H}\cdot(\gamma_{\rm i}^2\vec{v}_{\rm i}) - \overleftrightarrow{H}\cdot(\gamma_{\rm e}^2\vec{v}_{\rm e})| \ll |\overleftrightarrow{H}\cdot(\gamma_{\rm i}^2\vec{v}_{\rm i})|$. If the MHD-assumption of "synchronized accelerations" is not made, Ohm's law contains further current acceleration terms, inertial terms and gravitomagnetic terms (Khanna 1998a), which may be important for collisionless reconnection and particle acceleration along magnetic fields. This topic will be discussed elsewhere.

In the limit of quasi-neutral plasma ($Zn_{\rm i} \approx n_{\rm e}$) and $\gamma_{\rm e} \approx \gamma_{\rm i} \approx \gamma$, Eq. (5) reduces to

$$\vec{j} \approx \sigma\gamma(\vec{E} + \vec{v}\times\vec{B}) - \frac{\sigma}{en_{\rm e}}(\vec{j}\times\vec{B}) + \frac{\sigma}{en_{\rm e}\alpha_{\rm g}}\vec{\nabla}(\alpha_{\rm g}p_{\rm e}^{\rm e}) , \tag{7}$$

which contains all the terms, familiar from the non-relativistic generalized Ohm's law, but no gravitomagnetic terms.

2.2 *The gravitomagnetic battery*

The generation of magnetic fields by a plasma battery was originally devised by Biermann (1950) for stars. He showed that, if the centrifugal force acting on a rotating plasma does not possess a potential, the charge separation owing to the electron partial pressure cannot be balanced by an electrostatic field, and thus currents must flow and a magnetic field is generated.

In Khanna (1998b) I have re-formulated Biermann's theory in the Kerr metric. The base of this battery theory is Ohm's law of Eq. (7). Assuming that electrons and ions have non-relativistic bulk velocities in the plasma rest frame, superscripts $i, e, '$ can be dropped. With $p = p_{\rm i} + p_{\rm e} = (n_{\rm i} + n_{\rm e})kT$, the *impressed electric field* (IEF), $\vec{E}^{(i)} = \vec{\nabla}(\alpha_{\rm g}p_{\rm e})/en\gamma\alpha_{\rm g}$, can be re-expressed with the aid of the equation of motion for a 'cold' quasi-neutral plasma to yield

$$\vec{E}^{(i)} = \frac{m_{\rm i}}{(Z+1)e}\left(\gamma\vec{g} + \overleftrightarrow{H}\cdot(\gamma\vec{v}) - \frac{d(\gamma\vec{v})}{d\tau}\right) + \frac{Z\left(\vec{j}\times\vec{B} + (\vec{v}\cdot\vec{j})\vec{E}\right)}{(Z+1)en\gamma} . \tag{8}$$

τ is the proper time in a FIDO frame; i.e. $d/d\tau_{\rm p} = \gamma d/d\tau$. The criterium for magnetic field generation is that $\vec{\nabla}\times\alpha_{\rm g}\vec{E}^{(i)} \neq 0$. Here I restrict the discussion to the gravitomagnetic IEF $\vec{E}^{(i)}_{\rm gm}$. The function part of $\alpha_{\rm g}\vec{E}^{(i)}_{\rm gm}$ is

$$\begin{aligned}\left(\vec{\beta}\cdot\vec{\nabla} + \vec{\nabla}\vec{\beta}\cdot\right)(\gamma\vec{v}) &= \left(\beta^i(\gamma v^j)_{|i} + \gamma\beta^{i|j}v_i\right)\vec{e}_j \\ &= -\gamma v^\phi\tilde{\omega}^2\vec{\nabla}\omega - \omega\left((\gamma v^r)_{,\phi}\vec{e}_r + (\gamma v^\phi)_{,\phi}\vec{e}_\phi\right) ,\end{aligned} \tag{9}$$

where $\tilde{\omega} = (h_{\phi\phi})^{1/2}$ and $_|$ denotes the covariant derivative in 3-space. In axisymmetry $\alpha_{\rm g}\vec{E}^{(i)}_{\rm gm}$ is clearly rotational, unless some freak γ should manage to make $\gamma v^\phi\tilde{\omega}^2$ a function of ω alone. Thus the gravitomagnetic force drives a poloidal current and generates a toroidal magnetic field. Only if v^ϕ is non-axisymmetric, the gravitomagnetic IEF drives toroidal currents. The total IEF $(\alpha_g \vec{E}^{(i)}_{gm} + \alpha_g \vec{E}^{(i)}_{class})$ is likely to rotational in general. This will be clarified for specific velocity fields elsewhere.

In presence of a weak poloidal magnetic field the Biermann battery is limited due to modifications of the rotation law by the Lorentz force, rather than by ohmic dissipation. Then the contribution of the centrifugal force to the IEF becomes irrotational already at weak toroidal fields (Mestel & Roxburgh 1962). The gravitomagnetic battery term, on the other hand, is only linearly dependent on $\vec{v}$. The equilibrium field strength should therefore be higher than for the Biermann battery.

3 The MHD induction equation in the 3+1 split of the Kerr metric

In this section I review the axisymmetric dynamo equations in the 3+1 split of the Kerr metric (Khanna & Camenzind 1996a). Ohm's law is assumed to be of the standard form for a quasi-neutral plasma; Hall-term and IEF are neglected. Combining Maxwell's equations (Thorne et al. 1986) with Ohm's law yields the MHD induction equation

$$\frac{\partial \vec{B}}{\partial t} = \vec{\nabla} \times \left((\alpha_g \vec{v} \times \vec{B}) - \frac{\eta}{\gamma} \left(\vec{\nabla} \times (\alpha_g \vec{B}) + (\vec{E}_p \cdot \vec{\nabla}\omega) \tilde{\omega} \vec{e}_{\hat{\phi}} \right) \right) + (\vec{B}_p \cdot \vec{\nabla}\omega) \tilde{\omega} \vec{e}_{\hat{\phi}} \,. \quad (10)$$

The term standing with the magnetic diffusivity η is the current density, which, via Ampère's law, contains the coupling of the gravitomagnetic field with the electric field. In axisymmetry this is simply the shear of the poloidal electric field in the differential rotation of space, ω. Another induction term is the shear of the poloidal magnetic field by ω. This generates toroidal magnetic field out of poloidal magnetic field even in a zero-angular-momentum flow.

3.1 The gravitomagnetic dynamo

Introducing the flux Ψ of the poloidal magnetic field and the poloidal current T

$$\Psi = \frac{1}{2\pi} \int \vec{B}_p \cdot d\vec{A} = \tilde{\omega} A^{\hat{\phi}} \qquad T = 2 \int \alpha_g \vec{j}_p \cdot d\vec{A} = \alpha_g \tilde{\omega} B^{\hat{\phi}} \,, \quad (11)$$

where $A^{\hat{\phi}}$ is the toroidal component of the vector potential, eq. (10) splits into

$$\begin{aligned} &\frac{\partial \Psi}{\partial t} + \alpha_g (\vec{v}_p \cdot \vec{\nabla}) \Psi - \frac{\eta \tilde{\omega}}{\gamma} v^{\hat{\phi}} (\vec{\nabla}\omega \cdot \vec{\nabla}\Psi) - \frac{\eta \tilde{\omega}^2}{\gamma} \vec{\nabla} \cdot \left(\frac{\alpha_g}{\tilde{\omega}^2} \vec{\nabla}\Psi \right) \\ &\quad = \frac{\eta \tilde{\omega}}{\gamma \alpha_g} \left[\left(T \vec{v}_p - \frac{\eta}{\gamma} \vec{\nabla} T \right) \times \vec{e}_{\hat{\phi}} \right] \cdot \vec{\nabla}\omega \end{aligned} \quad (12)$$

$$\begin{aligned} &\frac{\partial T}{\partial t} + \alpha_g (\vec{v}_p \cdot \vec{\nabla}) T + \alpha_g \tilde{\omega}^2 T \left(\vec{\nabla} \cdot \frac{\vec{v}_p}{\tilde{\omega}^2} \right) - \alpha_g \tilde{\omega}^2 \vec{\nabla} \cdot \left(\frac{\eta}{\gamma \tilde{\omega}^2} \vec{\nabla} T \right) \\ &\quad = \alpha_g \tilde{\omega} (\vec{\nabla}\Psi \times \vec{e}_{\hat{\phi}}) \cdot \vec{\nabla}\Omega \,. \end{aligned} \quad (13)$$

These equations are the relativistic equivalent of the classical axisymmetric dynamo equations. It is important to note, however, that *no mean-field approach* was made,

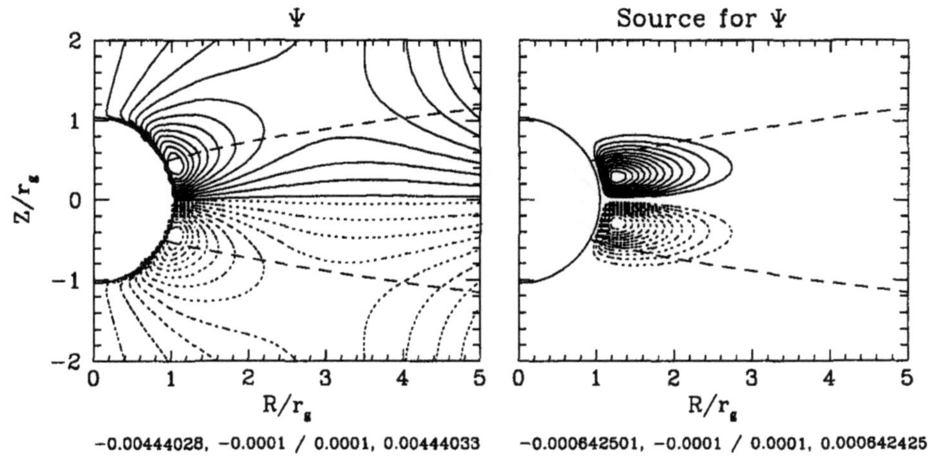

Figure 1. Left: Contours of magnetic flux Ψ showing a quadrupolar magnetosphere of an accreting, rapidly rotating black hole ($a = 0.998M$). Right: The gravitomagnetic current as source of Ψ. The disk is marked by long-dashed lines. Solid contours correspond to positive values, short-dashed contours indicate negative values. The range of contours is given below the boxes.

but Ψ has source terms anyway. They result from $\vec{E}_{\rm p} \cdot \vec{\nabla}\omega$. Obviously, Cowling's anti-dynamo theorem does not hold close to a rotating black hole. Growing modes of this gravitomagnetic dynamo were shown to exist for steep gradients of the plasma angular velocity Ω (Nùñez 1997). For simple accretion scenarios, growing modes could not be found in kinematic numerical simulations (Khanna & Camenzind 1996b). If, however, magnetic field is replenished by an outer boundary condition, the gravitomagnetic source terms generate closed loops around the black hole.

3.2 The magnetic field structure in the accretion disk close to the hole

In the accretion disk, magnetic field may be advected into the near-horizon area, where gravitomagnetic effects may become important. This can be simulated by advection/diffusion boundary conditions for $F = \Psi$, T

$$\frac{\partial F}{\partial n} + \frac{\gamma |v^{\hat{r}}|}{\eta} F = \frac{\partial F_{\rm out}}{\partial n} + \frac{\gamma |v^{\hat{r}}|}{\eta} F_{\rm out} \,, \tag{14}$$

where $\partial/\partial n$ is the derivative along the outer boundary normal. Figure 1 shows the stationary final state of a time-dependent simulation, in which $|B_{\rm p,out}|/|B_{\rm t,out}| = 1/50$. For such a dominantly toroidal magnetic field the gravitomagnetic source terms are strong enough to change the topology of Ψ. This may influence the efficiency of the electromagnetic extraction of rotational energy from the hole.

3.3 Black Hole driven currents in accretion flows

It was mentioned above that the shear of space does also induce a toroidal magnetic field (cf. Eq. [10]). In Eq. (13) this shear term is obscure, but still there, hidden

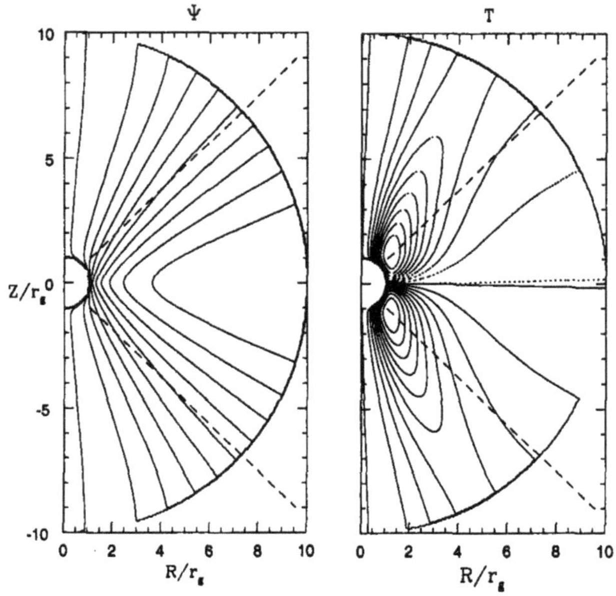

Figure 2. Poloidal magnetic field (left) and corresponding black hole generated current (right) in a zero-angular-momentum accretion flow. The specific angular memomentum of the black hole is $a = 0.998M$. Dotted contours have negative values. Dashed lines indicate the disk scale height H.

in $(\vec{\nabla}\Psi \times \vec{e}_{\hat{\phi}}) \cdot \vec{\nabla}\Omega \propto \vec{B}_{\rm p} \cdot \vec{\nabla}\Omega = \vec{B}_{\rm p} \cdot \vec{\nabla}(\alpha_{\rm g} v^{\phi} + \omega)$. In a zero-angular-momentum flow $v^{\phi} = 0$ (or, equivalently $\Omega = \omega$) and thus the current T is solely generated by the shear of space. Such a scenario can be used to assess, which fraction of the total current in a general accretion-ejection flow is driven by the hole. Figure 2 shows the steady state of an initially homogenous vertical magnetic field that has been dragged onto the hole by radial accretion. The major part of the black hole driven current is confined within the ergosphere. The toroidal field is of the same order as the poloidal field. The current's energy will be dissipated in the accretion flow and in the corona. In a realistic accretion-ejection flow, the plasma shear in the ergosphere of a rapidly rotating black hole will be similar to the shear of space. Thus the current system should have a structure similar to the current of Fig. 2. The energy extracted from the hole will therefore likely be deposited in the plasma close to the hole.

4 Conclusions

It was shown that a rotating black hole can generate magnetic fields in an initially un-magnetized plasma. In axisymmetry a plasma battery can only generate a toroidal magnetic field, but then the coupling of the gravitomagnetic potential with toroidal magnetic fields generates poloidal magnetic fields. Even an axisymmetric self-excited dynamo is theoretically possible, i.e. Cowling's theorem does not hold close to a Kerr black hole. Due to the joint action of gravitomagnetic battery and gravitomagnetic dynamo source term, a rotating black hole will always be surrounded by poloidal and toroidal magnetic fields (probably of low field strength, though). The gravitomagnetic dynamo source may generate closed poloidal magnetic field structures around the hole, which will influence the efficiency of the Blandford-Znajek mechanism.

The "shear-of-space" driven fraction of a global current can be assessed by a kinematic simulation of a zero-angular-momentum flow. The major part of the resulting current system is generated and closed in the corona near the hole. In a realistic accretion-ejection flow, the plasma shear in the ergosphere of a rapidly rotating black hole will be similar to the shear of space. The current system should therefore have a similar structure as in the example shown here, which means that the energy extracted from the hole is likely to be deposited in the disk corona.

Acknowledgments

This work was supported by the Deutsche Forschungsgemeinschaft (SFB 328).

References

1. J. Bicak, L. Dvorak, Gen. Relativ. Gravit. **7**, 959 (1976)
2. L. Biermann, Z. Naturforschg. **5a**, 65 (1950)
3. R.D. Blandford, R.L. Znajek, MNRAS **179**, 433 (1977)
4. R. Khanna, MNRAS **294**, 673 (1998a)
5. R. Khanna, MNRAS **295**, L6 (1998b)
6. R. Khanna, M. Camenzind, AA **307**, 665 (1996a)
7. R. Khanna, M. Camenzind, AA **313**, 1028 (1996b)
8. M. Núñez, Phys. Rev. Let. **79**, 796 (1997)
9. L. Mestel, I.W. Roxburgh, ApJ **136**, 615 (1962)
10. B. Punsly, ApJ **467**, 105 (1996)
11. B. Punsly, F.V. Coroniti, ApJ **350**, 518 (1990)
12. K.S. Thorne, R.H. Price, D.A. Macdonald, W.-M. Suen, X.-H. Zhang, in: *Black Holes: The Membrane Paradigm*, Thorne, K.S., Price, R.H., Macdonald, D.A., (Eds.), Yale Univ. Press, New Haven, pp. 67–120 (1986)
13. R.M. Wald, Phys. Rev. D **10**, 1680 (1974)

LENSE-THIRRING PRECESSION OF ACCRETION DISKS AND QUASI-PERIODIC OSCILLATIONS IN X-RAY BINARIES

D. MARKOVIC, F.K. LAMB
University of Illinois at Urbana-Champaign
Department of Physics, 1110 W. Green Street, Urbana, IL 61801, USA
E-mail: draza@physics.uiuc.edu, f-lamb@uiuc.edu

It has recently been suggested that gravitomagnetic precession of the inner part of the accretion disk, possibly driven by radiation torques, may be responsible for some of the 20–300 Hz quasi-periodic X-ray brightness oscillations (QPOs) observed in some low-mass binary systems containing accreting neutron stars and black hole candidates. We have explored warping modes of geometrically thin disks in the presence of gravitomagnetic and radiation torques. We have found a family of overdamped, low-frequency gravitomagnetic (LFGM) modes all of which have precession frequencies lower than a certain critical frequency $\omega_{\rm crit}$, which is ~1 Hz for a compact object of solar mass. A radiation warping torque can cause a few of the lowest-frequency LFGM modes to grow with time, but even a strong radiation warping torque has essentially no effect on the LFGM modes with frequencies $\gtrsim 10^{-4}$ Hz. We have also discovered a second family of high-frequency gravitomagnetic (HFGM) modes with precession frequencies that range from $\omega_{\rm crit}$ up to slightly less than the gravitomagnetic precession frequency of a particle at the inner edge of the disk, which is 30 Hz if the disk extends inward to the innermost stable circular orbit around a $2M_\odot$ compact object with dimensionless angular momentum $cJ/GM^2 = 0.2$. The highest-frequency HFGM modes are very localized spiral corrugations of the inner disk and are weakly damped, with Q values as large as 50. We discuss the implications of our results for the observability of Lense-Thirring precession in X-ray binaries.

1 Introduction

The possibility of observing gravitomagnetic (Lense-Thirring) precession of accretion disks around compact objects was considered early in the study of accretion onto black holes and neutron stars. The conclusion reached then[1,4,15–17] was that the combined action of gravitomagnetic and internal viscous torques forces the disk flow into the rotation equator of the compact object at ~100 gravitational radii and holds it there, if the outer disk is misaligned or the disk is driven by time-independent torques. As a result, it has been widely thought that because of this "Bardeen-Petterson effect" gravitomagnetic precession of the inner disk is unpromising as an explanation for X-ray oscillations.

Recently Ipser[6] investigated 'trapped' (i.e. localized[7]) pressure perturbations that warp the innermost parts of accretion disks, found that the frame dragging near a Kerr black hole causes precession of such perturbations, and suggested that this precession might be responsible for some of the low-frequency quasi-periodic oscillations (QPOs) observed in the X-ray brightness of black hole candidates.[25] The existence of these modes depends on the presence of specific (but not implausible) non-Keplerian angular velocity profiles. Ipser's general relativistic analysis did not take into account the viscosity of the gas in the disk and therefore did not address whether the Bardeen-Petterson effect suppresses such modes.

More recent suggestions[2,3,22,23] that gravitomagnetic precession of the inner disk, possibly excited by radiation torques, may be responsible for various QPOs and other features in the power spectra of some neutron star and black hole sources have renewed interest in the possibility of observing Lense-Thirring precession near compact objects.

In this paper we describe calculations of the properties of gravitomagnetic and radiation warping modes of geometrically thin, Keplerian disks in which the effects of viscosity are included. We have computed the linear warping modes of disks with a variety of radial temperature structures and for inner and outer boundary conditions appropriate to a variety of astrophysical situations.[10,11] We describe here only our results for isothermal disks with no driving torques acting on the inner and outer boundaries.

2 Solving the warp equation

In the work reported here we make several approximations to simplify the calculations and facilitate comparison with previous work. We treat the disk flow in the Newtonian approximation and include gravitomagnetic effects only to lowest post-Newtonian order. Therefore our treatment is accurate only outside the radius of the innermost stable circular orbit. We use an α model of the viscous shear stress in the accretion disk, setting $T_{r\phi} = \alpha P$ with α constant, and assume that the gas in the disk is fully ionized hydrogen.

Following Petterson,[15] we treat the disk as a family of rings. The orientation of each ring is specified by the unit vector $\mathbf{l}(R)$ orthogonal to the plane of the ring. For small tilt angles, $\mathbf{l} \approx (\beta\cos\gamma, \beta\sin\gamma, 1)$ and hence the disk warp is completely specified locally by the tilt β and twist γ. It is then convenient[4] to represent the warp by the real and imaginary parts of the complex variable $W \equiv \beta\cos\gamma + i\beta\sin\gamma = \beta e^{i\gamma}$.

We use Pringle's approximation[18] for the torque exerted on the disk by the radiation emitted from a central isotropic source with accretion efficiency ϵ. It is convenient to use as the radial variable[9]

$$x \equiv (\epsilon/\kappa)\,(R/R_{\rm g})^{1/2}\ , \tag{1}$$

where $R_g \equiv GM/c^2$ and $\kappa = \nu_2/\nu_1$ is the ratio (assumed constant) of the $R-\phi$ and $R-z$ viscosities,[10] and to look for global modes of the disk of the form $W(t,R) = e^{i\eta t}W(x)$. The warp equation in the presence of both radiation and gravitomagnetic torques can then be written as[10]

$$xW'' + (2 - isx)\,W' - i\left(A - \frac{B}{x^6}\right)W = 0, \tag{2}$$

where the prime denotes differentiation with respect to x,

$$A \equiv 2\eta/G_0 \tag{3}$$

is the complex mode frequency scaled by the characteristic angular frequency

$$G_0 \equiv \frac{\epsilon\alpha c k_B T_{\rm i}}{3GMm_{\rm p}} = 3.1\times10^{-4}\,\frac{\rm rad}{\rm s}\left(\frac{\epsilon}{0.2}\right)\left(\frac{\alpha}{0.05}\right)\left(\frac{T_{\rm i}}{10^7\,{\rm K}}\right)\left(\frac{2M_\odot}{M}\right) \tag{4}$$

associated with the radiation torque at the inner edge of the disk, and

$$B \equiv j \frac{4c}{G_0 R_g} \left(\frac{\epsilon}{\kappa}\right)^6 = \frac{1.7\times 10^4}{\kappa^6} \left(\frac{j}{0.2}\right) \left(\frac{\epsilon}{0.2}\right)^5 \left(\frac{0.05}{\alpha}\right) \left(\frac{10^7 \mathrm{K}}{T_i}\right) \tag{5}$$

is the gravitomagnetic frequency at $x = 1$ scaled by G_0, for a compact object with dimensionless angular momentum $j \equiv cJ/GM^2$. The constant s in Eq. (2) allows us to track the effect of the radiation torque: to include this torque we set $s = 1$; to neglect it, we set $s = 0$.

The strong dependence of B on the accretion efficiency ϵ and viscosity ratio κ is an artifact of our choice $x \propto \epsilon/\kappa$ for fixed R. If s is set to zero, so that the radiation torque is neglected, equation (2) becomes independent of ϵ, as it must; in this case ϵ is simply an arbitrary factor in the conversion from R to x and has no physical meaning.

Equation (2) is a linear, complex, second-order ordinary differential equation for the mode with complex frequency A. Specifying a solution of this equation, including the complex frequency, therefore requires six conditions. The tilt amplitude and twist angle at the inner edge of the disk are arbitrary, and hence only four conditions are physically meaningful. In general, the physics imposes two boundary conditions at the inner edge of the disk, at x_i, and two at the outer edge, at x_o. These boundary conditions restrict solutions to a countable set of "normal modes" $\beta(x)$ and $\gamma(x)$ corresponding to a discrete spectrum of precession frequencies $\omega = \mathrm{Re}\,A$ and growth rates $\sigma = -\,\mathrm{Im}\,A$.

Consider first the boundary conditions at the inner edge of the disk. The solution depends only on the ratio $W'(x_i)/W(x_i)$, because $W(x_i)$ is arbitrary. $W'(x_i) = 0$ corresponds to no external torque acting on the inner edge of the disk. $W(x_i) = 0$ corresponds to pinning of the inner edge of the disk in the rotation equator of the compact object and is only possible if there is a strong external torque. Similar considerations apply at the outer edge of the disk at x_o.

3 Results and discussion

Here we are particularly interested in gravitomagnetic precession of the disks around the neutron stars in the so-called atoll and Z sources.[25] These neutron stars are likely to have been spun up to their equilibrium spin rates, in which case $\dot{J}_z$, the z-component of the net flux of angular momentum through the disk, is zero. Therefore all the results reported here are for $\dot{J}_z = 0$. This assumption is not as restrictive as it might seem, because the tilt functions, precession frequencies, and damping rates of the disk warping modes are almost unaffected by the value of $\dot{J}_z$, unless it is almost equal to the flux being advected through the disk.[10,11]

When the gravitomagnetic torque is included, we have found two families of solutions: (1) a set of low-frequency modes—either the low frequency gravitomagnetic (LFGM) modes if the radiation torque is neglected, or the low frequency gravitomagnetic radiation (LFGMR) modes if the radiation torque is included—all of which have frequencies less than a certain critical frequency $\omega_{\rm crit}$, and (2) a set of high-frequency gravitomagnetic (HFGM) modes, all of which have frequencies greater than $\omega_{\rm crit}$. Using approximate analytical expressions valid for the high-order modes of an isothermal disk, we find,[10,11]

$$\omega_{\rm crit} \approx (x_{\rm i}/x_{\rm o})^{1/2} B/(11 x_{\rm i}^6) \,. \qquad (6)$$

The precession frequency of a gravitomagnetic mode is a weighted average of the single-particle precession frequency over the radial extent of the mode, and hence $\omega_{\rm crit}$ is roughly determined by the radius at which the mode function $\beta(x)$ peaks. The tilt functions of the highest-order LFGM and HFGM modes all peak at about the same place in the disk, so their frequencies are almost identical.

All the results reported here are for $\kappa = 1$, $G_0 = 3.1 \times 10^{-4}$ rad s^{-1}, $B = 1.7 \times 10^4$, $x_{\rm i} = 0.49$, and $x_{\rm o} = 50$. This value of $x_{\rm i}$ corresponds to $R_{\rm i} = 6\,GM/c^2$ for $\epsilon = 0.2$. If the outer radius of the disk is two or three times larger than assumed here, there would be a few additional modes with very low frequencies. The results presented here for the HFGM modes assume $W'(x_{\rm i}) = 0$ and $W'(x_{\rm o}) = 0$; the results presented for the LFGM and LFGMR modes assume only $W'(x_{\rm o}) = 0$.

Figure 1 shows the growth rates and precession frequencies of the lowest 18 LFGM modes. The $n = 0$ ($\omega = \sigma = 0$) mode corresponds to the static solution found by Bardeen & Petterson.[1] As n increases, the relative spacing $\Delta\omega/\omega$ between adjacent modes decreases and the spectrum therefore becomes quasi-continuous at large n. The spectrum of the precession frequencies and damping rates of the higher-order LFGM modes is plotted as a continuous curve in Figure 2. The slight

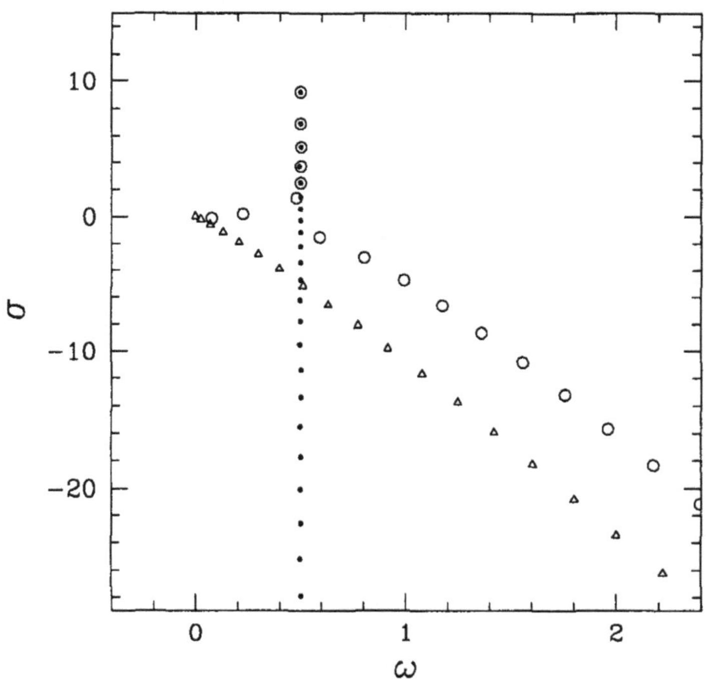

Figure 1. Spectrum of dimensionless growth rates σ and precession frequencies ω of the lowest 18 low-frequency gravitomagnetic (LFGM) modes (*open triangles*), radiation-warping (R) modes[10] (*dots*), and low-frequency gravitomagnetic radiation-warping (LFGMR) modes (*open circles*) of the accretion disk.

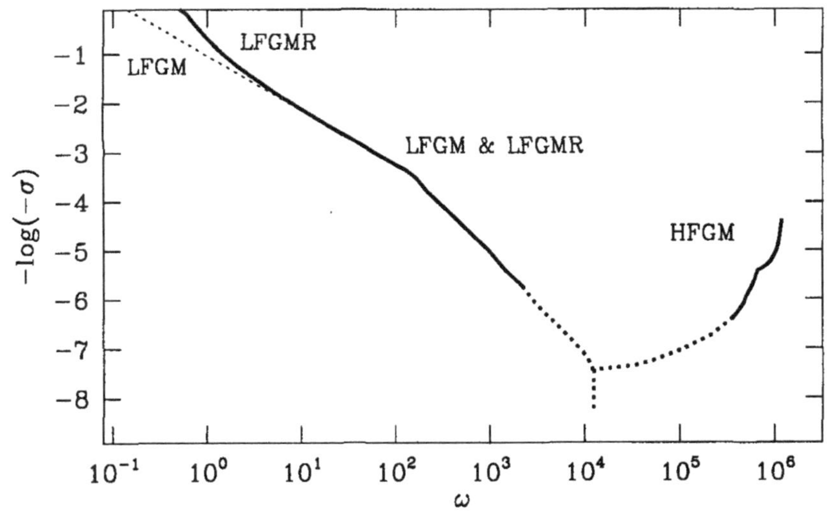

Figure 2. Spectrum of dimensionless damping rates $-\sigma$ and precession frequencies ω of low-frequency gravitomagnetic (LFGM) modes (*dotted line*), low-frequency gravitomagnetic radiation-warping (LFGMR) modes (*left-hand solid line*), and high-frequency gravitomagnetic (HFGM) modes (*right-hand solid line*) plotted as continuous curves. The damping rates of the LFGM and LFGMR modes with ω greater than $\sim$10 are indistinguishable on this scale. The LFGM and HFGM spectra become almost vertical at the critical frequency $\omega_{\rm crit}$ discussed in the text. This happens even if the disk is driven at the inner edge. The HFGM spectrum is essentially unaffected by any radiation warping torque. The heavy dotted portions of the curves show where the radial distance between corrugations becomes smaller than $10^{-2}\,r$.

break in the spectrum at $\omega \sim 10^2$ is where the damping rate becomes comparable to $\omega_{\rm crit}$.

If the radiation and gravitomagnetic torques are both included, generally all of these LFGMR modes are time-dependent (see Fig. 1). In particular, the static $n = 0$ LFGM mode is perturbed by the radiation-warping torque and becomes weakly time-dependent. The lowest-order LFGMR modes are unaffected by the gravitomagnetic torque, whereas the high-order LFGMR modes are essentially the same as the LFGM modes with similar frequencies. In between there are a few hybrid modes that are affected by both the radiation-warping torque and the gravitomagnetic torque. The HFGM modes are essentially unaffected by any radiation warping torque.

The vast majority of the modes that lie between the lowest-frequency LFGMR modes and the highest-frequency HFGM modes are highly overdamped short wavelength corrugations of the disk. In particular, the modes with frequencies in the interval $2 \times 10^3 < \omega < 4 \times 10^5$ have corrugation wavelengths $\lesssim$1% of the radius at which the mode peaks. When the wavelength becomes less than the vertical thickness of the disk, our treatment of the disk as infinitesimal is no longer valid.[14] In this regime, modes of the type we have found may be even more strongly damped or even nonexistent.

The LFGMR modes that have precession frequencies $\gtrsim 10^{-4}$ Hz are essentially unaffected by even a strong radiation-warping torque and have damping rates $\gtrsim$

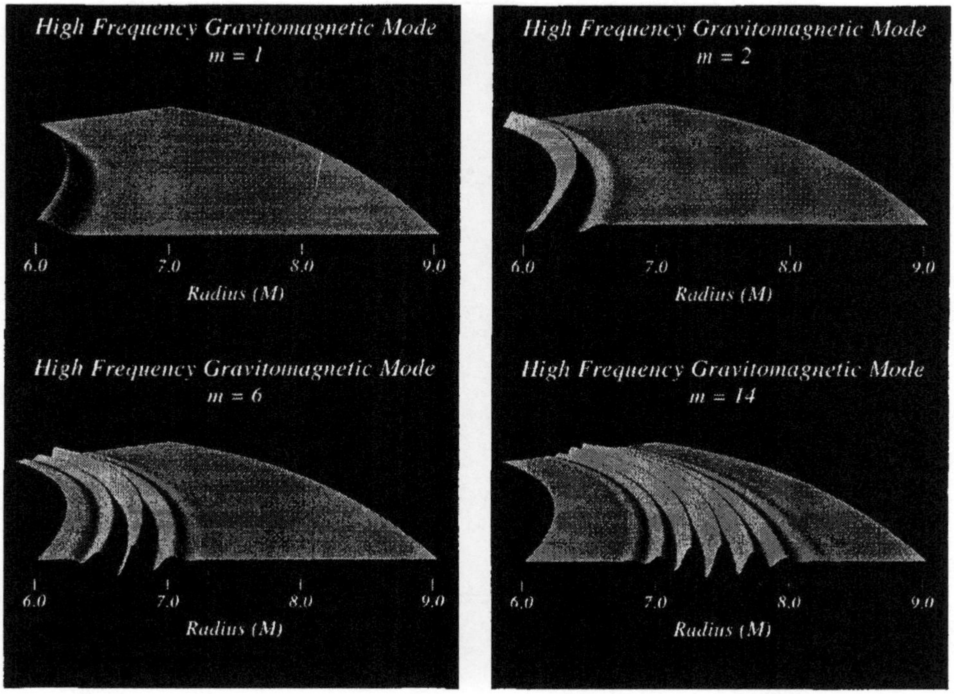

Figure 3. Four low-order undriven high-frequency gravitomagnetic (HFGM) modes. For $m = 1$, $\omega = -49\sigma = 0.97\omega_{\rm gm,i}$; for $m = 2$, $\omega = -15\sigma = 0.89\omega_{\rm gm,i}$; for $m = 6$, $\omega = -5.2\sigma = 0.73\omega_{\rm gm,i}$; whereas for $m = 14$, $\omega = -2.3\sigma = 0.53\omega_{\rm gm,i}$.

10 times their frequencies. The highest-frequency LFGMR modes are very short wavelength spiral corrugations that extend from the radius $r_{\rm i}$ of the inner edge of the disk only to $\sim 10\,r_{\rm i}$. These modes have precession frequencies $\sim \omega_{\rm crit}$, which is $\sim$ 1 Hz for a compact object of solar mass, and are very strongly damped. They therefore appear unpromising as an explanation of the 6–300 Hz QPOs observed in some accreting neutron stars and black hole candidates.

4 The lowest-order HFGM modes

The tilt functions $\beta[x(R)]$ of the lowest-order (highest-frequency) undriven HFGM modes have a single narrow peak near the inner edge of the disk (see Fig. 3). These modes are very localized spiral corrugations of the inner disk. The lowest-order undriven HFGM modes precess in the prograde direction with precession frequencies comparable to the gravitomagnetic precession frequency $\omega_{\rm gm,i} = B/x_i^6$ at the inner edge of the disk.

The damping rate of a mode is determined by the radial spacing δ between the disk corrugations. The damping rates $-\sigma$ of the highest-frequency undriven HFGM modes are given fairly accurately by $-\sigma \approx \pi^2 x_{\rm max}/\delta^2$, where $x_{\rm max}$ is the position of the maximum of the tilt function for the mode. Our numerical solutions show that for the dozen or so highest-frequency HFGM modes, δ decreases with increasing

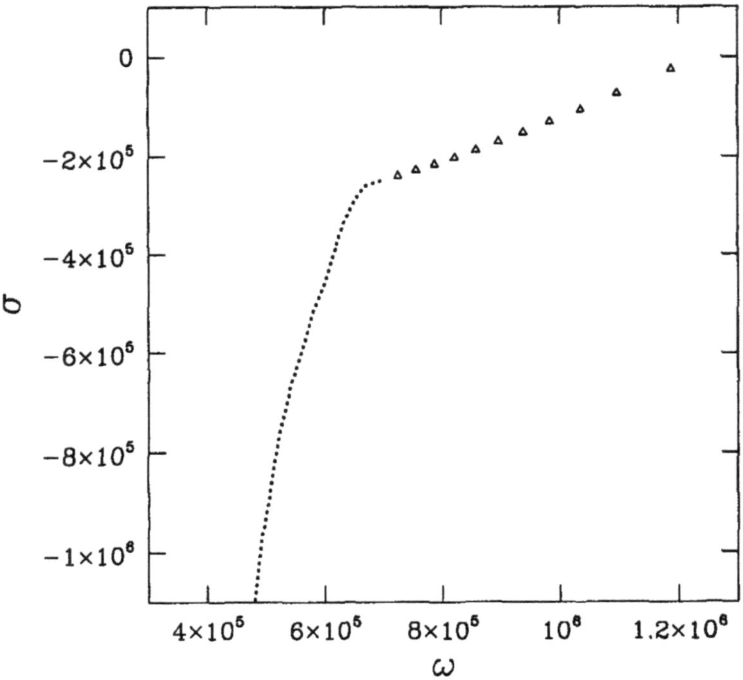

Figure 4. Spectrum of dimensionless growth rates σ and precession frequencies ω of undriven high-frequency gravitomagnetic (HFGM) modes of the accretion disk. The frequencies and damping rates of the 11 lowest-order (highest-frequency) HFGM modes are plotted individually as open triangles. The dense spectrum formed by the precession frequencies and damping rates of the higher-order HFGM modes is indicated by the dotted line.

mode number, but not uniformly, whereas the radial width Δ of the tilt function increases approximately linearly with increasing $x_{\max}$ from $m = 2$ up to at least $m \sim 14$.

With increasing mode number, the frequencies of the HFGM modes decrease (see Fig. 4). The dozen or so lowest-order HFGM modes have precession frequencies that range from $\approx \omega_{\rm gm,i}$ for the $m = 1$ mode to $\sim 0.5\,\omega_{\rm gm,i}$ for the $m = 14$ mode. These modes are underdamped, with Q values that range from 49 for the $m = 1$ mode to 2.3 for the $m = 14$ mode. The damping rates of the HFGM modes are sensitive to the viscosity in the disk. If the viscosity parameter α (see § 2) is 0.5 rather than 0.05, for example, the Q of the fundamental HFGM mode becomes 22 rather than 49. With further increases in the mode number, the HFGM modes become overdamped and their precession frequencies eventually approach $\omega_{\rm crit}$ asymptotically from above.

The warp functions, precession frequencies, and damping rates of the modes in the HFGM family depend on the boundary condition at the inner edge of the disk: a torque that oscillates with a given frequency and acts on the inner edge of the disk can always resonantly excite the appropriate driven HFGM mode that has the same frequency, if the torque is strong enough. The properties of the HFGM modes are only slightly affected if the Newtonian gravitational potential is replaced by a pseudopotential that mimics the steeper effective potential at small radii in general relativity.[10,11]

5 Implications for QPOs

The high-frequency HFGM modes may be involved in some of the rapid X-ray variability and high-frequency QPOs observed in some accreting neutron stars and black hole candidates. The highest possible frequency of the fundamental ($m = 1$) HFGM mode is the gravitomagnetic precession frequency of gas at the radius $R_{\rm isco}$ of the innermost stable circular orbit around the compact object. For a compact object with mass M and dimensionless angular momentum $j \ll 1$, $R_{\rm isco} \approx 6GM/c^2$ and hence the Lense-Thirring precession frequency of the innermost stable circular orbit is

$$\nu_{\rm gm,isco} \approx 30 \left(\frac{j}{0.2}\right)\left(\frac{2M_\odot}{M}\right) \text{ Hz} . \tag{7}$$

The largest value of j expected for neutron stars with spin frequencies ~ 300 Hz is ~ 0.2. Hence, for a $M = 2M_\odot$ neutron star with a radius smaller than that of the innermost stable circular orbit and a spin frequency of ~ 300 Hz, the *highest possible* frequency of the fundamental HFGM mode would be ~ 30 Hz, whereas for a $6M_\odot$ black hole with $j \sim 0.2$, the highest possible frequency would be ~ 10 Hz.

Stella & Vietri have recently proposed[23] that the 15–60 Hz features observed in the power spectra of some neutron-star sources are produced by gravitomagnetic precession of gas at the same radius as the gas in circular Keplerian orbit that is thought to generate the kilohertz QPOs,[13,24] which have frequencies ranging from 900 to 1200 Hz in the sources Stella & Vietri considered. The gravitomagnetic precession frequency $\nu_{\rm gm}$, the frequency $\nu_{\rm K}$ of the Keplerian-frequency kilohertz QPO, and the dimensionless angular momentum j of the star should then be related by

$$\nu_{\rm gm} = (GM4\pi/c^3)\, j\nu_{\rm K}^2 \,. \tag{8}$$

Solving this relation for I_{45}, the moment of inertia of the star in units of 10^{45} g cm^2, yields

$$I_{45}\left(\frac{M_\odot}{M}\right) = 2\left(\frac{\nu_{\rm gm}}{40\text{ Hz}}\right)\left(\frac{300\text{ Hz}}{\nu_{\rm spin}}\right)\left(\frac{1.2\text{ kHz}}{\nu_{\rm K}}\right)^2 . \tag{9}$$

Computations of the *maximum* values of $I_{45}M_\odot/M$ for modern neutron star equations of state are very close to unity. For example, for the FPS equation of state the maximum value of $I_{45}M_\odot/M$ is 0.87 and occurs for a $1.55M_\odot$ star; for the UU equation of state the maximum is 1.05 and occurs for a $2.1M_\odot$ star (M. C. Miller, 1998, personal communication; for a discussion of these equations of state, see Psaltis *et al.*[19]). Even for the obsolete and unrealistically stiff equations of state L and M, the maximum values of $I_{45}M_\odot/M$ are $\lesssim 2$.

These results show that the highest precession frequencies predicted by modern neutron star models are 2–4 times lower than the frequencies of the bumps seen in the power spectra of the atoll sources. Hence, in order to explain even the lower frequency bumps seen in some atoll sources by gravitomagnetic precession of the inner disk, the precessing HFGM corrugation would have to generate X-ray brightness oscillations at the second harmonic or I_{45}/M for these neutron stars would have to be twice as large as the largest values given by modern equations of state.[19,22,23] The maximum frequencies of the horizontal branch QPOs in the Z sources are $\sim 50\%$ higher than the frequencies of the bumps in the atoll source power spectra and are therefore at least three times higher than the highest precession frequencies predicted by modern neutron star equations of state.[19,23] Also, the frequencies of the horizontal branch QPOs do not vary with kilohertz QPO frequency as required by Eq. (8).[19]

In addition to resolving whether the frequencies predicted by the gravito- magnetic precession model can be reconciled with the frequencies of the features observed in the power spectra of neutron stars and black hole candidates, the excitation, damping, and visibility in X-rays of the low-order HFGM modes will have to be explored further before they can be regarded as a promising explanation for these features. Our results show that these modes are not excited by the radiation-warping torque. The two lowest-order HFGM precession modes involve vertical motions of the inner edge of the disk and this offers a possible way of exciting them. However, in order to excite steadily either of these two modes, a torque acting on the inner edge of the disk would have to match the variation of their phases with azimuth as well as their frequencies. On the other hand, coupling of the vertical motion to the radiation field, magnetic field, or surface of a neutron star may instead damp these modes.

The two lowest HFGM modes might also be excited by repetitive impulsive disturbance of the inner disk. If, however, such disturbances affect the inner portion of the disk as well as its innermost edge, they may excite the broad noise-like spectrum of underdamped HFGM modes, rather than one or two such modes, producing a broad hump rather than a QPO.

The highest order HFGM modes do not appear promising as a mechanism for producing quasi-periodic oscillations in X-ray brightness, because they are tightly wound spiral corrugations within a narrow annulus centered at a radius ~ 5 times larger than the radius of the inner edge of the accretion disk and do not affect the accretion disk at its inner edge. On their face, the lowest order HFGM modes appear more capable of causing oscillations in the X-ray emission, for example by modulating the accretion flow onto the compact object.

This work was supported in part by NSF grants AST 93-15133 and AST 96-18524 and by NASA grant NAG 5-2925.

References

1. Bardeen, J. M., & Petterson, J. A., ApJ 195, L65 (1975).
2. Cui, W., Zhang, S. N., & Chen, W., ApJ 492, L53 (1998).
3. Cui, W., this volume (1998).
4. Hatchett, S. P., Begelman, M. C., & Sarazin, C. L., ApJ 247, 677 (1981).
5. Iping, R. C., & Petterson, J. A., A&A 239, 221 (1990).

6. Ipser, J. R., ApJ 458, 508 (1996).
7. Kato, S. & Honma, F., Publ. Astr. Soc. Japan 43, 95 (1991).
8. Lamb, F. K., Shibazaki, N., Alpar, M. A., & Shaham, J., Nature 317, 681 (1985).
9. Maloney, P. R., Begelman, M. C., & Pringle, J. E., ApJ 472, 582 (1996).
10. Markovic, D. & Lamb, F. K. ApJ 507, 326 (1998).
11. Markovic, D. & Lamb, F. K., in preparation (1998).
12. Miller, M. C., & Lamb, F. K., ApJ 470, 1033 (1996).
13. Miller, M. C., Lamb, F. K., & Psaltis, D., ApJ 508, in press (1998).
14. Papaloizou, J.C.B., & Pringle, J. E., MNRAS 202, 1181 (1983).
15. Petterson, J. A., ApJ 214, 550 (1977a).
16. ———, ApJ 216, 827 (1977b).
17. ———, ApJ 226, 253 (1978).
18. Pringle, J. E., MNRAS 281, 357 (1996).
19. Psaltis, D., et al. 1998, ApJ, submitted
20. Stella, L., talk presented at the Symposium on The Active X-ray Sky: Results from BeppoSAX and Rossi-XTE, Accademia dei Lincei, Rome, October 1997.
21. ———, talk presented at the meeting of the High Energy Astrophysics Division of the AAS, Estes Park, Colorado, November 1997.
22. ———, this volume (1998).
23. Stella, L., & Vietri, M., ApJ 492, L59 (1998).
24. Strohmayer, T., et al., ApJ 469, L9 (1996).
25. van der Klis, M., in *X-Ray Binaries*, ed. W.H.G. Lewin, J. van Paradijs, & E.P.J. van den Heuvel (Cambridge University Press, Cambridge), 252 (1995).

Classical Papers

General Relativity and Gravitation, Vol. 16, No. 8, 1984

On the Gravitational Effects of Rotating Masses: The Thirring-Lense Papers[1]

BAHRAM MASHHOON, FRIEDRICH W. HEHL, and DIETMAR S. THEISS

Institut für Theoretische Physik, Universität zu Köln, D-5000 Cologne 41, Federal Republic of Germany

Received May 30, 1982

Abstract

The purpose of this work is to provide a critical analysis of the classical papers of H. Thirring [*Phys. Z.*, **19**, 33 (1918); *Phys. Z.*, **22**, 29 (1921)] and J. Lense and H. Thirring [*Phys. Z.*, **19**, 156 (1918)] on rotating masses in the relativistic theory of gravitation and to render them accessible to a wider circle of scholars. An English translation of these papers is presented which follows the original German text as closely as possible. This is followed by a concise account of the significance of the results of these papers as well as the possibility of measuring the gravitational effects of rotating masses.

§(1): *Introduction*

Hans Thirring (1888-1976) was born and educated in Vienna. He studied mathematics and physics at the University of Vienna and received his Ph.D. in 1911. His habilitation thesis of 1915 was on the theory of the specific heat of solids. His major fields of interest were the theory of relativity and applied physics; moreover, he had wide-ranging interests in psychology, politics, and in issues related to world energy resources. He became professor of physics in Vienna in 1921. For political reasons he had to leave the University in 1938. In 1945 he was reinstated and became the director of the Institute for Theoretical Physics. Among the books he authored there is one entitled *Die Idee der Relativitätstheorie*.

Josef Lense (b. 1890) was born and educated in Vienna. He studied astronomy at the University, obtaining his Ph.D. in 1914 [cf. Die jovizentrische Bewe-

[1] Dedicated to Professor Josef Lense on the occasion of his 92nd birthday.

0001-7701/84/0800-0711$03.50/0 © 1984 Plenum Publishing Corporation gung der kleinen Planeten, *Astron. Nachrichten*, **196**, 341-346 (1913)]. His habilitation thesis of 1921 on the integration of differential expressions, as well as his subsequent work on quadratic forms, was motivated by the geometric aspects of the relativity theory. He became a lecturer in Vienna in 1921, and a professor of mathematics at the Technical University of Munich in 1927. In addition, from 1946 until his retirement, he was the director of the Mathematical Institute in Munich. He has authored many books in mathematics, a subject in which his major fields of interest have been differential geometry and mathematical physics. He presently makes his home in Munich.

The collaboration of Lense and Thirring began in Vienna after Thirring, aware of Lense's background in astronomy, inquired whether Lense would be willing to carry through, in perturbation theory, the integration of the equations of motion of a test body in the field of a rotating mass together with the application of the results to the orbits of the planets and moons [1]. This resulted in a joint publication in the *Physikalische Zeitschrift* of 1918.

§(2): *English Translation of the Thirring-Lense Papers*

On the Effect of Rotating Distant Masses in Einstein's Theory of Gravitation

HANS THIRRING

The considerations of this study can best be made clear by a quotation from Einstein's fundamental paper of 1914.[1] He says the following in the introduction:

"At first it seems that such an extension of the theory of relativity has to be rejected for physical reasons. Namely, let K be a permissible coordinate system in the Galilei-Newtonian sense, K' a uniformly rotating coordinate system with respect to K. Then centrifugal forces act on masses which are at rest relative to K', while not on masses at rest relative to K. Already Newton considered this as proof that the rotation of K' has to be interpreted as "absolute," and that one cannot thus treat K' as "at rest" with the same justification as K. This argument is not valid, however, as explained by E. Mach in particular. That is, we need not attribute the existence of those centrifugal forces to the motion of K'; rather, we can attribute them as well to the average rotational motion of the ponderable distant masses in the surrounding relative to K', whereby we treat K' as at rest. If the Newtonian laws of mechanics do not admit such a conception, this could well be caused by the defects of this theory. . . ."

Since Einstein's theory seems to have been brought to completion in the

[1] A. Einstein, *Berl. Ber.*, 1914, p. 1030; see also *Ann. Phys. (Leipzig)*, **49**, p. 769 (1916).

publications of 1915, the question suggests itself: Is the new theory devoid of the defects of the Newtonian theory insofar as the rotation of distant masses, according to their equations, indeed produces a gravitational field equivalent to a "centrifugal field"? Perhaps one could regard a discussion of this question futile by stating that the required equivalence is guaranteed by the general covariance of the field equations. But things are not quite as simple since the boundary conditions for $g_{\mu\nu}$ at spatial infinity play a role as well. These questions of principal interest have been treated in the papers of De Sitter[1] and Einstein.[2] We will not go into these general questions, rather we will calculate and study the field of rotating distant masses using a specific and concrete example. For this purpose, the method of approximate integration of the field equations given by Einstein[3] is perfectly appropriate. This method will serve as the foundation for the following calculations. We choose as an example the field inside a uniformly rotating infinitely thin hollow sphere endowed with a constant mass density.

In the first section of this paper (which can be skipped without affecting the understanding of the rest) the $g_{\mu\nu}$ inside the spherical shell will be calculated approximately; in the second section the motion of a point mass in this field will be discussed.

A. Calculational Part: Calculation of the $g_{\mu\nu}$ for the Neighborhood of the Center of the Rotating Hollow Sphere

Notation:

a, radius of the hollow sphere;
M, its mass;
ω, its angular velocity;
x, y, z, rectangular coordinates of a point on the surface of the sphere;
x_0, y_0, z_0, coordinates of the point under consideration;
χ, gravitational constant;
ρ_0, naturally measured space density of matter.

In connection with the approximations used for calculating the field, the following should be mentioned in advance: The field in the neighborhood of the center of the sphere will be considered to be so weak that in the field equations only terms of first order with respect to $\gamma_{\mu\nu}$ are taken into account ($\gamma_{\mu\nu}$ is defined by $g_{\mu\nu} = -\delta_{\mu\nu} + \gamma_{\mu\nu}$). Since the higher-order terms are neglected, it is possible to use Einstein's method for the approximate integration of the field

[1] W. de Sitter, *Amsterdam Proc.*, 19, 527, 1917.
[2] A. Einstein, *Berl. Ber.*, 1917, p. 142.
[3] A. Einstein, *Berl. Ber.*, 1916, p. 688.

equations. According to the second type of approximation often used, the components of the velocity of the ponderable matter will be considered as small with respect to unity (velocity of light), so that in this crudest approximation, which yields Newton's theory, already the first powers can be neglected. Let us apply this approximation—which is totally independent of the first one—insofar as we drop the terms of third and higher orders in the velocities in comparison to 1. Finally, our calculations relate to the neighborhood of the center of the sphere; let r be the distance between the point under consideration and the center of the sphere and let R be the distance between the point under consideration and the integration element. Then we will expand $1/R$ in a power series in r/a up to second order.

Einstein's method of approximate integration yields for the computation of the $g_{\mu\nu}$ the following formulas:

$$g_{\mu\nu} = -\delta_{\mu\nu} + \gamma_{\mu\nu}, \qquad \delta_{\mu\nu} = 1, \quad \mu = \nu \tag{1}$$

$$= 0, \quad \mu \neq \nu$$

$$\gamma_{\mu\nu} = \gamma_{\mu\nu}' - \frac{1}{2}\sum_{\alpha} \gamma_{\alpha\alpha}' \tag{2}$$

$$\gamma_{\mu\nu}' = -\frac{\chi}{2\pi}\int \frac{T_{\mu\nu}(x, y, z, t-r)}{R}\, dV_0 \tag{3}$$

Here $T_{\mu\nu}$ denotes the covariant energy tensor of matter and dV_0 the spatial volume element of integration space,

$$R^2 = (x - x_0)^2 + (y - y_0)^2 + (z - z_0)^2$$

The coefficients $g_{\mu\nu}$ of the line element refer to the coordinates

$$x_1 = x, \quad x_2 = y, \quad x_3 = z, \quad x_4 = it$$

According to the first type of approximation, we are allowed to substitute the covariant energy tensor by the contravariant one. Neglecting the stresses, it is given by

$$T_{\mu\nu} = T^{\mu\nu} = \rho_0 \frac{dx_\mu}{ds}\frac{dx_\nu}{ds}$$

$$= \rho_0 \frac{dx_\mu}{dx_4}\frac{dx_\nu}{dx_4}\left(\frac{dx_4}{ds}\right)^2 \tag{4}$$

Let the hollow sphere rotate with angular velocity ω about the z axis, then for one of its points with polar coordinates, a, ϑ, φ, we have

$$\frac{dx_1}{dx_4} = -i\frac{dx}{dt} = ia\omega \sin\vartheta \sin\varphi$$

$$\frac{dx_2}{dx_4} = -i\frac{dy}{dt} = -ia\omega \sin\vartheta \cos\varphi \tag{5}$$

$$\frac{dx_3}{dx_4} = 0$$

These values upon substitution into (4) yields the following scheme for $T_{\mu\nu}$:

$$T_{\mu\nu} = \rho_0 \left(\frac{dx_4}{ds}\right)^2$$

$$\cdot \begin{Bmatrix} -a^2\omega^2 \sin^2\vartheta \sin^2\varphi & +a^2\omega^2 \sin^2\vartheta \sin\varphi\cos\varphi & 0 & ia\omega \sin\vartheta \sin\varphi \\ +a^2\omega^2 \sin^2\vartheta \sin\varphi\cos\varphi & -a^2\omega^2 \sin^2\vartheta \cos^2\varphi & 0 & -ia\omega \sin\vartheta\cos\varphi \\ 0 & 0 & 0 & 0 \\ ia\omega\sin\vartheta\sin\varphi & -ia\omega \sin\vartheta\cos\varphi & 0 & 1 \end{Bmatrix} \tag{6}$$

Since we take ρ_0 to be the naturally measured matter density, one has to insert the naturally measured spatial volume element for dV_0 as well in order to ensure the tensorial character of the integral (3). For this the following formula[1] holds:

$$dV_0 = \sqrt{g}\, i \frac{dx_4}{ds} dV \tag{7}$$

For the integration we introduce polar coordinates. Thus

$$\sqrt{g}\, dV = a^2 da \sin\vartheta\, d\vartheta\, d\varphi \tag{8}$$

Finally, one still has to express $1/R$ in terms of the integration variables. We choose the coordinate system such that the point under consideration is situated in the Z-X plane. Its coordinates are then

$$x_0 = r\sin\vartheta_0, \quad y_0 = 0, \quad z_0 = r\cos\vartheta_0$$

Then we have

$$\begin{aligned} R^2 &= (a\sin\vartheta\cos\vartheta - r\sin\vartheta_0)^2 + a^2\sin^2\vartheta\sin^2\varphi + (a\cos\vartheta - r\cos\vartheta_0)^2 \\ &= a^2\left[1 - \frac{2r}{a}(\sin\vartheta\cos\varphi\sin\vartheta_0 + \cos\vartheta\cos\vartheta_0) + \frac{r^2}{a^2}\right] \end{aligned}$$

Neglecting the terms mentioned at the beginning, the expansion in a binomial series yields

[1] See A. Einstein, *Berl. Ber.*, 1914, 1058, equation (47a).

$$\frac{1}{R}=\frac{1}{a}\left\{1+\frac{r}{a}(\sin\vartheta\cos\varphi\sin\vartheta_0+\cos\vartheta\cos\vartheta_0)-\frac{1}{2}\frac{r^2}{a^2}\right.$$

$$\left.+\frac{3}{2}\frac{r^2}{a^2}(\sin\vartheta\cos\varphi\sin\vartheta_0+\cos\vartheta\cos\vartheta_0)^2\right\} \tag{9}$$

We denote the expression inside the curly bracket by K and write

$$\frac{1}{R}=\frac{K}{a} \tag{9a}$$

Substitution of (6), (7), (8), and (9a) into (3) yields

$$\gamma_{11}'=\frac{i\chi}{2\pi}\rho_0 a^3\omega^2\,da\int_0^{2\pi}d\varphi\int_0^{\pi}d\vartheta\left(\frac{dx_4}{ds}\right)^3\sin^3\vartheta\sin^2\varphi K$$

$$\gamma_{22}'=\frac{i\chi}{2\pi}\rho_0 a^3\omega^2\,da\int_0^{2\pi}d\varphi\int_0^{\pi}d\vartheta\left(\frac{dx_4}{ds}\right)^3\sin^3\vartheta\cos^2\varphi K$$

$$\gamma_{44}'=-\frac{i\chi}{2\pi}\rho_0 a\,da\int_0^{2\pi}d\varphi\int_0^{\pi}d\vartheta\left(\frac{dx_4}{ds}\right)^3\sin\vartheta K$$

$$\gamma_{12}'=-\frac{i\chi}{2\pi}\rho_0 a^3\omega^2\,da\int_0^{2\pi}d\varphi\int_0^{\pi}d\vartheta\left(\frac{dx_4}{ds}\right)^3\sin^3\vartheta\sin\varphi\cos\varphi K$$

$$\gamma_{14}'=\frac{\chi}{2\pi}\rho_0 a^2\omega\,da\int_0^{2\pi}d\varphi\int_0^{\pi}d\vartheta\left(\frac{dx_4}{ds}\right)^3\sin^2\vartheta\sin\varphi K$$

$$\gamma_{24}'=-\frac{\chi}{2\pi}\rho_0 a^2\omega\,da\int_0^{2\pi}d\varphi\int_0^{\pi}d\vartheta\left(\frac{dx_4}{ds}\right)^3\sin^2\vartheta\cos\varphi K$$

$$\gamma_{13}'=\gamma_{23}'=\gamma_{33}'=\gamma_{43}'=0 \tag{10}$$

The absolute value of the quantity dx_4/ds differs from unity only by terms of order $\omega^2 a^2$; it appears as a factor in the small first-order terms $\gamma_{\mu\nu}'$; hence it is sufficient to calculate them starting with the expression for the line element in the "zeroth" approximation:

$$ds^2=-dx_1^2-dx_2^2-dx_3^2-dx_4^2$$

$$\frac{ds^2}{dx_4^2}=-1-\frac{dx_1^2+dx_2^2+dx_3^2}{dx_4^2}$$

$$=-1+\omega^2a^2\sin^2\vartheta$$

$$\frac{ds}{dx_4} = i\left(1 - \frac{\omega^2 a^2}{2}\sin^2 \vartheta\right)$$

$$\left(\frac{dx_4}{ds}\right)^3 = i\left(1 + \frac{3}{2}\omega^2 a^2 \sin^2 \vartheta\right) \qquad (11)$$

Because the accuracy of our calculation extends only to terms of order $\omega^2 a^2$, we can put $(dx_4/ds)^3 = i$ for all those $\gamma_{\mu\nu}'$ which already contain the factor ωa; merely for γ_{44}' will we use the expression (11). Furthermore, we put

$$\rho_0 da = \sigma$$

and then equations (10) become

$$\gamma_{11}' = -\frac{\chi}{2\pi}\sigma a^3 \omega^2 \int_0^{2\pi} d\varphi \int_0^{\pi} d\vartheta \sin^3 \vartheta \sin^2 \varphi K$$

$$\gamma_{22}' = -\frac{\chi}{2\pi}\sigma a^3 \omega^2 \int_0^{2\pi} d\varphi \int_0^{\pi} d\vartheta \sin^3 \vartheta \cos^2 \varphi K$$

$$\gamma_{44}' = \frac{\chi}{2\pi}\sigma a \int_0^{2\pi} d\varphi \int_0^{\pi} d\vartheta \sin \vartheta K \left(1 + \frac{3}{2}\omega^2 a^2 \sin^2 \vartheta\right)$$

$$\gamma_{12}' = \frac{\chi}{2\pi}\sigma a^3 \omega^2 \int_0^{2\pi} d\varphi \int_0^{\pi} d\vartheta \sin^3 \vartheta \sin \varphi \cos \varphi K$$

$$\gamma_{14}' = \frac{i\chi}{2\pi}\sigma a^2 \omega \int_0^{2\pi} d\varphi \int_0^{\pi} d\vartheta \sin^2 \vartheta \sin \varphi K$$

$$\gamma_{24}' = -\frac{i\chi}{2\pi}\sigma a^2 \omega \int_0^{2\pi} d\varphi \int_0^{\pi} d\vartheta \sin^2 \vartheta \cos \varphi K$$

$$\gamma_{13}' = \gamma_{23}' = \gamma_{33}' = \gamma_{34}' = 0 \qquad (12)$$

If one substitutes the value of K from (9) into these expressions and calculates the integrals, one obtains

$$\gamma_{11}' = -\frac{\chi}{2\pi}\frac{M}{3a}a^2\omega^2\left(1 - \frac{r^2}{5a^2}\right)$$

$$\gamma_{22}' = -\frac{\chi}{2\pi}\frac{M}{3a}a^2\omega^2\left\{1 - \frac{r^2}{5a^2}\left(1 - 3\sin^2 \vartheta_0\right)\right\}$$

$$\gamma_{44}' = \frac{\chi}{2\pi}\frac{M}{a}\left\{1 + a^2\omega^2\left[1 - \frac{r^2}{5a^2}\left(1 - \frac{3}{2}\sin^2\vartheta_0\right)\right]\right\}$$

$$\gamma_{24}' = -\frac{i\chi}{2\pi}\frac{M}{3a}\,\omega r\sin\vartheta_0$$

$$\gamma_{12}' = \gamma_{14}' = \gamma_{13}' = \gamma_{23}' = \gamma_{33}' = \gamma_{43}' = 0 \qquad (13)$$

From these expressions one obtains the $\gamma_{\mu\nu}$ and subsequently the $g_{\mu\nu}$ by using the equations (1) and (2). Furthermore, we substitute in place of the polar coordinates r and ϑ_0 of the point under consideration its rectangular coordinates and replace the Einstein gravitational constant χ by the conventional one: $k = \chi/8\pi$ (velocity of light = 1). Then one gets

$$g_{11} = -1 - \frac{2kM}{a}\left\{1 + a^2\omega^2 - \frac{\omega^2}{10}(2z_0^2 + x_0^2)\right\}$$

$$g_{22} = -1 - \frac{2kM}{a}\left\{1 + a^2\omega^2 - \frac{\omega^2}{10}(2z_0^2 - 3x_0^2)\right\}$$

$$g_{33} = -1$$

$$g_{44} = -1 + \frac{2kM}{a}\left\{1 + \frac{5a^2\omega^2}{3} - \frac{\omega^2}{6}(2z_0^2 - x_0^2)\right\}$$

$$g_{24} = -i\,\frac{4kM}{3a}\,\omega x_0 \qquad (14)$$

all remaining $g_{\mu\nu}$ vanish.

Now we want to free ourselves also from the special choice of the coordinate system. (We had placed the point under consideration in the Z-X plane.) For this purpose, we perform the transformation:

$$x_1' = x_1\cos\alpha + x_2\sin\alpha$$

$$x_2' = -x_1\sin\alpha + x_2\cos\alpha$$

$$x_3' = x_3$$

$$x_4' = x_4 \qquad (15)$$

Then by means of the transformation law of a covariant tensor of second rank

$$g_{\sigma\tau}' = \frac{\partial x_\mu}{\partial x_\sigma'}\frac{\partial x_\nu}{\partial x_\tau'}g_{\mu\nu}$$

the coefficient scheme becomes

$$g_{\mu\nu}=\begin{pmatrix} -1-\frac{2kM}{a}\left[1+a^2\omega^2-\frac{\omega^2}{10}(2z^2+x^2-3y^2)\right], & +\frac{2kM}{a}\frac{\omega^2}{5}xy, & 0, & +i\frac{4kM}{3a}\omega y \\ +\frac{2kM}{a}\frac{\omega^2}{5}xy, & -1-\frac{2kM}{a}\left[1+a^2\omega^2-\frac{\omega^2}{10}(2z^2-3x^2+y^2)\right], & 0, & -i\frac{4kM}{3a}\omega x \\ 0, & 0, & -1, & 0, \\ +i\frac{4kM}{3a}\omega y, & -i\frac{4kM}{3a}\omega x, & 0, & -1+\frac{2kM}{a}\left[1+\frac{5a^2\omega^2}{3}+\frac{\omega^2}{6}(2z^2-x^2-y^2)\right] \end{pmatrix} \tag{16}$$

The index 0 in the coordinates is dropped here; in the following x, y, z denote the coordinates of the point under consideration.

B. Physical Part: The Motion of a Point Mass inside the Rotating Hollow Sphere. We want to set up the equations of motion of a point mass situated near the center of our rotating hollow sphere. The field in this neighborhood is characterized by the coefficient scheme of the $g_{\mu\nu}$ [equation (16) of the first section].

As is well known, the law of motion for a point mass in Einstein's theory is given by the condition $\delta \int ds = 0$, or, if one carries through the variation,[1]

$$\frac{d^2x_\tau}{ds^2}=\Gamma^\tau_{\mu\nu}\frac{dx_\mu}{ds}\frac{dx_\nu}{ds}, \qquad \tau=1\ldots4 \tag{17}$$

We have, according to the first type of approximation, for the "field components" $\Gamma^\tau_{\mu\nu}$

$$\Gamma^\tau_{\mu\nu}=-\begin{Bmatrix}\mu\nu\\ \tau\end{Bmatrix}=\begin{bmatrix}\mu\nu\\ \tau\end{bmatrix}=\frac{1}{2}\left(\frac{\partial g_{\tau\nu}}{\partial x_\mu}+\frac{\partial g_{\mu\tau}}{\partial x_\nu}-\frac{\partial g_{\mu\nu}}{\partial x_\tau}\right) \tag{18}$$

We merely want to consider motions of the point mass which are small with respect to the velocity of light, so that we can neglect the squares and products of the components of the velocity. Then one can cancel all those terms on the right-hand side of the equations (17) in which the index 4 does not appear, and, besides, we can substitute the derivatives with respect to s by the derivatives with respect to t. Taking into account that $dx_4/dt = i$, the equations (17) become

$$\frac{d^2x_\tau}{dt^2}=2i\left(\Gamma^\tau_{14}\frac{dx_1}{dt}+\Gamma^\tau_{24}\frac{dx_2}{dt}+\Gamma^\tau_{34}\frac{dx_3}{dt}\right)-\Gamma^\tau_{44} \tag{19}$$

Hence in what follows, only those of the field components $\Gamma^\tau_{\mu\nu}$ are taken into

[1] Indices appearing twice are to be summed over from 1 to 4.

consideration which contain the index 4 at least once. These are 16 quantities, which (though *they are not tensorial components !*) can be arranged in our case according to the scheme of an antisymmetric tensor of second rank. Since the partial derivatives with respect to x_4 vanish altogether in the stationary field, the quantities $\Gamma^{\tau}_{\sigma 4}$ can be written in the following way:[2]

$$
\begin{array}{ll}
\Gamma^1_{14} = 0, & \Gamma^1_{24} = \frac{1}{2}\left(\frac{\partial g_{14}}{\partial x_2} - \frac{\partial g_{24}}{\partial x_1}\right), \\[2ex]
\Gamma^1_{34} = \frac{1}{2}\left(\frac{\partial g_{14}}{\partial x_3} - \frac{\partial g_{34}}{\partial x_1}\right), & \Gamma^1_{44} = -\frac{1}{2}\frac{\partial g_{44}}{\partial x_1} \\[2ex]
\Gamma^2_{14} = \frac{1}{2}\left(\frac{\partial g_{24}}{\partial x_1} - \frac{\partial g_{14}}{\partial x_2}\right), & \Gamma^2_{24} = 0, \\[2ex]
\Gamma^2_{34} = \frac{1}{2}\left(\frac{\partial g_{24}}{\partial x_3} - \frac{\partial g_{34}}{\partial x_2}\right), & \Gamma^2_{44} = -\frac{1}{2}\frac{\partial g_{44}}{\partial x_2} \\[2ex]
\Gamma^3_{14} = \frac{1}{2}\left(\frac{\partial g_{34}}{\partial x_1} - \frac{\partial g_{14}}{\partial x_3}\right), & \Gamma^3_{24} = \frac{1}{2}\left(\frac{\partial g_{34}}{\partial x_2} - \frac{\partial g_{24}}{\partial x_3}\right), \\[2ex]
\Gamma^3_{34} = 0, & \Gamma^3_{44} = -\frac{1}{2}\frac{\partial g_{44}}{\partial x_3} \\[2ex]
\Gamma^4_{14} = \frac{1}{2}\frac{\partial g_{44}}{\partial x_1}, & \Gamma^4_{24} = \frac{1}{2}\frac{\partial g_{44}}{\partial x_2}, \\[2ex]
\Gamma^4_{34} = \frac{1}{2}\frac{\partial g_{44}}{\partial x_3}, & \Gamma^4_{44} = 0
\end{array}
\tag{20}
$$

If one substitutes here the special values of the $g_{\mu\nu}$ from (16), one gets the following scheme:

[1] Note that this scheme corresponds to the six-vector $\mathfrak{M}$ of the electromagnetic field. The analogy between electrodynamics and (approximated) gravitational theory goes still further when one bears in mind that by the approximate integration the quantities $g_{14}, g_{24}, g_{34}, g_{44}$ are calculated in the same way in terms of the density and velocity of matter as the potentials $\mathfrak{A}_x$, $\mathfrak{A}_y$, $\mathfrak{A}_x$, Φ in terms of the electric four-current, and furthermore, that in our case the right-hand sides of equations (19) correspond altogether, except for numerical factors, to the components of the ponderomotive force $\mathfrak{E} + [\mathfrak{v}\mathfrak{H}]$!

$$\begin{matrix} 0 & i\dfrac{4kM}{3a}\omega & 0 & -\dfrac{kM}{3a}\omega^2 x \\ -i\dfrac{4kM}{3a}\omega & 0 & 0 & -\dfrac{kM}{3a}\omega^2 y \\ 0 & 0 & 0 & \dfrac{2kM}{3a}\omega^2 z \\ \dfrac{kM}{3a}\omega^2 x & \dfrac{kM}{3a}\omega^2 y & -\dfrac{2kM}{3a}\omega^2 z & 0 \end{matrix} \tag{21}$$

Then, we obtain the equations of motion for our specific problem from (19) to (21):

$$\begin{aligned} \ddot{x} &= -\frac{8kM}{3a}\omega\dot{y} + \frac{kM}{3a}\omega^2 x \\ \ddot{y} &= \frac{8kM}{3a}\omega\dot{x} + \frac{kM}{3a}\omega^2 y \\ \ddot{z} &= -\frac{2kM}{3a}\omega^2 z \end{aligned} \tag{22}$$

The right-hand sides of the equations represent the components of the force which our field exerts on the point mass with mass 1. As can be seen, the first terms of the X and Y components correspond to the Coriolis force and the second terms to the centrifugal force. At first sight, the third equation yields the surprising result that this "centrifugal force" possesses an axial component as well. Its occurrence in the field of the rotating sphere can be elucidated in the following way: From the viewpoint of the observer at rest, those surface elements of the hollow sphere situated near the equator have a larger velocity, and consequently a larger apparent (inertial and gravitating) mass, than those close to the poles. Hence the field of a rotating hollow sphere with a constant surface density corresponds to the field of a hollow sphere at rest with a surface density which increases with the polar angle ϑ. In the latter case, it is readily understandable that points lying outside the equatorial plane are dragged into it.

(Incidentally, one can also easily imagine that forces analogous to the centrifugal force occur in the interior of a hollow sphere endowed with such a nonuniform mass density. As is well known, in potential theory one can demonstrate the vanishing of the force field inside a hollow sphere with constant surface charge density as follows: The attractive force of the surface elements, which are seen from P under the angle $d\omega$ (see figure 1), is equal and opposite

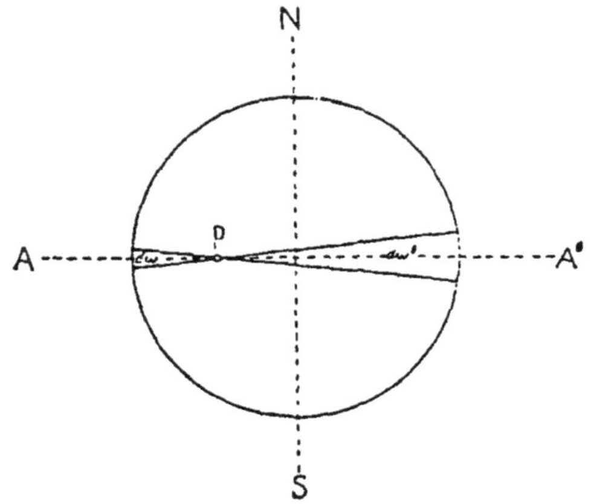

Fig. 1.

to the force exerted by those surface elements which lie within the opposite angle $d\omega'$. Of course, this no longer applies in the case of nonuniform surface density. Let AA' be the equatorial plane; then, with respect to the position of the point P as drawn in the figure, the surface elements lying inside the angle $d\omega$ are closer to the equator on the average, and are therefore specifically heavier than those of $d\omega'$. Hence there results a force in the direction of $A'A$, i.e., a tension that is perpendicular to the rotation axis, points away from it, and becomes smaller the closer the point P is shifted to the center.)

That we merely recognize a radial but never an axial component of the centrifugal force in nature could be reconciled with the results found here as follows: The approximation of the firmament by means of an infinitely thin hollow sphere is just incorrect. But even if we want to improve our approximation (say, by means of a spatial distribution of masses) we would never obtain a field equivalent to a real centrifugal field when using this method of integration. We would obtain such a field if we imagined all masses in the outer space (galaxies, etc.) as rotating and calculated their gravitational effect. But the solution for the retarded potentials [equation (3)] assumes the boundary condition $\lim \gamma_{\mu\nu} = 0$ for spatial infinity. As Einstein has shown in his cosmological paper,[1] these boundary conditions are approximately fulfilled for a coordinate system with respect to which the fixed stars are at rest on the average. Therefore, our solution (16) does not represent the field of a single rotating hollow sphere representing the Universe, rather the field inside of such a hollow sphere outside of which there are masses at still larger distances from the origin that are at rest on the average with respect to the chosen system of reference. For ex-

[1] *Berl. Ber.*, 1917, p. 142.

ample, the field as represented by equation (16) is that which would hold at the center of the Sun if instead of the Sun and all the planets a large hollow sphere existed, say, with the radius of Neptune's orbit and which rotated with an angular velocity ω with respect to the fixed stars. If there were observers at the center of this sphere, situated on a heavenly body, the gravitational field of which can be neglected and which rotates about the same axis with the angular velocity ω', the observers would perceive centrifugal and Coriolis forces consisting of the effect of their own rotation and the effect of the rotation of the hollow sphere. In the following, the influence of the field of the hollow sphere on the centrifugal force caused by its own rotation will be studied.

For this purpose, we introduce a coordinate system tied to the reference body rotating with the angular velocity ω'. This is performed by means of the transformation

$$\left.\begin{aligned} x' &= x \cos \omega' \frac{x_4}{i} + y \sin \omega' \frac{x_4}{i}, \qquad z' = z \\ y' &= -x \sin \omega' \frac{x_4}{i} + y \cos \omega' \frac{x_4}{i}, \qquad x_4' = x_4 \end{aligned}\right\} \tag{23}$$

By this transformation, the quantities of interest $g_{\mu 4}$ become

$$\begin{aligned} g_{14}' &= -iy'\left[\omega'\left(1 + \frac{2kM}{3a}\right) - \omega \frac{4kM}{3a}\right] \\ g_{24}' &= ix'\left[\omega'\left(1 + \frac{2kM}{3a}\right) - \omega \frac{4kM}{3a}\right] \\ g_{44}' &= -1 + \frac{2kM}{a}\left[1 + \frac{5a^2\omega^2}{3} - \frac{\omega^2}{3} z^2\right] + (x'^2 + y'^2) \\ &\quad \cdot \left\{\omega'^2\left(1 + \frac{2kM}{a}\right) - \omega\omega' \frac{4kM}{3a} + \omega^2 \frac{kM}{3a}\right\} \end{aligned} \tag{24}$$

If one constructs the equations of motion out of these quantities according to (19) and (20), one will get

$$\begin{aligned} \ddot{x} &= 2\left[\omega'\left(1 + \frac{2kM}{a}\right) - \omega \frac{4kM}{3a}\right]\dot{y} + \left\{\omega'^2\left(1 + \frac{2kM}{a}\right) - \omega\omega' \frac{4kM}{3a} + \omega^2 \frac{kM}{3a}\right\} x \\ \ddot{y} &= -2\left[\omega'\left(1 + \frac{2kM}{a}\right) - \omega \frac{4kM}{3a}\right]\dot{y} + \left\{\omega'^2\left(1 + \frac{2kM}{a}\right) - \omega\omega' \frac{4kM}{3a} + \omega^2 \frac{kM}{3a}\right\} y \\ \ddot{z} &= -\frac{2kM}{3a}\omega^2 z \end{aligned} \tag{25}$$

After setting $M = 0$ here, one obtains the usual centrifugal-Coriolis field:

$$\begin{aligned} \ddot{x} &= 2\omega'\dot{y} + \omega'^2 x \\ \ddot{y} &= -2\omega'\dot{x} + \omega'^2 y \\ \ddot{z} &= 0 \end{aligned} \tag{26}$$

After setting $M \neq 0$, $\omega = 0$, it becomes

$$\begin{aligned} \ddot{x} &= 2\omega'\left(1 + \frac{2kM}{a}\right)\dot{y} + \omega'^2\left(1 + \frac{2kM}{a}\right)x \\ \ddot{y} &= -2\omega'\left(1 + \frac{2kM}{a}\right)\dot{x} + \omega'^2\left(1 + \frac{2kM}{a}\right)y \\ \ddot{z} &= 0 \end{aligned} \tag{27}$$

wherefrom one can see how the inertial effects are influenced by the presence of the surrounding masses M. The centrifugal and Coriolis forces are multiplied by the factor $(1 + 2kM/a)$.

Finally, one can see from equation (25) that if the hollow sphere rotates in the same sense, this results in diminishing the centrifugal and Coriolis forces. When one sets

$$\omega' = \omega \frac{4kM}{3(2kM + a)} \tag{28}$$

then the Coriolis force vanishes. One could call the quantity $4kM/3(2kM + a)$ the dragging coefficient of the hollow sphere with respect to the Coriolis force. The centrifugal force cannot be made to vanish, since the expressions inside the curly brackets of equation (25), when set equal to zero, do not yield real roots for ω. In the reference system "at rest" ($\omega' = 0$), the expression for the centrifugal force was

$$\frac{kM}{3a}\omega^2(x^2 + y^2)^{1/2}$$

If one now allows the reference system to rotate in the same sense as the hollow sphere, the centrifugal force will at first decrease for small values of ω' and will reach a minimum when ω'/ω is equal to half the value of the "dragging coefficient.[1] Thereupon, it increases again and reaches the original value again which it had for $\omega' = 0$ as soon as ω'/ω is equal to the dragging coefficient. Then it will continue to increase along with ω', and for large ω' will reach a value hardly

[1] One can immediately convince oneself by differentiating the term inside the parenthesis.

different from the value it had in the absence of the hollow sphere [namely, $\omega'^2(x^2+y^2)^{1/2}$], since according to our assumptions $2kM/a$ is small compared to 1.

At first sight, it seems contradictory to the spirit of a theory of relativity that the right-hand sides of the equations of motion (25) do not depend on the difference $\omega - \omega'$ alone. One should not forget, however, that we do not deal with only two bodies (point mass and hollow sphere) in the problem treated here; rather, owing to the boundary conditions $\lim \gamma_{\mu\nu} = 0$, still further distant masses at rest relative to the original reference system are introduced as a third element determining the field.

Summary

Using a concrete example, it is shown that forces occur in the (Einsteinian) gravitational field of distant rotating masses which are analogous to the centrifugal and the Coriolis forces, respectively. The peculiarities associated with this special case are discussed.

Vienna, December 1917, Institute for Theoretical Physics of the University.
(Received 21 December 1917.)

Correction to My Paper: "On the Effect of Rotating Distant Masses in Einstein's Theory of Gravitation"[1]

HANS THIRRING

Herr Laue and Herr W. Pauli kindly brought to my attention the following errors in my above-mentioned paper: In the approximate integration of the Einsteinian field equations, the quantity dV in the formulas for the retarded potentials $\gamma_{\mu\nu}$ [equation (3) *l.c.*] simply denotes the usual spatial volume element of integration space (in polar coordinates: $r^2 dr \sin\vartheta\, d\vartheta\, d\varphi$) and not, as erroneously claimed by me, the naturally measured volume element

$$i\,\frac{dx_4}{ds}\,r^2 dr \sin\vartheta\, d\vartheta\, d\varphi$$

[1] This journal [*Phys. Z.*], 19, 33, 1918.

Therefore, one has to drop a factor $i(dx_4/ds)$ everywhere in the equations (10) and (12). Furthermore, in going from (12) to (13) we incorrectly set $\int \rho_0 dV = M$. Instead, the correct formula is:

$$\int \rho_0 dV_0 = i \int \rho_0 \frac{dx_4}{ds} dV = M$$

or taking (11) into account,

$$M = 4\pi\sigma a^2 \left(1 + \frac{\omega^2 a^2}{3}\right)$$

By eliminating these errors, one obtains for the coefficients $g_{\mu\nu}$ of the line element, in place of (10), the scheme presented below [equation (16)]*:

$$g_{\mu\nu} = \begin{Bmatrix} -1 - \frac{2kM}{a}\left[1 + \frac{a^2\omega^2}{3} - \frac{2\omega^2}{15}(z^2 + x^2 - 2y^2)\right], & + \frac{2kM}{a}\frac{\omega^2}{5}xy, & 0, & +i\frac{4kM}{3a}\omega y \\ + \frac{2kM}{a}\frac{\omega^2}{5}xy, & -1 - \frac{2kM}{a}\left[1 + \frac{a^2\omega^2}{3} - \frac{2\omega^2}{15}(z^2 + y^2 - 2x^2)\right], & 0, & -i\frac{4kM}{3a}\omega x \\ 0, & 0, & 0, & 0 \\ +i\frac{4kM}{3a}\omega y, & -i\frac{4kM}{3a}\omega x, & 0, & -1 + \frac{2kM}{a}\left[1 + a^2\omega^2 - \frac{2\omega^2}{15}(2z^2 - x^2 - y^2)\right] \end{Bmatrix} \tag{16}$$

The equations of motion for the point mass then read as follows:

$$\ddot{x} = -\frac{8kM}{3a}\omega\dot{y} + \frac{4kM}{15a}\omega^2 x$$

$$\ddot{y} = +\frac{8kM}{3a}\omega\dot{x} + \frac{4kM}{15a}\omega^2 y \tag{22}$$

$$\ddot{z} = -\frac{8kM}{15a}\omega^2 z$$

Hence the Coriolis force remains unchanged as compared to my original formula (22); however, for the centrifugal term one needs an additional factor of $\frac{4}{5}$.

Furthermore, the equations of motion with respect to the reference system rotating with angular velocity ω' read [see equation (25)]:

$$\ddot{x} = 2\left[\omega'\left(1 + \frac{2kM}{a}\right) - \omega\frac{4kM}{3a}\right]\dot{y} + \left\{\omega'^2\left(1 + \frac{2kM}{a}\right) - \omega\omega'\frac{8kM}{3a} + \omega^2\frac{4kM}{15a}\right\}x$$

*Translator's note: A typographical error should be corrected in equation (16): The entry for g_{33} must be "−1" instead of "0."

$$\ddot{y} = -2\left[\omega'\left(1 + \frac{2kM}{a}\right) - \omega\frac{4kM}{3a}\right]\dot{x} + \left\{\omega'^2\left(1 + \frac{2kM}{a}\right) - \omega\omega'\frac{8kM}{3a} + \omega^2\frac{4kM}{15a}\right\}y$$

$$\ddot{z} = -\frac{8kM}{3a}\omega^2 z \tag{25}$$

Also here the correction of the error simply yields an additional factor of $\frac{4}{5}$ in the terms with ω^2. Because of an additional mistake in the transformation to the rotating coordinate system, there originally occurred a factor of $\frac{4}{3}$ instead of $\frac{8}{3}$ in the term with $\omega\omega'$. Thus, two sentences in the next-to-last paragraph of my paper have still to be corrected as follows: "If one now allows the reference system to rotate in the same sense as the hollow sphere, the centrifugal force will at first decrease for small values of ω' and will reach a minimum when ω'/ω is equal to the 'dragging coefficient'. Thereupon, it increases again and reaches the original value which it had for $\omega' = 0$ as soon as ω'/ω is equal to twice the dragging coefficient."

The principal result of my paper (occurrence of centrifugal and Coriolis forces in the gravitational field of distant rotating masses) remains completely unchanged.

Vienna, 15 October 1920.

(Received 25 October 1980.)

On the Influence of the Proper Rotation of Central Bodies on the Motions of Planets and Moons According to Einstein's Theory of Gravitation

J. LENSE and H. THIRRING

In a paper published recently,[1] one of us has calculated approximately the field inside a rotating hollow sphere according to Einstein's theory of gravitation. This example seemed to be of principal interest in answering the question whether the rotation of distant masses according to Einstein's theory really produces a gravitational field equivalent to a "centrifugal field." Now it appears interesting, from another point of view, to consider the integration of the field

[1] Hans Thirring; this journal [*Phys. Z.*], **19**, 33, 1918; hereafter referred to as paper I.

equations for a rotating solid sphere, which can be simply carried out using the same methods. For, so long as one stood on the basis of Newton's theory, one could exactly substitute the field outside a sphere charged with uniform volume density (at rest or rotating) by the field of a material point of the same mass. Likewise, according to Einstein's theory, the field of an incompressible fluid sphere *at rest* is equivalent to that of a point mass,[1] but this is no longer valid for rotating spheres. As will be shown in what follows, additional terms appear here which correspond to centrifugal and Coriolis forces. Now since the planets move in the field of the Sun, which itself rotates, and the moons in turn in the field of the rotating planets, it does not seem excluded from the outset that one can obtain a novel confirmation of Einstein's theory by observing the perturbations caused by these additional terms. In the following, we calculate numerically the perturbations in the planetary orbital elements, which turn out to be below observational limits. For the moons of Jupiter, however, one obtains relatively large secular perturbations which should still lie within the errors of measurement.

§1. *The Calculation of the $g_{\mu\nu}$ for the Field of the Rotating Solid Sphere*

Notation:

l, radius of the sphere;
M, its mass;
ω, its angular velocity;
x', y', z', rectangular coordinates of a point of the integration space;
x, y, z, coordinates of the point under consideration;
k, gravitational constant;
ρ_0, naturally measured space density of matter.

The calculation is performed in analogy with that in the paper cited at the beginning: Einstein's method of approximate integration[2] is used, but this time, when forming the energy tensor of matter, the velocity of the masses producing the field is regarded as so small compared to 1 (velocity of light) that one can neglect the squares and products of the components of the velocity. (This is in contrast to the example treated in the previous paper. The terms of the centrifugal force, which are proportional to[*] w^2, are omitted and only the Coriolis terms appear.) Neglecting such terms is completely justified when one considers that $l\omega$ for the Sun and all the planets is very small in a system of units in which the velocity of light = 1. Besides, in the case treated here we consider the field at a larger distance from the surface of the sphere. Let r denote the distance be-

[1] K. Schwarzschild, *Berl. Ber.*, 1916, p. 424.
[2] A. Einstein, *Berl. Ber.*, 1916, p. 688.
[*] Translator's note: A typographical error should be corrected here. Read "ω^2" instead of "w^2."

tween the point under consideration and the center of the sphere, r' the distance between the center and the integration element, and R the distance between the point under consideration and the integration element, then we expand $1/R$ in a series in r'/r up to the second order.

Then, exactly as in paper I, we start from the approximate solution given by Einstein[1]:

$$g_{\mu\nu} = -\delta_{\mu\nu} + \gamma_{\mu\nu}, \qquad \delta_{\mu\nu} \begin{matrix} =1, & \mu = \nu \\ =0, & \mu \neq \nu \end{matrix}$$

$$\gamma_{\mu\nu} = \gamma'_{\mu\nu} - \frac{1}{2}\delta_{\mu\nu}\sum_{\alpha}\gamma'_{\alpha\alpha} \tag{1}$$

$$\gamma'_{\mu\nu} = -\frac{k}{2\pi}\int\frac{T_{\mu\nu}(x',y',z',t-R)}{R}dV_0$$

Thereafter, we form the energy tensor of stress-free matter

$$T_{\mu\nu} = T^{\mu\nu} = \rho_0\frac{dx_\mu}{ds}\frac{dx_\nu}{ds} = \sigma_0\frac{dx_\mu}{dx_4}\frac{dx_\nu}{dx_4}\left(\frac{dx_4}{ds}\right)^2 \tag{2}$$

with the expressions for the components of velocity:

$$\begin{aligned} \frac{dx_1}{dx_4} &= -i\frac{dx'}{dt} = ir'\omega\sin\vartheta'\sin\varphi' \\ \frac{dx_2}{dx_4} &= -i\frac{dy'}{dt} = -ir'\omega\sin\vartheta'\cos\varphi' \\ \frac{dx_3}{dx_4} &= 0 \end{aligned} \tag{3}$$

(r', ϑ', φ' polar coordinates of a point of the sphere; the rotation is performed about the Z axis), and by neglecting the terms with ω^2 we obtain

$$T_{\mu\nu} = \rho_0\left(\frac{dx_4}{ds}\right)^2 \cdot \left\{\begin{matrix} 0, & 0, & 0, & ir'\omega\sin\vartheta'\sin\varphi' \\ 0, & 0, & 0, & -ir'\omega\sin\vartheta'\cos\varphi' \\ 0, & 0, & 0, & 0 \\ ir'\omega\sin\vartheta'\sin\varphi', & -ir'\omega\sin\vartheta'\cos\varphi', & 0, & 1 \end{matrix}\right. \tag{4}$$

[1]In the corresponding equation (2) of paper I, the factor $\delta_{\mu\nu}$ has been inadvertently omitted.

According to equations (7) and (8) of paper I, we have to set for dV_0

$$dV_0 = i\,\frac{dx_4}{ds}\,r'^2\,dr'\,\sin\vartheta'\,d\vartheta'\,d\varphi' \tag{5}$$

In order to express $1/R$ in terms of the integration variables, we choose the coordinate system again such that the point under consideration is placed in the Z-X plane. Then, by introducing polar coordinates,

$$x = r\sin\vartheta, \qquad y = 0, \qquad z = r\cos\vartheta$$

hold and one has

$$R^2 = (r'\sin\vartheta'\cos\varphi' - r\sin\vartheta)^2 + r'^2\sin^2\vartheta'\sin^2\varphi' + (r'\cos\vartheta' - r\cos\vartheta)^2$$
$$= r^2\left[1 - \frac{2r'}{r}(\sin\vartheta'\cos\varphi'\sin\vartheta + \cos\vartheta'\cos\vartheta) + \frac{r'^2}{r^2}\right]$$

Upon expanding in binomial series up to second term,

$$\frac{1}{R} = \frac{1}{r}\left\{1 + \frac{r'}{r}(\sin\vartheta'\cos\varphi'\sin\vartheta + \cos\vartheta'\cos\vartheta) - \frac{1}{2}\frac{r'^2}{r^2}\right.$$
$$\left. + \frac{3}{2}\frac{r'^2}{r^2}(\sin\vartheta'\cos\varphi'\sin\vartheta + \cos\vartheta'\cos\vartheta)^2\right\} \tag{6}$$

We again denote the expression inside the curly bracket by K and write

$$\frac{1}{R} = \frac{K}{r} \tag{6a}$$

If we now substitute (4), (5), and (6a) into the last equation (1), then we obtain

$$\gamma'_{44} = -\frac{i\chi}{2\pi}\frac{\rho_0}{r}\int_0^l r'^2 dr'\int_0^{2\pi} d\varphi'\int_0^{\pi} d\vartheta'\left(\frac{dx_4}{ds}\right)^3 \sin\vartheta'\,K$$

$$\gamma'_{14} = \frac{\chi}{2\pi}\frac{\rho_0}{r}\,\omega\int_0^l r'^3 dr'\int_0^{2\pi} d\varphi'\int_0^{\pi} d\vartheta'\left(\frac{dx_4}{ds}\right)^3 \sin^2\vartheta'\sin\varphi'\,K$$

$$\gamma'_{24} = -\frac{\chi}{2\pi}\frac{\rho_0}{r}\,\omega\int_0^l r'^3 dr'\int_0^{2\pi} d\varphi'\int_0^{\pi} d\vartheta'\left(\frac{dx_4}{ds}\right)^3 \sin^2\vartheta'\cos\varphi'\,K$$

$$\gamma'_{11} = \gamma'_{22} = \gamma'_{33} = \gamma'_{12} = \gamma'_{13} = \gamma'_{23} = \gamma'_{34} = 0 \tag{7}$$

By neglecting the ω^2 terms and using the first type of approximation, we have

$$\left(\frac{dx_4}{ds}\right)^3 = i$$

[See equation (11) of paper I.] When one substitutes this value for $(dx_4/ds)^3$ as well as the expression for K from (6) and (6a) into (7), one obtains after evaluation of the integrals

$$\begin{aligned}
\gamma'_{44} &= \frac{\chi}{2\pi}\frac{M}{r} \\
\gamma'_{14} &= 0 \\
\gamma'_{24} &= -i\,\frac{\chi}{2\pi}\frac{M}{r}\frac{l}{5r}\,\omega l \sin\vartheta \\
\gamma'_{11} &= \gamma'_{22} = \gamma'_{33} = \gamma'_{12} = \gamma'_{13} = \gamma'_{23} = \gamma'_{34} = 0
\end{aligned} \tag{8}$$

Moreover, it follows from (1) that once rectangular coordinates are introduced and the Einsteinian gravitational constant is replaced by the Newtonian one $k = \chi/8\pi$,

$$\begin{aligned}
g_{11} &= g_{22} = g_{33} = -1 - \frac{2kM}{r} \\
g_{44} &= -1 + \frac{2kM}{r} \\
g_{24} &= -i\,\frac{4kM}{r}\frac{lx}{5r^2}\,\omega l \\
g_{12} &= g_{13} = g_{23} = g_{14} = g_{34} = 0
\end{aligned} \tag{9}$$

If now by a rotation of the system one frees oneself from the special choice of the coordinates in which the point under consideration is situated in the Z–X plane, one obtains the final scheme for the coefficients:

$$g_{\mu\nu} = \left\{ \begin{array}{cccc}
-1 - \dfrac{2kM}{r}, & 0, & 0, & i\,\dfrac{4kM}{5r}\dfrac{ly}{r^2}\,\omega l \\
0, & -1 - \dfrac{2kM}{r}, & 0, & -i\,\dfrac{4kM}{5r}\dfrac{lx}{r^2}\,\omega l \\
0, & 0, & -1 - \dfrac{2kM}{r}, & 0 \\
i\,\dfrac{4kM}{5r}\dfrac{ly}{r^2}\,\omega l, & -i\,\dfrac{4kM}{5r}\dfrac{lx}{r^2}\,\omega l, & 0, & -1 + \dfrac{2kM}{r}
\end{array} \right. \tag{10}$$

§2. *The Equations of Motion of a Point Mass in the Field of a Rotating Solid Sphere*

In the following, the equations of motion of a point mass in the field of a rotating solid sphere will be set up and we will assume that its velocity is so small that we can neglect the squares and the products of the components of velocity with respect to 1. Thereby, we emphasize from the outset that we are only interested in finding those *perturbation terms* of the planetary motion which are due to the rotation of the central bodies. In order to obtain a sufficiently exact solution of the planetary problems in accordance with Einstein's theory, one still has to add to the perturbation terms calculated here those which lead to the known perihelion precession.[1] The terms due to the proper rotation of the central bodies already emerge from the first approximation of Einstein's theory, while the perihelion perturbation just mentioned was only obtained in the second approximation. Nevertheless, it is still not permissible to take the former into account and to neglect the latter. The reason why one is not allowed to do that results from the following considerations: The ansatz for the force developed below and the Newtonian one differ by additional terms which are proportional to $\omega l v$, where v represents the velocity of the planets or the moons, respectively, while ωl represents the velocity of a point on the equator of the central body. But now the inequality

$$v > \omega l \tag{11}$$

is valid for both the Sun-planets system as well as the planets-moons systems that will be considered. If we take the terms with $\omega l v$ into account, we have all the more reason to consider those terms in the equations of motion which contain the squares and products of the components of velocity of the point mass. But if we do that, then we are not at all allowed to calculate with the first approximation alone, for those terms which are added in the second approximation to the Newtonian terms are in ratio to the former as $\alpha/r:1$ ($\alpha = 2kM$). But the square of the velocity of a planet is also of the same order as α/r; thus taking the quadratic terms in the velocities into account logically requires the consideration of terms emerging from the second approximation. It follows therefrom that the calculations performed here would not actually make sense owing to the validity of the inequality (11). Nevertheless, they can be used in practice if one keeps in mind that all perturbations considered here are so small as to allow us to superpose them linearly. One bases the computation on the equations of motion as given by Einstein in his Mercury paper and adds the perturbation terms calculated in what follows. In this way, one arrives at the desired result which is the calculation of an orbit including relativistic effects.

[1]A. Einstein, *Berl. Ber.*, 1915, p. 831.

As shown in paper I, by neglecting the above-mentioned terms and with the coordinates $x_1 = x, x_2 = y, x_3 = z, x_4 = it$ the general equations of motion

$$\frac{d^2 x_\tau}{ds^2} = \Gamma^\tau_{\mu\nu} \frac{dx_\mu}{ds}\frac{dx_\nu}{ds}$$

become

$$\frac{d^2 x_\tau}{dt^2} = 2i\left(\Gamma^\tau_{14}\frac{dx_1}{dt} + \Gamma^\tau_{24}\frac{dx_2}{dt} + \Gamma^\tau_{34}\frac{dx_3}{dt}\right) - \Gamma^\tau_{44} \tag{12}$$

The 16 quantities $\Gamma^\tau_{\sigma 4}$ that appear here are, according to the first type of approximation for a stationary field, given by

$$\Gamma^1_{14} = 0, \qquad \Gamma^1_{24} = \frac{1}{2}\left(\frac{\partial g_{14}}{\partial x_2} - \frac{\partial g_{24}}{\partial x_1}\right),$$

$$\Gamma^1_{34} = \frac{1}{2}\left(\frac{\partial g_{14}}{\partial x_3} - \frac{\partial g_{24}}{\partial x_1}\right), \qquad \Gamma^1_{44} = -\frac{1}{2}\frac{\partial g_{44}}{\partial x_1}$$

$$\Gamma^2_{14} = \frac{1}{2}\left(\frac{\partial g_{24}}{\partial x_1} - \frac{\partial g_{14}}{\partial x_2}\right), \qquad \Gamma^2_{24} = 0,$$

$$\Gamma^2_{34} = \frac{1}{2}\left(\frac{\partial g_{24}}{\partial x_3} - \frac{\partial g_{34}}{\partial x_2}\right), \qquad \Gamma^2_{44} = -\frac{1}{2}\frac{\partial g_{44}}{\partial x_2}$$

$$\Gamma^3_{14} = \frac{1}{2}\left(\frac{\partial g_{34}}{\partial x_1} - \frac{\partial g_{14}}{\partial x_3}\right), \qquad \Gamma^3_{24} = \frac{1}{2}\left(\frac{\partial g_{34}}{\partial x_2} - \frac{\partial g_{24}}{\partial x_3}\right),$$

$$\Gamma^3_{34} = 0, \qquad \Gamma^3_{44} = -\frac{1}{2}\frac{\partial g_{44}}{\partial x_3}$$

$$\Gamma^4_{14} = \frac{1}{2}\frac{\partial g_{44}}{\partial x_1}, \qquad \Gamma^4_{24} = \frac{1}{2}\frac{\partial x_{44}}{\partial x_2},$$

$$\Gamma^4_{34} = \frac{1}{2}\frac{\partial g_{44}}{\partial x_3}, \qquad \Gamma^4_{44} = 0 \tag{13}$$

For our field, as given by the equations (10), this scheme becomes*

*Translator's note: The entry for Γ^3_{14} in (14) must have "r^2" in the denominator instead of "z^2."

$$
\begin{matrix}
0, & -i\,\frac{2kM}{5r^2}\,\frac{\omega l^2}{r}\,\frac{x^2+y^2-2z^2}{r^2}, & -i\,\frac{6kM}{5r^2}\,\frac{\omega l^2}{r}\,\frac{yz}{r^2}, & \frac{kM}{r^2}\,\frac{x}{r} \\[2ex]
+i\,\frac{2kM}{5r^2}\,\frac{\omega l^2}{r}\,\frac{x^2+y^2-2z^2}{r^2}, & 0, & +i\,\frac{6kM}{5r^2}\,\frac{\omega l^2}{r}\,\frac{xz}{r^2}, & \frac{kM}{r^2}\,\frac{y}{r} \\[2ex]
+i\,\frac{6kM}{5r^2}\,\frac{\omega l^2}{r}\,\frac{yz}{z^2}, & -i\,\frac{6kM}{5r^2}\,\frac{\omega l^2}{r}\,\frac{xz}{r^2}, & 0, & \frac{kM}{r^2}\,\frac{z}{r} \\[2ex]
-\frac{kM}{r^2}\,\frac{x}{r}, & -\frac{kM}{r^2}\,\frac{y}{r}, & -\frac{kM}{r^2}\,\frac{z}{r}, & 0
\end{matrix}
\tag{14}
$$

By substituting these values for the $\Gamma^{\tau}_{\sigma 4}$ into (12), we obtain the desired equations of motion*:

$$
\begin{aligned}
\ddot{x} &= \frac{kM}{r^2}\,\frac{\omega l^2}{r}\left[\frac{4}{5}\,\frac{x^2+y^2-2z^2}{r^2}\,\cdot + \frac{12}{5}\,\frac{yz}{r^2}\,\dot{z}\right] - \frac{kM}{r^2}\,\frac{x}{r} \\
\ddot{y} &= -\frac{kM}{r^2}\,\frac{\omega l^2}{r}\left[\frac{4}{5}\,\frac{x^2+y^2-2z^2}{r^2}\,\dot{x} + \frac{12}{5}\,\frac{z}{r^2}\,\dot{z}\right] - \frac{kM}{r^2}\,\frac{y}{r} \\
\ddot{z} &= \frac{kM}{r^2}\,\frac{\omega l^2}{r}\,\frac{12}{5}\,\frac{\underline{z}}{r}\,\frac{x\dot{y}-y\ddot{x}}{r} - \frac{kM}{r^2}\,\frac{z}{r}
\end{aligned}
\tag{15}
$$

The last terms on the right-hand side represent the Newtonian force; as discussed above, one has to substitute in its place the force components according to Einstein's Mercury paper. The first terms on the right-hand side are the perturbation terms of interest caused by the proper rotation of the central bodies.

§3. *Calculation of the Perturbations due to the Proper Rotation of the Central Body*

The perturbation terms appearing in the equations (15) have to be regarded as components X, Y, Z of the perturbing force caused by the proper rotation of the central bodies. We decompose them into three other mutually perpendicular components S, T, W, where S denotes the radial one, T the transversal one, and W the orthogonal one (i.e., normal to the plane of the planetary orbit) and we introduce the following conventional astronomical notation:

*Translator's note: In the first of equations (15), a "y" is missing in "$\dot{y}$", in the second, an "x" is missing in "xz/r^2", and in the last one, "$z/\!/r$" and "$\ddot{x}$" must be replaced by "z/r" and "$\dot{x}$," respectively.

a, semimajor axis;
e, eccentricity;
$p = a(1 - e^2)$, semilatus rectum;
$i = \sphericalangle\, y\Omega\Pi$, inclination;
$\Omega = \sphericalangle\, XO\Omega$, longitude of the node;
$\tilde{\omega}$ = broken $\sphericalangle\, XO\Pi$, longitude of the pericenter;
L_0, mean longitude of the epoch, i.e., mean longitude of the planet or satellite at time $t = 0$ (also a broken angle, measured from the X axis);
$v = \sphericalangle\, \Pi OP$, true anomaly;
$u = \sphericalangle\, \Omega OP = v + \tilde{\omega} - \Omega$, argument of the latitude;
U, period of revolution of the planet or satellite in days;
$n = 2\pi/U = (kM/a^3)^{1/2}$, mean daily motion;
$C = r^2\dot{v} = na^2(1 - e^2)^{1/2}$, twice the areal velocity.

Furthermore, we abbreviate the constant appearing in the equations (15) by setting $4kM\omega l^2/5 = K$.

Now we have*

$$x = r(\cos u \cos \Omega - \sin u \sin \Omega \cos i)$$

$$y = r(\cos u \sin \Omega + \sin u \cos \Omega \cos i)$$

$$z = r \sin u \sin i$$

$$r = \frac{p}{1 + e \cos v}$$

$$x\dot{y} - y\ddot{x} = C \cos i$$

Moreover,

$$S = X(\cos u \cos \Omega - \sin u \sin \Omega \cos i) + Y(\cos u \sin \Omega + \sin u \cos \Omega \cos i) + Z \sin u \sin i$$

$$T = -X(\sin u \cos \Omega + \cos u \sin \Omega \cos i) - Y(\sin u \sin \Omega - \cos u \cos \Omega \cos i) + Z \cos u \sin i$$

$$W = X \sin \Omega \sin i - Y \cos \Omega \sin i + Z \cos i$$

If one substitutes the values of X, Y, Z obtained from equations (15) into these formulas for S, T, W by using the given relations and notations, one finds after a longer intermediary calculation

*Translator's note: In the fifth formula below, read "$\dot{x}$" instead of "$\ddot{x}$."

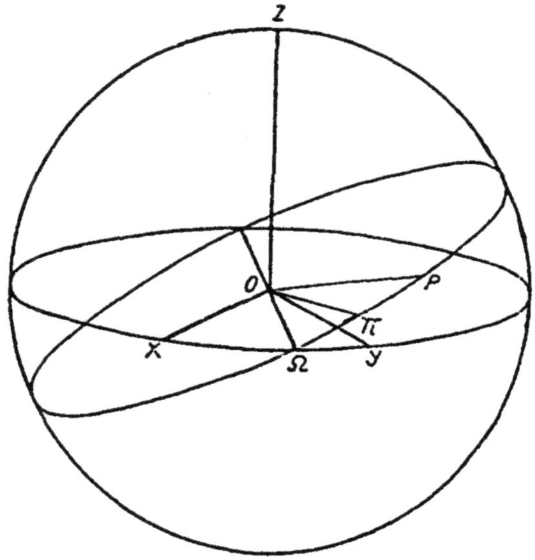

Fig. 2. Π and *P* denote the positions of the pericenter and of the planet or satellite, respectively, which are projected from the middle point *O* of the central body onto the sphere.

$$
\begin{aligned}
S &= \frac{KC\cos i}{r^4} \\
T &= -\frac{K\dot{r}\cos i}{r^3} = -\frac{KCe\cos i\sin v}{pr^3} \\
W &= \frac{K\sin i}{r^4}(2C\sin u + r\dot{r}\cos u) \\
&= \frac{KC\sin i}{r^4}\left(\frac{re\sin v\cos u}{p} + 2\sin u\right)
\end{aligned}
\tag{16}
$$

The variation of the orbital elements caused by the perturbing force is given by the equations

$$
\begin{aligned}
\frac{da}{dt} &= \frac{2}{n(1-e^2)^{1/2}}\left(Se\sin v + T\frac{p}{r}\right) \\
\frac{de}{dt} &= \frac{(1-e^2)^{1/2}}{na}\left[S\sin v + T\left(e + \frac{r+a}{a}\cos v\right)\right] \\
\frac{di}{dt} &= \frac{1}{C}Wr\cos u
\end{aligned}
$$

$$\frac{d\Omega}{dt} = \frac{1}{C \sin i} Wr \sin u$$

$$\frac{d\tilde{\omega}}{dt} = \frac{(1-e^2)^{1/2}}{nae}\left[- S\cos v + T\left(1+\frac{r}{p}\right)\sin v\right] + 2\sin^2\frac{i}{2}\,\frac{d\Omega}{dt}$$

$$\frac{dL_0}{dt} = -\frac{2}{na^2} Sr + \frac{e^2}{1+(1-e^2)^{1/2}}\frac{d\tilde{\omega}}{dt} + 2(1-e^2)^{1/2}\sin^2\frac{i}{2}\,\frac{d\Omega}{dt}$$

which, by substituting the values (16), can be represented in the following form:

$$\frac{da}{dt} = 0$$

$$\frac{de}{dt} = \frac{K\cos i}{Ca}\sin v \cdot \dot{v}$$

$$\frac{di}{dt} = \frac{K\sin i}{Cp}\cos u\ [e\sin v\cos u + 2(1+e\cos v)\sin u]\ \dot{v}$$

$$\frac{d\Omega}{dt} = \frac{K}{Cp}\sin u[e\sin v\cos u + 2(1+e\cos v)\sin u]\ \dot{v}$$

$$\frac{d\tilde{\omega}}{dt} = -\frac{K\cos i}{Ca}\left(2+\frac{1+e^2}{e}\cos v\right)\dot{v} + 2\sin^2\frac{i}{2}\,\frac{d\Omega}{dt}$$

$$\frac{dL_0}{dt} = -\frac{2K\cos i}{na^2p}(1+e\cos v)\,\dot{v} + \frac{e^2}{1+(1-e^2)^{1/2}}\frac{d\tilde{\omega}}{dt} + 2(1-e^2)^{1/2}\sin^2\frac{i}{2}\,\frac{d\Omega}{dt}$$

In the sense of perturbation theory, we regard the orbital elements on the right-hand side as constants due to the extraordinarily small factors K. We only integrate over v, i.e., we calculate the perturbations to first order in K, and observe that $u = v + \tilde{\omega} - \Omega$. Moreover, if we introduce $K_1 = K/na^3$, we obtain

$$\Delta a = 0$$

$$\Delta e = -\frac{K_1\cos i}{(1-e^2)^{1/2}}\cos v$$

$$\Delta i = -\frac{K_1\sin i}{2(1-e^2)^{3/2}}(\cos 2u + 2e\cos v\cos^2 u)$$

$$\Delta\Omega = \frac{K_1}{(1-e^2)^{3/2}}\left[v - \frac{1}{2}\sin 2u + e\left(\sin v - \frac{1}{2}\sin 2u\cos v\right)\right]$$

$$\Delta\tilde{\omega} = -\frac{K_1\cos i}{(1-e^2)^{3/2}}\left(2v + \frac{1+e^2}{e}\sin v\right) + 2\sin^2\frac{i}{2}\,\Delta\Omega$$

$$\Delta L_0 = -\frac{2K_1 \cos i}{1-e^2}(v + e \sin v) + \frac{e^2}{1+(1-e^2)^{1/2}} \Delta\tilde{\omega} + 2(1-e^2)^{1/2} \sin^2 \frac{i}{2} \Delta\Omega$$

There follows the interesting result that the perturbations in the semimajor axis vanish exactly. While only periodic terms appear in Δe and Δi, additional secular terms occur in the perturbation of the other elements; namely, because of $v = nt$ + periodic terms,

$$\Delta\Omega = \frac{K_1}{(1-e^2)^{3/2}} nt$$
$$\Delta\tilde{\omega} = \Delta L_0 = -\frac{2K_1}{(1-e^2)^{3/2}}\left(1 - 3\sin^2 \frac{i}{2}\right) nt \tag{17}$$

§4. *Numerical Results*

The numerical evaluation shows that these secular perturbations on the Sun-planets system even for the period of one century are beyond any observational possibility; for they reach a maximum of 0.01″ (for the perihelion of Mercury). Things are different in the planet-moons systems. Here somewhat larger numbers are encountered. It is better for the numerical calculation to transform the formulas (17). Let us introduce the following notation:

l, radius of the planet in centimeters;
τ, period of rotation of the planet in days;
a, semimajor axis of the satellite in centimeters;
a_1, semimajor axis of the planetary orbit in centimeters;
U, orbital period of the satellite in days;
U_1, orbital period of the planet in days;
J, number of days in a year;
c, velocity of light in cm $\sec^{-1}$.

Then the formulas resulting from (17)

$$2\Delta\Omega = -\Delta\tilde{\omega} = -\Delta L_0 = \frac{\pi^2 J}{9c^2}\frac{l^2}{\tau U^2} \tag{18}$$

yield, in units of arc seconds per century, the perturbations in the elements of the satellite orbit due to the rotation of the planet. We have set $e^2 = i^2 = 0$, which is allowed for the moons involved within the desired accuracy.

In addition, the perturbations are superposed additively in the sense of Section 2 with those discussed by Einstein in his Mercury paper, which are partly due to the direct effect of the planet and partly due to the perturbing effect of the Sun. The former are given by

$$\Delta\Omega = 0, \qquad \Delta\tilde{\omega} = \Delta L_0 = \frac{5\pi^2 J}{24c^2}\frac{a^2}{U^3(1-e^2)} \tag{19}$$

and the latter[1] by*

$$4\Delta\Omega = \Delta\tilde{\omega} = \Delta L_0 = \frac{5\pi^2 J}{12c^2}\frac{a_1{}^2}{U_1{}^3} \tag{20}$$

both in arc seconds per century. In equation (20), the eccentricity and the inclination of the planetary and satellite orbits were neglected again due to the fact that these terms are extremely small as shown in Table I. They are considerably smaller for all remaining moons.

The perturbations due to the proper rotation of the planet are contained in Table II. They remain below 0.5″ for all other satellites.

The largest terms are analogous to the Einsteinian perihelion motion of Mercury [formulas (19)], as emerges from Table III.† They are smaller than 0.5″ for the moons that are left out.

If we now add the three types of terms in order to obtain all relativistic effects, the following has to be taken into account: The correction of Newton's

Table I

	$\Delta\Omega$	$\Delta\tilde{\omega} = \Delta L_0$
Moon	+1.9″	+7.7″
Both moons of Mars	+0.7″	+2.7″

Table II

	Jupiter			Saturn				
	V	I	II	1	2	3	4	5
$\Delta\Omega$	+1′53″	+ 9″	+2″	+20″	+10″	+ 5″	+2″	+1″
$\Delta\tilde{\omega} = \Delta L_0$	−3′46″	−18″	−4″	−41″	−19″	−10″	−5″	−2″

[1] W. de Sitter, Planetary motion and the motion of the moon according to Einstein's theory, *Amsterdam Proc.*, 19, 1916. In the formulas (20), XY plane denotes the plane of planetary orbit. In de Sitter's paper, on p. 379 formula (38), the factor $\frac{1}{4}$ is missing in front of $\delta\tilde{\omega}$.

*Translator's note: The result of de Sitter cited in footnote 1 is, in fact, correct. Hence equation (20) gives values of $\Delta\tilde{\omega}$ and ΔL_0 too large by a factor of 4. Tables I and IV must be corrected accordingly.

†Translator's note: The existence of a satellite of Saturn, mentioned as "Saturn 10" or "Themis" in Tables III and IV, was announced by Pickering in 1905. This has not been confirmed by subsequent observations.

Table III[a]

		$\Delta\tilde{\omega} = \Delta L_0$		$\Delta\tilde{\omega} = \Delta L_0$
Mars	1	22″	Jupiter I	4′28
	2	2	II	1 24
Saturn	1	5′46″	III	26
	2	3 03	IV	6
	3	1 47	V	36 37
	4	59	Uranus 1	22
	5	25	2	10
	6	3	3	3
	7	2	4	1
	10	2	Neptune's moon	5

[a] $\Delta\Omega = 0$.

law treated in Einstein's Mercury paper is caused by a perturbing force acting along the radius vector. Its components, according to the cited paper, amount to

$$S = -\frac{3n^2a^3C}{2c^2}\frac{\dot{v}}{r^2}, \qquad T = W = 0$$

so they are independent of the choice of the coordinate system. Consequently, the corresponding perturbations [formulas (19) and Table III] can be related to an arbitrary XY plane. However, the variations of the elements contained in the formulas (20) and caused by the perturbing effect of the Sun (representing the deviation from the classical form) are referred to the orbital plane of the planet as already mentioned. The same holds for the numbers in Table I, whereas in Table II, which contains the perturbation terms caused by the rotation of the planets, everything refers to the equatorial plane of the central body in accordance with the choice of the coordinate system in the present paper. Therefore, the following is valid for Table IV, which summarizes all the relativistic effects: Terms (19) and (20) only appear in connection with the Moon and the two moons of Mars; hence, the plane of reference is the orbital plane of the planet. In the case of the satellites of Jupiter and Saturn, however, the reference plane is the equatorial plane of the corresponding central body since here again only terms (18) and (19) appear. The perturbations of the moons of Uranus and the moon of Neptune contain only terms (19); consequently, the plane of reference can be chosen arbitrarily.

The following can be stated about the column "Δt": The secular perturbation in the mean longitude causes a variation in the mean daily motion. That is, in the absence of relativistic effects a certain correction should be made for the time that elapses between two definite events, such as the darkenings of the moons of Jupiter. This correction for a time span of a hundred years is given in

Table IV

	$\Delta\Omega$	$\Delta\tilde{\omega} = \Delta L_0$	Δt
Moon	2"	8"	13.9s
Mars 1. Phobos	1	25	0.5
2. Deimos	1	5	0.4
Jupiter I	9	4' 10"	29.5
II	2	1 20	18.9
III	0	26	12.5
IV	0	6	7.1
V	1' 53"	32 51	1m 5.4s
Saturn 1. Mimas	20	5 05	19.2
2. Enceladus	10	2 44	15.0
3. Tethys	5	1 37	12.2
4. Dione	2	54	9.2
5. Rhea	1	23	6.9
6. Titan	0	3	3.3
7. Hyperion	0	2	2.7
10. Themis	0	2	2.9
Uranus 1. Ariel	0	22	3.7
2. Umbriel	0	10	2.7
3. Titania	0	3	1.5
4. Oberon	0	1	1.0
Neptune's moon	0	5	2.1

the last column of Table IV and can be obtained by means of the formula

$$\Delta t = \frac{1}{15} U \Delta L_0$$

Summary

The perturbation terms in the orbits of the planets and moons which, according to Einstein's theory, are caused by the proper rotation of the central body turn out to be smaller than those which emerge from the second approximation of the theory and lead to the perihelion motion of Mercury. They need not at all be taken into consideration in comparison with the latter terms in the case of the planetary orbits; however, for the orbits of the moons of Jupiter and Saturn they have to be taken into account. The secular perturbations due to all relativistic effects involved were calculated for the moons of the exterior planets. Though in some cases—for example, for the 5th moon of Jupiter—they reach a considerable magnitude, the present observations should not be precise enough as to allow a test of the theory using the perturbations of the orbits of the moons.

Vienna, February 1918. Institute for Theoretical Physics of the University.
(Received 21 February 1918.)

§(3): *Discussion of Thirring's Papers*

Hans Thirring investigated the gravitational "magnetic" field generated by a rotating hollow sphere. However, Mach's ideas on the origin of inertia provided the main impetus for his work. Mach criticized Newton's formulation of the laws of mechanics since the motion of matter was referred to an absolute space. This notion is related to an ensemble of global inertial frames each in uniform motion with respect to the others. A body in uniform motion persists in that state and offers "resistance" when accelerated with respect to absolute space. This characteristic property of matter is embodied in the notion of inertial mass. An observer can determine if matter is accelerated with respect to inertial frames since extra inertial forces (proportional to the inertial mass) appear in an accelerated frame. The success of Newtonian mechanics has to be reconciled with the fact that only the motion of matter with respect to other "matter" is physically detectable.

Newton considered the introduction of the abstract notion of absolute space necessary for the mathematical formulation of the laws of mechanics. On the contrary, Mach considered all motion to be relative. In rejecting the notion of absolute space Mach had predecessors in Leibniz and Berkeley, among others. If only relative motion has significance, the inertial frames must be determined by matter.

To give these vague ideas a more definite formulation, one may extend the principle of relativity to accelerated motion and postulate that inertial forces are due to the gravitational *field* generated by all the matter in the universe. According to Einstein's relativistic theory of gravitation (which has observational support for macroscopic phenomena), however, these notions must be rejected since they imply the *global equivalence* of inertial and certain gravitational forces in contrast to Einstein's principle of equivalence which is purely local. To illustrate this point, consider a variant of Newton's bucket experiment in which the bucket is uniformly accelerated. Other than forces of electromagnetic origin (such as viscosity), the fluid in the bucket is also subject to a *uniform* inertial force field (relative to the bucket). A contradiction arises, however, if the bucket is now treated as freely falling in the gravitational field generated by all the matter in the universe in accelerated motion, since according to Einstein's theory the only external gravitational forces that affect the motion of the fluid relative to the bucket are *tidal* forces.[1] This example also illustrates why the case of rotational acceleration is interesting: The (first-order) tidal and centrifugal forces have similar functional forms. It is clear that these arguments are based on the tensor character of the gravitational field in Einstein's theory and are completely independent of any boundary conditions on the space-time metric at spatial infinity.

[1] This argument against the relativity of accelerated motion can be formulated equally well within the framework of Newton's theory of gravitation.

Moreover, the gravitational interaction of the distant matter in the universe cannot be responsible for the inertial mass if Einstein's principle of equivalence is valid [2].[2]

Let us return to Mach's criticism of Newton's absolute space: Matter moves with respect to absolute space but cannot affect it. Mach's solution was that the configuration of distant matter in the universe determines the characteristics of this inertial space. The local metrical properties of space-time are related to the distribution of matter according to Einstein's theory of gravitation. Furthermore, this theory transforms the question of the inertial properties of our local space-time into a problem of modern cosmology. An important requirement for our local frame to be inertial is the local isotropy of space. This has been tested by several null experiments of great precision [4, 5]. The precise relationship between our local inertial frame and the rest frame determined by distant matter is a difficult problem of observational astronomy. It follows from the study of the motions of the inner planets of the solar system that the two frames do not rotate with respect to each other with an uncertainty of ~1 second of arc per century [6-8]. The discoveries of Hubble expansion and the cosmic microwave background radiation have opened up new possibilities for the discussion of this problem. Severe restrictions may be placed on the rotation of the universe from the recent studies of the isotropy of the microwave background radiation. Moreover, the dipole temperature anisotropy of this radiation may be used to determine the linear motion of our local frame with respect to the rest frame of the radiation [9].

The major contribution of the work of Thirring (and of Lense and Thirring) was the demonstration of how the local inertial frames are affected by the motion of nearby matter ("dragging of the inertial frames"). It is interesting to elucidate this issue in a coordinate invariant manner in relation to the work of Thirring. Consider first a freely falling "laboratory" frame in an external field and assume that a Fermi frame [10] is set up along the geodesic path of a representative point in the laboratory ("center of mass"). Let τ be the proper time along this path and let $\mathbf{\Omega}_D(\tau)$ be the rate of rotation of the locally non-rotating frame with respect to a certain local fiducial frame (e.g., the rest frame of the laboratory). If it is established that this latter frame does not rotate with respect to the rest frame of inertial observers at spatial infinity, $\mathbf{\Omega}_D$ corresponds to the dragging of the local inertial frames with respect to those of infinity. The motion of free particles with respect to the Fermi frame may be described to first order by

[2] In the theories of Newton and Einstein, the strength of the gravitational interaction is constant in space and time. On the basis of Mach's ideas regarding the origin of inertia, a decrease in the inertial mass may, however, be expected as a result of the Hubble expansion of the universe. This may be interpreted as a decrease in the "constant" of gravitation on a cosmological time scale. There is no firm evidence at present for such a variation [3].

$$\frac{d^2X^i}{d\tau^2} + K^i{}_j(\tau)\,X^j = 0$$

where X^i are the (Fermi) coordinates of a particle in the neighborhood of the representative point. Here $K_{ij}(\tau)$ is a symmetric traceless matrix denoting certain ("electric") components of the Riemann tensor evaluated in the Fermi frame. To express this equation of motion with respect to the fiducial frame which rotates with frequency $\mathbf{\Omega} = -\mathbf{\Omega}_D(\tau)$ with respect to the Fermi frame, let $x_i = M_{ij}(\tau)X^j$ be the relevant coordinate transformation and

$$\frac{dM_{ij}}{d\tau} = e_{ikl}\Omega_l M_{kj}$$

where e_{ijk} is the alternating symbol with $e_{123} = 1$. The equation of motion of a free particle with respect to the fiducial frame is then

$$\frac{d^2x_i}{d\tau^2} + 2e_{ijk}\Omega_j\frac{dx_k}{d\tau} + \left(\Omega_i\Omega_j - \delta_{ij}\Omega^2 + e_{ilj}\frac{d\Omega_l}{d\tau} + k_{ij}\right)x_j = 0$$

where $k_{ij} = M_{im}M_{jn}K_{mn}$.

Thirring considers a hollow sphere of mass M and radius a rotating uniformly with angular frequency ω. The shell has uniform mass density so that its angular momentum is $J = \frac{2}{3}Ma^2\omega$. The effect of rotation on the gravitational field in the interior of the sphere is considered to first order in the Newtonian constant k and to second order in ω. A test particle at rest at the center of the sphere follows a geodesic C in this field. The Fermi frame along C rotates uniformly in the same sense as the hollow sphere with a frequency

$$\Omega_D = \frac{2kJ}{a^3} = \frac{4kM}{3a}\,\omega$$

with respect to the rest frame of the inertial observers at infinity. In a local frame that rotates with frequency $\mathbf{\Omega} = -\mathbf{\Omega}_D$ with respect to the Fermi frame, the equation of motion of a free test particle near the center of the hollow sphere can be written as above with k_{ij} determined simply by

$$k_{ij}(\tau) = -\frac{1}{2}\left(\frac{\partial^2 g_{44}}{\partial x_i\partial x_j}\right)_C$$

Thirring's calculation of the metric tensor to second order in ω implies that k_{ij} is diagonal with $k_{11} = k_{22} = -(4kM/15a)\,\omega^2$ and $k_{33} = -2k_{11}$. The equation of motion comprising the dragging of the inertial frames and the gravitational tidal force of the rotating mass is the same as that given by Thirring, as expected.[3] Thirring neglected any stresses in the shell which, however, is not in general permissible due to the conservation laws of energy and momentum [11, 12]. When due account is taken of elastic stresses in the rotating shell, only terms in

[3]Hence Thirring's interpretation of the ω^2-terms as analogous to centrifugal forces cannot be maintained, even apart from the fact that gravitational forces can only be locally identified with inertial forces.

the metric tensor proportional to ω^2 turn out to be different from those given by Thirring; consequently, the proper tidal matrix is in fact just $\frac{1}{2}$ of that obtained from Thirring's results.

The weak-field result of Thirring for the dragging of the inertial frames is valid in the limit $kM \ll a$; its generalization to strong fields has been discussed by a number of authors [13-17]. These studies provide further insight into the dependence of Ω_D/ω on the nature of the matter distribution.

To summarize, Mach's ideas provided strong stimuli for the development of the relativistic theory of gravitation by subordinating the existence of inertia to the matter distribution and by suggesting the possible equivalence of inertia and gravitation. Following Einstein [18], attempts have been made to use extensions of Mach's ideas to solve the problem of boundary conditions in relativistic cosmology. The gravitational field equations are local differential equations for the metrical properties of space-time. Once the matter content has been specified, boundary conditions are in general necessary to provide a unique solution for the cosmic gravitational field. To avoid this problem, it is tempting to postulate, for instance, that the universe is spatially closed. It appears, however, that a much more extensive knowledge of the matter and radiation content throughout the universe is necessary before the problem of the cosmic field can be adequately tackled.[4]

§(4): *Discussion of the Paper of Lense and Thirring*

Josef Lense and Hans Thirring investigated the influence of the gravitational "magnetic" field of a rotating body on the motion of test masses in orbit around it. The gravitational field of a rotating body is characterized by a "vector potential," which in the first approximation corresponds to the gravitational "magnetic" dipole due to the mass current. The vector potential is given by[5,6]

$$\mathbf{G} \approx -2k \frac{\mathbf{J} \times \hat{\mathbf{r}}}{r^2}$$

for $r \gg 2kM$ and $r \gg A$, where $A = J/M$ is the angular momentum per unit mass of the body. The precession of a test gyroscope (i.e., a test gravitational "magnetic" dipole) at rest outside a matter distribution indicates the dragging of the inertial frames or, equivalently, the presence of the gravitational "magnetic"

[4] Review papers on Mach's ideas [19-21] may be consulted for more extensive discussions and further references to the literature.

[5] For a localized matter distribution, the "vector" potential is defined by $G_i = -g_{i4}/g_{44}$ in a system of Schwarzschild-like coordinates.

[6] Lense and Thirring calculated the gravitational "vector" potential for a *rotating solid sphere* in the weak-field approximation. In this procedure, deviations of the source from spherical symmetry are naturally reflected in the vector potential in a form similar to the multipole expansions of electrodynamics [22, 23].

field $-2\mathbf{\Omega}_D$,

$$\mathbf{\Omega}_D = -\frac{1}{2}\nabla \times \mathbf{G} = \frac{kJ}{r^3}[3(\hat{\mathbf{r}} \cdot \hat{\mathbf{J}})\hat{\mathbf{r}} - \hat{\mathbf{J}}]$$

The weak-field analogy with electrodynamics provides insight into the gravitational field of a rotating body.[7] This analogy is incomplete, however, in contrast to the notion of the dragging of the inertial frames which is of general validity. To illustrate this point, consider a test gyroscope with its center of mass at rest in the stationary field of a Kerr black hole [26]. Let **S** be the spin vector of the gyroscope with respect to a local orthonormal tetrad frame whose spatial axes correspond to the spherical coordinate axes of static observers at infinity. The gyroscope precesses according to the formula

$$\frac{d\mathbf{S}}{d\tau} = \mathbf{\Omega}_D \times \mathbf{S}$$

where the precession frequency is given by

$$\mathbf{\Omega}_D = f(r, \vartheta)\,\nabla \times \left(\frac{k\mathbf{J} \times \hat{\mathbf{r}}}{r^2}\right) + g(r, \vartheta)\,\frac{kM}{r} \cdot \frac{(k\mathbf{J} \times \hat{\mathbf{r}}) \times \hat{\mathbf{r}}}{r^3}$$

in Boyer–Lindquist coordinates [27]. The functions f and g are determined from

$$\Sigma^{3/2}(\Sigma - 2kMr)f = r^4 \Delta^{1/2}$$

$$kMg = (\Delta^{-1/2}\Lambda - r)f$$

where $\Sigma = r^2 + A^2\cos^2\vartheta$, $\Lambda = r^2 - A^2\cos^2\vartheta$, and $\Delta = r^2 + A^2 - 2kMr$. The radial coordinate ranges from the static limit

$$r = kM + (k^2M^2 - A^2\cos^2\vartheta)^{1/2}$$

to infinity. The weak-field results are obtained for $r \gg 2kM$,

$$f(r, \vartheta) = 1 + \frac{kM}{r} + \cdots$$

$$g(r, \vartheta) = 1 + \frac{5}{2}\left[1 - \frac{1}{5}\frac{A^2}{k^2M^2}(1 + 2\cos^2\vartheta)\right]\frac{kM}{r} + \cdots$$

while f and g both diverge as the static limit surface is approached. At this surface, the assumption that the gyroscope's center of mass is at rest as determined by the inertial observers at infinity breaks down. Despite the dramatic increase

[7]It is interesting to note that the formal analogy with electrodynamics, pointed out by Einstein [24], was explored by Thirring [25], who confirmed that in the weak-field limit the geodesic equation may be written in the form of the Lorentz force law in terms of certain gravitational "electric" and "magnetic" fields.

in the magnitude of precession frequency close to the static limit surface, the general direction of this vector is similar to its weak-field limit: At the poles, $\mathbf{\Omega}_D$ is parallel to **J** and is given by $2kr^{-3}f(r, 0)\,\mathbf{J}$, but as ϑ varies, its direction gradually deviates away from **J** until it reverses at the equator, where $\mathbf{\Omega}_D = -kr^{-3}(1 - 2kM/r)^{-1}\,\mathbf{J}$. Thus, even in the strong-field region, the gravitational influence of a rotating mass is characterized by the dragging of the inertial frames.

In the weak-field approximation a test body follows a perturbed Keplerian orbit about a central rotating mass. To characterize the motion, one may use instead of the position and velocity of the test mass at any given moment of time the six orbital parameters of the osculating ellipse which the test particle would follow if the perturbing force were turned off at that instant. In terms of these parameters, the equations of motion then reduce to the Lagrange "planetary" equations. Lense and Thirring found that the perturbing effect of the gravitational "magnetic" field produced no change in the semimajor axis (a) of the osculating ellipse, while the eccentricity (e) and the inclination (i) of the orbit were affected only in a periodic manner, so that they remained unchanged when averaged over time. The only secular effects appeared in the longitude of the ascending node, the longitude of the pericenter, and the mean longitude. They discovered that the average rate of advance of the ascending node was given by

$$\mathbf{\Omega}_{LT} = \frac{2k\mathbf{J}}{a^3(1 - e^2)^{3/2}}$$

while the longitude of the pericenter, as well as the mean longitude, varied on the average at a rate $(1 - 3\cos i)\,\Omega_{LT}$. It follows from these results that the average rate of change of the longitude of the pericenter relative to the longitude of the node is given by

$$-3\Omega_{LT}(\hat{\mathbf{J}}\cdot\hat{\mathbf{L}})\,\hat{\mathbf{L}}$$

where **L** is the orbital angular momentum and $\hat{\mathbf{J}}\cdot\hat{\mathbf{L}} = \cos i$. Thus the frequency of pericenter precession is on the average given by

$$\boldsymbol{\omega}_{LT} = \Omega_{LT}[\hat{\mathbf{J}} - 3(\hat{\mathbf{J}}\cdot\hat{\mathbf{L}})\,\hat{\mathbf{L}}]$$

which also characterizes the precession of the orbit as a whole. This result may be expressed as follows [28]: For an unperturbed orbit, the orbital angular momentum and the Lenz vectors are constants of the motion that characterize the orientation of the orbit. The rotation of the central mass produces on the average a precessional motion in these vectors with a frequency $\boldsymbol{\omega}_{LT}$, which therefore corresponds to the average precession frequency of the elliptical orbit.[8]

The pericenter precession due to the rotation of the central mass was first

[8]Note that the average frequency of precession of the orbital angular momentum is given, equally well, by $\mathbf{\Omega}_{LT}$.

considered for an equatorial orbit by de Sitter [29]. Lense and Thirring provided a general treatment valid for any orbital inclination.[9] This precession effect has been utilized in the theoretical determination of the angular momentum of the Kerr black hole [31, 32].

The observation of the Lense-Thirring precession in the orbits of the planets and moons could provide a new significant test of Einstein's theory in the solar system. This possibility was investigated in considerable detail by Lense and Thirring, who pointed out that the effects were too small to be measurable. This conclusion is equally valid today, although new possibilities have been opened up by the advent of space exploration.

To measure the Lense-Thirring precession, an orbit must have considerable eccentricity or inclination; furthermore, the observations must be so accurate as to allow the extraction of this small effect from the competing—and generally more significant—Newtonian and relativistic precessions. The multipole moments of the central masses and many-body effects account for the main Newtonian component, while the relativistic part is due to the purely spherical post-Newtonian fields of the Sun and the planets. The main effects of this post-Newtonian field are Einstein's pericenter precession and de Sitter's three-body effect due to the tidal influence of the Sun on a planet-moon orbit. Lense and Thirring discussed the relativistic perturbations in detail and provided estimates of the known significant effects, namely, Einstein's pericenter precession, the Lense-Thirring orbital precession, and de Sitter's three-body effect.[10] Moreover, they drew attention to the fact that for a given period of time (e.g., a century), the most significant relativistic orbital precession effect in the solar system is the Einstein precession for the moons of Jupiter and Saturn. The prospects for the observation of this effect have been reexamined recently [34]. A long-range observing program with improved accuracy is, in any case, necessary. The recent discovery of a new satellite of Jupiter and several new inner moons of Saturn by the Voyager spacecraft [35, 36] is encouraging, since the relativistic effects are expected to be even greater for these satellites. It is a cherished hope that, as space exploration continues, the various relativistic orbital effects can become measurable in the not-too-distant future.

The artificial satellites provide a new tool in attempts to test the predictions of Einstein's theory. Ginzburg [37] considered the possibility of measuring the perigee shift of an artificial satellite due to the rotation of the Earth, thereby providing a test of the Lense-Thirring pericenter precession. On the other hand, an interesting experiment has been suggested by Van Patten and Everitt [38, 39]

[9]In spite of a claim to the contrary [30], the Lense-Thirring results correctly give the frequency of the pericenter motion for any inclination angle i, including $i = \pi$.

[10]A summary of the various relativistic effects in the orbits of the planetary satellites has been given by Lense [33]. In terms of the general order of magnitude, the Einstein precession is the largest for these orbits, followed by the Lense-Thirring and de Sitter effects, respectively.

for the measurement of the Lense–Thirring precession of the orbital angular momentum. In this experiment, the dragging of the line of nodes is measured by placing two counter-orbiting drag-free spacecraft in polar orbits about the Earth.[11]

An alternative approach to the problem of determining the gravitational "magnetic" field of a rotating mass was considered by Schiff [41] and independently by Pugh. It consists in measuring the precession rate of a gyroscope in Earth orbit due to the coupling of the gyroscope's angular momentum to the rotation of the Earth. An experiment of this type using a superconducting gyroscope has been under active development at Stanford University by Fairbank, Everitt and their collaborators [42].

The Stanford gyroscope experiment and the Van Patten–Everitt counter-orbiting satellites experiment are the only two significant "laboratory" experiments thus far proposed to measure the dragging of the inertial frames due to a rotating mass [3].

Note Added in Proof

In a recent paper, V. B. Braginsky and A. G. Polnarev [*JETP Lett.*, **31**, 415 (1980)] have suggested a new experiment to measure the gravitational influence of a rotating mass. In this experiment, the contribution of the gravitational "magnetic" field of the Earth to the relative tidal acceleration experienced by masses at the ends of a spring in orbit around the Earth is measured. This idea has been discussed in detail by B. Mashhoon and D. S. Theiss [*Phys. Rev. Lett.*, **49**, 1542 (1982)], and the various general relativistic effects have been calculated. To distinguish the effect of the gravitational magnetic field from inertial effects, a local inertial frame along the orbit must be precisely defined. In fact, local gyroscopes must satisfy essentially the same performance criteria as in the Stanford gyroscope experiment.

Furthermore, a new gravitational effect of a rotating mass has recently been found by B. Mashhoon and D. S. Theiss. See B. Mashhoon, *Gen. Rel. Grav.*, March, 1984.

Acknowledgments

It is a pleasure to thank Professor J. Lense for helpful comments. The work of one of the authors (B.M.) was supported by the Alexander von Humboldt Foundation.

[11] In addition, a preliminary analysis [40] indicates that the Van Patten–Everitt experiment may also be used, by means of satellite-to-satellite range-rate measurements, to search for gravitational waves that are incident on the Earth and have a frequency in the range 10^{-4} Hz $\lesssim \nu \ll 10$ Hz.

References

1. Lense, J. (1981). Private communication.
2. Brans, C. H. (1977). *Phys. Rev. Lett.*, **39**, 856.
3. Shapiro, I. I. (1980). In *General Relativity and Gravitation: One Hundred Years After the Birth of Albert Einstein*, Vol. 2, Held, A., ed. (Plenum Press, New York), p. 469.
4. Hughes, V. W., Robinson, H. G., and Beltran-Lopez, V. (1960). *Phys. Rev. Lett.*, **4**, 342.
5. Drever, R. W. P. (1961). *Philos. Mag.*, **6**, 683.
6. Clemence, G. M. (1947). *Rev. Mod. Phys.*, **19**, 361.
7. Clemence, G. M. (1957). *Rev. Mod. Phys.*, **29**, 2.
8. Schiff, L. I. (1964). *Rev. Mod. Phys.*, **36**, 510.
9. Weiss, R. (1980). *Ann. Rev. Astron. Astrophys.*, **18**, 489.
10. Mashhoon, B. (1977). *Astrophys. J.*, **216**, 591.
11. Bass, L., and Pirani, F. A. E. (1955). *Philos. Mag.*, **46**, 850.
12. Hönl, H., and Maue, A. W. (1956). *Z. Phys.*, **144**, 152.
13. Brill, D. R., and Cohen, J. M. (1966). *Phys. Rev.*, **143**, 1011.
14. Cohen, J. M., and Brill, D. R. (1968). *Nuovo Cimento*, **B56**, 209.
15. De la Cruz, V., and Israel, W. (1968). *Phys. Rev.*, **170**, 1187.
16. Lindblom, L., and Brill, D. R. (1974). *Phys. Rev. D*, **10**, 3151.
17. Orwig, L. P. (1978). *Phys. Rev. D*, **18**, 1757.
18. Einstein, A. (1955). *The Meaning of Relativity* (Princeton University Press, Princeton, New Jersey).
19. Goenner, H. (1970). In *Ernst Mach: Physicist and Philosopher*, Boston Studies in the Philosphy of Science, Vol. VI, Cohen, R. S., and Seeger, R. J., eds. (Reidel, Dordrecht).
20. Reinhardt, M. (1973). *Z. Naturforsch.*, **28a**, 529.
21. Raine, D. J. (1981). *Rep. Prog. Phys.*, **44**, 1151.
22. Teyssandier, P. (1977). *Phys. Rev. D*, **16**, 946.
23. Teyssandier, P. (1978). *Phys. Rev. D*, **18**, 1037.
24. Einstein, A. (1913). *Phys. Z.*, **14**, 1261.
25. Thirring, H. (1918). *Phys. Z.*, **19**, 204.
26. Mashhoon, B. (1971). *J. Math. Phys.*, **12**, 1075.
27. Boyer, R. H., and Lindquist, R. W. (1967). *J. Math. Phys.*, **8**, 265.
28. Landau, L. D., and Lifshitz, E. M. (1971). *The Classical Theory of Fields*, 3rd ed. (Pergamon Press, Oxford).
29. De Sitter, W. (1916). *Mon. Not. Roy. Astron. Soc.*, **76**, 699.
30. Kalitzin, N. St. (1958). *Nuovo Cimento*, **9**, 365.
31. Kerr, R. P. (1963). *Phys. Rev. Letters*, **11**, 237.
32. Boyer, R. H., and Price, T. G. (1965). *Proc. Cambridge Philos. Soc.*, **61**, 531.
33. Lense, J. (1918). *Astron. Nachr.*, **206**, 117.
34. Hiscock, W. A., and Lindblom, L. (1979). *Astrophys. J.*, **231**, 224.
35. Voyager 2 Jupiter results: (1979). *Science*, **206**, 925.
36. Voyager 2 Saturn results: (1982). *Science*, **215**, 499.
37. Ginzburg, V. L. (1959). *Sci. Am.*, **200**, 149.
38. Van Patten, R. A. and Everitt, C. W. F. (1976). *Celest. Mech.*, **13**, 429.
39. Van Patten, R. A., and Everitt, C. W. F. (1976). *Phys. Rev. Letters*, **36**, 629.
40. Mashhoon, B. (1979). *Astrophys. J.*, **227**, 1019.
41. Schiff, L. I. (1960). *Phys. Rev. Lett.*, **4**, 215.
42. Everitt, C. W. F. (1973). In *Experimental Gravitation*, Bertotti, B., ed. (Academic Press, New York), p. 331.

REVIEWS OF MODERN PHYSICS VOLUME 21, NUMBER 3 JULY, 1949

An Example of a New Type of Cosmological Solutions of Einstein's Field Equations of Gravitation

KURT GÖDEL
Institute for Advanced Study, Princeton, New Jersey

1. THE MAIN PROPERTIES OF THE NEW SOLUTION

ALL cosmological solutions with non-vanishing density of matter known at present[1] have the common property that, in a certain sense, they contain an "absolute" time coordinate,[2] owing to the fact that there exists a one-parametric system of three-spaces everywhere orthogonal on the world lines of matter. It is easily seen that the non-existence of such a system of three-spaces is equivalent with a rotation of matter relative to the compass of inertia. In this paper I am proposing a solution (with a cosmological term $\neq 0$) which exhibits such a rotation. This solution, or rather the four-dimensional space S which it defines, has the further properties:

(1) S is homogeneous, i.e., for any two points P, Q of S there exists a transformation of S into itself which carries P into Q. In terms of physics this means that the solution is stationary and spatially homogeneous.

(2) There exists a one-parametric group of transformations of S into itself which carries each world line of matter into itself, so that any two world lines of matter are equidistant.

(3) S has rotational symmetry, i.e., for each point P of S there exists a one parametric group of transformations of S into itself which carries P into itself.

(4) The totality of time-like and null vectors can be divided into +- and −-vectors in such a way that: (a) If ξ is a +-vector, $-\xi$ is a −-vector, (b) a limit of +- (or −-) vectors, if $\neq 0$, is again a +- (or −-) vector. That is, a positive direction of time can consistently be introduced in the whole solution.

After a direction of time has been introduced in this way, a temporal orientation is defined for the world line of every (real or possible) particle of matter or light, i.e., it is determined for any two neighboring points on it which one is earlier. On the other hand, however, no uniform temporal ordering of *all* point events, agreeing in direction with all these individual orderings, exists. This is expressed in the next property:

(5) It is *not possible to assign a time coordinate* t *to* each space-time point in such a way that t always increases, if one moves in a positive time-like direction; and this holds both for an open and a closed time coordinate.

(6) Every world line of matter occurring in the solution is an open line of infinite length, which never approaches any of it's preceding points again; but there also exist closed time-like lines.[3] In particular, if P, Q are any two points on a world line of matter,[4] and P precedes Q on this line, there exists a time-like line connecting P and Q on which Q precedes P; i.e., it is theoretically possible in these worlds to travel into the past, or otherwise influence the past.

(7) There exist no three-spaces which are everywhere space-like and intersect each world line of matter in one point.

(8) If Σ is any system of mutually exclusive three-spaces, each of which intersects every world line of matter in one point,[5] then there exists a transformation which carries S and the positive direction of time into itself, but does *not* carry Σ into itself; i.e., an *absolute* time does not exist, even if it is not required to agree in direction with the times of all possible observers (where "absolute" means: definable without reference to individual objects, such as, e.g., a particular galactic system).

(9) Matter everywhere rotates relative to the compass of inertia with the angular velocity: $2(\pi\kappa\rho)^{\frac{1}{2}}$, where ρ is the mean density of matter and κ Newton's gravitational constant.

2. DEFINITION OF THE LINEAR ELEMENT AND PROOF THAT IT SATISFIES THE FIELD EQUATIONS

The linear element of S is defined by the following expression:[6]

$$a^2(dx_0^2 - dx_1^2 + (e^{2x_1}/2)dx_2^2 - dx_3^2 + 2e^{x_1}dx_0dx_2),$$

[1] See, for example, H. P. Robertson, Rev. Mod. Phys. 5, 62 (1933).

[2] As to the philosophical consequences which have been drawn from this circumstance see J. Jeans, "Man and the Universe," Halley Stewart Lecture (1935), and my article forthcoming in the Einstein volume of the Library of Living Philosophers.

[3] If the tangent of a line is discontinuous, the line is to be considered as time-like only if the corners can be so rounded off, that the resulting line is everywhere time-like.

[4] "World line of matter" without further specification always refers to the world lines of matter occurring as such in the solution under consideration.

[5] Another hypothesis about Σ under which the conclusion holds is that Σ is one-parametric and oriented (where the orientation refers to the space whose points are the elements of Σ).

[6] This quadratic form can also be written thus

$$a^2\left[(dx_0+e^{x_1}dx_2)^2 - dx_1^2 - \frac{e^{2x_1}}{2}dx_2^2 - dx_3^2\right],$$

which makes it evident that, as required, its signature is everywhere −2. The three-space obtained by leaving out the term $-dx_3^2$ has a simple geometric meaning (see below). Essentially the where a is a positive number. The matrices of the g_{ik} and g^{ik}, therefore, are the two:

$$a^2 \cdot \begin{Vmatrix} 1 & 0 & e^{x_1} & 0 \\ 0 & -1 & 0 & 0 \\ e^{x_1} & 0 & e^{2x_1}/2 & 0 \\ 0 & 0 & 0 & -1 \end{Vmatrix}, \quad \frac{1}{a^2} \cdot \begin{Vmatrix} -1 & 0 & 2e^{-x_1} & 0 \\ 0 & -1 & 0 & 0 \\ 2e^{-x_1} & 0 & -2e^{-2x_1} & 0 \\ 0 & 0 & 0 & -1 \end{Vmatrix}.$$

Owing to the fact that only two of the forty $\partial g_{ik}/\partial x_l$ are $\neq 0$, namely $\partial g_{22}/\partial x_1$ and $\partial g_{02}/\partial x_1$, the $\Gamma_{i,kl}$ and $\Gamma_{kl}{}^i$ can very easily be computed. One obtains the values:

$$\Gamma_{0,12} = -\Gamma_{1,02} = \Gamma_{2,01} = (a^2/2)e^{x_1},$$
$$\Gamma_{1,22} = -\Gamma_{2,12} = -(a^2/2)e^{2x_1},$$
$$\Gamma_{01}{}^0 = 1, \quad \Gamma_{12}{}^0 = \Gamma_{02}{}^1 = e^{x_1}/2,$$
$$\Gamma_{22}{}^1 = e^{2x_1}/2, \quad \Gamma_{01}{}^2 = -e^{-x_1}.$$

These $\Gamma_{i,kl}$ and $\Gamma_{kl}{}^i$, and those obtained from them by interchanging the last two (or the lower two) indices are the only ones that do not vanish.

Using for R_{ik} the formula[7]

$$R_{ik} = \frac{\partial}{\partial x_\sigma}\Gamma_{ik}{}^\sigma - \frac{1}{2}\frac{\partial^2 \log g}{\partial x_i \partial x_k} + \tfrac{1}{2}\Gamma_{ik}{}^\sigma \frac{\partial \log g}{\partial x_\sigma} - \Gamma_{\sigma i}{}^\rho \Gamma_{\rho k}{}^\sigma,$$

and taking account of the fact that $\partial/\partial x_i$, except for $i=1$, vanishes for every magnitude of the solution, and that $g = (a^8/2)e^{2x_1}$, we obtain

$$R_{ik} = \frac{\partial}{\partial x_1}\Gamma_{ik}{}^1 + \Gamma_{ik}{}^1 - \Gamma_{\rho i}{}^\sigma \Gamma_{\sigma k}{}^\rho.$$

This yields the values for the R_{ik}

$$R_{00} = 1, \quad R_{22} = e^{2x_1}, \quad R_{02} = R_{20} = e^{x_1};$$

all other R_{ik} vanish. Hence,

$$R = 1/a^2.$$

The unit vector u in the direction of the x_0-lines has the contravariant components $1/a$, 0, 0, 0 and, therefore, the covariant components a, 0, ae^{x_1}, 0.

Hence, we obtain:

$$R_{ik} = 1/a^2 \cdot u_i u_k.$$

Since, furthermore, R is a constant, the relativistic field equations (with the x_0-lines as world lines of matter), i.e., the equations[8]

$$R_{ik} - \tfrac{1}{2}g_{ik}R = 8\pi\kappa\rho u_i u_k + \lambda g_{ik}$$

are satisfied (for a given value of ρ), if we put

$$1/a^2 = 8\pi\kappa\rho, \quad \lambda = -R/2 = -1/2a^2 = -4\pi\kappa\rho.$$

The sign of the cosmological constant here is the opposite of that occurring in Einstein's static solution. It corresponds to a positive pressure.

3. PROOFS FOR THE PROPERTIES ENUMERATED

That there exists no-one parametric system of three-spaces orthogonal on the x_0-lines follows immediately from the necessary and sufficient condition which a vector field v in a four-space must satisfy, if there is to exist a system of three-spaces everywhere orthogonal on the vectors of the field. This condition requires that the skew symmetric tensor

$$a_{ikl} = v_i\left(\frac{\partial v_k}{\partial x_l} - \frac{\partial v_l}{\partial x_k}\right) + v_k\left(\frac{\partial v_l}{\partial x_i} - \frac{\partial v_i}{\partial x_l}\right) + v_l\left(\frac{\partial v_i}{\partial x_k} - \frac{\partial v_k}{\partial x_i}\right)$$

should vanish identically. The components of the corresponding vector

$$w^j = \frac{\epsilon^{jikl}}{6 \cdot \sqrt{g}} a_{ikl},$$

however, in our case (i.e., for $v_i = u_i$) have the values 0, 0, 0, $\sqrt{2}/a^2$. The non-vanishing of w^3 shows, moreover, that there exist no surfaces orthogonal on the x_0-lines in the subspaces $x_3 = \text{const.}$

If v is the unit vector representing the velocity of matter, the vector w (which evidently is always orthogonal to v) is twice the angular velocity of matter in a local inertial system in whose origin matter is at rest at the moment considered.[9] Hence, property (9) follows at once.

The properties (1) and (2) follow from the directly verifiable fact that the space S admits the following four systems of transformations into itself,

$$\begin{array}{llll} \text{(I)} & x_0 = x_0' + b & \text{(II)} & x_2 = x_2' + b \\ & x_i = x_i' \text{ for } i \neq 0 & & x_i = x_i' \text{ for } i \neq 2 \\ \text{(III)} & x_3 = x_3' + b & \text{(IV)} & x_1 = x_1' + b \\ & x_i = x_i' \text{ for } i \neq 3 & & x_2 = x_2' e^{-b} \\ & & & x_0 = x_0' \\ & & & x_3 = x_3' \end{array}$$

where b is an arbitrary real number.

A division of the time-like and null vectors into +- and −- vectors as required by (4) can be effected by defining ξ to be a +- or a −-vector according as to whether the inner product $(\xi u) = g_{ik}\xi^i u^k$ is $>$ or <0.

In order to prove (3) we introduce new coordinates r, φ, t, y (where r, φ, t are cylindrical coordinates in the subspaces $x_3 = \text{const.}$, and y, up to a constant factor, is $= x_3$) by the following formulas of transformation,

same three-space, but with the signature +3 and with more general values of the constants, has been investigated in connection with the theory of continuous groups, without any reference to relativity theory. See, for example, L. Bianchi, *Lezioni sulla teoria dei gruppi continui finiti di transformazioni* (Pisa, 1918), p. 565.

[7] Note that physicists frequently denote with $-R_{ik}$ what is here denoted with R_{ik}, with a corresponding change of sign in the field equations.

[8] The linear element is supposed to give time-like distances in seconds and space-like distances in light seconds. Therefore, the coefficient of $u_i u_k$ differs from the usual one by a factor c^2.

[9] This is an immediate consequence of the definition of a local inertial system, which requires that $g_{ik} = \pm\delta_k{}^i$ and $\partial g_{ik}/\partial x_l = 0$ for every i, k, l.

which are easily solvable with respect to the x_i,

$$e^{x_1}=ch2r+\cos\varphi sh2r$$

$$x_2e^{x_1}=\sqrt{2}\sin\varphi sh2r$$

$$tg\left(\frac{\varphi}{2}+\frac{x-2t}{2\sqrt{2}}\right)=e^{-2r}tg\frac{\varphi}{2},\quad\text{where}\quad\left|\frac{x_0-2t}{2\sqrt{2}}\right|<\frac{\pi}{2}$$

$$x_3=2y.$$

This leads[10] to the expression for the linear element,

$$4a^2(dt^2-dr^2-dy^2+(sh^4r-sh^2r)d\varphi^2+2\sqrt{2}sh^2rd\varphi dt),$$

which directly exhibits the rotational symmetry, since the g_{ik} do not depend on φ.

Property (6) now follows easily: If c is defined by $shc=1$ (i.e., $c=\log(1+\sqrt{2})$), then for any $R>c$ we have $sh^4R-sh^2R>0$; hence, the circle defined by $r=R$, $t=y=0$ is everywhere time-like (the positive direction of time, by the above definition, being that of increasing φ). Hence, the line defined by

$$r=R,\quad y=0,\quad t=-\alpha\varphi\quad(0\leq\varphi\leq 2\pi)$$

for sufficiently small α also will be everywhere time-like. However, the initial point Q of this line (i.e., the point corresponding to $\varphi=0$) and the end point P (i.e., that corresponding to $\varphi=2\pi$) are situated on the t-line: $r=R$, $y=\varphi=0$, and P precedes Q on this line if $\alpha>0$. Repeating this procedure any point preceding Q on its t-line can be reached, and because of the homogeneity of the solution the same can be done for every point.

Property (7), in view of (2) and (4), is an immediate consequence of (6). For, a three-space satisfying the two conditions stated in (7) in conjunction with time measured along the world lines of matter in their positive direction would yield a coordinate system with the property that the 0th coordinate always increases if one moves in a positive time-like direction, in contradiction to (6), which implies that all coordinates of the initial and the end point of a time-like line are equal in certain cases.

Property (5), for an open time coordinate, is an immediate consequence of the existence of closed time-like lines; for a closed time coordinate it follows from the fact that the subspaces $t=$const. would contradict property (7) (as can easily be shown owing to the simple connectivity of S).

In order to prove property (8), let U be an element of $\sum$; then U intersects the subspace S_0 of S defined by $x_3=0$ in a surface V (for it has one point in common with each x_0-line situated on S_0). Now, according to what was proved, V cannot be orthogonal on all x_0-lines in S_0. So let l be an x_0-line in S_0 on which it is not orthogonal, and P the point of intersection of V and l. Then by rotating S_0 around l (and every S_b defined by $x_0=b$ by the same angle around the x_0-line obtained from l by the translation $x_3'=x_3+b$), U goes over into a three-space different from U, but passing through P, hence not contained in $\sum$, since the elements of $\sum$ were assumed to be mutually exclusive. Hence $\sum$ goes over into a system different from $\sum$.

4. SOME ADDITIONAL THEOREMS AND CONSIDERATIONS ABOUT THE SOLUTION

I am mentioning without proof that, disregarding the connectivity in the large (which can be changed by identifying the points of certain point sets with each other), the solution given and Einstein's static universe are the only spatially homogeneous cosmological solutions with non-vanishing density of matter and equidistant world lines of matter.[11]

The space S has a simple geometric meaning. It is the direct product of a straight line and the three-space S_0, defined by $x_3=0$; and S_0 is obtained from a space R of constant positive curvature and signature $+--$ by stretching the metric[12] in the ratio $\sqrt{2}:1$ in the direction of a system of time-like Clifford parallels.[13]

This definition of S_0 also leads to an elegant representation of its group of transformations. To this end we map the points of R on the hyperbolic quaternions $u_0+u_1j_1+u_2j_2+u_3j_3$ of positive absolute value[14] by means of projective coordinates[15] $u_0u_1u_2u_3$ so chosen, that Klein's *fundamental quadric* takes *on the form*

---

[10] This computation is rather cumbersome. It is simpler to derive both forms of the linear element independently from each other from the geometrical meaning of S given below. The first form is obtained by taking for the x_1x_2-space of the coordinate system the point set corresponding to any two-parametric subgroup of the multiplicative group of the hyperbolic quaternions as defined in footnote 14.

[11] There exist stationary homogeneous solutions in which the world lines of matter are not equidistant. They lead, however, into difficulties in consequence of the inner friction which would arise in the "gas" whose molecules are the galaxies, unless the irregular motion of the galaxies is zero and stays so.

[12] By "stretching the metric in the ratio μ in the direction of the lines of a system π "I mean that a new distance PQ' of neighboring points is introduced by the equation $(PQ')^2=PR^2+(\mu\cdot RQ)^2$, where R is the foot of the perpendicular drawn from P on the line of π passing through Q; or, in other terms: $(ds')^2=ds^2+(\mu^2-1)(v_idx_i)^2$, where v is the field of the tangent vectors of unit length of the lines of π.

[13] That is, a system of pairwise equidistant straight lines which for each point of space contains exactly one line passing through it.

[14] Here the u_i are real numbers and the units j_n are defined by $j_1=i_1$, $j_2=i\cdot i_2$, $j_3=i\cdot i_3$, where the i_n are the units of the ordinary quaternions and i the imaginary unit, which is assumed to commute with all i_n. The term "hyperbolic quaternions" occurs in the literature in a different sense, but the number system just defined evidently is what should be so-called. For: norm$(u)=u\cdot\bar{u}$ $=u_0^2+u_1^2-u_2^2-u_3^2$, and moreover, the multiplicative group of these quaternions, if quaternions differing by a real factor are identified, is isomorphic with the group of transformations of the Lobatchefskian plane into itself. That the metric of R remains invariant under the transformations given in the text follows immediately from the equation norm$(uv)=$norm$(u)\cdot$norm(v).

[15] It is to be noted, however, that there exist different topological forms of spaces of constant positive curvature and signature -1, and that that form which can be represented in projective coordinates in a one-to-one manner does not lead exactly to the space S defined before, but rather to a space obtained from S by identifying any two points which are situated on the same line of the system π and whose distance on that line is equal to a certain constant. A corresponding difference subsists for the groups of transformation.

$u_0^2+u_1^2-u_2^2-u_3^2$. Then, any motion $u\rightarrow u'$ of R into itself can be represented in the form $u'=p\cdot u\cdot q$, where p and q are hyperbolic quaternions of positive norm. A system π of Clifford parallels can be represented by $\sigma^{\alpha}\cdot u$, where σ is a hyperbolic quaternion depending on π alone, and the individual lines of π are obtained by assigning a fixed value to u and varying α from $-\infty$ to $+\infty$. It follows that those motions of R into itself which leave π (and the orientation of its lines) invariant are represented by $u'=\sigma^{\beta}\cdot u\cdot q$, where β varies over all real numbers and q over all hyperbolic quaternions of positive norm. These motions, however, evidently form the four-parametric continuous group of transformations which carry S_0 into itself. The lines of π, of course, are the world lines of matter.

Evidently, in whatever ratio μ the metric of R is stretched in the direction of the lines of π, the resulting space R' has rotational symmetry. Therefore, the contracted Riemann tensor of $R'\times l$ (l being a straight line), if the coördinate system in the point considered is orthonormal, and its first basis vector $e^{(0)}$ has the direction of the π-lines, its last one $e^{(3)}$ the direction of l, has the form $\begin{pmatrix} a & & & \\ & b & 0 & \\ & 0 & b & \\ & & & 0 \end{pmatrix}$, where a and b are functions of μ. Computation shows that $\mu=\sqrt{2}$ yields $b=0$, i.e., $R_{ik}=a\cdot e_i^{(0)}e_k^{(0)}$, which makes it possible to satisfy the field equations in the manner described above.

As the the physical meaning of the solution proposed in this paper, it is clear that it yields no red shift for distant objects. For, by using the transformation (I) defined in the proof of the properties (1) and (2), one proves immediately that light signals sent from one particle of matter (occurring in the solution) to another one arrive with the same time intervals in which they are sent. For the period of rotation one obtains $2\cdot 10^{11}$ years, if for ρ the value of 10^{-30} g/cm^3 is substituted. Assuming galactical systems were formed by condensation of matter originally distributed uniformly, and taking for the ratio of contraction 1:200 (which is suggested by the observed average ratio of 1:200 between diameter and distance of galaxies), one obtains (using the law of conservation of angular momentum) for the average period of rotation of galactic systems $5\cdot 10^6$ years. This number is of the correct order of magnitude, but, in view of the fact that this would have to be approximately the period of rotation in the outer parts of the nebulae, the observed value is found to be considerably larger.[16] Of course such comparison with observation has very little significance before an expansion has been combined with the rotation. Moreover, an explanation would have to be found for the apparent irregularity of the distribution of the axes of rotation of the galaxies. This, however, is perhaps not impossible, since there exist various circumstances which would tend to blur the original order, or make it appear blurred, especially if the axes of rotation of matter in different places (unlike in the solution described above) were not parallel with each other. The radius of the smallest time-like circles, in the solution given in this paper, is of the same order of magnitude as the world radius in Einstein's static universe.

[16] From the numerical data which E. Hubble, Astrophys. J. 79, 74 (1934) gives about two galaxies of medium size follow periods of rotation of $2\cdot 10^7$ and $7\cdot 10^7$ years at a distance of about half the radius from the center. The period of rotation of the Andromeda Nebula in the central region is estimated at 1, $5\cdot 10^7$ years.

Spinning test-particles in general relativity. I

By A. Papapetrou
Physics Department, University of Manchester

(*Communicated by R. E. Peierls, F.R.S.—Received* 3 *May* 1951)

A method for the derivation of the equations of motion of test particles in a given gravitational field is developed. The equations of motion of spinning test particles are derived. The transformation properties are discussed and the equations of motion are written in a covariant form.

1. Introduction

The problem of the equations of motion in general relativity demands for its solution the determination, at the same time, of the gravitational field of the moving bodies. If one remembers how difficult it is to determine the gravitational field of a given distribution of matter—even in the simplest case of one body at rest—one sees immediately that the problem of the equations of motion will be one of extreme complexity when considered in its general form.

An important simplification occurs when one of the moving bodies is very small compared with the other bodies, so small that its influence on the motion of the other bodies can be neglected. Such a small body is usually called a *test particle*. The problem of the motion of test particles is simpler than the general one because it splits into two separate problems. First, one has to determine the basic gravitational field which will be there when the presence of the test particle is ignored. Then one has to find the equations of motion of the test particle in this basic field, which now is assumed to be already known. This second problem—which, in a restricted sense, may be called the problem of the motion of test particles—is the subject of the present paper.

Historically, what was discussed first was the problem of the motion of bodies with comparable masses, under the restrictive assumption that their velocities are small compared with that of light. The first solution of this problem was given by Einstein and co-workers (1938, 1940, 1949). The method used by those authors has two characteristic features: use is made of the field equations only, and any consideration of the interior of the bodies is avoided. A second method has been proposed by Fock (1939) and further developed by the present author (1951). Its main features are, first, that the interiors of the bodies are also considered, and secondly, that besides the field equations use is made of the 'dynamical equation'

$$\mathfrak{T}^{\mu\nu}{}_{;\nu} = 0.$$

This second method is essentially simpler than the first. It turns out that in order to determine the equations of motion to a given approximation the second method requires the calculation of the quantities $g_{\mu\nu}$ to an approximation by one step lower than the first method.

The motion of test particles has been discussed for the first time by Infeld & Schild (1949). The method used by those authors is that of Einstein and co-workers. They are considering the simplest kind of a test particle which we might call the *single-pole particle.* Their result is that the orbits of such particles are the geodesics of the basic field $g_{\mu\nu}$.

In the present paper the motion of test particles will be discussed from the point of view of the method of Fock. The results are more striking than with the slowly moving bodies of comparable masses. It is found that when the basic field is known, the motion of test particles is determined by the dynamical equation only. The single-pole particle will be treated here as an illustrative example; the main problem will be the motion of test particles whose internal structure requires for their description not only single-pole but also dipole terms, briefly the motion of *pole-dipole particles.* The method can also be used for particles with higher multipoles, so that it can be said that it solves the problem of the motion of test particles in its general form.

The very use of the word dipole in the theory of gravitation will cause distrust among many physicists. The reason is that such a dipole immediately suggests a mass density assuming positive and negative values, a case which does not occur in macroscopic physics. However, this argument has a meaning only so long as we consider the Newtonian theory of gravitation. In general relativity all ten components $\mathfrak{T}^{\mu\nu}$ will be of importance. Now a rotating particle has, necessarily, non-vanishing moments of the components $\mathfrak{T}^{i4}$ $(i = 1, 2, 3)$.* It is concluded, therefore, that in general relativity a spinning particle is a pole-dipole particle. This result will be sufficient to show that the motion of pole-dipole particles is not at all a question of a purely mathematical character. On the contrary, it constitutes an important physical problem.

It will be useful to say a word about the accuracy of the equations of motion which will be derived. If it is assumed that the mass and size of the test particle tend to zero, the equations of motion will be rigorous in all respects but one, namely, the neglect

* Throughout this paper Latin indices will take the values 1, 2, 3 only, while Greek indices will take the values 1, 2, 3 and 4.

of the gravitational radiation emitted by the test particle in its motion. The situation is analogous to that of the motion of a charged test particle in a given electromagnetic field, when the electromagnetic radiation emitted by the particle is disregarded. But in view of the extreme smallness of the gravitational radiation it may be expected that the equations of motion will be more than adequate for most physical problems.

2. Description of the method: application to the single-pole particle

In this section the method will be described in detail, and then applied to derive the equations of motion of a single-pole particle.

It is assumed that the dimensions of the test particle are very small compared with the characteristic length of the basic field (e.g. the distance from the central body in the case of the Schwarzschild field). In order to attach a rigorous meaning to the calculation it should be demanded that the dimensions of the particle tend to zero. Thus the particle will describe a narrow tube in the four-dimensional space. Inside this tube a line L is chosen which will 'represent' the motion of the particle. The co-ordinates of the points of L will be denoted by X^{α}; they may be considered as functions either of $t\,(X^4 = t)$ or of the proper time s on L. The demand that the particle be very small can be formulated in the following way. The tensor $\mathfrak{T}^{\mu\nu}$ describing the particle must be different from zero only within a sphere whose centre is the point with co-ordinates X^i and the radius a very small length R (this being true for any t). When the limiting case $R \to 0$ is assumed, the arbitrariness in the choice of L disappears.

Put
$$\delta x^{\alpha} = x^{\alpha} - X^{\alpha}, \tag{2·1}$$
and consider integrals of the form
$$\int \mathfrak{T}^{\mu\nu}\, dv, \quad \int \delta x^{\rho}\, \mathfrak{T}^{\mu\nu}\, dv, \quad \int \delta x^{\rho}\, \delta x^{\sigma}\, \mathfrak{T}^{\mu\nu}\, dv, \quad \ldots,$$
the integration being carried over the three-dimensional space for $t = \text{const}$. The structure of a test particle can now be defined quantitatively. A *single-pole* particle is one which has at least some of the integrals $\int \mathfrak{T}^{\mu\nu}\, dv \neq 0$, while all integrals with one or more factors δx^{ρ} are equal to zero. A *pole-dipole* particle has at least some of the integrals $\int \mathfrak{T}^{\mu\nu}\, dv$ and $\int \delta x^{\rho}\mathfrak{T}^{\mu\nu}\, dv \neq 0$, while all integrals with more than one factor δx^{ρ} are equal to zero. In the same way particles with higher multipoles can be defined.*

* In practical cases these results will be used as successive steps of an approximation procedure. For example, consider a small rotating planet moving around the sun. To the first approximation it will be treated as a single-pole particle. In the next step account is taken of the spin, treating the planet as a pole-dipole particle; the arbitrariness in the choice of the line L may, in this case, be avoided by picking up the centre of mass of the planet, in which case the components S^{i4} of the spin-tensor will vanish. There are also higher multipole terms which can be considered in a similar way. The justification of this procedure lies in the fact that the importance of higher multipoles is decreasing rapidly when the particle is very small; this is so because the additional terms corresponding to a multipole of order n will contain the factor $(R/r)^n$.

The equations of motion of a single-pole particle are now derived. The dynamical equation is written in the form

$$\frac{\partial \mathfrak{T}^{\alpha\beta}}{\partial x^{\beta}} + \Gamma^{\alpha}_{\mu\nu} \mathfrak{T}^{\mu\nu} = 0. \tag{2·2}$$

It follows from (2·2) that

$$\frac{\partial}{\partial x^{\gamma}} (x^{\alpha} \mathfrak{T}^{\beta\gamma}) = \mathfrak{T}^{\beta\alpha} - x^{\alpha} \Gamma^{\beta}_{\mu\nu} \mathfrak{T}^{\mu\nu}. \tag{2·3}$$

Integrating (2·2) and (2·3) over the three-dimensional space for t = const., then

$$\frac{d}{dt}\int \mathfrak{T}^{\alpha 4} dv = -\int \Gamma^{\alpha}_{\mu\nu} \mathfrak{T}^{\mu\nu} dv, \tag{2·4}$$

$$\frac{d}{dt}\int x^{\alpha} \mathfrak{T}^{\beta 4} dv = \int \mathfrak{T}^{\alpha\beta} dv - \int x^{\alpha} \Gamma^{\beta}_{\mu\nu} \mathfrak{T}^{\mu\nu} dv. \tag{2·5}$$

Inside the test particle $\Gamma^{\alpha}_{\mu\nu}$ can be developed in power series:

$$\Gamma^{\alpha}_{\mu\nu} = {}_0\Gamma^{\alpha}_{\mu\nu} + {}_0\Gamma^{\alpha}_{\mu\nu,\sigma} \delta x^{\sigma} + \ldots, \tag{2·6}$$

the subscript 0 meaning values at the point X^{α}. Introduce (2·6) in (2·4) and (2·5), and keep in mind that the particle is a single-pole. Then, omitting the subscript 0 which is no longer necessary,

$$\frac{d}{dt}\int \mathfrak{T}^{\alpha 4} dv + \Gamma^{\alpha}_{\mu\nu} \int \mathfrak{T}^{\mu\nu} dv = 0, \tag{2·4a}$$

$$\int \mathfrak{T}^{\alpha\beta} dv = \frac{dX^{\alpha}}{dt} \int \mathfrak{T}^{\beta 4} dv. \tag{2·5a}$$

Now introduce the quantities $M^{\alpha\beta}$ by the relation

$$M^{\alpha\beta} = u^4 \int \mathfrak{T}^{\alpha\beta} dv, \tag{2·7}$$

where $u^4 = \dfrac{dX^4}{ds}$, $ds^2 = g_{\mu\nu} dX^{\mu} dX^{\nu}$. Equations (2·4$a$) and (2·5$a$) take, then, the form

$$\frac{d}{ds}\left(\frac{M^{\alpha 4}}{u^4}\right) + \Gamma^{\alpha}_{\mu\nu} M^{\mu\nu} = 0, \tag{2·8}$$

$$M^{\alpha\beta} = u^{\alpha} \frac{M^{\beta 4}}{u^4}, \tag{2·9}$$

where $u^{\alpha} = \dfrac{dX^{\alpha}}{ds}$. Putting $\beta = 4$ in (2·9) then

$$M^{\alpha 4} = u^{\alpha} \frac{M^{44}}{u^4},$$

and introducing this in the right-hand side of (2·9)

$$M^{\alpha\beta} = m u^{\alpha} u^{\beta}, \tag{2·9a}$$

where $m = M^{44}/(u^4)^2$. With (2·9a) one gets from (2·8)

$$\frac{d}{ds}(m u^{\alpha}) + \Gamma^{\alpha}_{\mu\nu} m u^{\mu} u^{\nu} = 0. \tag{2·10}$$

This is the equation of motion of the single-pole particle. It contains not only the differential equation of the orbit, but also an equation for the quantity m. This last equation can be separated if (2·10) is multiplied by u_α. Taking into account the general relation

$$u_\alpha \frac{du^\alpha}{ds} + \Gamma^\alpha_{\mu\nu} u_\alpha u^\mu u^\nu = 0$$

then

$$\frac{dm}{ds} = 0. \tag{2·11}$$

With this result equation (2·10) reduces to

$$\frac{du^\alpha}{ds} + \Gamma^\alpha_{\mu\nu} u^\mu u^\nu = 0, \tag{2·12}$$

which is identical with the result of Infeld & Schild: The orbits of a single-pole particle are the geodesics of the field $g_{\mu\nu}$.

The quantity m is the rest-mass of the test particle. From (2·11) it is seen that the rest-mass of a single-pole particle remains constant throughout the motion. Further, it is possible to prove by a direct calculation that the quantity m is a *scalar* for general co-ordinate transformations. The proof is not given here because the same result will follow from the more general invariance properties which will be proved for the pole-dipole particle (§ 4).

3. The pole-dipole particle

The equations of motion of the pole-dipole particle may now be derived. Besides (2·2) and (2·3) the equation

$$\frac{\partial}{\partial x^\delta}(x^\alpha x^\beta \mathfrak{T}^{\gamma\delta}) = x^\beta \mathfrak{T}^{\gamma\alpha} + x^\alpha \mathfrak{T}^{\gamma\beta} - x^\alpha x^\beta \Gamma^\gamma_{\mu\nu} \mathfrak{T}^{\mu\nu} \tag{3·1}$$

must be used, which is again a direct consequence of (2·2). Now integrate equations (2·2), (2·3) and (3·1) over the three-dimensional space for t = const., considering (2·6) and (2·1) and remembering that the test particle is now a pole-dipole particle (i.e. one for which all integrals $\int \delta x^\rho \ldots \mathfrak{T}^{\mu\nu} dv$ with more than one factor δx^ρ vanish). Writing $\Gamma^\alpha_{\mu\nu}$ and $\Gamma^\alpha_{\mu\nu,\sigma}$ instead of ${}_0\Gamma^\alpha_{\mu\nu}$ and ${}_0\Gamma^\alpha_{\mu\nu,\sigma}$, then from (2·2)

$$\frac{d}{dt}\int \mathfrak{T}^{\alpha 4} dv + \Gamma^\alpha_{\mu\nu}\int \mathfrak{T}^{\mu\nu} dv + \Gamma^\alpha_{\mu\nu,\sigma}\int \delta x^\sigma \mathfrak{T}^{\mu\nu} dv = 0. \tag{3·2}$$

Then from (2·2), taking into account (3·2),

$$\int \mathfrak{T}^{\alpha\beta} dv = \frac{dX^\alpha}{dt}\int \mathfrak{T}^{\beta 4} dv + \frac{d}{dt}\int \delta x^\alpha \mathfrak{T}^{\beta 4} dv + \Gamma^\beta_{\mu\nu}\int \delta x^\alpha \mathfrak{T}^{\mu\nu} dv. \tag{3·3}$$

And finally from (3·1), taking into account (3·2) and (3·3),

$$\frac{dX^\alpha}{dt}\int \delta x^\beta \mathfrak{T}^{\gamma 4} dv + \frac{dX^\beta}{dt}\int \delta x^\alpha \mathfrak{T}^{\gamma 4} dv = \int \delta x^\alpha \mathfrak{T}^{\beta\gamma} dv + \int \delta x^\beta \mathfrak{T}^{\alpha\gamma} dv. \tag{3·4}$$

Equations (3·2) to (3·4) contain the equations of motion of the pole-dipole particle, though in a disguised form and mixed with other relations. The following calculations will be necessary for bringing the equations of motion into their usual form.

Introduce the quantities $M^{\lambda\mu\nu}$ by the relation

$$M^{\lambda\mu\nu} = -u^4 \int \delta x^\lambda \mathfrak{T}^{\mu\nu} dv. \tag{3·5}$$

Note that

$$M^{4\mu\nu} = 0, \tag{3·5a}$$

since all integrals refer to the hyperplane t = const. and therefore $\delta x^4 = 0$. Furthermore, introduce the components of the spin of the particle

$$S^{\alpha\beta} = \int \delta x^\alpha \mathfrak{T}^{\beta 4} dv - \int \delta x^\beta \mathfrak{T}^{\alpha 4} dv. \tag{3·6}$$

Because of (3·5) one also writes

$$u^4 S^{\alpha\beta} = -(M^{\alpha\beta 4} - M^{\beta\alpha 4}). \tag{3·6a}$$

Equation (3·4) takes, now, the form

$$u^4(M^{\alpha\beta\gamma} + M^{\beta\alpha\gamma}) = u^\alpha M^{\beta\gamma 4} + u^\beta M^{\alpha\gamma 4}. \tag{3·7}$$

This relation enables $M^{\lambda\mu\nu}$ to be expressed in terms of u^α and $S^{\alpha\beta}$. Two other relations of the form (3·7) may be written by taking the cyclic permutations of the indices α, β, γ. Adding the first and third of these relations and subtracting the second, then

$$2u^4 M^{\alpha\beta\gamma} = u^\alpha(M^{\beta\gamma 4} + M^{\gamma\beta 4}) + u^\beta(M^{\alpha\gamma 4} - M^{\gamma\alpha 4}) + u^\gamma(M^{\alpha\beta 4} - M^{\beta\alpha 4}).$$

In this relation the last four terms can be expressed in terms of $S^{\alpha\beta}$ by means of (3·6a). For the transformation of the remaining two terms one writes (3·7) with $\gamma = 4$:

$$u^4(M^{\alpha\beta 4} + M^{\beta\alpha 4}) = u^\alpha M^{\beta 44} + u^\beta M^{\alpha 44}.$$

According to (3·6a) and (3·5a)

$$u^4 S^{\alpha 4} = -M^{\alpha 44}, \tag{3·6b}$$

and

$$M^{\alpha\beta 4} + M^{\beta\alpha 4} = u^\alpha S^{4\beta} + u^\beta S^{4\alpha}.$$

Hence

$$2M^{\alpha\beta\gamma} = -(S^{\alpha\beta}u^\gamma + S^{\alpha\gamma}u^\beta) + \frac{u^\alpha}{u^4}(S^{4\beta}u^\gamma + S^{4\gamma}u^\beta). \tag{3·8}$$

Next consider equation (3·3). With the help of (2·7) and (3·5) equation (3·3) is written in the form

$$M^{\alpha\beta} = u^\alpha \frac{M^{\beta 4}}{u^4} - \frac{d}{ds}\left(\frac{M^{\alpha\beta 4}}{u^4}\right) - \Gamma^\beta_{\mu\nu} M^{\alpha\mu\nu}. \tag{3·9}$$

This relation enables one to express $M^{\alpha\beta}$ in terms of M^{44}, u^α and $S^{\alpha\beta}$. First, it is found that, if $\beta = 4$ is put in (3·9),

$$M^{\alpha 4} = u^\alpha \frac{M^{44}}{u^4} - \frac{d}{ds}\left(\frac{M^{\alpha 44}}{u^4}\right) - \Gamma^4_{\mu\nu} M^{\alpha\mu\nu}, \tag{3·9a}$$

and then, introducing (3·9a) in (3·9),

$$M^{\alpha\beta} = \frac{u^\alpha}{u^4}\left[\frac{u^\beta}{u^4} M^{44} - \frac{d}{ds}\left(\frac{M^{\beta 44}}{u^4}\right) - \Gamma^4_{\mu\nu} M^{\beta\mu\nu}\right] - \frac{d}{ds}\left(\frac{M^{\alpha\beta 4}}{u^4}\right) - \Gamma^\beta_{\mu\nu} M^{\alpha\mu\nu}. \tag{3·10}$$

Besides this result equation (3·9) contains another important relation. It is found immediately if one remembers that, according to the definition (2·7), it is $M^{\alpha\beta} = M^{\beta\alpha}$. Taking into account (3·6a), then

$$\frac{u^\alpha}{u^4}M^{\beta 4} - \frac{u^\beta}{u^4}M^{\alpha 4} + \frac{dS^{\alpha\beta}}{ds} + \Gamma^\alpha_{\mu\nu}M^{\beta\mu\nu} - \Gamma^\beta_{\mu\nu}M^{\alpha\mu\nu} = 0. \tag{3·11}$$

Or, with the help of (3·9a) and (3·6b),

$$\frac{dS^{\alpha\beta}}{ds} + \frac{u^\alpha}{u^4}\frac{dS^{\beta 4}}{ds} - \frac{u^\beta}{u^4}\frac{dS^{\alpha 4}}{ds} + \left(\Gamma^\alpha_{\mu\nu} - \frac{u^\alpha}{u^4}\Gamma^4_{\mu\nu}\right)M^{\beta\mu\nu} - \left(\Gamma^\beta_{\mu\nu} - \frac{u^\beta}{u^4}\Gamma^4_{\mu\nu}\right)M^{\alpha\mu\nu} = 0. \tag{3·12}$$

This relation contains time derivatives of the quantities $S^{\alpha\beta}$ only, i.e. it is what may be called the equation of motion of the spin.

Finally, it is necessary to transform equation (3·2). This can be written

$$\frac{d}{ds}\left(\frac{M^{\alpha 4}}{u^4}\right) + \Gamma^\alpha_{\mu\nu}M^{\mu\nu} - \Gamma^\alpha_{\mu\nu,\sigma}M^{\sigma\mu\nu} = 0, \tag{3·13}$$

with $M^{\alpha 4}$ and $M^{\mu\nu}$ defined by (3·9a) and (3·10). Equation (3·13) is the ordinary equation for the motion of the particle in the space.

The total number of the unknowns entering in these equations is ten: M^{44}, three independent components u^α and six components $S^{\alpha\beta}$. A quick glance at equations (3·12) and (3·13) would seem to show that there are exactly as many equations. This is, however, not the case, because three of equations (3·12)—those corresponding to $\alpha = 1, 2, 3$ and $\beta = 4$—are trivial identities. Hence the number of the equations of motion is *not sufficient* for the determination of all unknowns, i.e. for the complete determination of the motion.

This is nothing really new. It has been noticed already when the general pole-dipole particle has been discussed in special relativity (Mathisson 1937; Lubanski 1937; Hönl & Papapetrou 1939a, b). The results of that discussion can be obtained immediately from (3·11) and (3·13) by putting $\Gamma^\alpha_{\mu\nu} = 0$. One finds from (3·13)

$$\frac{dP^\alpha}{ds} = 0, \quad \left(P^\alpha = \frac{M^{\alpha 4}}{u^4}\right), \tag{3·13a}$$

and then from (3·11)

$$\frac{dS^{\alpha\beta}}{ds} + u^\alpha P^\beta - u^\beta P^\alpha = 0. \tag{3·11a}$$

Both equations can, in this case, be integrated directly. The integral of (3·13a) expresses the conservation of energy and momentum,

$$P^\alpha = \text{const.},$$

while equation (3·11a) leads to the conservation of angular momentum,

$$S^{\alpha\beta} + X^\alpha P^\beta - X^\beta P^\alpha \equiv J^{\alpha\beta} = \text{const.}$$

Furthermore, one finds that only when $S^{\alpha\beta} = 0$ it will follow from (3·11a) and (3·13a) that $P^\alpha \propto u^\alpha$ and therefore $u^\alpha = \text{const.}$, i.e. X^α will be linear functions of t (or s). When $S^{\alpha\beta} \neq 0$ the co-ordinates X^α may be arbitrary functions of t. More characteristically, in the frame in which $P^i = 0$ the pole-dipole particle will in general have

$u^i \neq 0$. This fact can be expressed by saying that a pole-dipole particle can have an arbitrary 'internal motion'.

The simplest possible internal motion—uniform motion on a circle—has been discussed in detail. It turns out that this internal motion gives a most successful classical analogue of the microscopic motion, the 'Zitterbewegung', of the electron (Hönl & Papapetrou 1940). One might, therefore, expect that the internal motion will be of importance in the study of the elementary particles. On the contrary, for macroscopic particles the internal motion must be avoided: The motion must be entirely determined by the equations of motion. For this purpose one has to reduce the number of independent components $S^{\alpha\beta}$ to 3. The simplest possibility consists, for example, in assuming that in a certain system of co-ordinates it will be $S^{i4} = 0$ (equivalent to the assumption that the energy density $\mathfrak{T}^{44}$ is positive-definite).

4. Transformation properties

What remains to be shown is that the equations of motion (3·12), (3·13) can be written in a covariant form. For this purpose one has to prove first that the quantities $S^{\alpha\beta}$ are the components of a tensor, and secondly that the quantity

$$m = \frac{1}{u^4}(M^{\alpha 4} + \Gamma^{\alpha}_{\mu\nu} S^{\mu 4} u^\nu)\, u_\alpha \tag{4·1}$$

is a scalar. These two theorems can be proved by making use of the transformation formulae of the quantities $M^{\lambda\mu\nu}$ and $M^{\mu\nu}$. The calculations leading to those formulae are quite simple but rather lengthy. For this reason only a description of the method which has been used will be given here, without the details of the calculations.

It is convenient to consider two types of transformations separately:

$$\text{(i)} \quad x^i = f^i(x'^\mu), \qquad x^4 = x'^4;$$

$$\text{(ii)} \quad x^i = x'^i, \qquad x^4 = f^4(x'^\mu).$$

An arbitrary transformation will be a superposition of one transformation (i) and one (ii).

The relation between the values $M^{\lambda\mu\nu}$ (in the system of co-ordinates x^μ) and $M'^{\lambda\mu\nu}$ (in the system x'^μ) can be found in the case (i) very easily, this being due to the fact that in both frames the integration extends over the same space. The hyperplane $t =$ const. is identical with $t' =$ const. The result is

$$M^{\lambda\alpha\beta} = \frac{\partial x^\lambda}{\partial x'^\rho}\frac{\partial x^\alpha}{\partial x'^\mu}\frac{\partial x^\beta}{\partial x'^\nu} M'^{\rho\mu\nu}. \tag{4·2}$$

In the case (ii) a more extensive calculation is needed, mainly because the hyperplanes $t =$ const. and $t' =$ const. are now different. The simplest transition from the first hyperplane to the second is by correlating those points of the two hyperplanes which have the same values of the co-ordinates x'^i and differ only in the values of x'^4. Furthermore, one must consider, in the first instance, infinitesimal transformations only. The result is

$$M^{\lambda\alpha\beta} = \frac{\partial x^\alpha}{\partial x'^\mu}\frac{\partial x^\beta}{\partial x'^\nu}\left(\frac{\partial x^\lambda}{\partial x'^\rho} - \frac{u^\lambda}{u^4}\frac{\partial x^4}{\partial x'^\rho}\right) M'^{\rho\mu\nu}. \tag{4·3}$$

One sees now that equation (4·2) is contained in (4·3). The additional terms in (4·3) vanish for a transformation of the type (i), since in this case $\partial x^4/\partial x'^i = 0$, while according to (3·5a) $M'^{4\mu\nu} = 0$. Furthermore, one can verify by a direct calculation that equation (4·3) has the group property: if it is valid for each of the transformations $x^\mu \to x'^\mu$ and $x'^\mu \to x''^\mu$, then it is valid also for the transformation $x^\mu \to x''^\mu$. One concludes that equation (4·3) is first valid for *any* infinitesimal transformation, and secondly, since a finite transformation can be decomposed into a series of infinitesimal ones, that equation (4·3) will be valid also for finite transformations. Thus equation (4·3) is the general transformation formula of the quantities $M^{\lambda\mu\nu}$.

In a similar way one can derive the transformation formula of the quantities $M^{\mu\nu}$. The final result, valid for infinitesimal as well as for finite transformations, is

$$M^{\alpha\beta} = \frac{\partial x^\alpha}{\partial x'^\mu}\frac{\partial x^\beta}{\partial x'^\nu}M'^{\mu\nu} - \left(\frac{\partial^2 x^\alpha}{\partial x'^\mu \partial x'^\rho}\frac{\partial x^\beta}{\partial x'^\nu} + \frac{\partial x^\alpha}{\partial x'^\mu}\frac{\partial^2 x^\beta}{\partial x'^\nu \partial x'^\rho}\right)M'^{\rho\mu\nu} + \frac{d}{ds}\left(\frac{\partial x^\alpha}{\partial x'^\mu}\frac{\partial x^\beta}{\partial x'^\nu}\frac{\partial x^4}{\partial x'^\rho}\frac{1}{u^4}M'^{\rho\mu\nu}\right). \tag{4·4}$$

Putting $\beta = 4$ in (4·3) one finds

$$M^{\lambda\alpha 4} = \frac{\partial x^\alpha}{\partial x'^\mu}\frac{\partial x^4}{\partial x'^\nu}\left(\frac{\partial x^\lambda}{\partial x'^\rho} - \frac{u^\lambda}{u^4}\frac{\partial x^4}{\partial x'^\rho}\right)M'^{\rho\mu\nu}.$$

Substituting for $M'^{\rho\mu\nu}$ the value given by (3·8), a number of terms cancel and what remains is

$$M^{\lambda\alpha 4} = -\tfrac{1}{2}u^4\frac{\partial x^\lambda}{\partial x'^\rho}\frac{\partial x^\alpha}{\partial x'^\mu}S'^{\rho\mu} + \frac{1}{2}\left(u^\lambda\frac{\partial x^\alpha}{\partial x'^\mu} + u^\alpha\frac{\partial x^\lambda}{\partial x'^\mu}\right)\frac{\partial x^4}{\partial x'^\rho}S'^{\rho\mu}.$$

On the right-hand side the first term is antisymmetric in λ, α, while the second is symmetric in the same indices. Hence

$$M^{\lambda\alpha 4} - M^{\alpha\lambda 4} = -u^4\frac{\partial x^\lambda}{\partial x'^\rho}\frac{\partial x^\alpha}{\partial x'^\mu}S'^{\rho\mu}.$$

Or, because of (3·6a),

$$S^{\lambda\alpha} = \frac{\partial x^\lambda}{\partial x'^\rho}\frac{\partial x^\alpha}{\partial x'^\mu}S'^{\rho\mu}. \tag{4·5}$$

Thus it has been proved that $S^{\lambda\alpha}$ is a tensor.

Another conclusion drawn from (4·3) is worth mentioning. A particle cannot be a pole-dipole in one frame and a single-pole in another. If some of the quantities $M^{\lambda\alpha\beta}$ do not vanish in one frame, then the same will happen in any other frame. This result is a special case of the following general theorem. The order of the highest non-vanishing multipole of a particle is invariant against co-ordinate transformations.

Writing (4·4) with $\beta = 4$ one finds

$$M^{\alpha 4} = \frac{\partial x^\alpha}{\partial x'^\mu}\frac{\partial x^4}{\partial x'^\nu}M'^{\mu\nu} - \left(\frac{\partial^2 x^\alpha}{\partial x'^\mu \partial x'^\rho}\frac{\partial x^4}{\partial x'^\nu} + \frac{\partial x^\alpha}{\partial x'^\mu}\frac{\partial^2 x^4}{\partial x'^\nu \partial x'^\rho}\right)M'^{\rho\mu\nu} + \frac{d}{ds}\left(\frac{\partial x^\alpha}{\partial x'^\mu}\frac{\partial x^4}{\partial x'^\nu}\frac{\partial x^4}{\partial x'^\rho}\frac{1}{u^4}M'^{\rho\mu\nu}\right). \tag{4·6}$$

In the right-hand side one can substitute $M'^{\mu\nu}$ according to (3·9). Then, considering (3·8) and the transformation formula for $\Gamma^\mu_{\rho\sigma}$,

$$\Gamma'^\mu_{\rho\sigma} = \frac{\partial x'^\mu}{\partial x^\lambda}\frac{\partial x^\epsilon}{\partial x'^\rho}\frac{\partial x^\zeta}{\partial x'^\sigma}\Gamma^\lambda_{\epsilon\zeta} + \frac{\partial x'^\mu}{\partial x^\lambda}\frac{\partial^2 x^\lambda}{\partial x'^\rho \partial x'^\sigma},$$

one finds

$$-\frac{\partial x^\alpha}{\partial x'^\mu}\frac{\partial x^4}{\partial x'^\nu}\Gamma'^\mu_{\rho\sigma}M'^{\nu\sigma\rho} = \Gamma^\alpha_{\epsilon\zeta}u^\epsilon S^{4\zeta} + \frac{\partial^2 x^\alpha}{\partial x'^\rho\partial x'^\sigma}\frac{\partial x^4}{\partial x'^\nu}S'^{\nu\rho}u'^\sigma - \frac{\partial x^\alpha}{\partial x'^\mu}\Gamma'^\mu_{\rho\sigma}\frac{u^4}{u'^4}u'^\sigma S'^{4\rho}.$$

Similarly, because of (3·8),

$$\frac{\partial x^4}{\partial x'^\nu}\frac{\partial x^4}{\partial x'^\sigma}\frac{M'^{\sigma\mu\nu}}{u^4} = \frac{1}{2}\frac{\partial x^4}{\partial x'^\nu}\left(-S'^{\nu\mu}+\frac{u'^\nu}{u'^4}S'^{4\mu}+\frac{u'^\mu}{u'^4}S'^{4\nu}\right) = \frac{\partial x^4}{\partial x'^\nu}\frac{M'^{\nu\mu 4}}{u'^4}.$$

Introducing these results in (4·6) one finally finds

$$\frac{1}{u^4}(M^{\alpha 4}+\Gamma^\alpha_{\rho\sigma}u^\rho S^{\sigma 4}) = \frac{\partial x^\alpha}{\partial x'^\mu}\frac{1}{u'^4}(M'^{\mu 4}+\Gamma'^\mu_{\rho\sigma}u'^\rho S'^{\sigma 4}); \tag{4·7}$$

i.e. the quantity on the left-hand side is a vector.

From the last result it follows immediately that the quantity m defined by (4·1) is a scalar.

5. The covariant equations of motion

It is now possible to write the equations of motion in a covariant form. Introducing the quantities

$$\frac{DS^{\alpha\beta}}{Ds} = \frac{dS^{\alpha\beta}}{ds} + \Gamma^\alpha_{\mu\nu}S^{\mu\beta}u^\nu + \Gamma^\beta_{\mu\nu}S^{\alpha\mu}u^\nu, \tag{5·1}$$

which are the components of a tensor,* equation (3·12) is written in the form

$$\frac{DS^{\alpha\beta}}{Ds} + \frac{u^\alpha}{u^4}\frac{DS^{\beta 4}}{Ds} - \frac{u^\beta}{u^4}\frac{DS^{\alpha 4}}{Ds} = 0. \tag{5·2}$$

Multiplying (5·2) by u_β, then

$$\frac{1}{u^4}\frac{DS^{\alpha 4}}{Ds} = u_\beta\frac{DS^{\alpha\beta}}{Ds} + \frac{u^\alpha}{u^4}u_\beta\frac{DS^{\beta 4}}{Ds}.$$

Using this result one writes (5·2) in the final form

$$\frac{DS^{\alpha\beta}}{Ds} + u^\alpha u_\rho\frac{DS^{\beta\rho}}{Ds} - u^\beta u_\rho\frac{DS^{\alpha\rho}}{Ds} = 0. \tag{5·3}$$

This is the covariant formulation of the equation of motion of the spin.

To transform equation (3·13) one rewrites (3·9*a*) with the help of (3·6*b*) and (5·2):

$$M^{\alpha 4}+\Gamma^\alpha_{\mu\nu}S^{\mu 4}u^\nu = \frac{u^\alpha}{u^4}(M^{44}+\Gamma^4_{\mu\nu}S^{\mu 4}u^\nu) + \frac{DS^{\alpha 4}}{Ds}.$$

* Generally, when a tensor $A::$ is given on a line L, the quantity $\frac{DA::}{Ds}$ will be defined as follows. $A::$ is extended (arbitrarily) to a tensorfield in a small tube surrounding L. Then, denoting by $(A::)_{;\nu}$ the covariant derivative of this field, one writes

$$\frac{DA::}{Ds} = (A::)_{;\nu}u^\nu.$$

This formula shows directly that $\frac{DA::}{Ds}$ will be a tensor.

Multiplying by u_α/u^4 one gets, according to (4·1),

$$m = \frac{1}{(u^4)^2}(M^{44} + \Gamma^4_{\mu\nu}S^{\mu4}u^\nu) + \frac{1}{u^4}u_\rho\frac{DS^{\rho4}}{Ds}.$$

Hence

$$\frac{1}{u^4}M^{\alpha4} = mu^\alpha - \Gamma^\alpha_{\mu\nu}S^{\mu4}\frac{u^\nu}{u^4} + u_\beta\frac{DS^{\alpha\beta}}{Ds}.$$

Introducing this in (3·13), it is found, after some elementary calculations,

$$\frac{d}{ds}\left(mu^\alpha + u_\beta\frac{DS^{\alpha\beta}}{Ds}\right) + \Gamma^\alpha_{\mu\nu}u^\nu\left(mu^\mu + u_\beta\frac{DS^{\mu\beta}}{Ds}\right) + S^{\mu\nu}u^\sigma(\Gamma^\alpha_{\nu\sigma,\mu} + \Gamma^\alpha_{\mu\rho}\Gamma^\rho_{\nu\sigma}) = 0. \quad (5·4)$$

This relation is already covariant. To see it one has to consider first that if A^α is a vector, then

$$\frac{dA^\alpha}{ds} + \Gamma^\alpha_{\mu\nu}u^\nu A^\mu \equiv \frac{DA^\alpha}{Ds} \quad (5·5)$$

is also a vector; and secondly, that the curvature tensor of the basic field $g_{\mu\nu}$ is given by

$$R^\alpha_{\nu\sigma\mu} = \Gamma^\alpha_{\nu\sigma,\mu} - \Gamma^\alpha_{\mu\sigma,\nu} + \Gamma^\alpha_{\mu\rho}\Gamma^\rho_{\nu\sigma} - \Gamma^\alpha_{\nu\rho}\Gamma^\rho_{\mu\sigma}. \quad (5·6)$$

With (5·5) and (5·6) equation (5·4) takes the evidently covariant form

$$\frac{D}{Ds}\left(mu^\alpha + u_\beta\frac{DS^{\alpha\beta}}{Ds}\right) + \tfrac{1}{2}S^{\mu\nu}u^\sigma R^\alpha_{\nu\sigma\mu} = 0. \quad (5·7)$$

Equation (5·7) is a generalization of (2·10) in the sense that it reduces to (2·10) when $S^{\alpha\beta} = 0$. It is worth stressing that only in this simplest case the orbits will be geodesics of the field $g_{\mu\nu}$. When the particle has a non-vanishing spin *the orbits will differ from the geodesics.*

A detailed discussion of equations (3·12), (3·13) for a certain class of test particles moving in the Schwarzschild field will be published in a subsequent paper.

References

Einstein, A. & Infeld, L. 1940 *Ann. Math.* **41**, 455.
Einstein, A. & Infeld, L. 1949 *Canad. J. Math.* **1**, 209.
Einstein, A., Infeld, L. & Hoffmann, B. 1938 *Ann. Math.* **39**, 65.
Fock, V. A. 1939 *J. Phys. U.S.S.R.* **1**, 81
Hönl, H. & Papapetrou, A. 1939*a* *Z. Phys.* **112**, 512.
Hönl, H. & Papapetrou, A. 1939*b* *Z. Phys.* **114**, 478.
Hönl, H. & Papapetrou, A. 1940 *Z. Phys.* **116**, 153.
Infeld, L. & Schild, A. 1949 *Rev. Mod. Phys.* **21**, 408.
Lubánski, J. 1937 *Acta Phys. Polon.* **6**, 356.
Mathisson, M. 1937 *Acta Phys. Polon.* **6**, 167.
Papapetrou, A. 1951 *Proc. Phys. Soc.* **64**, 57.

Spinning test-particles in general relativity. II

By E. Corinaldesi and A. Papapetrou
Physics Department, University of Manchester

(*Communicated by R. E. Peierls, F.R.S.—Received* 3 *May* 1951)

The equations of motion for spinning test particles are discussed for particles characterized by the condition $S^{i4} = 0$.

1. Introduction

In the present paper we shall discuss some applications of the equations of motion for spinning test particles derived in part I. The Schwarzschild field produced by a large massive body will be assumed as the basic field. As for the test particles, we shall make the assumption that they are of the macroscopic type, that is to say, that the centre of mass always lies inside the (three-dimensional) space occupied by the particle at a given time, the spin being the result of an ordinary rotation. The motion of such particles should be determined entirely by the equations of motion expressed by (5·3) and (5·4) of part I.

These, however, are not sufficient for the determination of all unknowns. In fact, the equations for the spin components are three, while the independent components of $S^{\alpha\beta}$ are six. Therefore, it is necessary to reduce the number of independent components by imposing suitable conditions to the $S^{\alpha\beta}$.

The condition

$$S^{i4} = 0 \tag{1·1}$$

suggests itself because of its simplicity. It must be pointed out, however, that an equation like (1·1) has no meaning, unless we specify the frame of reference in which it holds. In special relativity, where all frames are fully equivalent, one would be inclined to demand (1·1) in the rest-frame of the test particle. Then (1·1) could be generalized unambiguously to any frame of reference,

$$u_\beta S^{\alpha\beta} = 0. \tag{1·1a}$$

In general relativity, however, the choice of the frame in which (1·1) will be demanded is rather ambiguous. Let us consider, for example, a test particle which moves in the Schwarzschild field of an incomparably larger body. In the rest-frame of the test particle the central body is in motion, and it is not at all evident that this frame will be fully equivalent to the frame in which the central body is at rest. On the contrary, it has been suggested that this second frame—which we may call the rest-frame of the Schwarzschild field—would be of primary importance (Rosen 1940; Papapetrou 1948). Thus, we arrive at the alternative possibility, which consists in demanding that the condition (1·1) be fulfilled in the rest-frame of the Schwarzschild field. The first consequence of this formulation of the condition (1·1) is a considerable simplification in the calculations, since three of the spin components vanish identically just in the frame in which the calculations will be performed. Furthermore, this assumption, far from being arbitrary and abstract, can be justified by the following physical argument. Let us suppose for a moment that the reduction of the independent components $S^{\alpha\beta}$ has been obtained by using the condition (1·1*a*). Then, in the frame of the Schwarzschild field we will have $S^{i4} \neq 0$. The physical meaning of this result is that the point X^{α} chosen to 'represent' the particle would not coincide with the centre of mass of the test particle in that frame (Papapetrou 1939). If so, however, the most simple procedure would be to replace the point X^{α} by the mass-centre X'^{α} of the particle. Then, automatically, the equations $S'^{i4} = 0$ would hold for the new components of the spin. This can be done for all values of t. Therefore, it all amounts to abandoning the representation of the test particle by the world line L, locus of X^{α}, and adopting instead that provided by the world line L' of the centre of mass. This argument proves that the condition (1·1*a*) corresponds to a fictitious description of the test particle.

Therefore, we shall assume that *the condition* (1·1) *holds in the rest-frame of the Schwarzschild field.*

As for the calculations presented in this paper, their purpose will be twofold. In the first place, the study of the motion in the Schwarzschild field will have a theoretical interest in itself. The integrals of motion representing the conservation of energy and angular momentum of spinning test particles will be derived. By comparing them with the corresponding expressions deduced from the geodesic equations, the implications of the general theory presented in part I will be greatly clarified. On the other hand, we shall give an estimate of the influence of the spin on the planetary motion and on the deflexion of light rays in a gravitational field.

2. Equations of motion in polar co-ordinates

In polar co-ordinates the Schwarzschild field can be described by the line element*

$$ds^2 = e^{\mu}\,dt^2 - e^{-\mu}\,dr^2 - r^2(d\theta^2 + \cos^2\theta\,d\phi^2) \tag{2·1}$$

with

$$e^{\mu} = 1 - \frac{2r_0}{r}, \tag{2·2}$$

where r_0 is the gravitational radius of the central body. The non-vanishing Christoffel symbols are:

$$\left.\begin{aligned}
&\Gamma^1_{11} = -\frac{\mu'}{2}, \quad \Gamma^1_{22} = -r\,e^{\mu}, \quad \Gamma^1_{33} = -r\,e^{\mu}\cos^2\theta, \quad \Gamma^1_{44} = \frac{\mu'}{2}e^{2\mu},\\
&\Gamma^2_{12} = \Gamma^2_{21} = \frac{1}{r}, \quad \Gamma^2_{33} = \cos\theta\sin\theta,\\
&\Gamma^3_{13} = \Gamma^3_{31} = \frac{1}{r}, \quad \Gamma^3_{23} = \Gamma^3_{32} = -\tan\theta, \quad \Gamma^4_{14} = \Gamma^4_{41} = \frac{\mu'}{2},
\end{aligned}\right\} \tag{2·3}$$

where $\mu' = d\mu/dr$. As usual, the indices 1, 2, 3, 4 will refer to the co-ordinates r, θ, ϕ and t, respectively.

The definition of $DS^{\alpha\beta}/Ds$ given by (5·1) of part I, together with (1·1), gives†

$$\frac{DS^{i4}}{Ds} = \Gamma^4_{k\nu}S^{ik}u^{\nu}. \tag{2·4}$$

* See, for instance, Bergmann (1947, p. 212). The angle θ is chosen in such a way that the value $\theta = 0$ corresponds to the plane $z = 0$.

† As in part I, Latin indices will take the values 1, 2, 3, only, while Greek indices will run from 1 to 4.

With the help of this relation, the equations for the spin components

$$\frac{DS^{\alpha\beta}}{Ds}+\frac{u^\alpha}{u^4}\frac{DS^{\beta 4}}{Ds}-\frac{u^\beta}{u^4}\frac{DS^{\alpha 4}}{Ds}=0 \tag{2·5}$$

reduce to

$$\frac{DS^{ik}}{Ds}+\Gamma^4_{l\lambda}\frac{u^\lambda}{u^4}(u^iS^{kl}-u^kS^{il})=0. \tag{2·6}$$

Using the expressions (2·3) for the Christoffel symbols, these equations, after a simple calculation, go over into the form

$$\left.\begin{aligned}
&\dot{S}^{12}+\left(\frac{1}{r}-\mu'\right)\dot{r}S^{12}+r\,e^\mu\cos^2\theta\dot{\phi}S^{23}-\cos\theta\sin\theta\dot{\phi}S^{31}=0,\\
&\dot{S}^{23}+\left(\frac{\mu'}{2}-\frac{1}{r}\right)\dot{\phi}S^{12}+\left(\frac{2}{r}\dot{r}-\tan\theta\dot{\theta}\right)S^{23}+\left(\frac{\mu'}{2}-\frac{1}{r}\right)\dot{\theta}S^{31}=0,\\
&\dot{S}^{31}+\tan\theta\dot{\phi}S^{12}+r\,e^\mu\dot{\theta}S^{23}+\left[\left(\frac{1}{r}-\mu'\right)\dot{r}-\tan\theta\dot{\theta}\right]S^{31}=0;
\end{aligned}\right\} \tag{2·7}$$

the dot indicates the differentiation with respect to s.

The equations of motion

$$\frac{d}{ds}\left(mu^\alpha+u_\beta\frac{DS^{\alpha\beta}}{Ds}\right)+\Gamma^\alpha_{\mu\nu}u^\nu\left(mu^\mu+u_\beta\frac{DS^{\mu\beta}}{Ds}\right)+S^{\mu\nu}u^\sigma(\Gamma^\alpha_{\nu\sigma,\mu}+\Gamma^\alpha_{\mu\rho}\Gamma^\rho_{\nu\sigma})=0 \tag{2·8}$$

must now be written for the Schwarzschild field. Inserting the values (2·3) of the Christoffel symbols into (2·8), one finds after some elementary calculations

$$\left.\begin{aligned}
&\frac{d}{ds}[\dot{t}(m+m_s)]+\lambda^4(m+m_s)=0,\\
&\frac{d}{ds}[\dot{r}(m+m_s)]+\lambda^1(m+m_s)=0,\\
&\frac{d}{ds}[\dot{\theta}(m+m_s)]+\lambda^2(m+m_s)+\frac{3\mu'}{2r}\dot{r}S^{12}+\frac{3\mu'}{2}r\,e^\mu\cos^2\theta\dot{\phi}S^{23}=0,\\
&\frac{d}{ds}[\dot{\phi}(m+m_s)]+\lambda^3(m+m_s)-\frac{3\mu'}{2}r\,e^\mu\dot{\theta}S^{23}-\frac{3\mu'}{2r}\dot{r}S^{31}=0.
\end{aligned}\right\} \tag{2·9}$$

Here

$$\lambda^\alpha=\Gamma^\alpha_{\mu\nu}u^\mu u^\nu \tag{2·10}$$

(which is *not* a vector) has the following components:

$$\left.\begin{aligned}
&\lambda^4=\mu'\dot{r}\dot{t},\\
&\lambda^1=\frac{\mu'}{2}e^{2\mu}\dot{t}^2-\frac{\mu'}{2}\dot{r}^2-r\,e^\mu\dot{\theta}^2-r\,e^\mu\cos^2\theta\dot{\phi}^2,\\
&\lambda^2=\frac{2}{r}\dot{r}\dot{\theta}+\cos\theta\sin\theta\dot{\phi}^2,\\
&\lambda^3=\frac{2}{r}\dot{r}\dot{\phi}-2\tan\theta\dot{\theta}\dot{\phi}.
\end{aligned}\right\} \tag{2·10a}$$

The quantity m_s is defined by the formula

$$m_s=\frac{r^2\mu'}{2}(\cos^2\theta\dot{\phi}S^{31}-\dot{\theta}S^{12}). \tag{2·11}$$

It represents a sort of interaction energy describing a spin-orbit coupling, and will be the object of a more complete discussion in the following sections. In the meantime, we want only to point out that $m+m_s$ will play the role of an 'effective mass' throughout.

Multiplying the equations (2·9) by u_α, and taking into account the relation

$$u_\alpha \dot{u}^\alpha + u_\alpha \lambda^\alpha = 0,$$

we obtain
$$\frac{d}{ds}(m+m_s) = \frac{3\mu'}{2} r\dot{r}(\dot{\theta}S^{12} - \cos^2\theta\dot{\phi}S^{31}). \tag{2·12}$$

Using (2·12) the equations (2·9) go over into the form

$$\left.\begin{aligned}
&(m+m_s)(\ddot{t}+\lambda^4) - \frac{3}{r}\dot{r}\dot{t}m_s = 0,\\
&(m+m_s)(\ddot{r}+\lambda^1) - \frac{3}{r}\dot{r}^2 m_s = 0,\\
&(m+m_s)(\ddot{\theta}+\lambda^2) - \frac{3}{r}\dot{r}\dot{\theta}m_s + \frac{3\mu'}{2r}\dot{r}S^{12} + \frac{3r\mu'}{2}e^\mu\cos^2\theta\dot{\phi}S^{23} = 0,\\
&(m+m_s)(\ddot{\phi}+\lambda^3) - \frac{3}{r}\dot{r}\dot{\phi}m_s - \frac{3\mu'}{2r}\dot{r}S^{31} - \frac{3r\mu'}{2}e^\mu\dot{\theta}S^{23} = 0,
\end{aligned}\right\} \tag{2·13}$$

which is more suitable for the applications.

Using (2·9) or (2·13) it is possible to derive first integrals expressing the conservation of energy and angular momentum. The procedure is analogous to that followed in the study of spinless particles. The first of (2·9) combined with the first of (2·10*a*) gives

$$e^\mu \dot{t}(m+m_s) = E = \text{const.}, \tag{2·14}$$

which is the integral of energy. On the other hand, the last of (2·9), combined with the equations (2·7), gives

$$r^2\cos^2\theta\left[(m+m_s)\dot{\phi} + \frac{\mu'}{2}S^{31}\right] + S_z = I_z = \text{const.} \tag{2·15}$$

Here
$$S_z = -r^2\cos\theta\sin\theta S^{23} - r\cos^2\theta S^{31}$$

is the component of the spin in the z-direction of a system of Cartesian co-ordinates,* and I_z represents the z-component of the total angular momentum.

There are two other relations, of a structure similar to (2·15), which express the conservation of the x- and y-components of the angular momentum. These are usually disregarded in the discussion of the motion of spinless particles, the reason being that such particles are always moving on plane orbits; one can assume that the motion takes place on the plane $\theta = 0$, in which case I_x and I_y vanish identically. On the contrary, the motion of spinning particles is, in general, not a plane one, and consequently the full expression of the conservation of angular momentum will be needed. We shall give it in § 4 in Cartesian co-ordinates.

* Cf. the formulae (4·1) and (4·3).

3. Plane motion

We shall now discuss the question whether the equations (2·7) and (2·9) admit any solution representing a motion in a plane passing through the central body. Without any loss of generality, this plane can be taken as the equatorial plane $\theta = 0$. Apart from some special cases of doubtful physical significance, the general condition for a motion in the plane $\theta = 0$ is found to be

$$S^{23} = S^{12} = 0, \tag{3·1}$$

with only $S^{31} \neq 0$. The meaning of this condition can be clarified by noticing that the Cartesian components of the spin in this case are (cf. (4·3))

$$S_x = S_y = 0, \quad S_z = rS^{13}.$$

Thus the spin must be perpendicular to the plane of motion $z = 0$.

On the assumption (3·1), the equations (2·7) reduce to

$$\dot{S}^{31} + \left(\frac{1}{r} - \mu'\right)\dot{r}S^{31} = 0. \tag{3·2}$$

This can be easily integrated,

$$r\,e^{-\mu}S^{31} \equiv -e^{-\mu}S_z = \text{const.} \tag{3·3}$$

Therefore, in the present case (3·3) stands besides (2·14) and (2·15), as a first integral for the spin only.

The formula (3·3) has a remarkable, though qualitative analogy with the gravitational red shift of spectral lines. In fact, let us assume that S_z be proportional to the angular velocity of the internal rotation of the test particle; the period of such a rotation is then proportional to $e^{-\mu} = \left(1 - \frac{2r_0}{r}\right)^{-1}$, i.e. it increases with decreasing r according to a law similar to that describing the influence of the gravitational potential on the wave-length of light.

Besides the strict plane motion based on the condition (3·1), the motion will be very nearly plane in all cases when the spin is much smaller than the orbital angular momentum. This condition is fulfilled in the planetary motions. An interesting question is whether the introduction of the spin in the planetary motion will lead to any appreciable effects. To answer this question we shall have to make an estimate of the order of magnitude of the spin terms for the planetary motion.

By using the integral (2·14), and noticing that

$$\frac{d^2}{ds^2}(\ldots) = \dot{t}^2\frac{d^2}{dt^2}(\ldots) + \ddot{t}\frac{d}{dt}(\ldots)$$

the variable s can be eliminated from (2·13). Thus we obtain

$$\left.\begin{aligned}
&\frac{d^2r}{dt^2} - \frac{3\mu'}{2}\left(\frac{dr}{dt}\right)^2 - r\,e^{\mu}\left(\frac{d\theta}{dt}\right)^2 - r\,e^{\mu}\cos^2\theta\left(\frac{d\phi}{dt}\right)^2 + \frac{\mu'}{2}e^{2\mu} = 0,\\
&\frac{d^2\theta}{dt^2} + \left(\frac{2}{r} - \mu'\right)\frac{dr}{dt}\frac{d\theta}{dt} + \cos\theta\sin\theta\left(\frac{d\phi}{dt}\right)^2 + \frac{3\mu'}{2E}e^{\mu}\left[r\,e^{\mu}\cos^2\theta\frac{d\phi}{dt}S^{23} + \frac{1}{r}\frac{dr}{dt}S^{12}\right] = 0,\\
&\frac{d^2\phi}{dt^2} + \left(\frac{2}{r} - \mu'\right)\frac{dr}{dt}\frac{d\phi}{dt} - 2\tan\theta\frac{d\theta}{dt}\frac{d\phi}{dt} - \frac{3\mu'}{2E}e^{\mu}\left[r\,e^{\mu}\frac{d\theta}{dt}S^{23} + \frac{1}{r}\frac{dr}{dt}S^{31}\right] = 0.
\end{aligned}\right\} \tag{3·4}$$

For comparison, the Newtonian equations

$$\frac{d^2\mathbf{r}}{dt^2} = -\frac{r_0\mathbf{r}}{r^3}$$

will be written in polar co-ordinates:

$$\left.\begin{aligned}
&\frac{d^2r}{dt^2}+\frac{r_0}{r^2}-r\left(\frac{d\theta}{dt}\right)^2-r\cos^2\theta\left(\frac{d\phi}{dt}\right)^2=0,\\
&\frac{d^2\theta}{dt^2}+\frac{2}{r}\frac{dr}{dt}\frac{d\theta}{dt}+\cos\theta\sin\theta\left(\frac{d\phi}{dt}\right)^2=0,\\
&\frac{d^2\phi}{dt^2}+\frac{2}{r}\frac{dr}{dt}\frac{d\phi}{dt}-2\tan\theta\frac{d\theta}{dt}\frac{d\phi}{dt}=0.
\end{aligned}\right\} \tag{3·4a}$$

For vanishing spin, the difference between corresponding equations of the two systems is a perturbation of the order of magnitude r_0/r compared with the Newtonian terms. It is responsible, for example, for the rotation of the perihelion of the planet. We shall refer to such a perturbation as the 'ordinary relativistic terms'. Let us now consider one typical spin-dependent term, e.g. the last term in the third of the (3·4). Its order of magnitude, compared with the Newtonian terms, is

$$\frac{S_z}{E}\frac{r_0}{r^2}\left(r\frac{d\phi}{dt}\right)^{-1}.$$

Since

$$S_z \approx \frac{ER^2}{\tau}, \quad \left(r\frac{d\phi}{dt}\right) \approx \frac{r}{T},$$

R being the radius of the planet, τ the period of the rotation of the planet around its axis, and T the period of the revolution around the sun, the term in question is seen to be of the order of magnitude $\frac{r_0}{r}\frac{R^2}{r^2}\frac{T}{\tau}$ compared with the Newtonian terms, and therefore of the order $\frac{R^2}{r^2}\frac{T}{\tau}$ compared with the ordinary relativistic terms. The ordinary relativistic perturbation, however, is itself very small, and is observable for Mercury alone. In the case of this planet, $T = \tau$ and $R^2/r^2 \approx 10^{-8}$. Similar numerical results would be obtained for the other spin-dependent terms in (3·4). Therefore, the effect of the spin is exceedingly small in the case of the planetary motion.

4. Equations of motion in Cartesian co-ordinates

In this section the equations (2·6) and (2·8) will be written in Cartesian co-ordinates. The use of these has the advantage of making possible a better understanding of the meaning of m_s; it also simplifies the discussion of deflexion problems.

Therefore, let us introduce a spatial frame of reference x, y, z, related to the system of polar co-ordinates used in § 2 by the equations

$$x^1 \equiv x = r\cos\theta\cos\phi, \quad x^2 \equiv y = r\cos\theta\sin\phi, \quad x^3 \equiv z = r\sin\theta. \tag{4·1}$$

For the new system of co-ordinates the non-vanishing Christoffel symbols are*

$$\left.\begin{aligned}\Gamma^l_{ik} &= \mu'\frac{x^l}{r}\left[\mathrm{e}^{\mu}\,\delta_{ik}-\frac{x^i x^k}{2r^2}(1+2\mathrm{e}^{\mu})\right], \quad \Gamma^l_{44} = \frac{\mu'}{2r}\mathrm{e}^{2\mu}x^l,\\ \Gamma^4_{i4} &= \Gamma^4_{4i} = \frac{\mu'}{2r}x^i, \quad \Gamma^4_{44} = 0.\end{aligned}\right\} \tag{4·2}$$

Let us introduce a three-dimensional vector $\mathbf{S}$, whose components S_x, S_y, S_z are the values of S^{23}, S^{31}, S^{12} in the Cartesian frame. Remembering that $S^{\alpha\beta}$ is a tensor, we find that the relation between S_x, S_y, S_z and the components S^{23}, S^{31}, S^{12} in polar co-ordinates reads

$$\left.\begin{aligned}r^2\cos\theta S^{23} &= -\cos\theta\cos\phi S_x-\cos\theta\sin\phi S_y-\sin\theta S_z,\\ r\cos\theta S^{31} &= \sin\theta\cos\phi S_x+\sin\theta\sin\phi S_y-\cos\theta S_z,\\ rS^{12} &= \sin\phi S_x-\cos\phi S_y.\end{aligned}\right\} \tag{4·3}$$

We can now write the equations (2·6) and (2·8) in Cartesian co-ordinates either directly by using the expressions (4·2) instead of (2·3), or by transforming the equations (2·7) and (2·13) with the help of (4·1) and (4·3). The equations for the spin in Cartesian co-ordinates are found to be:

$$\dot{\mathbf{S}} = \mu'\left[\dot{r}\mathbf{S}+\frac{1}{r}\mathrm{e}^{\mu}(\mathbf{r}\mathbf{S})\,\dot{\mathbf{r}}-\frac{1}{2r}(\dot{\mathbf{r}}\mathbf{S})\,\mathbf{r}-\frac{1}{2r^2}(1+2\mathrm{e}^{\mu})\,\dot{r}(\mathbf{r}\mathbf{S})\,\mathbf{r}\right], \tag{4·4}$$

where $\mathbf{r}\equiv(x,y,z)$.

By the equation

$$r^2\dot{t}\boldsymbol{\omega} = (\mathbf{r}\times\dot{\mathbf{r}}), \tag{4·5}$$

we define a quantity $\boldsymbol{\omega}$, which can be interpreted as the angular velocity, referring to the motion of the test particle on its orbit. Then, after some elementary calculations, we find that the co-ordinates x, y, z of the test particle must satisfy the equation

$$(m+m_s)(\ddot{\mathbf{r}}+\boldsymbol{\lambda})-\frac{3m_s}{r}\dot{r}\dot{\mathbf{r}}-\frac{3\mu'}{2r}\left(\mathrm{e}^{\mu}\dot{t}(\mathbf{r}\mathbf{S})\,\boldsymbol{\omega}+\frac{\dot{r}}{r}(\mathbf{r}\times\mathbf{S})\right) = 0. \tag{4·6}$$

Here the vector $\boldsymbol{\lambda}$ is defined by the equation

$$\boldsymbol{\lambda} = \frac{\mu'}{2r}(\mathrm{e}^{2\mu}\dot{t}^2+2\mathrm{e}^{\mu}\dot{\mathbf{r}}^2-(1+2\mathrm{e}^{\mu})\,\dot{r}^2)\,\mathbf{r}. \tag{4·7}$$

The equation for t is not altered by a purely spatial transformation, and is the same as the first of (2·13).

Finally, equation (2·11) for the definition of the quantity m_s, which also appears in (4·6), takes the form

$$m_s = -\frac{\mu' r}{2}\dot{t}(\boldsymbol{\omega}\mathbf{S}). \tag{4·8}$$

Here m_s has assumed the characteristic form of a spin-orbit interaction energy. In fact, it is proportional to the scalar product of the spin with the vector $\boldsymbol{\omega}$, which is connected with the orbital angular momentum.

The energy integral retains its form (2·14). The integral expressing the conservation of the angular momentum, on the other hand, can be derived either from (4·6) or by transforming (2·15) to Cartesian co-ordinates: once the equation for the

* See, for instance, Weyl (1922, p. 253).

component I_z has been written in these co-ordinates, the other two components can be derived very easily by making use of symmetry considerations. Thus we obtain the vector equation

$$\mathbf{S}+E\,e^{-\mu}r^2\boldsymbol{\omega}+\frac{\mu'}{2r}[(\mathbf{rS})\,\mathbf{r}-r^2\mathbf{S}]=\mathbf{I}=const. \tag{4·9}$$

It is remarkable that in the right-hand side of (4·9), besides the first and second term, which are, respectively, the spin and what can be interpreted as the orbital angular momentum, there is an additional term, which causes a drastic departure from the ordinary Newtonian mechanics.

The existence of the spin-orbit coupling expressed by (4·8) seems to suggest that a precession of the orbital angular momentum and of the spin around **I** might take place.

Here we shall confine ourselves to the case of the planetary motion. In this case an approximate treatment is possible, based on the two assumptions: (i) $|\mathbf{S}|\ll|\mathbf{I}|$, (ii) $r\gg r_0$. Because of (i) the motion is very nearly plane (the plane of motion changing slowly with the time, and always passing through the central body). Because of (ii) the orbit is very nearly an ellipse (with a superimposed slow rotation of the perihelion). Let $\mathbf{S}_0$ be the value of the spin at a certain time, say $t=0$. At a later time, the spin will have assumed the value

$$\mathbf{S}=\mathbf{S}_0+\delta\mathbf{S}. \tag{4·10}$$

If t is not very large compared with the period T of the orbital motion, then $|\delta\mathbf{S}|\ll|\mathbf{S}_0|$, and the equation (4·4) can be approximated as follows:

$$\dot{\mathbf{S}}\approx\mu'\left[\dot{r}\mathbf{S}_0-\frac{3\dot{r}}{2r^2}(\mathbf{rS}_0)\,\mathbf{r}+\frac{1}{r}\mathbf{S}_0\times(\dot{\mathbf{r}}\times\mathbf{r})\right]. \tag{4·11}$$

Averaging (4·11) over one period T (of the orbital motion), we obtain

$$\left(\frac{d\mathbf{S}}{dt}\right)_{\text{av.}}=\left(\frac{2r_0}{r}\right)_{\text{av.}}(\boldsymbol{\omega}\times\mathbf{S}_0), \tag{4·12}$$

since the average value of the first two terms vanishes. Being $|\boldsymbol{\omega}|\approx 2\pi/T$, it is evident from (4·12) that **S** describes a precession around $\boldsymbol{\omega}$ with the period

$$T_{\text{prec.}}\approx\frac{T}{(2r_0/r)_{\text{av.}}}. \tag{4·13}$$

As the orbits of the planets are very nearly circles, $(1/r)_{\text{av}}$ will not be very sensitive to the way in which the averaging process is carried out. For the earth we find

$$T_{\text{prec.}}\approx 5\times10^7\,T=5\times10^7\text{ years}. \tag{4·14}$$

For comparison, we might mention that the non-spherical shape of the earth has the consequence of inducing a precession with a period of 26000 years.

We would like to end this section with a general remark. If the spin precession is connected with any energy-consuming secondary effects, e.g. because of gravitational radiation, one should expect the motion to tend towards a final state characterized by a minimum of the spin-orbit interaction. According to (4·8), m_s will have a minimum when the spin is orientated in the same direction as the orbital angular momentum (in which case the motion will be plane, cf. § 3). This remark might be of interest for the theories of the origin of the planetary system.

5. The hyperbolic motion

Besides the elliptic motion, the Newtonian equations admit solutions representing the motion of particles on hyperbolic orbits. Such orbits, on the other hand, will be practically rectilinear, with a very small deflexion, in all cases when the particle is very energetic and the 'impact parameter', b, is much larger than the gravitational radius r_0. (By 'impact parameter' we mean the distance between the particle and the central body at the position of closest approach.)

The 'hyperbolic motion' has been already treated in general relativity by using the geodesic equations, this corresponding to the motion of spinless particles. In the present section we shall study the problem of the deflexion of very energetic spinning particles in the Schwarzschild field. Thus we shall be able to see whether the influence of the spin could alter appreciably the numerical results deduced by using the equations of the geodesics. In particular, we shall give an estimate of the influence that the spin of the photon might have on the deflexion of light.

In the literature, the discussion of deflexion problems is based on the use of polar co-ordinates. Here, on the contrary, Cartesian co-ordinates will be introduced. Their use makes the calculations much shorter and more compact, as it will be shown.

Using the energy integral (2·14), the variable s can be eliminated from the equation (4·6). Thus we obtain

$$\frac{d^2\mathbf{r}}{dt^2}+\frac{\mu'}{2r}\left[e^{2\mu}+2e^{\mu}\left(\frac{d\mathbf{r}}{dt}\right)^2-(1+2e^{\mu})\left(\frac{dr}{dt}\right)^2\right]\mathbf{r}$$
$$-\mu'\frac{dr}{dt}\frac{d\mathbf{r}}{dt}-\frac{3\mu' e^{\mu}}{2rE}\left[\frac{1}{r}\frac{dr}{dt}(\mathbf{r}\times\mathbf{S})+e^{\mu}(\mathbf{r}\mathbf{S})\,\boldsymbol{\omega}\right]=0. \qquad (5\cdot1)$$

Let the equations
$$x=b,\quad y=\beta t,\quad z=0 \qquad (5\cdot2)$$
with $b \gg r_0$ and $\beta \to 1$, represent the 'unperturbed' motion of a spinning test particle in the Schwarzschild field. As for the spin, arbitrary constant values (S_x, S_y, S_z) can be attributed to its components in the unperturbed motion, with the only restriction that they should not be too large, so that the following perturbation method be permissible.

In all small terms of the equation (5·1) we shall replace $\mathbf{r}$ and $\mathbf{S}$ by the unperturbed values. Thus we can write the first-order equations

$$\left.\begin{aligned}
&\frac{d^2x}{dt^2}+\frac{3r_0}{(b^2+t^2)^{\frac{5}{2}}}\left(b^3-\frac{S_z t^2}{E}\right)=0,\\
&\frac{d^2y}{dt^2}+\frac{r_0 t}{(b^2+t^2)^{\frac{5}{2}}}\left(b^2-2t^2+\frac{3bS_z}{E}\right)=0,\\
&\frac{d^2z}{dt^2}-\frac{3r_0}{E(b^2+t^2)^{\frac{5}{2}}}\left[(b^2-t^2)S_x+2bS_y t\right]=0.
\end{aligned}\right\} \qquad (5\cdot3)$$

Similarly, from (4·4) we could write the first order equations for the spin components.

We shall restrict ourselves to the discussion of the following cases:

(i) $\mathbf{S} = (0, 0, S)$. On this assumption the motion will be strictly in the plane $z = 0$ (cf. (3·1) et seq.). Integrating the first two equations (5·3) we obtain

$$\left.\begin{aligned}\frac{dx}{dt} &= -\frac{br_0 t}{\sqrt{(b^2+t^2)}}\left(\frac{2}{b^2}+\frac{1}{b^2+t^2}\right)+\frac{r_0 S t^3}{Eb^2(b^2+t^2)^{\frac{3}{2}}}\\ \frac{dy}{dt} &= 1-\frac{r_0}{\sqrt{(b^2+t^2)}}\left(2-\frac{b^2}{b^2+t^2}\right)+\frac{r_0 b S}{E(b^2+t^2)^{\frac{3}{2}}}.\end{aligned}\right\} \tag{5·4}$$

The values of the integration constants have been fixed by assuming that $dx/dt = 0$ and $dy/dt = 1$ for $t = 0$. The angle of deflexion $\delta\phi$ can now be obtained very easily from the asymptotic value of the ratio $(dx/dt):(dy/dt)$ for $t \to \pm\infty$. We find

$$\delta\phi \approx \frac{2r_0}{b}\left(1-\frac{S}{2bE}\right). \tag{5·5}$$

The spin-independent part of this deflexion agrees with the value given in the literature for the bending of light rays in the Schwarzschild field. (As for the spin-dependent term, we notice that it vanishes when E tends to infinity.)

(ii) If the unperturbed values of the spin are $(0, S, 0)$ (spin parallel to the velocity), or $(S, 0, 0)$ (spin perpendicular to the velocity and in the xy-plane), the perturbed motion will not be plane. In both these cases, however, the first two of (5·3) are independent of S, and, consequently, the deflexion will be identical with that which is found for spinless particles. Also the corrections to the spin can be evaluated very easily. It does not, however, seem worthwhile to give the details of the calculation.

These results could be applied tentatively to the problem of the deflexion of light. One finds then that the spin of the photon has practically no influence on the value of the deflexion. Assuming, namely, the spin to be orientated as in (i), and taking $S \approx \hbar$, $E = h\nu = h\dfrac{c}{\lambda}$, one finds for the second term in the bracket of (5·5) the extremely small value λ/b. Actually, one should expect the spin of the photon to be orientated as in (ii) (first case); then, however, it would not give any contribution to the deflexion.

References

Bergmann, P. G. 1947 *Introduction to the theory of relativity*. New York: Prentice-Hall.
Papapetrou, A. 1939 *Prakt. Akad. Athen.* **14**, 540.
Papapetrou, A. 1948 *Proc. R. Irish Acad.* **52**, 11.
Rosen, N. 1940 *Phys. Rev.* **57**, 147.
Weyl, H. 1922 *Space-time-matter*. London: Macmillan.

WSEG
Research Memorandum
No. 11

PROPOSAL FOR A SATELLITE TEST OF THE CORIOLIS PREDICTION OF GENERAL RELATIVITY*

Table of Contents

Abstract
Introduction
Measurement of Orientation
Significance of Experiment
Feasibility of the Experiment
  Major Atmospheric Perturbations
  A Technique to Minimize the Atmospheric Perturbations
  Remaining Atmospheric Perturbations
  The Gravitational Perturbation
  The Magnetic Perturbation
Useful By-Products of the Experiment
Summary
Acknowledgments
Table I Summary of Parameters

George E. Pugh
12 November 1959

*This article has been reset with errors corrected by editors of this volume.

PROPOSAL FOR A SATELLITE TEST OF THE CORIOLIS PREDICTIONS OF GENERAL RELATIVITY[a]

Abstract

A technique is proposed to use the gyroscopic stability of a spinning artificial satellite to measure the apparent Coriolis force (Lense–Thirring effect) predicted by general relativity. It is established that the primary satellite ought to be shielded from atmospheric buffeting by a secondary or tender satellite that would accompany the primary test vehicle.

Other perturbations considered are interaction with the magnetic field of the earth, and interaction of any gravitational quadrupole moment of the satellite with the gradient of the earth's gravitational field. These interactions are sufficiently small that design of a vehicle to avoid excessive interaction should be possible.

The spinning shielded satellite technique proposed here could make possible a rather accurate measurement of the curvature of space near the earth. This curvature manifests itself in a precession of the spin (the de Sitter–Fokker effect), which is about a factor of 12 larger than the Lense–Thirring effect and thus could provide a logical first step for such a program.

Development of the exterior, or tender, satellite alone would offer a number of subsidiary advantages. It could be used with a dummy interior satellite ot obtain drag-free trajectories and thus provide a very accurate tool to study air drag, as well as gravitation and geodetics. By using an interior satellite with very different nuclear composition the tender could be used to establish the equivalence of gravitational and inertial mass to greater accuracy.

WSEG RESEARCH MEMORANDUM NO. 11 PROPOSAL FOR A SATELLITE TEST OF THE CORIOLIS PREDICTION OF GENERAL RELATIVITY

Introduction

1. The purpose of this proposal is to call attention to the fact that the development of satellite technology will soon allow an experimental verification of the "induced Coriolis force," or Lense–Thirring effect predicted by general relativity. In discussing this prediction for the theory, Einstein remarked that the magnitude of this effect (which amounts to about 2×10^{-6} radians per year near the surface of the earth) "is so small that confirmation by laboratory experiments is not to

[a]Views contained in this paper are the authors' and do not necessarily reflect those of the Weapons Systems Evaluation Group or the Department of Defense.

be thought of." However, artificial satellites can provide an almost force-free environment that cannot be approached in surface laboratories. This new environment makes the experimental confirmation of the prediction a distinct possibility.

2. Conceptually, the experiment can be described very simply. A satellite is launched into a nearly circular polar orbit which is high enough to be almost free of the earth's atmosphere, but close enough to the earth that induced Coriolis forces will not be too strongly attenuated. The satellite is then given a large angular momentum or spin about an axis nearly perpendicular to the plane of the orbit. The direction of this axis of angular momentum relative to the fixed stars is accurately determined and recorded for a period of a year or more. Neglecting minor perturbations, the prediction of classical physics is that this angular momentum axis will not rotate relative to the fixed stars. However, relativity predicts a well-defined rotation or precession of this axis in the direction of rotation of the earth. In effect, space itself near the earth should appear to rotate very slowly with the earth.

3. If, on the other hand, the satellite is launched again in a polar orbit but with its spin axis in the plane of the orbit and parallel to the spin axis of the earth, then the Lense–Thirring effect on the orientation of the spin axis should vanish, and another relativistic effect known as the de Sitter–Fokker effect should be observed.[b]

4. This second effect which appears to be a factor of twelve larger than the first is a result of the curvature of space near a massive object. As a consequence of this curvature, a vector, such as the angular momentum vector when displaced parallel to itself around a Schwartzschild field, will be found to be no longer parallel to its initial orientation ofter the circu is complete.

Measurement of Orientation

5. In order to get useful results from the experiment it is essential to have a technique for measuring very accurately the orientation of the spin axis with respect to the distant stars. It seems likely that a number of ways for accomplishing this could be devised. However the following very simple method is offered to demonstrate that the problem of measurement in the experiment is basically solvable.

6. A telescope is mounted in the satellite looking along the angular momentum axis. As the satellite spins, each star in the field of view traces a circle on the recording apparatus. The totality of circles thus generated provides a characteristic signature indicating with high accuracy the exact orientation of the angular momentum axis. The resolution angle α of a 25 cm telescope using visible light is given by

$$\alpha = 1.22\frac{\lambda}{D} = \frac{1.22 \times 5.5 \times 10^{-5}}{25} = 2.7 \times 10^{-6}$$

[b]The author is grateful to Prof. P. G. Bergmann for calling his attention to the importance of this effect in the experiment.

However, with good optics it is normally not difficult to measure differences in angle equal to about one third the nominal resolution. The de Sitter–Fokker effect at the end of one month is about equal to the telescope resolution so a measurable effect should be available in two weeks. As the duration of the experiment increases the percentage accuracy should increase so that theoretically at the end of one year an answer could be available to 5 per cent accuracy. However, improvements in measuring technique or a larger telescope could reduce this time.

7. It is worth noting that this method of determining the spin axis does not require that the spin axis be aligned with the telescope axis with high accuracy. It is only required that the spin axis be included in the field of view of the telescope.

8. There remain, of course, a number of difficult engineering problems. The telescope must be focused, and the image must be recorded and transmitted. However, it is too early to become deeply involved in the engineering problems of the experiment.

Significance of Experiment

9. Quite aside from the implications to any particular theory, the history of science has demonstrated that when experiments are conducted which either open new fields, or greatly increase accuracy in old familiar fields, the chance of learning things new and basic is very great. The experiment proposed here represents an improvement of about a factor of 10^3 in the determination of the locally non-rotating frame of reference near the earth. It also provides a first feasible technique to measure the curvature of space near the earth.

10. Actually some implications of the curvature of space near massive objects have already been observed. The bending of light near the sun and the precession of the perihelion of mercury have botah been determined experimentally to an accuracy that requires this curvature or some equivalent effect in the theory to account for the observations.

11. However, the experiment proposed here offers an independent confirmation at this effect and should provide higher accuracy than was obtained in the previous experiments.

12. However, no implication of the rotation of space near a heavy rotating massive object has ever been observed. Thus, the measurement of this rotation (the Lense–Thirring effect) could be of crucial theoretical importance to the theory of relativity.

13. While the curvature of space or the equivalent of it is contained in almost any theory that can patch up the apparent internal inconsistencies of special relativity, the rotation effect is a higher order prediction of relativity and alternative theories making different predictions for this effect are still experimentally allowed.

14. Yilmaz has recently exhibited several alternative formulations of relativity which predict the three classical experiments and appear to preserve general covariance but do not predict the Lense–Thirring effect or predict it in lesser degree.

(Phys. Rev. III, 1417 (1958)). At the 1959 New York meeting of the American Physical Society, he suggested a hypothetical test of these predictions using the orbit of a satellite appear feasible as a definitive test because of the extremely large magnitude of perturbations that are not accurately known, such as the quadrupole and higher order moments of the earth's gravitational field. However, the spin of a satellite is much less influenced by such effects and may provide a feasible technique for the experiment.

15. The origin of the different predictions is fundamental in our understanding of physical laws. When Einstein devised his General Theory, he was forced by the evidence of the Michaelson Moreley experiment to construct the theory in such a way that physical laws were invariant to a translation of coordinates. On the basis of mathematical simplicity, and as a logical generalization of this invariance, he also assumed that the laws should be independent of a *rotation* of coordinates and in fact any other coordinate transformation. This latter assumption is deeply embedded in the structure of the theory and represents an additional departure from classical theory not specifically required by the experimental evidence.

16. Once this postulate is accepted, rotation has meaning only when defined *relative* to other massive bodies. Consequently, a spinning satellite will measure its rotation relative to both the earth and the distant stars; i.e., both will contribute to the determination of a "non-rotating" angular momentum axis. To the extent that the earth contributes to this determination, the local "non-rotating axis" will actually rotate relative to the distant stars. Thus the Coriolis prediction in a qualitative way follows directly from the basic assumptions of the theory. (On the other hand, the formal mathematics of the theory appears to be consistent with the existence of a "preferred" or non-rotating coordinate system, which can be imposed through boundary conditions at infinity.)

17. The practical significance of the theory of relativity in the field of physics is much greater than would be implied by the very small size of the physical effects it predicts for microscopic objects of limited size. The predictions of relativity become crucial in the interpretation of cosmological problems. Moreover, the structure of relativity provides a powerful guide in the selection of new theories concerning the basic structure of matter. Consequently, it is very important that the theory of relativity itself be provided with as rigorous an experimental foundation as is possible.

Feasibility of the Experiment

18. Aside from data collection, the prime requirement for feasibility of the experiment is that the external torques acting about the center of mass of the satellite with a component *perpendicular* to the spin axis must be very small. The following three interactions appear to be the most important factors that might contribute to a change in the diction of the angular momentum axis:

(a) The earth's atmosphere acting through the translation of the satellite to produce torques with a component perpendicular to the spin.
(b) The gradient of the gravitational field interacting with the mass moments of the satellite.
(c) The magnetic field of the earth interacting with electric currents on or in the satellite or with any magnetization of the satellite.

19. The relative importance of these effects depends in large degree on the specific for of the experiment that is to be selected. The final choice of an experiment (involving actual selection of the orbit and spin axis; the mass, volume and moments of inertia for the satellite; and the external shield and internal stabilization mechanism (if any); together with recording and telemetering techniques) will depend on a large number of specific harware considerations that cannot be discussed now. Consequently, one cannot attempt to calculate perturbations on a realistic basis. However, to provide a basis for the discussion we will consider a specific idealized version of the experiment.

20. Table I summarizes the parameters assumed for these calculations.

21. The satellite mass and radius were selected to be consistent with a reasonable launching capability. The moment of inertia, $1/2mr^2$, falls between the moments for a hollow or solid sphere. The spin was chosen to give a maximum internal acceleration not exceeding 10 G.

22. With appropriate theoretical interpretation the orbit could be equatorial, polar, or intermediate; however, we will discuss only the equatorial and polar orbits because it simplifies visualization and calculations. The orbit radius is selected in round numbers to give a reasonable altitude where air drag would not be excessive. Higher altitudes might well produce significantly less interaction; however, data on air densities above 600 km are too incomplete to be very useful.

23. The Lense–Thirring precession on the surface of the earth is shown together with an estimate of the effect at the chosen orbit radius. The precession of 6×10^{-14} rad/s corresponds to an applied "torque" on the satellite discussed here of 1.0×10^{-10} N · m. This pseudo-torque will be compared with other actual torques that might alter the spin axis.

24. The corresponding parameters for the de Sitter–Fokker effect are also shown. However, the experimental perturbations will be compared primarily against the more stringent requirements of the Lense–Thirring effect. The requirements for the de Sitter–Fokker effect are, of couse, always about a factor of twelve less severe.

Major Atmospheric Perturbations

25. Classically, air drag at high altitudes (where mean free paths are sufficient that collective air motion can be ignored) is divided into two components:

(a) A specular reflection component in which incident particles bounce elastically like light from a mirror.

Table 1. Summary of Parameters

Mass of Statellite	
Mass of Statellite	$= 10^3$ kg $= m$
Radius of Satellite	$= 0.5$ m $= r$
Moment of Inertia	$= \frac{1}{2}\, mr^2 = I$
Spin	$= 14.14$ rad/s $= \omega$
Max Acceleration at Rim	$= 100$ m/s$^2 \cong 10$ **G**
Instrumentation	= one telescope of 25.4 cm of diameter and associated recording and transmitting equipment
	Orbit Data
Radius of Orbit	$= 7 \times 10^6$ m
Altitude	= 632 km
Acceleration (g)	$= 8.15$ m/s^2
Period	= 97 min
Velocity	= 7560 m/s
Atmospheric Density	$= 10^{-15}$ g/cm^3
	Relativistic Precession Lense–Thirring Effect
$\frac{d\phi}{dt} = 2.7 \times 10^{-6}$ rad/year on surface of earth[a] $\cong 2.0 \times 10^{-6}$ rad/year at orbit altitude or 6×10^{-14} rad/s	
Equivalent Torque $= I\omega \frac{d\phi}{dt} = 1.0 \times 10^{-10}$ N · m	
	de Sitter–Fokker Effect
$\frac{d\phi}{dt} = 3.3 \times 10^{-5}$ rad/year for indicated orbit[b] or $\cong 1.0 \times 10^{-12}$ rad/s	
Equivalent Torque $= I\omega \frac{d\phi}{dt} = 1.7 \times 10^{-9}$ N · m	

[a] Based on a calculation by H. Yilmaz

[b] Based on a calculation by F. Pirani

(b) A thermalized component in which incident particles stop to be thermalized and then either rebound with a thermal distribution of velocities or remain attached to the surface.

26. It is clear that neither component of itself could give rise to a net torque (if we neglect the spin of he satellite). Specular reflections produce a change in momentum which is by definition perpendicular to the surface; consequently, in bouncing from a spherical surface each individual reflection results in zero net torque. For thermalizing collisions, on the other hand, while the average torque or rebound

is zero, the torque associated with the initial stopping of the incident particles is finite. If only thermalizing collisions occured this net torque for the collision would be cancelled by similar collisions on the opposite side of the satellite. However, the ratio of thermalizing to specular collisions is normally temperature dependent. Thus if one hemisphere of the satellite were warmer than the other, (as the side facing the sun almost surely would be) one would exect a number of thermalizing collisions in one hemisphere which would not be balanced by thermalizing collisions in the other hemisphere. For a satellite moving at high velocity perpendicular to its spin axis, this could result in a significant torque perpendicular to the spin axis.

27. To estimate the magnitude of this effect we consider an air density ρ of about 10^{-15} g/cm^3 at 600 km. (This is just about a factor of two higher than the estimate by T. S. Stern, Physics of Fluids, **1**, 165 (1958)).

28. Then the total material μ swept out by the satellite in orbit is given by

$$\mu = \rho v A = 10^{-9} \times 7560 \times \frac{\pi}{4} = 5.9 \times 10^{-6} \text{ g/s}$$

If all particles bounced elastically this would result in change of velocity of

$$2v\frac{\mu}{m} = 9 \times 10^{-8} \text{ m/s}$$

which corresponds to a total drag force of 9×10^{-5} N. Of course, the direct drag is far too small to be of concern in the experiment. Even after ten years the change in orbit radius would be negligible.[c]

29. However, it is large compared to the size torques allowed. For this force to produce a torque equal to the Lense–Thirring relativistic torque of 1.0×10^{10} N · m. it is only necessary that the center of effort for the torque be displaced by $(1.0 \times 10^{-10}) \times (9 \times 10^{-5})$ or 1×10^{-7} m. An effect of this magnitude might be expected if as much as one collision in 50000 on one hemisphere was not matched by a corresponding collision on the opposite hemisphere.

30. It is, of course, possible that the temperature coefficient for thermalized vs. specular reflection could be made very low or that the temperature could be kept very uniform. However, where possible in an experiment of this type it is better to remove perturbations that might tend to throw doubt on the results.

A Technique to Minimize the Atmospheric Perturbations

31. As it turns out, it is probably not very difficult to shield the satellite from the larger atmospheric forces. Moreover, such shielding has a number of subsidiary benefits associated with other perturbations. To provide the shield, the primary experimental satellite can be encased in a larger hollow sphere or "tender" satellite

[c]At these high altitudes the ionization is considerable and electro-magnetic contributions to drag are probably more important (especially for a slightly charged body) than the direct Newtonian collisions. However, in the calculation of torques these electro-magnetic effects do not appear to be very important.

having the center of mass. The tender satellite could be equipped with light beams or other sensing mechanism to monitor the position of the primary satellite *without* exerting significant forces or torques on the primary satellite. The use of *external* vernier rocket jet would allow minor repositioning of the tender with respect to the primary satellite so that the two would not collide during the course of the experiment.

32. Two additional advantages of the use of the "tender" are immediately apparent:

(a) It provides a shield against micro-meteorites that might otherwise defeat the experiment.
(b) it shields the satellite against the direct rays of the sun, thus allowing better temperature control and minimizing difficulties with thermal expansion.

Other advantages of the tender will become clear when we take up other perturbations.

33. There would also be a number of useful by-products from the development of a shielded satellite. Such a satellite would provide means for obtaining very accurate gravitational data since atmospheric to keep the tender in position might provide very accurate data on atmospheric drag. It is even possible that data resulting from such orbits might allow one to correct for classical perturbations in orbital-type relativistic experiments.

Remaining Atmospheric Perturbations

34. Having now disposed of the high-relative velocity of the satellite and the atmosphere by placing the satellite in a container so that the local atmosphere moves with the satellite, we now consider the remaining atmospheric perturbations. These perturbations will be the result of transfer of angular momentum from the tender to the satellite wia the residual gas contained in the tender.

35. Again using a density of 10^{-5} g/cm^3 the total net torque resulting from this effect assuming the worst situation with 100 per cent thermalizing collisions is about 1×10^{-9} N $\cdot$ m. This torque is about ten times the Lense–Thirring relativistic effect. However, it should have no net effect on the experiment because the torque involved has no component perpendicular to the spin. One practical problem in the design of the tender will be to provide an adequate opening to the outside (probably out the back of the rocket) to allow sufficient "pumping speed" to maintain the good vacuum in the tender, despite outgassing problems, and nevertheless keep the primary satellite shielded from high-velocity molecules outside the tender. On the other hand, it is possible that the speed of the tender could be utilized as a vacuum pump to attain an even better vacuum in the satellite than exists in the exterior atmosphere.

36. We turn now to a consideration of other perturbations that might influence a shielded satellite.

The Gravitational Perturbation

37. As we noted earlier, the interaction of the gravitational gradient with the mass moments of the satellite can príduce torques that would alter the angular momentum axis. Of course, if the satellite were perfectly spherical the quadrupele and all higher order moments would vanish and there would be no perturbation. To estimate the magnitude of the effect, however, let us assume that a small mass ring (total mass ℓ) is placed on the equator of the satellite in addition to the intrinsic mass of the spherical satellite. (In some concepts such a ring might be considered desid=rable as a technique to fix the principal axis and stabilize the angular momentum axis in the body of the satellite.) If we let θ represent the angle between the axis of the mass ring and the gravitational gradient, then the torque produced will be given by:

$$\text{Torque} = \frac{4}{\pi}\left(\frac{\ell}{1g}\right)\frac{r}{R}\sin\theta\cos\theta$$

Thus, if the resulting precession is to be small compared to the Lense–Thirring relativistic effect, we have the requirement that

$$\frac{\ell}{m}\sin\theta\cos\theta \ll \frac{1.0\times10^{-10}}{3.7\times10^{-4}} = 0.27\times10^{-6}\,.$$

Consequently, if no restriction is placed on θ we have the requirement that

$$\ell \ll 0.4\text{ g}$$

38. However, the limitation may not be so severe. For instance, in polar orbit, if the spin axis is always perpendicular to the plane of the orbit, the value of $\sin\theta\cos\theta$ is zero. Of course, any error in the initial alignment of the spin angle would result in some net torque, but this effect should be easily calculable, to the accuracy that the satellite moments are known.

39. Fortunately, it should be possible to measure the ratio of the satellite's moments of inertia very accurately in flight. It is only necessary to record with the telescope the period of the classical precession of the spin axis in the body of the satellite. Once the period is known, classical corrections for the gravitational interaction of the moments with the earth's gravitational field should be possible with high accuracy.

40. There is, of course, no essential reason for using an intentionally nonspherical satellite. It is probable that the drift of the rotation axis in the body of a spherical satellite would be so slight that it would not interfere with measurements. Indeed, in a truly spherical satellite the drift rate is zero. However, any irregularities in the body resulting from uneven temperatures or other causes would define a new major axis, so that in practice some drift would exist even if it were almost vanishingly small. This problem is, of curse, considerably simplified by the use of the tender satellite to stabilize temperatures and shield from the direct rays of the sun.

41. One strategy to overcome any drift would be to use a number of adjustable moment-shims that would allow one to define the major axis at will. These shims

could then be used toward the end of the experiment to realign the spin axis in the body of the satellite.

The Magnetic Perturbations

42. The magnetic field of the earth can interact with magnetic moments of the satellite. Such magnetic moments might result from induced currents in the satellite, electronic current loops in the instrumentation of the satellite, magnetization of materials in the satellite, or rotation of the satellite as a charged sphere.

43. It is clear that the use of ferro-magnetic materials in the satellite must be avoided, and that other materials should be as non-magnetic as possible. However, if these precautions are followed on would not expect serious trouble from magnetic materials.

44. Any electrical circus in the satellite should be designed to minimize the effective area of D. C. current loops. Indeed, where possible the circus should be turned off during the course of the experiment. To estimate the magnitude of these interactions conside a current loop around the equator of the satellite. The resulting torque τ is given by

$$\tau = I\vec{B} \times \vec{A} = 4 \times 10^{-5} \times \frac{\pi}{4} I = 3.2 \times 10^{-5} I$$

Thus, if the resulting torque is to be less than 1×10^{-10} the current in such a loop would have to average less than 0.3×10^{-5} A or alternatively, a one-ampère circu would have to be restrictd to an effective area of less than 0.03 cm. Of course, it is the component of the circu area-vector parallel to the spin axis that counts. The component of area perpendicular to the spin axis does not result in a net torque unless the current is very large.

45. Probably induced currents in the satellite resulting from rotation in the earth's magnetic field will be one of the more serious problems. The forces resulting from these currents can be made to vanish either by making the satellite a perfect conductor or a perfect insulator. Barring major breakthroughs in superconductors, the latter alternative appears the easiest to approximate.

46. To get a feeling for the size of effects one can expect from induced currents, let us consider a conducting loop on the surface of the satellite that passes through both poles. The induced voltage in such a loop is given by $E = d(\vec{B} \cdot \vec{A})/dt$ or $E = \omega BA \cos\theta = 4.4 \times 10^{-4}$ V. Thus, if the resistance of the loop were less than 50 Ω, the torque would exceed the Lense–Thirring relativistic "torque".

47. However, in any proposed experiment, most of the torque so generated would primarily be directed toward slowing the rotation. Because of time integration over spin and orbit, the net effect perpendicular to the spin axis will tend to be quite small.

48. In addition, one must avoid large conducting bodies in the satellite. The use of laminated and high resistance components is desirable. Probably even the telescope mirrors should be laminated. This is illustrated by fact that the maximum

torque resulting from a solid rectangular piece of material is given approximately by

$$\tau \cong \omega B^2 \frac{V}{\rho} W^2 .$$

where V is volume of the material, ρ is the resistivity, and W is the width or second largest dimension of the rectangle. For example, a copper bar 2 cm $\times$ 1 cm $\times$ 1 cm would produce a torque just about equal to the relativistic torque. It is clearly advantageous keep W as small as possible. Again, the tender can simplify the problem by making it possible to put all or almost all electronic equipment in the tender.

49. A negligible torque will result from the rotation of the satellite even if one assumes it to be charged to about 12 Volt (which seems very unlikely in a shielded satellite). The total charge on such a satellite is about 3×10^{-9} C.

Thus, with a rotation about 2.3 cycles per second, the total "current" is less than 10^{-8} A. This is a factor of 10^{-3} smaller than would be required for a serious perturbation.

Useful By-Products of the Experiment

50. Development of the technique required to use the tender satellite concept would also open the door to a number of other useful experiments which will be mentioned briefly.

51. The fact that the interior satellite is almost completely shielded from external forces implies that the trajectory of the entire satellite will be essentially a pure geodesic without any atomspheric perturbation. The trajectory, therefore, might be useful to improve international geodetic data. A modified version to provide higher sustained thrust for the tender could be used to investigate gravitational anomalies at low altitudes where air drag would usually preclude such measurements entirely.

52. On the other hand, if a technique can be developed to monitor the total thrust required to maintain the tender in position, then this information might be useful to provide a minute-by-minute evaluation of the drag on the tender.

53. If such a satellite and tender were constructed using very different materials having different "packing fractions" (such as copper vs. beryllium) then the technique could be used to provide a check of the equivalence of gravitational and inertial mass. For this purpose it would probably be good to use a high orbit where the air drag is exceptionally low.

54. The accuracy obtainable in this experiment depends entirely on the accuracy of the tender sensing and control mechanism. For instance, discrepancy of one part in 10^6 in the ratio for inertial to gravitational mass would result in a seven meter difference in the radius for a stable orbit. If instrumentation were accurate enough to permit detection of a 7 cm difference in radius then a discrepancy of one part in 10^8 might be observable.

Summary

55. Assuming no major interactions have been ignored, it appears that there is no single perturbation of such magnitude that it would make the experiment non-feasible. The physical design of a satellite to avoid all interactions with the desired level of confidence may be difficult, but can probably be done within a few years. Considerable work should be done both to improve the theoretical understanding of the relativistic problem and to determine more accurately the size of the classical interactions and the difficulties of practical design within the limitations imposed.

56. A reasonable program to investigate the relativistic effects would probably begin by checking out techniques in the simpler classical gravitation and inertial mass could probably be carried out in the process of checking out the gyroscopic techniques. The first real gyroscopic experiment should probably be measurement of the large de Sitter–Fokker effect. Measurement of the Lense–Thirring effect, while of greater interest, should probably be postponed until all techniques are fully established.

57. The experiments is an important one which appears to be within the responsibilities of the NASA. It would be, in the most literal sense a direct measurement of space itself.

Acknowledgments

58. The author wihes to acknowledge many helpful discussions with his colleagues Hugh Everett III and R. Schwartz, as well as some recent useful discussions with P. G. Bergmann, F. Pirani and H. Yilmaz. However, these individuals are not responsible for errors or omissions that may remain in this proposal.

MOTION OF A GYROSCOPE ACCORDING TO EINSTEIN'S THEORY OF GRAVITATION*

BY L. I. SCHIFF

INSTITUTE OF THEORETICAL PHYSICS, DEPARTMENT OF PHYSICS, STANFORD UNIVERSITY

Communicated April 19, 1960

The Experimental Basis of Einstein's Theory.—Einstein's theory of gravitation, the general theory of relativity, has been accepted as the most satisfactory description of gravitational phenomena for more than forty years. It is a theory of great conceptual and structural elegance, and it is designed so that it automatically agrees in the appropriate limits with Galileo's observation of the equality of gravitational and inertial mass, with Newton's mechanics of gravitating bodies, and with Einstein's special theory of relativity. Leaving aside the very important matter of elegance, we wish in this section to examine the experimental basis of the theory. This basis consists of the three points of limiting agreement with earlier results just mentioned, together with certain astronomical evidence.

The equality of gravitational and inertial mass was originally formulated in terms of equal accelerations for all freely falling test particles, regardless of mass or chemical composition. In this form, it is a consequence of general relativity theory insofar as test particles move in accordance with the geodesic equations for a Riemannian metric. However, the experimental evidence on freely falling particles is not of very great accuracy. Much more precise experiments were performed about half a century ago by Eötvös and collaborators,[1] and are now being repeated with improved technique by Dicke.[2] Since they make use of particles that are not in free fall but are subjected to nongravitational constraints, the relation with Einstein's theory is not quite so simple as just indicated. On the other hand, we can regard these experiments as establishing with great confidence the *principle of equivalence*, which we express in the following way: all observations made locally on a system in a static, uniform gravitational field in the absence of local background matter agree with corresponding observations made on the same system when it is subjected to an equivalent acceleration in the absence of the field.

This statement of the equivalence principle goes beyond the direct evidence of the Eötvös experiments. For one thing, the Eötvös experiments do not compare observations made in the presence and absence of a gravitational field, but rather compare observations made with an acceleration in one direction and a gravitational field in another. More important, the observations made are not perfectly general, but consist of mass comparisons. However, there is a great deal of physical content to a precise mass measurement, since many of the phenomena known in physics enter into it with sufficient effect to be noticeable;[3–5] this occurs through the Hamiltonian of the system, of which the mass is essentially the ground-state eigenvalue. It would be remarkable if the equivalence principle were to apply to the ground states of the Hamiltonians of physics, and not also to the excited states that determine, for example, the transition frequencies. Thus, while this formulation of the equivalence principle is an extrapolation from the direct evidence of the Eötvös experiments, it is not so great an extrapolation as might at first be supposed.

The other points of contact with Einstein's theory of gravitation are most readily discussed in terms of the formalism of the theory. In the simplest interesting case, that of the gravitational field about a stationary, spherically symmetric mass m, the metric that represents the solution of the field equations can be written in the Schwarzschild standard form[6]

$$ds^2 = (1 - 2m/r)dt^2 - dr^2/(1 - 2m/r) - r^2(d\theta^2 + \sin^2\theta\, d\phi^2), \quad (1)$$

where units have been chosen so that the speed of light and the Newtonian gravitational constant are equal to one. Then special relativity is valid whenever m/r may be neglected in comparison with unity. The metric (1) may be supplemented by the geodesic equation of motion for a test particle, and the null-geodesic equation of motion for a light ray; alternatively, these equations of motion can be obtained from the field equations themselves.

Suppose now that we wish to verify the structure of equation (1) by comparison with observation. We may then write the metric in the form

$$ds^2 = (1 + \alpha m/r + \beta m^2/r^2 + \ldots)dt^2 - (1 + \gamma m/r + \delta m^2/r^2 + \ldots)dr^2 - r^2(d\theta^2 + \sin^2\theta\, d\phi^2), \quad (2)$$

where $\alpha, \beta, \gamma, \delta, \ldots$ are of order unity. This implies an expansion in powers of the quantity m/r, which is very small in all cases of observational interest. It also implies spherical symmetry, in which case $d\theta$ and $d\phi$ appear in the combination shown and any multiplying series in powers of m/r is readily transformed away by a change of the radial variable. It then follows that the limiting case of Newtonian mechanics requires only that $\alpha = -2$. The gravitational red shift is also accounted for in the same way. The theory of the gravitational deflection of light passing close to the sun results from the null-geodesic equation for a light ray, together with the above value for α and the choice $\gamma = +2$. Finally, the theory of the precession of the perihelion of the orbit of the planet Mercury results from the geodesic equation of motion for a test particle, the foregoing values for α and γ, and the choice $\beta = 0$. Higher terms in the series of (2) have not been subjected to experimental test.[7]

We now ask to what extent the above numerical values of α, β, and γ, and the equations of motion, may be inferred without recourse to Einstein's theory of gravitation. It is argued elsewhere[8] that the values of α and γ, and the null-geodesic equation for a light ray, can be obtained correctly from the equivalence principle as formulated above, together with the special theory of relativity. On the other hand, the value of β, which depends in an essential way on the nonlinearity of the field equations, and the geodesic equation for a test particle, cannot be obtained in this way. Thus, the planetary orbit precession remains as the sole experimental basis for Einstein's theory. Recent terrestrial experiments on the gravitational red shift[9, 10] should be thought of as providing additional evidence for the equivalence principle as formulated above, rather than for the general theory of relativity.

There are at least three general ways in which one might look for new experimental verifications of Einstein's theory of gravitation to supplement the planetary orbit precession. The first is related to cosmological implications, such as variations of certain natural constants with time,[2] and the structure and evolution of the universe.[11] As an example of the latter, a search could be made for a minimum apparent size of extra-galactic nebulae, since the most distant visible nebulae might be expected to appear larger and redder than those at intermediate distances. The second approach consists in searching for gravitational radiation, either arising from extra-terrestrial sources, or possibly generated in the laboratory.[12] The third proposal is specifically designed to involve terms in the metric (2) beyond α and γ, and the equation of motion of matter of finite rest mass beyond the Newtonian approximation. This could be accomplished by measuring the precession of the spin axis of a torque-free gyroscope in the gravitational field of the rotating earth;[13, 14] although the β, δ, . . . terms in (2) are not involved, new off-diagonal space-time components of the metric tensor that arise from earth rotation become significant. The present paper is devoted mainly to a derivation and discussion of the results already published in a brief notice.[14]

Equations of Motion of a Spinning Test Particle.—Papapetrou[15] has derived covariant equations of motion for the center of mass and the spin angular momentum of a spinning test particle (torque-free gyroscope). These equations are incomplete in two respects. (1) They refer to free fall, that is, to motion in a pure gravitational field. Thus, while they describe a gyroscopic satellite, they do not describe a gyroscope in an earth-bound laboratory. Our first task, then, will be to generalize these equations by including a non-gravitational constraining force **F**. (2) There are only three independent equations for the six components of the angular momentum tensor, so that supplementary conditions must be imposed. These conditions may be chosen in either of two natural ways. Corinaldesi and Papapetrou[16] require the timelike components of the angular momentum tensor to vanish in the coordinate system in which the central attracting body is at rest. Pirani[17] requires these components to vanish in the rest-frame of the gyroscope. We shall consider both of these possibilities here.

Papapetrou uses the method of Fock[18] to obtain the equations of motion. This consists in starting with the "dynamical equation" for the energy-momentum tensor, according to which its covariant divergence is zero:

$$T^{\mu\nu}{}_{;\nu} = 0. \tag{3}$$

$T^{\mu\nu}$ represents the test particle, and is supposed to vanish outside of a narrow tube in four-dimensional space-time that surrounds the world line of some representative point X^μ of the particle. The space components of X^μ, X^i ($i = 1, 2, 3$), are regarded as functions of the time $X^4 = t$, or of the proper time s along the world line. This world line need not be a geodesic of the gravitational field, which is described by the metric tensor $g_{\mu\nu}$. The equations of motion are obtained by manipulating integrals of the form

$$\int T^{\mu\nu}\, dv, \qquad \int (x^\alpha - X^\alpha) T^{\mu\nu} dv, \qquad \ldots, \tag{4}$$

where the integrations extend over three-dimensional space for t = constant. A point test particle is one for which some of the components of the first integral of (4) fail to vanish, but the other integrals are always zero. A spinning test particle is one for which some of the components of both integrals fail to vanish, but an integral containing more than one factor $(x^\alpha - X^\alpha)$ is always zero.

The four-velocity of the representative point of the particle is $u^\mu = dX^\mu/ds$. We generalize the nongravitational constraining force **F** to a four-force F^μ such that $F^i = \mathbf{F}$ and $F^\mu u_\mu = 0$,[19] and assume that it is applied at the point X^μ. Equation (3) is still valid, but $T^{\mu\nu}$ must now be thought of as consisting of two parts, one representing the test particle and the other the force. We avoid a detailed specification of the second part by assuming that its covariant divergence is zero except at the point of application, and is there proportional to F^μ. Then if we denote just the first part by $T^{\mu\nu}$, equation (3) is to be replaced by

$$T^{\mu\nu}{}_{;\nu} = (F^\mu/u^4)\delta(x^1 - X^1)\delta(x^2 - X^2)\delta(x^3 - X^3), \tag{5}$$

where the right side is a product of Dirac δ functions. That equation (5) is actually covariant is most readily shown by using it to calculate the equation of motion of a point test particle after the manner of Papapetrou. The result is

$$m_0(du^\mu/ds + \Gamma^\mu{}_{\alpha\beta}u^\alpha u^\beta) = F^\mu, \tag{6}$$

which reduces to the geodesic equation when $F^\mu = 0$. The rest mass of the particle,

$$m_0 \equiv (1/u^4)\int T^{44}\, dv, \tag{7}$$

is a scalar,[15] and can be shown to be constant along the world line.

We may also apply Papapetrou's procedure to a spinning test particle, using, however, equation (5) in place of (3). The spin angular momentum is defined by

$$S^{\mu\nu} \equiv \int (x^\mu - X^\mu)T^{\nu 4}\, dv - \int (x^\nu - X^\nu)T^{\mu 4}\, dv, \tag{8}$$

which can be shown explicitly to have the transformation properties of a tensor.[15] The equation of motion of the spin is unaffected by inclusion of F^μ, and may be written in the covariant form[15]

$$(DS^{\mu\nu}/Ds) + u^\mu u_\alpha(DS^{\nu\alpha}/Ds) - u^\nu u_\alpha(DS^{\mu\alpha}/Ds) = 0, \tag{9}$$

where D/Ds represents covariant differentiation along the world line:

$$(DS^{\mu\nu}/Ds) \equiv (dS^{\mu\nu}/ds) + \Gamma^\mu{}_{\alpha\beta}S^{\alpha\nu}u^\beta + \Gamma^\nu{}_{\alpha\beta}S^{\mu\alpha}u^\beta. \tag{10}$$

The equation of motion of the representative point of the particle is modified from equations (6) and (7) by terms of order $S^{\mu\nu}$. However, this effect of the spin on the orbital motion of the particle is completely negligible in situations of current experimental interest.

It is sometimes convenient to rewrite equation (9) in the noncovariant form

$$(DS^{\mu\nu}/Ds) + (u^\mu/u^4)(DS^{\nu 4}/Ds) - (u^\nu/u^4)(DS^{\mu 4}/Ds) = 0. \tag{11}$$

This may be obtained from equation (9) in the following way:[20] set $\nu = 4$ in (9) and multiply by (u^ν/u^4); then set $u = 4$ in (9) and multiply by (u^μ/u^4); then add these two equations and substitute into (9). Since (11) is antisymmetric in u and ν, it seems at first to comprise six independent equations. However, if we put $\mu = i$, $\nu = 4$, we obtain a trivial identity. Thus only three of these equations are actually independent; the same remark applies to (9). It is therefore necessary to impose a supplementary condition.

The physical meaning of the supplementary condition is best seen by writing $S^{\mu\nu}$ in rectangular coordinates. Then since T^{i4} is the momentum density in the i direction, it follows from equation (8) that

$$\mathbf{S} = (S_x, S_y, S_z) = (S^{23}, S^{31}, S^{12}) \tag{12}$$

is the spin angular momentum vector with respect to the representative point

$$\mathbf{r} = (X^1, X^2, X^3). \tag{13}$$

Further, for $\mu = i$, $\nu = 4$, the second integral in (8) is zero since the integration extends over three-dimensional space for constant time, so that x^4 and X^4 are both equal to t. We thus obtain

$$S^{i4} = \int (x^i - X^i)T^{44}\, dv = m_0 u^4 \epsilon^i, \tag{14}$$

where m_0 is given by (7), and ϵ^i is the position of the center of mass of the particle with respect to X^i.

The Corinaldesi-Papapetrou (CP) supplementary condition is[16]

$$S^{i4} = 0 \tag{15}$$

in the rest-frame of the central attracting body, so that X^i is the center of mass in this coordinate system. On the other hand, Pirani's supplementary condition is[17]

$$S^{\mu\nu} u_\nu = 0. \tag{16}$$

In the rest-frame of the particle, $u^i = 0$ and u_i is zero or negligibly small,[21] so that $S^{i4} = 0$ and X^i is the center of mass in the particle rest-frame. As is well known,[22] the position of the center of mass of an object that possesses internal angular momentum is not a Lorentz-invariant quantity. It follows that the supplementary condition removes the ambiguity inherent in the choice of the representative point X^i in terms of which the motion of the spinning test particle is described and at which the constraining force is applied, by specifying that this point is the center of mass in one or another coordinate system.

In using the CP condition, it is more convenient to start from equation (11) than from (9). With the help of (15) and (10), equation (11) becomes

$$(DS^{ij}/Ds) = \Gamma^4{}_{k\mu}(u^\mu/u^4)\,(S^{ik}u^j - S^{jk}u^i). \tag{17}$$

We now rewrite (17) in terms of the rectangular coordinates (12) and (13), making use of the standard form (1) of the Schwarzschild metric. To lowest order, which requires knowledge only of α and γ in (2), we obtain[16]

$$(d\mathbf{S}/dt) = (m/r^3)\,[2\mathbf{S}(\mathbf{r}\cdot\mathbf{v}) + 2\mathbf{v}(\mathbf{r}\cdot\mathbf{S}) - \mathbf{r}(\mathbf{v}\cdot\mathbf{S}) - (3\mathbf{r}/r^2)(\mathbf{r}\cdot\mathbf{v})(\mathbf{r}\cdot\mathbf{S})], \tag{18}$$

where $\mathbf{v} = d\mathbf{r}/dt$. It is instructive also to find the analog of (18) when the isotropic form of the Schwarzschild metric is used;[6] to first order, this metric is

$$ds^2 = (1 - 2m/r)dt^2 - (1 + 2m/r)(dr^2 + r^2 d\theta^2 + r^2 \sin^2\theta\, d\phi^2), \tag{19}$$

and is the same to this order as the metric expressed in Fock's harmonic coordinates.[23] Equation (17) then becomes to lowest order

$$(d\mathbf{S}/dt) = (m/r^3)\,[3\mathbf{S}(\mathbf{r}\cdot\mathbf{v}) + \mathbf{v}(\mathbf{r}\cdot\mathbf{S}) - 2\mathbf{r}(\mathbf{v}\cdot\mathbf{S})]. \tag{20}$$

In using the Pirani condition (16), it is more convenient to start from equation (9) than from (11). We differentiate (16) and substitute into (9) to obtain

$$(DS^{\mu\nu}/Ds) = (u^\mu S^{\nu\alpha} - u^\nu S^{\mu\alpha})(du_\alpha/ds - u^\beta u_\rho \Gamma^\rho{}_{\alpha\beta}). \tag{21}$$

Since $u_\alpha = g_{\alpha\gamma}u^\gamma$, du_α/ds can be expressed in terms of the nongravitational constraining acceleration $f^\mu = F^\mu/m_0$ by means of equation (6). In doing this, terms of higher than first order in $S^{\mu\nu}$ on the right side of (21) are neglected. After some reduction, (21) becomes to this approximation

$$(DS^{\mu\nu}/Ds) = (u^\mu S^{\nu\alpha} - u^\nu S^{\mu\alpha})f_\alpha. \tag{22}$$

The components S^{i4} and f_4 may be eliminated from (22) by using the supplementary condition (16) and the relation $f_\mu u^\mu = 0$. The standard form of the Schwarzschild metric then gives for the spin equation of motion

$$(d\mathbf{S}/dt) = (3m/r^3)[\mathbf{v}(\mathbf{r}\cdot\mathbf{S}) - (\mathbf{r}/r^2)(\mathbf{r}\cdot\mathbf{v})(\mathbf{r}\cdot\mathbf{S})] + \mathbf{S}(\mathbf{v}\cdot\mathbf{f}) - \mathbf{f}(\mathbf{v}\cdot\mathbf{S}); \tag{23}$$

the isotropic form leads to

$$(d\mathbf{S}/dt) = (m/r^3)[\mathbf{S}(\mathbf{r}\cdot\mathbf{v}) + 2\mathbf{v}(\mathbf{r}\cdot\mathbf{S}) - \mathbf{r}(\mathbf{v}\cdot\mathbf{S})] + \mathbf{S}(\mathbf{v}\cdot\mathbf{f}) - \mathbf{f}(\mathbf{v}\cdot\mathbf{S}). \tag{24}$$

Transformation to the Gyroscope Rest-Frame.—The four equations of motion—(18), (20), (23), and (24)—are quite different from each other, and yet presumably all describe the same physical system. They may be reconciled by transforming each to the rest-frame of the gyroscope. Physically, this corresponds to the fact that measurements on the gyroscope are most readily interpreted as being made by a co-moving observer, who may then transmit these measurements to the outside world. If, for example, the gyroscope is in a satellite, it may be observed by a human being who travels with it, or by a device which telemeters information to the earth below. One of the questions that is most important from an observational point of view is whether or not there is a change in the frequency of rotation of the gyroscope, since this would be the simplest quantity to measure with precision. Corinaldesi and Papapetrou[16] argued on the basis of equation (18), which is the only one of the four that they obtained, that since $(d\mathbf{S}/dt)$ has a component parallel to $\mathbf{S}$, its magnitude will change, and hence the rotation frequency might be expected to change. On the other hand, the four equations predict quite different values for the change in magnitude of $\mathbf{S}$.

The physical situation is as follows. The co-moving observer (human or otherwise) measures the rotation frequency by comparison with a standard clock that is carried along, and interprets this as being the ratio of the magnitude of the angular momentum to the moment of inertia, both measured in the gyroscope rest-frame. The moment of inertia is determined by the dimensions of the gyroscope, and hence by comparison with a standard measuring rod that is also carried along by the observer. Thus, if we transform $\mathbf{S}$ to the co-moving system, we have a consistent set of observations. As we shall see, it turns out that the magnitude of the transformed angular momentum $\mathbf{S}_0$ is constant. This means that if the dimensions do not change, the rotation frequency is constant, and the gyroscope behaves like a clock which can be set to any desired frequency. The frequency will of course exhibit Doppler and gravitational shifts when observed from the outside, just like that of a more conventional clock. This will occur, for example, if the gyroscope is in a satellite and its rotation frequency is telemetered to the earth below.

The transformation to the gyroscope rest-frame consists of two parts, a coordinate transformation involving changes of order m/r, and a Lorentz transformation involving changes of order v^2. Since we need work only to first order in m/r

and in v^2, we can deal with the two transformations separately, and combine their effects later.

The first transformation recognizes the fact that a coordinate length δx^i in the original system corresponds to a proper length $\delta s = (-g_{ii})^{1/2}\delta x^i$, and that it is this δs that is measured by the standard rod of the co-moving observer. Thus when the isotropic metric (19) is used, all coordinate lengths are to be multiplied by $(1 + m/r)$, to first order, to convert them to the co-moving system. From (12), we see that each component of **S** transforms like the product of two coordinates, so that **S** must be multiplied by $(1 + 2m/r)$ to obtain $\mathbf{S}_0$, again to first order:

$$\mathbf{S}_0 = (1 + 2m/r)\mathbf{S}. \tag{25}$$

With the standard metric (1) the situation is slightly more complicated. Here, radial length intervals are to be multiplied by $(1 + m/r)$, and tangential length intervals are to be left unchanged. It then follows from (12) that the tangential components of **S** are to be multiplied by $(1 + m/r)$, and the radial component is to be left unchanged:

$$\mathbf{S}_0 = \mathbf{S} + (m/r)[\mathbf{S} - (\mathbf{r}/r^2)(\mathbf{r}\cdot\mathbf{S})]. \tag{26}$$

The Lorentz transformation is most conveniently written down if **v** is chosen to be along one of the rectangular axes, say x:

$$\begin{array}{lll} S_0^{12} = \gamma S^{12} + v\gamma S^{24}, & S_0^{23} = S^{23}, & S_0^{31} = \gamma S^{31} - v\gamma S^{34}, \\ S_0^{14} = S^{14}, & S_0^{24} = \gamma S^{24} + v\gamma S^{12}, & S_0^{34} = \gamma S^{34} - v\gamma S^{31}; \end{array} \tag{27}$$

here, $\gamma = (1 - v^2)^{-1/2}$. The CP condition (15) requires that $S^{14} = S^{24} = S^{34} = 0$, so that equations (27) become, with the help of (12):

$$\begin{array}{lll} S_{0x} = S_x, & S_{0y} = \gamma S_y, & S_{0z} = \gamma S_z, \\ S_0^{14} = 0, & S_0^{24} = v\gamma S_z, & S_0^{34} = -v\gamma S_y. \end{array} \tag{28}$$

Thus, in this case, the component of **S** parallel to **v** is left unchanged, but the perpendicular components are multiplied by $(1 + {}^1/_2v^2)$, to first order. This may be written in a rotation-covariant manner as follows:

$$\mathbf{S}_0 = \mathbf{S} + {}^1/_2[v^2\mathbf{S} - \mathbf{v}(\mathbf{v}\cdot\mathbf{S})]. \tag{29}$$

The Pirani condition (16) requires, when **v** is along x, that

$$S^{14} = 0, \qquad S^{24} = vS^{12}, \qquad S^{34} = -vS^{31}.$$

Here, effects of order m/r have been neglected, since we are permitted to separate the coordinate and Lorentz transformations in lowest order. Equations (27) then become

$$\begin{array}{lll} S_{0x} = S_x, & S_{0y} = \gamma^{-1}S_y, & S_{0z} = \gamma^{-1}S_z, \\ S_0^{14} = 0, & S_0^{24} = 0, & S_0^{34} = 0. \end{array} \tag{30}$$

The last three of equations (30) are in agreement with the remark made just after equation (16), that when the Pirani condition is used, $S^{i4} = 0$ and hence X^i is the center of mass in the gyroscope rest-frame. The first three of equations (30) show that the component of **S** parallel to **v** is left unchanged, while the perpendicular components are multiplied by $(1 - {}^1/_2 v^2)$, to first order. In rotation-covariant form this becomes

$$\mathbf{S}_0 = \mathbf{S} - {}^1/_2[v^2\mathbf{S} - \mathbf{v}(\mathbf{v}\cdot\mathbf{S})]. \quad (31)$$

Equation (18) was obtained by using the standard form of the Schwarzschild metric together with the CP supplementary condition. Thus, in order to transform it to the gyroscope rest-frame we combine the transformations (26) and (29) to first order:

$$\mathbf{S}_0 = \mathbf{S} + (m/r)[\mathbf{S} - (\mathbf{r}/r^2)(\mathbf{r}\cdot\mathbf{S})] + {}^1/_2[v^2\mathbf{S} - \mathbf{v}(\mathbf{v}\cdot\mathbf{S})]. \quad (32)$$

The time derivative of (32) is

$$(d\mathbf{S}_0/dt) = (d\mathbf{S}/dt) - (m\dot{r}/r^2)\mathbf{S} + (3mr/\dot{r}^4)\mathbf{r}(\mathbf{r}\cdot\mathbf{S}) - (m/r^3)\mathbf{v}(\mathbf{r}\cdot\mathbf{S}) - (m/r^3)\mathbf{r}(\mathbf{v}\cdot\mathbf{S}) + v\dot{v}\mathbf{S} - {}^1/_2\dot{\mathbf{v}}(\mathbf{v}\cdot\mathbf{S}) - {}^1/_2\mathbf{v}(\dot{\mathbf{v}}\cdot\mathbf{S}). \quad (33)$$

Here, dots denote time derivatives, and terms of order $(m/r)(d\mathbf{S}/dt)$ and $v^2(d\mathbf{S}/dt)$ have been neglected in comparison with $(d\mathbf{S}/dt)$. Also, the difference between the differential time intervals dt in the two coordinate systems has been neglected, since it is of fractional order m/r and v^2. Several substitutions can be made on the right side of equation (33). It is evident that $\dot{r} = (\mathbf{r}\cdot\mathbf{v})/r$ and $\dot{v} = (\mathbf{v}\cdot\dot{\mathbf{v}})/v$. The acceleration $\dot{\mathbf{v}}$ can be obtained from the equation of motion (6). Since we require $\dot{\mathbf{v}}$ only to first order in m/r and v^2, spin corrections to (6) can be neglected and the Newtonian approximation used:

$$\dot{\mathbf{v}} = -(m/r^3)\mathbf{r} + \mathbf{f}, \quad (34)$$

where $\mathbf{f} = \mathbf{F}/m_0$ is the acceleration that arises from the nongravitational constraint. Finally, $(d\mathbf{S}/dt)$ is to be taken from (18). With all these substitutions, equation (33) becomes

$$(d\mathbf{S}_0/dt) = (3m/2r^3)[\mathbf{v}(\mathbf{r}\cdot\mathbf{S}) - \mathbf{r}(\mathbf{v}\cdot\mathbf{S})] + \mathbf{S}(\mathbf{v}\cdot\mathbf{f}) - {}^1/_2\mathbf{f}(\mathbf{v}\cdot\mathbf{S}) - {}^1/_2\mathbf{v}(\mathbf{f}\cdot\mathbf{S}). \quad (35)$$

Since according to (32), $\mathbf{S}$ differs from $\mathbf{S}_0$ only by terms of order m/r and v^2, it may be replaced by $\mathbf{S}_0$ on the right side of (35).

Equation (20), which arises from the isotropic metric and the CP condition, may be transformed in just the same way. The transformation is the combination of (25) and (29):

$$\mathbf{S}_0 = \mathbf{S} + (2m/r)\mathbf{S} + {}^1/_2[v^2\mathbf{S} - \mathbf{v}(\mathbf{v}\cdot\mathbf{S})]. \quad (36)$$

Differentiation of (36), followed by substitution from (20) and (34), again leads to equation (35).

When the Pirani condition is used, equation (23) with the standard metric must be transformed by means of (26) and (31), and equation (24) with the isotropic metric must be transformed by means of (25) and (31). In both cases, the result analogous to equation (35) is

$$(d\mathbf{S}_0/dt) = (3m/2r^3)[\mathbf{v}(\mathbf{r}\cdot\mathbf{S}) - \mathbf{r}(\mathbf{v}\cdot\mathbf{S})] + {}^1/_2\mathbf{v}(\mathbf{f}\cdot\mathbf{S}) - {}^1/_2\mathbf{f}(\mathbf{v}\cdot\mathbf{S}). \quad (37)$$

Again, $\mathbf{S}$ may be replaced by $\mathbf{S}_0$ on the right side.

The difference between equations (35) and (37) evidently arises from the difference between the CP and Pirani supplementary conditions rather than from the difference between the standard and isotropic forms of the metric. It is therefore related to the specification of the representative point (13), at which $\mathbf{F}$ is applied, as being the center of mass in one or another coordinate system. The specification of this point has no physical significance so long as $\mathbf{F} = 0$, and indeed we note that equations (35) and (37) agree in this case. However, when $\mathbf{F} \neq 0$, it will result in a torque $-\boldsymbol{\epsilon} \times \mathbf{F}$ about the center of mass, where $\boldsymbol{\epsilon}$ is the vector position of the center of mass with respect to the representative point. Since measurements are to be made by a co-moving observer, such a torque in his coordinate system is instrumental, and to be avoided by applying any constraining force at the center of mass in the rest-frame of the gyroscope. This corresponds precisely to the Pirani supplementary condition. We therefore conclude that (37) gives $(d\mathbf{S}_0/dt)$ as measured by a co-moving observer when instrumental torques have been eliminated.

Even when an instrumental torque is present, however, a correction can be made for it. When the CP supplementary condition is used, the vector $\boldsymbol{\epsilon}$ in the gyroscope rest-frame can be obtained by substituting the S_0^{i4} given by the last three of equations (28) into equation (14). This gives to first order

$$\epsilon_x = 0, \qquad \epsilon_y = vS_z/m_0, \qquad \epsilon_z = -vS_y/m_0.$$

Since $\mathbf{v}$ is here along x, this may be written in a rotation-covariant manner as

$$\boldsymbol{\epsilon} = -(\mathbf{v} \times \mathbf{S})/m_0.$$

Thus when the CP condition is used, we must subtract the instrumental torque

$$-\boldsymbol{\epsilon} \times \mathbf{F} = [(\mathbf{v} \times \mathbf{S}) \times \mathbf{F}]/m_0 = \mathbf{S}(\mathbf{v}\cdot\mathbf{f}) - \mathbf{v}(\mathbf{f}\cdot\mathbf{S})$$

from the rate of change of the gyroscope angular momentum given by the right side of (35). When this is done, equations (35) and (37) agree.

Effects of Earth Rotation and Extra-Terrestrial Objects.—Equation (37) gives the rate of change of angular momentum measured by a co-moving observer when the central attracting body (the earth) is spherically symmetrical and at rest. The effect of the axial rotation of the earth can be included by adding off-diagonal space-time components to the metric tensor. These components were computed by de Sitter[24] and by Lense and Thirring.[25] In the isotropic metric, with the vector angular velocity $\boldsymbol{\omega}$ of the earth directed along the positive z axis, the additional components are ($x^1 = x, x^2 = y, x^3 = z$):

$$g_{14} = -2I\omega y/r^3, \qquad g_{24} = 2I\omega x/r^3, \qquad g_{34} = 0, \tag{38}$$

where $I = 2mR^2/5$ is the moment of inertia of the earth of radius R, assumed to be homogeneous.

It is not difficult to see that the contribution of (38) to $(d\mathbf{S}/dt)$ is the same for the four spin equations of motion (18), (20), (23), and (24), and that it is not affected by the transformation to the gyroscope rest-frame. When this contribution is added to the right side, equation (37) may be written

$$(d\mathbf{S}_0/dt) = \boldsymbol{\Omega} \times \mathbf{S}_0, \tag{39}$$

where

$$\boldsymbol{\Omega} = {}^1/_2(\mathbf{f} \times \mathbf{v}) + (3m/2r^3)(\mathbf{r} \times \mathbf{v}) + (I/r^3)[(3\mathbf{r}/r^2)(\boldsymbol{\omega}\cdot\mathbf{r}) - \boldsymbol{\omega}]. \tag{40}$$

Simple order of magnitude estimates based on equations (39) and (40) suffice to show that the effects of the moon, the sun, and the galaxy are negligible in comparison with the effect of the earth.

Interpretation of the Spin Equation of Motion.—The form of equation (39) shows that the magnitude of the spin angular momentum measured by a co-moving observer is constant. Thus, if the moment of inertia of the gyroscope does not change, the rotation frequency remains constant. As remarked above, the gyroscope then behaves like a clock that can be set to any desired frequency, and this frequency will exhibit Doppler and gravitational shifts when observed from outside.

It also follows from equation (39) that if a number of gyroscopes are traveling together with various magnitudes and directions for their angular momentum vectors, the angles between these vectors, as measured by a co-moving observer, remain constant. The spin axes of all the gyroscopes precess with the common vector angular velocity $\mathbf{\Omega}$ given by (40). This precession takes place with respect to the inertial frame, which is generally believed to be defined by the distant extragalactic nebulae, the so-called "fixed stars." Thus, an experimental verification of equations (39) and (40) will consist in essence of a series of comparisons of the direction of the spin axis of a gyroscope and particular "fixed stars," made at different times. If the gyroscope is in motion when a comparison is made, the well-known correction for aberration must be made.[26]

The first term on the right side of equation (40) is the Thomas precession,[27] which is a special relativity effect. It is independent of gravitation, and is present even when m is set equal to zero in (34) and (40). The second term is a consequence of Einstein's theory of gravitation that goes beyond the equivalence principle. Even though this term involves only the values of α and γ in (2), which can be obtained correctly from the equivalence principle, it also involves the equation of motion of matter of finite rest mass beyond the Newtonian approximation. It is interesting to note that if $\dot{\mathbf{v}}$ is infinitesimally small, so that $\mathbf{f}$ given by (34) is equal to $(m/r^3)\mathbf{r}$, the gravitational precession is three times the Thomas precession, and of the same sign. The effects produced by this second term in various special cases have been discussed by several authors, notably de Sitter,[28] Fokker,[29] and Pirani.[17] However, in none of these papers is the differential equation (39) or the full expression (40) for the time-varying precession angular velocity $\mathbf{\Omega}$ exhibited, nor is the relation between the different possible coordinate systems and supplementary conditions discussed.

The third term on the right side of (40) arises from the rotation of the earth, and is present even when the gyroscope is at rest. It has the interesting property that it is parallel to $\boldsymbol{\omega}$ at the poles ($\mathbf{r}$ parallel or antiparallel to $\boldsymbol{\omega}$), and antiparallel to $\boldsymbol{\omega}$ at the equator ($\mathbf{r}$ perpendicular to $\boldsymbol{\omega}$). This is physically plausible if we think of the moving earth as "dragging" the metric with it to some extent. At the poles, there is a tendency for the metric to rotate with the earth, and hence to cause the spin to precess in the direction of rotation of the earth. At the equator, we note particularly that the gravitational field, and hence also the dragging of the metric, falls off with increasing radial distance. If, then, we imagine the gyroscope oriented so that its axis is perpendicular to that of the earth, the side of the gyroscope nearest the earth is dragged with the earth more than the side away from the earth, so that the spin precesses in the opposite direction to the rotation of the earth. This third term is also a consequence of general relativity theory that goes beyond the equivalence principle, partly because the non-Newtonian equation of motion is required, and partly because the off-diagonal space-time components of the metric tensor cannot be inferred from the equivalence principle.

Experimental Consequences.—Two experimental arrangements especially commend themselves: a gyroscope in a satellite, and a gyroscope at rest in a laboratory fixed with respect to the earth.[30]

The satellite gyroscope is in free fall, so that $\mathbf{f} = 0$. Then for an orbit in the earth's equatorial plane, for example,

$$\mathbf{\Omega} = (3m/2r)\boldsymbol{\omega}_0 - (I/r^3)\boldsymbol{\omega}, \tag{41}$$

where $\boldsymbol{\omega}_0 = (\mathbf{r} \times \mathbf{v})/r^2$ is the instantaneous orbital angular velocity vector of the gyroscope. If m and r are to be expressed in cgs units, m/r must be replaced by Gm/c^2r, where G is the Newtonian gravitational constant and c is the speed of light. It is convenient to replace Gm by gR^2, where g is the acceleration of gravity at the surface of the earth. Then (41) may be written

$$\mathbf{\Omega} = (gR/c^2)[(3\boldsymbol{\omega}_0/2)(R/r) - (2\boldsymbol{\omega}/5)(R/r)^3], \tag{42}$$

where $gR/c^2 = 7.0 \times 10^{-10}$. It is easily seen that the second term of (42) is never more than a per cent or two of the first; thus $\mathbf{\Omega}$ is roughly parallel to $\boldsymbol{\omega}_0$ even when the orbit is not equatorial. For a satellite of moderate altitude, the precession is approximately 6×10^{-9} radians per revolution when the gyroscope spin axis is in the plane of the orbit.

The laboratory gyroscope is constrained to remain at rest with respect to the rotating earth, so that

$$\mathbf{v} = \boldsymbol{\omega} \times \mathbf{r}, \qquad d\mathbf{v}/dt = \boldsymbol{\omega} \times \mathbf{v}. \tag{43}$$

The nongravitational constraining acceleration $\mathbf{f}$ may then be obtained from equations (34) and (43), and substituted into (40). The only experimental parameter in (40) is then the latitude λ of the laboratory. It is convenient to write $\mathbf{\Omega}$ in the form

$$\mathbf{\Omega} = (2gR\cos^2\lambda/c^2)[1 - (\omega^2R/4g)]\boldsymbol{\omega} + (2g\sin\lambda/c^2\omega)(\boldsymbol{\omega} \times \mathbf{v}) + (2gR/5c^2)(3\sin^2\lambda - 1)\boldsymbol{\omega} - (6g\sin\lambda/5c^2\omega)(\boldsymbol{\omega} \times \mathbf{v}). \tag{44}$$

The first line represents the first two terms of (40), and the second line the third term. Each line has been divided into a part proportional to $\boldsymbol{\omega}$, which results in a secular precesion, and a part proportional to $\boldsymbol{\omega} \times \mathbf{v}$, which contributes nothing to the time integral of $\mathbf{\Omega}$ over a whole number of days. Thus the secular precession arising from the effect of earth rotation on the metric vanishes when $\sin\lambda = 3^{-1/2}$, or $\lambda = 35°16'$. Since $\omega^2R/4g$, the ratio of centripetal to gravitational acceleration, is very small compared to unity, equation (44) shows that the secular part of $\mathbf{\Omega}$ is given to good approximation by

$$\mathbf{\Omega}_s = (4gR/5c^2)(1 + \cos^2\lambda)\boldsymbol{\omega}.$$

If the gyroscope spin axis is perpendicular to the earth's axis, the precession is approximately 3.5×10^{-9} $(1 + \cos^2\lambda)$ radians per day.

In comparing the two experimental arrangements, it should be noted that while the secular precessions per revolution are roughly the same, the period of a satellite of moderate altitude is much shorter than a day, so that the precession per unit time of the satellite gyroscope will be about 15 times that of the earth-bound gyroscope. Moreover, most of the experimental difficulties that seem to rise with a high-precision gyroscope, especially instrumental torques, are greatly reduced if the gyroscope does not have to be supported against gravity. On the other hand, it is much simpler at present to monitor a gyroscope in the laboratory than in a satellite.

* Supported in part by the U. S. Air Force through the Air Force Office of Scientific Research.

[1] For a summary of these experiments see v. Eötvös, R., D. Pekár, and E. Fekete, *Ann. Physik*, **68**, 11 (1922).

[2] Dicke, R. H., *Science*, **129**, 621 (1959).

[3] Wapstra, A. H., and G. J. Nijgh, *Physica*, **21**, 796 (1955).

[4] Dicke, R. H., *Revs. Mod. Phys.*, **29**, 355 (1957).

[5] Schiff, L. I., these PROCEEDINGS, **45**, 69 (1959).

[6] Møller, C., *The Theory of Relativity* (Oxford: Oxford University Press, 1952), p. 326.

[7] The remarks in this paragraph have been made with particular force by Eddington, A. S., *The Mathematical Theory of Relativity* (Cambridge: Cambridge University Press, 1924), p. 105. It should be noted that β is no longer zero if the isotropic form of the metric is used.

[8] Schiff, L. I., *Am. J. Physics*, **28**, 340 (1960). A part of this argument has also been given by Balazs, N. L., *Z. Physik*, **154**, 264 (1959).

[9] Pound, R. V., and G. A. Rebka, Jr., *Phys. Rev. Lett.*, **4**, 337 (1960).

[10] Cranshaw, T. E., J. P. Schiffer, and A. B. Whitehead, *Phys. Rev. Lett.*, **4**, 163 (1960).

[11] Wheeler, J. A., *Onzième Conseil de Physique Solvay* (Brussels: Editions Stoops, 1958); *Nuovo cimento supp.* (to be published 1960).

[12] Weber, J., *Phys. Rev.*, **117**, 306 (1960).

[13] Pugh, G. E., WSEG Research Memorandum No. 11 (Weapons Systems Evaluation Group, The Pentagon, Washington 25, D. C., November 12, 1959). A spinning satellite contained within a protecting satellite is proposed here; theoretical results are quoted for the precession in free fall, but are not quite correct so far as earth-rotation effects are concerned. A similar, more qualitative, proposal was also made in 1959 by R. A. Ferrell (unpublished).

[14] Schiff, L. I., *Phys. Rev. Lett.*, **4**, 215 (1960). The full expression for the precesion of a torque-free gyroscope, either in free fall or constrained in an earth-bound laboratory, is quoted here without proof.

[15] Papapetrou, A., *Proc. Roy. Soc.*, **A209**, 248 (1951).

[16] Corinaldesi, E., and A. Papapetrou, *Proc. Roy. Soc.*, **A209**, 259 (1951).

[17] Pirani, F. A. E., *Acta Phys. Polonica*, **15**, 389 (1956).

[18] Fock, V. A., *J. Phys. U.S.S.R.*, **1**, 81 (1939).

[19] Møller, C., *op. cit.*, p. 295.

[20] Actually, Papapetrou (reference 15) derived the noncovariant equation (11) first, and then obtained the covariant form (9) from it.

[21] u_i will not precisely vanish in the particle rest-frame for a rotating central body, but can be neglected.

[22] Møller, C., *op. cit.*, p. 170.

[23] Fock, V., *The Theory of Space Time and Gravitation* (New York: Pergamon Press, 1959), p. 194.

[24] de Sitter, W., *M. N. Roy. Astron. Soc.*, **76**, 699 (1916).

[25] Lense, J., and H. Thirring, *Phys. Zeits.*, **19**, 156 (1918).

[26] Bergmann, P. G., *Introduction to the Theory of Relativity* (New York: Prentice-Hall, 1946), p. 37.

[27] Thomas, L. H., *Phil. Mag.*, (7), **3**, 1 (1927); Møller, C., *op. cit.*, p. 56.

[28] de Sitter, W., *M. N. Roy. Astron. Soc.*, **77**, 155, 481 (1916).

[29] Fokker, A. D., *Kon. Akad. Weten. Amsterdam, Proc.*, **23**, 729 (1920).

[30] The second type of experiment was suggested to the author by W. M. Fairbank.

The Gyroscope Experiment (*).

I: General Description and Analysis of Gyroscope Performance.

C. W. F. Everitt

W. W. Hansen Laboratories of Physics, Stanford University - Stanford, Cal.

1. – Introduction.

The idea of testing Einstein's general theory of relativity by observations on gyroscopes in space was suggested by Schiff and independently though rather less completely by Pugh in 1960 [1], three years after the launch of the first sputnik. Many years earlier Eddington had mentioned the possibility of making a gyroscope test of general relativity; during the 1930's Blackett examined the practical problems but concluded that to try the experiment then on Earth was hopeless. The analyses given below show that this is still so. However space research opens new possibilities, especially if it is combined with another area of research from the last fifteen years: large-scale low-temperature technology. Work on the gyroscope program has been carried on jointly by members of the Department of Physics and Department of Aeronautics and Astronautics at Stanford, with support from the National Aeronautics and Space Administration since 1963.

For a gyroscope moving in free fall about a massive sphere, the formula found by Schiff for the precession rate in the framework of the fixed stars is

$$\boldsymbol{\Omega} = \frac{3GM}{2c^2R^3}(\boldsymbol{R}\wedge\boldsymbol{v}) + \frac{GI}{c^2R^3}\left[\frac{3\boldsymbol{R}}{R^2}(\boldsymbol{\omega}\cdot\boldsymbol{R}) - \boldsymbol{\omega}\right], \tag{1}$$

where $\boldsymbol{R}$ and $\boldsymbol{v}$ are the co-ordinate and velocity of the gyroscope and M, I and $\boldsymbol{\omega}$ are the mass, moment of inertia and angular velocity of the central body. The first term in eq. (1) gives the geodetic precession Ω_G due to the motion of the gyroscope through the gravitational field, the second term is the mass-current precession Ω_M due to the rotation of the central body itself. In

(*) Work supported in part by NASA Grant No. NGR 05-020-019.

the scalar-tensor theory the geodetic term is multiplied by $(4 + 3\omega)/(6 + 3\omega)$ and the mass-current term by $(3 + 2\omega)/(4 + 2\omega)$ where ω is the Brans-Dicke parameter. For a gyroscope in a 500 km polar orbit about the Earth the two effects are at right angles as illustrated in Fig. 1 and have (in general relativity) integrated values equal approximately to 6.9 arc·s/y for the geodetic term and 0.05 arc·s/y for the mass-current term. In the scalar-tensor theory the values are approximately 6.3 and 0.043 arc·s/y. Corrections for various higher-order terms have been worked out by O'CONNELL and BARKER and by WILKINS [2].

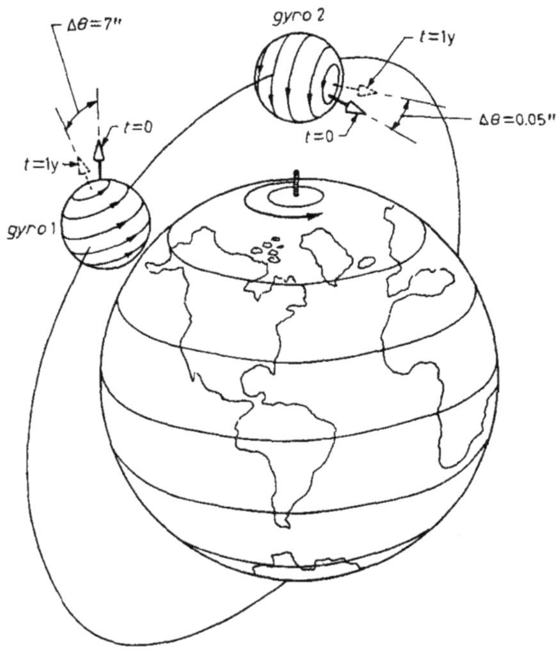

Fig. 1. – Relativistic motions of gyroscopes.

The most important is that due to the quadrupole mass-moment of the Earth. This affects the relativity term directly and also indirectly through its influence on the shape of the orbit. It causes an additional precession of the order of 0.01 arc·s/y. WILKINS has made a detailed study of the requirements on knowledge of orbit parameters to determine the geodetic effect with a precision of 1 part in 10000. Errors can be kept below this level by integrating the terms over each orbit and piecewise fitting of data from successive orbits. The orbit information needed is within the capacity of existing satellite tracking networks.

An experiment to observe either of the relativistic effects consists in measuring the precession rate of one or more gyroscopes with respect to the axis of a telescope pointing at a suitable fixed star. A polar orbit is chosen because it simplifies the separation of the effects and also because it helps in averaging certain error terms in the gyroscope. Other orbits are not ruled out: it is possible in principle to separate the two effects unambigously in any inclined orbit by taking advantage of the modulation of the geodetic term caused by the recession of the modes due to the Earth's quadrupole mass-moment, although with the existing design of the experiment the data processing scheme sets practical limits to such a plan. In an orbit that is nearly but not quite polar gyroscope 2 which measures the mass-current term is disturbed also by a component of the geodetic precession. Some writers have supposed that this mixture of terms restricts the accuracy of the experiment. Actually the term can be calculated out in two distinct ways: 1) by evaluating the geodetic precession from gyro 1 using the known orbit parameters, 2) by observing that whereas the angular change due to the mass-current precession varies linearly with time t, the geodetic component affecting gyro 2 varies as t^2 because of the slow recession of the orbit plane. Indeed a nonpolar orbit has one potential advantage, in that it partially eliminates the uncertainty due to the unknown proper motion of the reference star. Consider an orbit inclined at about 10° to the polar axis. During the course of a year the orbit plane turns through roughly three half-revolutions. The primary measurement of the geodetic effect is then a sinusoidal variation of angle in gyro 1 with an amplitude of about 2 arc·s and a period of eight months. The proper motion causes a linear drift of the star which can be subtracted out. Unfortunately this argument applies only to the geodetic precession. To eliminate proper motion from the measurement of the mass-current precession, one must either make observations with two satellites in different orbits or improve the knowledge of proper motion by other means. The new telescope technology which has been developed for the present experiments offers some interesting ways of making such improvements, which we are investigating.

The design goal of the experiment has been to measure the two effects predicted by SCHIFF with an accuracy approaching 0.001 arc·s/y. This would determine the geodetic precession to 1 part in 7000 and the mass-current precession to about 2 percent. If one accepts the parametrized post-Newtonian formalism due to EDDINGTON and others as a suitable framework for testing relativistic theories then since the geodetic precession depends on $1+2\gamma$ in that formalism, an experiment with the hoped for accuracy would measure γ to 1 part in 5000 [3]. On a broader view the special interest of the experiment lies in extending relativistic measurements into a new domain, especially through the direct observation of the effects of off-diagonal terms in the metric obtained from the mass-current effect. To measure Ω_M to 2 percent a gyroscope is needed with residual drift errors from extraneous sources less than $1.6\cdot10^{-16}$ rad/s. This desired accuracy will be called Ω_0. It appears to be just within the capabilities of existing technology.

2. – Description of the experiment.

The gyroscope designed for the experiment is illustrated in Fig. 2. It consist of a ball 4 cm in diameter made from optically selected fused quartz, coated with a thin film of superconductor. The ball is electrically suspended within a spherical quartz housing by voltages applied to mutually perpendicular sets of condenser plates. It is spun up initially to a speed of about 200 revolutions per second by a gas jet system designed by BRACKEN and EVERITT [4], after which the gas is pumped out and the ball is allowed to run freely in a vacuum. The entire device is surrounded by a spherical superconducting magnetic shield. *Illustrations of the actual parts are reproduced in the next paper.*

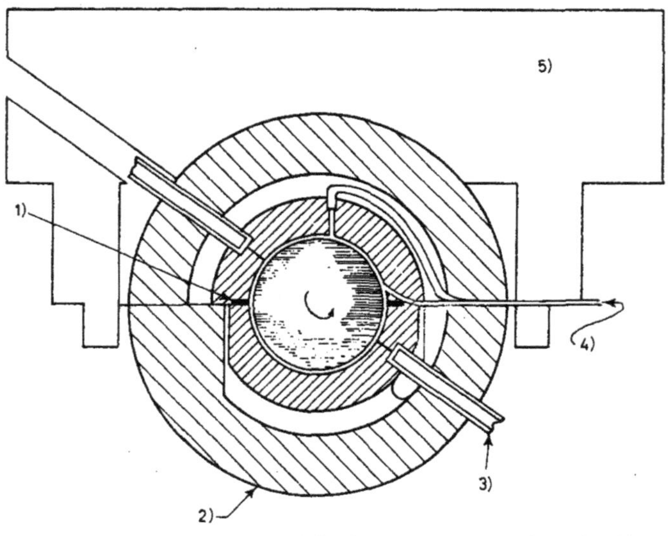

Fig. 2. – Gyroscope for laboratory model of relativity experiment: 1) readout ring, 2) superconducting magnetic shield, 3) leads to support electrodes, 4) inlet and exhaust for gas spin-up, 5) mounting ring.

The basic ideas of the electrically suspended gyroscope was developed in 1953 by the late ARNOLD NORDSIECK and his colleagues at the University of Illinois. The arrangement we have adopted is a modification of the version built for guidance purposes by Honeywell, Inc. Its chief novelty lies in exploiting the possibilities of superconducting shielding and readout techniques. The shield shown in Fig. 2 has two distinct functions. First, it provides a region of extremely *low* magnetic field ($5 \cdot 10^{-7}$ G or less) for the gyroscope, eliminating the torques due to eddy currents which prevent existing mechanical gyroscopes of any kind, electrically suspended or otherwise, from reaching the required drift performance. Second, it provides a region of extremely *constant* magnetic field, in which a magnetic readout of gyro angle, based on observations of the London moment in the spinning superconducting sphere, can operate without disturbance from changes in the external field. The advantages of the new readout will be explained shortly. Techniques for obtaining ultra low magnetic field in superconductors are covered in the paper by FAIRBANK.

In a rotating superconductor there is, according to the London equations, a magnetic moment aligned parallel to the instantaneous spin axis. The existence of this effect was deduced in 1953 by LONDON, following an earlier investigation from a different theory of superconductivity due to BECKER, SAUTER and HELLER [5]. It was detected experimentally in 1963 [6]. In a spinning sphere the moment M_L is a dipole of magnitude

$$M_L = \frac{mc}{3e} r^3 \omega_s = 3\cdot 10^{-8} r^3 \boldsymbol{\omega}_s , \tag{2}$$

where r is the radius and $\boldsymbol{\omega}_s$ is the spin angular velocity. Figure 3 shows the proposed readout. The gyro rotor is surrounded by a superconducting loop. Since the resistence of the loop is zero, any change in flux due to change of orientation of M_L will generate a cancelling current which may be used to measure the direction of the spin axis. Two alternative readout circuits are available and are being developed in parallel for the experiment. One depends on modulating the inductance of a second superconducting loop in series with the first, and hence converting the direct current into an alternating current proportional in amplitude to the gyro angle. The other circuit is based on quantum interference effects in a Josephson junction. The field sensitivity

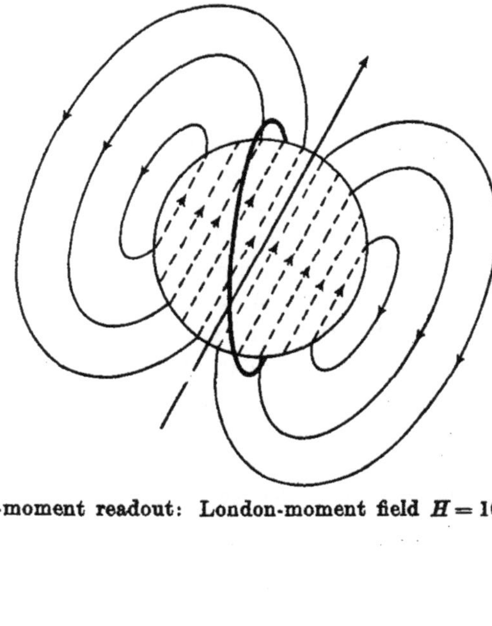

Fig. 3. - London-moment readout: London-moment field $H = 10^{-7}\boldsymbol{\omega}_s$ G.

of existing Josephson junction magnetometers corresponds to a resolution of about 1 arc s in 1 s of time. By integrating for longer periods the full 0.001 arc s accuracy is obtained. The paper by FAIRBANK contains further details about the two techniques and methods of improving their resolution.

The London-moment readout has several advantages for the present experiment. The fact that it measures the instantaneous spin axis directly is of very great importance. Most but not all of the other readout schemes that have been developed or proposed for mechanical gyroscopes depend on having reference marks of some kind on the surface of the ball. For such methods to work there must be a sufficient difference between the moments of inertia of the gyro-rotor in different axes to define a preferred axis about which it will spin. To reach the required limit on Ω_0, however, the rotor must (as is proved in Sect. **3**) be made spherical to 5 parts in 10^7 and homogeneous to 3 parts in 10^7. Dynamical arguments show that it is difficult, if not impossible, to align and maintain the axis of a rotor with such extreme homogeneity. Compared with those methods that do look directly at the spin axis the London-moment readout has advantages in sensitivity and null stability, which come from the special properties of the superconducting circuits.

Figures 4 and 5 illustrate the final spacecraft. The chief structural element is a superinsulated Dewar vessel, containing about 800 litres of liquid helium, designed to maintain cryogenic temperatures for about two years. Solar paddles, electronics and telemetry are mounted on a girth ring attached to the outer shell of the Dewar, as is an eight-foot long sun shield which prevents stray light from entering the star-tracking telescope. Within the Dewar is a separate chamber open at one end to space, in which is contained an experimental package comprising the telescope, four gyroscopes and a proof mass for a drag-free control system. For mechanical stability, all these parts are made from fused quartz, optically contacted together and maintained at liquid-helium temperatures to eliminate thermal distortion. The reference is a bright star near the Earth's equatorial plane; most probably Rigel. The satellite orbit is chosen such that the boresight of the telescope lies in the orbit plane: in this way one readout axis of the telescope provides reference for Ω_G and the other for Ω_M. To check and average any null shifts between the telescope and gyro readouts, and also to improve gyro drift performance, the spacecraft is rolled slowly about the line of sight to the star with period of about five hours. Considerations of this roll maneuver incidentally determined the design of the sunshield and the peculiar arrangement of solar paddles illustrated in Fig. 5. Two of the four gyroscopes are arranged with their axes respectively paralled and antiparallel to the roll axis: in them the relativity signals Ω_G and Ω_M appear as orthogonal components in the two readout planes. The other two gyroscopes are arranged orthogonally: one nearly parallel to the Earth's axis (not quite so since Rigel is a few degrees from the equatorial plane) and, therefore, susceptible only to Ω_G, the other normal to the orbit plane and susceptible only to Ω_M.

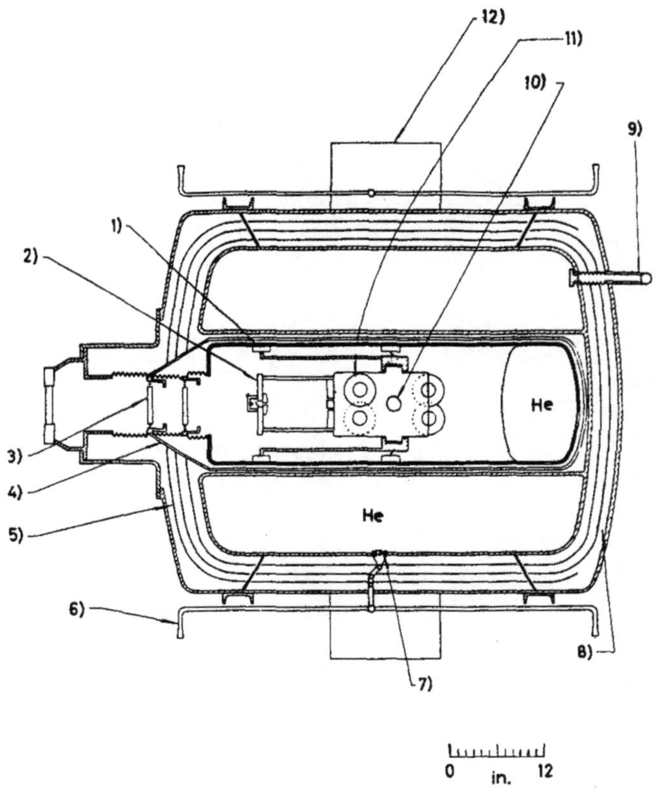

Fig. 4. – Flight apparatus: 1) cryogenic actuators, 2) telescope, 3) gold-coated windows, 4) Dewar within Dewar, 5) heat exchangers, 6) attitude control jet (1 of 12), 7) superfluid plug, 8) superinsulation, 9) fill line, 10) zero-g proof mass, 11) gyro, 12) instrumentation.

There is thus triple redundancy in the measurement of each effect. The two perpendicular gyroscopes also provide an accurate roll reference, which is needed for demodulating the data from the parallel gyros. Attitude control of the satellite is accomplished by gas jets supplied by the helium gas evaporating from the Dewar. The natural boil-off rate is enough to allow the use of proportional thrusters which give a very smooth pointing control system. The same thrusters are utilized in the translational control system. Making the satellite drag-free helps in two ways: it improves averaging of the residual accelerations on the gyroscopes and it reduces errors in the orbit determinations needed in analysing relativity data. Design of the control systems has been due to members of the Stanford Department of Aeronautics and Astronautics, who were responsible for the first drag-free satellite, successfully launched and operated by the U.S. Navy in July 1972.

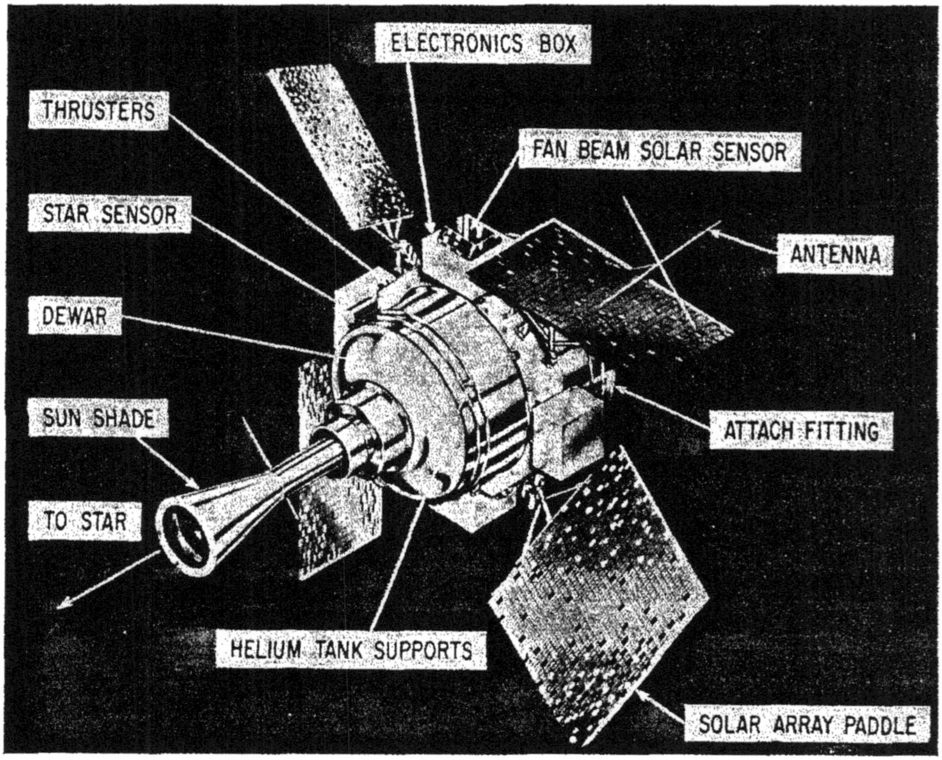

Fig. 5. – General view of spacecraft.

Several features in the design of the apparatus may best be grasped after considering the form of the experimental data. The final outputs are difference signals between the gyroscope and telescope readouts, whose content may be divided into six parts:

1) relativity drifts Ω_G and Ω_M, the measurement of which is the goal of the experiment;

2) proper motion of the reference star;

3) annual aberration of starlight, which gives a sinusoidal displacement between gyro and telescope outputs *in the plane of the ecliptic*, with amplitude $\pm 20{,}116$ arc·s and period one year;

4) orbital aberration of starlight: a sinewave with amplitude *in the plane of the orbit* ± 5 arc·s and period 90 minutes. Since the telescope boresight lies in the orbit plane, the star is occulted by the Earth for half of each orbit and only one side of the wave is observed;

5) disturbances due to attitude motions of the satellite;

6) intrinsic noise and errors in the gyro and telescope readouts.

If Ω_G is indeed 6.9 arc·s/year, the total range of motion (G-S) between the gyro axes and the line of sight to the star during the course of a year is, then, roughly 25 arc·s in the plane of the orbit and up to 50 arc·s in the plane of the ecliptic. The signals in the parallel gyroscopes are modulated at satellite roll arte, and also include constant components (G'-T') from any misalignment between the gyro and telescope readouts and (G_0-S) from any initial misalignment between the spin axes of the gyros and the line of sight to the star. To keep the range (G-S) as small as possible, the experiment should be designed so that (G'-T') and (G_0-S) are no larger than one or two arc·s. Also, if possible, the alignement error (G'-T') should be held constant to better than 0.001 arc·s over the year; for even though the rolling of the satellite allows certain errors to be removed in data analysis, it is safest not to rely on that except as a check.

In principle, the 50 arc·s displacement might be allowed to appear in either the gyroscope readout or the telescope readout, or even in both. In practice it is a far less forbidding problem to obtain wide linear range from the gyroscope, which has a readout derived from linear changes in a magnetic field, than from the telescope, which depends on motions of a star image. Thus the telescope is maintained as close to null as possible, and the displacements due to relativity drift and aberration of starlight are taken up in the gyro readouts.

The telescope, illustrated in Fig. 6, is a folded Schmidt-Cassegrain system, of 150 in. focal length and 5.5 in. aperture, designed in conjunction with DAVIDSON. The light is divided by a beam splitter near the focal plane to give two star images, one for each readout axis. Each image then falls on the sharp edge of a roof prism, where it is again subdivided and passed to a light chopper and photodetector at ambient temperature. In this way a reference accuracy better than 0.001 arc·s is obtained even though the determination is far beyond the diffraction limit, since the location of the center of the star image is limited only by the sharpness of the knife edge and by photon-counting statistics.

Much attention has been paid to the mechanical and optical stability of the telescope. It is made entirely of fused quartz: the parts are held together in molecular adhesion by the technique of « optical contacting » rather than cement, for since the displacement over the diameter of the primary mirror (16 cm) corresponding to a tilt of 0.001 arc·s is only 8 Å, differential aging in even the thinnest film of epoxy adhesive (say 2000 Å thick) may be expected to cause appreciable error. The helium environment is equally critical to stability. Thermal distortion is proportional to the transverse heat flux through the structure times the ratio of coefficient of expansion to thermal conductivity. At ambient temperature, to avoid distortions greater than 0.001 arc·s, the solar

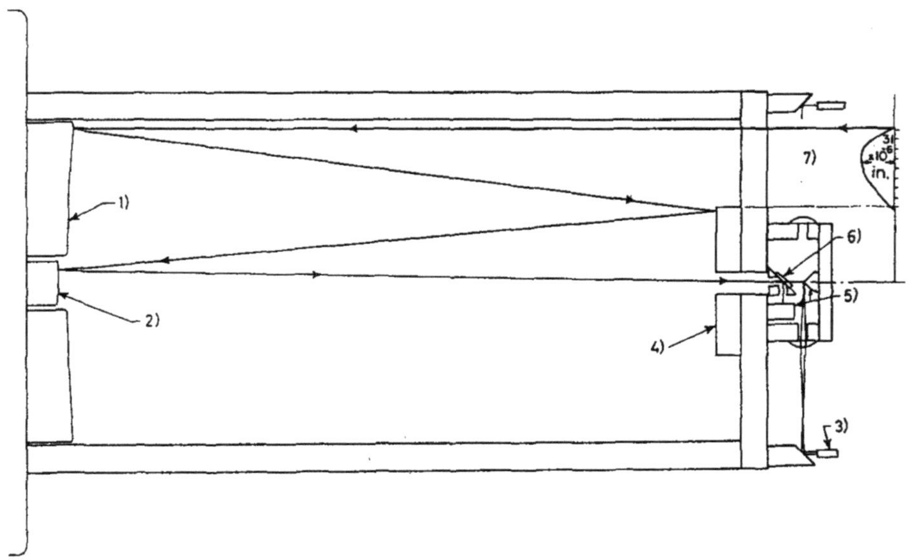

Fig. 6. – Star-tracking telescope: 1) primary, 2) tertiary, 3) light pipe, 4) secondary, 5) image dividers, 6) beam splitter, 7) correction in plates (2×10^4); effective focal length 150 in., physical length 13 in., radius primary 46 in., radius secondary 70.050 in., radius tertiary 7.9517 in., aperture 5.6 in.

radiation falling on the spacecraft would have to be cancelled at the telescope to one part in 10^5. At low temperatures, on the other hand, expansion coefficients decrease by several orders of magnitude and thermal conductivities for some materials increase, with the result that the allowable heat flux may be as much as two orders of magnitude greater than the total heat input into the Dewar. The light-chopping technique eliminates optical null shift due to changes in sensitivity of the photomultipliers. Some other sources of error such as atmospheric turbulence, and sag and creep of the structure under its own weight, which makes a long-term null stability of 0.001 arc·s for the telescope almost unthinkable on Earth, vanish in space. Probably the greatest null shift comes from the gradual relaxation of strains in the quartz after its formation on Earth: these are minimized by annealing the structure in the vertical position for several days at about 1100 °C, then only the focus is affected by dimensional changes. An analysis by VAN PATTEN and myself has also given criteria for effects of ultraviolet darkening, tarnishing of the mirrors, aging of the light pipes and other errors. An important general conclusion is that it is advantageous for several reasons to use diffraction-limited optics. To do so, however, places stringent requirements on the pointing control system.

The ultimate limit to sensitivity of the telescope is set by photon noise from the starlight. The noise equivalent angle $\langle\theta_t\rangle$ for each readout axis is calculated from the fluctuation in intensity of the light falling on the photocell during each signal period. For a star of magnitude M and color temperature $\bar{\lambda}$ the result is approximately

$$\langle\theta_i\rangle \simeq 2\cdot 10^{-6}\frac{\delta}{D}\sqrt{\frac{2.51^{-M}\Delta\nu}{\bar{\lambda}\,\varepsilon\,\eta}}\,, \tag{3}$$

where δ is the image diameter, D the telescope aperture, ε the effective light loss in each channel, η the quantum efficiency of the detector and $\Delta\nu$ the bandwidth. For diffraction-limited optics, δ may be taken as 1.22 λ/D. A telescope of 5.5 in. aperture, having $\varepsilon\eta$ about 0.01, resolves the direction of a first magnitude star to 0.01 arc·s in 0.1 s of time. Greater accuracy is achieved by integrating for longer periods. Resolutions approaching the limit set by (3) have been obtained on an experimental star simulator and detector in the laboratory. Appreciable nonlinearities occur when the angular-displacement center of the Δ at the image across the prism edge exceeds

$$\Delta < 1.15\,\delta^{\frac{2}{3}}\,\theta_0^{\frac{1}{3}}\,, \tag{4}$$

where θ_0 is the allowed error (less than 0.001 arc·s). For the 5.5 in. aperture, Δ is about ± 0.05 arc·s. More precise computer calculations by Pondrom jr. have confirmed the general results of eqs. (3) and (4) and provide design criteria for the telescope pointing system. Numerical substitution into (3) shows that the noise begins to be unacceptable in a telescope of this efficiency for stars fainter than third magnitude. Rigel is of 0.16 magnitude.

Errors occur in the telescope if the roof prisms have curvature, or small niches on the dividing edge. Then the motion Δ of the star image in one axis may shift the null point of the readout for the other axis. This effect was first investigated by Nidey. An analysis similar in principle to his but slightly different in detail yields a criterion for the maxium allowable chip width w:

$$w < \Delta^{\frac{1}{2}}\,\theta_0^{\frac{1}{2}}\,f\,, \tag{5}$$

where f is the focal length. From (5) the maximum nick diameter allowable in a 150 in. focal length telescope is $35\cdot 10^{-6}$ cm. Prisms have already been made with no nick greater than $15\cdot 10^{-6}$ cm.

The foregoing indicates that a telescope may be designed capable of resolving angles of 0.001 arc·s after about 10 s integration time, having very good null stability and a range of linear readout of about ± 0.05 arc·s. Such a telescope has been built and delivered to Stanford by Davidson Optronics Company for use in the laboratory model of the gyro relativity experiment discussed in the next paper. Two things are then needed to obtain full accuracy relativity data: a pointing control system capable of keeping the telescope aligned within the ± 0.05 arc·s linear range, and a data instrumentation system to subtract and process the gyro and telescope signals for transmission to Earth.

With the compliances expected in the spacecraft structure, the main attitude control system may be expected to maintain pointing to 0.5 arc·s, or about a factor of ten less precise than is needed. Consideration of this problem led VAN PATTEN and myself to suggest the addition of an inner fine pointing servo loop, acting directly on the gyro-telescope package, utilizing cryogenic magnetic forces as illustrated in Fig. 4. A detailed analysis and laboratory simulation of such a two loop system has been performed by BULL and DEBRA. They have established that adequate pointing accuracy can be achieved in the presence of telescope noise and normal satellite disturbances. Sensitivity to parameter changes has been investigated and model helium thrusters have been built and tested in conjunction with the analogue simulation. The problem of pointing seems to be effectively solved.

Figure 7 illustrates the principles of the data instrumentation system, conceived by VAN PATTEN. The heavy lines represent an integrating data loop which supplies continually updated relativity information in digital form,

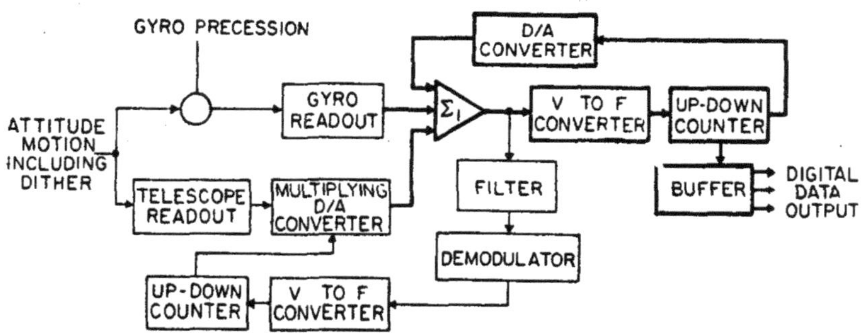

Fig. 7. – Relativity data instrumentation system.

after subtracting and summing the gyroscope and telescope signals with the final signal in the precision summing amplifier Σ_1. The output of Σ_1 consists of an amplitude-modulated suppressed-carrier alternating current signal. This signal is processed in a sampling demodulator and filter to obtain a direct current output with extremely low zero offset, and then integrated by means of a 17-bit up-down binary counter, which contains the readout signal for storage and telemetry. The integrating loop is closed via a 17-bit digital to analogue converter summed into Σ_1. Its operation may be understood as follows. Call the gyro output G, the telescope output T and the signal in the up-down counter R. The summing amplifier provides the function $(T-G+R)$, which is maintained at null, making the final signal R equal to $(G-T)$, the quantity of interest in the experiment.

To avoid error, the scale factors of the gyro and telescope readouts have to be carefully matched; for differences in scale factor will lead in the presence of a pointing error Δ' to a null shift identical in form with the relativity signal. If the nominal scale factor of each readout is k but there is a difference Δk between them, the limit on $\Delta k/k$ is θ_0/Δ', which means, for example, that if the pointing servo has an offset of 0.03 arc·s the maximum acceptable scaling difference to keep the error below 0.001 arc·s is about 1%. The calibration technique depends on introducing a low-frequency dithering motion into the inner pointing servo to make the entire gyrotelescope package swing back and forth across the line of sight to the star with an amplitude of about 0.03 arc·s at 0.1 Hz. If the scale factors of the two readouts are not equal, a signal appears at the output of the summing amplifier Σ_1, where it is synchronously detected, and integrated by means of a voltage to frequency converter and bidirectional counter. The resulting digital signal drives a multiplying digital to analogue converter which scales the telescope readout. A detailed analysis shows that the system may be designed to maintain the 1% scaling accuracy with reasonable settling time in the presence of gyro noise.

Signals may be derived from the instrumentation system for use in attitude control electronics. Since the telescope lies in the orbit plane, the star is occulted during half of each orbit. During this and other interruptions such as the occasional visual flare or meteorite impact, control logic is activated to inhibit the telescope output T and refer the controller to the equivalent function $(G-R)$ which is available from the instrumentation system. Thus the telescope remains pointing in the correct line of sight by reference to the gyroscope. Reacquisition and resumption of measurement occur automatically when the telescope is turned on again. The only loss is the immediate loss of data for the final statistical analysis.

In the actual rolling satellite appropriate encoders and resolvers are added to the instrumentation system to unroll and process the data for both parallel and perpendicular gyroscopes. These modifications are described in detail elsewhere [7]. All portions may be made with solid-state integrated circuitry for maximum reliability. The instrumentation system provides enough filtering to process the data and store it in a memory of reasonable size, without the complexities and reliability problems of an onboard computer. The choice of time constant for the integrating loop depends on several factors. The optimum figure appears to be about 20 s. This is long enough to give substantial noise filtering while allowing recovery from interruptions within one or two minutes. The output of the instrumentation loop is sampled every few seconds and stored in a core memory containing 10^4 words of 17 bits, from which it is transmitted once per orbit for further processing on the ground. With a time constant of 20 s the output of the data loop consists of: 1) full relativity signal, 2) full annual aberration in ecliptic plane, 3) 99.98% orbital aberration in orbit plane and 4) residual noise. A beautiful and important feature of the experiment is that the aberration signal, being known with great precision, provides an inherent calibration of the data. Thus the telescope scale factor (which might be expected to change through effects such as aging of the photomultiplier) is automatically driven to the same value as the gyroscope by the dithering technique, while the gyroscope scale factor (which should remain constant) is automatically checked and calibrated in space by the aberration signals.

Experimental work at Stanford on the laboratory model of the gyroscope, on a method for controlling liquid helium in space, and on the design and construction of long hold time helium Dewars, is described in the next paper. The rest of this paper deals with the problem of gyro errors.

3. – Analysis of gyro errors.

The drift rate of a gyroscope due to some extraneous nonrelativistic torque may be written

$$\Omega_r = \frac{\Gamma_r}{I\omega_s}, \tag{6}$$

where I is the moment of inertia, ω_s the spin angular velocity and Γ_r the resolved component of the torque perpendicular to the spin axis. For a gyroscope to be capable of performing the experiment either the sum of all the Ω_r's must be made less than Ω_0, or their magnitudes must be exactly known. With one partial exception, which will be explained shortly, it is not possible to calculate out the Ω_r's exactly, so the gyroscope has to be designed to make their effects negligible.

The moment of inertia of a sphere of density ϱ and radius r is $(8\pi/15)\,\varrho r^5$. Replacing ω_s in (3) by v_s/r where v_s is the peripheral velocity of the ball, one has

$$\Omega_r = \frac{15}{8\pi}\,\frac{\Gamma_r}{\varrho r^4 v_s}. \tag{7}$$

There is an upper limit to the peripheral velocity of any gyro-rotor set ultimately by the bursting speed of the material and more immediately by the elastic distortion due to centrifugal forces. For a spinning elastic sphere, the difference $\Delta r/r$ between polar and equatorial radii is given by

$$\frac{\Delta r}{r} = \frac{2}{7}\,\frac{\varrho v_s^2}{E}\left(1 + \frac{11}{14}\sigma\right), \tag{8}$$

where E is Young's modulus and σ Poisson's ratio. Hence, if a certain limit is assigned to $\Delta r/r$, the maximum peripheral velocity is proportional to the combination of properties for the material of the ball

$$\sqrt{E/\varrho}(1-11\sigma/28)\,.$$

Several alternative methods exist of classifying the torques acting on the gyroscope. Here they are grouped according to their effects on a number of distinct design problems as follows:

1) size and choice of material for an ideal gyro-rotor;

2) homogeneity of rotor and the problem of an Earth-bound experiment;

3) torques due to suspension and residual electric charge, and their effect on sphericity and centering accuracy;

4) optimum spin speed;

5) requirements on residual accelerations on the satellite as determined by 1) and 2);

6) magnetic torques and readout reaction;

7) gas torques and the spin-up problem;

8) ultimate limits.

3·1. *Size and choice of material for an ideal gyro-rotor.* – The extraneous torques may be divided into two broad categories: those related to the surface area of the ball and those related to its volume. Each surface-dependent torque Γ_σ will be proportional to (area) $\times$ (radius) $\times$ $\sigma(r)$, where $\sigma(r)$ is a function which in some instances is constant and in others depends on deviations from perfect sphericity in the shape of the ball. Similarly each volume-dependent torque is proportional to (density) $\times$ (volume) $\times$ (radius) $\times$ $\varphi(r)$, where $\varphi(r)$ is a function measuring the deviations from perfect homogeneity of the ball. Over a fair range of radii $\sigma(r)$ may be taken as proportional to r^s, where $0<s<1$; and (r) as proportional to r^v where $0<v<1$. Thus Γ_σ varies as $r^{(3+s)}$. and Γ_φ as $r^{(4+v)}$. Substituting into (6) one finds that the drift rates Ω_σ and Ω_φ due to all torques in these categories vary as $r^{(s-1)}$, and r^v. Thus some errors increase and some decrease with increasing rotor size, but in neither case is there much advantage to a change in diameter. In fact, the diameter of 4 cm chosen for the experiment was fixed more by the design of readout than the drift performance.

When the functional expressions for Γ_σ and Γ_φ are combined with eqs. (7) and (8) one finds for a given distortion $\Delta r/r$ of the rotor that the drift errors due to volume-dependent torques vary as $\sqrt{\varrho/E}\,[1+(11/28)\sigma]$ and those due to surface-dependent torques are proportional to $\sqrt{1/E\varrho}\,[1+(11/28)\sigma]$. The first expression is virtually constant for all materials, the second tends to decrease with increasing density. There are also certain kinds of error for example those due to the action of a residual magnetic field on the London moment, which do not fit into either category but are proportional to $1/\varrho$. The choice of material for the rotor, therefore, depends on two distinct questions. To minimize the surface-dependent torques and those due to the London moment and differential damping effects, the material should be selected for lowest $1/\sigma$ and $1/\sqrt{\varrho E}$. To minimize volume-dependent torques it should be selected for extreme homogeneity. The correct choice depends which category of torque is dominant. Fused quartz was chosen for the present experiment chiefly for homogeneity. The density variations of Schlierren-quality quartz have been determined by making optical interference measurements in three axes through cubes of the material with accurately parallel sides. The homogeneity obtained so far is about 1 part in 10^6.

3·2. *Homogeneity of the rotor and the problem of an Earth-bound experiment.* – Consider a perfectly spherical gyro-rotor which is not quite homogeneous. The support will act through the center of geometry and the inertia through the center of mass. Let the residual acceleration acting on the system be f. Then there will be a *mass unbalance* error Ω_u. Evaluating the error for a rotor with uniform density gradient parallel to the spin axis and then introducing a factor c_1 between 0 and 1 to represent the symmetry of the actual density distribution one finds

$$\Omega_u < \frac{c_1}{2}\frac{\Delta\varrho}{\varrho}\frac{f}{v_s}. \tag{9}$$

Taking $\Delta\varrho/\varrho$ as $3\cdot10^{-7}$, c_1 as 0.5 and v_s as 2500 cm/s, one finds that the average value of f needed to make Ω_u less than $0.25\,\Omega_0$ is $10^{-9}\,g$. Conversely, if f were made equal to g, the acceleration at the surface of the earth, the variations in $\Delta\varrho/\varrho$ would have to be reduced about nine orders of magnitude below the value currently obtained with quartz.

Equation (9) shows immediately the problems of doing an experiment on Earth. Some torque averaging is obtained in experiments with a vertical gyroscope in an observatory near the equator. For this f in eq. (9) is replaced not by g but by $g\sin\lambda$ where λ is the latitude of the observatory. One may also try to improve the balance of the gyro. A method of doing this is to deposit small quantities of matter of different density (gold, for example), on the surface of the ball to bring the center of mass and center of geometry together. The mass balance may be checked by measuring the pendulum period of the ball before spin up. It is easy to show that if Ω_0 is the desired accuracy and ω_s the spin speed of the gyro, the pendulum period has to exceed

$$T > 2\pi\sqrt{\frac{\sin\lambda}{\Omega_0\omega_s}}. \tag{10}$$

For an observatory within a few arc·seconds of the equator, the period required is about one day, which is long but not quite absurd. In reality the suspension errors discussed in the next Section make the problem much more difficult. For a space experiment $\sin \lambda$ in eq. (10) is replaced by f/g and the period becomes about 20 min.

Another entirely different source of error which places a restriction on the homogeneity of the gyro-rotor is caused by the action of the Earth's gravitational field on the quadrupole mass moment of the gyroscope. The secular drift rate Ω_g due to this effect integrated around the satellite orbit is

$$\Omega_g = \frac{3}{2}\frac{\Delta I}{I}\frac{1}{\omega_s}\frac{GM}{R^3}\sin 2\beta\,, \tag{11}$$

where ΔI is the difference between principal moments of inertia, β is the angle between the gyroscope axis and the orbit plane, and G, M and R have the same meanings as in eq. (1). The magnitude of the effect is practically identical in space and on Earth. Two cases have to be considered. The difference in moment of inertia may arise from inhomogeneities in the rotor; then $\Delta I/I$ is replaced by $c_2(\Delta\varrho/\varrho)$ where c_2 is another symmetry factor different and smaller than c_1. Alternatively the difference may arise from centrifugal distortion of the ball; then $\Delta I/I$ is replaced by $\Delta r/r$ from eq. (8). According to (11) the gravity gradient error vanishes when the gyroscope is parallel or perpendicular to the orbit plane, as is to be expected from symmetry. The first result may be transformed into a second condition on density. For a gyroscope aligned within 1° of the orbit plane, which is reasonable for a near-polar orbit, $c_2(\Delta r/r)$ has to be less than 1 part in 10^6. Thus the two restrictions on density are of the same order.

The error Ω_g associated with centrifugal distortion of the ball differs from every other disturbance acting on the gyroscope in that it depends on quantities whose magnitude can be calculated exactly. However the correction is not likely to be necessary in a polar orbit.

The conclusion is then that the density needs to be uniform to about 3 parts in 10^7, with the variations somewhat symmetrically distributed; and that the average residual acceleration on the satellite should be $10^{-9}\,g$ or less.

3'3. *Torques due to the suspension system and residual electric charge, and their effects on sphericity and centering accuracy.* – The discussion of suspension torques in my earlier paper on the subject contained some mistakes which are put right here.

The electrical suspension system utilizes three mutually perpendicular pairs of condenser plates deposited on the spherical inner surface of the gyro housing. The plates are circular with a diameter of about 2 cm and are about $5\cdot 10^{-3}$ cm from the rotor. Twenty kHz alternating voltages are applied to each plate; when the gyroscope experiences an acceleration parallel to an electrode axis, the voltage is raised on one plate and lowered on the other to keep the ball centered. A 1 MHz sensing voltage is also applied to the electrodes to measure the ball position, and this too exerts forces, and therefore torques, on the gyroscope.

Almost any scheme that may be conceived of for supporting a massive body against gravity—wheter it is based on electric, magnetic or hydrodynamic forces—depends ultimately on creating differential pressures over the body's surface. If the body is nearly, but not quite, spherical, a change in orientation results in work being done against these pressures. Conversely the pressures exert a torque which tends to rotate the body to the minimum energy point, having a magnitude equal to the differential of the energy with respect to angular position. For a true sphere, the torque evidently vanishes: it depends, therefore, on the size and form of the deviations from perfect sphericity. Disregarding for the moment extraneous effects, such as the interaction of a magnetic suspension with residual magnetic moment in the ball or the turbine torques in a gas bearing, we may expect the torques on a specific body to be identical, or at least have close analytic similarities, whatever the method of suspension; for the pressure that has to be exerted over a given area to balance a particular acceleration is always the same. Only if the system can be designed so that the energy put into it at one point can be taken out elsewhere, does this statement require qualification. In an electrically suspended gyroscope that may in principle be done by having what is known as « sum of the energies » control, *i.e.* by adjusting simultaneously the voltages on all the electrodes to make $\sum_i C_i V_i^2$ constant, where C_i and V_i are the capacitances and voltages on the i-th electrode.

Let us begin by considering what may be called the plain man's suspension system, in which a pressure is exerted on one area of the sphere to counteract gravity. To fix our ideas we may think of a lightweight sphere floating on a pool of mercury. If the total deviation from sphericity is Δr, then the difference between maximum and minimum energy is $Mf\Delta r$, where again f is the residual acceleration. Since the gyroscope spins at high speed, it may be treated as an axially symmetric body. Upper bounds on the torque are obtained by expanding the shape of the rotor in a series of Legendre polynomials and identifying the deviation from sphericity Δr with each polynomial in turn. The resultant expression for the drift rate Ω_s for an acceleration in the most disadvantageous possible direction with respect to the gyro spin axis is

$$\Omega_s < \frac{5}{2} n\chi \left(\frac{\Delta r}{r}\right) \frac{f}{v_s}, \tag{12}$$

where n is the order of the polynomial and χ is a quantity lying between 0 and 1 which depends on the size of the support pads and the scale of the deviations from sphericity. If the support were a localized point and the gyro-rotor had surface corrugations represented by a harmonic of high order, the drift rate for that term could become very large. In a real gyroscope, however, the pads are extended and the deviations from sphericity are mostly in the lower harmonics, so the product $\chi(\Delta r/r)$ is vanishingly small for all except the lowest-order terms. Satisfactory upper limits are thus obtained by considering the first even and first odd harmonics: that is ellipsoidal and pear-shaped rotors for which n is 2 and 3 respectively. It is interesting before going further to notice the similarity between eq. (12) and eq. (9) for the mass unbalence error. Taking $(\Delta r/r)$ as 10^{-6}, f in space as having an average of about $10^{-9}\,g$, one finds that the drift rate Ω_s to be expected with the plain man's suspension system is $2\cdot 10^{-15}$ or about a factor of 12 higher than the desired value for Ω_0.

Various mechanizations have been proposed to obtain electrical suspensions with performance better than the plain man's suspension system. Consider first briefly one where the suspension forces are switched off for long periods, allowing the ball to drift freely in the housing over a deadband b, after which a sharp pulse is applied to recenter it. This is known as the pulse-on-demand system. Such a support scheme has considerable attractions for the design of low-level accelerometers and at first sight seems to promise an improvement in gyro-performance also. In fact, it does not. For the velocity acquired by the ball over the distance b is $v_b^2 = 2fb$ where f is again the ambient acceleration. In order to recenter the ball, a much larger acceleration f' must be applied at the limit of the deadband. Let t' be the period of application of f' and t the free floating time, then extending (10), the average drift rate Ω'_s for the pulse-on-demand system is

$$\Omega'_s = 5n\chi\left(\frac{\Delta r}{r}\right)\frac{f'}{v_s}\frac{t'}{t}. \tag{13}$$

But $v_b = 2f't' = 2ft$, so $f't'/t$ is simply equal to f and a gyroscope mechanized with pulse-on-demand has a drift rate equal to one with the plain man's suspension system.

We have seen that the quantity χ tends to get smaller the greater the area of the support pad. The ideal limit is a neutral density body floating immersed in an incompressible fluid. There the support pressure extends over the entire surface and the torque vanishes absolutely. An electrical suspension can be arranged to mimic this characteristic of flotation by applying voltages to all six electrodes at once. The torque level is actually reduced by raising the voltages above the minimum. A suspension system mechanized in this way is said to be *preloaded*; its characteristics are expressed in terms of a preload acceleration h, which is defined as the acceleration parallel to an electrode axis required to send the voltage on one plate to zero. With h equal to $10^{-6}\,g$, which is reasonable for the space experiment, the normal operating voltage is about 0.7 V. Practical considerations in the design of the position-sensing bridge dictate that the one megacycle measuring signal cannot be much less than 0.5 V; so its influence on gyro-performance must also be thoroughly studied. On Earth h has to be comparable to or greater than g and the operating voltage is about 2000 V.

Analysis of the errors with a preloaded suspension has been greatly helped by an investigation of conventional electrically suspended gyroscopes due to MATCHETT of Honeywell Incorporated. We consider a situation where the electrode surfaces are symmetrical and lie on a perfect sphere and examine three terms only: 1) an ellipsoid tilted with respect to the suspension axes, 2) an ellipsoid aligned with the electrodes but displaced from the center of the cavity and 3) a pear-shaped body. There are also terms arising from asymmetries in the electrode configuration, which are small compared with the worst cases of misaligned or miscentered bodies, but become significant if advantage is taken of the reductions in the main terms due to symmetry. Various other terms up to the seventh order harmonic have been computed by DER MATCHETT and by Mr. P. EBY of NASA Marshall Center.

Define axes x, y, z in the gyro-housing, through the centers of the support electrodes, and axes x', y', z' in the rotor, z' being the direction of spin. Take the potentials of the electrode pairs as V_2, V_4 (x-axis) V_3, V_5 (y-axis) V_1, V_6 (z-axis), then from the force equation the difference in potentials along the x-axis due to an acceleration f_x is

$$V_2^2 - V_4^2 = \frac{8aMd^2}{Ar^2}\, f_x \equiv kf_x\,, \tag{14}$$

with equivalent expressions for the y and z axes, d being the distance between rotor and electrode and A the ratio of the area a of an electrode to its area projected on a plane, that is, in this instance 1.1. The preload h_x is for a suspension mechanized to keep $(V_2 + V_4)$ constant

$$(V_2 + V_4)^2 = kh_x\,, \tag{15}$$

where k has the same meaning as in eq. (12). Define $V'_x \equiv V'_y + V'_z$, then from (14) and (15)

$$V_x^2 = \frac{kh_x}{2}[1 + (f_x/h_x)^2]\,, \tag{16}$$

with similar definitions for V_y and V_z. The torque acting on a tilted ellipsoidal rotor, expressed in rotor coordinates may be shown to be

$$ {}_1\Gamma_e = 0.065\,\frac{a_2 r^2}{d^2}\sin\theta_0 \left\{ i'_x \sin 2\varphi_0\,\frac{(V_y^2 - V_x^2)}{2} + \right. \tag{17}$$
$$\left. + i'_y \cos\theta_0 (V_z^2 - V_y^2\cos^2\varphi_0 - V_x^2 \sin^2\varphi_0)\right\},$$

where a_2 is the coefficient of the spherical harmonic, $i'_x i'_y i'_z$ are the unit vectors in rotor coordinates, θ_0 φ_0 are the Euler angles, and the units are Gaussian. Before going further, it should be noticed that if the suspension is mechanized in such a way as to maintain $V_x^2 = V_y^2 = V_z^2$ the torque on a misaligned ellipsoidal rotor vanishes. A suspension with this feature is said to have « sum of the squares » control. The more general case of « sum of the energies » control, where $\sum_i C_i V_i^2$ is held constant, has already been referred to. In practice, neither control technique is fully effective since there are errors due to 1) incomplete compensation, 2) imperfections in electrode forms, 3) the action of the measuring voltages.

The errors are reduced by designing the suspension system to maintain the preload accelerations nearly equal on all three axes. They may also be reduced by exploiting symmetries in the gyro housing, for example by pointing the spin axis approximately parallel to an electrode axis, making $\theta_0 \to 0$, or by rolling the satellite. Defining a preload compensation factor ζ equal to $(h_x - h_y)/h_z$ etc., one finds the following upper limit on the drift rate, ${}_1\Omega_e$ due to a misaligned elliptic rotor

$$(18) \qquad {}_1\Omega_e < 1.35\left(\frac{\Delta r}{r}\right)\left(\frac{1}{v_s}\right)\left[\frac{f^2}{h} + \zeta h\right]\theta_0\,,$$

where θ_0 is now the small residual misalignment from perfect symmetry. For a gyroscope mechanized with « sum of the squares » control, θ_0 should be replaced by the « sum of the squares » control factor $\xi \equiv (V_x^2 - V_y^2)/V_z^2$. In theory, one might exploit both « sum of the squares » and symmetry; but there is no reason to, for at that level the error is dominated by terms omitted from the calculation. The other errors are evaluated in a similar way, though the analysis is more tedious. The limit for a miscentered ellipsoid is

$$(19) \qquad {}_2\Omega_e < 3.28\left(\frac{\Delta r}{r}\right)\left(\frac{t}{d}\right)\left(\frac{f}{v_s}\right),$$

where t is the displacement, d the rotor-electrode gap. The limit for a pear-shaped rotor is

$$(20) \qquad {}_3\Omega_e < 0.8\left(\frac{\Delta r}{r}\right)\left(\frac{f}{v_s}\right).$$

The analytical forms of eqs. (18) to (20) should be compared with eq. (12). Several other points should also be noticed. First, the deviation $\Delta r/r$ for the ellipsoid terms of eqs. (18) and (19) is the combined result of polishing errors and centrifugal distortion. In practice, centrifugal distortion is dominant and

$\Delta r/r$ may be evaluated from eq. (8). The deviation $\Delta r/r$ in eq. (20) on the other hand, is due solely to polishing errors. Next, the form of the acceleration term $[f^2/h + \zeta h]$ for the residual drift rate of the misaligned ellipsoid should be noticed. It implies that there is an optimum preload for a given average level of residual acceleration f. Physically its meaning may be understood as follows. In the absence of external accelerations, the gyro tends to drift towards the minimum energy position (where its axis is aligned with the electrode having the lowest preload) with a rate proportional to the second term ξh. The first term represents the effect of external acceleration: it is reduced by increasing the preload because then the suspension more closely resembles a hydrostatic system: the percentage change in voltage to compensate the acceleration is less. Since the acceleration enters as f^2, the torque due to a sinusoidally varying acceleration on a misaligned ellipsoid rectifies. The importance of this result will be discussed under Subsect. 3·5.

Evidently the critical factors for gyroscope performance are the preload compensation factor ζ, the centering accuracy t, and either the misalignment angle θ_0 or the « sum of the squares » control factor ζ. Typically, the combination of electronic and mechanical limitations gives values of ζ as $2\cdot 10^{-3}$, t/d as $3\cdot 10^{-3}$ and θ_0 or ξ as $5\cdot 10^{-3}$. For a gyroscope operating on Earth the resultant best drift rate is about 10^5 arc·s/y. In space the same numerical values give a limiting drift rate of about $0.9\cdot 10^{-3}$ arc·s/y; and, if the satellite is rolled about the line of sight to the star, as proposed, θ_0 and t/d both average to around 10^{-4}, reducing the error still further. However, the analysis omits the difficult higher-order terms.

In addition to the suspension erros, there is a torque acting on the gyro when a residual electric charge on the ball induces charge on the inner surface of the gyro-housing. This term evidently vanishes if the ball is exactly centered and either it or the cavity is perfectly spherical. The magnitude of the effect is

$$(21)\qquad {}_4\Omega_e < \frac{Q^2}{4r^4}\left(\frac{\Delta r}{r}\right)\left(\frac{1}{v_s}\right)\left[\frac{5t}{d'} + \frac{3\Delta r'}{d'}\right]\left[\frac{A'd}{A'd + Ad'}\right],$$

where Q is the total charge in electrostatic units, t the displacement of the ball from the center of the cavity, A the electrode area, d the rotor-electrode distance, and A', d' and r' are the area, distance and polishing error of the raised conducting regions which surround the spin-up channels. Numerical substitution gives an upper limit of 10^8 electrons for the charge on a ball of 4 cm diameter to keep ${}_4\Omega_e$ below $0.25\Omega_0$ for the nonrolling satellite $7\cdot 10^9$ electrons for the rolling satellite. The corresponding static potential between ball and case in the absence of roll is 0.03 V which may be contrasted with the potential of 0.7 V applied to the support. The difference comes about through the absence of symmetry and compensation factors for the static torque. For the rolling satellite, the allowed static potential is 2.1 V.

3'4. *Optimum spin speed.* – Evidently drift rates due to constant sources decrease in proportion to v_s as the spin speed of the gyro is raised. Those that depend on the centrifugal distortion of the ball increase in proportion to v_s. There is, therefore, an optimum spin speed. Roughly speaking, this is reached when the deviations from sphericity due to centrifugal distortion become comparable with the polishing errors. In the present system it is about 200 Hz.

3'5. *Magnitude of residual accelerations on the gyroscope.* – The residual acceleration acting on the gyroscope may be written

$$f = f_1 + f_2 + f_3 ,$$

where f_1 is the acceleration arising from air drag, solar radiation and other external forces on the satellite, f_2 is the action of the self gravitation of the satellite on the gyroscope, f_3 is the gravity gradient acceleration arising from the Earth's field because the gyroscope is some distance from the mass-center of the satellite. Combined with f_3 is a smaller factor caused by the slight difference between the in-track velocity of the gyroscope from that of a free body the same distance from the center of the Earth. For a typical satellite in a 500 mile orbit

$$\left\{\begin{array}{l} f_1 \sim 6\cdot 10^{-8}\, g , \\ f_2 \sim 10^{-9} g , \\ f_3 \sim 1.5\, l \;(\boldsymbol{a}\sin\alpha + 2\boldsymbol{b}\cos\alpha)\cdot 10^{-9}\, g , \end{array}\right. \tag{21}$$

where l is the distance between the gyroscope and the mass center, α the angle between l and the orbit plane, and $\boldsymbol{a}$ and $\boldsymbol{b}$ are unit vectors defining the plane through l perpendicular to the orbit.

The value of f_1 is reduced to a value $\bar{f}$ equal to f_2 by making the satellite drag free. Since the averaging per orbit in a conventional satellite is about 20%, the improvement in gyro performance gained by making the satellite drag free is about a factor of 10. In a rolling satellite the value of f_3 is averaged, except in its effect on ${}_1\Omega_s$, which rectifies in the manner already explained. Making allowances for departures from circularity of the orbit, the averaging of f_3 is probably good to about 1% and taking l as 20 cm. $\bar{f}_3$ is about $5\cdot 10^{-10}\, g$ or somewhat less than $\bar{f}$. The rectified value of ${}_1\Omega_s$ is proportional to $[f_0^2/2h + \zeta_0 h]$ where f_0 is the peak of f_3 and ζ_0 is the effective average of the compensation factor in the rolling satellite. Given f_0 and ζ_0, the error is minimized by making the preload equal to $f_0/\sqrt{2\zeta_0}$ that is about $30 f_0$. The suspension is made automatically adjustable to higher preloads for emergencies such as the impact of micrometeorites on the satellite.

In addition to the accelerations already mentioned, the gyroscope is subject to disturbances due to noise in the pointing servos. Their magnitude may be as high as $10^{-7}\,g$ in some instances. However, the effect may probably be made negligible by choosing the bandwidths of the pointing servo and gyrosuspension in such a way that the high-frequency disturbances do not affect the gyroscope. It is possible to do this because their amplitude of motion is very small.

3·6. *Magnetic torques and readout limitations.* – The first magnetic disturbance to consider is the action on the London moment of any residual trapped flux in the magnetic shield. Its magnitude is

$$ {}_1\Omega_m = 2\cdot 10^{-8}\,\frac{H_s}{\varrho r^2}\,, \tag{22} $$

where H_s is the component of trapped flux perpendicular to the spin axis. It is independent of the spin speed of the ball. If the gyro-shields is maintained in a fixed orientation, eq. (20) sets an upper limit on H_s of 10^{-7} G. In a rolling satellite with the gyroscopes aligned parallel to the roll axis, the effect averages; there are then other factors, both torques and readout errors, which determine an upper limit on the field as follows.

One error of interest is the reaction of the gyro readout loop itself on the London moment due to the readout current. Its magnitude is

$$ {}_2\Omega_m = 2\cdot 10^{-15}\,\frac{\sin 2\beta}{K\varrho r^2}\,, \tag{23} $$

where K is the gain of the feedback servo used in the readout circuit and β is the readout angle. The error is completely negligible in all real situations; even for quite large readout angles. Much more important is the effect of rotating trapped flux in the ball. If the readout circuit has any dissipation, the component of spin in the plane of the loop will be slowed down, but the transverse component remains unaffected. This is an example of a significant class of disturbances which may be called differential damping torques. The error has the form

$$ {}_3\Omega_m = \frac{15}{8}\,\frac{R_L H_b^2}{\varrho r L \omega_s}\sin 2\beta\,, \tag{24} $$

where H_b is the field trapped in the ball (which is usually comparable with H_s), L is the inductance of the readout circuit, and R_L is a complicated resistive term, different in the two cases of Josephson junction and modulated inductance magnetometers, which may also in some circumstances be a function of ω_s. Typically, with a modulated-inductance magnetometer, the upper limit on the trapped flux in the ball set by equation (24) is $5\cdot10^{-6}$ G.

Trapped flux introduces errors which affect the readout as well as the gyroscope. It does so in several distinct ways. Rotating trapped flux may, if large enough, saturate the readout amplifiers. It also causes a rectification signal, associated with phase shifts in the readout circuit, which may change and produce a zero error as the ball slows down. With a Josephson-junction magnetometer, the trapped flux may cause field changes too rapid for the readout to follow, in which case the magnetometer will lose lock unless steps are taken to cancel the signal. With a modulated inductance magnetometer, trapped flux in the ground plane causes an error if the modulation amplitude changes. Proper design can make all these terms negligible below about 10^{-6} G.

The conclusion is that the field should be kept below 10^{-6} G and preferably below 10^{-7} G. There is also a stringent limit on the constancy of the field. Even though readout measurements based on superconducting circuits have in principle absolute null stability, errors may arise from changes in the external field through the gyro-loop. Much of the external flux is excluded by the presence of the superconducting ball, but still the maximum change in field allowed is 10^{-12} G. Field stabilities of this level may be assured by using a superconducting shield as described in the next paper.

A source of readout error of a different kind, which is important to the design of the gyroscope, is the displacement of the ball from the center of the housing. Defining the displacement axes with x in the direction of spin and z perpendicular to the plane of the loop, the error angle θ_d is approximately

$$\theta_d = \frac{\pi}{3}\frac{xz}{r^2}\,. \tag{25}$$

The resultant maximum allowed displacement is about $2\cdot10^{-5}$ cm, which may be compared with the limit $8\cdot10^{-6}$ cm from the suspension torques. An equation analogus to (25) was derived by MAXWELL in 1863 during his interesting analysis of the errors in the British Association experiment to determine the ohm in absolute units [8].

3·7. *Gas torques and the spin-up problem.* – Residual gas in the cavity, gradually slows down the ball. Assuming diffuse reflection, the exponential spin-down time is given by

$$\tau = \frac{1}{5}\sqrt{\frac{2\pi k}{m}}\,\frac{\varrho r T^{\frac{1}{2}}}{p}\,, \tag{26}$$

where p is the pressure in the cavity, m the mass of the helium atom, k Boltzmann's constant and T the absolute temperature. For helium at 2 °K and 10^{-9} Torr, the time τ is about 300 y. Gas damping is much greater than trapped flux damping or anything else that contributes to slowing the ball. Intuitively then, one might expect serious precession errors due to departures from sphericity or centering of the rotor or the gyro-housing. Actually such errors vanish completely so long as the pressure is uniform. This rather surprising result comes about because the momentum transfer between two moving surfaces is proportional to (number of molecules) × (frequency of collision per molecule). The first factor is proportional to the distance d between the ball and the housing; the second factor is proportional to $1/d$. Thus τ is independent of d as eq. (26) shows. Errors may occur, however, if there are any pressure gradients in the cavity. If there is a minimum pressure axis inclined at an angle γ to the spin axis, the damping of spin components parallel and perpendicular to that axis will not be the same. The drift rate in a typical situation is

$$\Omega_p \sim \frac{5}{6}\sqrt{\frac{m}{2\pi k}}\,\frac{\Delta p}{\varrho r T^{\frac{1}{2}}}\sin 2\gamma\,, \tag{27}$$

where Δp is the total pressure variation in the cavity. At 10^{-9} Torr, Ω_p may be reduced below Ω_0 by making γ about 2 arc·min and $\Delta p/p$ about 5 percent.

Such limits on p, Δp and γ are extremely difficult to achieve. In fact, the gas-drag torque provides one of the most powerful arguments in favor of rolling the satellite, since that automatically introduces averaging for gyros parallel or perpendicular to the roll axis. The effective alignment obtained on the average by roll gives a γ of about 10 arc·s, which means that the requirements on Δp can be relaxed by about one order of magnitude. So $\Delta p/p$ might be 0.5 and p about $3\cdot 10^{-9}$ Torr. With a satellite rolled continuously in one direction, a new torque is introduced for the gyros with their axes aligned perpendicular to the roll axis, owing to gas drag, but its effects are negligible if the roll period exceeds not one hour.

The difficulty with gas torques focuses attention on one of the central problems in the design of the gyroscope: spin-up. One might suppose that some other method of spinning the ball besides gas jets would have residual torques that are more easily reduced. On the contrary, some quite general considerations exist to show that the difficulties of gas spin-up, severe as they are, are much less forbidding than almost any other system that can be devised.

To spin a gyro-rotor initially at rest, a torque Γ_s of some kind must be applied for a time t_s, after which Γ_s must be reduced to a level where the residual component Γ_r perpendicular to the spin axis does not cause significant drift errors. Neglecting drag torques $\Gamma_s = I\dot{\omega} = I\omega_s/t_s$. Substituting into eq (4), one finds.

$$\frac{\Gamma_r}{\Gamma_s} < \Omega_0 t_s \,. \tag{28}$$

Since it is operationally convenient to spin up in a time shorter than the half-period of the satellite orbit, t_s has to be 2000 s or less. Hence with $\Omega_0 \sim \sim 10^{-16}$ rad/s, the ratio Γ_r/Γ_s must be less than $2 \cdot 10^{-13}$. Gas forces are among the very few for which such a torque switching ratio is possible. Almost the only conceivable alternative is a mechanical clutch and drive motor.

But there is another consideration of great cogency. To obtain the full London moment, the gyro has to be spun up below its superconducting transition temperature. Now any spin-up system will dissipate some heat. At cryogenic temperatures, the only way of getting rid of the heat is by having some gas in the cavity. Let η be the mechanical efficiency of the spin-up system, so that $(1-\eta)$ represents the proportion of energy lost; and let T_c be the temperature of the cavity, T_r the maximum temperature rise allowed for the ball, and α the reduced mutual accommodation coefficient for the two surfaces. Then the minimum acceptable operating pressure during spin up for gas with specific-heat ratio γ is

$$p > \frac{2}{15}\sqrt{\frac{2m}{k}} \cdot \left(\frac{\gamma-1}{\gamma+1}\right)\left(\frac{1-\eta}{\alpha}\right)\sqrt{\frac{T_c+T_r}{T_c-T_r}}\,\frac{\varrho r v_s^2}{t_s} \,. \tag{29}$$

For helium gas the numerical factor reduces to $1 \cdot 10^{-5}$ CGS units. Taking α as 0.5, T_c as 2 °K, T_r as 6 °K, the minimum value for p is $2 \cdot 10^{-4}(1-\eta)$ Torr. Thus an efficiency of 99.9995 % would be needed to allow the gyro to be spin-up at the final operating pressure of 10^{-9} Torr. Actually that restriction can be relaxed slightly by taking advantage of the specific heat of the fused quartz, allowing the ball to warm up and then slowly cool down afterwards; but the restriction is still 99.993 %. There seems no escape with any conceivable system from having some gas in the cavity at higher than normal pressure during spin-up. Since nearly all the difficulty in reaching very low pressure lies in covering the last orders of magnitudes from 10^{-7} to 10^{-9} Torr, no other method is likely to be easier than gas spin up.

3·8. *Ultimate limits.* – The ultimate limits to gyro-performance are provided by random walk disturbances of various kinds. A gyroscope at a given temperature T may be expected on statistical mechanical grounds to experience fluctuations in the transverse components of its lattice angular momentum, due to exchange of energy with the molecular degrees of freedom, by an amount ΔJ of order $\sqrt{kTI}$. The root-mean-square fluctuation in direction $\langle\theta\rangle$

is then given by

$$\langle\theta\rangle \sim \frac{1}{\omega_s}\sqrt{\frac{kT}{I}}\,. \tag{30}$$

Equation (30) gives the jitter in direction and sets a limit on the precision of measurement. If Ω_0 is the desired drift rate, the minimum time for a significant observation is $\langle\theta\rangle/\Omega_0$ or about 6 hours for the existing experiment. This limit on *measurement* must be distinguished from the random walk *drifts* of the spin axis, which are evaluated by considering the transfer of angular momentum to the ball through impacts of gas mulecules, photons or cosmic rays. Much the largest of these effects is due to gas molecules. They give an effective drift rate

$${}_1\Omega_p \sim \frac{3}{4\pi}\left(N_0^{\frac{1}{2}} m^{\frac{1}{2}} k^{\frac{1}{2}}\right) \frac{p^{\frac{1}{2}}\, T^{\frac{1}{2}}}{\varrho r^2 v_s}\, t^{-\frac{1}{2}}\,, \tag{31}$$

where N_0 is Loschmidt's number and the remaining quantities have meanings already assigned. Numerical substitution makes ${}_1\Omega_p$ about $1.5\cdot 10^{-16} t^{-\frac{1}{2}}$ at a temperature of 2 °K and a pressure of 10^{-9} Torr. The effects of photons and cosmic rays are at least two orders of magnitude smaller. Thus the random-walk drifts are negligible after a few second, and the only significant effect of thermal fluctuations is to set a minimum time for observation through jitter. The reason for the drifts being so much smaller than the jitter was first explained to me by SCHIFF. It is that the relaxation times of the processes are very much longer than the observation time t_0.

4. – Summary and conclusion.

The error analysis provides design restrictions on the rotor, housing, suspension system, spin-up system and environment, needed to attain gyro drift performance of 10^{-16} rad/s. Table I summarizes the results both for a rolling and a nonrolling satellite, assuming a 500 mile near-polar orbit, with a 4 cm diameter gyroscope spinning at 200 Hz. Some of the number quoted are restrictions, others, such as the Earth's gradient acceleration, are known fixed values. The various restrictions are not independent: an improvement of performance in one area would allow relaxation of restrictions elsewhere. The immense challenge that exists in achieving all the requirements at once needs no emphasis; nevertheless, each seems within the bounds of possibility. The specifications on the gyro-rotor, low magnetic fields, and drag-free satellite performance have already been met or surpassed. The three areas where most progress is needed are: 1) mechanizing the suspension electronics for space,

TABLE I. – *Design requirements for accuracy of* 10^{-16} rad/s.

				Fixed space craft	Rolling space craft
Gyro-rotor	Homogeneity		$\Delta\varrho/\varrho$	$\sim 3\cdot10^{-7}$	$\sim 3\cdot10^{-7}$
	Sphericity at rest		$\Delta r'/r$	$\sim 5\cdot10^{-7}$	$\sim 5\cdot10^{-7}$
Gyro-suspension	Sphericity of cavity		$\Delta r''/r$	$\sim 10^{-5}$	$\sim 10^{-5}$
	Centering accuracy		t/d	$\sim 3\cdot10^{-3}$	$\sim 10^{-4}$ (mean)
			t	$\sim 8\cdot10^{-6}$ cm	$\sim 8\cdot10^{-6}$ cm
	Preload	acceleration	h	$\sim 10^{-8}\,g$	$\sim 3\cdot10^{-8}\,g$
		voltage	V	~ 0.7 V	~ 2.1 V
	Preload compensation factor		ζ	$\sim 3\cdot10^{-3}$	$\sim 10^{-2}$
	Symmetry or sum of squares control factor		θ_0 / ξ	$\sim 5\cdot10^{-3}$	$\sim 10^{-4}$ (mean)
	Distance of gyros from drag-free proof mass		l	< 20 cm	< 20 cm
Environment	Acceleration due to self-gravitation		f	$\sim 10^{-9}\,g$	$\sim 10^{-9}\,g$
	Acceleration due to Earth's gradient		f_0	$\sim 2\cdot10^{-8}\,g$	$\sim 2\cdot10^{-8}\,g$
	Residual magnetic fields	best	H	$< 10^{-7}$ G	$< 10^{-7}$ G
		using symmetry	H	$< 4\cdot10^{-6}$ G	$< 4\cdot10^{-6}$ G
	Residual electrification on ball	charge	Q	$< 10^{8}$ electrons	$< 7\cdot10^{9}$ electrons
		voltage	V	< 0.03 V	< 2.1 V
	Residual gas pressure	magnitude	p	$\sim 10^{-9}$ Torr	$\sim 10^{-9}$ Torr
		gradients	$\Delta p/p$	$\sim 5\cdot10^{-2}$	~ 0.5
Spin-up system	Torque switching ratio		Γ_r/Γ_s	$\sim 2\cdot10^{-13}$	$\sim 2\cdot10^{-13}$

2) reaching satisfactory operating pressures, and 3) maintaining low residual charge on the ball. Work on these and other problems of the experiment is in progress at Stanford and at NASA Marshall Space Flight Center. Some details are given in the next paper.

* * *

All aspects of the experiment have been discussed with W. M. FAIRBANK, to whom I am indebted in many ways. In addition to colleagues mentioned in the text, special thanks are due to Dr. R. H. CANNON. Our debt to the late L. I. SCHIFF is incalculable.

REFERENCES

[1] L. I. SCHIFF: *Proc. Nat. Acad. Sci.*, **46**, 871 (1960); G. E. PUGH: *WSEG Research Memorandum*, No. 11, Weapons Systems Evaluation Group, The Pentagon, Washington 25 D. C. (12 November 1959).

[2] R. F. O CONNELL: this volume p. 496; B. M. BARKER and R. F. O'CONNELL: *Phys. Rev. Lett.*, **20**, 69 (1968); D. C. WILKINS: *Ann. of Phys.*, **61**, 277 (1960); and private communication.

[3] C. M. WILL: this volume p. 1.

[4] T. D. BRACKEN and C. W. F. EVERITT: *Adv. Cry. Eng.*, **13**, 168 (1968).

[5] F. LONDON: *Superfluids*, Vol. **1** (New York, 1960), p. 83.

[6] A. F. HILDEBRANDT: *Phys. Rev. Lett.*, **12**, 190 (1964); M. BOL and W. M. FAIRBANK: *Proceedings of the IX International Conference on the Low Temperature Physics* (New York, 1965), p. 451; A. KING jr., J. B. HENDRICKS and H. E. RORSCHACH: *Proceedings of the IX International Conference on the Low Temperature Physics* (New York, 1965), p. 466.

[7] *Mission definition study of the Stanford relativity satellite*, Ball Brothers Research corporation Report No. 71-07, NASA Contract NAS W 7284.

[8] J. C. MAXWELL: *Rep. Brit. Ass.*, **33**, 192 (1964), correction E.

PHYSICAL REVIEW D VOLUME 5, NUMBER 4 15 FEBRUARY 1972

Bound Geodesics in the Kerr Metric*

Daniel C. Wilkins
Institute of Theoretical Physics, Department of Physics, Stanford University, Stanford, California 94305
and Joseph Henry Laboratories, Princeton University, Princeton, New Jersey 08540
(Received 25 June 1971)

The bound geodesics (orbits) of a particle in the Kerr metric are examined. (By "bound" we signify that the particle ranges over a finite interval of radius, neither being captured by the black hole nor escaping to infinity.) All orbits either remain in the equatorial plane or cross it repeatedly. A point where a nonequatorial orbit intersects the equatorial plane is called a node. The nodes of a spherical (i.e., constant radius) orbit are dragged in the sense of the spin of the black hole. A spherical orbit near the one-way membrane traces out a helix-like path lying on a sphere enclosing the black hole.

I. INTRODUCTION

Do effects of general relativity proper play a central role in such celestial phenomena as the pulsars and quasars? This question merits serious consideration if only for the reason that there is no (macroscopic) device as effective as the gravitational field of a collapsed star for the release of energy.[1]

At present, two sorts of energy-releasing processes are known. In one, a particle falling into a black hole emits electromagnetic or gravitational waves; the hole does not participate actively. In the other, a possible mechanism for which has been proposed by Penrose, the black hole itself provides a source of energy. A particle near the one-way membrane breaks up into two particles. One can arrange that one of the fragments will escape to infinity with an energy larger than that possessed by the original particle provided the other fragment is captured by the black hole. It may be that this mechanism, more contrived than the radiation process, will have less direct importance for astrophysics.

The Penrose process does not violate conservation of energy. As shown by Christodoulou,[2] the extra energy of the escaping particle is taken from the rotational energy of the black hole. By repeating the Penrose process many times one may deplete a black hole of its entire rotational energy, which can amount for a charged black hole to as much as 50% of its rest mass.[3]

The energy release possible through radiation processes is equally remarkable. Christodoulou and Ruffini[3] have shown that a charged particle falling into a charged black hole can, under appropriate conditions, emit its entire rest-mass energy.

Preliminary to investigating the problem of radiation, one must understand the kinematics of test particles for which the effect of radiation is neglected. Darwin[4,5] has already considered the geodesics of a particle in the spherically symmetric Schwarzschild field. The Kerr-Newman field, which describes a body having charge, mass, and spin angular momentum, is richer in structure than the Schwarzschild field. In this paper, we study the bound geodesics or "orbits" in the charge-free Kerr metric.[6] We have restricted ourselves to the charge-free case both because this case can be a very relevant one for astrophysics, and because we would like to distinguish the purely relativistic effects of the gravitational field from the electrodynamic ones.

In Sec. II, we deduce a necessary condition for binding to occur. Subsequently, we specialize to spherical orbits, that is, orbits of constant radius. In Newtonian mechanics, one can understand much about the general orbit by studying the circular-orbit case. Here, too, one believes that most features of interest are present in the case of spherical orbits.

In the limit of large radius, a spherical orbit goes asymptotically to a Keplerian circle. Considering a sequence of orbits of ever smaller radius, we find that as the radius decreases, the line of the (ascending or descending) node is increasingly "dragged" in the sense of the spin of the black hole. (The line of the ascending node is a line of constant azimuth in the equatorial plane passing through that point at which the orbit, going from negative to positive latitudes, intersects the equatorial plane.)

At a certain radius, a horizon occurs. This horizon, known as the one-way membrane, has the property that anything penetrating the surface from outside will be unable to escape again. The most rapidly rotating Kerr particle for which causality does not break down[7] has an angular momentum equal to its mass squared (in geometrical units). Stable orbits are possible down to its one-way membrane. As the orbital radius approaches that of the horizon, the dragging of the nodes increases without limit. During the time that a particle makes one oscillation in latitude, it will be swept through many complete azimuthal revolutions. Consequently, an orbit near the horizon will have a helix-like shape with axis parallel to the spin of the black hole.

II. CONDITIONS FOR BINDING

Carter[7] has given the first integrals of the equations of motion of a particle in the Kerr-Newman field. For the sake of a clearer physical interpretation, we use his Eq. (9) to transform to Boyer-Lindquist coordinates; in these coordinates, the metric is symmetric under simultaneous inversion of the axial and stationary Killing vectors. The metric is

$$ds^2 = \rho^2\Delta^{-1}dr^2 + \rho^2 d^2\theta + \rho^{-2}\sin^2\theta[a\,dt - (r^2+a^2)d\phi]^2 - \rho^{-2}\Delta(dt - a\sin^2\theta\, d\phi)^2,$$

where

$$\rho^2 = r^2 + a^2\cos^2\theta, \qquad \Delta = r^2 - 2mr + a^2 + e^2, \tag{1}$$

a, m, and e are, respectively, the specific angular momentum, mass, and charge of the black hole. Specializing to the charge-free case, the equations of motion take the form[8]

$$\rho^2\dot{r} = \pm\sqrt{R}, \tag{2a}$$

$$\rho^2\dot{\theta} = \pm\sqrt{\Theta}, \tag{2b}$$

$$\rho^2\dot{\phi} = (\Phi\sin^{-2}\theta - aE) + a\Delta^{-1}P, \tag{2c}$$

$$\rho^2\dot{t} = a(\Phi - aE\sin^2\theta) + (r^2+a^2)\Delta^{-1}P, \tag{2d}$$

with

$$\Theta = Q - \cos^2\theta[a^2(\mu^2 - E^2) + \Phi^2\sin^{-2}\theta],$$
$$P = E(r^2+a^2) - \Phi a, \tag{3}$$
$$R = P^2 - \Delta[\mu^2 r^2 + Q + (\Phi - aE)^2].$$

The dot denotes differentiation with respect to a parameter λ, defined in terms of the proper time by

$$\tau = \mu\lambda. \tag{4}$$

The signs in (2a) and (2b) can be chosen independently. The constant a denotes the angular momentum per unit mass of the central body. E, Φ, and Q are three constants of the particle's motion. E and Φ refer, respectively, to the energy and to the z component of angular momentum; Q is related to the θ velocity, $\dot{\theta}$.

We rewrite $R(r)$ in a form independent of the masses. Writing $R(r)$ out in detail,

$$R(r) = (E^2 - \mu^2)r^4 + 2\mu^2 m r^3 + [a^2(E^2-\mu^2) - \Phi^2 - Q]r^2 + 2m[(aE - \Phi)^2 + Q]r - a^2 Q.$$

Dividing through by $\mu^2 m^4$,

$$R(r)/\mu^2 m^4 = (\hat{E}^2 - 1)\hat{r}^4 + 2\hat{r}^3 + [\hat{a}^2(\hat{E}^2 - 1) - \hat{\Phi}^2 - \hat{Q}]\hat{r}^2 + 2[(\hat{a}\hat{E} - \hat{\Phi})^2 + \hat{Q}]\hat{r} - \hat{a}^2\hat{Q}, \tag{5}$$

where

$$\hat{E} = E/\mu, \qquad \hat{\Phi} = \Phi/m\mu, \quad \hat{Q} = Q/m^2\mu^2, \qquad \hat{r} = r/m, \quad \hat{a} = a/m. \tag{6}$$

We shall henceforth take

$$\mu = m = 1,$$

which is equivalent to using the caret variables. Since $\mu^2 = 1$ is greater than zero, the geodesics described by (1), (2), and (3) are timelike.

One can determine the values of E, Φ, and Q for which binding occurs by using the effective potentials.

We first discuss the effective radial potential $V(r)$. $V(r)$ is defined as that value of E such that $\dot{r} = 0$ at radius r; by Eq. (2a) it follows that $R(r) = 0$. Rewriting $R(r)$ in a more convenient form,

$$R(r) = [r^4 + a^2(r^2+2r)]E^2 - 4a\Phi rE + a^2\Phi^2 - (r^2 + Q + \Phi^2)(r^2 - 2r + a^2), \tag{7}$$

a quadratic in E. $R(r)$ has not one but two roots:

$$V_\pm(r, \Phi, Q) = \frac{4a\Phi r \pm \sqrt{D}}{2[r^4 + a^2(r^2+2r)]}. \tag{8}$$

The discriminant D depends on Φ only through Φ^2. In general, the reality of the radial velocity leads by (2a) to the requirement that $R(r)$ be non-negative. If D is negative, this requirement is fulfilled

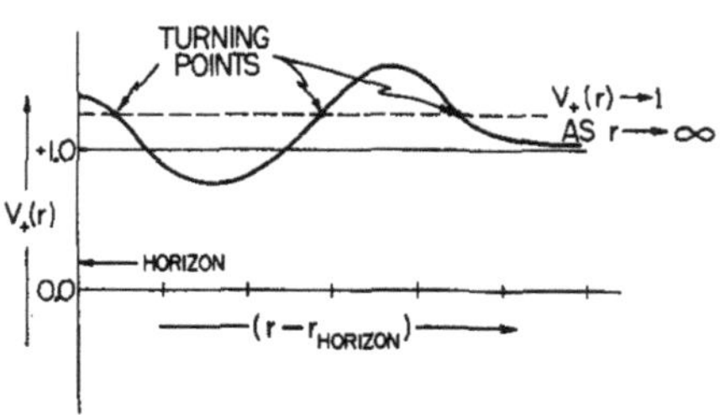

FIG. 1. Sketch of an effective radial potential which would bind a particle of energy greater than unity (dashed line). There must be at least three turning points. We show that such a potential is impossible by proving that there may be at most two turning points for $E > 1$.

for all values of the energy. If, however, D is non-negative, the energy must satisfy either of the following conditions:

$$E \geq V_+(r, \Phi, Q) \tag{9a}$$

or

$$E \leq V_-(r, \Phi, Q). \tag{9b}$$

Whether or not the effective potential is real, we may always make the one-to-one correspondence,

$$E, \Phi, Q \leftrightarrow -E, -\Phi, Q. \tag{10}$$

Inspection of the equations of motion, (2) and (3), shows that with proper choice of sign, the 4-velocities of two such corresponding motions will differ only in sign.

We now prove that if $E^2 \geq 1$, and if the specific angular momentum of the black hole lies in the causality-preserving range $0 \leq a \leq 1$, then the motion is unbound.

The proof depends on setting an upper limit on the number of turning points. Consider a conceivable $V_+(r)$ sketched in Fig. 1. Notice that by (7) and (8), $V_+(r)$ goes asymptotically to unity at large radii. As indicated by the figure a bound state with $E \geq 1$ is only possible if there is a range of energies for which three or more turning points occur. We will prove in fact, that there are at most two. $E \leq -1$ also satisfies $E^2 \geq 1$. By the correspondence (10), however, if our theorem is proven for $E \geq 1$, it will necessarily also hold for $E \leq -1$.

Let us reexpress $R(r)$ in terms of the new coordinate x, defined by

$$r = x + 1.$$

Substituting this into (5) yields

$$R(x) = (E^2-1)x^4 + (4E^2-2)x^3 + [(6+a^2)E^2 - a^2 - \Phi^2 - Q]x^2 + [(4+2a^2)(E^2-1) + 6 + 2a^2E^2 - 4a\Phi E]x + [(2aE-\Phi)^2 + (E^2+1+Q)(1-a^2)]. \tag{11}$$

In the language of classical algebra, whenever a term in the polynomial is followed by one of the opposite sign, that is described as a "variation of sign."[9] A polynomial (with real coefficients) cannot have more positive roots than there are variations of sign.[10] If $E \geq 1$, the first two terms are non-negative. Write the last three terms as

$$c_2x^2,\ c_1x,\ c_0.$$

As many as three variations of sign can occur if

$$c_2, c_0 < 0, \quad c_1 > 0. \tag{12}$$

From (12),

$$c_1 - (c_0 + c_2) > 0,$$

or by (11),

$$Qa^2 > 3E^2 - 1 > 0. \tag{13}$$

If $a = 0$ (Schwarzschild), (13) is impossible. If $0 < a \leq 1$, (13) implies a positive Q, whence c_0 is non-negative; but c_0 non-negative contradicts (12). Since (12) cannot be satisfied, there may be at most two variations of sign and hence two positive roots. We have shown this for $r \geq 1$. It must be true *a fortiori* outside the horizon, since the latter occurs at

$$r = 1 + (1-a^2)^{1/2} \geq 1.$$

When $E^2 < 1$, the first term is negative. Four variations would be possible only if the succeeding terms had the signs +, −, +, −. But if the second term is positive, then

$$E^2 > \tfrac{1}{2}.$$

Equation (13) continues to hold, showing that the last three terms cannot have the supposed signs. Hence, not more than three variations of sign are possible. One sees then that for given Q, Φ, and $|E| < 1$, there may be at most one region of binding.

By considering the motion in latitude, and using the preceding theorem, we will prove that Q must be ≥ 0 for binding.

By analogy with $V(r)$ the effective θ potential, $V^2(\theta)$, is defined as that value of E^2 which makes $\dot\theta = 0$ when the polar angle $= \theta$. By (2b) and (3),

$$0 = \Theta(\theta) = Q - \cos^2\theta\{a^2[1 - V^2(\theta)] + \Phi^2 \sin^{-2}\theta\}$$

or

$$V^2(\theta) = 1 + a^{-2}(\Phi^2 \sin^{-2}\theta - Q\cos^{-2}\theta). \tag{14}$$

The significance of $V^2(\theta)$ can be seen by writing

$$\Theta(\theta) = a^2\cos^2\theta[E^2 - V^2(\theta)]. \tag{15}$$

A particle at polar angle θ can only satisfy

$$E^2 < V^2(\theta) \tag{16}$$

if it is confined to the equatorial plane and has $Q = 0$. That Q is not negative follows from (14) according to which $Q > 0$ implies

$$E^2 > V^2(\pi/2) = -\infty.$$

For all cases other than $Q = 0$,

$$E^2 \geq V^2(\theta). \tag{17}$$

But (14) shows that if $Q<0$, then

$$V^2(\theta)>1, \tag{18}$$

By our earlier theorem, (17) and (18) together imply that the particle is unbound. Thus, for binding we require

$$Q\geq 0. \tag{19}$$

It follows from (2b), (3), and the theorem that an orbit has $Q=0$ if and only if it is confined to the equatorial plane.

Typical examples of $V^2(\theta)$ with $Q>0$ are shown in Fig. 2.[11] A particle can only reach the axis ($\theta=0, \pi$) if $\Phi=0$. With the help of the figure we draw the conclusion: Every orbit either remains in the equatorial plane ($Q=0$) or crosses it repeatedly ($Q>0$).

III. SPHERICAL ORBITS

From here on we will treat the most interesting case, the extreme Kerr case:

$$a=1.$$

In addition, we restrict ourselves to spherical orbits. The particle's radial coordinate will be stable at some value r_0 if $R(r)$ vanishes at $r=r_0$

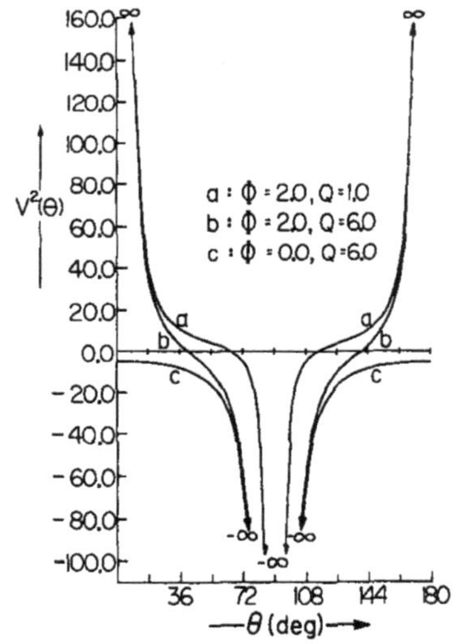

FIG. 2. Examples of the effective θ potential with $Q>0$. Only when $\Phi=0$ can particles of finite energy reach the axis ($\theta=0°$, 180°). In this and subsequent figures, the spin parameter, a, equals unity.

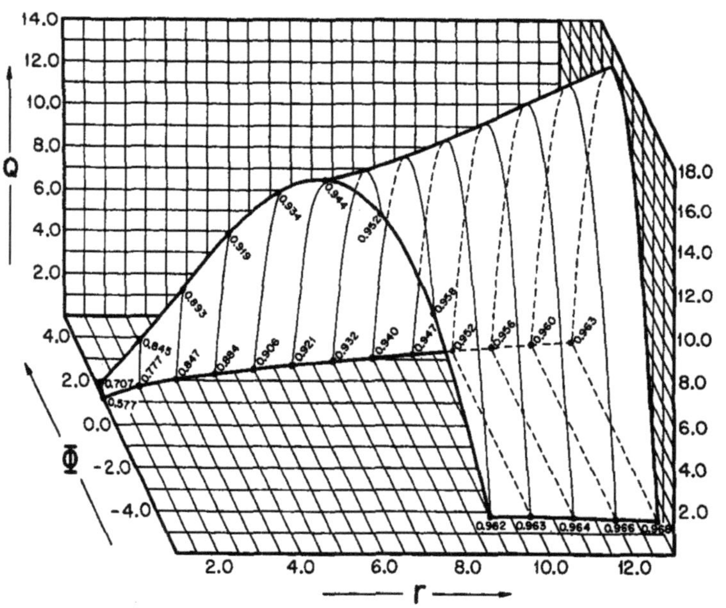

FIG. 3. Portion of the surface of stable spherical orbits. The energies of the orbits of least and greatest binding are given for integral values of the radius. For small radii, only orbits revolving in the sense of the black hole (those with $\Phi>0$) are stable.

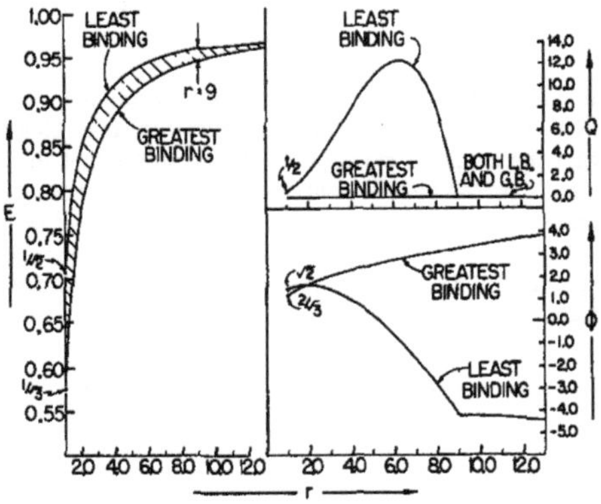

FIG. 4. Constants of motion of spherical orbits of least and greatest binding. As seen in the plot of Φ versus r and in Fig. 3, all orbits with radius $\lesssim 5.3$ are co-revolving ($\Phi > 0$). The discontinuity of slope at $r=9$ for the orbits of least binding is not puzzling in view of the same discontinuity evident in Fig. 3.

and goes negative nearby. This will be the case if

$$R(r_0)=0, \tag{20a}$$

$$\left.\frac{\partial R}{\partial r}\right|_{r=r_0}=0, \tag{20b}$$

$$\left.\frac{\partial^2 R}{\partial r^2}\right|_{r=r_0}<0. \tag{20c}$$

Replacing (20c) by

$$\left.\frac{\partial^2 R}{\partial r^2}\right|_{r=r_0}>0 \tag{20c$'$}$$

yields instead the conditions for an unstable spherical orbit.

In the following we shall only be interested in the plus-root solutions, that is, those satisfying (9a). The behavior of the minus-root solutions is easily determined with the help of (10).

Solving Eqs. (20a) and (20b) simultaneously eliminates two of the four unknowns, E, Q, Φ, and r. Imposing the stability requirement (20c) and the boundedness requirement (19) determines a subset of this two-parameter set of trajectories.[12] In Fig. 3, the set of stable spherical orbits of small radius is represented as a two-dimensional surface in (r, Q, Φ) space. The intersections of the surface with the $Q=0$ plane are the equatorial orbits. For $r \geq 9$, there are two such lines, standing for the co-revolving[13] ($\Phi > 0$) and counterrevolving ($\Phi < 0$) equatorial orbits.

For a fixed radius the energy varies monotonically along the surface. The co-revolving equatorial orbits have the largest binding energy, $1-E$. For a given radius ≥ 9, the counterrevolving equatorial orbit has the smallest binding energy. For a radius <9, such an orbit is unstable (and thus not shown in Fig. 3). Instead, the orbit situated at an inflection point of the effective radial potential, that is, with

$$\frac{\partial^2 R}{\partial r^2}=0,$$

is the least tightly bound (for a given radius <9); the set of these orbits constitutes the curved edge of the opening in the surface. (See Fig. 4.)

For large radii, the surface goes to the Schwarzschild limit:

$$Q(r)=\frac{r^2}{r-3}-\Phi^2 \geq 0,$$
$$E^2(r)=1-\frac{r-4}{r(r-3)}. \tag{21}$$

Stable orbits occur all the way to the horizon, at $r=1$. The one-parameter family of horizon-skimming orbits is

$$2/\sqrt{3} \leq \Phi \leq \sqrt{2}, \tag{22a}$$

$$Q=\tfrac{3}{4}\Phi^2-1, \tag{22b}$$

$$E=\tfrac{1}{2}\Phi. \tag{22c}$$

The lower limit on Φ results from the restriction to $Q \geq 0$. To understand the upper limit one must consider an orbit just outside the horizon. Setting $r=1+\lambda$, with $0<\lambda \ll 1$, one can show that

$$\left.\frac{\partial^2 R}{\partial r^2}\right|_{r=1+\lambda}=2(\Phi^2-2)\lambda+O(\lambda^2). \tag{23}$$

Applying (20c) yields the upper limit, $\Phi=\sqrt{2}$.

Equations (22b) and (23) do not of themselves rule out the alternative range

$$-\sqrt{2} \leq \Phi \leq -2/\sqrt{3}. \tag{22a$'$}$$

Consideration of $V_+(r)$, however, shows that (22a) describes a solution of (9a), and (22a′) a solution of (9b).

From (21), observe that in the Schwarzschild case, the squared angular momentum, Φ^2+Q, depends only on the radius. By contrast, it follows from Eqs. (22) that, for orbits at the horizon, Φ^2+Q varies almost by a factor of 2 in the extreme Kerr case.

IV. DRAGGING OF THE NODES

Lense and Thirring[14] have shown that in the weak-field limit, the nodes of a circular orbit are dragged in the sense of the spin by an angle

$$\Delta\Omega \cong 2(a/m)(m/r)^{3/2} \tag{24}$$

per revolution. An exact expression for $\Delta\Omega$, correct for all distances, can be given in the Kerr case.

Assume, say, that θ is decreasing. Dividing (2c) by (2b) then gives

$$\frac{d\phi}{d\theta}=-\frac{\Phi\sin^{-2}\theta-E+P\Delta^{-1}}{[Q-\cos^2\theta(1-E^2+\Phi^2\sin^{-2}\theta)]^{1/2}}. \tag{25}$$

Setting

$$z=\cos^2\theta \tag{26}$$

and assuming $\theta\leq\frac{1}{2}\pi$, we integrate to obtain

$$\phi=\tfrac{1}{2}\Phi\int\frac{dz}{(1-z)Y(z)}+\frac{P\Delta^{-1}-E}{2}\int\frac{dz}{Y(z)}, \tag{27}$$

where

$$Y(z)=[\beta z^3-(\alpha+\beta)z^2+Qz]^{1/2}, \tag{28}$$

with $\alpha=\Phi^2+Q$ and $\beta=1-E^2$. Turning points in θ occur when the denominator of (25) vanishes or, equivalently, when

$$\beta z^2-(\alpha+\beta)z+Q=0. \tag{29}$$

Since α, β, and Q are all non-negative and since $\alpha\geq Q$, the roots, $z_\pm$, of (29) are real and non-negative. The range of z for which the motion takes place includes the equatorial value, $z=0$:

$$0\leq z\leq z_- . \tag{30}$$

Equation (30) corresponds to one-quarter of a complete oscillation in latitude. From (2c) it is clear that the azimuth changes by the same amount in each quarter oscillation; that is, one gets the same change whatever the signs of $\dot\theta$ and $\theta-\frac{1}{2}\pi$.

Using a standard table of integrals,[15] the change of azimuth during one-quarter oscillation of latitude is cast into a more intelligible form:

$$\Delta\phi=\frac{1}{(\beta z_+)^{1/2}}[\Phi\Pi(-z_-,k)+(P\Delta^{-1}-E)K(k)], \tag{31}$$

where

$$k^2=z_-/z_+$$

and

$$K(k)=\int_0^{\pi/2}\frac{dx}{(1-k^2\sin^2x)^{1/2}},$$

$$\Pi(n,k)=\int_0^{\pi/2}\frac{dx}{(1+n\sin^2x)(1-k^2\sin^2x)^{1/2}}$$

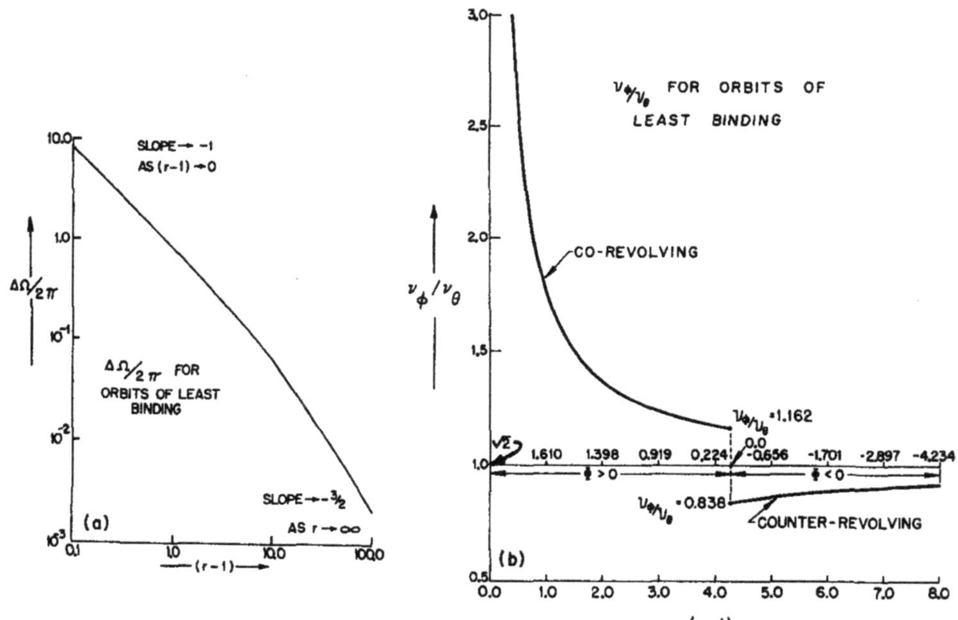

FIG. 5. (a) Angle of dragging of nodes per revolution versus distance from the one-way membrane. $\Delta\Omega$ increases continuously as the orbital radius decreases. (b) Ratio of the ϕ and θ frequencies versus distance from the one-way membrane. This changes discontinuously from a value less than unity to a value greater than unity, in going from counterrevolving to co-revolving orbits.

are elliptic integrals of the first and third kinds, respectively.

An orbit is called co-revolving if $\Delta\phi$ is positive. The first term on the right-hand side of (31) has the sign of Φ; one can show that the second term is always positive for a spherical orbit satisfying (9a). When Φ is negative the first term dominates the second; even when Φ approaches zero, the integral by which it is multiplied blows up so that the first term remains dominant. It follows that the sign of Φ determines that of $\Delta\phi$ and hence whether an orbit is co-revolving.

If the θ and ϕ frequencies were equal, $\Delta\phi$ would equal $\frac{1}{2}\pi$. Thus the ratio of the frequencies is given in general by

$$\nu_\phi/\nu_\theta = |\Delta\phi|/\tfrac{1}{2}\pi, \tag{32}$$

Substituting values of E, Φ, and Q for various orbits, one finds that

$$\begin{aligned} \nu_\phi/\nu_\theta &< 1 \quad \text{for} \quad \Phi < 0 \\ &> 1 \quad \text{for} \quad \Phi > 0 . \end{aligned} \tag{33}$$

Equation (33) signifies that the nodes are always dragged in the sense of the spin.

The angle of advance of the nodes per nodal period is

$$\Delta\Omega = 2\pi\,|\nu_\phi/\nu_\theta - 1| . \tag{34}$$

One finds, as one would expect, that $\Delta\Omega$ always varies continuously when the constants of motion change continuously. ν_ϕ/ν_θ, however, undergoes a finite discontinuity when Φ passes through zero. Figure 5 displays the contrasting behavior of $\Delta\Omega$ and ν_ϕ/ν_θ for the least tightly bound orbits. The plot of ν_ϕ/ν_θ for the most tightly bound orbits does not show the discontinuity since such orbits are all co-revolving.

For large radii, one may replace E, Φ, and Q by their Schwarzschild values. From (21), (31), (32), and (34), the leading term of $\Delta\Omega$ is just the Lense-Thirring result, (24), with $a = m = 1$.

Contrast (24) with the effect in Newtonian theory of a mass quadrupole moment:

$$\Delta\Omega \propto \frac{\cos i}{r^2}, \tag{35}$$

where i is the angle of inclination. The $\Delta\Omega$ of (35) differs from that of (24) in these respects: (a) higher-order dependence on the radius; (b) dependence on the inclination of the orbit, e.g., the nodes do not regress for a polar orbit; and (c) the sense of rotation of the nodes depends on the motion of the orbiting particle – the nodes move contrary to the azimuthal velocity.

For small radii, only the term Δ^{-1} in (31) diverges. Using Eq. (8), the effective potential at $r = 1 + \lambda$ (expanded about $r = 1$) is

$$V(r) \simeq \tfrac{1}{2}\Phi + \frac{\partial V}{\partial r}\lambda + \frac{1}{2}\frac{\partial^2 V}{\partial r^2}\lambda^2 .$$

Since the slope of $V(r)$ vanishes at $r = 1 + \lambda$,

$$\lambda \simeq -\frac{\partial V}{\partial r}\Big/\frac{\partial^2 V}{\partial r^2} .$$

Thus

$$E = V(r) \simeq \tfrac{1}{2}\Phi - \frac{1}{2}\frac{\partial^2 V}{\partial r^2}\lambda^2 = \tfrac{1}{2}\Phi + O(\lambda^2).$$

Using this to evaluate P to order λ, one finds

$$P\Delta^{-1} = \Phi\lambda^{-1} + O(1). \tag{36}$$

There results the asymptotic formula

$$\frac{\nu_\phi}{\nu_\theta} \simeq \frac{1}{r-1}\,\frac{2K(k)}{\pi(\beta z_+)^{1/2}} . \tag{37}$$

Substitution of the values (22) for orbits at the horizon reveals that the coefficient of $(r-1)^{-1}$ varies from about 0.817 for $\Phi = 2/\sqrt{3}$ to about 0.835 for $\Phi = \sqrt{2}$.

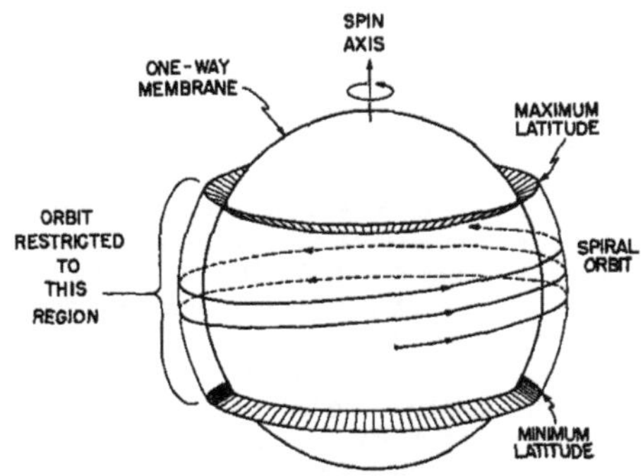

FIG. 6. Sketch of path of spherical orbit near one-way membrane (not drawn to scale.)

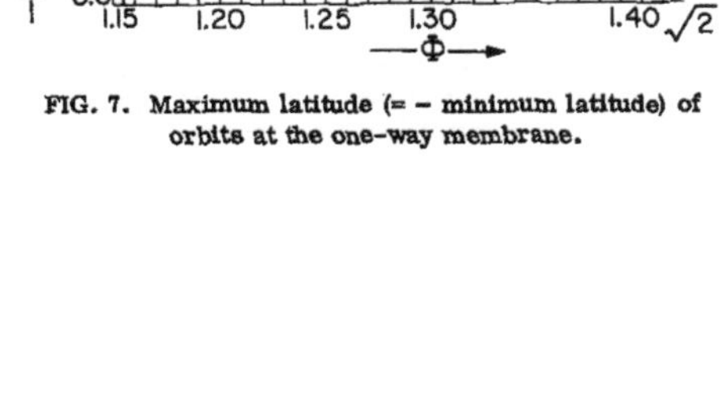

FIG. 7. Maximum latitude (= – minimum latitude) of orbits at the one-way membrane.

An orbital path close to the horizon is sketched in Fig. 6. The particle traces out a kind of helix lying on a sphere. As the particle approaches the maximum latitude, the angular separation between successive loops of the helix decreases. Reaching the maximum latitude, the particle begins the winding descent to the minimum latitude, located below the equatorial plane symmetrically to the maximum. Using Eqs. (22), (26), (29), and (30), we have plotted in Fig. 7 the maximum latitudes of orbits at the horizon.

It is known that there is a close resemblance between the linearized gravitational theory and classical electromagnetism. This prompts one to ask whether there occur orbits analogous to the spirals of electrons in the earth's magnetic field. The best place to look for such exotic behavior is near the horizon. From Eqs. (3), (9a), and (8) it follows that $P\Delta^{-1}$ always diverges at the horizon. Divide Eq. (2c) by Eq. (2d) to obtain

$$\frac{d\phi}{dt} \approx \tfrac{1}{2} \text{ for } r \approx 1.$$

One can see from this that the retrograde motion required for any kind of looping is impossible: Dragging forces a particle near the horizon to revolve always in the same sense as the black hole.

Unlike the weak-field region it is not true in general that $\Delta\Omega$ depends only on the radius. For example, for $r = 9$, $\Delta\Omega/2\pi$ decreases from 0.0814 for the counterrevolving equatorial orbit to 0.0607 for the co-revolving equatorial orbit.[16]

V. PERIODS

The proper θ period is obtained by integration of (2b). Squaring (2b), multiplying through by $\cos^2\theta \sin^2\theta$, and making the change of variables (26), one obtains

$$\dot{z}^2 = \frac{4}{r^2 + z^2} Y(z)$$

with $Y(z)$ as in (27). Hence, apart from a sign,

$$d\tau = \frac{(r^2 + z)dz}{2Y(z)} .$$

Integrating,

$$\tfrac{1}{4}\tau_{\theta,\phi} = \int_0^{z_-} \frac{(r^2 + z)dz}{2Y(z)}$$

or

$$\tau_{\theta,\phi} = \frac{4}{(\beta z_+)^{1/2}} (r^2 + z_+) K(k) - 4\left(\frac{z_+}{\beta}\right)^{1/2} E(k), \quad (38)$$

where

$$E(k) = \int_0^{\pi/2} (1 - k^2 \sin^2 x)^{1/2} dx$$

is the complete elliptic integral of the second kind.

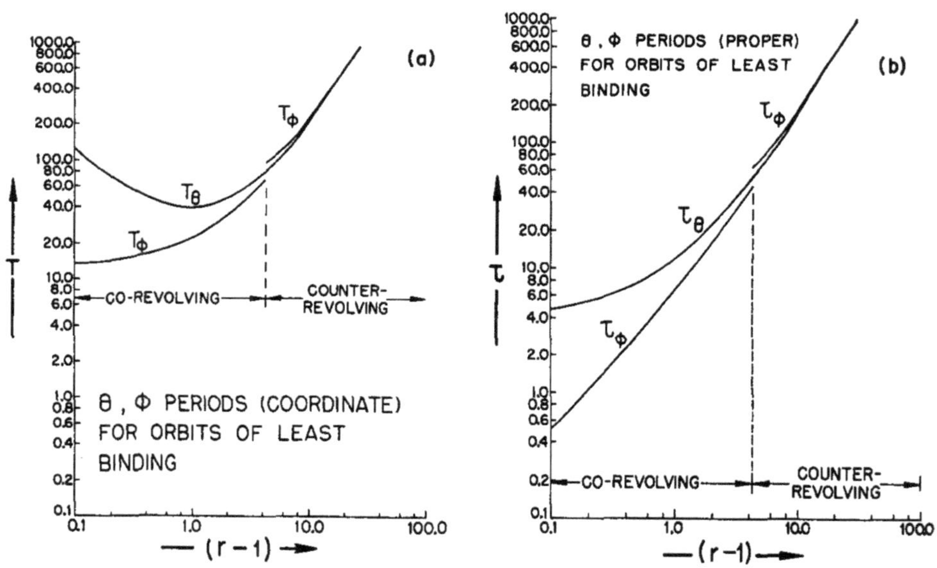

FIG. 8. The θ, ϕ periods (coordinate and proper) for orbits of least binding. The ϕ period is greater (less) than the θ period for counterrevolving (co-revolving) orbits. For the most tightly bound orbits (not shown), which are all co-revolving and equatorial, the ϕ period is everywhere less than the θ period.

To obtain the coordinate period, divide Eq. (2d) by Eq. (2b). Integration then yields

$$\tau_{\theta,c}=4\left[E\left(\frac{z_+}{\beta}\right)^{1/2}+\frac{1}{(\beta z_+)^{1/2}}[\Phi+P\Delta^{-1}(r^2+1)-E]\right]K(k)$$

$$-4E\left(\frac{z_+}{\beta}\right)^{1/2}E(k). \tag{39}$$

To find the proper azimuthal period divide $\tau_{\theta,\phi}$ by

$$\nu_\phi/\nu_\theta=\tau_\theta/\tau_\phi,$$

which is given by (31) and (32); likewise for the coordinate period.

The periods determined above refer to $\mu=m=1$. The scaling law for times is obtained in similar manner to Eqs. (6). Equation (2b) is to be cast into the mass-independent form

$$\hat{\rho}^2\frac{d\hat{\theta}}{d\hat{\lambda}}=(\hat{\Theta})^{1/2}, \tag{40}$$

where $\hat{\Theta}$ is the function Θ expressed in terms of caret variables. The polar angle is already scale free. From (1), (3), and (6) one sees that

$$\hat{\rho}^2=m^{-2}\rho^2,$$

$$\hat{\Theta}=(m\mu)^{-2}\Theta.$$

It follows that (2b) can be cast into the form (40) by setting

$$d\hat{\lambda}=\mu m^{-1}d\lambda=m^{-1}d\tau.$$

The second equality results from (4). Putting $m=1$,

$$d\hat{\lambda}=d\hat{\tau}$$

or

$$d\hat{\tau}=m^{-1}d\tau. \tag{41}$$

Similarly, from (2d), (3), and (6),

$$d\hat{t}=m^{-1}dt. \tag{42}$$

Equations (41) and (42) together with Eq. (6) constitute a general rule applicable to geodesics with $\mu^2>0$.

Figure 8 shows the periods for the least-bound orbits with radii ≤ 30. Here we see that the discontinuity in ν_ϕ/ν_θ mentioned earlier is due entirel to a discontinuity in ν_ϕ.

Note added in proof. One can show from (7) that

$$D=4r\Delta\{\Phi^2r^3+(r^2+Q)[r^3+a^2(r+2)]\}.$$

Equations (8) and (19) then imply that for an orbit the effective radial potential is everywhere real outside the horizon.

ACKNOWLEDGMENTS

Without the encouragement, guidance, and infectious enthusiasm of Professor R. Ruffini, this work would not have been accomplished. I am also indebted to Professor J. A. Wheeler and Dr. Demetrios Christodoulou for helpful discussions. To those at Stanford University and at Princeton who enabled me to spend some months in the stimulating atmosphere of Princeton's Department of Physics, I express my heartfelt gratitude.

---

*Preparation for publication assisted in part by NSF Grants No. GP14361 and No. GP7669.

[1]R. Ruffini and J. A. Wheeler, in *The Significance of Space Research for Fundamental Physics*, edited by A. F. Moore and V. Hardy (European Space Research Organization, Paris, 1970).

[2]D. Christodoulou, Phys. Rev. Letters 25, 1596 (1970).

[3]D. Christodoulou and R. Ruffini, Phys. Rev. D 4, 3552 (1971).

[4]C. Darwin, Proc. Roy. Soc. (London) A249, 180 (1958).

[5]C. Darwin, Proc. Roy. Soc. (London) A263, 39 (1961).

[6]By "bound" we mean that the particle remains within a finite radial distance of the black hole but is never captured.

[7]B. Carter, Phys. Rev. 174, 1559 (1968).

[8]We have corrected the signs of the terms containing P in Eqs. (2c) and (2d).

[9]The polynomial must be written in the order of decreasing (or increasing) powers. Terms with zero coefficients are *not* included.

[10]R. Descartes, *The Geometry of René Descartes*, translated by D. E. Smith and M. L. Latham (Open Court, Chicago, 1925). Descartes merely states the rule. For a proof see, e.g., H. E. Buchanan and L. C. Emmons, *A Brief Course in Advanced Algebra* (Houghton-Mifflin, Boston, 1937).

[11]The symbol $V^2(\theta)$ does not stand for the square of a real-valued function; rather it is meant to suggest a function having the dimensions of squared energy. That it can take on negative values is consequently not problematical.

[12]The artificiality of considering the θ and r motions separately here reveals itself in the curious circumstance that a trajectory may be "stable" and yet unbound.

[13]By co-revolving we mean that during one complete oscillation in latitude, the particle's azimuth changes in the sense of the spin of the black hole, that is to say, it increases. In Sec. IV we explain why an orbit is co-revolving if and only if $\Phi>0$.

[14]J. Lense and H. Thirring, Physik. Z. 19, 156 (1918).

[15]W. Grobner and N. Hofreiter, *Integraltafel* (Springer, Vienna, 1949), Vol. 1.

[16]To give a meaning to $\Delta\Omega$ for equatorial orbits, one imagines the motion to deviate infinitesimally from the equatorial plane.

2324 10

PHYSICAL REVIEW D VOLUME 10, NUMBER 8 15 OCTOBER 1974

Generalized Wilkins effect and selected orbits in a Kerr-Newman geometry*

Mark Johnston and Remo Ruffini
Joseph Henry Laboratories, Princeton University, Princeton, New Jersey 08540
(Received 8 April 1974)

Wilkins has pointed out that, in addition to the periodicities associated with circular orbits in the equatorial plane of a Kerr geometry, there exist periodicities connected with the longitudinal motion of particles. We extend this result to spherical orbits of charged particles in a Kerr-Newman geometry. We give explicit examples of nonspherical orbits, illustrating dragging of inertial frames and these longitudinal periodicities. The relevance of these results for the physics of collapsed objects is discussed.

I. INTRODUCTION

Recent observations of structure down to 50 msec in x-ray signals from binary x-ray sources[1] and the possible identification of these sources with totally collapsed objects[2] have made the analysis of orbits of test particles in the generalized Kerr metric of the greatest relevance, the more so since it has been shown[3] from purely energetic arguments that a collapsed object is quite likely to possess an electromagnetic field. The aim is to find a "signature" in the x-ray signal that could be eventually related to periodic motion of particles around a collapsed object, and from these periodicities to infer some of its properties (mass, electromagnetic fields, angular momentum). The shorter the periods, the closer are the orbits to the surface of the collapsed object. At the same time a clear cutoff should exist in the observed x-rays corresponding to the minimum possible periodicities connected with orbital motion around a collapsed object. From this we see the need for an analysis of orbits in the most general metric expected to exist around a collapsed object.

This is the motivation for our generalization of the Wilkins results for the motion of a charged particle in the field of a black hole with arbitrary values of the angular momentum and charge within the limits fixed by the existence of a regular horizon. We also examine a few numerical examples of nonspherical orbits of charged and uncharged particles in the field of an extreme charged and spinning black hole.

II. GEODESIC EQUATIONS AND EFFECTIVE POTENTIALS

The motion of a charged particle in the background field of metric $g_{\mu\nu}$ and electromagnetic vector potential A_μ is described by a Lagrangian

$$\mathcal{L} = \tfrac{1}{2} g_{\alpha\beta}\dot{x}^\alpha \dot{x}^\beta - \frac{\epsilon}{\mu} A_\alpha x^\alpha , \qquad (1)$$

where ϵ and μ are respectively the charge and the mass of a test particle and the dot denotes ordinary differentiation with respect to an affine parameter λ which is chosen to be related to the proper time τ of the particle by

$$\tau = \mu\lambda . \qquad (2)$$

We then have the normalization condition for $u^\alpha = \dot{x}^\alpha$

$$g_{\alpha\beta}u^\alpha u^\beta = -\mu^2, \qquad (3)$$

and the equation of motion of the particle

$$\frac{du^\alpha}{d\tau} + \Gamma^\alpha_{\beta\gamma}u^\beta u^\gamma = \frac{\epsilon}{\mu}F^\alpha{}_\beta u^\beta. \qquad (4)$$

Here $F^\alpha{}_\beta$ and $\Gamma^\alpha_{\beta\gamma}$ are the electromagnetic tensor and the Christoffel symbols of the background geometry. We assume for the Kerr-Newman[4] metric the Boyer-Lindquist[5] coordinates

$$\begin{aligned} ds^2 &= \rho^2\Delta^{-1}dr^2 + \rho^2 d\theta^2 \\ &\quad + \rho^{-2}\sin^2\theta\,[a\,dt - (r^2+a^2)d\phi]^2 \\ &\quad - \rho^{-2}\Delta[dt - a\sin^2\theta\, d\phi]^2, \end{aligned} \qquad (5)$$

where as usual

$$\Delta = r^2 - 2mr + a^2 + e^2 \qquad (6a)$$

and

$$\rho^2 = r^2 + a^2\cos^2\theta. \qquad (6b)$$

Here a, e, and m give respectively the angular momentum, the charge, and the mass (in geometrical units) of the given geometry. Carter[6] has shown how Eqs. (4) can be reduced to a set of total differential equations

$$\dot\theta = \rho^{-2}\sqrt{\Theta}, \qquad (7a)$$

$$\dot r = \rho^{-2}\sqrt{R}, \qquad (7b)$$

$$\dot t = -a\rho^{-2}(aE\sin^2\theta - \Phi) + \rho^{-2}(r^2+a^2)\Delta^{-1}P, \qquad (7c)$$

$$\dot\phi = -\rho^{-2}(aE - \Phi\sin^{-2}\theta) + a\rho^{-2}\Delta^{-1}P, \qquad (7d)$$

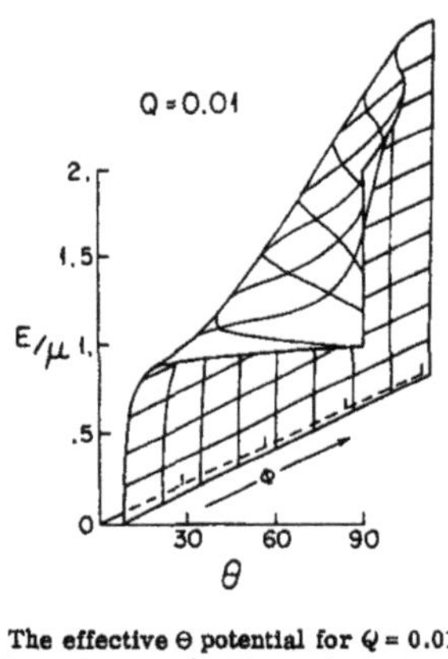

FIG. 1. The effective Θ potential for $Q = 0.01$ and $a = 1$ (extreme Kerr black hole). The angle θ, measured in degrees out of the equatorial plane, indicates where $\Theta = 0$. On the vertical axis is indicated the energy of the particle in units of the particle rest mass as measured by a far observer at infinity. On the third axis is the value of the ϕ component of the angular momentum of the particle. Units along the Φ axis are the same in Figs. 1, 2, and 3, and are indicated in Fig. 3.

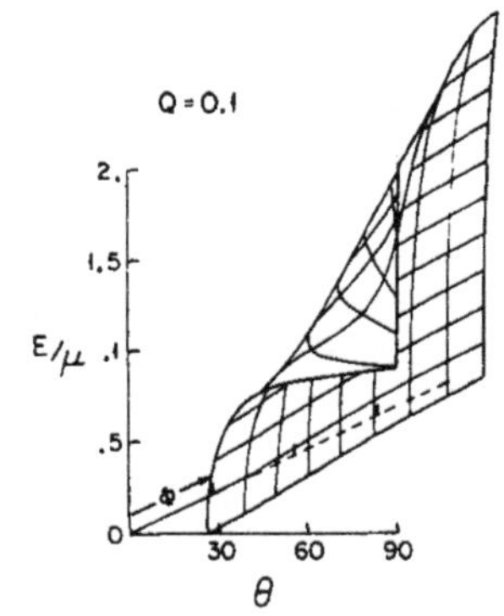

FIG. 2. The effective Θ potential for $Q = 0.1$ and $a = 1$.

with

$$\Theta = Q - \cos^2\theta[a^2(\mu^2 - E^2) + \Phi^2\sin^{-2}\theta], \qquad (8a)$$

$$P = E(r^2+a^2) - \Phi a - \epsilon e r, \qquad (8b)$$

$$R = P^2 - \Delta(\mu^2 r^2 + K), \qquad (8c)$$

$$Q = K - (\Phi - aE)^2, \qquad (8d)$$

where K, E, and Φ are constants of the motion on the trajectory of the particle. E and Φ denote respectively the energy and the angular momentum of the particle in the ϕ direction as measured by an observer at rest at infinity. For interpretation of the variable K see, e.g., Christodoulou and Ruffini[7] or Hughston *et al.*[8]

Making the transformations

$$\hat r = r/m,\quad \hat a = a/m,\quad \hat e = e/m,\quad \hat t = t/m,$$

$$\hat E = E/\mu,\quad \hat\epsilon = \epsilon/\mu,\quad \hat\Phi = \Phi/m\mu,\quad \hat\tau = \tau/m,$$

$$\hat\Theta = \Theta/m^2\mu^2,\quad \hat K = K/m^2\mu^2,\quad \hat R = R/\mu^2 m^4$$

is equivalent to setting $m = \mu = 1$; henceforth we make that assumption and omit the caret.

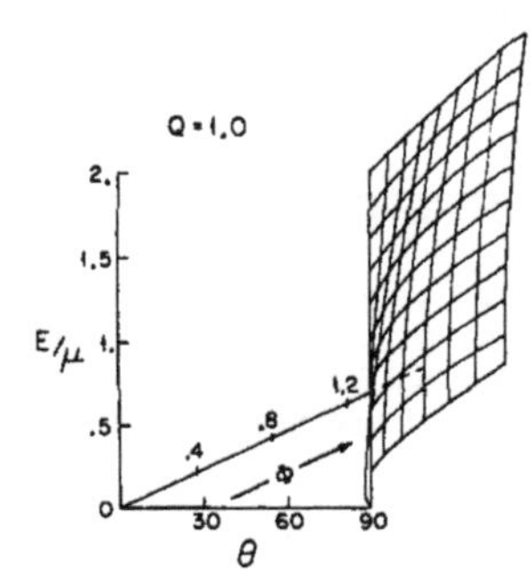

FIG. 3. The effective Θ potential for $Q = 1$ and $a = .1$.

Following Ruffini and Wheeler[9] we define the effective potential for the motion of a particle in the radial and θ direction as given by the equations

$$\rho^2\dot{r} = \sqrt{R} = 0 \qquad (9a)$$

and

$$\rho^2\dot{\theta} = \sqrt{\Theta} = 0, \qquad (9b)$$

or simply $R = 0$ and $\Theta = 0$. Using the equations given by Carter, we have for the effective potential in the θ direction

$$Q - \cos^2\theta[a^2(1-E^2) + \Phi^2\sin^{-2}\theta] = 0. \qquad (10)$$

We can then conclude that if $Q = 0$ and the motion crosses the equatorial plane then the motion has to be confined to the equatorial plane, and if $Q > 0$ the motion is limited to a maximum angle out of the equatorial plane depending on the angular momentum and energy of the test particle. The motion may reach the symmetry axis, 90° of the equatorial plane, only if $\Phi = 0$ and if $Q \geq -a^2(E^2 - 1)$. Three effective potentials are given in Figs. 1–3 for selected values of Q, and a value of $a = 1$. These effective potentials are functions only of a and Q and are independent of the charge of the particle and the charge of the black hole.

The effective potential in the radial direction is given by

$$E(r) = [-B \pm (B^2 - 4AC)^{1/2}]/2A, \qquad (11)$$

where

$$A = (r^2 + a^2)^2 - a^2\Delta, \qquad (12a)$$

$$B = 2a\Phi\Delta - (r^2 + a^2)(\Phi a + \epsilon e r), \qquad (12b)$$

$$C = (\Phi a + \epsilon e r)^2 - \Delta(r^2 + Q + \Phi^2). \qquad (12c)$$

The positive or negative sign corresponds to the positive or negative root solution in the sense of Christodoulou and Ruffini.[7] Wilkins has shown that solutions with $E^2 > 1$ for black holes with $a^2 < 1$ are unbound in the radial direction.

III. GENERALIZED WILKINS EFFECT

We now focus our attention on spherical orbits. We have from Eqs. (7)

$$\frac{d\phi}{d\theta} = \frac{-\Phi/\sin^2\theta - aE + aP/\Delta}{\{Q - \cos^2\theta[(1-E^2)a^2 + \Phi^2/\sin^2\theta]\}^{1/2}}.$$

We integrate ϕ as θ goes from $\pi/2$ to one of the turning points defined by Eq. (9a). Substituting for $z = \cos^2\theta$ we obtain for the change in ϕ, $\Delta\phi$, as θ goes through $\frac{1}{4}$ of a complete oscillation,

$$\Delta\phi = \tfrac{1}{2}\Phi\int_0^{z_-}\frac{dz}{(1-z)Y(z)} + a\,\frac{P\Delta^{-1} - E}{z}\int_0^{z_-}\frac{dz}{Y(z)}, \qquad (14)$$

where

$$Y(z) = [z(\beta z^2 - (\alpha+\beta)z + Q]^{1/2},$$

$$\alpha = \Phi^2 + Q,$$

$$\beta = a^2(1-E^2). \qquad (15)$$

Here $z_\pm$ corresponds to the turning point in the θ motion ($\Theta = 0$). We then have

$$\Delta\phi = \frac{1}{(\beta z_+)^{1/2}}[\Phi\Pi(-z_-, k) + a(P\Delta^{-1} - E)F(k)]. \qquad (16)$$

Here

$$k^2 = \frac{z_-}{z_+},$$

and

$$F(k) = \int_0^{\pi/2}(1 - k^2\sin^2 x)^{-1/2}dx \qquad (17)$$

is the complete elliptic integral of first kind and

$$\Pi(n, k) = \int_0^{\pi/2}(1 + n^2\sin^2 x)^{-1}(1 - k^2\sin^2 x)^{-1/2}dx \qquad (18)$$

is the complete elliptic integral of third kind.

Following Wilkins[10] we can define the ratio between the frequencies in the φ and in the θ direction by

$$\frac{\nu_\phi}{\nu_\theta} = \frac{|\Delta\varphi|}{\pi/2}, \qquad (19)$$

and the period in the θ coordinate, as seen by an observer of infinity, by a direct integration of $dt/d\theta$ obtained from Eqs. (7). This gives

$$\tau_\theta = 4\left[a^2E(z_+/\beta)^{1/2} + \frac{(r^2 + a^2)P\Delta^{-1} + a\Phi(1 - aE)}{(\beta z_+)^{1/2}}\right]F(k)$$

$$- 4a^2E(z_+/\beta)^{1/2}E(k). \qquad (20)$$

Here

$$E(k) = \int_0^{\pi/2}(1 - k^2\sin^2 x)^{1/2}dx$$

is the complete elliptic integral of second kind. From Eqs. (20) and (19) we obtain for the period in the ϕ direction (still measured by an observer at infinity) the relation

$$\tau_\phi = \tau_\theta(\nu_\theta/\nu_\phi). \qquad (21)$$

IV. NUMERICAL INTEGRATIONS FOR SELECTED NONRADIAL ORBITS

To explicitly visualize some of the effects of dragging of inertial frames and precession of the

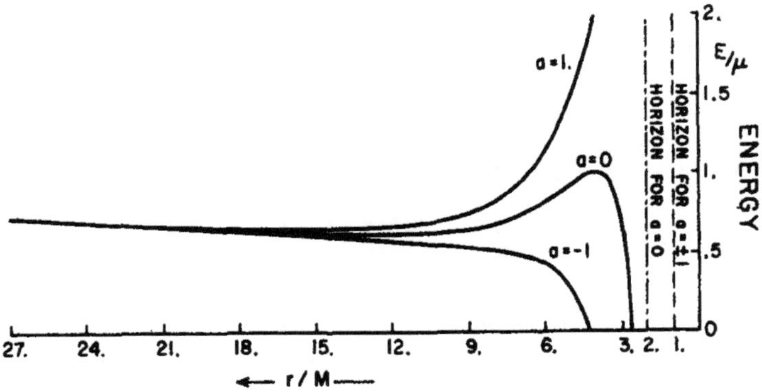

FIG. 4. Radial effective potential for the motion shown in Figs. 5, 6, and 7. The angular momentum of the particle in the ϕ direction has been chosen to be $\Phi = 4$. The other constant of the motion, Q, is here set to zero. The three diagrams correspond to values of three different values of the parameter a of the black hole: $a = \pm 1$, $a = 0$.

perihelia we have numerically computed the trajectory of a bound orbit corresponding to an energy $E = 0.97$ and angular momentum $\Phi = 4$ in the equatorial plane ($Q = 0$). Compared and contrasted are the three cases in which the black hole is endowed with extreme values of the angular velocity $a = \pm 1$ and the case in which $a = 0$ (Schwarzschild black hole). We have given in Fig. 4 the three effective potentials for the cases under consideration. In Figs. 5-7 the actual orbits of the particles are given and contrasted.

Nonradial motions out of the equatorial plane are even more complicated. While in the Schwarzschild case the motion remains in a plane tilted with respect to the equatorial plane at all times, in the Kerr case the Wilkins effect analyzed in the previous paragraph forces the particle to oscillate between points with $\theta_- < \theta < \theta_+$. Christodoulou and Ruffini[11] have defined an angular velocity of the black hole

$$\Omega = \frac{a}{r_+^2 + a^2}, \tag{22}$$

where as usual by r_+ we indicate the radius of the horizon,

$$r_+ = 1 + [1 - (a^2 + e^2)]^{1/2}. \tag{23}$$

The reason for this definition follows from the fact that if we follow the geodesic motion of any particle falling into a generalized Kerr black hole, it will approach asymptotically with time (as seen by an observer at infinity) circular motion of constant latitude on the horizon. From Eqs. (7) it follows in fact that

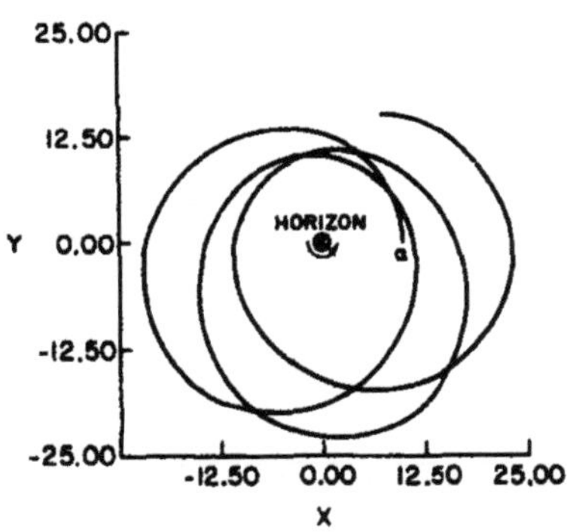

FIG. 5. Corotating orbit for extreme Kerr black hole with $a = 1$. The energy of the particle has been chosen $E = 0.97$, and angular momentum $\Phi = 4$. a indicates the starting point of the trajectory. The particle is uncharged ($\epsilon = 0$).

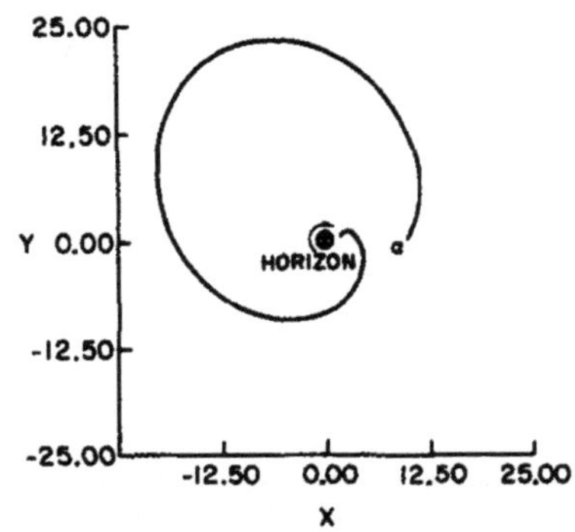

FIG. 6. Counterrotating orbit for extreme Kerr black hole with $a = -1$ and $E = 0.97$, $\Phi = 4$, $\epsilon = 0$.

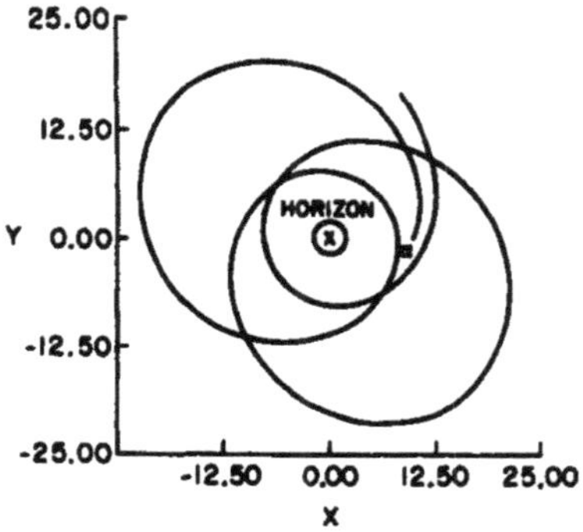

FIG. 7. Orbit of a particle with $E = 0.97$, $\Phi = 4$, $\epsilon = 0$, in the field of a Schwarzschild black hole ($a = 0$).

$$\lim_{r \to r_+} \frac{dr}{d\phi} = 0, \quad (24a)$$

$$\lim_{r \to r_+} \frac{d\theta}{d\phi} = 0, \quad (24b)$$

and

$$\lim_{r \to r_+} \frac{d\phi}{dt} = \Omega. \quad (24c)$$

To explicitly illustrate frame dragging and the precession of perihelia in the general case of non-equatorial orbits we have numerically integrated the motion of an uncharged cloud of particles, each particle with energy $E = 1$, neglecting all self-interactions, in the field of an extreme Kerr black hole with $a = 1$. In Figs. 8 and 9 the particles have $Q = 10$ and $\Phi = 0$, and the two cases are considered with $\dot{\theta} < 0$ and $\dot{\theta} > 0$. It is clear from this picture how all the particles, although with $\Phi = 0$, acquire the value of the angular velocity given by Eq. (22). Figure 10 shows a cloud of particles, revolving around an extreme Kerr-Newman black hole with $a^2 = e^2 = \frac{1}{2}$, each particle having energy $E = 0.968$, $Q = 10$, and angular momentum $\Phi = 2$.

We have finally considered the motion of charged particles in the field of an extreme generalized Kerr solution.

A few examples of stable bound orbits are given in Table I, and the motion of a particle with bind-

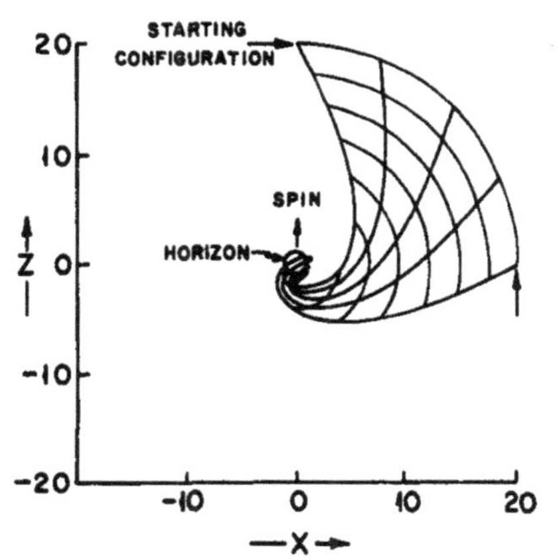

FIG. 8. Motion of an uncharged cloud of particles with no angular momentum around an extreme Kerr black hole with $a = 1$. The constants of the motion for the particle are $E = 1$, $\Phi = 0$, $Q = 10$. Initially $\dot{\theta} < 0$.

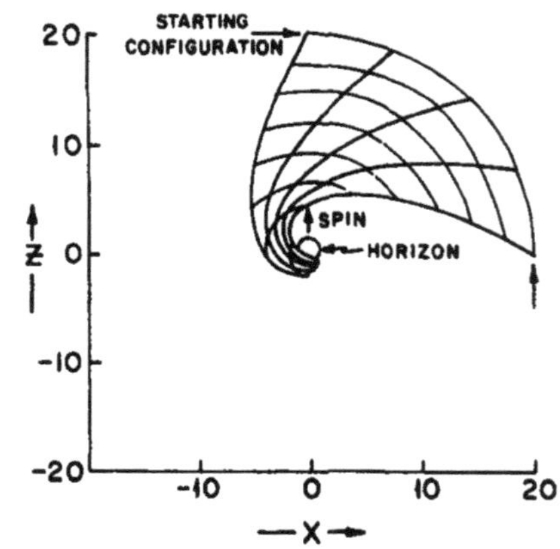

FIG. 9. Motion of an uncharged cloud of particles with no angular momentum around an extreme Kerr black hole with $a = 1$. The constants of the motion are $E = 1$, $\Phi = 0$, $Q = 10$. Initially $\dot{\theta} > 0$.

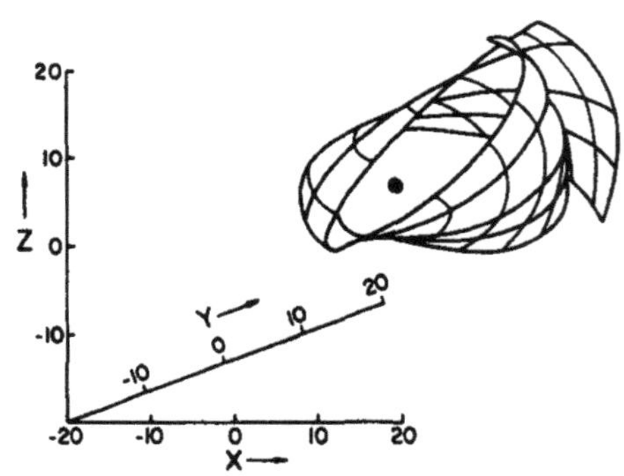

FIG. 10. Motion of uncharged cloud of particles corotating about an extreme spinning charged black hole. The orbits are stable; the constants of the motion are $E = 0.968$, $\Phi = 2$, $Q = 10$. The vertical lines indicate isochronous points (as seen from infinity).

TABLE I. Location of some circular tightly bound charged-particle orbits (the corotating cases). $a=e=\sqrt{2}/2$, $Q=0$.

ϵ	Φ	r	E
-0.2	1.7	1.2	0.707108
-1	2.5	2.00	0.707107
-100	75.3	15.35	0.287
-1000	716.6	70.28	0.139

ing energy of the order of 70% of the rest mass of the particle shown in Fig. 11. It is interesting to notice that, since the strength of the electric field is always much larger at the surface of a charged black hole than the value of its magnetic field,[12,13] no orbits spiraling around the magnetic lines of force are possible. The situation will be drastically different, however, if the black hole is endowed with a magnetosphere with a charge excess opposite to the charge of the black hole. In this case the electric field of the black hole will be partly screened and looping motions should, indeed, be possible.[14]

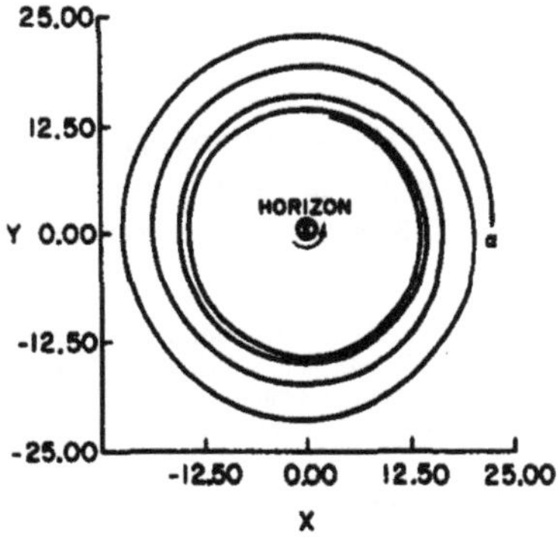

FIG. 11. Orbit of a highly charged particle ($\epsilon=-100$) near a black hole with $a^2=e^2=\frac{1}{2}$. The constants of the motion are $E=0.3$, $\Phi=75.5$, $Q=0$. The particle circles the hole more than four times in passing from apastron to periastron. A continuation of the motion would show the particle spiraling back along a similar trajectory to the outer turning point, then repeating the process. As the charge becomes more negative the orbits become more and more bound, but larger values of the angular momentum are required to maintain stable orbits, and the location of such orbits moves outward.

*Work partially supported by the N.S.F. under Grant No. GP30799X to Princeton University.

[1]S. Rappaport, R. Doxsey, and W. Zaumen, Astrophys. J. 168, L43 (1971).
[2]R. Ruffini, Bull. Am. Phys. Soc. 18, 109 (1973).
[3]R. Ruffini and A. Treves, Astrophys. Lett. 13, 109 (1973).
[4]E. T. Newman, E. Couch, R. Chinnapared, A. Exton, A. Prakash, and R. Torrence, J. Math. Phys. 6, 918 (1965).
[5]R. H. Boyer and R. W. Lindquist, J. Math. Phys. 8, 265 (1967).
[6]B. Carter, Phys. Rev. 174, 1559 (1968).
[7]D. Christodoulou and R. Ruffini, Phys. Rev. D 4, 3552 (1971).
[8]L. Hughston, R. Penrose, P. Sommers, and M. Walker, Commun. Math. Phys. 27, 303 (1972).
[9]R. Ruffini and J. A. Wheeler, in *The Significance of Space Research for Fundamental Physics*, edited by A. F. Moore and V. Hardy (European Space Research Organization, Paris, 1970).
[10]D. Wilkins, Phys. Rev. D 5, 814 (1972).
[11]D. Christodoulou and R. Ruffini, Gravity Research Foundation essay, 1971 (unpublished).
[12]For a detailed analysis of orbits of maximum binding in extreme Kerr geometry see Ref. 7 and L. Hively and R. Ruffini (unpublished).
[13]D. Christodoulou and R. Ruffini, in *Black Holes*, 1972 Les Houches Lectures, edited by C. DeWitt and B. S. DeWitt (Gordon and Breach, New York, 1973).
[14]Computations on a plasma corotating around a rotating black hole are currently under examination by J. Wilson at Lawrence Livermore Laboratory, Livermore, California.

C. R. Acad. Sc. Paris, t. 279 (23 décembre 1974) Série A — 971

RELATIVITÉ. — *Sur certaines vérifications nouvelles de la Relativité générale rendues possibles par la découverte d'un pulsar membre d'un système binaire.* Note (*) de MM. **Thibaut Damour** et **Remo Ruffini**, presentee par M. André Lichnerowicz.

Cette Note montre comment la récente découverte d'un pulsar membre d'un système binaire pourrait fournir des informations très importantes pour la vérification de la Relativité générale ainsi que pour la connaissance de la structure et des processus d'émission des pulsars. On présente des estimations des principaux effets relativistes que l'on peut espérer observer dans un tel système.

Récemment J.H. Taylor [1] a découvert une nouvelle radiosource pulsante (*cf.* tableau pour les paramètres) membre d'un système binaire (comme l'atteste la variation régulière de la période de pulsation P sur des temps $\tau \sim 8$ h). Il est important de vérifier (*a*) que cet objet est bien un pulsar [2], c'est-à-dire que sa période P doit croître lentement avec le temps et satisfaire de plus l'inégalité $(dE/dt)_{em} \leqq I\,(4\,\pi^2/P^3)\,(dP/dt)$, où I est le moment d'inertie de l'étoile à neutron et $(dE/dt)_{em}$ la puissance émise par le pulsar, (*b*) et ensuite que le compagnon de cet objet est une étoile compacte : naine blanche, étoile à neutron ou black-hole. Si le compagnon n'est pas une étoile compacte les « marées » induites le doteraient d'un moment quadrupolaire qui se signalerait par une rapide précession du périastre (360°/an) [3].

Si ces deux conditions sont remplies, alors on disposera dans cet objet du meilleur outil depuis l'observation des sources de rayons X membres d'un système binaire pour l'étude de la structure des corps compacts plongés dans des champs gravitationnels intenses.

L'effet Doppler ordinaire (y compris la partie « transversale »), le rougissement gravitationnel (observables grâce à la grande excentricité de l'orbite), ainsi que le changement de fréquence dû au retard pris par la lumière si elle passe près du compagnon sont décrits par la formule [4] :

$$\frac{P_{obs}}{P_{em}} = \frac{(u^\alpha k_\alpha)_{em}}{(u^\alpha k_\alpha)_{obs}},$$

où P_{obs} est la période observée sur Terre, P_{em} la période propre du pulsar, u^α la quadri-vitesse de l'observateur ou de l'émetteur et $k^\alpha = dx^\alpha/d\lambda$ le vecteur tangent à la géodésique de longueur nulle dotée du paramètre affine λ qui joint l'observateur à l'émetteur. Le vecteur k^α étant transporté parallèlement à lui-même on en déduit :

$$k_\alpha^{em} = k_\alpha^{obs} + \frac{1}{2}\int_{obs}^{em} g_{\beta\gamma,\alpha}\, k^\beta\, k^\gamma\, d\lambda.$$

Sur la Terre on peut choisir $k_0 = +1$ et $k_i = n_i$ où n_i est un 3-vecteur unité définissant la « direction radiale ». Alors on a

$$(u^\alpha k_\alpha)^{obs} = (1 + v^i n_i)$$

(effet Doppler dû à la vitesse $v^i = dx^i/dx^0$ de la Terre) et

$$(u^\alpha k_\alpha)^{em} = (1 + v^i n_i)(-g_{00} - 2\, g_{0i}\, v^i - g_{ik}\, v^i v^k)^{-1/2} + \frac{1}{2} u^\alpha \int_{obs}^{em} g_{\beta\gamma,\alpha}\, k^\beta\, k^\gamma\, d\lambda.$$

où le deuxième membre est évalué au point d'émission. Le terme $(1 + v^i n_i)$ donne l'effet Doppler ordinaire qui est ici modifié entre autres par l'effet Einstein $(-g_{00})$ et par l'effet transverse $(-g_{ik}\, v^i\, v^k)$. Le dernier terme enfin, usuellement plus faible que les autres peut devenir important si le signal lumineux passe près du compagnon de masse M_2 et de vitesse v_2 (il faut que l'inclinaison $i \simeq 90°$) on peut alors l'estimer à $u^0_{em}\,(4\,M_2/b_0)\,(v^b_2 - v^b_{em})$ où b_0 est le paramètre d'impact mesuré sur un axe b abaissé du compagnon sur le rayon.

Le périastre précesse avec la vitesse angulaire

$$\Omega_p \simeq 3\,\Omega_L (M_1 + M_2)/a\,(1 - e^2),$$

où $\Omega_L \simeq (M_1 + M_2)^{1/2}\, a^{-3/2}$ est la vitesse angulaire moyenne sur l'orbite, M_1 est la masse du pulsar, M_2 celle du compagnon et a le demi-grand axe de l'orbite relative. Le tableau montre que cet effet est appréciable. Appréciable aussi (tableau) sera la précession du moment cinétique propre du pulsar induite par le couplage spin-orbite :

$$\Omega_{SL} \simeq 3\,\Omega_L\, M_2/\{2\,a\,(1 - e^2)\}.$$

C'est précisément cet effet (ainsi que celui dont on parle plus bas) que cherche à mesurer une expérience terrestre dont la conception est due à Schiff. Cet effet est du plus grand intérêt non seulement pour la vérification de la Relativité générale mais encore pour la connaissance de la structure des étoiles à neutron. En effet pour la première fois il sera possible d'observer les processus d'émission d'un pulsar à des angles variables. Si comme cela a été théoriquement postulé, la radiation est émise dans des cônes issus des pôles magnétiques l'intensité observée sera modulée par la fréquence Ω_{SL}. Elle pourra même s'annuler périodiquement si le faisceau émis est assez étroit.

TABLEAU

Paramètres du pulsar binaire correspondant à une masse de l'étoile à neutron de 0,7 $M_\odot$ en fonction de l'inclinaison i de l'orbite

L'excentricité de l'orbite est $e = 0{,}615$ et la période du système binaire est 27 908 s, $M_2/M_\odot$ indique la masse du compagnon, v_1 la vitesse du pulsar déduite de l'effet Doppler, v_2 la vitesse du compagnon, a le semi-grand axe de l'orbite relative, Φ_p la contribution relativiste à la précession du périastre et Φ_{LS} la précession due au couplage, spin-orbite. $\dot{E}$ et τ sont respectivement le flux et la vie moyenne du système pour l'émission gravitationnelle.

i	$M_2/M_\odot$	v_1 (km/s)	v_2 (km/s)	a (10^3 km)	Φ_p (deg/an)	Φ_{LS} (deg/an)	$\dot{E}$ (erg/s)	$\tau = (-E/\dot{E})$ (10^9 an)
90........	0,605	323	373	1 501	2,51	0,58	$5{,}7.10^{30}$	2,0
50........	0,908	422	325	1 609	2,88	0,82	$1{,}1.10^{31}$	1,5
20........	4,37	944	151	2 361	6,19	2,67	$1{,}2.10^{32}$	0,45

Si de plus le compagnon est doué d'un moment angulaire J_2, il y a lieu de tenir compte de l'interaction des deux moments angulaires (interaction « spin-spin ») due à l'effet Lense-Thirring. La précession correspondante sera $\Omega_{SS} \sim J_2/a^3$. Si le compagnon est une naine blanche on trouve que $\Omega_{SS} \sim \Omega_{SL}$ en revanche s'il est une étoile à neutron ou un black-hole Ω_{SS} sera bien plus petit que Ω_{SL} [5].

C. R. Acad. Sc. Paris, t. 279 (23 décembre 1974) Série A — 973

Signalons aussi qu'un tel système ouvre la possibilité d'étudier l'action d'un corps collapsé en rotation sur la polarisation des radiations qui passent dans son voisinage. En particulier Mashoon ([6]) a étudié la déflection différentielle des radiations à polarisation circulaire droite ou gauche. Pour un black-hole de Kerr extrême cet effet peut atteindre environ une minute d'arc pour une fréquence de 4.10^8 Hz. Mais cet effet ainsi que celui dû au retard de la lumière peut être difficile à détecter en raison du court temps d'éclipse des objets en question : environ 15 s pour une naine blanche et une trentaine de millisecondes pour une étoile à neutron ou un black-hole.

Enfin il est intéressant de calculer l'énergie gravitationnelle émise par un tel système (tableau). Le flux sur Terre correspondant s'élèverait à environ 10^{-15} erg/cm^2.s bien trop faible pour être mesuré. Il est donc clair que l'effet d'une telle radiation sur le système est négligeable.

Concluons en disant que, même si le système découvert par J.H. Taylor ne permet pas d'observer tous les effets décrits ici, il est probable que d'autres systèmes semblables le permettront, vue la très petite région de la Galaxie observée par ce dernier.

(*) Séance du 25 novembre 1974.

([1]) J. H. Taylor et R. A. Hulse, *Discovery of a Pulsar in a Binary System* (soumis à *Ap. J. Letters*).

([2]) *Cf.* par exemple, H. Gursky et R. Ruffini, *Neutron Stars, Black Holes and Binary X-Ray Sources*, éditeur H. D. Reidel, Dordrecht, 1975.

([3]) B. Flannery et E. P. J. van den Heuvel (soumis pour publication).

([4]) On utilise des unités géométriques : $G = c = 1$, la signature $+2$, les indices grecs vont de 0 à 3 et les indices latins de 1 à 3.

([5]) Pour détails et bibliographie *voir* T. Damour et R. Ruffini (soumis pour publication à *Phys. Rev. letters*).

([6]) B. Mashoon, *Nature*, 250, 1974, p. 316.

Department of Physics,
Princeton University,
Princeton N. J., U. S. A.;
and
Institute for Advanced Study,
Princeton N. J., U. S. A.

PHYSICAL REVIEW LETTERS

VOLUME 36 — 22 MARCH 1976 — NUMBER 12

Possible Experiment with Two Counter-Orbiting Drag-Free Satellites to Obtain a New Test of Einstein's General Theory of Relativity and Improved Measurements in Geodesy*

R. A. Van Patten and C. W. F. Everitt

Department of Aeronautics and Astronautics and W. W. Hansen Laboratories of Physics, Stanford University, Stanford, California 94305

(Received 29 October 1975)

In 1918, Lense and Thirring calculated that a moon orbiting a rotating planet would experience a nodal dragging effect due to general relativity. We describe an experiment to measure this effect to 1% with two counter-orbiting drag-free satellites in polar earth orbit. In addition to tracking data from existing ground stations, satellite-to-satellite Doppler ranging data are taken near the poles. New geophysical information is inherent in the polar data.

A satellite orbiting a massive body such as Earth may be regarded as a gyroscope. Some time ago, it occurred to one of us that a new test of general relativity might be attempted by examining data from a polar-orbiting drag-free satellite for evidence of a relativistic frame-dragging effect analogous to the motional precession of an ordinary gyroscope predicted by Schiff. The initial plan was to examine data already available after the successful launch in September 1972 of the U. S. Navy's Triad Satellite, with the DISCOS[1] control system developed by the Stanford University Guidance and Control Laboratory, which reduces external nongravitational accelerations on the satellite orbit to below $10^{-11}g$ This opens the possibility of searching for a small relativistic effect on the orbit plane.

Lense and Thirring[2] calculated in 1918 that a rotating central body drags the inertial frame such that the right ascension of the ascending node of an orbiting satellite advances at a rate given by

$$\dot{\vec{\Omega}}_{LT} = \frac{2GI\omega}{c^2a^3(1-e^2)^{3/2}}\,\vec{n}^{(2)}, \qquad (1)$$

where G is the gravitational constant, $I\omega$ is the magnitude of the central body's angular momentum, $\vec{n}^{(2)}$ is the unit vector parallel to the central body's spin axis, and a and e are the semimajor axis and eccentricity of the satellite's orbit. Lense and Thirring gave numbers for bodies orbiting the major planets. For a satellite 800 km above Earth, $\dot{\vec{\Omega}}_{LT}$ is 0.18 arcsec/yr, which amounts to a lateral displacement at the equator after 2.5 yr of 13.9 m (referred to Earth's surface).

Equation (1) should be compared with Schiff's result[3] for the motional precession Ω_M of a gyroscope in general relativity, which, when integrated around an orbit and multiplied into the gyro spin vector, yields a drift rate

$$\dot{\vec{n}}_M^{(1)} = \frac{GI\omega}{2c^2a^3(1-e^2)^{3/2}} \times[\vec{n}^{(2)}\times\vec{n}^{(1)} + 3(\vec{n}^{(2)}\cdot\vec{n})(\vec{n}^{(1)}\times\vec{n})], \qquad (2)$$

where $\vec{n}$, $\vec{n}^{(1)}$, and $\vec{n}^{(2)}$ are unit vectors defining the orbit axis, the gyro axis, and Earth's axis, respectively.[4] The two effects have the same functional dependence on I, ω, a, and e but, whereas the gyro precession depends on the orbit inclination i, reversing sign when $\cos 2i = -\frac{1}{3}$ or $i = 54°16'$, the nodal advance is independent of i. In identical polar orbits, $\vec{\Omega}_{LT} = 4\vec{n}_N^{(1)}$. Wilkins has pointed out that both effects originate in the off-diagonal terms g_{14}, g_{24} in the metric for the rotating Earth, but the center of mass of an orbiting body follows a geodesic and has timelike four-velocity, while the spin vector of a gyroscope obeys Fermi-Walker transport and is spacelike.[5] Thus an experiment to check Eq. (1), if feasible, would provide a separate and distinct test of gravitational frame dragging.

In addition to the Lense-Thirring drag, there is a small geodesic precession of the orbit plane due to the motion about the sun. This effect of the sun on a gyroscope was first pointed out by Fokker in 1920.[6] The precession in the plane of the ecliptic amounts to 0.021 arcsec/yr. This adds a component of 0.020 arcsec/yr to the nodal motion of the satellite and also causes a change of 0.0084 arcsec/yr in the inclination. The total relativistic displacement of the ascending node $(\Omega_{LT}+\Omega_G)$ is thus 13.9 + 1.51 m in 2.5 yr.

In-flight tracking data has confirmed that the DISCOS reduces along-track accelerations on the TRIAD satellite to below $10^{-11}g$. The corresponding cross-track deviations of the TRIAD satellite from an ideal geodesic path are no more than 4 cm/yr.[7]

For a satellite in orbit about Earth, the nonrelativistic regression $\vec{\Omega}_Q$ of the node due principally to Earth's quadrupole mass-moment is

$$\vec{\Omega}_Q = -\tfrac{3}{2}\bar{\omega}_0[R_e/a(1-e^2)]^2 J_2 \cos i\, \vec{n}^{(2)}, \qquad (3)$$

where R_e is Earth's equatorial radius, and $\bar{\omega}_0$ is the satellite's mean motion. The quantities J_2 and R_e are well known, and $\bar{\omega}_0$, a, and e for any particular satellite can be determined with sufficient accuracy; however, tracking uncertainties in the time history of the inclination angle i cause errors in determination of the time history of $\vec{\Omega}_Q$ with a single satellite about six times larger than the relativistic nodal motion $\dot{\Omega}_{LT}+\dot{\Omega}_G$.

We have conceived an experiment which appears to have the remarkable property of measuring $\Omega_{LT}+\Omega_G$ and, at the same time, yielding information about perturbations of orbital plane and altitude about two orders of magnitude more accurate than any previously obtained. The experiment depends on two counter-orbiting drag-free satellites in polar orbits, initially adjusted by in-flight corrections to be very nearly equal and opposite but planned in such a way that collisions are avoided. The orbit characteristics are such that encounters occur close to the north and south poles. At each polar passing, the distance between the two spacecraft is measured by satellite-to-satellite Doppler ranging to about 1 cm accuracy, while at lower latitudes, standard ground-based Doppler tracking data is obtained for each satellite independently. The satellite-to-satellite data yield a unique measurement of the angle 2α which is the sum of the coinclinations $(i_1'+i_2')$ for the two orbits. Recalling that the orbit inclination is defined from the ascending node so that counter-orbiting satellites in the same plane have opposite coinclinations, we have for the nodal rates from each satellite

$$\dot{\Omega}_{Q_1} + \Delta\dot{\Omega}_Q = KJ_2(i_1' + \Delta i'), \qquad (4)$$

$$\dot{\Omega}_{Q_2} - \Delta\dot{\Omega}_Q = KJ_2(i_2' - \Delta i'), \qquad (5)$$

where $\Delta i'$ is the angle from the true pole to an estimated pole location, and $\Delta\dot{\Omega}_Q$ is the nodal rate associated with $\Delta i'$. Combining (4) and (5) with the definition of α and integrating over the life of the experiment, we obtain a calculated sum of the two nodal regressions independent of the uncertainty in orbit inclinations:

$$[\Omega_{Q_1}+\Omega_{Q_2}]_{\rm calc} = 2KJ_2\int_{t_1}^{t_2}\alpha\, dt. \qquad (6)$$

The regression sum calculated from (6) *contains* only the effect of Earth's quadrupole moment; on the other hand, the measured regressions $\Omega_{1\,\rm meas}$ and $\Omega_{2\,\rm meas}$ of the two satellites, individually tracked from ground stations at lower latitudes, each also contain the relativistic term $\Omega_{LT}+\Omega_G$:

$$2(\Omega_{LT}+\Omega_G) = \Omega_{1\,\rm meas} + \Omega_{2\,\rm meas} - [\Omega_{Q_1}+\Omega_{Q_2}]_{\rm calc}. \qquad (7)$$

Thus, by combining polar-ranging and ground-tracking data, the relativistic effect can be determined.

The orbit planes are subject to solar and lunar gravity-gradient perturbations. These yield, in general, a combination of secular and periodic disturbance terms which, for matched counter-orbiting satellites, are equal in magnitude but opposite in direction. Each term has an associated indirect component due to the tides. Two secular terms exist, arising from the tilts of the orbit normal with respect to the planes of the ecliptic and the lunar orbit. The lunar effect also has a long-term variation since the moon's orbit vector moves about the ecliptic vector in a 5° radius circle with a 19-yr period. By choosin a suitable node for the satellite orbit, the combined secular and 19-yr terms can be very nearly balanced out for both satellites at any launch date such that the net long-term change in the polar separation $2a\alpha$ between the two satellites is held below ± 100 m over the $2\frac{1}{2}$-yr period.

With the removal of long-term effects, we find that the variations in $2a\alpha$ have a twice-yearly term of ± 1100 m and a twice-monthly term of ± 180 m due to the direct torques from the sun and moon. The corresponding tidal effects are each about one-sixth of the first two terms.

All ten effects (secular and periodic) just described propagate into the nodal motions of the two satellites through the regression equations (3) and (6). In addition, the gravity-gradient terms have nodal components that yield corresponding perturbations of the angles Ω_1 and Ω_2 for the two satellites, of amplitude 30% to 50% of the perturbations of the coinclination angles i_1' and i_2'. However, none of the nodal perturbations cause any significant error in determining $\Omega_{LT}+\Omega_G$ since all have very nearly equal and opposite effects on the two orbits.

The perturbations of the satellite orbits over the poles by the higher order mass moments of Earth are at frequencies which are integral multiples of Earth's rotational frequency. We are indebted to J. V. Breakwell and D. Schaechter of Stanford University for analyzing the effects. Harmonics J_{lm} with even m affect the lateral separation but not the vertical separation of the two satellites at the poles. Harmonics J_{lm} with odd m affect the vertical separation but not the lateral separation. For a near-circular 800-km orbit, the largest lateral perturbation (J_{22}) occurs twice daily at about ± 540 m, while the largest vertical perturbation (J_{31}) occurs once daily at ± 174 m. Harmonics like J_{41} with even l and odd m have a relatively minor effect on the vertical separation and introduce a small difference between the daily histories of the vertical separations at the two poles. Harmonics like J_{32} with odd l and even m have a similar small effect on the lateral separations. Analysis done with several recent earth gravity models leads to differences in perturbations far exceeding the 1 cm measurement accuracy expected. Thus, the experiment offers the prospect of making very precise checks on existing and future gravity models.

An additional vertical polar perturbation having a period of about 15 weeks arises from the difference in eccentricity of the two orbits. The

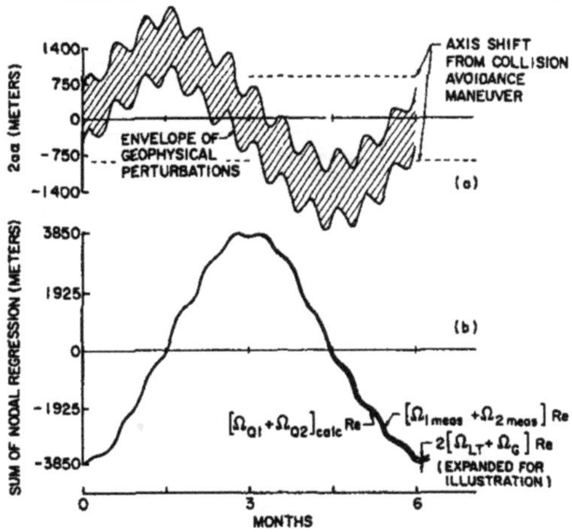

FIG. 1. (a) Form of lateral polar data. (b) Illustration of form of the sum of the nodal motions.

amplitude can be made small compared to the geophysical vertical perturbations by trimming the eccentricities to nearly zero. The solar, lunar, and geophysical lateral polar perturbations all sum in the lateral separation of the two satellites at polar encounter. This distance, $2a\alpha$, is accurately determined at polar encounters from the slant-range Doppler distance measurements. Under the assumption of certain orbital parameter constraints, the distance $2a\alpha$ will have, over a 6-month period, a form such as illustrated in Fig. 1(a). The daily lateral variations are contained in the envelope of data in Fig. 1(a). When the quantity α is integrated over the experimental period, the high-frequency geophysical terms all integrate to relatively small periodic components in $[\Omega_{Q_1}+\Omega_{Q_2}]_{calc}$ with the remaining solar and lunar terms having the form illustrated in the lower curve of Fig. 1(b). The quantity $\Omega_{1_{meas}}+\Omega_{2_{meas}}$ obtained from tracking data is illustrated in the upper curve of Fig. 1(b). The relativistic quantity $2(\Omega_{LT}+\Omega_G)$, which is a secular term, appears in Fig. 1(b) as the difference between the upper and lower curves. This simple illustration is representative of the more complex statistical estimation process which would be used in the actual data processing.

The error budget dictates the use of closely matched, very nearly polar orbits. Some of the orbit requirements are only feasible with drag-free satellites and call for in-flight corrections during the first phase of the mission. During the course of the experiment, occasional (i.e., every

VOLUME 36, NUMBER 12 PHYSICAL REVIEW LETTERS 22 MARCH 1976

TABLE I. Twin satellite experiment error budget.

	Error Source	Nodal Shift at Equator (meters)
1.	Measurement of mean lateral separation of orbits at poles	0.26
2.	Determination of nodal angle, Ω_1	0.11^a
	Determination of nodal angle, Ω_2	0.11^a
3.	Calculation of oblateness effects from polar data due to uncertainty in J_2.	0.09
4.	Nodal shift caused by orbital adjustments performed at the equator.	0.08
5.	Nodal shift cause by fluctuations in self gravity of the No. 1 satellite on its proof mass.	0.04
	Same as above for satellite No. 2.	0.04
	Composite error (root sum square)	0.33
	Relativistic effect = $2R_e(\Omega_{LT} + \Omega_G)$	30.8
	Probable experimental error	$1.1\%^a$ (1σ)

[a]Assuming UT 1 time-base correction from lunar-laser ranging data (Ref.9).

3 months) orbit adjustments are also needed to assure collision avoidance.[8] The resultant errors may be kept small by restricting the adjustments to impulses applied at the equatorial crossings.

We present in Table I the seven most important error sources in the measurement of $\Omega_{LT}+\Omega_G$, expressed in equivalent meters of $R_e\Delta\Omega$. When the seven estimated errors are statistically combined, the resulting estimated 1σ error $R_e\Delta(\Omega_{LT}+\Omega_G)$ becomes 0.33 m or about 1% of $2R_e(\Omega_{LT}+\Omega_G)$ for a $2\frac{1}{2}$-yr experiment. For a 1-yr experiment, the error is about 4%. A detailed discussion of each error source is given elsewhere.[8]

This experiment offers the possibility of making the first measurement of the Lense-Thirring nodal dragging effect and of obtaining important new information about orbital perturbations due not only to a rigid Earth but also to tidal-elastic deformation thereof. Measurement of the frame dragging or Lense-Thirring precession makes this experiment stand along with the Schiff gyro experiment[10] as the only two experiments thus far proposed that distinguish between the gravitational fields produced by matter at rest and by matter in rotation.

*This work was partially supported under National Aeronautics and Space Administration Grant No. NGR 05-020-019 through the Marshall Space Flight Center and National Aeronautics and Space Administration Contract No. 5-21960 through the Goddard Space Flight Center.

[1]Staff of the Space Department of the Johns Hopkins University Applied Physics Laboratory, Silver Spring, Maryland, and Staff of the Guidance and Control Laboratory, Stanford University, Stanford, California, Spacecr. Rockets 11, 637 (1974); J. Dassoulas, D. B. DeBra, and R. E. Jenkins, APL Tech. Dig. 12, No. 2, 2 (1973); K. Moe, D. B. DeBra, R. A. Van Patten, M. M. Moe, G. Oelker, and M. B. Ruggera, Jr., to be published.

[2]J. Lense and H. Thirring, Phys. Z. 19, 156 (1918).

[3]L. I. Schiff, Proc. Nat. Acad. Sci., U. S. A. 46, 871 (1960).

[4]B. M. Barker and R. F. O'Connell, Phys. Rev. D 6, 956 (1971).

[5]D. C. Wilkins, private communication.

[6]A. D. Fokker, Proc. K. Ned. Akad. Wet. 23, 729 (1920).

[7]A. W. Fleming and M. G. Tashker, Stanford University, Stanford, California, Report No. SUDAAR 445, 1972 (unpublished), p. 34.

[8]R. A. Van Patten and C. W. F. Everitt, to be published.

[9]J. G. Williams, private communication.

[10]C. W. F. Everitt, in *Experimental Gravitation*, edited by B. Bertotti (Academic, New York, 1973) p. 331; J. Lipa, W. M. Fairbank, and C. W. F. Everitt, *ibid.*, p. 361.

Contemporary Papers

VOLUME 56, NUMBER 4 PHYSICAL REVIEW LETTERS 27 JANUARY 1986

Measurement of the Lense-Thirring Drag on High-Altitude, Laser-Ranged Artificial Satellites

Ignazio Ciufolini

Center for Theoretical Physics, Center for Relativity, and Physics Department, University of Texas, Austin, Texas 78712

(Received 16 October 1984; revised manuscript received 19 April 1985)

We describe a new method of measuring the Lense-Thirring relativistic nodal drag using LAGEOS together with another similar high-altitude, laser-ranged satellite with appropriately chosen orbital parameters. We propose, for this purpose, that a future satellite such as LAGEOS II have an inclination supplementary to that of LAGEOS. The experiment proposed here would provide a method for experimental verification of the general relativistic formulation of Mach's principle and measurement of the gravitomagnetic field.

PACS numbers: 04.80.+z

In special and general relativity there are several precession phenomena associated with the angular momentum vector of a body. If a test particle is orbiting a rotating central body, the plane of the orbit of the particle is dragged by the intrinsic angular momentum J of the central body, in agreement with the general relativistic formulation of Mach's principle.[1]

In the weak-field and slow-motion limit the nodal lines are dragged in the sense of rotation, at a rate given by[2]

$$\dot{\Omega} = [2/a^3(1-e^2)^{3/2}]J, \tag{1}$$

where a is the semimajor axis of the orbit, e is the eccentricity of the orbit, and geometrized units are used, i.e., $G=c=1$. This phenomenon is the Lense-Thirring effect, from the names of its discoverers in 1918.[2]

In addition to this there are other precession phenomena associated with the intrinsic angular momentum or spin $\mathbf{S}$ of an orbiting particle. In the weak-field and slow-motion limit the vector $\mathbf{S}$ precesses at a rate given by[1] $d\mathbf{S}/d\tau = \dot{\mathbf{\Omega}} \times \mathbf{S}$ where

$$\dot{\mathbf{\Omega}} \equiv -\tfrac{1}{2}\mathbf{v}\times\mathbf{a} + \tfrac{3}{2}\mathbf{v}\times\nabla U + \frac{1}{r^3}\left[-\mathbf{J} + \frac{3(\mathbf{J}\cdot\mathbf{r})\mathbf{r}}{r^2}\right], \tag{2}$$

where $\mathbf{v}$ is the particle velocity, $\mathbf{a} \equiv d\mathbf{v}/d\tau - \nabla U$ is its nongravitational acceleration, $\mathbf{r}$ is its position vector, τ is its proper time, and U is the Newtonian potential.

The first term of this equation is the Thomas precession.[3] It is a special relativistic effect due to the noncommutativity of nonaligned Lorentz transformations. It may also be viewed as a coupling between the particle velocity $\mathbf{v}$ and the nongravitational forces acting on it.

The second (de Sitter[4]–Fokker[5]) term is general relativistic, arising even for a nonrotating source, from the parallel transport of a direction defined by $\mathbf{S}$; it may be viewed as spin precession due to the coupling between the particle velocity $\mathbf{v}$ and the static —$g_{\alpha\beta,0}=0$ and $g_{i0}=0$—part of the space-time geometry.

The third (Schiff[6]) term gives the general relativistic precession of the particle spin $\mathbf{S}$ caused by the intrinsic angular momentum $\mathbf{J}$ of the central body—$g_{i0}\neq 0$.

We also mention the precession of the periapsis of an orbiting test particle due to the angular momentum of the central body. This tiny shift of the perihelion of Mercury due to the rotation of the Sun was calculated by de Sitter in 1916.[7]

All these effects are quite small for an artificial satellite orbiting the Earth.

We propose here to measure the Lense-Thirring dragging by measuring the nodal precession of laser-ranged Earth satellites. We shall show that two satellites would be required; we propose that LAGEOS[8–10] together with a second satellite LAGEOS X with opposite inclination (i.e., with $I^X = 180° - I$, where $I \simeq 109.94°$ is the orbital inclination of LAGEOS) would provide the needed accuracy.

The major part of the nodal precession of an Earth satellite is a classical effect due to deviations from spherical symmetry of the Earth's gravity field —quadrupole and higher mass moments.[11] These deviations from sphericity are measured by the expansion of the potential $U(r)$ in spherical harmonics. From this expansion of $U(r)$ follows[11] the formula for the classical precession of the nodal lines of an Earth satellite:

$$\dot{\Omega}_{\text{class}} \simeq -\frac{3}{2}n\left(\frac{R_\oplus}{a}\right)^2\frac{\cos I}{(1-e^2)^2}\left\{J_2 + J_4\left[\frac{5}{8}\left(\frac{R_\oplus}{a}\right)^2(7\sin^2 I - 4)\frac{1+\frac{3}{2}e^2}{(1-e^2)^2}\right] + \dots\right\}, \tag{3}$$

where $R_\oplus$ is the mean equatorial radius, the J_{2n} are the even zonal harmonic coefficients of the gravitational field, $n = 2\pi/P$ is the orbital mean motion, and P is the period.

Clearly the orbital parameters n, a, e, and I must be very accurately known to identify the classical contribution to the precession. We investigate these points below, but the difficulty of measurement of the Lense-Thirring dragging is apparent from the following facts: LAGEOS has a classical precession of $\sim 126°$ per year, while its Lense-Thirring precession is only 31 arc msec/yr. Our principal difficulty arises from poor knowledge of certain moments of the potential of the Earth.

A way of overcoming this problem would be to measure accurately $J_2, J_4, J_6, \ldots$, by orbiting several high-altitude, laser-ranged satellites, plus LAGEOS to measure the Lense-Thirring effect. A better solution would be to orbit polar satellites. In 1976, Van Patten and Everitt[12] proposed the measurement of the Lense-Thirring effect with two counter-rotating drag-free satellites in polar orbit, i.e., counter-rotating in the same plane perpendicular to the equator. Because the classical precession is proportional to $\cos(I)$, which vanishes for polar orbits with $I = \pi/2$, the contribution to the nodal precession due to the quadrupole and higher moments of the Earth is zero for polar satellites. However, in order to measure the inclination with the technology then available to sufficient accuracy to extract the Lense-Thirring precession, it would have been necessary to use two counter-rotating drag-free polar satellites, measuring their distance with satellite-to-satellite Doppler ranging while the satellites were near the poles.

Today, however, the tracking accuracy of satellites has improved dramatically. The knowledge of the orbits of some artificial satellites using ground-based pulsed laser ranging systems will soon be sufficient to measure general relativistic perturbations.[13,14] Consequently, a general method[15,16] to measure the Lense-Thirring effect would be to have two laser-ranged satellites with any arbitrary pair of opposite (i.e., I and $180° - I$) inclinations to allow the separation of the classical and the relativistic precessions.

In particular LAGEOS, launched in 1976, has the best known orbit of any satellite because of its high altitude—semimajor axis = 12 270 km—and its design to minimize air drag and radiation[8-10] pressure. For this satellite the Lense-Thirring effect, Eq. (1), due to the Earth's intrinsic angular momentum $\mathbf{J}_\oplus \simeq 5.9 \times 10^{40}$ gm cm^2/sec, is $\dot{\Omega}_{LT} \simeq 31$ arc msec/yr, eastward. For an artificial satellite the measurable quantity is $\dot{\Omega} - \omega_e$, where ω_e is the Earth's time-dependent rotation rate. The estimates of UT1—in effect the Earth's rotation rate—with the use of lunar laser ranging (LLR) and very long-baseline interferometry (VLBI) have an accuracy of fractions of a millisecond[17] of time corresponding to an accuracy of a few arc milliseconds of rotation. Consequently, the nodal precession is measurable[18] with an accuracy of 1 or 2 arc msec/yr. Therefore, the relativistic precession is quite detectable, if other contributions can be accurately accounted for.

In addition to the Lense-Thirring effect it is also necessary to consider the Sun's de Sitter–Fokker effect, i.e., the spin precession of the system "Earth plus an orbiting satellite" considered as a gyroscope, due to the static field of the Sun. The average geodesic precession projected into the Earth's equatorial plane is, for LAGEOS,[19] $\dot{\Omega}_{DF} \simeq 17$ arc msec/yr, eastward. The relativistic node shift is dominated by the Lense-Thirring effect, and is in an eastward direction.

The orbital parameters n, a, and e appearing in Eq. (3) are measured with sufficient accuracy for our purpose via the LAGEOS laser ranging. However, the uncertainty in the inclination and thus in the classical part of the nodal precession may well be the dominant source of uncertainty in determining the relativistic part of the precession. Currently, a four-year span of LAGEOS orbital inclination residuals (from a dynamically consistent reference orbit) when fitted with a simple model of periods 1050, 560, and 280 days leads to residuals about this curve on the order of 2 arc msec.[18] These residuals are determined from independent 15-d intervals of laser tracking data. The Lense-Thirring effect is a secular effect, while most classical perturbations are of a periodic nature. Hence, it is the average (over a sufficiently long time) of the inclination error that concerns us. Under visual inspection the residuals appear to be Gaussian distributed. In one year, there are 24 such data points, so that after one year one expects $\sim 2/\sqrt{24} \sim 0.4$ arc msec of random error in the estimate of I for the span of that year, implying from Eq. (3) a contribution to the nodal precession error of ~ 3 arc msec/yr. This random error presumably can be reduced somewhat by an even longer period of data, though a careful study of systematic errors and of the statistics will be required to estimate the ultimate accuracy. Secondly, there are some correlations with our proposed second satellite which will lead to an improvement in the model, and a reduction in the residuals. For example, the effects of uncertainties in tidal parameters can be reduced by adjusting their values by use of data from the two satellites; changes in the inclination due to indeterminacy in the values of the J_{2n} are equal and opposite for the two satellites,[11] thereby, from Eq. (3), corresponding to equal and opposite nodal precessions. Any change in the inclination due to polar motion is already solved for in the experimental value of the inclination $I(t)$, and errors introduced should be relatively short period. A recent paper[20] shows 2-arc-msec residuals between VLBI and LAGEOS polar motion data, based on 5-d data intervals. This supports our expectation of average pole accuracy over longer periods at the 0.5-arc-msec level or better in the near future, and thus relatively small contributions to the uncertainty in the inclination.

We propose an experiment that is easier and less costly to implement than polar satellites would be. We propose that the forthcoming[21] satellite LAGEOS II (or some other future satellite LAGEOS X) be chosen to have the following orbital parameters:

$$a_{\text{LAGEOS } X} = a_{\text{LAGEOS}} \pm a_\epsilon,$$
$$I_{\text{LAGEOS } X} = \pi - I_{\text{LAGEOS}} \pm I_\epsilon, \quad (4)$$
$$e_{\text{LAGEOS } X} = e_{\text{LAGEOS}} \pm e_\epsilon,$$

where a_ϵ, I_ϵ, and e_ϵ are small. In other words, the second satellite must have almost the same semimajor axis, period, and eccentricity as LAGEOS but an almost supplementary inclination. Because the classical precession (3) is linearly dependent on $\cos(I)$, with this particular choice of orbital parameters the classical nodal precessions of LAGEOS and LAGEOS X will be nearly equal and opposite. On the other hand the Lense-Thirring and the geodesic precession—which are independent of I—will be the same for both satellites so that, apart from other perturbations, the sum of the two measured total precessions will be twice the relativistic precession of LAGEOS. The philosophy of this two-satellite proposal is that one uses external data to fix the even zonal harmonics of the Earth's potential. One must conservatively take an external estimate of the error in these terms. Unfortunately, the present uncertainty δJ_{2n} in the numerical value of the even zonal harmonic coefficients J_{2n} is of the order[18,22] $\delta J_{2n} \simeq 10^{-6} J_2$. Since the Lense-Thirring precession $\dot{\Omega}_{\text{LT}}$ is $\dot{\Omega}_{\text{LT}} \simeq 6.9\times 10^{-8}\dot{\Omega}_{\text{class}}$, the indeterminacy in the theoretical value of the classical nodal precession is more than 10 times the relativistic effect to be measured. The errors in these quantities limit the acceptable differences a_ϵ, I_ϵ, and e_ϵ between the two satellite orbits.

In order to have less than 3% individual contributions to the experimental uncertainty (based on expected improvements in the uncertainties $\delta J_{2n}/J_2$ to $\sim 3\times 10^{-7}$) the differences a_ϵ, I_ϵ, and e_ϵ between the orbital parameters of LAGEOS and LAGEOS X must be less than $a_\epsilon \simeq \pm 16$ km, $I_\epsilon \simeq \pm 0.13°$, and $e_\epsilon \simeq \pm 0.04$. Moreover, in order to calculate the total precession $\dot{\Omega}_{\text{total}}$ it is also necessary to consider nonlinear contributions from even zonal harmonics; contributions from nonzonal spherical harmonics; lunar, solar, and planetary gravitational perturbations; atmospheric drag, radiation pressure; and earth tides. All these disturbing forces cause periodic and secular orbital perturbations that are very small compared to perturbations due to the Earth's quadrupole moment. The nonlinear contributions from the J_{2n} coefficients are equal and opposite for LAGEOS and LAGEOS X[18] and anyway, since the precession associated with $(J_2)^2$ is of the order $\dot{\Omega}_{(J_2)^2} \simeq 10^{-4}\dot{\Omega}_{\text{class}}$, these contributions could be calculated with sufficient accuracy to allow the determination of the Lense-Thirring effect. Similarly the secular nodal precessions associated with the nonzonal spherical harmonics and lunar, solar, and planetary perturbations can be calculated with sufficient accuracy to allow their extraction.[18] Concerning the nodal precession associated with radiation pressure and earth tides, information on these perturbations is extracted from the time variation of $\dot{\Omega}$ when measured over a long enough period of time.[18,23] Because LAGEOS is a spherical satellite and has a very small orbital eccentricity, the error in the calculated secular rate of nodal precession due to direct solar radiation pressure is small[19] compared to $\dot{\Omega}_{\text{LT}}$. Regarding the direct solar radiation pressure, the main effect for a nearly symmetric satellite is due to eclipses of part of the orbit. The eclipse seasons for LAGEOS last roughly 90 days and are separated by periods about twice that long. When the Sun is at equal distances from the satellite orbital plane, but in opposite regions, that is, before and after the Sun crosses the orbital plane, the out-of-plane components of the solar radiation pressure force are nearly equal and opposite; the contributions of the satellite eclipse to the nodal precession are nearly equal and opposite. For this reason the error, over one year, in the calculated secular rate of nodal precession due to the satellite eclipses is small[19] compared to $\dot{\Omega}_{\text{LT}}$. Systematic effects due to the Earth's albedo and to small differences in the satellite albedo for changes in the direction of the solar radiation with respect to the satellite rotation axis have to be considered, particularly in terms of interaction with eclipse-season effects. However, the prospects for modeling the radiation pressure and the atmospheric drag effects on the node with small errors are good.[19] A detailed analysis of various perturbing forces will be the subject of a following paper.[19]

In summary we have

$$\dot{\Omega}^{\text{LAGEOS}}_{\text{exp}} = \dot{\Omega}^{\text{LAGEOS}}_{\text{class}} + \dot{\Omega}^{\text{LAGEOS}}_{\text{rel}} + \dot{\Omega}^{\text{LAGEOS}}_{\text{other}}, \quad (5)$$

where $\dot{\Omega}_{\text{exp}}$ is the experimentally measured value of the rate of nodal precession and $\dot{\Omega}_{\text{other}}$ is the small rate of advance, computable with sufficient accuracy, due to classical disturbing forces other than those caused by the even zonal spherical harmonics.

For the second satellite, we have

$$\dot{\Omega}^{\mathrm{LAGEOS}\,X}_{\mathrm{class}} = -\dot{\Omega}^{\mathrm{LAGEOS}}_{\mathrm{class}} + \Delta\dot{\Omega}_{\mathrm{class}}; \tag{6}$$

$$\Delta\dot{\Omega}_{\mathrm{class}} = \frac{3}{2}n\left(\frac{R_{\oplus}}{a}\right)^2 \frac{\cos I}{(1-e^2)^2}\left[\left(-\frac{7}{2}\frac{a_\epsilon}{a} - I_\epsilon \tan I + \frac{4ee_\epsilon}{1-e^2}\right)J_2 + \dots\right], \tag{6}$$

and the total nodal precession is

$$\dot{\Omega}^{\mathrm{LAGEOS}\,X}_{\mathrm{exp}} = \dot{\Omega}^{\mathrm{LAGEOS}\,X}_{\mathrm{class}} + \dot{\Omega}^{\mathrm{LAGEOS}\,X}_{\mathrm{rel}} + \dot{\Omega}^{\mathrm{LAGEOS}\,X}_{\mathrm{other}} = -\dot{\Omega}^{\mathrm{LAGEOS}}_{\mathrm{class}} + \Delta\dot{\Omega}_{\mathrm{class}} + \dot{\Omega}^{\mathrm{LAGEOS}}_{\mathrm{rel}} + \dot{\Omega}^{\mathrm{LAGEOS}\,X}_{\mathrm{other}}. \tag{7}$$

Finally, from (5) and (7), we have

$$2\dot{\Omega}^{\mathrm{LAGEOS}}_{\mathrm{rel}} = \dot{\Omega}^{\mathrm{LAGEOS}}_{\mathrm{exp}} + \dot{\Omega}^{\mathrm{LAGEOS}\,X}_{\mathrm{exp}} - \dot{\Omega}^{\mathrm{LAGEOS}}_{\mathrm{other}} - \dot{\Omega}^{\mathrm{LAGEOS}\,X}_{\mathrm{other}} - \Delta\dot{\Omega}_{\mathrm{class}}.$$

In conclusion we stress that because of the increased tracking accuracy of the orbits of artificial satellites using laser ranging systems, the Lense-Thirring and de Sitter–Fokker relativistic nodal precessions are now significant and should be taken into account in the theoretical determination of orbital perturbations of high-altitude satellites such as LAGEOS. These relativistic nodal precessions should also be taken into account in geodesy calculations for a more accurate determination of the irregular gravitational field of the Earth, i.e., quadrupole and higher mass moments.

The experiment described above will provide an alternative verification, together with the Everitt-Fairbank experiment,[24] of Mach's principle and will give at last a measurement of the Earth's gyrogravitational or magnetogravitational field. Finally, we observe that orbiting a satellite with an inclination opposite to that of LAGEOS will also give a better determination of UT1.[18,19] For these reasons I propose that the forthcoming high-altitude laser-ranged satellite LAGEOS II or a similar future satellite have an inclination I of 70.06°.

I thank Professor P. Bender, Professor B. Bertotti, Professor M. Demianski, Professor B. DeWitt, Professor L. Shepley, and Professor S. Weinberg for helpful discussions and particularly Professor R. Matzner, Professor B. E. Schutz, Professor B. D. Tapley, Professor J. A. Wheeler, and Professor R. J. Eanes. This work was supported in part by National Science Foundation Grants No. PHY82-05717 and No. PHY8404931.

[1]C. W. Misner, K. S. Thorne, and J. A. Wheeler, *Gravitation* (Freeman, San Francisco, 1973), p. 1066.

[2]J. Lense and H. Thirring, Phys. Z **19**, 156 (1918); English translation by B. Mashhoon *et al.*, Gen. Relativ. Gravit. **16**, 711 (1984).

[3]L. H. Thomas, Philos. Mag. 3, 1 (1927).

[4]W. de Sitter, Mon. Not. Roy. Astron. Soc. **77**, 155 (1916).

[5]A. D. Fokker, Proc. K. Ned. Akad. Wet. **23**, 729 (1920).

[6]L. I. Schiff, Proc. Natl. Acad. Sci. U.S.A. **46**, 871 (1960).

[7]W. de Sitter, Mon. Not. Roy. Astron. Soc. **76**, 699 (1916).

[8]D. E. Smith and P. J. Dunn, Geophys. Res. Lett. **7**, 437 (1980).

[9]C. F. Yoder *et al.*, Nature (London) **303**, 757 (1983).

[10]R. J. Eanes, B. E. Schutz, and B. D. Tapley, to be published.

[11]W. M. Kaula, *Theory of Satellite Geodesy* (Blaisdell, Waltham, 1966).

[12]R. A. VanPatten and C. W. F. Everitt, Phys. Rev. Lett. **36**, 629 (1976).

[13]For a discussion of the measurement of the perigee precession of LAGEOS, see N. Ashby and B. Bertotti, Phys. Rev. Lett. **52**, 485 (1984).

[14]A theoretical method to measure the Riemann curvature tensor with a generalized version of the geodesic deviation equation [I. Ciufolini, to be published] and using four laser-ranged test particles is described in I. Ciufolini and M. Demianski, to be published.

[15]I. Ciufolini, dissertation, University of Texas at Austin, 1984 (unpublished).

[16]I. Ciufolini, Bull. Am. Phys. Soc. **6**, 1169 (1985).

[17]W. E. Carter, D. S. Robertson, J. E. Pettey, B. D. Tapley, B. E. Schutz, R. J. Eanes, and Miao Lufeng, Science (Washington, D.C.) **224**, 957 (1984).

[18]B. D. Tapley, B. E. Schutz, and R. J. Eanes, private communication.

[19]I. Ciufolini, to be published.

[20]D. S. Robertson, W. E. Carter, B. D. Tapley, B. E. Schutz, and R. J. Eanes, Science (Washington, D.C.) **229**, 1259 (1985).

[21]National Aeronautics and Space Administration/Piano Spaziale Nazionale LAGEOS II Study Group, Final Report, 1983 (unpublished).

[22]F. J. Lerch *et al.*, J. Geophys. Res. A **84**, 3879 (1979).

[23]The experiment should be done over a period of a few years.

[24]The experiment to measure the Schiff precession of the spin axis of a gyroscope orbiting the Earth is described by C. W. F. Everitt, in *Experimental Gravitation*, edited by B. Bertotti (Academic, New York, 1973), p. 331.

VOLUME 58, NUMBER 11 PHYSICAL REVIEW LETTERS 16 MARCH 1987

New Test of General Relativity: Measurement of de Sitter Geodetic Precession Rate for Lunar Perigee

Bruno Bertotti[(a)]

Institute of Astronomy, University of Cambridge, Cambridge, United Kingdom

Ignazio Ciufolini

Center for Space Research, University of Texas at Austin, Austin, Texas 78712

and

Peter L. Bender[(b)]

Joint Institute for Laboratory Astrophysics, University of Colorado and National Bureau of Standards, Boulder, Colorado 80309

(Received 12 January 1987)

According to general relativity, the calculated rate of motion of lunar perigee should include a contribution of 19.2 msec/yr from geodetic precession. We show that existing analyses of lunar-laser-ranging data confirm the general-relativistic rate for geodetic precession with respect to the planetary dynamical frame. In addition, the comparison of Earth-rotation results from lunar laser ranging and from very long-baseline interferometry (VLBI) shows that the relative drift of the planetary dynamical frame and the extragalactic VLBI reference frame is small. The estimated accuracy is about 10%.

PACS numbers: 04.80.+z, 95.10.Jk, 96.20.−n

General relativity predicts that a gyroscope attached to a test body orbiting around the sun precesses with respect to the distant standard of rest—distant galaxies—with the (prograde) angular velocity

$$\mathbf{\Omega}_G = -\tfrac{3}{2}\mathbf{v}\times\mathbf{r}M_\odot/r^3, \tag{1}$$

where $\mathbf{v}$ is the velocity of the body, $\mathbf{r}$ the radius vector from the sun, and $M_\odot$ the gravitational radius of the sun.[1,2] This is called the geodetic (or de Sitter) precession. On the basis of the theory of Fermi transport[3,4] the spin vector $\mathbf{S}$ of a gyroscope moves according to the formula

$$d\mathbf{S}/dt = \mathbf{\Omega}\times\mathbf{S}. \tag{2}$$

For a general post-Newtonian metric and for a body which moves with proper (i.e., nongravitational) acceleration $\mathbf{a}$, Eq. (1) generalizes to[5-7]

$$\mathbf{\Omega} = -\tfrac{1}{2}\mathbf{v}\times\mathbf{a} + (\gamma+\tfrac{1}{2})\mathbf{v}\times\nabla U; \tag{1'}$$

here U is the Newtonian potential. The first term is present also when $U=0$ and space-time is flat; it describes the Thomas precession, well known in special relativity. The Lense-Thirring effect[8-10] and geodetic precession are the main examples of the peculiar (and Machian!) influence of mass motions on the inertial frames, and have never been tested.

A space experiment to measure the relativistic precessions is in preparation in the United States.[11] It involves flying superconductive gyroscopes on a low Earth satellite. The rotation axes are measured with respect to a stellar reference system. The experiment is based on the principle of equivalence, according to which the dynamical laws governing a system of laboratory size are those of ordinary electromagnetism and special relativity; hence the gyroscope behavior determines parallel transported directions along a world line.

We shall use here instead a gravitationally bound system, the earth-moon system, small with respect to the rest of the planetary system (and we shall use the world "local" in this sense). Similarly to the case of laboratory experiments, the motion due only to interactions within the system determines a privileged local frame; for example, the frame of ideal Keplerian motion. In this frame one can expect that the gravitational effects of the sun and of planets other than the Earth are essentially reduced to their tidal forces and show up only through the curvature tensor. One can say that this is a generalized principle of equivalence. Because of the nonlinear character of gravitation, however, this is by no means a trivial statement.

The work by Ashby and Bertotti[12,13] is essentially a proof, in a suitable approximation, of this principle. They extend Fermi's construction of a local inertial frame to the neighborhood of the (gravitating) Earth. A local metric is obtained in which the Earth—whose center is at rest—appears solely through its own Schwarzschild solution to the appropriate order and its corrections due to higher harmonics and its rotational gravitomagnetic field. The sun shows up in the tidal terms, with very small relativistic corrections, and in a minute nonlinear interaction with the Earth. Both these corrections are negligible to the level required in the present paper. They also compute explicitly the coordi-

 © 1987 The American Physical Society

nate transformation from the usual, post-Newtonian (PN) frame, in which the center of gravity of the solar system is at rest, to this "Ptolemaic" frame. In an expansion with respect to the distance r from the earth, it turns out, as expected, that the linear part of this coordinate transformation corresponds to geodetic precession; in other words, the spatial base vectors $\mathbf{\Lambda}_{(j)}$ of the new frame precess with the angular velocity $\mathbf{\Omega}_G$ with respect to the old "Copernican," post-Newtonian frame; there is no Thomas precession. In the new frame the dynamics of the Earth and its satellites are governed by the usual, classical laws, corrected with the locally generated relativistic effects, if needed. These are the advance of the perigee and the Lense-Thirring precession due to the Earth's angular momentum. The Lense-Thirring effect for the moon is much smaller than the geodetic precession.

If we introduce the angular momentum per unit mass, $\mathbf{L}=\mathbf{r}\times\mathbf{v}$, of the Earth in its solar orbit, Eq. (1) shows that $\mathbf{\Omega}_G$ is directed along the pole of the ecliptic, in the direction of $\mathbf{L}$. $\mathbf{\Omega}_G$ has a constant part

$$\Omega_0=\tfrac{3}{2}\,\omega_\odot M_\odot/a_\oplus=19.2\text{ msec/yr}, \qquad (3)$$

and a correction with a 1-yr period due to the eccentricity $e_\oplus$ of the Earth's orbit,

$$\Omega_1\cos\omega_\odot t=\tfrac{9}{2}\,(\omega_\odot M_\odot e_\oplus/a_\oplus)\cos\omega_\odot t$$
$$=0.96\cos\omega_\odot t\text{ msec/yr}. \qquad (4)$$

$\omega_\odot$ is the sidereal frequency and t is reckoned from a perihelion passage; $a_\oplus$ is the semimajor axis of the Earth.

If geodetic precession were not present, the sidereal mean motion of the moon and the lunar perigee and node rates would all be changed by the same amount. de Sitter, in his classical paper of 1916,[1] predicted this effect for the node and the perigee and found that at that time the observations were not accurate enough. In principle we could determine the geodetic precession rate by comparing in the PN frame the observed and calculated rates for any of these motions. We choose to consider the 8.85-yr-period sidereal rate of perigee motion, for which the nonrelativistic part of the calculated rate is determined with high accuracy from the observed mean motions of the moon and Earth and other known quantities. The observed rate of perigee motion $\dot{\tilde\omega}$ minus the mean motion rate $\dot L'$ for the sun can be determined accurately from variations in the Earth-moon distance.

Currently, numerically integrated ephemerides for the moon are calculated with the use of programs developed at the Jet Propulsion Laboratory and Massachusetts Institute of Technology which include all known general relativistic effects of significant magnitude.[14,15] A recent lunar-ephemeris fit to lunar-laser-range observations from August 1969 to July 1986 is described by Newhall *et al.*[16] The lunar motion is tied in to the PN planetary dynamical frame[17] via other kinds of solar-system observations, including particularly microwave tracking data for the Viking landers on Mars. The mean motion of the Earth with respect to the planetary dynamical frame (PN) is essentially determined by its motion with respect to the perihelion of Mars plus the calculated perihelion precession rate for Mars. Since the general relativistic contribution to the perihelion precession rate for Mercury has been verified to 0.5% accuracy[15] and the corresponding rate for Mars is much less, the uncertainty in the rate is small. The overall uncertainty in the Earth's mean motion, or equivalently in the mean motion of the sun $\dot L'$, is 0.25 msec/yr.[18]

The accuracy of the observed value of $\dot{\tilde\omega}-\dot L'$ can be determined from how well the lunar-range data fit the calculated emphemeris. We consider the quantity

$$\dot l-\dot D=\dot L'-\dot{\tilde\omega}, \qquad (5)$$

where L is the lunar mean longitude, $l\equiv L-\tilde\omega$, and $D\equiv L-L'$. From the leading terms in the known expressions[19,20] for the partial derivatives of the lunar range with respect to l and D, we can find the change in range $\Delta\rho$ which would correspond to perturbations $\Delta(\dot l-\dot D)$ and $\Delta(\dot l+\dot D)$:

$$\Delta\rho\simeq at\{[0.0272\sin l+0.0144\sin(l-2D)-0.0077\sin 2D]\Delta(\dot l-\dot D)$$
$$+[0.0272\sin l-0.0048\sin(l-2D)+0.0077\sin 2D]\Delta(\dot l+\dot D)\}. \qquad (6)$$

Here a is the lunar semimajor axis, t is the time, and the periods of the three periodic terms are 27.55, 31.81, and 14.77 d. Thus the signature of an error in $\dot l-\dot D$ is an oscillating error in range which has an envelope increasing linearly in time. We have not been able to find any other parameter in the lunar problem which would give error terms that mimic the $t\sin(l-2D)$ and $t\sin(2D)$ terms from Eq. (6).

The present Jet Propulsion Laboratory lunar ephemeris fits the range normal points for the McDonald Observatory from the end of 1975 to early 1982 with an rms residual of about 18 cm.[21] Similarly, the rms residuals for the McDonald and Haleakala Observatories since December 1985 are about 6 cm.[21] We have calculated the maximum value of $\Delta(\dot l-\dot D)$ which would give no more than twice the observed rms range residuals for the first halves of 1976 and 1986, and find

$$\delta|\dot l-\dot D|\lesssim 1.5\text{ msec/yr}. \qquad (7)$$

This appears to be a reasonable estimate for the present uncertainty in the difference between the observed and calculated values for $\dot{\tilde\omega}-\dot L'$.

The uncertainty in the calculated value of $\dot{\tilde{\omega}}$ based on data through May 1982 was given by Dickey *et al.*[22] as 0.45 msec/yr. The single largest source of error was the uncertainty in the inclination of the lunar orbit to the ecliptic. Combining this with the uncertainties in the observed values of $\dot{\tilde{\omega}}-\dot{L}'$ and $\dot{L}'$ gives

$$\delta[\dot{\tilde{\omega}}_{\rm obs}-\dot{\tilde{\omega}}_{\rm calc}] \lesssim 1.6 \text{ msec/yr}. \qquad (8)$$

Since the general-relativistic rate of 19.2 msec/yr for geodetic precession was included in the calculated value, this indicates that an accuracy of $\pm 10\%$ for the general-relativistic rate appears to be achievable with presently available data. However, detailed computer studies of the data are needed in order to determine the actual accuracy, as discussed later.

The agreement between the calculated post-Newtonian motion of the lunar perigee and the observed motion is an indirect verification of the generalized equivalence principle[12,13]. Without such agreement it would not be possible to transform away locally the relativistic effects of the sun with a geodetic rotation. It is also important, in the light of Mach's principle, to connect the local dynamical frame with the extragalactic very long-baseline-interferometry (VLBI) frame. According to the standard theory the local, parallel-propagated Fermi axes—with respect to which the local motion is inertial—rotate with respect to the VLBI frame with the geodetic precession $\mathbf{\Omega}_G$. Hence *the rotational angular velocity $\mathbf{\Omega}^{(E)}$ of the Earth relative to the VLBI frame is the sum of its angular velocity with respect to the local Fermi axes and the geodetic precession velocity $\mathbf{\Omega}_G$*:

$$\mathbf{\Omega}^{(E)}_{\rm VLBI}=\mathbf{\Omega}^{(E)}_{\rm local}+\mathbf{\Omega}_G. \qquad (9)$$

If we assume, as usual, that the distant standard of rest of the PN planetary frame—connected to the Fermi frame with the "geodetic" rotation—agrees with the VLBI frame, Eq. (9) is equivalent to the agreement between the kinematical (VLBI) and planetary (PN) determinations of the angular velocity of the Earth.

This can be approximately checked by comparing values for Earth rotation time UT1 as determined by VLBI with values with respect to the PN planetary dynamical frame determined from lunar laser ranging. The Polaris-IRIS VLBI results[23] are available from late 1980 to the present, and results from the NASA Deep Space Network also are available. The UT1 results from lunar laser ranging are available up through May 1982[22] and from early 1985 until July 1986.[16] Although a thorough comparison including the latest data has not been done recently, the UT1 rates from the two techniques appear to agree to better than 1.5 msec/yr.[24] If we estimate the uncertainty in the lunar-laser-ranging results since late 1980 as being roughly 1 msec/yr and the VLBI uncertainty as considerably less, the uncertainty in the relative rotation rate between the planetary dynamical frame and the extragalactic VLBI frame would be about 2 msec/yr. Thus an accuracy for the contribution from geodetic precession to the mean motion of perigee with respect to the VLBI frame of about 10% of the geodetic precession rate appears to be achievable.

It also would be possible in principle to determine the geodetic precession rate by use of other observable quantities besides the mean motion of lunar perigee. The general precession of the Earth's rotation axis has been determined by observations of stellar proper motions,[25] by astrometric observations of the planets and the sun,[26] by lunar laser ranging,[16] and by VLBI.[27] However, comparison with a calculated value is not possible because of our poor knowledge of the mean moment of inertia of the Earth. Similarly, the contribution of geodetic precession to the mean motion of the moon cannot be determined accurately because we do not know the mass of the Earth-moon system and the lunar semimajor axis accurately enough to calculate the classical contribution to the mean motion. For the node of the lunar orbit, the contribution of geodetic precession to the mean motion is poorly determined because the motion of the node has relatively little effect on the lunar range.

Another possibility would be to use an artificial earth satellite such as LAGEOS[28,29] instead of the moon. The node of the LAGEOS orbit, for example, can be determined accurately enough, but, it precesses as a result of the Earth's oblateness and other perturbations. Its classical precession is determined by the even zonal harmonic coefficients of the Earth's gravity field, and the uncertainty[9] in these coefficients is far too large. For example, an uncertainty in J_2 of a part in a million produces an uncertainty of 450 msec/yr in the nodal precession. The moon, which is 30 times farther away than LAGEOS, is very little affected by the uncertainty in J_2 or in higher terms.

As explained in Refs. 9 and 10, the uncertainties in the Earth's gravitational field could be overcome by means of a new LAGEOS-type satellite, coupled with the present LAGEOS. They must have the same semimajor axis a and inclinations which add up to 180°. With this orbital configuration, the point lying half way between the two nodes provides a locally inertial direction. The sum of the geodetic and Lense-Thirring precessions for this locally inertial direction with respect to the extragalactic VLBI reference frame can be determined by combining satellite laser ranging and VLBI measurements of UT1.

In summary, it has been shown that the data are consistent with the general relativistic rate for the geodetic precession with respect to both the planetary dynamical frame and the extragalactic VLBI frame. The estimated accuracy is about 10%.

Since submitting this Letter, we have learned that detailed studies to determine the geodetic precession rate from lunar-laser-ranging data are being carried out at the Harvard-Smithsonian Center for Astrophysics.[30] The preliminary results indicate worse accuracy than our estimates in this Letter. The results are sensitive to the set of parameters which is solved for and to the editing of the data. We therefore caution readers that independent determinations of the accuracy based upon full computer studies are needed, in which the geodetic precession rate is specifically parametrized and all other relevant parameters are solved for simultaneously.

We wish to thank N. Ashby, R. J. Eanes, R. A. Matzner, J. Ries, B. E. Schutz, B. D. Tapley, and M. A. Vincent for helpful discussions. We particularly wish to thank J. G. Williams for extensive discussions concerning the accuracy of determining the lunar perigee motion, and T. M. Eubanks for information on the comparison of UT1 rates. Earlier discussions with R. W. King, J. O. Dickey, and X. X. Newhall concerning the analysis of lunar-laser-ranging data and with R. D. Reasenberg and R. W. Hellings concerning the Viking lander tracking results also are appreciated. And finally, we are indebted to R. D. Reasenberg and I. I. Shapiro for the information they provided about the unpublished Center for Astrophysics results concerning geodetic precession. One of us (I.C.) was supported by National Academy of Sciences under Grant No. NAS 5-28192. Another of us (P.L.B.) was partially supported by NASA under Order No. S-56150-D.

[a]Permanent address: Departimento di Fisica, Nucleare e Teorica, Università di Pavia, 27100 Pavia, Italy.

[b]Quantum Physics Division, National Bureau of Standards.

[1]W. De Sitter, Mon. Not. R. Astron. Soc. **77**, 155 (1916).

[2]A. D. Fokker, Proc. K. Ned. Akad. Wet. **23**, 729 (1920).

[3]E. Fermi, Atti Accad. Naz. Lincei, Cl. Sci. Fis. Mat. Nat., Rend. **31**, 21, 51 (1922).

[4]A. G. Walker, Proc. Roy. Soc. Edinburgh **52**, 345 (1932).

[5]C. W. Misner, K. S. Thorne, and J. Wheeler, *Gravitation* (Freeman, San Francisco, 1973), p. 1066.

[6]C. M. Will and K. Nordtvedt, Jr., Astrophys. J. **177**, 757 (1972).

[7]C. M. Will, *Theory and Experiment in Gravitational Physics* (Cambridge Univ. Press, Cambridge, 1981).

[8]J. Lense and H. Thirring, Phys. Z **19**, 156 (1918) [English translation by B. Mashhoon *et al.*, Gen. Rel. Gravit. **16**, 711 (1984)].

[9]I. Ciufolini, Phys. Rev. Lett. **56**, 278 (1986).

[10]I. Ciufolini, Celest. Mech. (to be published).

[11]J. P. Turneaure *et al.*, in *Proceedings of the Fourth Marcel Grossmann Meeting on General Relativity, Rome, Italy, 1985*, edited by R. Ruffini (Elsevier, Amsterdam, 1986), p. 411; D. Bardas *et al.*, in *Cryogenic Optical Systems and Instruments II*, edited by R. K. Melugin, SPIE Technical Publication No. 619 (SPIE, Bellingham, Washington, 1986), p. 29; see also related papers in this volume.

[12]N. Ashby and B. Bertotti, Phys. Rev. Lett. **52**, 485 (1984).

[13]N. Ashby and B. Bertotti, Phys. Rev. D **15**, 2246 (1986).

[14]J. G. Williams *et al.*, Phys. Rev. Lett. **36**, 551 (1976).

[15]I. I. Shapiro *et al.*, Phys. Rev. Lett. **36**, 555 (1976).

[16]X. X. Newhall *et al.*, in *Proceedings of the International Astronomical Union Symposium Number 128, Earth Rotation and Reference Frames for Geodesy and Geodynamics, Berkeley, West Virginia, 1986* (Reidel, Dordrecht, The Netherlands, 1986).

[17]J. Kovalevsky, in *Proceedings of the International Astronomical Union Colloquial Number 26, Reference Coordinate Systems for Earth Dynamics, Warsaw, Poland, 1975*, edited by B. Kolaczek and G. Weiffenbach (Polish Academy of Science, Warsaw, 1975), p. 123.

[18]J. G. Williams, Icarus **57**, 1 (1984).

[19]E. W. Brown, Mem. R. Astron. Soc. **53**, 39 (1987), and **57**, 51 (1905).

[20]J. P. Hauser, Ph.D. thesis, University of Colorado, 1974 (unpublished).

[21]J. G. Williams, private communication, based on results of Newhall *et al.* (Ref. 16).

[22]J. O. Dickey *et al.*, J. Geophys. Res. **90**, 9353 (1985).

[23]D. S. Robertson and W. E. Carter, in *Proceedings of the International Conference on Earth Rotation and the Terrestrial Reference Frame, Columbus, Ohio, 1985*, edited by I. Mueller (Department Geodetic Science and Surveying, Ohio State University, 1985), p. 296.

[24]T. M. Eubanks, private communication.

[25]W. Fricke, Astron. J. **72**, 1368 (1967); W. Fricke, in *Reference Coordinate Systems for Earth Dynamics*, edited by E. M. Gaposchkin and B. Kolaczek (Reidel, Dordrecht, The Netherlands, 1981), p. 331.

[26]J. H. Lieske *et al.*, Astron. Astrophys. **58**, 1 (1977).

[27]T. A. Herring *et al.*, J. Geophys. Res. **91**, 4745 (1986).

[28]D. E. Smith and P. J. Dunn, Geophys. Res. Lett. **7**, 437 (1980).

[29]S. C. Cohen and D. E. Smith, J. Geophys. Res. **90**, 9217 (1985).

[30]R. D. Reasenberg and I. I. Shapiro, private communication.

VOLUME 61, NUMBER 23 PHYSICAL REVIEW LETTERS 5 DECEMBER 1988

Measurement of the de Sitter Precession of the Moon: A Relativistic Three-Body Effect

I. I. Shapiro, R. D. Reasenberg, J. F. Chandler, and R. W. Babcock
Harvard-Smithsonian Center for Astrophysics, Cambridge, Massachusetts 02138
(Received 27 July 1988)

We analyzed lunar laser-ranging data, accumulated between 1970 and 1986, to estimate the deviation of the precession of the Moon's orbit from the predictions of general relativity. We found no deviation from this predicted de Sitter precession rate of nearly 2 angular sec per century (sec/cy), to within our estimated standard error of 0.04 sec/cy. This standard error, 2% of the predicted effect, incorporates our assessment of the likely contributions of systematic errors, and is about threefold larger than the statistical standard error.

PACS numbers: 95.10.Jk, 96.20.−n

One of the dynamical consequences of Einstein's theory of general relativity, noted first by de Sitter,[1] is a non-Newtonian precession of the Moon as it orbits the Earth in a system freely falling in the gravitational field of the Sun. de Sitter noted, too, that the magnitude of this "geodetic precession"[2] would be approximately 2 angular sec per century (sec/cy). About 50 yr later, in the late 1960's, one of us (I.S.), with Martin Slade and others, undertook the task of reanalyzing the world's accumulation of optical observations of the Moon, about 350000 in all, spanning the period from 1750 to almost 1970. Also included in the analysis were radar observations of the Moon and tracking data of the Surveyor spacecraft. The goal was to detect the predicted geodetic precession reliably. We failed: The result obtained,[3] 1.5 ± 0.6 sec/cy, was of only marginal reliability. We also looked into the possibility of detecting the contribution of the geodetic precession to the total precession of the Earth's spin axis. For this purpose, we considered using a combination of seismic models of the Earth and data from the tracking of artificial satellites to determine accurately the Newtonian contribution, which depends most importantly on the fractional difference of the principal moments of inertia of the Earth. The total precession was already known with sufficient accuracy and would be known even better from the accumulation of very-long-baseline interferometry (VLBI) observations of compact extragalactic radio sources. This approach, too, was unavailing: The Newtonian contribution to the precession could not be determined with the required accuracy. Another possibility considered by one of us (I.S.) was to determine the fractional difference in the Earth's moments of inertia from its effect on the Earth's principal, 18.6-yr, nutation. The amplitude of this nutation term could be measured with the requisite accuracy from the VLBI observations of the extragalactic sources. A covariance analysis,[4] and a consideration of the relevant uncertainties in other aspects of the needed model of the Earth, showed that this approach might be feasible. VLBI data over a sufficient time span and of sufficient accuracy will have been gathered within a few more years for a useful attempt to be made at this determination. On a somewhat longer time scale, the planned National Aeronautics and Space Administration gyroscope experiment holds promise of far higher accuracy.[5]

The main hope for early success lay in laser ranging to the optical corner reflectors[6] placed on the Moon by the Apollo astronauts in the late 1960's and early 1970's. It was clear that an approximately 18-yr accumulation of such ranging data would be required to separate the de Sitter precession reliably from any contribution due to the uncertainty of the amplitude (and phase) of the 18.6-yr period "classical" term in the motion of the lunar node. In the early 1980's, as this accumulation interval drew to a close, the present authors began a special analysis of the lunar laser-ranging data to try to measure this geodetic precession. Meanwhile, as we were completing this analysis, Bertotti, Ciufolini, and Bender[7] (BCB) inferred the presence of the geodetic precession from a lack of any known problems in the comparisons of (1) lunar laser-ranging (LLR) data with the general-relativistic theory which predicts this precession, and (2) the results, deduced from the LLR data, for the time dependence of the orientation of the Earth with independent determinations of this dependence. Their estimated uncertainty for each method was 10%; i.e., they imply that any larger deviation between the predicted geodetic precession and that affecting the observations would have been noticeable from such comparisons.

In the remainder of this paper, we (1) discuss our parametric model of the Moon's motion, with emphasis on the aspect that allows us to distinguish, as directly as feasible, the contribution of any geodetic precession; (2) describe the data sets we analyzed to determine the geodetic precession; (3) present the result we obtained for the geodetic precession, with special emphasis on the contributing source to its overall uncertainty; and (4) comment on the paper by BCB, in particular demonstrating that they significantly underestimated the uncertainties accompanying their approaches. In this last dis-

© 1988 The American Physical Society

cussion, we show why the uncertainties for their approaches exceed those in ours by at least thirtyfold for the first of their two methods and by about tenfold for the second.

Our model of the Moon's motion consists of two coupled sets of differential equations, one set describing the orbit and the other the rotation. The coupling is, of course, supplied by the nonspherical figure of the Moon, which affects both motions. The orbital equations are those of a general relativistic three-body system in a parametrized post-Newtonian (PPN) framework.[8] We include the perturbations due to the (1) leading terms in the spherical-harmonic expansion of the gravitational fields of the Earth and Moon, (2) Newtonian effects of the other planets, and (3) drag from tides on Earth. Similarly, the rotational equations include the torques produced by the Earth and the Sun on the leading moments of the Moon's figure and the effect of the anelasticity of the Moon.[3,9] The Earth's torque on the Moon includes the effect of the second zonal harmonic of the Earth's gravitational field.

Our model also includes the equations for the partial derivatives of the motions with respect to the (large) set of relevant adjustable parameters. We integrate the whole ensemble of equations simultaneously by numerical methods[8] to obtain a tabulation as a function of time for the positions and velocities of the relevant bodies, and for the partial derivatives with respect to these parameters. These latter include the mass of the Moon and of the Earth-Moon system; six initial conditions each for the heliocentric motion of the Earth-Moon system, for the Moon's geocentric motion, and for the Moon's rotation; a coefficient for the tidal drag on the Moon; the Moon's moment-of-inertia ratios and its gravitational harmonic coefficients through degree and order three; the PPN metric parameters β and γ; the rate $\dot{G}$ of a possible secular variation in the gravitational coupling constant G; and an *ad hoc* parameter h related to any extra precession of the Moon's orbit about the ecliptic pole that is not included in, or cancels part of, the predicted general-relativistic geodetic precession.

The partial derivatives for other relevant parameters, those that either do not affect at all, or do not affect significantly, the motions of bodies, are easier to compute. These latter parameters include instrumental biases; the coordinates of the laser-ranging sites and of the retroreflector locations; Love numbers for the Earth and Moon[9]; corrections to universal time (UT1) and to the Earth's pole position; corrections to the International Astronomical Union (IAU) values for the constant of general precession and for the coefficients of the series expression for the Earth's nutation[10]; and the scale of the solar system in light travel time units (i.e., the astronomical unit in seconds).

The parameter h is the key to our study since it is a measure of the (dis)agreement between the observed motion of the Moon and the de Sitter precession. We formulated the equations of motion[8] in a coordinate system rotating in the instantaneous ecliptic at a (variable) rate $h\Omega$, where Ω is the de Sitter precession rate. Any departure in the lunar motion from the de Sitter rate is characterized by the numerical factor h, where $h=0$ would be consistent with the prediction from general relativity and $h=1$ would imply a 100% error in this prediction.

The de Sitter precession may be thought of as having contributions from two sources: The first is the effect of mass on the curvature of space, which results in locally measured angles differing from those measured with respect to the fixed stars. The second source, which contributes half as much as the first, is the gravitational analog of the spin-orbit coupling of an electron in an atom. In the PPN formalism, the sum of these two contributions is proportional to $\gamma+\frac{1}{2}$.

Our primary data set was the 1970-1986 collection of LLR data, about 4400 echo-delay measurements for laser signals sent from various sites on the Earth to the various corner reflectors on the Moon. To aid in distinguishing the effects of lunar orbit precession from those of Earth nutation, we included the results of VLBI measurements in the form of (loose) *a priori* constraints on the corrections[10] to some of the coefficients of the shorter-period terms (annual, semiannual, and fortnightly) in the IAU series[11] representing the Earth's nutations. In part of our analysis, the "local" part, we also used the orbits and masses of the other planets and of relevant asteroids as obtained from our separate analysis of radar and spacecraft-tracking data. In the "global" part, we analyzed virtually all of the solar-system data simultaneously. In no case, however, did we include the optical observations of the Moon, since their standard errors are so large as to outweigh the advantage of the long time span they cover.[12]

We started our local analysis with a nominal value of zero for h and appropriate values for the other parameters, and applied our weighted-least-squares filter[13] to obtain estimates for all the relevant parameters, as well as to obtain the corresponding standard deviations and normalized correlation matrix. Of course, in this approach, we set the PPN parameters γ and β to unity and estimated neither. We found no significant departure of the geodetic precession of the Moon from that predicted by general relativity: $h=0.019\pm0.010$ (statistical standard errors are given here and hereafter unless otherwise noted). We also repeated this analysis, with $h\equiv0$, and estimated in its place an *ad hoc* coefficient λ, multiplying the relativistic terms in the equations describing the Moon's orbital motion[8]; we obtained $\lambda=1-0.010\pm0.011$, consistent with general relativity. In both analyses we estimated a total of 335 parameters, including 250 representing corrections to the temporal behavior of the Earth's orientation, and utilized all the LLR observations. We also estimated h and λ simultaneously and found the magnitude of their normalized correlation to be approximately 0.5.

These local analyses, as indicated, utilized separately determined values for such other solar-system parameters as the masses and orbital initial conditions for the other planets. We therefore also carried out a global analysis, to be described in detail elsewhere,[8] using all of our data—LLR as well as radar and spacecraft-tracking data for the planets—and solving for all relevant parameters simultaneously: a total of about 20000 observations and about 1300 parameters, including ~900 to represent the topography on Mercury and Venus. Although this latter approach is philosophically preferable, in practice it is not distinguishable: We obtained $h = 0.002 \pm 0.006$, not significantly different from the result from the local analysis. However, because of possible systematic errors, discussed below, neither of these statistical standard errors can be accepted as, or simply related to, the true uncertainty.

We attempted to account for the systematic errors in the global analysis by considering first the changes in our estimate of h which would result from large, but not "unreasonable," imposed changes in our estimates of other parameters. The largest effects come from those parameters that are highly correlated with h or that have "realistic" uncertainties significantly larger than the statistical standard errors, because of correspondingly large sensitivities either to systematic errors in the data or to omission from the model. This study showed that the largest of the effects on h was under 0.01 and virtually all were well under this value.

Second, we repeated our weighted-least-squares global analysis, but with different subsets of the data and of the parameters. For example, we split the LLR data into two "halves" according to time of year: approximately January-June (about 1900 observations) and July-December (about 2500 observations), and included one half at a time in otherwise identical analyses. The respective estimates for h were -0.023 ± 0.010 and 0.016 ± 0.008. We also investigated the effect of the use of progressively shorter spans of LLR data in both the global and the local analyses. The results[8] demonstrated that having LLR data over a significant fraction of 18 yr is crucial to detecting the de Sitter precession reliably.

We attempted to quantify any systematic effects that might be present in the LLR postfit residuals from the local and global analyses by (1) forming a histogram of the normalized residuals (each residual divided by the "final" standard error[8] of the observation), and (2) Fourier analyzing the residuals. The results show (1) very nearly Gaussian distributions, with (2) no significant periodicities present. The interpretation of the Fourier transforms, however, is complicated by the data having been taken at irregular intervals, and with fortnightly gaps. In addition, the fortnightly frequency is strongly absorbed by the lunar model, as are the monthly and yearly frequencies.

On the basis of all these studies, we conclude that a reliable estimate of the "true" standard error of our measurement of the de Sitter precession is 2% of the predicted effect, i.e., $h \simeq 0.00 \pm 0.02$.

We turn now to a discussion of the BCB paper. Their argument consisted of two parts; we deal with each in turn. The final argument held that any error in the general-relativistic prediction of geodetic precession would appear as an increase in the LLR postfit residuals, since the model used to analyze the data was consistent with general relativity. In particular, BCB considered the difference between the Moon's predicted mean perigee rate and the Sun's mean motion, and calculated the error in the difference that would increase twofold the rms of the residuals for two selected half-year periods.[7] This approach runs the risk that the full effect of masking (see below) on the estimate of the parameter of interest may be far more severe than anticipated. In this case, considering the full data and parameter set, we find that the cumulative effect of all other parameters leads to the statistical standard error for h being 24 times as large in our global analysis as it would be were only h being estimated. This "masking factor" represents the extent to which a linear combination of changes in the other parameters can be made to reproduce the signature of h and has the following consequence in the analysis of the LLR data: The mean-square residual R^2, which is the sum square of the postfit weighted residuals per degree of freedom, has a quadratic dependence on h: $R^2 \sim 1 + 2h^2$ for the LLR data set. One might expect R to be unity. However, the *a priori* knowledge of the standard errors in the LLR measurements, primarily from instrumental effects, is not precise; only an increase in R above the present level by more than 30% would be a reasonable indication of a problem[14]; in fact, an independent study of the residuals[15] shows that, for different large subsets of the data, R varies, but not monotonically with time, from 0.7 to 1.5. Our own earlier analysis showed variations in R, for somewhat different subsets, from about 0.6 to 2.0, also exceeding $\pm 30\%$. Systematic trends in the residuals would not be discernible easily, in view of the very uneven temporal spacing of the observations. An increase in R of more than 30% corresponds to a value of h greater than 0.6, implying, therefore, an uncertainty of *at least* 0.6 in h if determined *via* this method. BCB indicated that their result had an uncertainty equivalent to 0.1, but, as they stated, this value was not based on any detailed analysis. BCB did not examine any residuals; as noted above, the only statistic they considered was the rms of residuals for certain portions of the data. It would appear that the main problem with their assignment of a standard error was their implicit assumption that the effect of masking would be no more than a factor of 2. Moreover, a method based, as is BCB's, on the rms of residuals could not take significant advantage of a "square root of n"

VOLUME 61, NUMBER 23 PHYSICAL REVIEW LETTERS 5 DECEMBER 1988

reduction factor in the statistical standard error of the result, even were systematic errors not a factor, because of the limit set by the imprecise knowledge of the measurement standard errors.

The second BCB argument held that any (large) departure from the de Sitter precession would have been evident in the comparison between the mean rates of change of UT1, determined separately from LLR and from VLBI data. This idea is basically sound, but the estimated accuracy suffers in the same way, though to a lesser extent, from the lack of data analysis. To test this second method, we performed an analysis that attempted to match the scenario described in BCB. We carried out a series of analyses, corresponding to our local solution, but with h constrained to different values in each. Since we estimate corrections to UT1 values for epochs spread throughout the time span of the LLR observations, we found the weighted-least-squares estimate of the slope of these corrections in each case. Comparison of the results showed that the estimated UT1 rates reflected only about half of the imposed offset ($h \neq 0$) in the geodetic precession rate, thereby yielding a masking factor of about 2. We then estimated the standard deviation of h as that value needed to distinguish, at the 1-standard-deviation level, the difference between the UT1 rates estimated from the LLR data and that estimated from the VLBI data. We obtained $\sigma(h) = 0.2$, rather than the equivalent 0.1 given by BCB. Thus, their second approach yields a result still short by about tenfold of the accuracy achievable by actual analysis of the LLR data.

We thank Peter L. Bender for several stimulating discussions, especially on the paper of BCB. We also thank the National Science Foundation, Grant No. PHY-84-09671, and the National Aeronautics and Space Administration (NASA), Grant No. NAGW-967, for partial support.

[1]S. de Sitter, Mon. Not. Roy. Astron. Soc. **77**, 155 (1916).

[2]See, for example, S. Weinberg, *Gravitation and Cosmology: Principles and Applications of the General Theory of Relativity* (Wiley, New York, 1972), p. 237. This precession is also sometimes referred to as "geodesic."

[3]M. A. Slade, Ph.D. thesis, Massachusetts Institute of Technology, 1971 (unpublished).

[4]T. Stevenson, B.A. thesis, Massachusetts Institute of Technology, 1979 (unpublished).

[5]C. F. W. Everitt, private communication.

[6]C. O. Alley *et al.*, Science **167**, 458 (1970).

[7]B. Bertotti, I. Ciufolini, and P. L. Bender, Phys. Rev. Lett. **58**, 1062 (1987).

[8]J. F. Chandler, R. W. Babcock, R. D. Reasenberg, and I. I. Shapiro, to be published.

[9]R. J. Cappallo, C. C. Counselman, III, R. W. King, and I. I. Shapiro, J. Geophys. Res. **86**, 7180 (1981); A. J. Ferrari, W. S. Sinclair, W. L. Sjogren, J. G. Williams, and C. L. Yoder, J. Geophys. Res. **85**, 3939 (1980).

[10]T. A. Herring, C. Gwinn, and I. I. Shapiro, J. Geophys. Res. **9**, 11 416 (1986).

[11]P. A. Seidelmann, Celestial. Mech. **27**, 79 (1982).

[12]The LLR data, on the other hand, are so sensitive that we were able to show via a numerical experiment, with the appropriate partial derivatives, that, remarkably, the mass of Jupiter can be estimated from LLR data alone with a statistical standard error barely more than an order of magnitude greater than that obtained from all relevant spacecraft, Jupiter-satellite, and asteroid observations.

[13]See, for example, M. E. Ash, Massachusetts Institute of Technology Lincoln Laboratory Technical Note No. 1972-5, 1972 (unpublished).

[14]R. W. King, private communication.

[15]A. L. Whipple, private communication.

THE ASTROPHYSICAL JOURNAL, 509:856–860, 1998 December 20
© 1998. The American Astronomical Society. All rights reserved. Printed in U.S.A.

DETERMINATION OF THE GEOMETRY OF THE PSR B1913+16 SYSTEM BY GEODETIC PRECESSION

MICHAEL KRAMER
Astronomy Department, 601 Campbell Hall, University of California, Berkeley, California 94720; michael@astro.berkeley.edu[1]
Received 1998 May 22; accepted 1998 July 27

ABSTRACT

New observations of the binary pulsar B1913+16 are presented. Since 1978 the leading component of the pulse profile has weakened dramatically by about 40%. For the first time, a decrease in component separation is observed, consistent with expectations of geodetic precession. Assuming the correctness of general relativity and a circular hollow-cone–like beam, a fully consistent model for the system geometry is developed. The misalignment angle between pulsar spin and orbital momentum is determined, giving direct evidence for an asymmetric kick during the second supernova explosion. It is argued that the orbital inclination angle is 132°.8 (rather than 47°.2). A prediction of this model is that PSR B1913+16 will no longer be observable after the year 2025.

Subject headings: binaries: close — pulsars: individual (B1913+16) — relativity — stars: neutron

1. INTRODUCTION

Following the discovery of the first binary pulsar, B1913+16 (Hulse & Taylor 1975), it was shown that its unseen companion is another neutron star (Taylor, Fowler, & McCulloch 1979). Although four other neutron star binary systems are now known (see e.g., Fryer & Kalogera 1997 for a recent discussion), PSR B1913+16 remains a unique system in the sense that it exhibits a measurable amount of geodetic precession (Damour & Ruffini 1974; Barker & O'Connell 1975a, 1975b; Esposito & Harrison 1975). The effect occurs if the pulsar spin axis is misaligned with respect to the orbital angular momentum vector. In this case, a coupling of the vectors causes a precession of the pulsar spin around the orbital momentum. As a result, the angle between spin axis and our line of sight will change with time, so that different portions of the emission beam are observed. Consequently, one expects changes in the measured pulse shape as a function of time. Weisberg, Romani, & Taylor (1989, hereafter WRT), reported a change in the relative amplitude of the prominent leading and trailing components in the pulse profile of PSR B1913+16. Analyzing observations made at 1408 MHz in a time interval from 1981 to 1987, they determined a change in the amplitude ratio of $1.2\% \pm 0.2\%\ \mathrm{yr}^{-1}$, leading to a secular weakening of the leading component. If the emission beam exhibits an overall hollow-cone shape, one would also expect a change in the *separation* of the two components rather than only a change in relative intensity. This was, however, not observed within an upper limit of 0°.06 for the given time interval. As an interpretation, WRT argue that geodetic precession alters the observer's cut through an irregular and patchy emission beam structure, similar to that proposed by Lyne & Manchester (1988).

Cordes, Wasserman, & Blaskiewicz (1990, hereafter CWB) used polarization data of PSR B1913+16 to compare waveforms and position-angle (PA) swings that were obtained at slightly different frequencies between 1397 and 1416 MHz from 1985 to 1988. In the classical polar cap model, the PA swing depends on the impact parameter of our line of sight (Radhakrishnan & Cooke 1969), so that geodetic precession should alter the swing. While CWB detected only a possible change in the waveform, they could not find any significant change in the PA swing. In order to reconcile their observations with those of WRT, they argue that while geodetic precession is acting, the existence of a core component, which is more prominent at lower frequencies (see Taylor, Fowler, & McCulloch 1979), is affecting the otherwise uniform hollow-cone structure and the stability of the component ratio.

In the following we present new observations of PSR B1913+16 and propose a model giving an estimate of the misalignment angle between pulsar spin axis and orbital momentum. Such an angle is particularly useful to constrain the progenitor system and the asymmetry of the second supernova explosion (e.g., Cordes & Wasserman 1984; Burrows & Woosley 1986). In § 2 we describe our observations and the analysis of the data that we model in § 3. We discuss our results in § 4 and draw our conclusions in § 5.

2. OBSERVATIONS AND DATA ANALYSIS

The observations were made on six epochs between 1994 April and 1998 April using the 100 m Effelsberg radio telescope of the Max Planck Institut für Radioastronomie. We used a tunable 1.3/1.7 GHz HEMT receiver installed in the prime focus, yielding a system temperature of 25–35 K on the cold sky with a telescope gain of 1.5 K/Jy. The system provided left- and right-hand circularly polarized signals that were centered on 1410 MHz. Full details of the observing system can be found in Kramer et al. (1998).

Apart from a variation in observing time (30–60 minutes) depending on interstellar scintillation, each of the incoherently dedispersed data sets was obtained and analyzed in a homogeneous way. Total power profiles were produced by gain calibrating and summing the two circularly polarized intensities. These profiles were smoothed to match the dispersion smearing across one filter-bank channel, resulting in an effective time resolution of 337 μs compared to a sampling time of 57.6 μs (see Fig. 1). The effective time resolution of our data is a factor of about 2.5 worse than that of the data presented by WRT, but a factor of about 1.6 better than CWB. Since our data are at the same center frequency (i.e., within 0.1%) and have time resolution similar to WRT and CWB, we can accurately compare our data set to the previous observations.

[1] Also: Max Planck Institut für Radioastronomie, Auf dem Hügel 69, 53121 Bonn, Germany

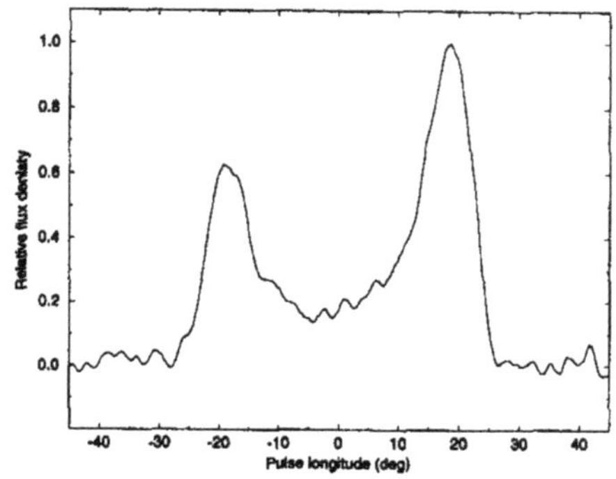

FIG. 1.—Pulse profile of B1913+16 observed in 1995 April

At first, we measured the amplitude ratio of the leading and trailing components using the off-pulse rms as an uncertainty estimator. These are presented in Figure 2, where we also plot data from WRT. It is seen that the new data are in good agreement with the earlier measured trend. Performing a weighted least-squares fit, we obtained a ratio change of $-1.9\% \pm 0.1\%$ per year. It is interesting to note that our fit is also in agreement with the component ratio from 1978 March (Fig. 2), which was measured using the profile published by Taylor & Weisberg (1982). The amplitude ratio between profile midpoint and trailing component seems to remain constant at a value of about 0.17 over a time span of 20 yr within the errors.

Measuring the separation between the leading and trailing components, we followed the approach of WRT by fitting Gaussian curves to the central portion of each component. The formal accuracy determined by the fit error can be up to an order of magnitude better than the sampling time, similar to a standard template matching procedure. We note, however, that the components are asymmetric in shape, so that the resulting centroids might be affected by

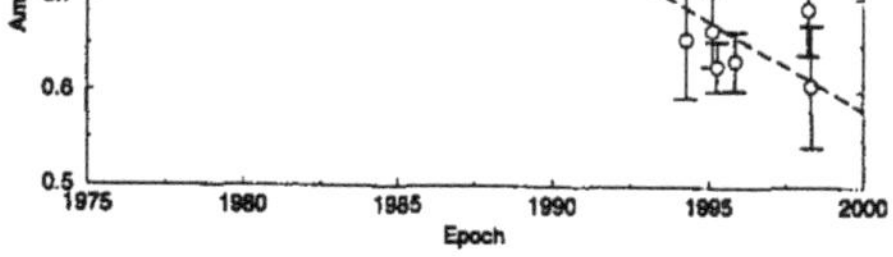

FIG. 2.—Amplitude ratio of leading and trailing components as a function of time. Filled circles were adapted from WRT, while open circles represent Effelsberg measurements. The dashed line represents a linear fit to the data with a slope of 1.9% yr^{-1}. The amplitude ratio adapted from Taylor & Weisberg (1982) is shown as a filled square.

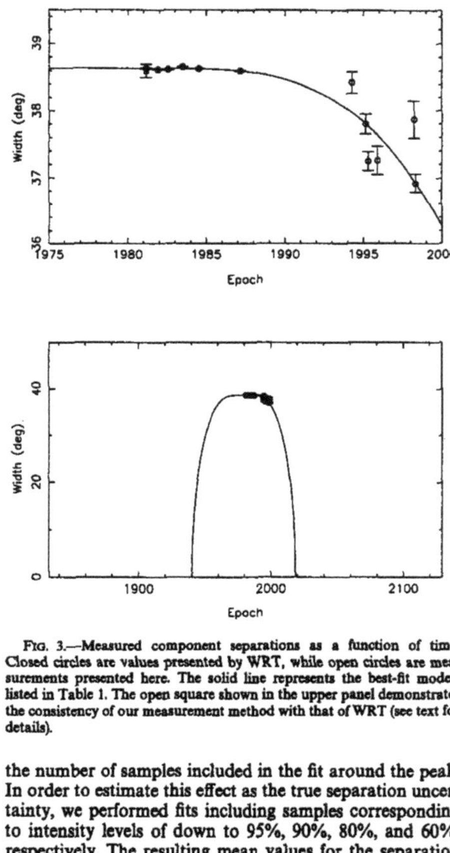

FIG. 3.—Measured component separations as a function of time. Closed circles are values presented by WRT, while open circles are measurements presented here. The solid line represents the best-fit models listed in Table 1. The open square shown in the upper panel demonstrates the consistency of our measurement method with that of WRT (see text for details).

the number of samples included in the fit around the peak. In order to estimate this effect as the true separation uncertainty, we performed fits including samples corresponding to intensity levels of down to 95%, 90%, 80%, and 60%, respectively. The resulting mean values for the separation and their corresponding errors are presented in Figure 3 with values adapted from WRT. In order to verify this approach, we measured the component separation of the 1981 profile presented by WRT at corresponding intensity levels. We obtained a value of $38^\circ \pm 0\overset{\circ}{.}1$, in very good agreement with their value. This gives confidence in the consistency between our measurements and those of WRT.

3. MODELING GEODETIC PRECESSION

In what follows we will make two assumptions: (1) general relativity is the correct theory of gravitation within the uncertainties of our measurements, and (2) the emission beam exhibits an overall circular hollow-cone–like shape, with intensity possibly depending on magnetic longitude. The first assumption is obviously well justified, given the excellent agreement of the measured orbital decay with the prediction from general relativity (Taylor & Weisberg 1982, 1989). While the assumption of an intensity dependence on magnetic longitude accounts for the component ratio change, the hollow-cone–like shape follows the arguments given by CWB, which can now be based on even more data presented for slowly rotating pulsars (e.g., Rankin 1993; Gil, Kijak, & Seiradakis 1993; Kramer et al. 1994; Gould 1994). These data, derived from obviously random cuts through emission beams, indicate that the opening angle of the beam, ρ, follows a period relation as expected for a circular beam in a dipolar field structure. Recent results suggest that this is not true for very fast rotating pulsars ($P \lesssim 40$ ms) but that it is still applicable to PSR B1913+16 (Kramer et al. 1998).

Assumption (1) and the timing parameters presented by Taylor (1992) imply an orbit inclination angle, $i = 47^\circ\!.2$, or $i = 180^\circ - 47^\circ\!.2 = 132^\circ\!.8$ and an expected precession rate of $\Omega_p = 1^\circ\!.21\ \mathrm{yr}^{-1}$ (Barker & O'Connell 1975a, 1975b). Modeling the effects of precession, we use the same notation and coordinate system as introduced by CWB but define the precession phase

$$\phi(t) = \Omega_p \cdot (T_0 - t) \qquad (1)$$

such that T_0 describes the closest approach of the pulsar spin axis, $\boldsymbol{\Omega}$, to the line of sight, $\boldsymbol{n}$ (see Fig. 4). Note that our coordinate system represents the usual observer's practice (i.e., pulsar-centered with z-axis antiparallel to the angular momentum vector) and thus differs from that described by Damour & Taylor (1992), which is Earth-centered with the primary axis along the line of sight. For a comparison, in Figure 4 we refer to Damour & Taylor's definition of $\boldsymbol{K}$, $\boldsymbol{k}$ and $\boldsymbol{s}_1$ (see their Fig. 1). Despite this difference, the misalignment angle, λ, between pulsar spin and orbital momentum is the same in both their convention and ours. The angle between pulsar spin and line of sight, β, will change with time according to

$$\cos\beta(t) = \cos\lambda\cos i + \sin\lambda\sin i\cos\phi(t)\,, \qquad (2)$$

affecting also the impact parameter, $\sigma(t) = \beta(t) - \alpha$, i.e., the angle between the magnetic axis, $\boldsymbol{m}$, and the line of sight at their closest approach. If the inclination angle between magnetic axis and pulsar spin axis is denoted by α and the emission beam exhibits an opening angle, ρ, the observed pulse width (or component separation if ρ is defined accordingly) changes with time as

$$W(t) = 4\sin^{-1}\left\{\sqrt{\frac{\sin^2(\rho/2) - \sin^2[\sigma(t)/2]}{\sin\alpha\sin\beta(t)}}\right\} \qquad (3)$$

using assumption (2). Obviously, only four free parameters describe the system discussed here, i.e., α, λ, T_0, and ρ. In order to determine them, we fitted the measured component separations to the equations above. First we started a robust Simplex least-squares fit algorithm (e.g., Caceci & Cacheris 1984) on grid positions covering the whole parameter space (i.e., $0 \le \alpha,\ \lambda \le 180$, $1997 - 150 \le T_0 \le 1997 + 150$, $0 \le \rho \le 90$), treating ρ as a free parameter. However, because of WRT's very precise measurements, ρ can be essentially fixed after calculating β and σ for chosen values of α, λ, T_0 and a component separation measured in 1981. The χ^2 hypersphere was thus searched on a finer grid in the remaining three dimensions. Both methods yield the same results with a best reduced $\chi^2_{red} = 6.94$ and 8 degrees of freedom. This relatively large χ^2value suggests that some of the separation uncertainties are underestimated. Nevertheless, the minimum is well constrained, as shown by the 68%, 90%, and 99% *confidence contours* (Fig. 5) and *confirmed* by Monte-Carlo simulations. However, since the orbit inclination angle can be either $i = 47^\circ\!.2$ or $i = 132^\circ\!.8$, four equivalent model solutions exist, which are listed in Table 1 (note that the angles are related to each other by a 180° ambiguity). Quoted uncertainties represent 99% confidence

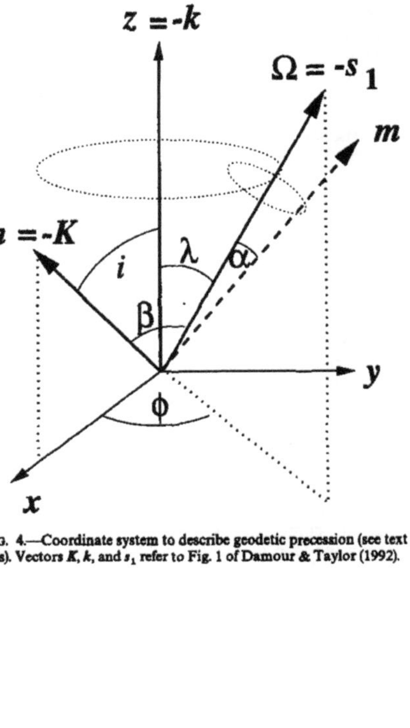

FIG. 4.—Coordinate system to describe geodetic precession (see text for details). Vectors $\boldsymbol{K}$, $\boldsymbol{k}$, and $\boldsymbol{s}_1$ refer to Fig. 1 of Damour & Taylor (1992).

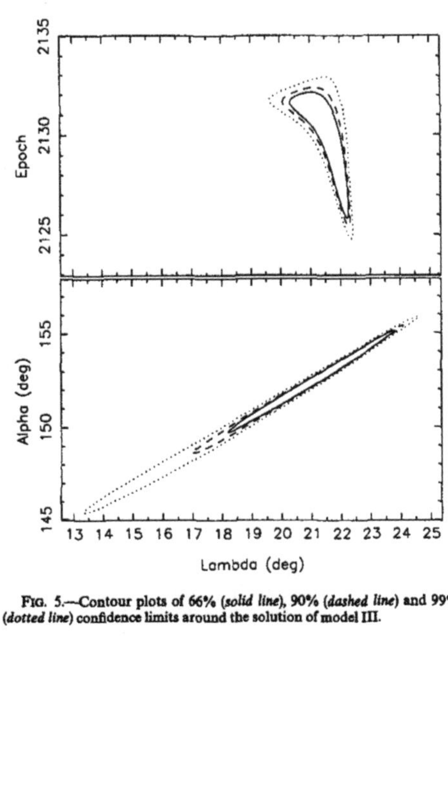

FIG. 5.—Contour plots of 66% (*solid line*), 90% (*dashed line*) and 99% (*dotted line*) confidence limits around the solution of model III.

TABLE 1

BEST MODEL SOLUTIONS

Model	i (deg)	α (deg)	λ (deg)	T_0 (yr)	ρ (deg)
I.	47.2	27^{+8}_{-3}	22^{+3}_{-8}	1980 ± 4	9.0^{+4}_{-3}
II.	47.2	153^{+3}_{-8}	158^{+8}_{-3}	2128 ± 4	9.0^{+4}_{-3}
III.	132.8	153^{+3}_{-8}	22^{+3}_{-8}	2128 ± 4	9.0^{+4}_{-3}
IV.	132.8	27^{+8}_{-3}	158^{+8}_{-3}	1980 ± 4	9.0^{+4}_{-3}

limits. As we discuss in detail below, we consider model solution III to be the correct one.

4. DISCUSSION

We observe a gradual decrease in the amplitude of the leading component, which is apparently caused by geodetic precession (c.f. WRT, CWB). While the amplitude ratio of the trailing component to the profile midpoint seems to remain unchanged, flux density variations due to interstellar scintillation prevent concluding whether or not the profile has weakened as a whole. Nevertheless, the observed amplitude change strongly suggests a latitudinal dependence of the emission strength. This is not too surprising as, e.g., dependencies of the spectral index on the distance to the magnetic axis have been observed before (e.g., Lyne & Manchester 1988; Kramer et al. 1994). The case of PSR B1913+16 is unique, however, as such dependence can be seen for a single pulsar. Assuming the model parameters in Table 1, the impact parameter has changed by about 2° between 1981 and 1998. This corresponds to about 20% of the determined opening angle, which seems a realistic value for the given amplitude change. Although the beam is thus not uniformly filled, this does not contradict the assumption about a circular beam shape. In fact, CWB showed that the core component, which lags the profile midpoint significantly, can disturb a possible uniformity of the beam. Taylor et al. (1979) argued similarly, based on reported profile changes at 430 MHz. Later, however, they attributed the effect to a comparison of data obtained with a different number of polarization channels (Taylor & Weisberg 1982). We thus believe that the simple assumption of a circular hollow-cone–like beam shape is well justified.

Since we have determined the system geometry without using polarization data, we can actually compare our results with those determined using polarization observations by CWB and Blaskiewicz, Cordes, & Wassmerman (1991, hereafter BCW). Using the same observer's convention as CWB and BCW, we can make this comparison directly, without the transformation of the involved angles that would be necessary using Damour & Taylor's (1992) convention. Computing the impact parameter σ for the observing epochs of CWB and BCW, we note that model solutions I and IV yield a negative value, while model solutions II and III give a positive impact angle. Both CWB and BCW present PA curves that are well defined over a wide range of longitudes ($\gtrsim 60°$), exhibiting a positive slope with a flattening at pulse longitudes separated from the profile midpoint by more than 15°. This behavior clearly indicates a positive impact parameter, i.e., an outer line of sight (Narayan & Vivekanand 1982; Lyne & Manchester 1988), so that we can exclude the first and fourth model solutions. The remaining ones differ in particular in the value of the misalignment angle. While solution III indicates an almost aligned system, solution IV implies almost antiparallel spins. Such a retrograde spin requires an asymmetric kick during the second supernova explosion that has to be much greater than the presupernova orbital velocity of the pulsar (e.g., Hills 1983; Kaspi et al. 1996). Following the calculations of Hills (1983) or Freyer & Kalogera (1997), this would imply a kick velocity of about 500 km s^{-1}, which is in conflict with Freyer & Kalogera's estimate of about 200 km s^{-1} as the most probable value. Such large kicks are also hard to reconcile with the (transverse) system velocity of PSR B1913+16 of about 130 km s^{-1} (Taylor & Weisberg 1989). Hence, we reject this solution and consider model solution III to be the one that best describes the data.

Although we are confident in the consistency between our data and WRT's, we now discuss the possibility that the comparison of data that were obtained with different observing systems can bear some uncertainties. Obviously, we minimized possible effects by using very similar observing parameters and an identical data analysis procedure as WRT. We can nevertheless study remaining uncertainties, assuming even larger measurement errors than estimated. The best model solutions listed in Table 1 predict that the PA swing steepens with time, while a sign change in the impact parameter will occur in the near future (or has already occurred recently). With increased measurement errors, solutions are allowed (e.g., $\lambda \approx 22°$, $\alpha \approx 112°$), where the impact parameter has changed sign prior to the observations of CWB, so that it will remain positive with less steep PA swings. Studies of various χ^2 hyperspheres, however, indicate that the most interesting parameter, λ, will most likely remain in a range $13° \lesssim \lambda \lesssim 25°$.

Using model III to compute the change in the impact angle, σ, between CWB's observing epochs, we derive a change of $\Delta\sigma = \sigma(1988.5) - \sigma(1984.5) = 0°.8 - 1°.1 = -0°.3$, which is only slightly higher than their estimated detection limit. Following also the discussion by CWB that the core component affects the PA curve, it is thus not surprising that they were not able to detect a change. In fact, the authors argue that the impact parameter should be much smaller than implied by their measured slope of the PA swing, which is also consistent with our result. Following the arguments of CWB, BCW excluded the central portion of the PA curve, when they fitted a (relativistic) Radhakrishnan-Cooke (1969) model to measured PAs of PSR B1913+16. At 1403 MHz they obtained $\alpha = 171° \pm 64°$ and $\sigma = 0°.2 \pm 1°$, which is indeed in good agreement with model III.

Our results imply a misalignment of the pulsar spin to the orbital momentum vector of $\lambda = 22^{+3}_{-8}{}^{\circ}$. Such value is in agreement with early simulations made by Bailes (1988), who found 20° as a typical value for systems like PSR B1913+16. The model also implies that the component separation remains almost unchanged for about 60 yr, which corresponds to a 20% likelihood of observing the pulsar in that phase given a precession period of about 300 yr. Thus, it is easy to understand why WRT did not detect any change in the component separation.

5. CONCLUSIONS

We have analyzed new observations of PSR B1913+16, showing that the first component has weakened by about 40% since 1978. For the first time we detected a small change in component separation, which we modeled assuming geodetic precession and a circular hollow-cone–

like beam. If our model is correct, the change of the component separation should be easily measurable in future high signal-to-noise ratio observations, as it will decrease with a higher rate. A test is also provided by monitoring the PA curve whose slope will possibly change sign in the near future.

Due to geodetic precession, the pulsar will not beam toward Earth for a significant fraction of its lifetime. In fact, our proposed model predicts that PSR B1913+16 will not be observable after the year ~2025, in good agreement with results by Istomin (1991). Shortly before this, the leading component will disappear if it continues to weaken with the measured rate. Reappearing again around the year 2220, PSR B1913+16 will, in total, be observable only for about a third of the precession period (see Fig. 3). While this does not affect birth rate calculations of double neutron star systems, the relatively small inferred (two-pole) beaming fraction of about 0.35 should be considered in future calculations.

So far, we have assumed that general relativity is the correct theory of gravitation in order to calculate the precession rate. With more and better data, we can treat the precession rate as a free parameter, so that the performance of another test for general relativity will be possible.

While timing observations are not able to decide between the "true" orbital inclination angle, our final model suggests that its value is $i = 132^\circ\!.8$ rather than $i = 47^\circ\!.2$. While polarization data provide information of the system in the two dimensions of the projected sky, geodetic precession and its observed effects yield information in direction of our line of sight. Combining all the available information allows us to draw a three-dimensional picture and to the remove the ambiguity in the orbital inclination angle of PSR B1913+16 for the first time.

It is clear that a misalignment angle of $\lambda = 22^{+3}_{-8}{}^\circ$ requires an asymmetric kick during the second supernova explosion. An asymmetric kick has been already inferred for J0045−7319 (Kaspi et al. 1996) and has been argued for B1913+16 on the basis of evolutionary models (e.g., Wijers et al. 1992; van den Heuvel & van Paradijs 1997; Fryer & Kalogera 1997). Our observations provide direct evidence for this hypothesis and give a natural explanation for the large inferred birth velocities of pulsars (e.g., Lyne & Lorimer 1993).

I am indebted to all people involved in the project to monitor millisecond pulsars in Effelsberg, in particular to Kiriaki Xilouris, Christoph Lange, Dunc Lorimer, and Axel Jessner. During this work I enjoyed stimulating discussions with Norbert Wex and Duncan Lorimer, who also read the manuscript with great care. It is a particular pleasure to thank Don Backer for the countless fruitful and helpful discussions. I am grateful for the constant support of Richard Wielebinski, and acknowledge the receipt of the Otto-Hahn Prize, during whose tenure this paper was written.

REFERENCES

Barker, B. M., & O'Connell, R. R. 1975a, Phys. Rev. D, 12, 329
———. 1975b, ApJ, 199, L25
Bailes, M. 1988, A&A, 202, 109
Blaskiewicz, M., Cordes, J. M., & Wasserman, I. 1991, ApJ, 349, 546 (BCW)
Burrows, A., & Woosley, S. 1986, ApJ, 308, 680
Caceci, M. S., & Cacheris, W. P. 1984, Byte, 5, 340
Cordes, J. M., Wasserman, I. 1984, ApJ, 279, 798
Cordes, J. M., Wasserman, I., & Blaskiewicz, M. 1990, ApJ, 349, 546 (CWB)
Damour, T., & Ruffini, R. 1974, Phys. Rev. D, Acad. Sci. Paris, 279, série A, 971
Damour, T., & Taylor, J. H. 1992, Phys. Rev. D, 45, 1840
Esposito, L. W., & Harrison, E. R. 1975, ApJ, 196, L1
Fryer, C., & Kalogera, V. 1997, ApJ, 489, 244
Gil, J. A., Kijak, J., & Seiradakis, J. H. 1993, A&A, 272, 268
Gould, D. M. 1994, Ph.D. thesis, Univ. Manchester
Hills, J. G. 1983, ApJ, 267, 322
Hulse, R. A., & Taylor, J. H. 1975, ApJ, 195, L51
Istomin, Ya. N. 1991, Soviet Astron. Lett., 17, 301
Kaspi, V. M., Bailes, M., Manchester, R. N., Stappers, B. W., & Bell, J. F. 1996 Nature, 381, 584
Kramer, M., Wielebinski, R., Jessner, A., Gil, J. A., & Seiradakis, J. H. 1994, A&AS, 107, 515
Kramer, M., Xilouris, K. M., Lorimer, D. R, Doroshenko, O., Jessner, A., Wielebinski, R., Wolszczan, A., & Camilo, F. 1998, ApJ, 501, 270
Lyne, A. G., & Lorimer, D. R. 1993, Nature, 369, 127
Lyne, A. G., & Manchester, R. N. 1988, MNRAS, 234, 477
Narayan, R., & Vivekanand, M. 1982, A&A, 113, L3
Radhakrishnan, V., & Cooke, D. J. 1969, Astrophys. Lett., 3, 225
Rankin, J. M. 1993, ApJ, 405, 285
Taylor, J. H. 1992, Phil. Trans. R. Soc. Lond. A, 341, 117
Taylor, J. H., Fowler, L. A., & McCulloch, P. M. 1979, Nature, 277, 437
Taylor, J. H., Weisberg, J. M. 1982, ApJ, 253, 908
———. 1989, ApJ, 345, 434
van den Heuvel, E. P. J., & van Paradijs, J. 1997, ApJ, 483,
Weisberg, J. M., Romani, R., & Taylor, J. H. 1989, ApJ, 347, 1029 (WRT)
Wijers, R. A. M. J., van Paradijs, J., & van den Heuvel, E. P. J. 1992, A&A, 261, 145

www.ingramcontent.com/pod-product-compliance
Ingram Content Group UK Ltd.
Pitfield, Milton Keynes, MK11 3LW, UK
UKHW052113170926
8624IPUK00026B/373

9 789812 38347